Ultrasound in Peripheral, Neuraxial and Perineuraxial Regional Anaesthesia

Eryk Eisenberg • Elisabeth Gaertner
Editors

Ultrasound in Peripheral, Neuraxial and Perineuraxial Regional Anaesthesia

Springer

Editors
Eryk Eisenberg
Département d'anesthésie et réanimation
Clinique Pôle Santé République
Clermont-Ferrand, France

Elisabeth Gaertner
Département d'anesthésie et réanimation chirurgicale
Hôpitaux Civils de Colmar
Colmar, France

ISBN 978-3-031-08803-2 ISBN 978-3-031-08804-9 (eBook)
https://doi.org/10.1007/978-3-031-08804-9

This Springer imprint is published by the registered company Springer Nature Switzerland AG
The registered company address is: Gewerbestrasse 11, 6330 Cham, Switzerland

To my sons, Sasha and Solal.

May I never disappoint you.

"We are a biological accident which does what it may"—Jacques Brel

"Without art, a gift is nothing other than a dirty habit"—Georges Brassens (Le mauvais sujet repenti)

"Two intellectuals seated will go less far than an imbecile who is walking"—from Michel Audiard

"I do not wish to achieve immortality by means of my work. I wish to achieve immortality by not dying!"—Woody Allen

Foreword

"Ultrasound is now making so much noise that we are beginning to hear it".—Dr Daniel Lichtenstein

This book is the fruition of a long-term work started in 2004, for which the first result was the first edition published in 2007. I am happy to state that 7 years after its publication, its content has continued to be fully of current interest. Nevertheless, a few updates are now required:

Technical advances in ultrasound imaging have made it necessary to create an updated iconography. Although a few images in the work are the result of high-end imaging equipment (3D/4D ultrasound images and CT scan ultrasound fusion imaging, MRI ultrasound), almost the totality however has been carried out with ultrasound imagers which are affordable to the majority of anaesthesia departments. The technical performance of this equipment does not cease to increase, image quality is being refined in an astonishing manner, but however it is prudent to not have "blind faith in our ability to see".

By echoing (no pun intended) the adage "*Give a small boy a hammer and he will find that everything he encounters needs a pounding*" (Abraham Kaplan), the daily use of ultrasound is being extended to new medical techniques: anaesthetics, analgesics, diagnostics, etc. Therefore, it was appropriate to offer the reader a description of some of them.

Lastly, notwithstanding the citation according to which "experience is a lantern hooked onto one's back and which only illuminates the path travelled" (Confucius), of course it has been inescapable to allow "the light of experience" to mix somewhat the present and future enlightenment of our practices.

From the Same Publisher

Principes et protocoles en anesthésie pédiatrique (3rd edition), F. Duflo, S. Combet, M. de Queiroz Siqueira, 2014.

Le blessé de guerre, S. Mérat, 2014.

Anesthésiste-Réanimateur/Chirurgien: un seul bloc, V. Travers, H. Cuche, E. Gaertner, 2013.

La communication dans le soin. Hypnose médicale et techniques relationnelles, F. Bernard, H. Musellec, E. Gaertner, 2013.

Médicaments en anesthésie (*3rd edition*), B. Dalens, 2013.

Hépatologie aiguë en anesthésie, réanimation, urgence, F. Aubrun, S. Duperret, 2013.

Kétamine (2nd edition), G. Mion *et al.*, 2012.

Cœur et anesthésie (2nd edition), P. Coriat, Y. Le Manach *et al.*, 2012.

Les Monitorages des paramètres physiologiques en situation critique, J.-J. Lehot, M. Cannesson *et al.*, 2012.

Guide pratique d'anesthésie locorégionale 3rd edition), X. Paqueron and Y. Cimino, 2012.

Fiches techniques en salle de surveillance post-interventionnelle, C. Péraldi *et al.*, 2012.

Intubation. De l'oxygénation à l'intubation difficile, C. Erb, H. Menu, É. Wiel, 2012.

Protocoles d'anesthésie-réanimation obstétricale (2nd edition), P. Dailland *et al.*, 2011.

Procédures anesthésiques liées aux techniques chirurgicales, S. Mérat *et al.*, 2011.

Guide pratique d'anesthésie locorégionale pédiatrique: clinique et échographie, S. Combet, M. De Queiroz Siqueira and F. Duflo., 2011.

Anesthésie et analgésie locorégionale en chirurgie abdominale et périnéale, M. Beaussier, P. Niccolaï *et al.*, 2010.

Anesthésie en chirurgie orthopédique et traumatologique, E. Gaertner *et al.*, 2010.

Mémo Cardio en urgence, F. Lapostolle and J.-P. Torres, 2010.

Hypnose, douleurs aiguës et anesthésie, Cl. Viroe and F. Bernard, 2010.

Comprendre les émotions en anesthésie réanimation, J.-C. Lleu, 2010.

Hygiène en anesthésie (4th edition), A.-M. Saire-Mauffrey, E. Gaertner, H. de Montclos *et al.*, 2010.

Les Produits de l'anesthésie (4th edition), X. Sauvageon *et al.*, 2010.

Anesthésie et réanimation du patient obèse, J.-E. Bazin, P. Coriat *et al.*, 2009.

Nouveaux risques en anesthésie réanimation, J.-J. Lehot, M. Canesson *et al.*, 2009.

Principes et protocoles en anesthésie pédiatrique (2nd edition), F. Duflo *et al.*, 2009.

Manuel d'échographie en réanimation et service d'urgence, V. Noble, B. Nelson, A.N. Suttingco, translated by E. Gaertner, 2009.

Iatrogénie et toxicologie en urgence, F. Baud and R. Bédry, 2009.

Douleur, inflammation et auto-immunité à l'usage des spécialités médicales, M. Sorel *et al.*, 2009.

Précis d'anesthésie clinique, P.G. Barash, B.F. Cullen, R.K. Stoelting, E. Gaertner, 2008.

Urgences et infections, C. Martin, 2008.

Anesthésie, analgésie et sédation du patient âgé, F. Aubrun, 2008.

Foreword to the 2014 French Edition

Seven years after the publication of the first edition of *Echographie en anesthésie régionale périphérique (Ultrasound in peripheral regional anaesthesia)*, edited by Dr. Eryk Eisenberg, we had the privilege to preview the second edition of this well-established, outstanding book. It is with pleasure that we welcome the second edition with a few words about the author and his collaborators. Dr. Eisenberg is one of the most fervent educators in the field of ultrasound-guided regional anaesthesia and peripheral nerve blocks. His must-go-to annual symposium, traditionally hosted in December in Clermont-Ferrand, is the most popular regional anaesthesia educational event in France, with rich tradition, well received and attended by delegates from all over Europe. The collaborators chosen by Dr. Eisenberg for this second edition are some of the best-known practitioners and educators in the field. It is thus no surprise that this team lead by Dr. Eisenberg has crafted probably the most comprehensive volume on ultrasound in regional anaesthesia to date. The first edition (2007) that comprised 140 pages has been nearly tripled in volume and now features nearly 400 pages, 500 ultrasound images and nearly 50 anatomical dissections. The new edition is also accompanied by a dedicated website with videos of state-of-the-art techniques, richly accompanied by didactic and illustration material.

The scope of the book is uniquely comprehensive and logically organized. We particularly value the utmost didactic approach to illustrations with multiple diagrams, picture-in-picture samples and meticulous labelling of the images.

Aesthetic choice of transparent colours and miniature labels to avoid distraction from the relevant detail are not only effective but also artistic. The illustrations and images are all of top quality and consistent in style. The vignetting treatment of images to unclutter the unnecessary detail will also be appreciated by readers. The ultrasound images are outstanding and carefully selected for didactics and simplicity to demonstrate the most clinically relevant situations. Importantly, they have a real-life appearance, including actual needle paths and desired disposition of injectate during nerve block procedures; most are from the original database of Dr. Eisenberg. All the supplementary didactic material is authoritative and presented as an artful balance of years of clinical experience by the Editor and a summary of the peer-reviewed literature.

In summary, this second edition, *Ultrasound in peripheral, neuraxial and perineuraxial regional anaesthesia*, is truly an amazing text didactically and aesthetically. It is a must-have book for anyone interested in learning, advancing the existing knowledge or teaching ultrasound-guided regional anaesthesia. Sincere congratulations to the Editor and all of the contributors on producing such an educational masterpiece.

The New York School of Regional Anaesthesia (NYSORA) — Admir Hadzic
New York, USA

Faculty of Medicine — Manoj Kumar Karmakar
The Chinese University of Hong Kong
Hong Kong, Hong Kong

Postscript to the 2014 French Edition

Writing a postscript for the *Ultrasound in peripheral, neuraxial and perineuraxial regional anaesthesia* book coordinated by Dr Éryk Eisenberg is an honor and a privilege. This new edition arrives in a timely manner, 7 years after the first edition (*Ultrasound in peripheral regional anaesthesia*), to further strengthen the nationally recognized characteristic of the book as a reference work.

Dr. Éryk Eisenberg knew how to work with experts in the field, who succeeded in their project, in writing of each chapter of the volume, of combining an original presentation with an educational aspect and a very high level of pictorial and iconographic quality, without overlooking comprehensiveness in the clinical approach to peripheral, compartmental or spinal nerve blocks.

The vast majority of techniques in regional anaesthesia are perfectly discussed. Many techniques have been added to the first edition, and other chapters have been brilliantly updated. This work contains in fact 364 pages which makes it a pleasure to read, including over 500 original illustrations, enhanced by almost anatomical transparencies, the assumed and real trajectory of needle insertion and of catheters, and of discrete but basic annotations. The quality of the images and of the videos is unequalled and the references are complete.

Dr. Éryk Eisenberg is an expert clinician in ultrasound-guided regional anaesthesia. He is a recognized and respected as leading opinion-maker in the field. He worked in collaboration with other clinicians, all of whose names appear under the subject references and who have perfectly combined scientific rigorousness and clinical spirit. They jointly knew how to provide academic and practical meaning, in the innovative use of ultrasound in regional anaesthesia. This work, together with a user-friendly and very complete website, is without a doubt the ultimate French-language reference in this innovative field of anaesthesia and intensive care. Reading of the book *Ultrasound in peripheral, neuraxial and perineuraxial regional anaesthesia* undoubtedly has convinced you of this.

We extend our heartfelt congratulations and our accompanying support on this excellent teaching tool.

Department of Intensive Care-Anaesthesiology
CHU: University Hospital Centre
Montpellier, Lapeyronie Hospital
Montpellier, France

Xavier Capdevila

Department of Anaesthesia
University Regional Maternity of Nancy, CHU
Nancy, France

Hervé Bouaziz

Acknowledgements

To my friends and family, without whom nothing would be.

To my wonderful children, an essential, nourishing sap by each second, whose many attempts at making my "labour" less arduous have always been crowned by success!

"Here Dad, I made you some little computers to help you finish your book faster!"

To my friends (who know who they are) for being there to offer their support.

To all men who attempt to seek to push back the "sterile and mute abyss, which swallows up what we hardly are"—Fernando Pessoa (*The Book of Disquiet*)

To Jean-David who one day told me: "I*gnore things doesn't matter, what does matter, however, is not knowing where to look*".

I wish from the bottom of my heart that you managed to find the answers that you went so far to seek, so very far…

"Each man during his night goes off toward his light".—Victor Hugo

To our patients (from *patior:* to tolerate, to endure):

"Our life is a woven wind".—Joseph Joubert

For the iconography:

- PHOTOGRAPHS: Valentin UTA (www.valentinuta.com)
- Models of photographs: Laetitia Aurières, Jean Halberstadt
- ILLUSTRATIONS: Élisabeth FOUCHÉ (ADN) and Guillaume BLANCHET

I warmly thank each and every person who accepted to be associated with this work by providing his or her knowledge, professionalism, passion and talent:

- Christophe Aveline
- Rafaël Blanco
- Bénédicte Bouche
- Olivier Choquet
- Philippe Clavert
- Laurent Delaunay
- Mathilde De Queiroz
- Antoine Drizenko
- Bertrand Fabre (dissections and photographs)
- Virginie Dewé-Faucon (translations)
- Élisabeth Gaertner
- Philippe Gautier (iconographs)
- Admir Hadzic (iconographs)
- Ian Harper (translations)
- Arthur C Hertling (translations)
- Denis Jochum
- Manoj Kumar Karmakar
- Wing Hong Kwok
- Didier Morau
- Vincent Tubert
- Paul Joseph Zetlaoui

In particular, I would like to thank Dr Ian HARPER for his extensive translation work and his unwavering friendship.

Through the support granted by their firms, I would particularly like to thank for their trust and friendship:

- Élisabeth Fouché, Jean Christian Garnier (ADN)
- Albert Louys, Sandrine Aulagner (Fujifilm—SonoSite)
- Gilles Boutet, Fabien Jenn, Arnaud Tellier (SonoSite)
- Richard Darmon, Philippe Coruzzi, Christian Kouyoumdjian (Gamida)
- Simone Pajunk-Schelling, Martin Hauger (Pajunk)
- Laure Terrier, Jean Marc Degasperi, Élodie Goutteyron (Vygon)
- Corinne Ochoa, Frédérique Ranchon (Abbott)
- Manuel Rios (Baxter)
- Philippe Gendre (AstraZeneca)
- Gilles Recuyer, Olivier Blachère (Mepy Système)
- Martine Berdichevski, Paul Marcombes, Étienne Richard (Sonoscanner)
- Crystal Lin (Mindray)
- Charbel Abisafi (Fujifilm)
- Frédérique Nay, Hélène Demargne (BBraun)
- Charles El Koubi (EDM)

For their professionalism and their precious aid:Maud Thévenin, Anne Chevalier, Alix Thimiakis and Valérie Parroco, from Arnette Publications (John Libbey Eurotext)

Contents

List of Videos

Contributors

Christophe Aveline Hôpital Privé Sévigné, Cesson Sévigné, France

Rafael Blanco Anaesthetic Department, Corniche Hospital, Abu Dhabi, United Arab Emirates

Hervé Bouaziz Service d'anesthésie, Maternité régionale universitaire de Nancy, CHU, Nancy, France

Bénédicte Bouche Clinique Bretéché, Centre d'évaluation et traitement de la douleur chronique, Nantes, France

Xavier Capdevila Service d'anesthésie-réanimation, CHU de Montpellier, Hôpital Lapeyronie, Montpellier, France

Olivier Choquet Département d'anesthésie-réanimation A, CHU de Montpellier, Hôpital Lapeyronie, Montpellier, France

Philippe Clavert Service de chirurgie du membre supérieur, Centre de Chirurgie Orthopédique et de la Main (CCOM), Illkirch, France

Laurent Delaunay Département d'anesthésie-réanimation, Clinique Générale, Vivalto Santé, Annecy, France

Virginie Dewé-Faucon Service d'anesthésie-réanimation, Clinique Saint Charles, La Roche-sur-Yon, France

Eryk Eisenberg Département d'anesthésie et réanimation, Clinique Pôle Santé République, Clermont-Ferrand, France

Bertrand Fabre Institut Curie, Centre René Huguenin, Saint Cloud, France

Elisabeth Gaertner Département d'anesthésie et réanimation chirurgicale, Hôpitaux Civils de Colmar, Colmar, France

Philippe Gautier Clinique Sainte Anne-Saint Remi, Bruxelles, Belgium

Admir Hadzic The New York School of Regional Anesthesia (NYSORA), New York, USA

Ian Harper Department of Anaesthesia, Northumberland Healthcare NHS Trust, Newcastle upon Tyne, UK

Arthur C. Hertling Department of anesthesiology, Perioperative Care and Pain Medicine, NYU Grossman School of Medicine, New York, NY, USA

Denis Jochum Service d'anesthésie-réanimation, Groupe hospitalier du Centre Alsace, Hôpital Albert Schweitzer, Colmar, France

Manoj Kumar Karmakar Department of Anesthesia and Intensive Care, Prince of Wales Hospital, The Chinese University of Hong Kong, Sha Tin, Hong Kong

Wing Hong Kwok Department of Anesthesia and Intensive Care, Prince of Wales Hospital, The Chinese University of Hong Kong, Sha Tin, Hong Kong

Didier Morau Département d'anesthésie-réanimation A, CHU de Montpellier, Hôpital Lapeyronie, Montpellier, France

Mathilde De Queiroz Département d'anesthésie-réanimation, Hôpital Femme-Mère-Enfant, Bron, France

Vincent Tubert Clinique Pôle Santé République, Clermont-Ferrand, France

Paul-Joseph Zetlaoui Département d'anesthésie-réanimation, Hôpital de Bicêtre, Le Kremlin-Bicêtre, France

Physical Basis of Ultrasound

1

Eryk Eisenberg

Contents

E. Eisenberg (✉)
Département d'anesthésie et réanimation, Clinique Pôle Santé République, Clermont-Ferrand, France

E. Eisenberg, E. Gaertner (eds.), *Ultrasound in Peripheral, Neuraxial and Perineuraxial Regional Anaesthesia*,
https://doi.org/10.1007/978-3-031-08804-9_1

History

In 1794, by observing how bats fly, the Italian biologist Lazaro Spallanzani (1729–1799) suspected the existence of ultrasound waves. It was only nearly a century later that R. Kroning and then F. Galton studied this phenomenon.

In 1880, Jacques and Pierre Curie discovered **piezoelectricity**. This is the property of certain bodies (crystals) which can become polarised electrically whenever a mechanical force (direct effect) is applied to them and their deformation results from the effect of an electrical field (inverse effect): this is a reciprocal phenomenon. Piezoelectricity is the basis of **electromechanical transduction**, a technique used for production and detection of ultrasound.

By placing an ultrasound transducer on the skin of a person, we can use these two basic discoveries today to perceive what is hidden beneath the layer of the epidermis.

The idea of **ultrasound echo-location** methods was developed in 1912 due to the sinking of the *Titanic* and it is thanks to Paul Langevin and his first quartz transducers that, in 1916, the underwater detection of objects with ultrasound was born.

Karl Théodore Dussik, an Austrian psychiatrist, was the first person to have used ultrasound in a diagnostic context in humans. Towards the end of the 1930s, he and his younger brother Friedrich, a physicist, developed a method of detection of intracranial tumours by screening for a midline shift of brain structures. The results of their first studies were published in 1942 [1] and the first photos were presented in 1947. Introducing the notion of **hyperphonography**, the "ventriculogram" (Fig. 1.1) is reputed to represent the shape of the cerebral lateral ventricles. Unfortunately, it then was demonstrated by W. Güttner and his collaborators in Siemens laboratories that the technique used by the Dussik brothers (Fig. 1.2) could not allow them to obtain such an image because of the extent of absorption and of reflection of ultrasounds by the skull bones. Identical images were produced by using an empty skull, which made it possible to confirm the artefactual nature of the ventriculogram produced by the Dussik brothers [2].

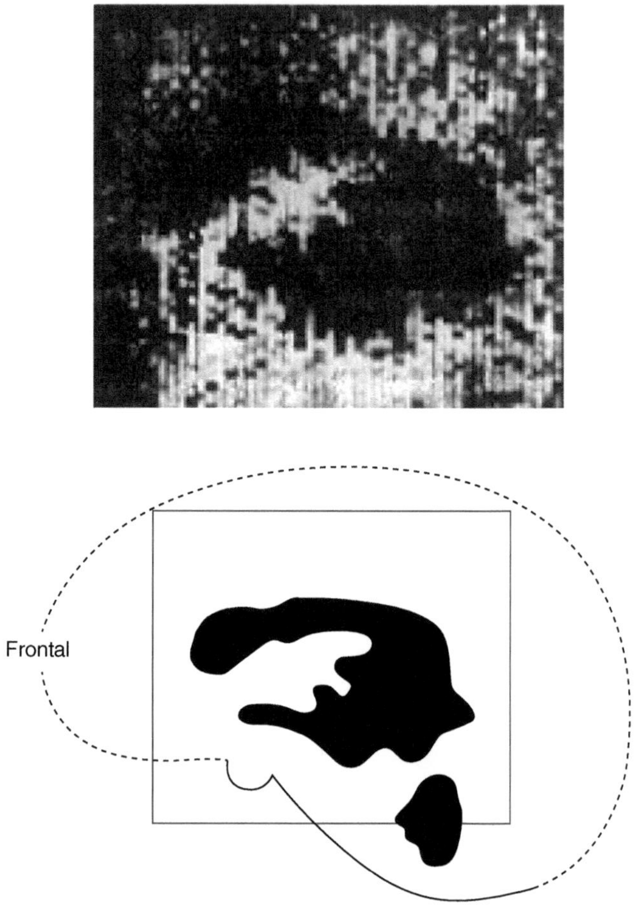

Fig. 1.1 The "ventriculogram" by the Dussik brothers

After the end of the Second World War, as the idea of clinical ultrasonography was becoming established [3], technical advances and applied research enabled the design and then the manufacture, in Japan and in the USA, of increasingly complex high-precision medical imaging equipment. Developments in the 1970s formed the image of ultrasound that we know today. Ultrasound imaging thrived thanks to transducers which produced images in shades of grey and the introduction of velocimetry by the Doppler Effect which enabled the study of blood flow.

More recently, the focusing of the ultrasound beam, important advances made in the field of high frequency and miniaturisation of probes have opened up to the era of intracavitary ultrasonography. Finally, by means of the remarkable development of digital techniques, it is now possible to use harmonic imaging, three-dimensional and four-dimensional imaging, needle electromagnetic guidance systems, or even fusion of ultrasound images with CT-scan or MRI-scan images, etc.

In clinical practice, inspired by interventional radiology techniques, the use of imaging (particularly ultrasound) applied to regional anaesthesia (RA) has recently been developed. As evidenced in Fig. 1.3, from an editorial by Rathmell [4], the number of published articles concerning imaging and anaesthesia (or pain) very quickly followed an eloquent growth curve.

In 2005, 29 articles were found in the same journals with *ultrasonography* and *peripheral nerve block* being the only items. In May 2013, a total of 362 articles emerged, involving the same research.

From the first description made in 1978 by La Grange et al. [5] to the exponential growth which we have witnessed during the last few years, steps were taken involving the nec-

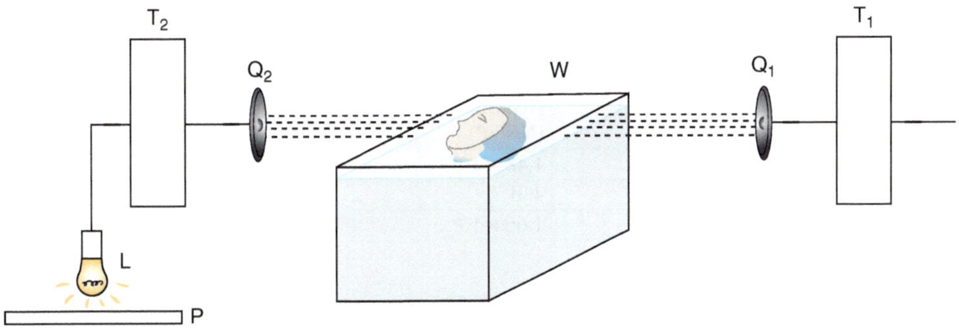

Fig. 1.2 T_1: ultrasound generator; Q_1: transmitter; Q_2: receiver; T_2: amplifier converter; W: water tank; L: light; P: film

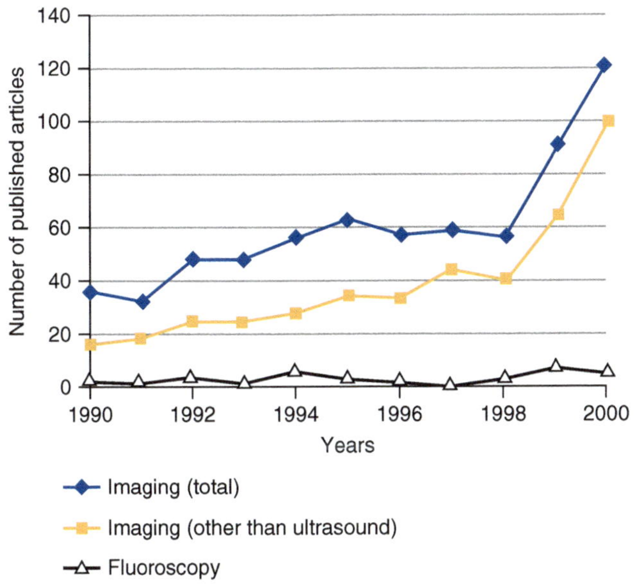

Fig. 1.3 Increase in the number of published articles in the field of anaesthesia and pain between 1990 and 2000. Results obtained from a search of MEDLINE with the keywords *"diagnostic imaging"*, *"echocardiography"* and *"fluoroscopy"* in the following journals: *Anesthesiology, Anesthesia and Analgesia, Regional Anesthesia and Pain Medicine* and *Pain*

essary technical development and the availability of increasingly more sophisticated functional equipment, increasingly smaller in size.

Physical Principles

Sound waves, which include **ultrasound waves**, are mechanical vibrations which are propagated in liquids and solids. Waves are considered to be **ultrasound** (and therefore inaudible by the human ear) when their frequency is between 20 kHz and 200 MHz. Above 200 MHz, the notion of **hypersound** occurs. Diagnostic medical ultrasonography covers the usual range of 2–15 MHz, although investigation at low depths of penetration is possible with higher frequencies (20 MHz for skin, 20–30 MHz for endovascular ultrasound and even 50 MHz for the cornea and the anterior segment of the eye).

Similar to waves generated on the surface of a body of water into which a stone is thrown when someone throws a stone into it, ultrasounds are propagated by resulting changes to pressure and vibration of within the material, transmitted increasingly closer in the form of an elastic wave.

The **Velocity** (***c***) of wave propagation (m/s) does not depend on its frequency but on the nature of the medium, and in particular its density and elasticity (Table 1.1). Velocity in air is slow (330 m/s), intermediate in water and soft tissue (about 1500 m/s) and rapid in cortical bone (3000–4000 m/s).

Frequency (***f***) and **wavelength** (***λ***) are related by the equation:

$$\lambda = c / f$$

Therefore, for f = 5 MHz, the wavelength in water at 37 °C (where c = 1540 m/s) is:

$$\lambda = 0.3\,\text{mm}$$

In any given medium, the velocity of propagation is constant. Interdependence exists between frequency, penetration and image resolution. **Penetration** is much better when an ultrasound wave is long (and therefore the frequency is low). The higher the frequency, the smaller the wavelength, the more shallow the penetration, the better the **spatial resolution** (axial, lateral and transverse). Resolution is the principal indicator of quality in ultrasound imaging.

Acoustic impedance which can be defined as the resistance of a medium to propagation of an ultrasound wave also depends on the density and elasticity of the tissue. In order for an ultrasound wave emitted by the transducer to be received by the latter, it must be reflected at the acoustic interface, the border between the two different media of acoustic impedance.

Therefore, when the wave passes from one medium to another, the **interface** which separates the tissues is at the origin of a "junction" of the native incident wave, which can occur by:

Table 1.1 Examples of velocity of propagation and of acoustic impedance in biological tissues for temperatures between 20 and 37 °C (from Bonnin et al. [7])

Tissue	Velocity of propagation of ultrasound *c* (m/s)	Acoustic impedance *Z* (N/s/m³)
Water	1480 (20 °C)	1.48×10^6
Air	340 (20 °C)	440
Blood	1566	1.66×10^6
Spongy bone	1450–1800	
Cortical bone	3000–4000	4×10^6 to 8×10^6
Fat	1450	1.38×10^6
Liver	1560	1.65×10^6
Muscle	1550–1630	1.65×10^6 to 1.74×10^6
Skin	1600	

- **Reflection:** part of the ultrasound beam "rebounds" (echoes) off the interface for which it is at the origin of the image.
- **Refraction:** the beam transmitted to the tissue located more deeply (whose direction can vary depending on the angle of the incident beam and on the difference in the velocity of propagation in the two media that it crosses) in turn enables image generation.

The intensity of ultrasound reflection, proportional to the difference in impedance of the two contiguous tissues, contributes to show the anatomical borders of organs [6]. The proportion of energy reflected is very low (<1%) for an interface between two soft tissues, while there is almost total **reflection** for a soft tissue/air interface, which makes ultrasound penetration impossible in air-bearing organs, such as the gut or lung tissue, which are normally aerated. On the contrary, the effects of refraction are not usually very great considering the small differences in the velocity of propagation between the different soft tissues (Table 1.1).

When the dimensions of the interface are smaller than the ultrasound wavelength, we no longer speak of refraction but of **diffusion** (in three spatial planes). This phenomenon contributes to the definition of the ultrasound structure of the tissue by multiple small diffusing heterogeneous structures which are located in the tissue being crossed.

In addition to reflection, refraction and diffusion, part of the **attenuation** that the ultrasound beam undergoes as it crosses the tissue is the result of **absorption** and of **deterioration** of its energy in the form of heat. Attenuation is greater at higher frequencies. This explains why high frequencies allow for investigation only in shallow depths of tissue, while low frequencies are required to visualise deeper structures.

Ultrasound generated by the transmitter therefore has three possible outcomes:

crossing tissue completely. This case does not produce any image. In fact, the ultrasound does not generate any echo since they do not encounter any reflecting structure: the tissue is **trans-sonorous** or **anechoic** (example: water).

can be partially reflected and refracted within the tissue. The images then depend on the acoustic impedance of the tissue crossed. If it is heterogeneous, the image represents the different acoustic impedances encountered using shades of grey, giving the tissue examined its own **echo-structure**.

not crossing the medium to which they are applied. They are totally reflected, enabling visualisation of the (**hyperechoic**) outline of the tissue but without providing any information on its content.

Therefore, we can summarise a few elementary rules for ultrasound imaging:

- At the junction of different tissues (interfaces), the waves are refracted and reflected.
- The higher the difference in acoustic impedance between the two tissues, the higher the reflection.
- The higher the difference in velocity of propagation of ultrasound, the higher the refraction.
- The highest frequency waves are more readily absorbed and diffused than those of low frequency. This is why they enable the study only of superficial structures, while use of low frequencies is required to visualise deeper anatomical structures.
- For B mode ultrasonography (Brightness), it is important for ultrasounds to approach the interface studied perpendicularly in order to obtain an optimum echo (maximum reflection from direction of the transducer).

Ultrasound Image (Fig. 1.4)

Transducers (Probes)

Electromechanical transduction enables the generation of ultrasound from electrical energy, and also the translation of the mechanical energy changes, which are changes to acoustic pressure, into an electrical signal. In the case of piezoelectricity, this involves a reversible process enabling the ultrasound transducer to be both the transmitter and receiver of ultrasounds, at the origin of image creation.

Transmission

Currently, PZT ceramic (lead zirconate titanate), composites or polymers have replaced quartz as the piezoelectric material. In practice, by applying an electrical impulse to them,

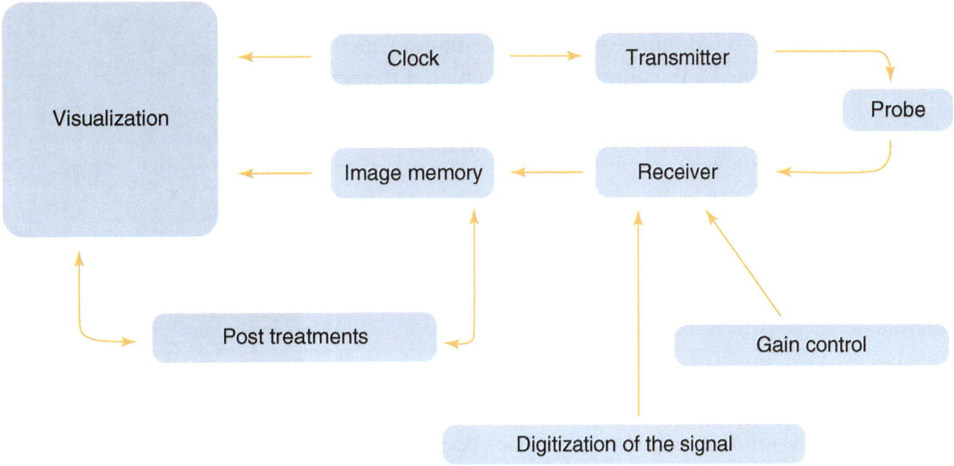

Fig. 1.4 Synoptic diagram of an ultrasound imager. (From Bonin et al. [7])

their vibration generates the production of ultrasounds. Their frequency is directly related to the duration of the electrical impulse. Short impulses produce a wide range of frequencies (a high band width), while long impulses produce nearer frequencies, used in particular to obtain a Doppler effect.

Reception

Ultrasound waves reflected by the tissues that the cross (or echoes) returning to the transducer, are then converted into an electrical signal. The latter, after its computer processing, gives rise to an ultrasound image.

It is necessary to select an appropriate transducer (in terms of frequency and shape/size) for the tissues that one wishes to study.

Principle of Image Formation

The analogue signal generated by receipt of ultrasound is converted into a digital signal. The varying intensity of the digital signal is then used to form a grey scale image on the US monitor: white/light grey pixels coding for stronger echoes, dark grey/black pixels for little or no echo detected. The grey scale image is then optimised by different post-treatment processes (adjustment of contrast, reinforcement of outlines, measurement of distances and surfaces, close-up, etc.).

The intensity of the reflected ultrasound signal is increasingly weaker from greater depths (signal attenuation). Attenuation occurs due to the combined effects of reflection, refraction, diffusion, absorption and degradation. To produce a more homogenous image US, machines compensate for attenuation by increasing the amplification of the return signal with increasing depth (time gain compensation, TGC).

A, B, M, 3D Mode, Tissue Harmonic Imaging, Real-Time Spatial Compound Imaging

A mode (as in Amplitude) is particularly used to measure distance in ophthalmology or in dermatology. **B mode** (as in Brightness) is most commonly used in medical practice. It involves real-time echotomography, since ultrasound data are collected in a plane of section by means of a large number of ultrasound waves sent whose axis is shifted each time. Each point on the screen is related both to time of transmission of the echo and the ultrasound beam axis.

M mode (as in motion) is reserved for cardiac applications, thanks to its one-directional characteristic which enables analysis of tissue motion (cardiac valves).

3D mode: Unlike two-dimensional ultrasonography where structures visualised lie in the plane of section scanned by ultrasound, three-dimensional ultrasonography performs volumetric ultrasound scanning. Real-time volumetric examination is called 4D ultrasonography. Based on summation of 2D planes, specific software makes it possible to obtain 3D imaging, where it is possible to select a plane of section of whatever direction.

Concerning **tissue harmonic imaging,** the tissue under examination is subjected to ultrasound with a "basic" frequency (for example 3.5 MHz) and only the harmonic frequency of the signal returning to the transducer (7 MHz) is used to generate the image [7], which improves image resolution. This is related, on one hand, to the higher ultrasound frequency, and on the other hand, to the fact that the harmonic component crosses the tissue only once (during return to the transducer). Tissue-harmonic imaging is less subject to signal degradation than the other modes. **Real-time spatial compound imaging:** *compound, cross-beam, multi-beam scanning, sonocity*, etc. correspond to technological tools specific to each type of ultrasound, but whose overall principle remains the same: summation of several images generated in the same plane by the ultrasound beam with different angulation, enabling to decrease artefacts significantly. The majority of modern equipment is equipped with this.

Main Artefacts

Acoustic Shadowing

Interfaces with a high reflection factor greatly reflect ultrasounds which “dazzle” the ultrasound scanner. Therefore, energy transmission is weak when it reaches deeper layers, and this causes a defect in information on the tissues in this area: this is the “shadow cone” (Fig. 1.5).

Posterior Enhancement

Fluid filled structures only minimally absorb the ultrasound beam. Due to the lack of beam attenuation (see above) tissues located deep to fluid containing structures have increased echogenicity and so appear brighter (Fig. 1.6).

Multiple Echoes

These echoes are related to multiple reflections between the transducer and a particularly reflective interface. They can also be generated by reverberations inside of a structure with high impedance, such as a needle (Fig. 1.7).

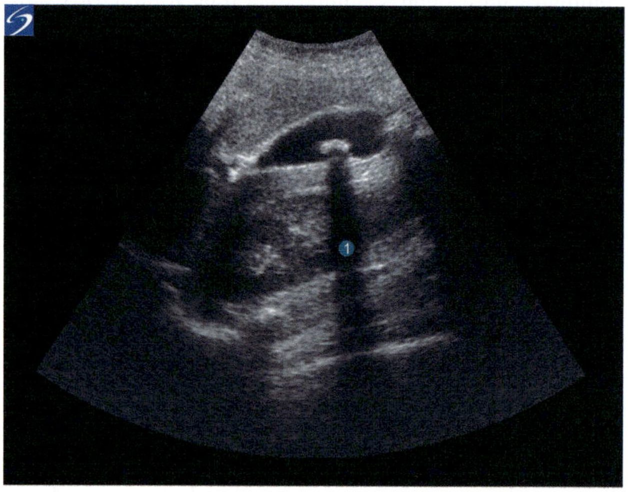

Fig. 1.5 Acoustic shadow (1)

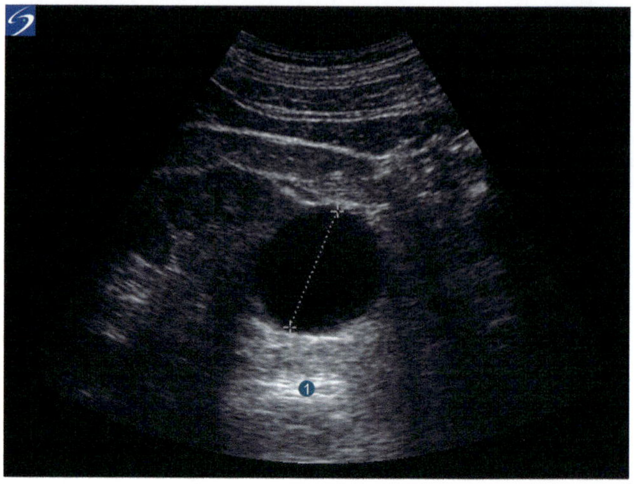

Fig. 1.6 Posterior enhancement (in 1)

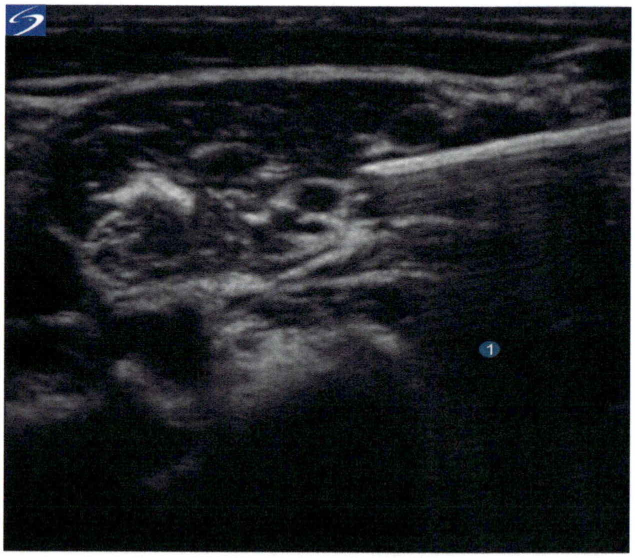

Fig. 1.7 Multiple reflexion artefacts (in 1)

Anisotropy (Figs. 1.8 and 1.9)

The echogenicity (and appearance) of a structure studied may change depending on the angle of the insonating beam. The maximum return echo occurs when the ultrasound beam is produced perpendicular to a structure. Altering the insonating angle may cause a structure to change from being brightly hyperechoic to darkly hypoechoic.

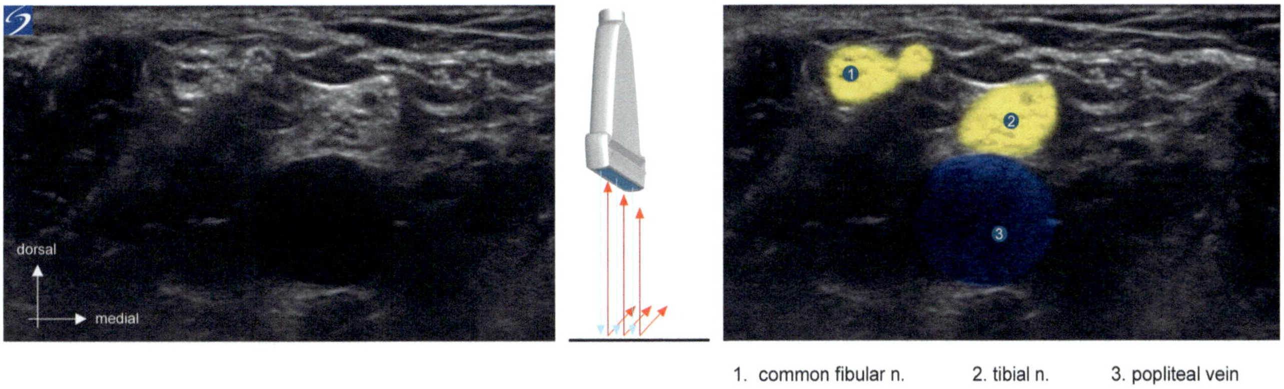

Fig. 1.8 Principle of anisotropy: ultrasound waves approaching their target perpendicularly and generating an optimal image

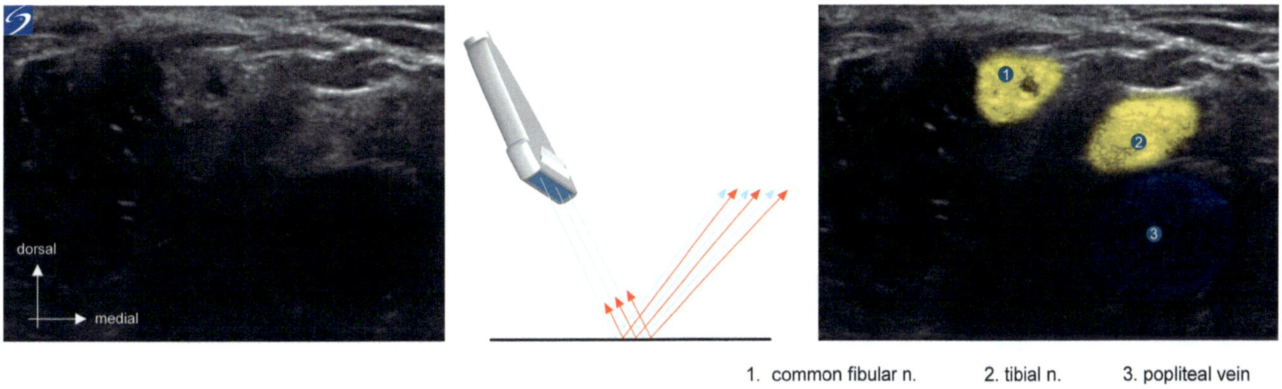

Fig. 1.9 Principle of anisotropy: ultrasound waves approaching their target according to an angle different from the perpendicular are reflected away from the probe; therefore, the image is of lower quality

Ultrasound Velocimetry

Ultrasound velocimetry enables the real-time study of the speed of flow. Primarily applied to blood flow, it is most often based on the Doppler Effect.

Continuous Doppler Ultrasound

The simplest process consists of using a continuous signal. Simultaneous to the continuous transmission of ultrasound with frequency f_i, the "receptor" part of the transducer captures the ultrasound waves reflected whose frequency, $f_i + \delta fn$, and the direction of flow are calculated by the system. The principal disadvantage of continuous Doppler is its inability to differentiate between two blood vessels located at different depths in the same ultrasound beam.

Pulsed Doppler Ultrasound (Fig. 1.10)

Similar to ultrasonography, transmission of a brief impulse by the transducer which immediately afterwards switches to receptor mode enables the selection of echoes in a temporal observation window "a Doppler window" corresponding to a given depth. In this case, the system is able to analyse a flow at a given depth, independently of other signals located along the pathway of the ultrasound of the same bundle, contrary to continuous Doppler.

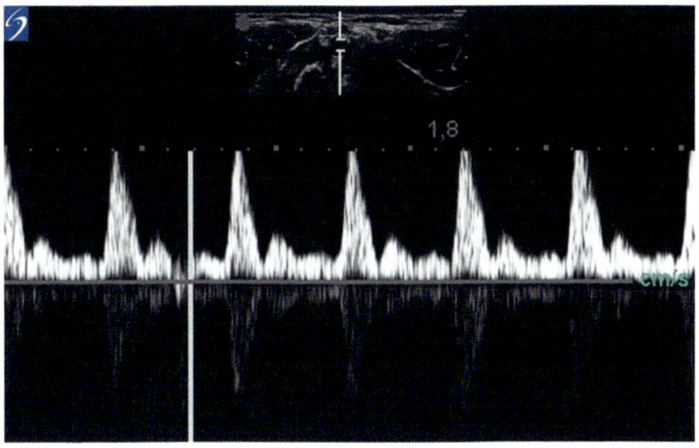

Fig. 1.10 Pulsed Doppler ultrasound waveform

Colour Doppler (Fig. 1.11)

A "morphological" ultrasound image is very well supplemented by "functional" ultrasound velocimetry. The "functional" velocity images are colour-coded and superimposed on "morphological" ultrasound images in shades of grey. Convention dictates that the displacement of fluids in the direction of the transducer should be coded red and that, conversely, displacement away from the transducer be coloured in blue. Furthermore, colour intensity is proportional to the speed measured. A flow strictly perpendicular to ultrasound waves comprises a limit of this function because colour coding is impossible.

Doppler Energy or Angiographic Mode (Fig. 1.12)

For this mode it is the energy of the echoes of mobile targets and not their speed of displacement which is colour-coded. It depends neither on speed of displacement of the targets nor the incidence of ultrasound compared to the direction of the flow. This mode is especially useful in the study of flows perpendicular to the ultrasound.

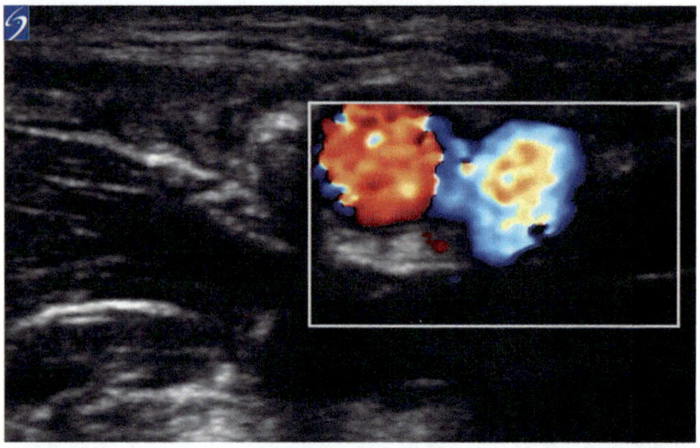

Fig. 1.11 Colour Doppler: femoral artery and vein side by side, whose opposite flows are displayed in red and blue, respectively

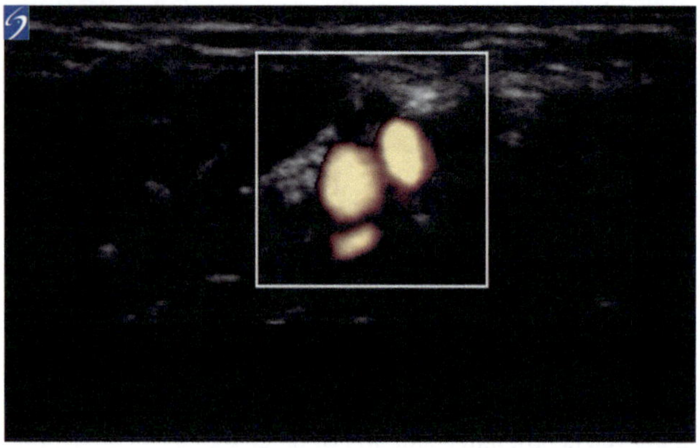

Fig. 1.12 Power Doppler: signals from contiguous vessels whose direction of flow cannot be determined

Biological Effects of Ultrasound

The biological effects of ultrasound are mechanical, chemical and thermal.

At **high** acoustic power (W/cm^2), the mechanisms of attenuation at the origin of biological effects are that of absorption by viscous relaxation (thermogenesis), a process of cavitation which produces mechanical cellular damage and then chemical damage, and direct or indirect oxidation by free radical formation.

For **low** acoustic power, **used in diagnostic imaging** (mW/cm^2), biological effects are represented by reflection/diffusion (formation of the ultrasound image) and absorption by viscous relaxation (thermogenesis). The critical fields of clinical application are obstetrics (imaging, pulsed Doppler, colour Doppler), neurology with intracranial Doppler and perhaps to a certain extent, paediatrics.

Although Doppler investigations use higher acoustic intensities and durations of exposure than ultrasonography, no harmful effect related to prolonged exposure to ultrasound has to date been demonstrated in the field of medical diagnostic imaging. However, it seems wise in principle to withhold some reservations regarding the field of obstetrics. Ultrasound investigation of the eye requires use of transducers with appropriate acoustic power in order to prevent harm to the retina.

The Ultrasonographer and the Quality of the Image Obtained

The quality of the examination greatly depends on the practitioner who is going to perform it. The latter will be confronted with a certain number of technical choices which will be determined based on the type of examination, the patient morphology and the available equipment. There is also a non-negligible human factor in the quality of the images generated: experience and dexterity play a primary role to ensure optimum quality of the examination. Three principal "decisive" choices can be differentiated.

Choice of Transducer

This choice is determined based on the patient's morphological parameters, the structures sought and the equipment available. Two criteria can be differentiated: the frequency of the transmitted beam and the type of wave (wave size, shape of the ultrasound field).

Frequency

The existence of broadband transducers (transmission in a range of frequencies, for example, between 2 and 5 MHz, 6 and 13 MHz) enables, with a single transducer, to assign priority to a certain range of frequencies in order to prioritise high frequency resolution ("RES") or lower frequency penetration ("PEN").

Type of Transducer

Structurally, transducers can be either linear or convex. **Linear** transducers, with high frequency and thus low penetration, are designed for the study of soft tissue, organs and relatively superficial structures (≤5–7 cm); they offer good spatial resolution. **Convex** transducers, with lower frequency, offer less optimum spatial resolution but enable the study of deeper structures; furthermore, the "trapezoidal" ultrasound field provided by the convex transducer enables a wider examination.

Adjustment of Settings

Transmission of the Ultrasound Beam and Gain

The acoustic transmission power corresponds to the energy transmitted to the patient. In order to prevent any biological effect, it must be adjusted to the lowest setting. Usually, this setting cannot be changed by the practitioner in ultrasound systems in general use. The "gain", whose value may be changed, corresponds to amplification of the signal generated by return of echoes to the transducer. Therefore increasing it, which lightens the image until it becomes completely white if done excessively, has no harmful biological effect and can be used without concern.

Adjustment of the Focal Points

Similar to focusing used in optics, focusing the ultrasound beam at a given depth makes it possible to obtain better image quality by means of "concentrating the ultrasound energy". Some ultrasound systems may allow positioning of one or more focal points within the imaging field. Other systems will automatically adjust the range of frequencies and position of the focal point based on whether you choose to work in the "resolution" mode (RES) for superficial study, "penetration" (PEN) for deep investigation, or in the "general" (GEN) for intermediate depths of study.

Reception of the Ultrasound Wave, Dynamic Range

The ultrasound machine attributes different shades of black and white to different levels of sound intensity received. This comprises the "grey scale". In addition to the possibility of adjusting the gain curve according to depth (TGC), adjustment of logarithmic compression changes the range of levels in the grey scale.

Pace of Imaging

This is an indirect adjustment because the pace of images is adapted to the parameters chosen, such as depth, width of sector, close-up, etc. It corresponds to the number of images updates acquired per second.

Choice of Techniques and Mode of Analysis of the Image

Two techniques will influence image quality in terms of shade of grey to supplement the conventional B mode ultrasound: tissue harmonic imaging mode (THI) and real-time spatial composite imaging. They may be combined if required.

How to Read an Ultrasound Image

Control of spatial location is a delicate point, undoubtedly confusing for inexperienced users. This stage precedes and conditions the ease with which the operator can control the method.

It is of fundamental importance to know on screen how the image is oriented.

B mode ultrasound imaging represents the tissue structures of a section whose shape varies depending on the type of transducer used. The thickness of the section obtained is generally approximately a millimetre. However, this varies slightly depending on depth and focalisation of the ultrasound beam.

Superficial and Deep Planes

Once the transducer is placed on the skin, by convention the screen always displays the superficial planes above and the deeper planes below (unless this order has been reversed by using the upper/lower button).

Cephalic, Caudal, Lateral, Medial, Ventral, Dorsal

There is a physical point of reference on one end of the transducer, represented on the screen by a symbol placed on one side of the image (a small coloured circle, the logo of the brand of ultrasound system, etc.). It is advisable with ultrasound-guided RA to place the marker of the transducer in a manner so that the image display is oriented for the ultrasound operator in the same way as the area examined (Fig. 1.13) and not conversely (Fig. 1.14). Consequently, during the procedure, the needle moves on the screen in the same direction as the anaesthetist sees on the patient.

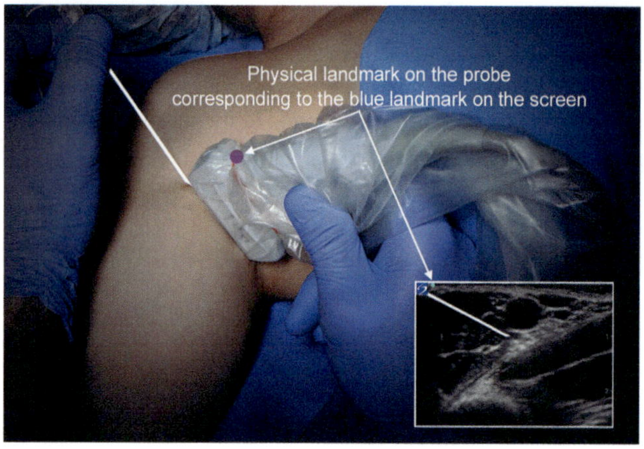

Fig. 1.13 Needle progressing on the screen in the same direction as external vision

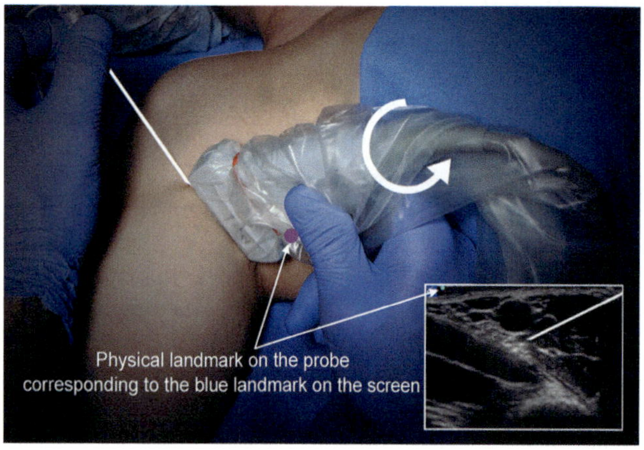

Fig. 1.14 Needle progressing on the screen in the opposite direction of external vision

Radiologists use a convention that consists of placing the most cephalic part of the image on the left side of the screen, but this has not been adopted by cardiologists, who have chosen to place the cephalic part of the image to the right of the screen in longitudinal sections.

Therefore, whatever the type of ultrasound section (transversal, sagittal, oblique, etc.), the image is oriented in the same way as you look at the patient, from the place where one is positioned to perform this procedure.

In total, in one section, we will always have the following:

- on the upper part of the screen, the most superficial structure (skin)
- on the lower part of the screen, the deeper structures
- on the left of the screen, the structures that the practitioner locates to his left in relation to the site of needle puncture
- on the right of the screen, the structures that the practitioner locates to his right in relation to the site of needle puncture

Depth Studied

A centimetric ruler generally indicates, on screen, the depth studied.

Ultrasound Physical Findings (Table 1.2)

Fluids in which there are no particles in suspension and which allow passage of ultrasounds, do not generate any image on screen (black); they are anechoic or trans-sonorous. The existence of particles, blood and/or mucous appears as more or less regular shades of grey.

Interfaces with a high factor of reflection (soft tissue/air, soft tissue/bone) reflect the majority of ultrasound waves (which "dazzle" the ultrasound transducer) and weakly transmit energy from deeper: they are at the origin of a hyperechoic outline, followed by a shadow cone (acoustic shadow).

Soft tissues are more or less echoic: fasciae are represented in the shape of "lines" (more or less hyperechoic) and muscles are overall hypoechoic, more or less heterogeneous.

In the **nerve roots**, the **nerves** are hypoechoic, surrounded by a clear border corresponding to the epineurium, to the perineurium and/or to the mesoneurium (a **"single bundle"** presentation). Further in the periphery, they take on a more heterogeneous presentation, like a "honeycomb", a "cluster of grapes", or also called a "follicular" presentation.

Table 1.2 Characteristics of structures encountered during ultrasound-guided peripheral nerve blocks

Structure	Ultrasonography	Doppler
Nerve	Hypoechoic, homogeneous in the nerve root, heterogeneous in the nerve trunk	No Doppler effect
Muscle	Hypoechoic, heterogeneous	
Tendon	Hyperechoic, ± homogenous	
Fascia	Hyperechoic, homogenous	
Fat	Hypoechoic, heterogeneous	
Bone	Very hyperechoic surface, with underlying acoustic shadow	
Artery	Echoic wall, anechoic content, little compressible, pulsatile	Measurable Doppler effect, colour coding
Vein	Echoic wall, anechoic content, compressible, calibre which can vary with ventilation	
Needle	Hyperechoic, but variable according to angulation in relation to the ultrasound beam (the more perpendicular the needle is to the ultrasound, the more it is hyperechoic, with "reverberation artefact")	No Doppler effect
Catheter	Some catheters are more echoic (often if there is a guide wire)	
Local anaesthetic	Anechoic	Doppler effect during injection

References

1. Dussik KT. On the possibility of using ultrasound waves as a diagnostic aid. Neurol Psychiatry. 1942;174:153–68.
2. Hueter TF, Bolt RH, Ballantine HT Jr. On the detection of brain tumors by ultrasonics. J Acoust Soc Am. 1950;22:686–7.
3. Dénier A. Les ultrasons, leur application au diagnostic. Presse Med. 1946;22:307–8.
4. Rathmell JP. Imaging in regional anesthesia and pain medicine: we have much to learn. Reg Anesth Pain Med. 2002;3:240–1.
5. La Grange P, Foster P, Pretorious L. Application of the Doppler ultrasound blood flow detector in supraclavicular brachial plexus block. Br J Anaesth. 1978;50:965–7.
6. Buddemeyer EU. The physics of diagnostic ultrasound. Radiol Clin N Am. 1975;13:391–402.
7. Bonnin A, Broussouloux C, Convard JP, Legmann P, Seguin G, Blery M. Échographie. Paris: Masson; 2004.

Ultrasound in Regional Anaesthesia: Why and When?

2

Eryk Eisenberg and Paul-Joseph Zetlaoui

Contents

Whatever the era or the technique, regional anaesthesia (RA) has always been an exercise in applied anatomy. Knowing where a nerve is, knowing how to approach it without risk to the adjacent organs, knowing how to inject the local anaesthetic close enough to the nerve without touching it nor injecting into a blood vessel is always a delicate exercise, which was performed either totally blindly (by eliciting paraesthesiae), or with the kind of *blind's white stick* which is the neurostimulator. By providing direct anatomical vision, ultrasonography has made it possible to simplify the conduct of RA. The rapid adoption and popularity of ultrasonography in anaesthesia is not merely a passing fashion but because it offers two main benefits: that of **quality** and of **safety**.

E. Eisenberg (✉)
Département d'anesthésie et réanimation, Clinique Pôle Santé République, Clermont-Ferrand, France

P.-J. Zetlaoui
Département d'anesthésie-réanimation, Hôpital de Bicêtre, Le Kremlin-Bicêtre, France
e-mail: paul.zetlaoui@bct.aphp.fr

Limits of Neurostimulation

Basis of Neurostimulation

Neurostimulation was developed to compensate for the limitations of landmark-based techniques (lack of precision and nerve injury) and had enabled regional anaesthesia to pass from era of raw artistry to that of skilled work. Neurostimulation used by any anaesthetist, subject to specific training, enabled the majority of nerve blocks to be per-

E. Eisenberg, E. Gaertner (eds.), *Ultrasound in Peripheral, Neuraxial and Perineuraxial Regional Anaesthesia*, https://doi.org/10.1007/978-3-031-08804-9_2

formed successfully. Combining knowledge of surgical anatomy with neurophysiology, anaesthetists developed a deeper understanding about the site of needle insertion and the induced motor and sensory responses which resulted in successful RA.

The development of neurostimulation and of operating algorithms has made it possible to standardise the practice of RA [1]. However, some problems remain when locating nerves, sometimes related to the neurostimulator itself [2].

We used to operate with the belief that MIS (minimal intensity simulation) prevented direct contact between the nerve and the needle, and therefore reduced the possibility of intraneural injection. However, we did not answer the preliminary question of knowing if there is a limit to the level of neurostimulation intensity that can predict the position of the tip of the needle within the nerve, in contact with the nerve or at a distance from the nerve. An experimental study by Tsai et al. showed that minimal stimulation intensity cannot provide information on the exact position of the needle bevel. Only a response obtained for stimulation intensity below 0.2 mA per 100 μs would translate the intraneural position of the needle. Above this intensity, no value predicts the position of the needle (within the nerve, in contact with the nerve or 1 mm from the nerve) (Fig. 2.1) [3]. Therefore, the myth according to which neurostimulation, even correctly performed according to appropriate algorithms, can prevent intraneural injection must be questioned. This first argument alone justifies ultrasonography for RA.

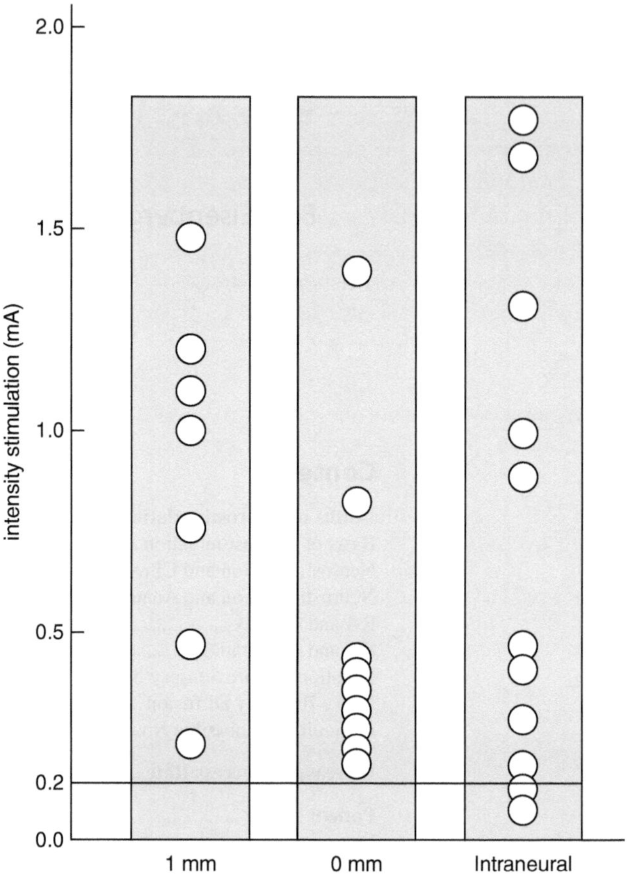

Fig. 2.1 Relation between minimal stimulus intensity and motor response based on position of the tip of the needle in relation to the nerve (at 1 mm, in direct contact or intraneural). (Redrawn from Tsai et al. [3])

Neurostimulation and Ultrasonography

Fundamentally, ultrasound guidance and neurostimulation are based on different concepts. The dogmas on which neurostimulation is based are the identification of skin or bony landmarks. From this, it is necessary to determine the site of the needle insertion, the direction and the angle of needle puncture, to consider the depth of the nerve, to obtain an expected motor/sensory response, to find the MIS to prevent intraneural injection or injecting too far from the nerve, and to inject the LA. Thereafter, to wait for an appropriate time for the block to establish to confirm that the procedure was correctly performed overall. In clinical practice, Dufour et al. showed that during a median nerve block at the elbow, performed via neurostimulation by experienced anaesthetists who had searched with MIS, an intraneural injection (then defined as an increase of over 75% in the diameter of the nerve) was found in 43% of cases [4]. Therefore, neurostimulation, even properly performed, can lead to almost 50% of intraneural injections.

In ultrasonography, we look for a nerve that we wish to block by exploring the area in which it is usually located. Whatever the actual position of the nerve (at times very different from the theoretical position), it will eventually be located. Subsequently, depending on local anatomy, a needle puncture site is determined so that the skin-nerve pathway avoids a "vulnerable" structure. The puncture site and direction of the needle can be very different from what they would have been with neurostimulation. When visualisation is *optimum*, it is generally possible to ascertain that the bevel of the needle is in fact next to the target nerve and not in intraneural position. The injection then is administered under direct vision. If visualisation is *optimum*, an intraneural injection will be detected with the first millilitre, and the needle can then be repositioned. The homogenous spread of the local anaesthetic around the nerve ensures success of the nerve block in almost 100% of cases. As the needle is removed, there is no longer much doubt regarding the block's success.

Ultrasonography has demonstrated the limits of neurostimulation. Ultrasound guidance provides information about the position of the tip of the needle in relation to the nerve; but is it useful to *always* use neurostimulation and ultrasound guidance in combination? Probably not, but insofar as conditions of *optimum* visualisation are not *always* encountered, the combination of neurostimulation and of ultrasound guidance retains a **logical** place at least in situations of imperfect visualisation (which is very often the case). And if it is likely that ultrasonography improves the performance of neurostimulation [5], then it is necessary to refer to new algorithms which combine neurostimulation and ultrasound guidance; the one presented by Jochum et al. can be recommended (Fig. 2.2) [6]. The other question is whether neurostimulation improves the performance of ultrasound. The study by Perlas et al.—which shows that when the needle is in contact with the nerve in ultrasonography, paraesthesia is triggered in only 38% of cases and there is a neurostimulation response in only 74.5% of cases [7]—was not conceived for this. It discredits some of the dogmas of neurostimulation but not the utility of some of its aspects. Dingemans et al. even develop the concept that the combination of neurostimulation and ultrasound guidance may be less productive than ultrasound guidance alone [8]. In this study, ultrasound guidance alone at the 30th minute achieved 86% complete nerve blocks in 4 principal nerves of the forearm, while the combination of ultrasound guidance and neurostimulation only led to 57% complete nerve blocks. But this study again considers neurostimulation from the dogmatic standpoint, not mentioning the "archaic" aspect, which relates to MIS, the distance between the needle bevel and the contour of the nerve, and the assumed efficacy of the injection. In this case too, these results do not provide any reason to conclude a lack of utility of the combination of ultrasonography and neurostimulation.

The question which we must answer then, if we have an ultrasound system, is to determine if it is still possible to perform nerve blocks with neurostimulation alone. The recent editorial by Delaunay et al. clearly answers this question by raising another question: "Can we accept to continue to perform a procedure for which we do not control all of the effects even though an alternative exists?" [9]. The answer is probably no.

Neurostimulation and Anatomical Variations

Neurostimulation assumes that the location of the nerve can be predicted based on surface landmarks with low variability. The technique of sciatic nerve block in the buttock described by Labat, which had been improved by the contribution of Winnie, perfectly illustrates this principle, since these landmarks are sufficient in the majority of cases. Yet these techniques are not infallible. Furthermore, when a motor response is obtained, it is sometimes difficult to interpret, such as in the presence of the Martin and Grüber anastomosis[1] which is a traditional pitfall in neurostimulation for the uninformed anaesthetist.

Variations in location of the musculocutaneous nerve, or even its absence as a separate nerve, may complicate the task of locating it with neurostimulation in the axillary area [10, 11]. Ultrasound mapping before needle puncture makes it possible to specifically locate the usual or unusual position of the musculocutaneous nerve (Fig. 2.3) and its relations to the median nerve, and therefore to choose the best needle puncture technique and the best site for injection of a local anaesthetic.

Locating the median nerve in the wrist, by obtaining flexion of the fingers with neurostimulation during needle puncture in the depression between the two palmar tendons, is usually easy. But the median nerve may divide into two branches in the forearm, one of which (the palmar branch) may be located outside of the intertendinous space. Therefore, approaching the nerve using traditional markers can result in an incomplete nerve block in about 10% of cases. Furthermore, the median artery, an embryological remnant, may complicate the approach to the median nerve in the wrist. The existence of a persistent median artery is usually associated with a high division of the median nerve [12]. Therefore, locating the median artery in the wrist with ultrasound (Fig. 2.4) requires mapping of the median nerve in the forearm to look for an early separation of the palmar branch of the nerve, thus justifying a more proximal approach to the nerve in the forearm, or even the distal part of the arm. This relatively common variation (unilateral in 25% of cases, bilateral in 6% of cases) was never considered in nerve blocks with neurostimulation [13]. Also, the presence of a median artery requires changing the technique to avoid puncturing it.

These numerous anatomical variations, which are often responsible for repeated needle puncture, failure of nerve block and sometimes complications, are some of the major points when answering the question: "Why use ultrasound in RA?"

RA and Obesity

Obesity is a common cause of failure of nerve block, because of the difficulty in locating the cutaneous landmarks, the limitations of neurostimulation, and of the increased distance of the nerves from the skin surface [14]. In an obese subject, the risk of failure of RA is higher than that in a non-obese subject [15]. Performing a sciatic nerve block or a single axillary nerve block may be problematic solely using neurostimulation. Simply locating the axillary artery can be an achievement. The increase in incidence of obesity in the population will increasingly justify the use of ultrasonography, which is a valuable aid for nerve blocks that in principle are simple [16].

[1] Martin and Grüber anastomosis: anastomosis between the median and ulnar nerves in the proximal part of the forearm.

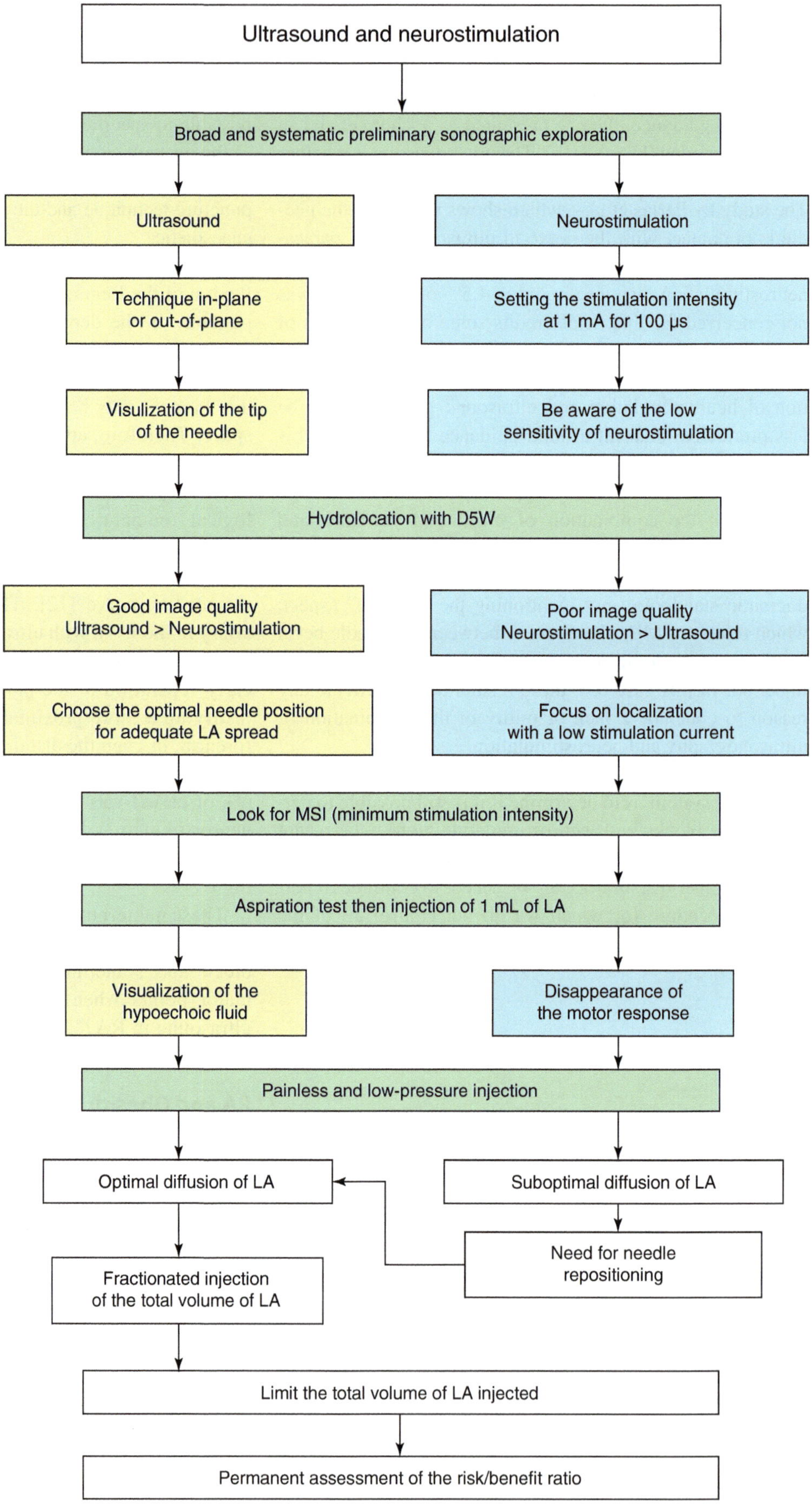

Fig. 2.2 Algorithm for dual neurostimulation and ultrasound guidance (from [6]). *LA* local anaesthesia; *D5W* dextrose 5% solution in water

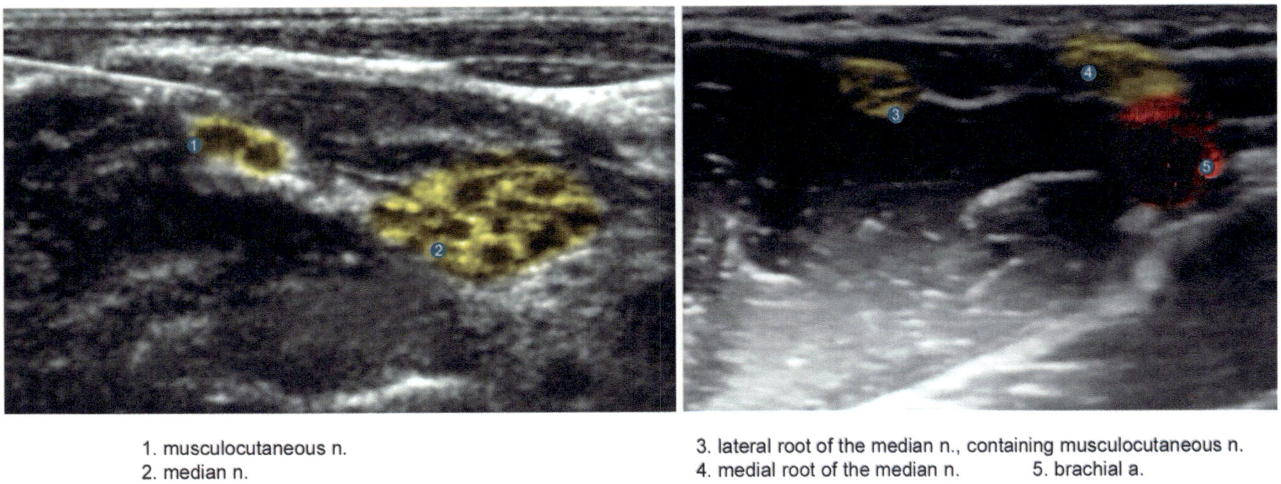

Fig. 2.3 Two different positions of the musculocutaneous nerve in the axillary area. Left side: The musculocutaneous nerve is in continuity with the median nerve. Neurostimulation can be puzzling, as the two responses cannot be separated. With ultrasound, a single injection is necessary to block both nerves. Right side: The musculocutaneous nerve is almost in a subcutaneous position, very distant from the brachial plexus in the axillary area. (Iconography: Paul Joseph Zetlaoui. From [10])

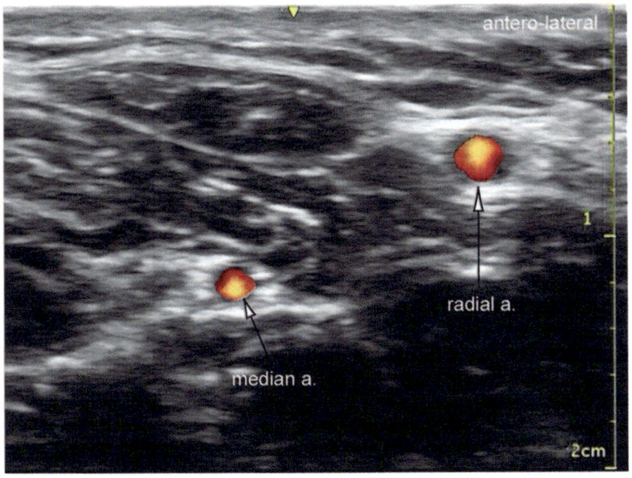

Fig. 2.4 Median artery. Ultrasound image of the median nerve in the distal part of the forearm showing the existence of a median artery (embryonic remainder) in the median nerve. The existence of a median artery is usually associated with early take-off of the palmar branch of the median nerve, which is located laterally to the flexor carpi radialis tendon. This anatomical configuration can be responsible for an incomplete block in a distal approach to the median nerve. (Iconography: Paul Joseph Zetlaoui)

RA and Amputation

In subjects who have suffered amputation of a limb, neurostimulation may not be feasible. However, such patients can often undergo procedures on the stump of the amputated limb. In these cases, in the absence of traditional landmarks and without the possibility of a response with neurostimulation, ultrasound can enable the injection of LA and can do so with very high success rate [17, 18].

Location of Pure Sensory Nerves

Neurostimulation locates the motor nerves; strategies have been devised in an attempt to locate the purely sensory nerves, with no motor response, by looking for paraesthesia synchronous with neurostimulation performed with high current flow rates. Ultrasound makes it possible to resolve these problems related to pure sensory nerves. The saphenous nerve is easily located along the thigh, near the sartorius muscle [19]. Similarly, the cutaneous branch of the radial nerve, conventionally blocked by ring infiltration around the cephalic vein at the distal end of the radius, is easy to locate (with an 18 MHz transducer) and to block using ultrasound (Fig. 2.5).

Block of the greater occipital nerve has particularly benefited from the use of ultrasound. A small and purely sensory nerve is anaesthetised in some cases of chronic headache [cervicogenic headache or headache due to hypotension of the cerebral spinal fluid (CSF)], and for anaesthesia-analgesia in surgery of the scalp. Its approach was carried out solely by blind infiltration; ultrasound has facilitated and popularised this technique, leading to improved management of patients [20].

Nerve Block by Diffusion

Nerve block by diffusion [fascia iliaca block, TAP block (transversus abdominis plane), etc.] traditionally has been described as a "blind" technique—there is no nerve that can be stimulated at the site of needle puncture—and is based on the tactile sensation of penetrating fascial layers. Although these techniques were reported to be effective in

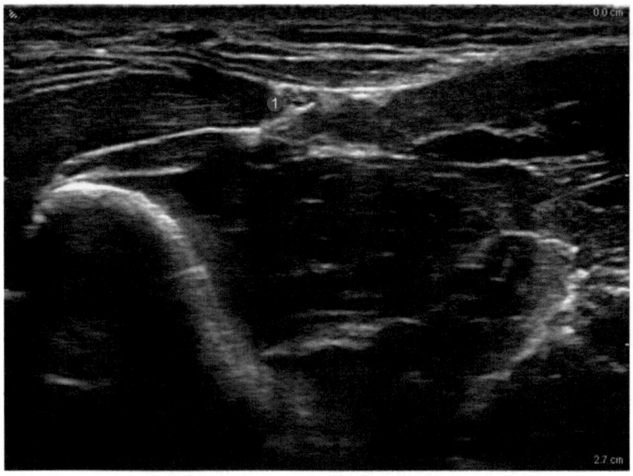

1. superficial sensory cutaneous branch of the radial n.

Fig. 2.5 Visualisation of the sensory cutaneous branch of the radial nerve in the distal 1/3 of the forearm

children, their results in adults were more random. However, even in children, ultrasound studies showed that performing a blind nerve block based solely on the technique of crossing fascial planes was not very effective. Weintraud et al. showed that when the blind technique was performed in the ilio-inguinal and ilio-hypogastric nerve block in children by experienced paediatric anaesthetists, the injection was performed in the proper anatomical plane in only 10% of cases, with effective analgesia (by diffusion spread) in less than 50% of cases [21]. Conversely, Willschke et al., in the same context, demonstrated that ultrasound guidance made it possible to quarter the dose of local anaesthetic necessary for an effective block by improving the precise location of the injection site [22].

Similarly, in adults, Dolan et al. showed that ultrasound made it possible to double the accuracy of a fascia iliaca nerve block, with the incidence of correctly located injection rising from 50 to 95% between blind needle puncture and ultrasound-guided needle puncture [23].

Therefore, for so-called diffusion nerve blocks, ultrasound considerably improves the accuracy of the injection site and therefore the **quality** of the block, and makes it possible to reduce the necessary LA dose, thus improving patient **safety.**

Difficult or Impossible Neurostimulation

Certain disorders that cause neuropathies have also raised problems for anaesthetists in peripheral nerve block. Diabetes is probably the most common of these clinical cases where neurostimulation loses its reliability more. Sites had demonstrated the utility of ultrasound guidance for a popliteal nerve block in a diabetic patient [24], and Minville et al. had shown for popliteal catheter insertion the utility of ultrasound in a patient with severe peripheral neuropathy in combination with arterial disease, in whom neurostimulation conducted according to conventional criteria was unable to provide a satisfactory location of the nerve. Thus, ultrasound becomes the only alternative to guide the tip of the needle as close as possible to the nerve and to allow the spread of the local anaesthetic in contact with the nerve. In the clinical case of Minville et al., even though the needle was in contact with the nerve, the minimum intensity of stimulation to obtain a motor response was 3.2 mA (for 100 μs), thus confirming the limits of neurostimulation [25].

Unlike neurostimulation, which requires movement in response to nerve stimulation, ultrasound guidance can achieve the same results without the need for a motor response. This difference can be appreciated in trauma cases when neurostimulation causes painful movement at the site of a fracture. Moreover, a plaster splint or a voluminous wound dressing can mask the most subtle motor responses, increasing the risk of failure of the nerve block. Therefore, in traumatology or in emergency medicine, ultrasound guidance has a natural indication to improve the overall quality of RA.

Ultrasonography can prove useful when it is necessary to manage failure of anterior primary block. This involves clinical cases reporting failure of nerve blocks with neurostimulation, where ultrasound makes it possible to locate the nerve or the plexus targeted and to perform an effective "rescue" nerve block [25, 26].

Increase in Success Rate

The aim of ultrasound guidance is to improve the success rate of RA. Starting in 1997, Marhofer et al. reported that ultrasound improves the quality of femoral nerve block compared to neurostimulation [27]. In the same study, the authors also reported that the onset time of the block is short (16 vs. 27 min) and that accidental vascular puncture was found only in the "neurostimulation" group. Since then, almost everything had been said; but it remained to be confirmed—which has now been done in many other studies. In 2009, a systematic review by Abrahams et al., based on 13 controlled clinical trials, confirmed the advantage of ultrasound in reducing the number of failures in RA, decreasing the time of conduct of a nerve block, decreasing the onset time of a block (−29%), prolonging the duration of the nerve block (+25%), and in decreasing the risk of vascular needle puncture [28]. These results have also been confirmed in children [29]. In a more recent review in 2011, Antonakakis et al. challenged these results showing that in 17 randomised studies, only 4 studies revealed a significant increase in the quality of the nerve block when using ultrasound, while the 13

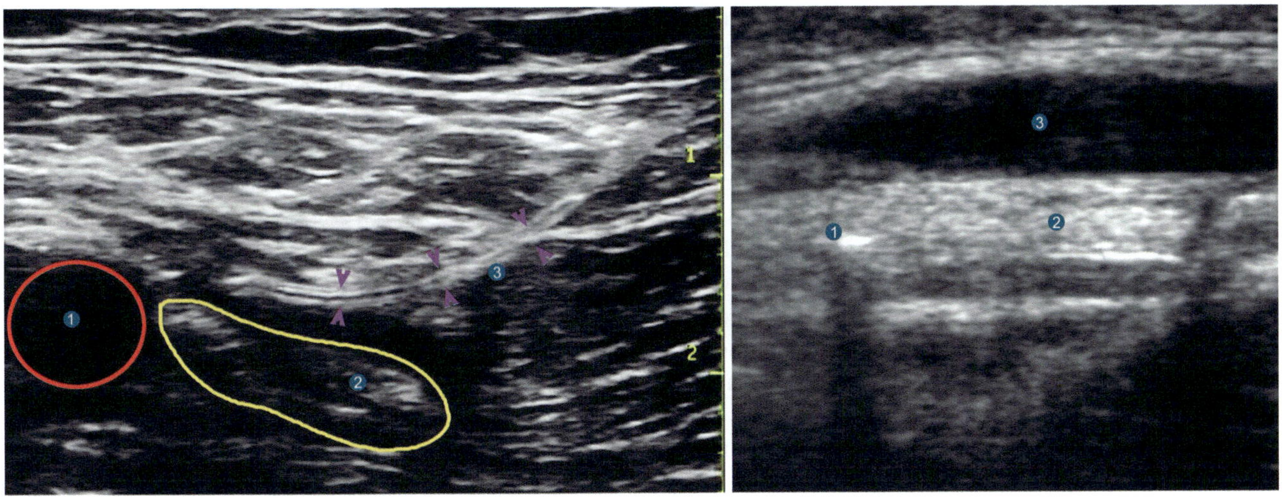

1. femoral a. 2. femoral n. branches 3. catheter 1. catheter tip 2. radial n. 3. branchial vein

Fig. 2.6 Visualisation of catheters with ultrasound imaging. On the left, standard catheter with no tip marker. Though the catheter is clearly visible, it is impossible to determine precisely the position of its tip. On the right, a catheter containing with a distal metallic marker (stimulating electrode); the catheter is not easily visualised but its tip is clearly identified. (Iconography: Paul Joseph Zetlaoui)

other studies did not find a difference. However, none of these 13 studies reported the superiority of other techniques for location of structures compared to ultrasound [30]. Lastly, the authors mention the difficulty of defining a "successful nerve block" or the "quality of a block" to explain these results. This increase in success rate can be explained by several factors. Although the primary factor is better positioning of the needle closer to the nerve, other factors, such as improvement in taking into account anatomical variations, particular aspects of patients or limits of neurostimulation, should also be considered.

The insertion of perineural catheters can be performed using three methods: neurostimulation with a needle followed by blind insertion of a catheter, use of a stimulating catheter or ultrasound guidance. Incidentally, these methods may also be combined. Studies comparing ultrasound guidance to other methods always show the superiority of ultrasound whether in terms of time of conduct of the procedure, pain related to it or puncture of a blood vessel. It is possible that a larger proportion of catheters can be placed properly with ultrasound [31]. Some studies have demonstrated better quality of analgesia when a catheter has been inserted using ultrasound [32]. However, when catheters are in the correct position, there does not appear to be any superiority or inferiority of ultrasound with respect to other techniques in terms of quality of analgesia obtained; then again, ultrasound is often associated with a decrease in the need for local anaesthetic or morphine. Ultrasound also makes it possible to locate the final position of the end of the catheter either by direct visualisation (Figs. 2.6 and 2.7)—all the more so since some catheters are fitted with a guide wire which makes it possible to locate them—or by the Doppler method [33]. Overall, ultrasound facilitates and improves the insertion of

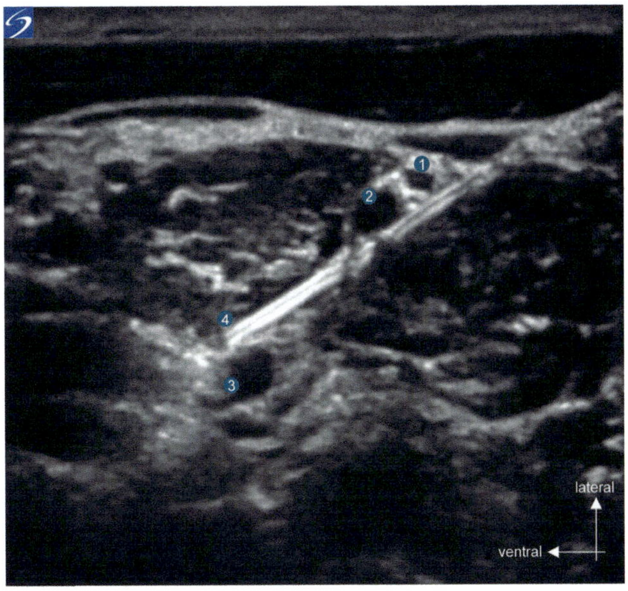

1. C5 n. root 2. C6 n. root 3. C7 n. root 4. catheter tip

Fig. 2.7 Interscalene catheter whose technical characteristics enable excellent ultrasound visualisation along its entire length

perineural catheters, and with ultrasound guidance it no longer seems necessary to resort to using stimulating catheters.

Patient Safety

Decrease of Local Anaesthetic Doses

One of the most unexpected benefits of ultrasound guidance in RA has been a decrease in the doses of LA necessary to obtain an effective nerve block. Following a seminal study by Willschke et al. showing that doses of local anaesthetic could be reduced drastically (−75%) to successfully perform ilio-inguinal and ilio-hypogastric nerve block in children [22], many studies have reported similar results. They have investigated the minimal effective volume necessary to obtain a nerve block with ultrasound guidance (MEV: 50 or 95). A study by Falcão et al. showed that an effective ultrasound-guided interscalene block is obtained with 0.95 mL of 5 mg/mL bupivacaine with adrenaline and that effective postoperative analgesia is obtained with 2.3–4.3 mL, with low risk of diaphragmatic paresis [34], which was confirmed by Gautier et al. who performed an effective block with 5 mL of ropivacaine [35]. For an axillary block, O'Donnell et al. showed that a volume of 1 mL per nerve (20 mg/mL lidocaine with adrenaline) was sufficient for a 180-min surgical block [36]. For the femoral nerve, Casati et al. reported that the MEV_{50} is 15 mL with ultrasound and 26 mL with neurostimulation. For MEV_{95}, the volumes are 22 and 41 mL, respectively [37]. Therefore, for almost all nerves, it is possible to show that ultrasound guidance enables a halving of the necessary volumes of local anaesthetic. In the literature, only nerve blocks around the clavicle appear to fail to demonstrate this phenomenon of dose reduction. For supraclavicular or infraclavicular approaches to the brachial plexus, MEV_{95} studies have listed volumes ranging between 30 and 40 mL, with no noteworthy difference between ultrasound-guided and nerve-stimulated techniques [38]. These results are a little surprising because in light of the especially favourable anatomy (i.e. the area where all components of the brachial plexus are grouped most closely), it seems logical to expect a decrease in the volumes of anaesthetic required.

Observed again on the interscalene level, however, this possible decrease in the dose injected has made it possible to reduce the incidence and extent of diaphragmatic dysfunction [34, 35].

Vascular Injection

Ultrasound significantly reduces the incidence of accidental intravascular injection in RA. This had been reported by Marhofer et al. in 1997 [27]. Other studies have confirmed this benefit [39–41]. This decrease in risk of vascular needle puncture has made it possible to perform nerve blocks in patients under anticoagulant treatment or who have coagulation disorders. Bigeleisen reported an infraclavicular block without complication in a patient under anticoagulant treatment [42]. In haemophiliac patients, the conduct of single superficial nerve blocks and catheter insertions are procedures that are performed regularly by trained practitioners. They also reflect our everyday activity. Dufour et al. showed the importance of ultrasound for conduct of nerve blocks of the upper limb in patients with an arterial venous fistula [43]. In these specific cases, ultrasound makes it possible to improve the safety of patients or to eliminate the need to exceed safe LA dose limits when the benefit-to-risk ratio appears favourable.

Decrease in Adverse Effects of Acute Systemic Toxicity

It should not be overlooked that the risk of needle puncture and intravascular injection continues to exist in spite of the use of ultrasound. There are several clinical cases which remind us of this fact, underlining that we should adhere to all safety precautions [44]. Visualisation of the wider needle puncture area with low pressure exerted on the probe (so as not to collapse blood vessels) and use of colour Doppler ultrasound make it possible to look for blood vessels and to evaluate the risk of vascular puncture and of intravascular injection. Visualisation of the needle tip and of the volume injected in the form of a hypoechoic area is a fundamental safety feature [45]. The decrease in the incidence of unidentified vascular puncture has led to a decrease in adverse effects from acute systemic toxicity. Orebaugh et al. [5], in a single centre study of over 14,000 nerve blocks, showed that the incidence of adverse effects from acute systemic toxicity is significantly higher with neurostimulation alone than with the combination ultrasound plus neurostimulation (6/5436 vs. 0/9069, $p = 0.0061$).

Delayed onset adverse effects from drug toxicity, secondary to systemic absorption of high doses of local anaesthetic are also probably less common thanks to the major reduction in the total doses injected, made possible by ultrasound. However, there does not appear to be any study reporting this effect.

Pneumothorax and Needle Puncture of an Adjacent Organ

Pneumothorax is a serious complication of infraclavicular, supraclavicular and paravertebral nerve blocks. Moreover, the latter had practically disappeared from contemporary practice for this reason. Ultrasound, by providing the possibility of visualising the pleura, has made it possible to dramatically reduce the incidence of this complication. Perlas et al., who reported a series of 500 supraclavicular blocks with no pneumothorax, promote the resurgence in use of this technique [46]. However, it should be kept in mind that this risk is still not zero as reported by a few clinical cases, and that pneumothorax occurring following an ultrasound-guided nerve block suggests existence of a technical problem in conduct of the block (imperfect technique, poor choice of the probe, needle or approach) [47].

Similarly, it can be assumed that the incidence of peritoneal or visceral needle puncture during a TAP block has considerably decreased since the systematic use of ultrasound.

Locating the Spine

Ultrasound location for neuraxial block is an important chapter in development of ultrasound in RA, where the limits of neurostimulation are no longer called into question. Firstly, ultrasound has made it possible to specifically locate the level of needle puncture with more accuracy than palpation of bony structures [48]. Secondly, the utility of ultrasound when identifying anatomical structures for epidural anaesthesia has been demonstrated. Ultrasonography makes it possible to locate the skin-epidural space distance with a high degree of accuracy. In 2010, in a review of the literature by Perlas [49], these last two parameters (identification of anatomical structures and assessment of skin-epidural space distance) were classified as IIb and Ia, respectively, with regard to their level of evidence; in patients with an anatomy considered to be normal, the ultrasound location allows for a decrease in the number of needle punctures, the number of changes in space and number of redirections of the needle to find the epidural space (level of evidence Ia), but with no significant improvement in anaesthetic efficacy.

Three years later, in 2013, Shaikh et al. showed different results, more in favour of ultrasound [50]. This meta-analysis, combining 14 randomised studies including 1334 patients (674 ultrasound examinations vs. 660 controls), showed that ultrasound improves the overall success rate (decrease in overall failure rate divided by 7). It also enables a decrease in the incidence of traumatic needle puncture, in the total number of needle punctures and in the number of redirections of the needle.

Based on these two reviews of the literature, it appears that ultrasound improves the conduct of spinal anaesthesia by decreasing the number of failures, the number of traumatic needle punctures, the number of needle punctures and of redirections of the needle, as well as the number of improperly inserted catheters.

In spinal anaesthesia, obesity may pose a real problem for the anaesthetist who in some cases is unable to locate the midline. Locating the interspinous space is sometimes illusory, and evaluating the skin to ligamentum flavum distance is often impossible. In such cases, ultrasound location or ultrasound guidance can prove to be very useful to locate the midline, the interspinous space and the distance between skin and the ligamentum flavum/dura mater complex. Balki et al. demonstrated in obese parturients the utility of ultrasound by showing a good correlation between distance measured with ultrasound and real distance between skin and the epidural space [51]. Furthermore, ultrasound made it possible to specifically locate landmarks for needle puncture.

In patients who have already undergone spinal surgery, ultrasound can be an appreciable aid in performing spinal anaesthesia [52].

In children, spinal ultrasound makes it possible to obtain especially remarkable images [53]. Performed by experienced anaesthetists, ultrasound (ultrasound located or ultrasound guidance) makes it possible to decrease the number of contacts with bone (Ib) or to reduce time for conduct of an epidural injection (Ib). Furthermore, changes to the diameter of the epidural space related to the volume injected may make it possible to predict extension of nerve block [54]. Again in children, colour flow Doppler ultrasonography can distinguish a caudal injection from an intrathecal injection, which is detected by a coloured *panache* in the CSF [55].

Therefore, we should clearly ask ourselves whether ultrasound is going to become the standard method for locating structures for spinal anaesthesia.

Teaching

Two different questions are combined in this title: "How to teach RA today?" and "How to teach ultrasound in RA?" It appears obvious we can no longer teach RA without teaching ultrasound guidance. During the era of neurostimulation, it was necessary to teach the landmarks for conduct of needle puncture, the sites of needle puncture and the desired motor/sensory responses for each nerve, at each characteristic site of puncture. Other factors are necessary to learn RA with ultrasound. Common recommendations from the American Society of Regional Anaesthesia (ASRA) and the European Society of Regional Anaesthesia (ESRA) offer a range of ten items for teaching [56]:

1. Visualise major landmarks, including blood vessels, fascia and bone structures.
2. Knowing how to identify nerves or plexuses in the short axis view.
3. Knowing normal anatomy and knowing how to identify a variation.
4. Planning a strategy for needle puncture to avoid needless tissue trauma.
5. Maintaining aseptic methods throughout the procedure, in particular with ultrasound system and the probe.
6. Following the needle in real time during its handling (or movement).
7. Using ultrasound in combination with an alternative technique, such as neurostimulation.
8. When the needle is assumed to be in place, injecting a small volume as a test dose. If this test dose is not visu-

alised, it should be considered that the tip of the needle is intravascular and outside of the field of the ultrasound.
9. Re-position the needle, if spread of the solution appears inadequate. Visualisation of the solution injected should be maintained throughout the duration of the injection to avoid an intravascular injection.
10. Complying with all usual safety recommendations (resuscitation equipment, repeated aspirations, test dose, usual monitoring, etc.).

In addition to these recommendations, other aspects of good practice must also be reinforced. The study by Sites et al. showed which precise items this practical training should focus on to improve performance of interns [57]. The "set of 7 errors" describes the following errors:

- absence of visualisation of the needle during its progression
- involuntary movements of the transducer
- inability to identify inadequate spread of the local anaesthetic
- inability to recognise an intramuscular injection
- fatigue
- an error in right to left or upper to lower orientation
- choice of an inappropriate needle puncture site based on position of the probe

Grau et al. had shown that interns inexperienced with RA acquire proficiency twice as fast if they had learned to identify anatomical structures, including the "ligamentum flavum/dura mater complex" under ultrasound, before starting to perform epidural anaesthesia [58]. Such published reports should help us propose a model for teaching ultrasound-guided RA in order to advance the safety and reliability of RA [58–60].

A Conclusion for the Future: 3D or 4D Ultrasound?

Ultrasonography has not solved all problems related to RA. A series of over 12,000 nerve blocks presented by Sites et al. showed that there are still a few cases of intravascular injections (0.18%), a few cases of paraesthesia persistent for more than 5 days (0.09%) and for more than 6 months (0.008%) and a few cases of pneumothorax (0.006%) [61]. This demonstrates that we still must know where the needle is during its advancement and know where the local anaesthetic is being injected.

Technological advances offer exceptional 3D images (Fig. 2.8) of nerves and of nerve blocks performed [62]. Apart from quality of these images, do we need them in clinical practice? Probably not. These reconstructed images post hoc are of little utility in the OR. In order for advances in technology in this area to be useful for us, we need 4D ultrasound, i.e. real-time 3D. We need to see, in real time, advancement of the needle towards the nerve, spread of the local anaesthetic as intended, and we need to be certain that injection of the local anaesthetic (a potentially fatal medicine) is neither intravascular nor intraneural. Current attempts show images which are not sufficiently accurate to provide clinical benefit. In this sense, 3D ultrasound is a mere step towards 4D, and the ability of ultrasound specialists to process images should be considerably increased to ensure greater safety for the patient during injection of a local anaesthetic. This provides a better answer to the question "why use ultrasound in RA?": to improve patient safety.

Fig. 2.8 3D image of the brachial plexus in the axillary area. (Iconography: Olivier Choquet)

References

1. Dupré LJ. Neurostimulateur en anesthésie locorégionale. Cahiers Anesthésiol. 1992;40:503–10.
2. Jochum D, Iohom G, Diarra DP, Loughnane F, Dupré LJ, Bouaziz H. An objective assessment of nerve stimulators used for peripheral nerve blockade. Anaesthesia. 2006;61:557–64.
3. Tsai TP, Vuckovic I, Dilberovic F, Obhodzas M, Kapur E, Divanovic KA, Hadzic A. Intensity of the stimulating current may not be a reliable indicator of intraneural needle placement. Reg Anesth Pain Med. 2008;33:207–10.
4. Dufour E, Cymerman A, Nourry G, et al. An ultrasonographic assessment of neurostimulation-guided median nerve block at the elbow: a local anesthetic spread, nerve size, and clinical efficacy study. Anesth Analg. 2010;111:561–7.

5. Orebaugh SL, Kentor ML, Williams BA. Adverse outcomes associated with nerve stimulator-guided and ultrasound-guided peripheral nerve blocks by supervised trainees: update of a single-site database. Reg Anesth Pain Med. 2012;37:577–82.
6. Jochum D, Bondàr A, Delaunay L, Egan M, Bouaziz H. One size does not fit all: proposed algorithm for ultrasonography in combination with neurostimulation for peripheral nerve blockade. Br J Anaesth. 2009;103:771–3.
7. Perlas A, Niazi A, McCartney C, Chan V, Xu D, Abbas S. The sensitivity of motor response to neurostimulation and paresthesia for nerve localization as evaluated by ultrasound. Reg Anesth Pain Med. 2006;31:445–50.
8. Dingemans E, Williams SR, Arcand G, et al. Neurostimulation in ultrasound-guided infraclavicular block: a prospective randomized trial. Anesth Analg. 2007;104:1275–80.
9. Delaunay L. Faut-il continuer à réaliser des blocs nerveux périphériques en neurostimulation seule? Ann Fr Anesth Réanim. 2013;32:217–9.
10. Dibiane C, Deruddre S, Zetlaoui PJ. A musculocutaneous nerve variation described during ultrasound-guided axillary nerve block. Reg Anesth Pain Med. 2009;34:617–8.
11. Orebaugh SL, Pennington S. Variant location of the musculocutaneous nerve during axillary nerve block. J Clin Anesth. 2006;18:541–4.
12. Gassner EM, Schocke M, Peer S, Schwabegger A, Jaschke W, Bodner G. Persistent median artery in the carpal tunnel: color Doppler ultrasonographic findings. J Ultrasound Med. 2002;21:455–61.
13. Del Rosario E, Kopka A. Persistent median artery at the forearm. Acta Anaesthesiol Scand. 2013;57:399.
14. Mariano ER, Brodsky JB. Comparison of procedural times for ultrasound-guided perineural catheter insertion in obese and nonobese patients. J Ultrasound Med. 2011;30:1357–61.
15. Nielsen KC, Guller U, Steele SM, Klein SM, Greengrass RA, Pietrobon R. Influence of obesity on surgical regional anesthesia in the ambulatory setting: an analysis of 9038 blocks. Anesthesiology. 2005;102:181–7.
16. Brodsky JB, Mariano ER. Regional anaesthesia in the obese patient: lost landmarks and evolving ultrasound guidance. Best Pract Res Clin Anaesthesiol. 2011;25:61–72.
17. Plunkett AR, Brown DS, Rogers JM, Buckenmaier CC III. Supraclavicular continuous peripheral nerve block in a wounded soldier: when ultrasound is the only option. Br J Anaesth. 2006;97:715–7.
18. Van Geffen GJ, McCartney CJ, Gielen M, Chan VW. Ultrasound as the only nerve localization technique for peripheral nerve block. J Clin Anesth. 2007;19:381–5.
19. Tsai PB, Karnwal A, Kakazu C, Tokhner V, Julka IS. Efficacy of an ultrasound-guided subsartorial approach to saphenous nerve block: a case series. Can J Anaesth. 2010;57:683–8.
20. Greher M, Moriggl B, Curatolo M, Kirchmair L, Eichenberger U. Sonographic visualization and ultrasound-guided blockade of the greater occipital nerve: a comparison of two selective techniques confirmed by anatomical dissection. Br J Anaesth. 2010;104:637–42.
21. Weintraud M, Marhofer P, Bosenberg A, et al. Ilioinguinal/iliohypogastric blocks in children: where do we administer the local anesthetic without direct visualization? Anesth Analg. 2008;106:89–93.
22. Willschke H, Marhofer P, Bösenberg A, et al. Ultrasonography for ilioinguinal/iliohypogastric nerve blocks in children. Br J Anaesth. 2005;95:226–30.
23. Dolan J, Williams A, Murney E, Smith M, Kenny GN. Ultrasound guided fascia iliaca block: a comparison with the loss of resistance technique. Reg Anesth Pain Med. 2008;33:526–31.
24. Sites BD, Gallagher J, Sparks M. Ultrasound-guided popliteal block demonstrates an atypical motor response to neurostimulation in 2 patients with diabetes mellitus. Reg Anesth Pain Med. 2003;28:479–82.
25. Minville V, Zetlaoui PJ, Fessenmeyer C, Benhamou D. Ultrasound guidance for difficult lateral popliteal catheter insertion in a patient with peripheral vascular disease. Reg Anesth Pain Med. 2004;29:368–70.
26. Howell SM, Unger MW, Colson JD, Serafini M. Ultrasonographic evaluation of neck hematoma and block salvage after failed neurostimulation-guided interscalene block. J Clin Anesth. 2010;22:560–1.
27. Marhofer P, Schröendorfer K, Koinig H, et al. Ultrasonographic guidance improves sensory block and onset time of three-in-one blocks. Anesth Analg. 1997;85:854–7.
28. Abrahams MS, Aziz MF, Fu RF, Horn JL. Ultrasound guidance compared with electrical neurostimulation for peripheral nerve block: a systematic review and meta-analysis of randomized controlled trials. Br J Anaesth. 2009;102:408–17.
29. Rubin K, Sullivan D, Sadhasivam S. Are peripheral and neuraxial blocks with ultrasound guidance more effective and safe in children? Paediatr Anaesth. 2009;19:92–6.
30. Antonakakis JG, Ting PH, Sites B. Ultrasound-guided regional anesthesia for peripheral nerve blocks: an evidence-based outcome review. Anesthesiol Clin. 2011;29:179–91.
31. Mariano ER, Loland VJ, Sandhu NS, et al. Comparative efficacy of ultrasound-guided and stimulating popliteal-sciatic perineural catheters for postoperative analgesia. Can J Anaesth. 2010;57:919–26.
32. Bendtsen TF, Nielsen TD, Rohde CV, Kibak K, Linde F. Ultrasound guidance improves a continuous popliteal sciatic nerve block when compared with neurostimulation. Reg Anesth Pain Med. 2011;36:181–4.
33. Dhir S, Ganapathy S. Use of ultrasound guidance and contrast enhancement: a study of continuous infraclavicular brachial plexus approach. Acta Anaesthesiol Scand. 2008;52:338–42.
34. Falcão LF, Perez MV, de Castro I, Yamashita AM, Tardelli MA, Amaral JL. Minimum effective volume of 0.5% bupivacaine with epinephrine in ultrasound-guided interscalene brachial plexus block. Br J Anaesth. 2013;110:450–5.
35. Gautier P, Vandepitte C, Ramquet C, DeCoopman M, Xu D, Hadzic A. The minimum effective anesthetic volume of 0.75% ropivacaine in ultrasound-guided interscalene brachial plexus block. Anesth Analg. 2011;113:951–5.
36. O'Donnell BD, Iohom G. An estimation of the minimum effective anesthetic volume of 2% lidocaine in ultrasound-guided axillary brachial plexus block. Anesthesiology. 2009;111:25–9.
37. Casati A, Baciarello M, Di Cianni S, et al. Effects of ultrasound guidance on the minimum effective anaesthetic volume required to block the femoral nerve. Br J Anaesth. 2007;98:823–7.
38. Tran de QH, Dugani S, Dyachenko A, Correa JA, Finlayson RJ. Minimum effective volume of lidocaine for ultrasound-guided infraclavicular block. Reg Anesth Pain Med. 2011;36:190–4.
39. Sauter AR, Dodgson MS, Stubhaug A, et al. Electrical neurostimulation or ultrasound guidance for lateral sagittal infraclavicular blocks: a randomized, controlled, observer blinded comparison study. Anesth Analg. 2008;106:1910–5.
40. Chan VW, Perlas A, McCartney CJ, et al. Ultrasound guidance improves success rate of axillary brachial plexus block. Can J Anaesth. 2007;54:176–82.
41. Martínez Navas A, De la Tabla González RO. Ultrasound-guided technique allowed early detection of intravascular injection during an infraclavicular brachial plexus block. Acta Anaesthesiol Scand. 2009;53:968–70.

42. Bigeleisen PE. Ultrasound-guided infraclavicular block in an anticoagulated and anesthetized patient. Anesth Analg. 2007;104:1285–7.
43. Dufour E, Laloë PA, Culty T, Fischler M. Ultrasound and neurostimulation-guided axillary brachial plexus block for resection of a hemodialysis fistula aneurysm. Anesth Analg. 2009;108:1981–3.
44. Zetlaoui PJ, Labbe JP, Benhamou D. Ultrasound guidance for axillary plexus block does not prevent intravascular injection. Anesthesiology. 2008;108:761.
45. Zetlaoui PJ, Labbe JP, Benhamou D. Accidental intravascular injection of local anesthetic? Anesthesiology. 2008;109:1144–5.
46. Perlas A, Lobo G, Lo N, Brull R, Chan VW, Karkhanis R. Ultrasound-guided supraclavicular block: outcome of 510 consecutive cases. Reg Anesth Pain Med. 2009;34:171–6.
47. Bhatia A, Lai J, Chan VW, Brull R. Case report: pneumothorax as a complication of the ultrasound-guided supraclavicular approach for brachial plexus block. Anesth Analg. 2010;111:817–9.
48. Furness G, Reilly MP, Kuchi S. An evaluation of ultrasound imaging for identification of lumbar intervertebral level. Anaesthesia. 2002;57:277–80.
49. Perlas A. Evidence for the use of ultrasound in neuraxial blocks. Reg Anesth Pain Med. 2010;35(2 Suppl):S43–6.
50. Shaikh F, Brzezinski J, Alexander S, et al. Ultrasound imaging for lumbar punctures and epidural catheterisations: systematic review and meta-analysis. BMJ. 2013;346:f1720.
51. Balki M, Lee Y, Halpern S, Carvalho JC. Ultrasound imaging of the lumbar spine in the transverse plane: the correlation between estimated and actual depth to the epidural space in obese parturients. Anesth Analg. 2009;108:1876–81.
52. Chin KJ, MacFarlane AJR, Chan V, Brull R. The use of ultrasound to facilitate spinal anesthesia in a patient with previous lumbar laminectomy and fusion: a case report. J Clin Ultrasound. 2009;37:482–5.
53. Tsui BC, Suresh S. Ultrasound imaging for regional anesthesia in infants, children, and adolescents: a review of current literature and its application in the practice of neuraxial blocks. Anesthesiology. 2010;112:719–28.
54. Triffterer L, Machata AM, Latzke D, et al. Ultrasound assessment of cranial spread during caudal blockade in children: effect of the speed of injection of local anaesthetics. Br J Anaesth. 2012;108:670–4.
55. Tsui B, Leipoldt C, Desai S. Color flow doppler ultrasonography can distinguish caudal epidural injection from intrathecal injection. Anesth Analg. 2013;116:1376–9.
56. Sites BD, Chan VW, Neal JM, et al. The American Society of Regional Anesthesia and Pain Medicine and the European Society of Regional Anaesthesia and Pain Therapy Joint Committee recommendations for education and training in ultrasound-guided regional anesthesia. Reg Anesth Pain Med. 2010;35(2 Suppl):S74–80.
57. Sites BD, Spence BC, Gallagher JD, Wiley CW, Bertrand ML, Blike GT. Characterizing novice behavior associated with learning ultrasound-guided peripheral regional anesthesia. Reg Anesth Pain Med. 2007;32:107–15.
58. Grau T, Bartusseck E, Conradi R, Martin E, Motsch J. Ultrasound imaging improves learning curves in obstetric epidural anesthesia: a preliminary study. Can J Anaesth. 2003;50:1047–50.
59. Dessieux T, Estebe JP, Bloc S, Mercadal L, Ecoffey C. Évaluation de la courbe d'apprentissage des internes pour l'échoguidage sur un fantôme. Ann Fr Anesth Réanim. 2008;27:797–801.
60. Ecoffey C, Estèbe JP. De la neurostimulation à l'échoguidage: une révolution pour la pratique clinique quotidienne et l'enseignement de l'anesthésie locorégionale. Ann Fr Anesth Réanim. 2008;27:795–6.
61. Sites BD, Taenzer AH, Herrick MD, et al. Incidence of local anesthetic systemic toxicity and postoperative neurologic symptoms associated with 12,668 ultrasound-guided nerve blocks: an analysis from a prospective clinical registry. Reg Anesth Pain Med. 2012;37:78–82.
62. Clendenen SR, Riutort K, Ladlie BL, Robards C, Franco CD, Greengrass RA. Real-time three-dimensional ultrasound-assisted axillary plexus block defines soft tissue planes. Anesth Analg. 2009;108:1347–50.

Ultrasound in Regional Anaesthesia: How?

3

Eryk Eisenberg, Laurent Delaunay, Denis Jochum, Mathilde De Queiroz, Didier Morau, Manoj Kumar Karmakar, and Wing Hong Kwok

Contents

E. Eisenberg (✉)
Département d'anesthésie et réanimation, Clinique Pôle Santé République, Clermont-Ferrand, France

L. Delaunay
Département d'anesthésie-réanimation, Clinique Générale, Vivalto Santé, Annecy, France

D. Jochum
Service d'anesthésie-réanimation, Groupe hospitalier du Centre Alsace, Hôpital Albert Schweitzer, Colmar, France
e-mail: jochumd@calixo.net

M. De Queiroz
Département d'anesthésie-réanimation, Hôpital Femme-Mère-Enfant, Bron, France

D. Morau
Département d'anesthésie-réanimation A, CHU de Montpellier, Hôpital Lapeyronie, Montpellier, France

M. K. Karmakar · W. H. Kwok
Department of Anesthesia and Intensive Care, Prince of Wales Hospital, The Chinese University of Hong Kong, Sha Tin, Hong Kong
e-mail: karmakar@cuhk.edu.hk

E. Eisenberg, E. Gaertner (eds.), *Ultrasound in Peripheral, Neuraxial and Perineuraxial Regional Anaesthesia*,
https://doi.org/10.1007/978-3-031-08804-9_3

Equipment

Ultrasound Systems

The ultrasound system has become an essential component in the technical arsenal of the anaesthetist-intensivist. It is indispensable in assessing internal structures, enabling safe vascular access, allowing non-invasive haemodynamic evaluation and in promoting the practice of safe and effective regional anaesthesia.

A modern, portable ultrasound system is a costly investment but clinicians must engage with hospital managers (and those with purchasing responsibility) to promote its utility and merit.

If one were to state the principal characteristics of the ideal apparatus, the terms "high performance", "mobile and light", "easy to handle", "functional", "simple to use", "robust and reliable" and "user-friendly" immediately come to mind. It is true that our mobile and versatile activity (operating room, post-surgical recovery room, anaesthesia induction room, intensive care department and, potentially, the emergency room receiving area, the hospital department or even the emergency ambulance) mean that a conventional ultrasound system, such as those dedicated to the activity of radiologists, would not be very practicable. Therefore, in choosing an **ultrasound system** intended for use in anaesthesia and in intensive care, it is appropriate to consider system that is **easy to transport** and which offers the required technical characteristics, i.e.:

- operates on both mains power and batteries
- relatively small in size
- robust
- easy and quick to adjust
- stable computer operating system
- possible archiving and export of images and videos
- computer and video connectors
- possible personalised pre-adjustment
- good quality screen (size and resolution)
- essential available modes:
  B-mode ultrasound
  colour Doppler and/or power (pulse Doppler is not essential in RA)
- available adjustments:
  overall gain and depth-related gain (TGC—time gain compensation).
  studied depth.
  number and position of focus.
  potential adjustment of frequency of the probe used.
  zoom function.
  dynamic range, compound, cross-beam, multi-beam, sonocity, harmonics, etc., which are a large selection of parameters that can significantly enhance the image. Compound, cross-beam, multi-beam and sonocity correspond to technological tools specific to each brand of ultrasound system but whose overall principle is equivalent. This involves the summation of several images generated in a given plane by ultrasound transmission with different angles, allowing for a significant reduction in artefacts. The majority of modern apparatus are equipped with this function.
  grey scale.
  screen brightness, contrast.
  a multiprobe selector.
- "upgradability" of the system
- moderate price
- etc.

Transducers

Ultrasound probes (transducers) (Fig. 3.1) are the "cornerstones" of the technique. As they are specific to each application of ultrasound, the probes must be selected appropriately. In fact, they must satisfy the following requirements:

- **Ergonomic requirements** for each application with their specific requirements for size, shape, and displacement: a surface or endocavity ultrasound system for adults or paediatric patients, etc.
- **Technical requirements** for each use (particularly related to the depth and shape of the field studied). The probe capabilities in respect of working frequency range, image resolution and maximum examination depth are all important considerations.

Choosing the probe correctly (based on the studied area) is a legibility pledge of the images generated. The depth of a

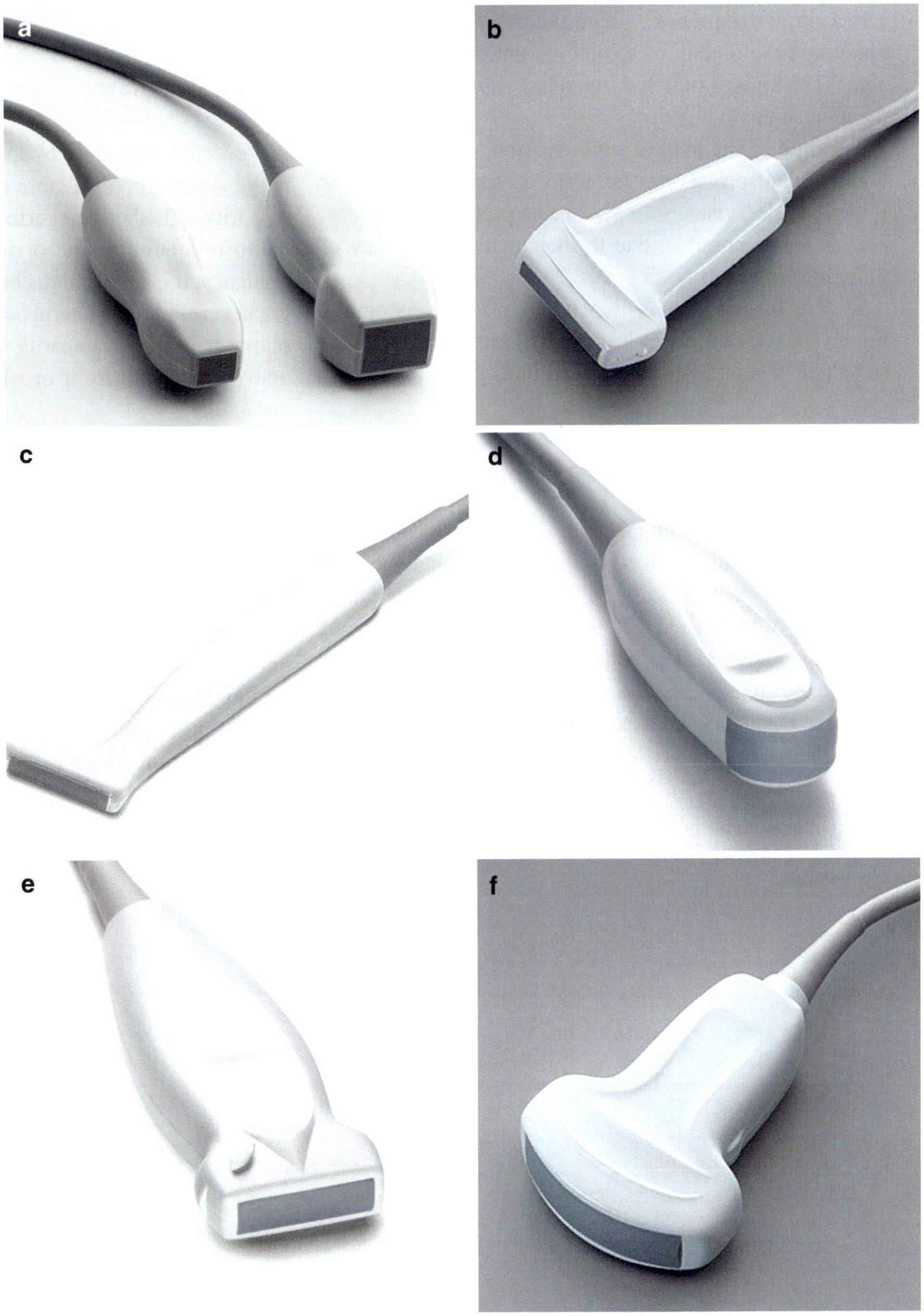

Fig. 3.1 A few possible probe types. (**a**) Cardiac probes; (**b**) 38 mm high frequency linear probe; (**c**) "hockey stick" high frequency linear probe—small footprint (25 mm); (**d**) intermediate frequency curvilinear probe; (**e**) 25 mm high frequency linear probe; (**f**) low frequency curvilinear probe

structure influences its quality of visualisation. Nevertheless, not all deep structures are difficult to see and, conversely, those which are superficial are not always perfectly visible. For example, by using a "low frequency" probe to visualise a sciatic nerve with a depth of 5–8 cm, the capacity of the ultrasound to reach the target and thus to generate the image of a nerve is increased. However, this is at the expense of spatial resolution, and therefore of the "clarity" of the outline and precision of the ultrasound structure of the nerve.

Therefore, for a given procedure, the characteristics of the probe will be chosen in three distinct fields:

- **Ultrasound frequency**, which determines the depth of study and the quality of spatial resolution (image precision). The highest possible frequency should be preferred for a given depth.
- **The shape of the field of investigation**, either rectangular for linear probes or "fan-shaped" for curved probes ("large angle" visualisation). Linear probes are preferable, as far as possible, due to their higher frequency of emission (Fig. 3.2).
- **The size of the probe**, depending on the anatomical region studied (a more or less limited area) and according to patient characteristics (adult or child).

This is why we mainly use, according to need, the following:

- A "high frequency" **linear transducer** (between 6 and 18 MHz) enabling a refined study of an area between 0.5 and 6 cm in depth (ideally, a 38 mm probe to obtain a wide field of vision and a 25 mm "hockey-stick" probe of small size for easier handling and for difficult-to-investigate areas)
- A lower frequency **convex transducer** (between 2 and 5 MHz) which enables easier study of tissues located beyond 6 cm and provides a broad field of vision but less refined details (ideally, a "conventional" convex probe and a "microconvex" probe for tighter spaces)

Needles

The same recommendations for use of needles with short bevels developed for peripheral nerve blocks performed with neurostimulation can be transferred to the practice of ultrasound guidance. However, care should be taken to not attribute overly high safety properties to such needle bevels [1].

Nerve Stimulator

As revealed in a study by Denis Jochum [2], the "range" of nerve stimulators available demonstrates a wide qualitative disparity. Whether you perform peripheral nerve blocks with neurostimulation alone or in combination with ultrasound guidance, it is essential to choose equipment which offers the greatest reliability in terms of signal quality, precision of the output current and linearity of intensity variation. This is an unavoidable priority if one follows a philosophy of quality and safety.

The Ultrasound Transmission Gel

Transmission of ultrasound between the ultrasound probe and the tissues under examination requires the use of a "coupling" medium on the skin surface. This allows for a reduction of the impedance between the air and the skin whose existence is essential to obtain a good quality image. A gel usually is preferred for its residual qualities, avoiding the almost continuous application which is necessary when using a liquid as a

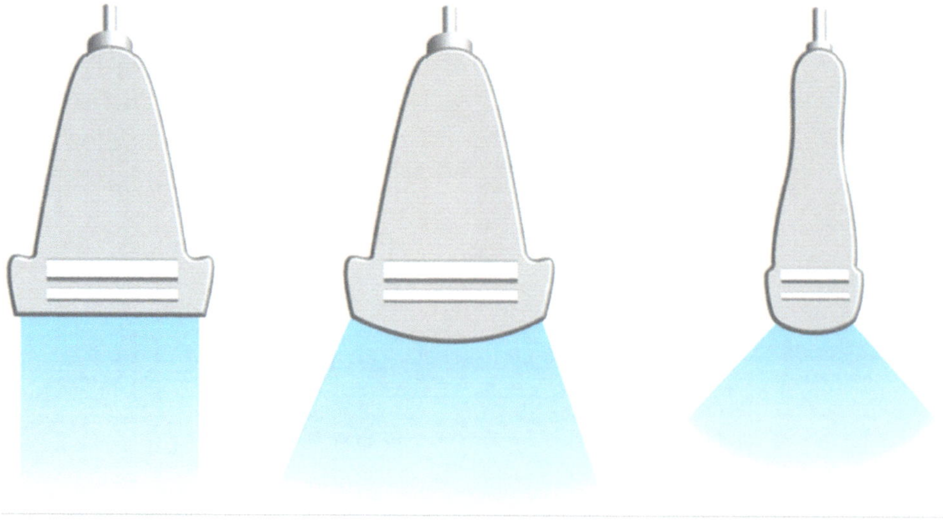

Fig. 3.2 Different forms of ultrasound field according to the type of probe

contact medium between the probe and the skin (some practitioners use saline or an available cutaneous antiseptic). It is essential to use a **sterile gel in individual packets**.

Although gels are in a very high proportion comprised of water, the manufacturers do not reveal their exact composition! Yet it has been shown that several of them contain largely propylene glycol and glycerol, substances whose toxicity for muscles and for nerves has been demonstrated [3]. During skin needle puncture through the gel, the needles can even introduce small quantities of it in contact with the nerves approached [4]. Certainly, this involves very small quantities, but an inflammatory reaction can be generated experimentally by placing a nerve in contact with the gel [5], also found when exposing the meninges to the gel [6]. The quantities of gel placed in contact with peripheral nerves or the central nervous system in anaesthetic practice are very likely much smaller than in the experimental studies which have been conducted. Similarly, the type of inflammatory reaction described in the aforementioned studies is not accompanied by histological lesions. However, "the absence of evidence of obvious risk does not mean proof of safety" [3]. **Therefore, it is recommended to not insert the needle through the gel**. It is sufficient to limit oneself to wiping away the excess gel from the needle puncture site when the probe is properly positioned to start the procedure.

Hygiene and Ultrasound-Guided RA

Extract from expert formalised recommendations (EFR) "Ultrasonography in locoregional anaesthesia"—SFAR 2011 [7].

"6. Conditions of conduct, hygiene

It is 'probably recommended' to perform an ultrasound-guided block in an awake, calm and cooperative patient. However, in cases where the benefit/risk ratio is favourable and justified, it is possible to perform a nerve block in a patient under anaesthesia (general or regional) or under sedation. In this case, ultrasound 'probably' provides additional safety. Due to the risk of cross-transmission and the need for a sterile environment required in RA, 'it is recommended' to comply with measures of aseptic practice for the ultrasound probe. 'It is recommended' before each procedure that probes and cables be wiped, cleaned and disinfected. The entire apparatus should be cleaned regularly. 'It is recommended', during use of an ultrasound probe, to use an appropriate single-use sterile protective sheath made for this purpose and a single-dose sterile gel. 'It is recommended' in the absence of perforation or tearing during removal of the protection that disinfection of the probe between each patient be the same as that corresponding to disinfection of the lower level. 'It is recommended' in case of a break in the sheath or staining of the probe that disinfection be of a higher level (Table 3.1). 'It is recommended' at the end of the operating list to clean the probe with a detergent, to rinse it, to dry it and to store it in a clean place. 'It is recommended' to seek validation of the different procedures for cleaning and disinfection from the Committee to Fight Nosocomial Infection (CLIN) and/or the department of hygiene."

Table 3.1 Excerpt from the EFR "Ultrasonography in locoregional anaesthesia"—SFAR 2011 [7]. Classification of medical devices and levels of treatment required

Definition of equipment	Class of equipment	Level of risk of infection	Treatment required
Introduction into the vascular system or into a cavity or sterile tissue whatever the approach	Critical	High risk	Sterilisation or sterile single-use if by default, high level disinfection
In contact with the mucosa or the superficially abraded skin	Semi-critical	Medium risk	Intermediate type disinfection
In contact with the patient's intact skin or without contact with the patient	Non-critical	Low risk	Low-level disinfection

During conduct of an ultrasound-guided peripheral nerve block, one should not overlook the usual recommendations required for nerve blocks with neurostimulation, which remain of current interest [7]. Rules on aseptic procedure, which must be scrupulously complied with, are complicated slightly by the presence of additional materials at the site of needle puncture: the probe and the gel.

The skin should be prepared rigorously at the site of needle puncture and in the area which is to undergo ultrasound examination, according to the so-called "4-phase ultrasound procedure":

1. Wiping of the skin with an antiseptic soap from the same range as the antiseptic chosen.
2. Rinsing with sterile water or sterile physiological saline.
3. Drying of the area with a clean cloth.
4. Application of the antiseptic; allowing time for drying before the invasive procedure.

Painting of the skin is performed with the aid of a "surgical" antiseptic, conventionally iodinated polyvidone or chlorhexidine, over an extensive area, well beyond the point of skin needle puncture, to be able to move the probe without desterilising it.

Whether it involves a peripheral nerve block with a "single shot" or placement of a perineural catheter, ultrasound guidance in RA requires taking specific hygiene precautions for the ultrasound probe. The probe is a reusable medical device

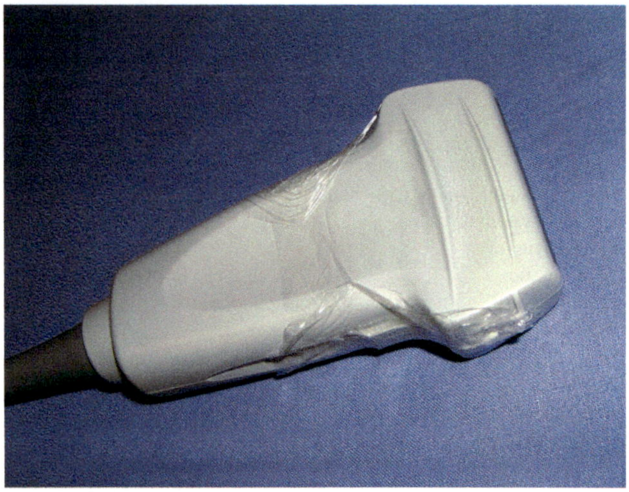

Fig. 3.3 Probe covered by a simple semipermeable occlusive dressing, not recommended

(RMD) which can be responsible for transmission of infection between patients by direct contact with blood or other biological fluids. Ultrasound probes are heat-sensitive devices and only one disinfection can be performed. It is also necessary to perform traceability of such disinfection. Recommendations concerning peripheral nerve blocks in adults continue to be of current interest [8]. The conduct of RA procedures should be performed with strict technique and according to aseptic procedures in conformity with recommendations of scientific societies and regulations. Since no specific text existed on use of ultrasound in RA, recommendations on disinfection of these medical devices were established [7, 9]. For good practice of asepsis with ultrasound probes, the aim is to do everything possible to avoid contamination of the probe in order to be able to carry out minimum low-level disinfection between each procedure. The low cost of a nonspecific device for protection of the probe, for example, a semipermeable transparent occlusive dressing, is not a sufficient argument to justify its use in this area (Fig. 3.3). When removing this transparent dressing, contamination of the probe is almost inevitable, which makes it no longer possible to perform this low-level disinfection and consequently makes this practice incompatible with activity of the OR.

Recommendations specify that for ultrasound performed during a perineural nerve block with a single or continuous injection, a sterile gel and a sterile protection are necessary. This sterile protection requires a dedicated sheath (EC marking) (Figs. 3.4 and 3.5) which should be adjusted to the size of the probe and have a sufficient length to maintain a sterile environment. It is preferable to have a telescopic sheath fitted for elastic protection and not a sticky band because of risk of tearing of the protection during removal of this fixation. **Disinfection of the probe between each patient must at least be the same as the disinfection conventionally used for non-critical devices** (Table 3.1). Transcutaneous probes

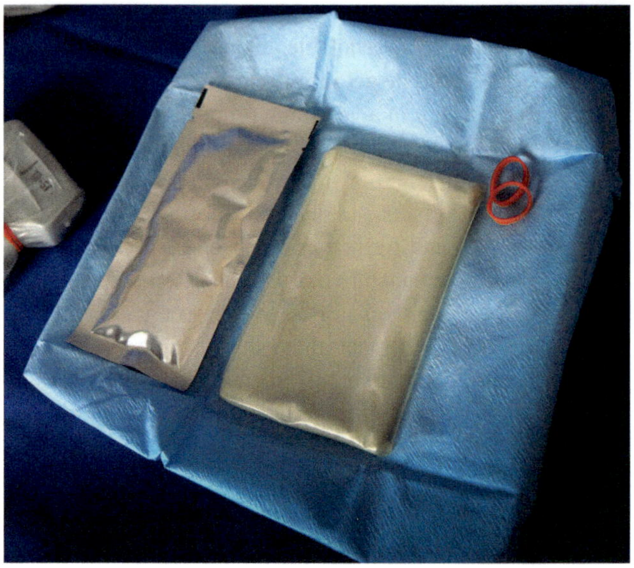

Fig. 3.4 Pack containing a protective sheath dedicated to ultrasound probes, a dose of sterile gel and two rubber bands for securing

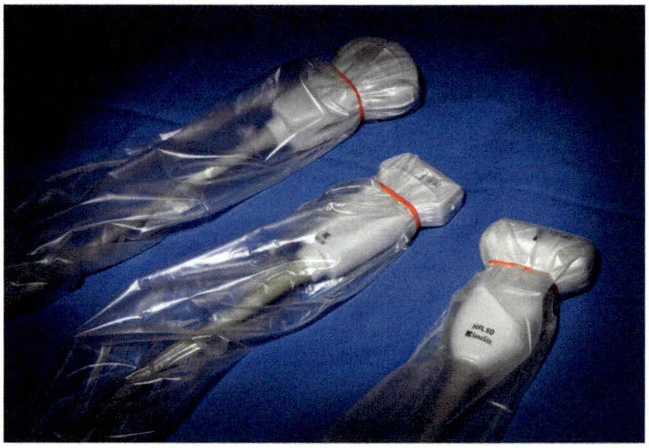

Fig. 3.5 Ultrasound probes fitted with their dedicated protective sheath

and cables should be wiped and cleaned with a product recommended by the manufacturer between each patient, for example, with a “non-woven” material soaked with detergent/disinfectant. All traces on the keyboard and the probe carried must be cleaned regularly.

There remains the problem of a break observed in the sheath, which requires use of higher level disinfection [10]. The majority of manufacturers do not offer a probe which can be completely immersed with the cable. The equipment which comprises the probe tolerates pre-disinfection with a detergent, but this is rarely the case for disinfectant products, for example, peracetic acid. It may be useful to investigate other means to facilitate this disinfection process. These alternatives can be automatic devices for rapid disinfection, either by physical disinfection with ultraviolet C radiation [11–13], or by chemical disinfection with a peracetic acid in spray form. However, even with these devices, the initial

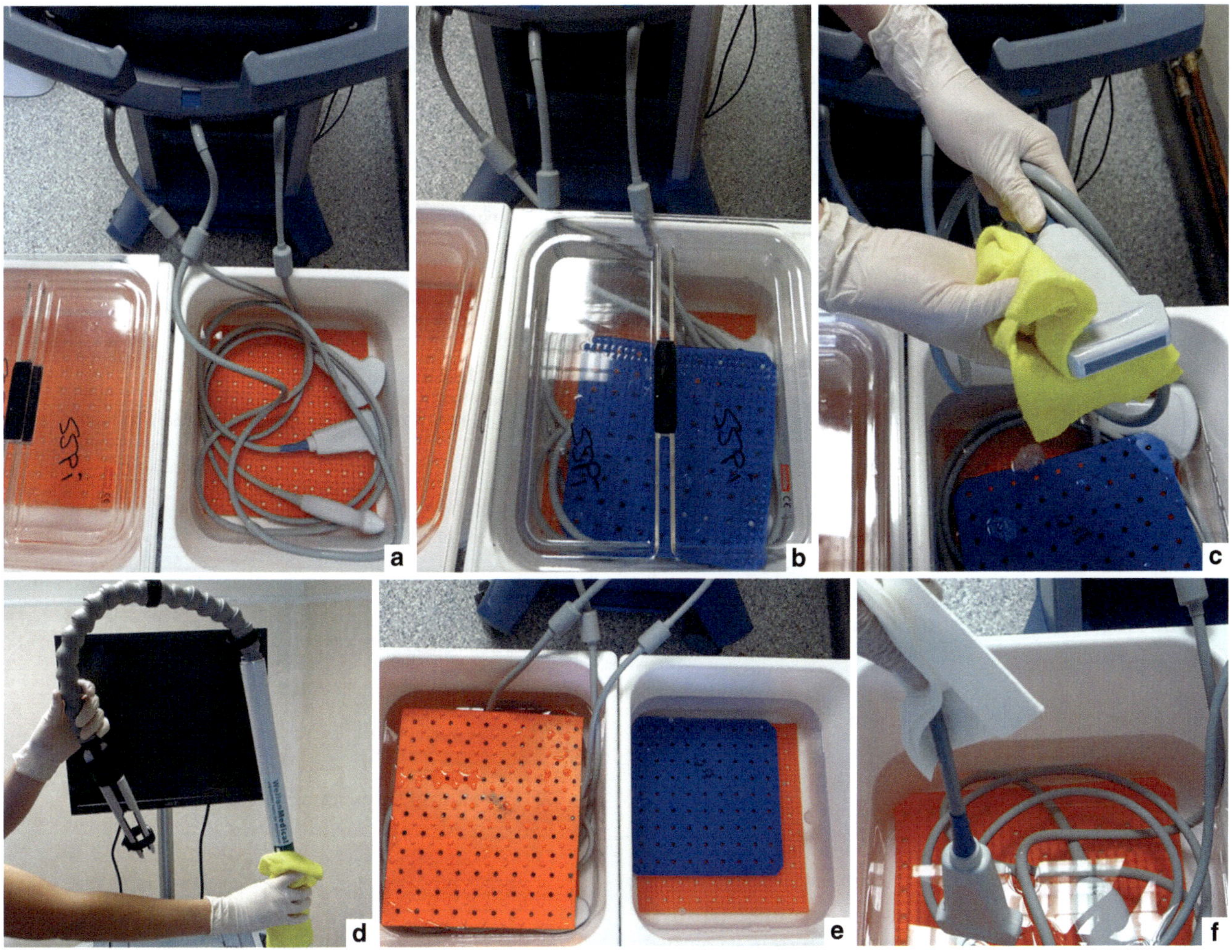

Fig. 3.6 (**a**) Positioning of probes and of cables in the soaking tub; (**b**) soaking of probes and of cables in detergent; (**c**) cleansing of probes; (**d**) cleansing of the ultrasound system; (**e**) rinsing tub; (**f**) Drying of probes and cables

phase of cleaning remains highly important to ensure better efficacy of disinfection. This is why the protective sheath remains very useful to facilitate this cleansing.

Moreover, even in the absence of a break in the protection, at the end of the operating list it is recommended to clean the probe with a detergent, to rinse it, to dry it (Fig. 3.6) and to store it in a clean place [14]. It is essential to establish procedures for cleaning and disinfection in each institution.

Expert Formalised Recommendations (EFR) Concerning Hygiene in Ultrasound-Guided RA [7]

- Due to the risk of cross-transmission and need for a sterile environment required in RA, it is recommended to comply with aseptic measures for the ultrasound probe.
- Before each procedure, it is recommended for probes and cables to be wiped, cleaned, and disinfected. The entire apparatus should be cleaned regularly.
- It is recommended, during use of an ultrasound probe, to use an appropriate single-use sterile protective sheath made for this use and a single-dose sterile gel.
- In the absence of perforation or a tear during removal of the protection, it is recommended that disinfection of the probe between each patient be the same as that corresponding to low level disinfection. In case of a break in the sheath or soiling of the probe, it is recommended that disinfection be of a higher level (Table 3.1). At the end of the operating list, it is recommended to cleanse the probe with detergent, to rinse it, to dry it, and to store it in a clean place.

- It is recommended to seek validation of the different procedures for cleaning and disinfection from the CLIN and/or the department of hospital hygiene.

In practice: Maintenance of immersible ultrasound probes in anaesthetic use.

- The morning before start of the surgical programme: careful wiping of all probes, cables, probe supports, keyboards, and ultrasound handles with a wipe pre-soaked with ready-to-use disinfectant detergent.
- **Before the procedure:** It is mandatory to protect the ultrasound probe with a sterile protection and to use a sterile gel.

Preparation of the probe before the procedure: placement of the protective sheath according to Fig. 3.7 following the detailed instructions in the logical operational chart.

- **After the procedure:** Removal of the sheath according to Fig. 3.8 following the instructions explained on the logical operational chart.
- **At the end of the operating list:** Bring ultrasound systems to the decontamination room, do not remove cables from the ultrasound machine, soak all probes (limit = visible swelling on the cable) in neutral detergent for 5 min, gently cleanse with a clean wipe, rinse with tap water, dry with a wiping square, clean the cable, the probe support as well as the entire ultrasound apparatus with a wipe soaked with detergent and taken in the soaking solution and thoroughly dried, rinse with tap water (when the wipe has been thoroughly wrung out), dry with a wiping square and fill out the traceability form (Fig. 3.6).
- **In case of a tear in the sheath or staining of the probe,** perform cleaning and intermediate level disinfection: Bring the ultrasound machine to the decontamination room, do not remove the cables from the ultrasound machine, soak the probe (limit = visible swelling on the cable) in disinfectant detergent for medical devices for 15 min (adjust duration of soaking according to product function), gently cleanse with a clean small cloth, rinse with tap water, soak the probe in the disinfectant for "10 min" (adjust duration of soaking depending on product function), rinse with tap water, dry with a wiping square and fill out the traceability form for the two phases of treatment: cleansing and disinfection.

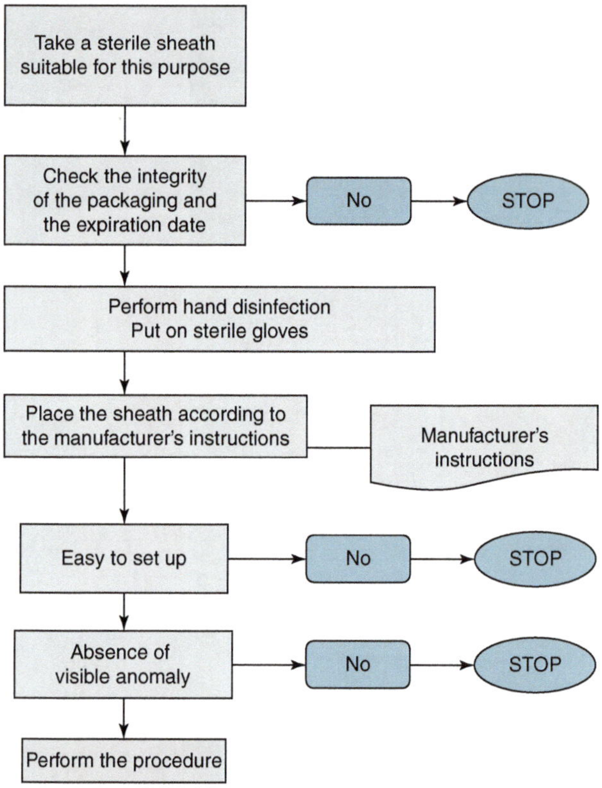

Fig. 3.7 Placement of the protective sheath for the probe

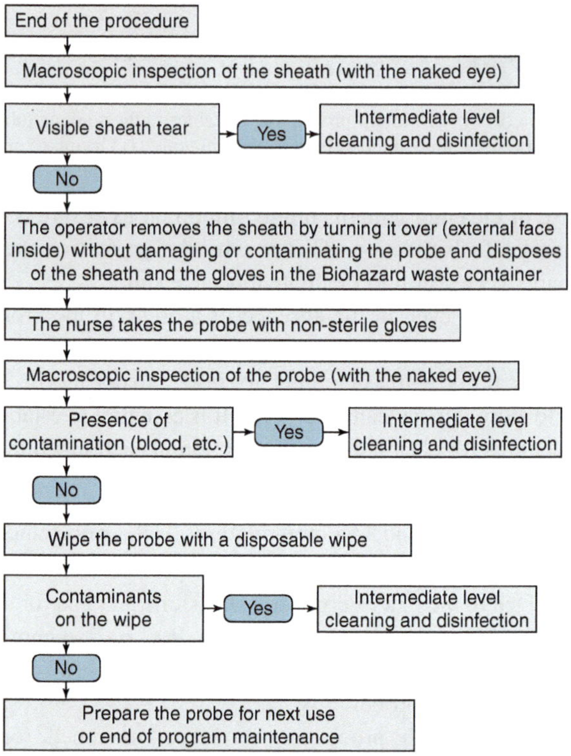

Fig. 3.8 Removal of the protective sheath after use. DASRI (in French): waste products from health care activities with infection hazards

Conduct of Ultrasound-Guided Peripheral Nerve Blocks

Excerpt from the EFR "Ultrasonography in locoregional anaesthesia"—SFAR 2011 [7]

"2. General Rules, Training and Procedure for Conduct"

"[...] Understanding the techniques of guiding of the needle 'in plane' and 'out of the plane' is a prerequisite for the safety and success in execution of RA. Due to the inter-individual variability in the rapidity of acquisition of the technique, 'it is recommended' to follow one's own learning curve. Additional measures 'are recommended' for conduct of a nerve block: neurostimulation and/or hydrolocalisation and/or hydrodissection and/or displacement of tissues with movements of the needle. In case of difficulty in visualisation of the ultrasound anatomy, 'it is recommended' to use neurostimulation in combination with ultrasound guidance."

"3. Equipment and Technical Aspect"

"'It is recommended' to have probes of an appropriate frequency and shape for the anaesthesia performed. 'It is recommended' to use the highest possible frequency to assign priority to spatial resolution and to improve precision of the image. Choice of a probe depends on type of block and depth of the target. 'It is recommended' to use different functions proposed by the ultrasound system and to adjust their settings to the native image and depth of the target: general and serial gain, the depth studied, number and position of focal points, multiple incidence imaging, Doppler mode, etc. 'It is recommended' to perform, before the anaesthetic procedure, wide dynamic visualisation of the anatomical elements, precisely looking for target and adjacent structures, using the functional assistance available on the ultrasound system. Compliance with this procedure makes it possible to plan the trajectory of the needle, to determine the plane of visualisation of the nerve (short and/or long axis) and technique of progression of the needle. 'It is probably recommended' to visualise target nerves in the 'short axis' for superficial and deep nerve blocks. The choice of the approach of the needle in plane or out of plane is independent of the depth of the target. It is recommended to use needles dedicated to RA. 'It is recommended' that unintentional movements of the probe be confirmed and corrected, to follow progression of the end of the needle and to visualise the distribution of the local anaesthetic."

"4. Technical Rules on Safety"

"In order to limit risk of intraneural injection, 'it is probably recommended' to approach the nerve tangentially and to verify before injection, by using small movements of the needle, that its tip is not stuck in the nerve. 'It is recommended' to discontinue the injection of the anaesthetic solution in the absence of real-time visualisation of the diffusion of the local anaesthetic and/or in case of pain, paraesthesia, resistance to the injection or swelling of the nerve. 'It is recommended' to remove the needle in case of an intraneural injection because it is impossible to demonstrate the safety of such an injection in spite of its often painless characteristic".

It can be considered that the vast majority of peripheral nerve blocks have been widely described. Many approaches have been considered for each of them. Some of them are very useful, others very "ingenious" but carry a potential risk, which means that they are rarely used, and lastly, sometimes it is found that they have been described seemingly to discourage their use. Some variants regularly give rise to new relevant published reports.

In medicine, as in many fields, **a close relationship exists between knowledge, tools, techniques and results.** All progress obtained in any of these fields has an impact on others. Fortunately, anaesthesia does not escape this rule; introduction of new materials or new techniques or advances in basic knowledge are often the basis of innovations in how to approach nerves.

Although the approach first used in neurostimulation has mostly been entirely satisfactory, it should not be forgotten that they have been described for this "blind" technique. In neurostimulation, they make it possible to limit needle puncture accidents and to obtain motor (or sensory) responses more readily than is sought, but not all are appropriate, useful and/or necessary for ultrasound guidance.

The advent and then the current popularity of this technique in RA have made it possible to add to the list of approaches for many peripheral nerve blocks:

- Most often, the operator is free of the surface markers by having a "visual" on a target nerve. The insertion point and trajectory of the needle with ultrasound often differ by this simple fact.
- It is important to have probes that are morphologically adapted to patients (from a paediatric patient to an obese patient) and to the different sites of needle puncture. The difference in overall probe size can sometimes make positioning of the needle and its visualisation difficult. Coordinating probe handling with needle movements can sometimes present difficulties; it is necessary in these cases to call upon ingenuity so that the needle is visible on the ultrasound screen while continuing to see the important structures clearly.

Understanding the need to change some technical aspects or to look for new approaches in ultrasound-guided RA is to recognise the fact that to be under **optimum** conditions of success and safety of a peripheral nerve block, it is always necessary to be able to see **simultaneously** three elements which are not always reflected in the methods used in neurostimulation:

- the target nerve(s)
- the tip of the needle
- the spread of the local anaesthetic

Target Nerves

Histology (Figs. 3.9 and 3.10)

Peripheral nerves consist of sets of nerve bundles enveloped by three distinct layers of tissue, which are (from superficial to deep): the epineurium, the perineurium and the endoneurium.

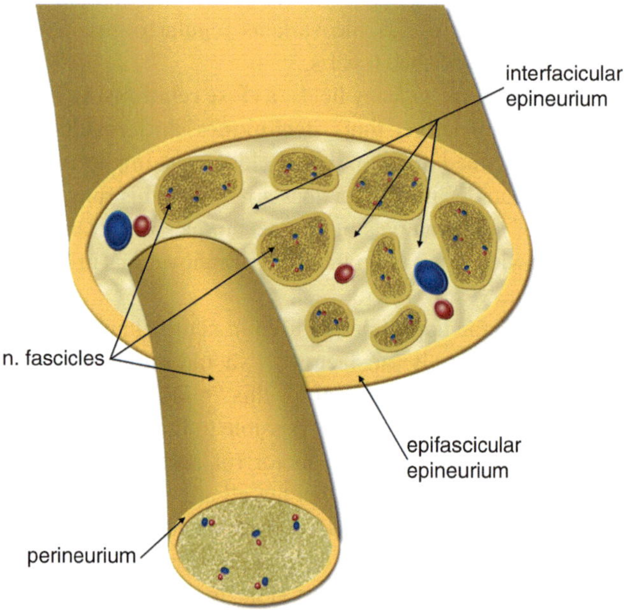

Fig. 3.9 Structural diagram of a nerve

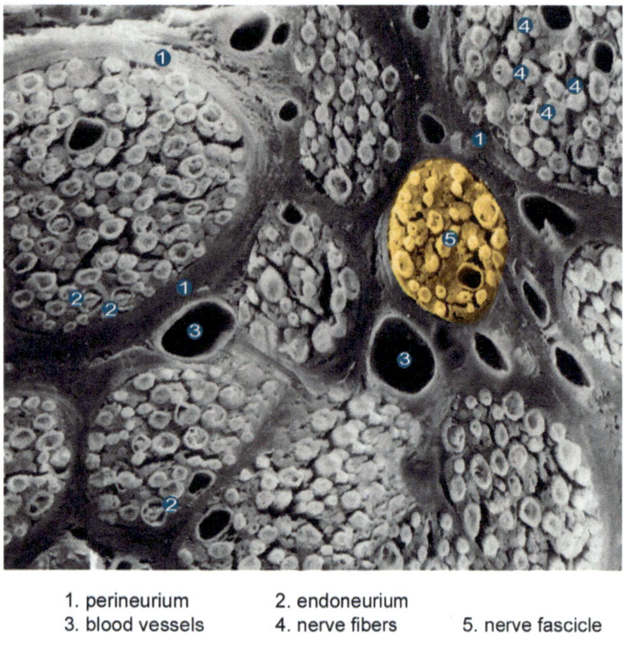

Fig. 3.10 Histological section of a nerve under electron microscopy

The **epineurium** has two components: the intrafascicular (or internal) epineurium which separates and surrounds the bundles, and the perifascicular (or external) epineurium which surrounds the entirety of the nerve (relatively loose fibrous envelope). Each of the bundles, surrounded by the **perineurium** (a dense resistant fibrous sheath), consists of a collection of juxtaposed "neural fibres", with the latter containing axonal tissue. Each neural fibre is covered by the **endoneurium**. The echoic component of nerve is represented by the perineurium and the epineurium, as well as by the connective tissue, lymphatic and vascular structures placed within it, whose density ranges depending on the level studied (emergence from the spine or a more peripheral nerve). Blood vessels and connective tissue are thus found between the fascicles (in the internal epineurium) but also between the neural fibres within each fascicle. In the periphery of the external epineurium, peripheral nerves are surrounded by connective tissue variably called mesoneurium, paraneurium, adventitia, etc., in variable proportions depending on the site studied. These envelopes have a plural role, both for nutrient supply (vascular supply), mechanical protection and to facilitate sliding in relation to the adjacent anatomical structures in the areas of great mobility (in particular the popliteal and axillary fossae) [15–17] (Fig. 3.11).

Ultrasound Image

Nerves are "filled" tubular structures for which it is relatively easy to follow the pathway with ultrasound, and more difficult with a CT or MRI scan. They are mobile and can vary in their position by impressing movements of the limb involved or by applying more or less pressure with the ultrasound probe. Nerves are not very compressible, unlike blood vessels which, additionally, are characteristically identified with Doppler ultrasound.

It is not possible to make them disappear by applying pressure with the probe, but it is possible to modify their shape a little.

In a transverse section ("short axis"), the nerve has a rounded, triangular or oval outline [18–20]. In it can be differentiated from small hypoechoic nodules (the fascicules),

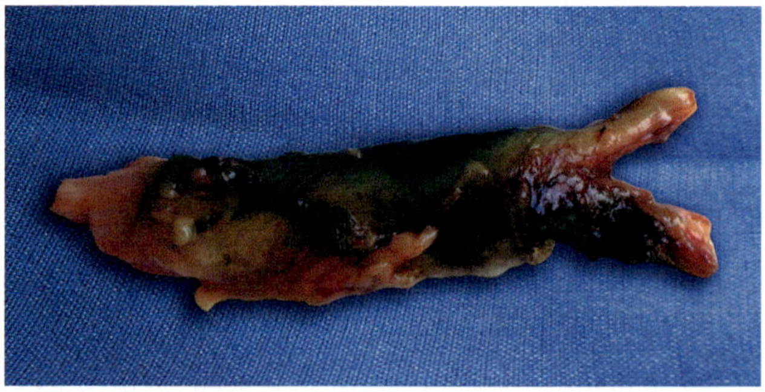

Fig. 3.11 Sciatic nerve bifurcation in the popliteal area: note the importance of the perineurium in which green ink has been injected

which are bathed in a hyperechoic environment (perineurium, interfascicular and perifascicular perineurium, paraneurium), producing a **follicular** presentation (Figs. 3.12 and 3.13).

In a longitudinal section ("long axis"), the nerve is striped with a "**fascicular**" structure (Fig. 3.14) characterised by multiple hypoechoic strips that are more or less discontinuous, separated by hyperechoic lines. It can sometimes be confused with a tendon whose "fibrillar" ultrasound presentation approaches it, although the parenchyma of the latter appears "thinner", comparable to the strings of the bow of a violin [21], with multiple continuous hyperechoic lines (Fig. 3.15).

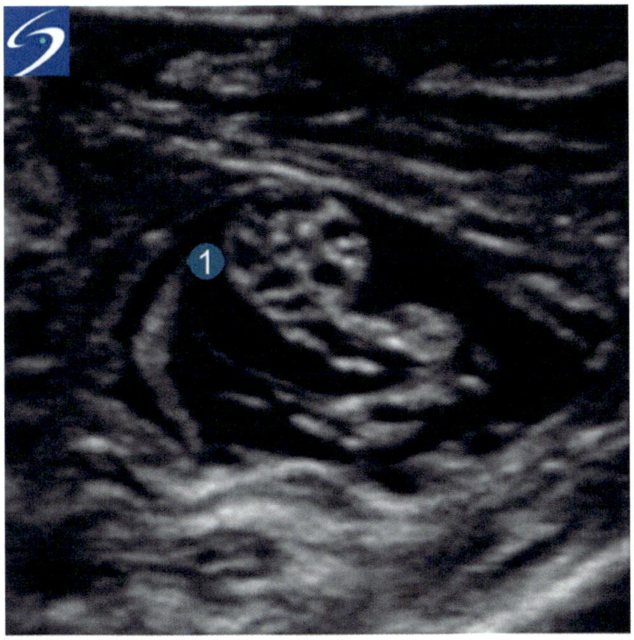

1. tibial n.

Fig. 3.12 Honeycomb appearance of the tibial nerve. Transverse section

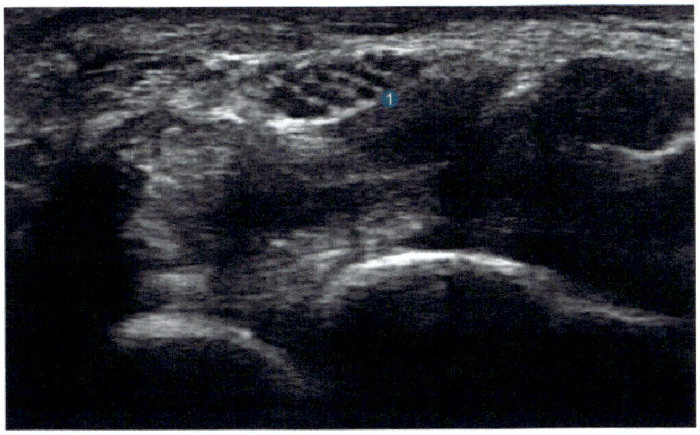

1. median n.

Fig. 3.13 Honeycomb appearance of the median nerve. Transverse section

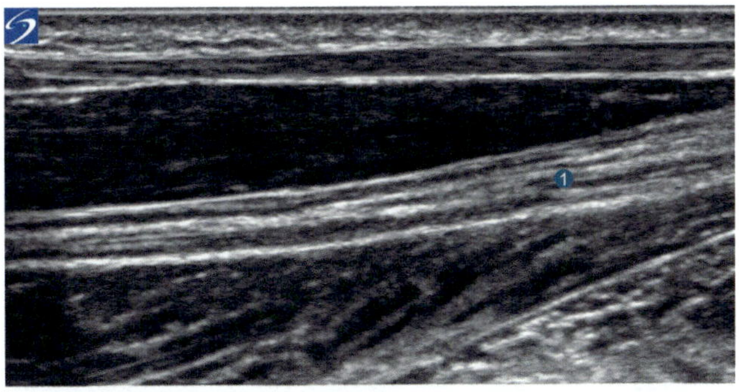

1. "fascicular" aspect of a peripheral n.

Fig. 3.14 Longitudinal section of a nerve

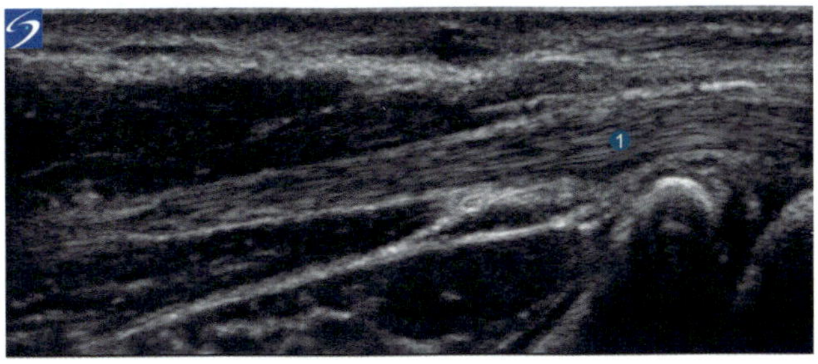

1. "fibrillar" aspect of a tendon

Fig. 3.15 Longitudinal section of a tendon

Since the bundles and the interfascicular and perifascicular connective tissue which comprise it have different echogenicity, the ultrasound presentation of the nerve varies depending on the level in which the following is observed:

- Insofar as a high fascicular density is found inside, the anterior branches of the cervical nerve routes of the brachial plexus, as can be observed in the interscalene space, have a totally hypoechoic ultrasound presentation of a "monofascicular" type [22], encircled only by a hyperechoic border (Figs. 3.16 and 3.17). This type of image can be differentiated from a vascular structure by the absence of posterior reinforcement (inconstant) and of the Doppler signal.
- The more it is studied distally, the more its specific multifascicular histology reveals it as a "follicular" heterogenous ultrasound structure in a 'honeycomb" shape (Figs. 3.12 and 3.13). However, note that even with a 15 MHz probe, ultrasound only allows for the appearance of a third of the fascicles visible with an optical microscope [23].

Even though the hyperechoic intraneural tissue (perineurium, interfascicular and perifascicular epineurium) may appear more abundant in heavier patients [24], the dimensions of nerves are invariable based on age and are relatively

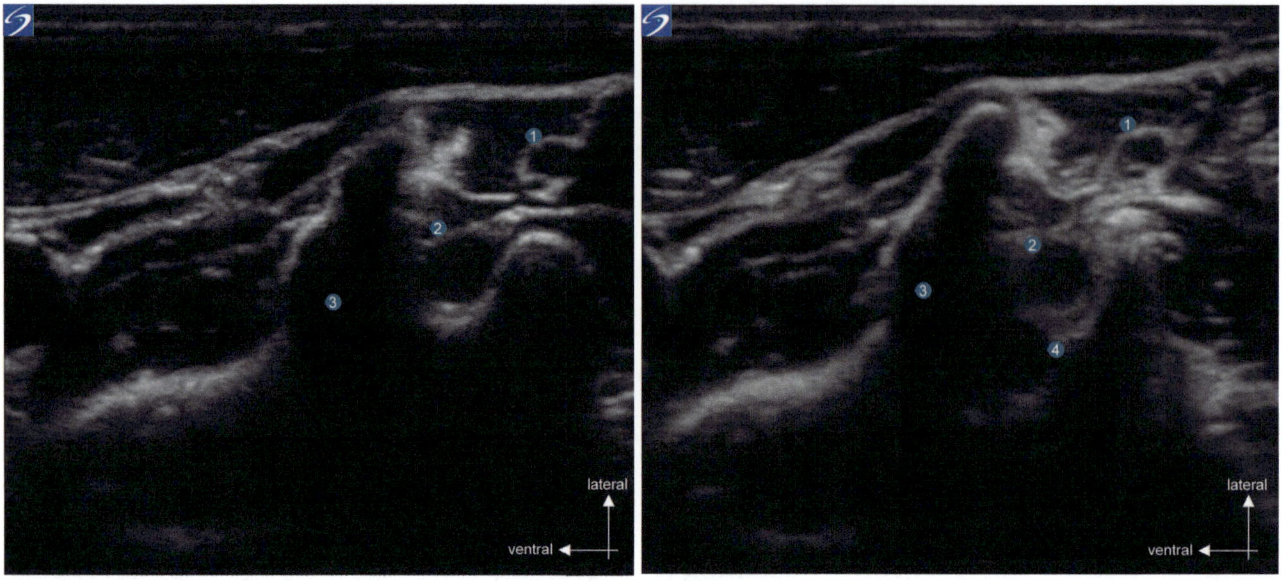

Fig. 3.16 "Mono-fascicular" appearance of cervical nerve roots

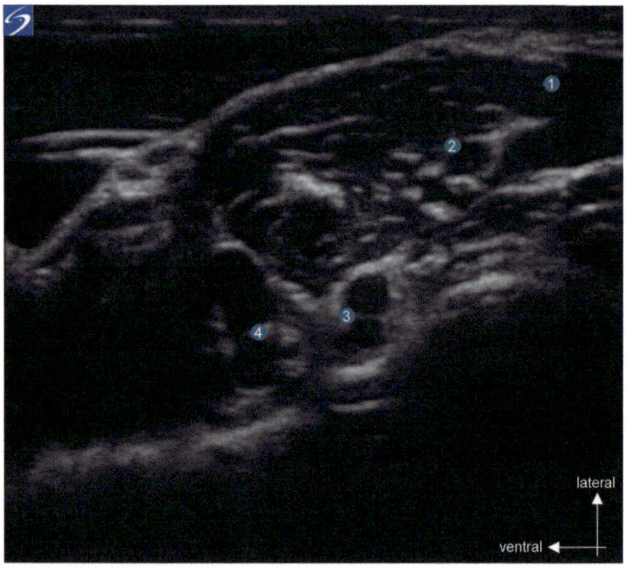

Fig. 3.17 Image illustrating risk of confusion between hypoechoic nerve (1–2–3) and vascular structures (4)

uninfluenced by the physical characteristics of individuals. The ultrasound structure of the nerve is not modified in case of motor or sensory hereditary neuropathy [25].

The size of a nerve hardly decreases at the end of its trajectory. When, for example, the median nerve is observed from its formation up to the wrist, in spite of different branches that it may give rise to, we can note that its size hardly varies. The axonal distribution with variable successive branches that it has given rise to was partially compensated by more abundant connective tissue. Figures 3.18, 3.19 and 3.20 show it in the brachial

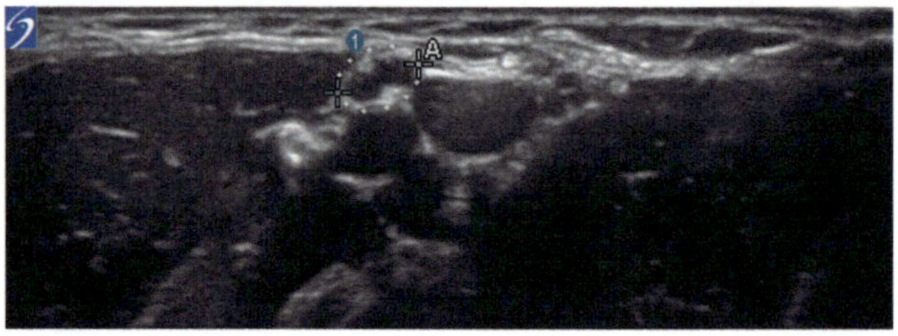

Fig. 3.18 Median nerve in the proximal brachial area

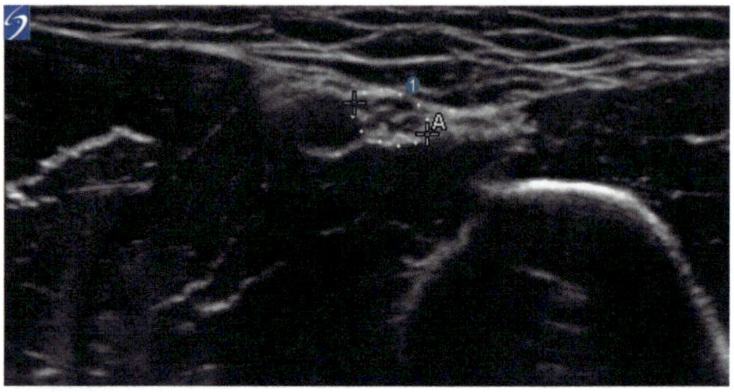

Fig. 3.19 Median nerve in proximity to the medial epicondyle

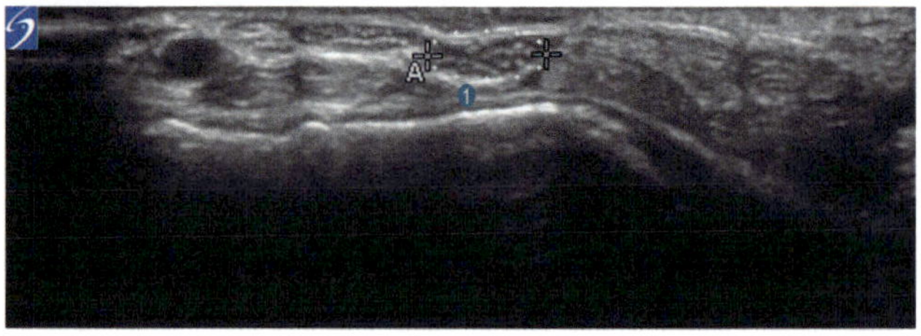

Fig. 3.20 Median nerve at the level of the distal end of the radius

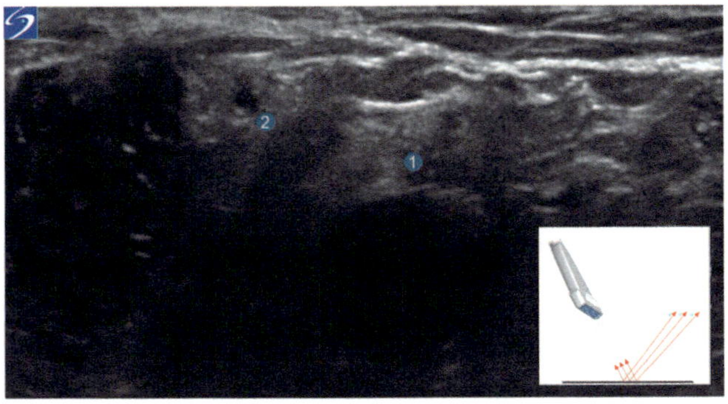

1. tibial n. 2. common fibular n.

Fig. 3.21 Tibial and peroneal nerves poorly visible as a result of unfavourable tilting of the probe

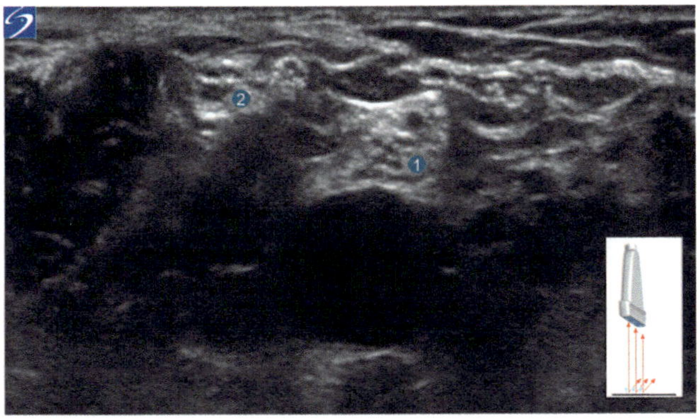

1. tibial n. 2. common fibular n.

Fig. 3.22 Tibial and peroneal nerves better identified thanks to a more favourable tilting of the probe

canal above the medial epicondyle and 1 cm above the flexion fold of the wrist, respectively.

Anisotropy is an artefact which consists of a variation in echogenicity of structures that are studied depending on the incidence of the ultrasound which approach them. Contrary to tendons whose sometimes similar appearance can be the source of confusion, the nerves are rarely subject to anisotropy. More precisely, they are the **bundles** which are rarely subject to the artefact of anisotropy. On the contrary, it is possible to vary the echogenicity of the **perineurium**, of the **epineurium** and the **paraneurium** by modifying the incidence of the ultrasound bundle. Therefore, this process becomes all the more evident as one is distanced from the spinal emergence of the nerve. It often permits, by allowing the appearance of hyperechoic bundles in the perineural and epineural connective tissue for which the echogenicity is varied, the confirmation of the nerve nature of the structure that may be difficult to identify.

Using anisotropy often proves very useful to improve the visibility of the nerves that we are looking for. This is true when they are located in a fatty cellular environment, such as in the popliteal or femoral area, where major heterogeneity in surrounding structures exists (fat, fasciae, etc.), as well as when they are in close contact with muscle structures, such as when we wish to study the brachial plexus in the axillary area. Figures 3.21 and 3.22 show the tibial and common fibular nerves as they appear within the popliteal fossa according to changes in the angle between the nerves and the ultrasound beam.

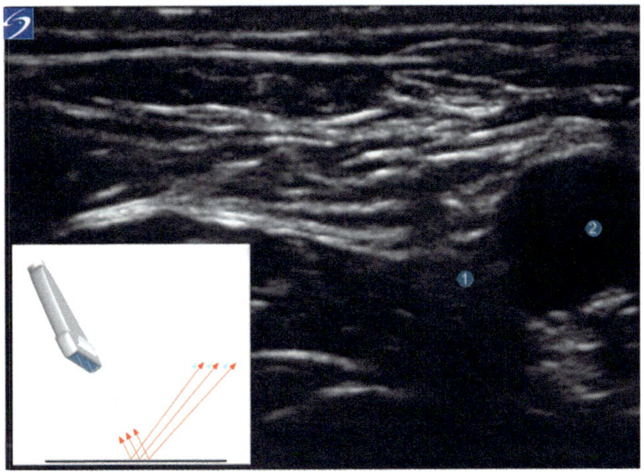

1. femoral n. 2. femoral a.

Fig. 3.23 Poorly visualised femoral nerve as a result of unfavourable tilting of the probe

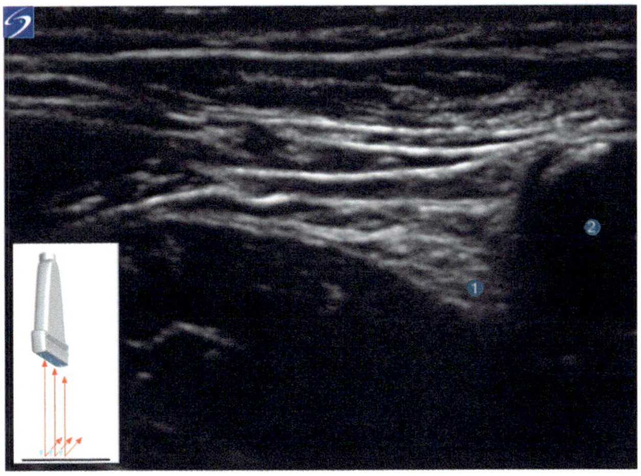

1. femoral n. 2. femoral a.

Fig. 3.24 Femoral nerve better visualised thanks to a more favourable tilting of the probe

The femoral nerve in the inguinal fold usually divides into multiple terminal branches in a "divergence". Its echogenicity is often poor and it is necessary to attempt to best visualise it by slightly tilting the probe on either side of the perpendicular to the skin [26] (Figs. 3.23 and 3.24).

For peripheral nerve blocks, in the majority of cases it is possible to visually locate the nerves sought. However, they can sometimes be only partially distinguished or not at all in spite of their presence. Generally, this relates to deep nerve blocks, in which the limitation is both technical (ultrasound performances) and anatomical. Whenever we know that a nerve lies in immediate proximity to another tissue structure that is visible with ultrasound, even though the nerve itself is not (because of its size, its position or its depth), the adjacent visible structure can be used as a marker for positioning of the needle and injection of the local anaesthetic. However, it is then more than ever necessary to use secondary guidance methods for assistance, particularly neurostimulation.

For example, in an infraclavicular approach to the brachial plexus, good visualisation of all of the nerve structures is obtained only in less than 30% of cases [27]. It is not always easy to locate the cords which surround the axillary artery and vein. The **lateral** cord is most often visible, but the **medial** cord (which—depending on the level—is of variable position with relation to the axillary artery) and the **posterior** cord (which is regularly confused with the posterior acoustic enhancement located deep to the artery) are much more difficult to locate. Therefore, during conduct of a nerve block, it is often necessary to be satisfied with visualisation:

- of the axillary artery.
- of the lateral cord, in lateral and cephalic position with relation to the axillary artery.

The end of the needle must then be positioned in the deep aspect of the axillary artery in order to inject the local anaesthetic, without damaging the posterior and medial cords which may particularly be found along the pathway of the needle. In this type of situation, **neurostimulation** remains a precious tool for:

- **aid in finding location**, via motor responses corresponding to the nerves that you think that you are approaching (by visual confirmation) or that you fear that you may encounter without being able to differentiate them;
- **safety**: it enables to modify the position of the needle if a motor response is obtained for a stimulation <0.2 mA, in order to avoid intrafascicular intraneural injection;
- and lastly, observation of the **appropriate spread** of the local anaesthetic: a crucial element in the successful conduct of the procedure.

It may appear aesthetically attractive to visualise the target nerve in its "long axis" (longitudinal section). However, firstly, it is often difficult to obtain this image in a stable manner considering the precision of position of the probe required and an often imperfectly rectilinear trajectory of nerves. Secondly, this limits the ability to position the needle around the nerve. It is much simpler and satisfactory to obtain the image of the nerve in its "short axis" (transverse section). Thus, greater freedom of movement of the probe is maintained, as well as a greater facility in positioning the needle in the different quadrants around the nerve and better vision of spread of the local anaesthetic in its contact.

Needle

The hazardous nature of RA is inherent to the local and systemic toxic potential of the local anaesthetic injected, but also to the risk of injury related to the needle with respect to the nerves involved and to the adjacent anatomical structures. These processes have been demonstrated by Selander et al. in 1979 [28], and the principles guiding the practice of RA have always aimed to avoid needle-to-nerve contact and intraneural injection. Nerve punctures do not invariably result in symptomatic lesions [1]. However, whatever the size and type of the needle, needle puncture invariably induces an inflammatory reaction. Available studies on lesions induced by **nerve puncture** suggest, firstly, that the bundles tend to move out of the path of the needle, and secondly, that short needle bevels perhaps carry less risk of damaging the fascicles but that fascicular injury that they generate probably are more severe than with long needle bevels [29–31]. In order to avoid puncturing a nerve, it is imperative to always know where the tip of the needle lies.

The principal factors of visibility of the needle with ultrasound are as follows:

- its size [32]
- its reflecting potential (smooth or textured lining)
- the configuration of the needle/probe used ("in plane" or "out of plane")
- the angle formed by the reflecting surface of the needle and direction of the ultrasound beam

Visibility of the needle increases:

- with its size
- with the angle that it forms with respect to the ultrasound beam
- with the existence, on or in the lining of the needle, of structures that reflect ultrasound

The tissue characteristics of the environment of the needle also condition part of its visibility. Structures which are not very echoic or anechoic (blood vessels, local anaesthetic) make it easier to locate. Conversely, the existence of heterogeneous anatomical elements with multiples interfaces "cloud" the image.

Choice of technique for insertion of the needle with respect to the ultrasound probe plays a role in the visibility of the images. It is possible to choose insertion "in plane" or "out of plane" of the ultrasound.

"Out of Plane" Insertion

In this approach, the needle tip is placed next to one of the two faces of the ultrasound probe. It is recommended to prick the skin close enough to the probe (1 cm) so that the distance between the puncture site and the target to reach is about equivalent in depth to the target in the plumb line of the probe (Fig. 3.25a). It is thus possible to determine in real time if the length of the needle inserted is consistent with the

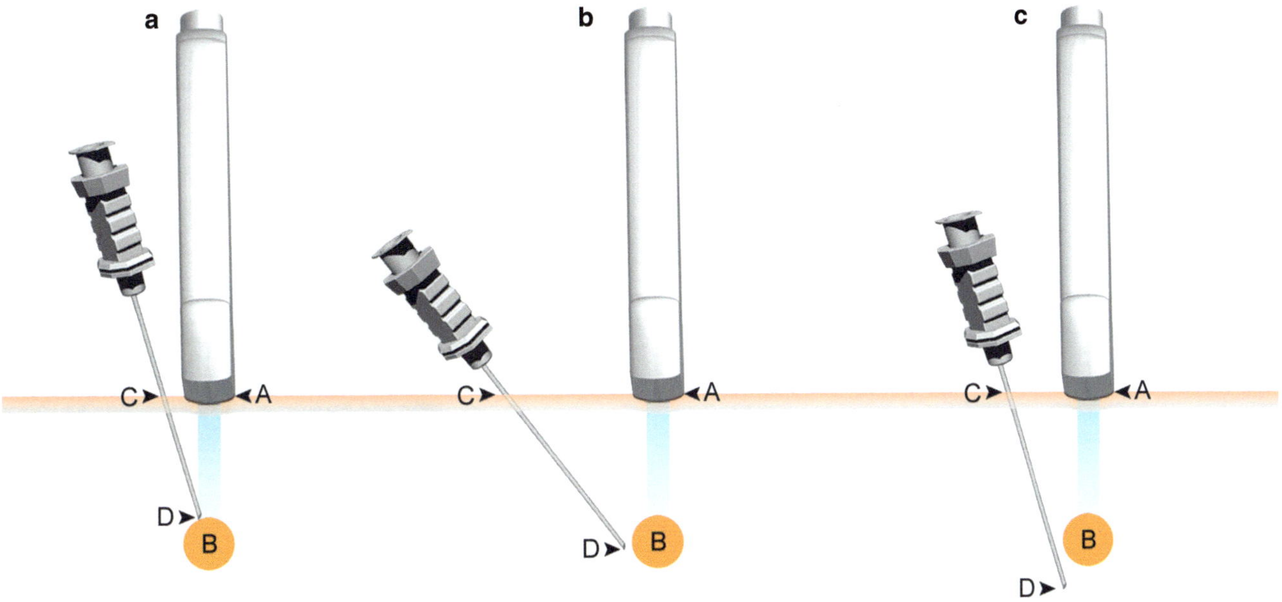

Fig. 3.25 Diagram showing the advantage of injection close to the probe: (**a**) length of insertion [CD] roughly equal to depth of the target on the screen [AB]. (**b**) length of insertion [CD] much greater than depth of the target on screen [AB] while tip of the needle is close to the target. (**c**) length [CD] greater than [AB] immediately indicates too deep insertion of the needle

depth of the target. If the length inserted is greater than the depth of the target, it can be immediately deduced without a trigonometric calculation that the needle has been introduced too deep (Fig. 3.25c). However, the more distant the chosen needle puncture site is from the probe, the less this deduction is simple and intuitive (Fig. 3.25b).

In the out-of-plane approach, the needle is represented on the screen by a more or less hyperechoic pinpoint image followed by a more or less pronounced acoustic shadow (Fig. 3.26). Depending on the nature of the tissue environment and of the needle, the latter is more or less visible (Fig. 3.27). Since it will not be possible to differentiate the image of a cross section of the **body** of the needle from that of its **tip**, it is necessary to use additional means to precisely determine the position of the tip. In an approach to a nerve in proximity to the pleura or to blood vessels (supraclavicular or infraclavicular block), making an error when positioning the needle tip clearly risks the puncture of a blood vessel or a pneumothorax. But without even considering the existence of vulnerable structures nearby nerves, the simple risk of needle puncture and/or intraneural injection in itself is a risk not to be neglected.

Additional methods for locating the needle are principally:

- ultrasound observation of tissue movements induced by movement of the needle
- special attention paid to sensations during advancement of the needle and their analysis (jerking movement, loss of resistance, crossing of fasciae, etc.) in the patient's perception (pain, paraesthesia, etc.)

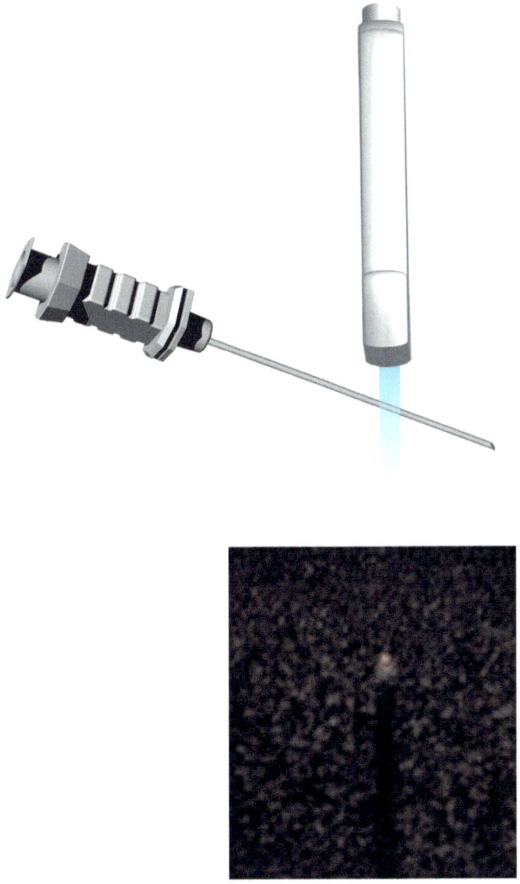

Fig. 3.26 Dot-like visualisation of the needle with an out-of-plane approach

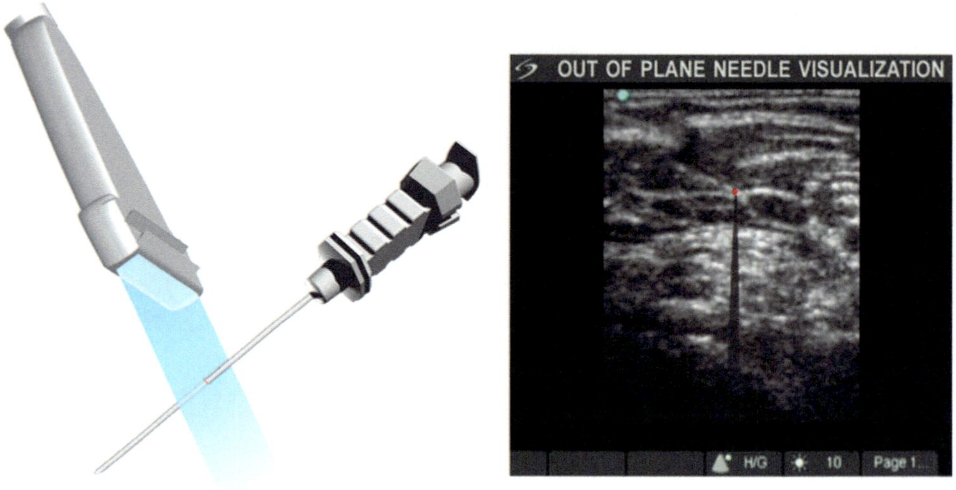

Fig. 3.27 Theoretical visualisation of a transverse section of the needle during out-of-plane insertion. Depending on the tissue studied, visualisation of the needle can be more or less difficult

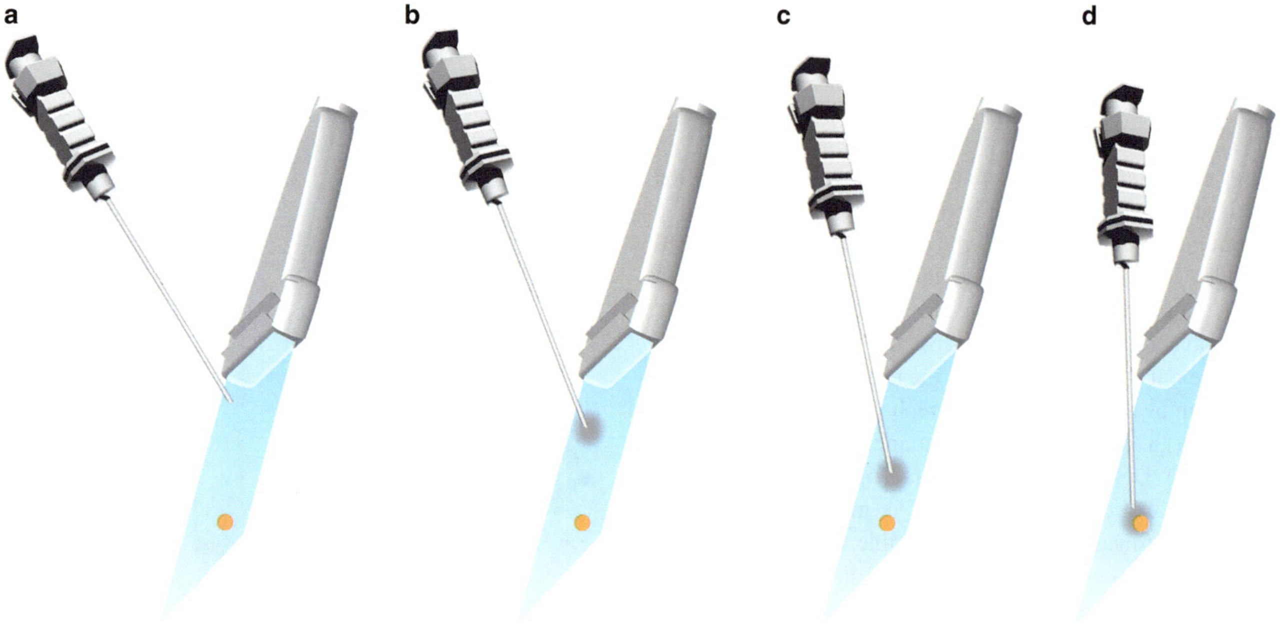

Fig. 3.28 Principle of hydrolocation

- neurostimulation and analysis of motor and/or sensory responses
- hydrolocalisation (Fig. 3.28)

The latter, as soon as the skin is crossed, consists of administering small successive injections of the liquid during advancement of the needle in order to observe on the ultrasound image the depth at which the tip of the needle lies. This is a simple and effective method of identifying the poorly visible pathway of the needle "out of plane", but which is sometimes also used when orientated "in plane". Ideally it is desirable to perform such "microinjections" with a dextrose 5% in water solution in order to be able to continue to perform neurostimulation simultaneously, to avoid muscle toxicity of local anaesthetics during intramuscular hydrolocalisation, and to eliminate an unnecessary increase in the dosage of the local anaesthetic injected.

Insertion "In Plane"

In this approach, the chosen needle puncture point lies at one end of the probe and the trajectory of the needle is directed into the ultrasound plane within which it is important to stay. In the case where the probe can absolutely not be moved without losing the ideal image, it is the needle which must not be removed from the plane. If, however, it is possible to slightly move the probe without compromising the informative nature of the image, the practitioner can perform small adjustments of the position of the probe in order to maintain a longitudinal view of the needle at all times during the procedure. In the majority of cases, maintaining the needle in the ultrasound plane requires simultaneous readjustments both of the needle and of the probe. In fact, the total immobility of the probe during the anaesthetic procedure is necessary in rare cases only (real-time ultrasound-guided spinal anaesthesia, epidural anaesthesia, paravertebral nerve blocks, etc.).

The advantage of the "in plane" approach is longitudinal visibility of the needle with easier determination of the position of its tip (Figs. 3.29 and 3.30). Bending of the needle (due to its flexibility, which is more pronounced when it is a narrower gauge) or a slight divergence between its path and the ultrasound plane, however, can result in a false impression concerning position of the tip (Fig. 3.31).

In an in-plane approach, the more the angle formed by the needle and direction of the ultrasound approaches 90° (needle is parallel to the surface of the skin on which the probe is applied), the more the needle is visible, but the more reverberation artefacts by the body of the needle itself are generated [33] (Fig. 3.32). In this approach fewer differences in echogenicity exist between the different models of needles.

Even though ultrasound guidance makes them probably less common than in all other techniques for conduct of peripheral nerve blocks, vascular punctures have been

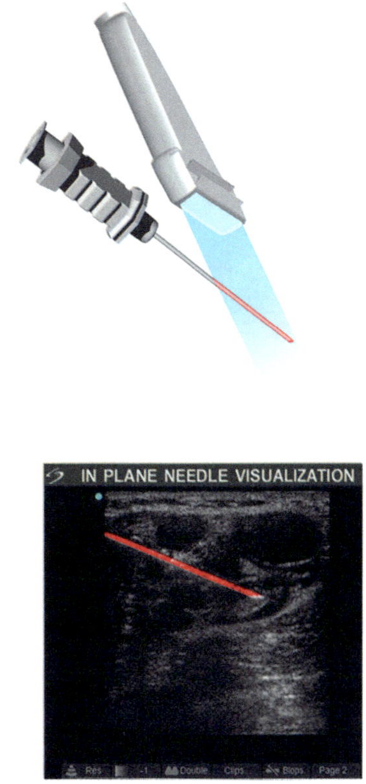

Fig. 3.30 Theoretical visualisation of the needle with an in-plane insertion

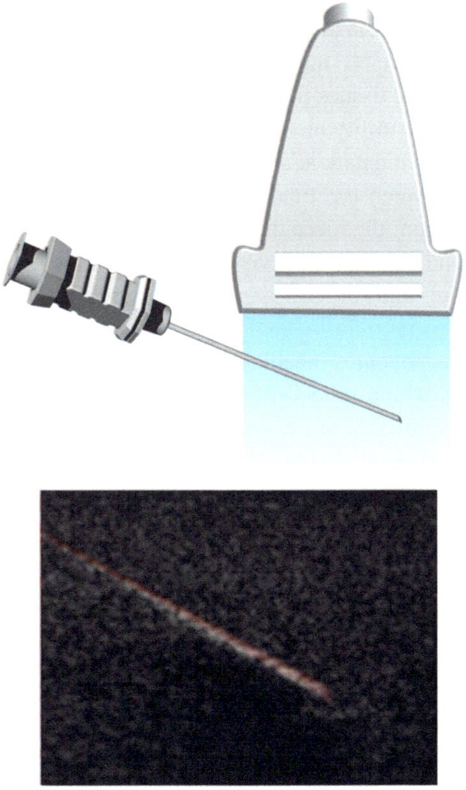

Fig. 3.29 Visualisation of the needle in an in-plane approach

Fig. 3.31 False impression of knowing the position of the bevel (tip) due to an oblique section of the needle by the ultrasound beam

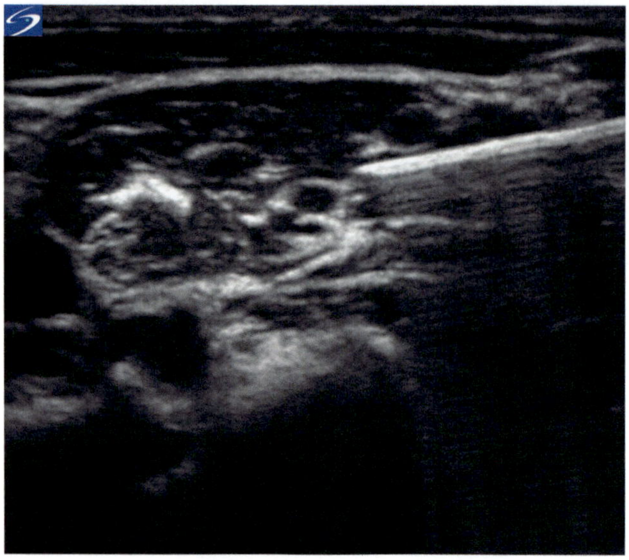

Fig. 3.32 Reverberation artefacts

described with ultrasound, in spite of an "in plane" needle orientation [34, 35], which emphasises the need to visualise the tip of the needle, the only "hazardous" item along with the local anaesthetic.

To date, the majority of needles which were used in RA and in neurostimulation were not designed to be especially echogenic. Even though the bevel sometimes is slightly more echogenic so than the body of the needle (due to the facets of the tip), it is not always easy to identify. Consequently, performing a peripheral nerve block with a needle "in plane", for a beginner (but also for an experienced anaesthetist), is an additional guarantee of safety in the sense that this enables better control of the position of the tip of the needle. The procedure, when adopting the "out of plane" approach and performing hydrolocalisation, creates an approximation because the needle is hardly visible. This "out of plane" technique (and also its concept) often involves not taking a needle tip position in immediate proximity of the nerves but rather slightly at a distance. The local anaesthetic is then injected into a favourable space of diffusion, allowing it to reach the targeted nerve(s).

The utility of visualising the needle "in plane" is reinforced during conduct of "multi-echolocation" nerve blocks, as for example an axillary nerve block. In this case, successively, without changing the skin needle puncture site, the anaesthetist redirects the needle towards several nerves located in the same ultrasound field. This procedure is easier when the needle can be seen longitudinally.

How to Improve Vision of the Needle?

One cannot deny the considerable utility of the **visual dimension** provided by ultrasound in RA. However, it can be achieved only by optimising the quality of the image. The intrinsic qualities of needles, catheters, but also of ultrasound systems and of their probes in themselves do not make it possible to spontaneously offer the best conditions of visualisation. Many factors play a role: some are human and others technical and material. Other than training the operator, which from the outset must be done in an appropriate manner for the procedure that he or she is preparing to perform, the methods for optimisation of visibility of the needle should be performed prior to the procedure everytime.

Training of the Practitioner

- Theoretical and practical knowledge: the physics of ultrasound, technical characteristics of the ultrasound system used, anatomy and RA, etc.
- Appropriate training supervised by competent instructors: on phantoms, computer simulator, the anatomy laboratory. These strategies train the novice practitioner to optimise their performance and to constantly know the position of the needle tip during anaesthetic procedures. Ergonomic parameters have an important place in improving the comfort of working conditions, and therefore performance: position of the screen opposite the practitioner, optimising muscle relaxation to avoid contortions and operator fatigue, the hand which holds the probe should at the same time be in contact with the patient to avoid accidentally sliding, etc.
- Observation, followed by performance of the procedure on patients, supervised by experienced practitioners.

Visibility Factors Related to the Needle

- The needle is much more visible when the acoustic impedance of the tissues in which it lies is low.
- Traditionally, visibility increases with the diameter of the needle (as does the patient's discomfort).
- The angle between the trajectory of the needle and the ultrasound beam; the more the needle is perpendicular to the ultrasound, the more it is visible. Choice of the needle puncture site and of the angle of approach thus is essential.
- So-called "echogenic" needles: produced with surface irregularities which reflect ultrasound in multiple directions (either particles in the surface lining or notches in the body of the needle) (Fig. 3.33).
- Shape and direction of the needle bevel.

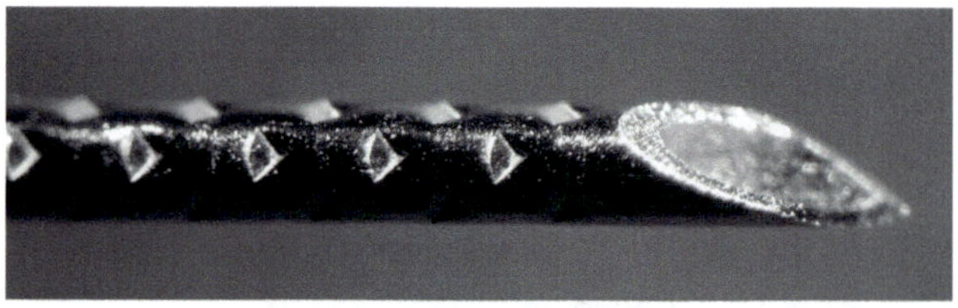

Fig. 3.33 Surface irregularities of the needle added in order to make them more echogenic: *cornerstones* (Pajunk®)

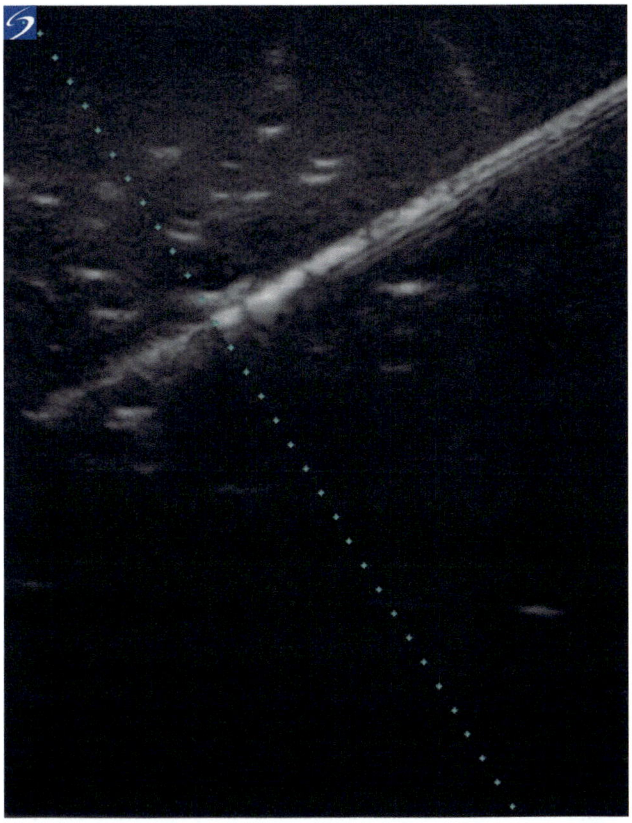

Fig. 3.34 Sonogram with use of the *beam steering* function (MBe, Sonosite®) in which we note better visibility of the needle in the area where it is active (to the right of the dotted line)

Visibility Factors Related to Artefacts

Anisotropy, reverberation, refraction, attenuation, distortion, etc.

Factors Related to Ultrasound Mode

- *Spatial compound* (summation of several images generated by ultrasound firing from different angles): This improves image clarity and resolution, decreases a certain number of artefacts and produces better definition of the outline of the needle.
- *Beam steering* (inclination of the ultrasound beam): Enables inclination of the ultrasound beam, which can be adjusted so that it can be more perpendicular to the axis of the needle. The signal generated by the latter, therefore, is improved (Fig. 3.34).
- *Frequency compound* (integration of several images generated with different ultrasound frequencies): Improvement of the tissue image but not of the needle.
- *Harmonic imaging:* Tendency to improve needle visibility.
- *Gain and TGC (time gain compensation):* Optimising these parameters improves needle visibility, in particular by decreasing the artefact generated by the skin in superficial layers.

Needle/Ultrasound Beam

- The in-plane approach generally enables better visibility of the needle, provided that it is correctly placed in the plane (which is only about 1 mm thick on average) and that the angle between the trajectory of the needle and the direction of the ultrasound is as close as possible to 90°.
- The out-of-plane approach shows the needle only in a transverse section image, which is generally of poor quality. Generally, an attempt is made to place the probe so that the needle is aligned to the mid-point of the probe. If the trajectory of the needle is not perpendicular to the ultrasound plane but is slightly oblique, its visibility may be substantially improved. Use of hydrolocalisation can improve the view of the section of the needle.
- Needle guide: For an in-plane approach, the use of a mechanical guide for needle puncture can greatly facilitate stability of the needle trajectory within the ultrasound beam. Guides with variable angulation are more appropriate for ultrasound-guided RA than those with fixed angulation.
- By holding the syringe in a "traditional" manner, the practitioner must repeatedly release and then take the needle back according to the injection phases of the LA. This may create difficulty in maintaining good visibility of the needle. An alternative technique for holding the syringe may make it possible to hold the needle throughout the procedure, and therefore to perform the procedure in a more fluid manner [36] (Figs. 3.35 and 3.36).

Electromagnetic Follow-Up of the Needle (GPS)

This is a spatial method to acquire the position and direction of the needle in real time. A signal generated by a device integrated in the needle is located in a controlled magnetic field, enabling the measuring system to calculate position and orientation. The parameters calculated are displayed on the screen of the ultrasound system. The SonixGPS™ system

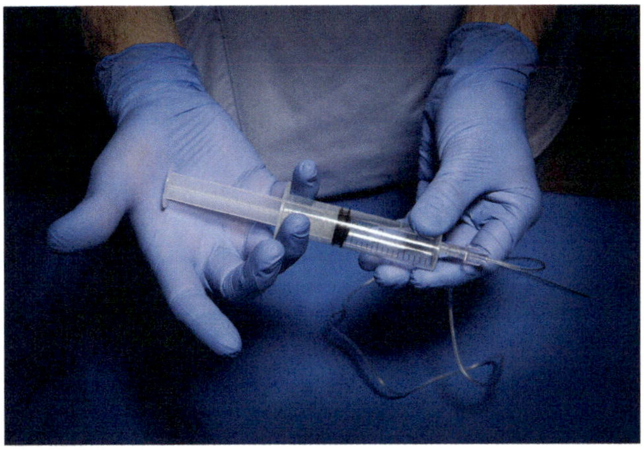

Fig. 3.35 Technique for holding the syringe: the syringe is placed in such a manner so that the plunger is located at the base of the thumb (thenar eminence)

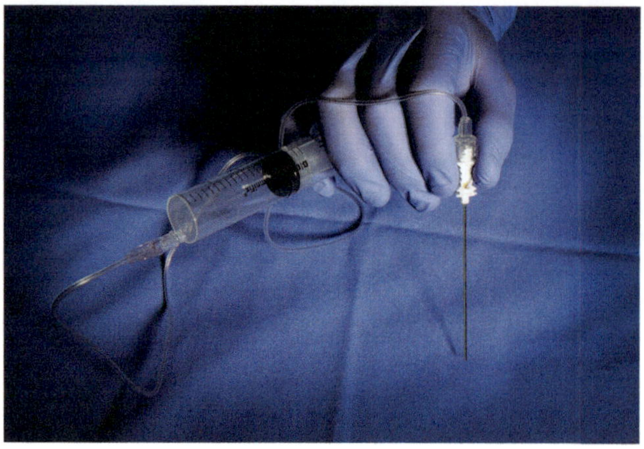

Fig. 3.36 Technique for holding the syringe; thumb and index are thus freed in order to hold the needle

(Ultrasonix, Richmond, BC, Canada), which was recently introduced on the market, makes it possible in particular to know the instantaneous position of the needle and its planned movement according to its direction. One of the principal limitations lies in the interference generated by the metal materials surrounding the electromagnetic field generator.

Various Techniques

- **Neurostimulation**: A visual technique can only be used under good visibility conditions! If this is not the case, then it is essential to use another technique to facilitate ultrasound guidance [37]. Neurostimulation makes it possible to reveal or to confirm position of the needle in contact with a nerve structure [38–40].
- **Colour Doppler**: Movements of the needle can produce Doppler signals, which aid in locating the needle under difficult conditions. A vibrating device that is placed on the base of the needle is commercially available. Vibrations induced in the body and at the tip of the needle generate a more or less visible Doppler signal [41].
- Whether it is intended to assess the depth to which the point is located in out-of-plane approaches *(hydrolocalisation)* or in demonstrating planar spread during needle puncture *(hydrodissection)*, **injection of the liquid** can variably improve visibility of the needle, at least by showing the position where the bolus appears. Injection of liquid containing micro-air bubbles can improve further visibility of the injected substance. However, the air is also a source of artefacts which can penalise the remainder of the procedure.

Method of Spread of the Local Anaesthetic

The method of spread of the fluid injected (local anaesthetic, dextrose 5% in water solution, etc.) is one of the **fundamental** observable parameters provided by ultrasound in RA, along with the trajectory of the needle. It predicts whether or not the

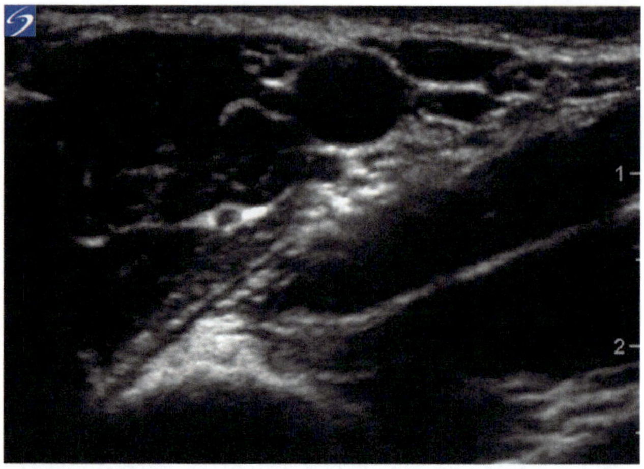

Fig. 3.37 Brachial plexus in the axillary area before injection of the local anaesthetic

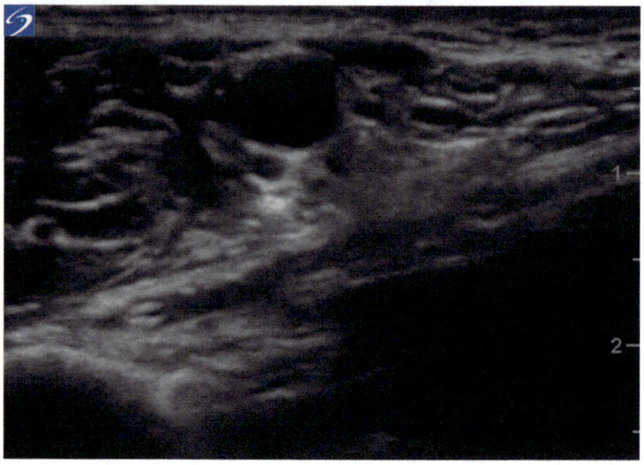

Fig. 3.38 "De-structuring" of the anatomical relations of the nerves of the brachial plexus after injection of the local anaesthetic

distribution of the local anaesthetic is appropriate (intra- or extraneural, in contact or too far away from the nerve, etc.) and provides the potential to correct the position to ensure optimum efficacy, which at present is **predictable**. Injection of the local anaesthetic produces major visual changes in the region studied. It reinforces the contrast between certain structures (nerve/cell and fatty tissue), often improves the visibility of the needle significantly, but also modifies anatomical relations (Figs. 3.37 and 3.38) and reduces visibility of blood vessels as they are surrounded by the anechoic local anaesthetic. Thus, it is recommended when administering multiple injections in a given site (for example, axillary nerve block) to start with the deepest nerves and those which are most distant from the needle puncture site, and finish with the most superficial and closest nerves in order to not have to cross an area which has already been injected. Care also must be taken to avoid accidental injection of air bubbles, since the latter are the source of artefacts which alter the accuracy of the procedure. Conversely, under some conditions, injection of a very low volume of air can make it possible to specify the position of the needle bevel when visibility is poor. In summary, injection of liquid (local anaesthetic or dextrose 5% in water solution) generally makes it possible to assess the position of the needle end or, in any case, the depth at which the point is located ("hydrolocalisation").

"Appropriate" Neurostimulation

Introduction

The ultrasound image is reconstructed, and is not a direct visual representation of the structures observed. Therefore, it is important to know the limits and the artefacts which may occur when the apparatus "creates" this image. In fact, recommendations on use of ultrasonography for RA, in North America and Europe [37], as well as national (French) recommendations [7], specify the use of ultrasound in combination with another method of nerve location. Neurostimulation intuitively is the most obvious. Hydrolocalisation, i.e. tissue movements during advancement of the needle and tactile sensations, which are presented by some as an alternative method, is, in practice, not incompatible with neurostimulation and can be used at any time during the procedure. The large number of studies comparing ultrasound and neurostimulation, which, it should be noted, are sometimes criticised in terms of methodology, has resulted in a contrast of the two techniques. Yet it must be recognised that the information that they provide is different and tends to be complimentary. After the fundamental first sonographic anatomical studies of different nerve blocks, it would have undoubtedly been more logical to look for the limits of ultrasound in terms of imaging rather than seeking to demonstrate by exhaustive methods its superiority in comparison compared to neurostimulation alone. Was it necessary to comprise an entire literature (including several meta-analyses) to demonstrate that to cross a room full of obstacles, the use of a white stick was less effective than switching on the light?

The purpose of these paragraphs is to offer an in-depth approach uniting the different techniques of location available and learning a method for this new way of administering peripheral nerve blocks.

Ultrasound-Guided Procedure: Place of Neurostimulation

Currently, we have ultrasonography and neurostimulation: two techniques to locate a nerve. It is necessary to be aware of the limits of these two methods but each one provides contributory information. Furthermore, their respective performance can be improved by simple tips that will be discussed below.

The joint use of ultrasonography and of neurostimulation provides the benefits of each technique while reducing the limits specific to each process. Functional information from neurostimulation may enable to reduce the limits of ultrasonography and to increase the safety aspect. For example, it makes it possible to correct errors in interpretation (for example, confusion between a nerve root and a blood vessel), or positioning, in particular, for practitioners who are not experienced in ultrasonography. It provides useful information in patients who are not very echogenic. It can be used as a safeguard, even for experienced operators, in counteracting the difficulty involved in constantly maintaining the tip of the needle in the narrow beam of the ultrasound. Even though the quality of imaging has considerably evolved during the last few years, the width (thickness) of the beam is measured in millimetres, or smaller units, with the latest generation matrix probes. The research which has studied the benefit of the combination use of ultrasonography and neurostimulation is all open to criticism at least on one essential point [42–45]. They use a nerve stimulator with basic intensities which are too low to systematically detect nerve structures. These studies have nevertheless determined the minimal intensities which must be used to warn of the presence nearby of nerve structures, i.e. 1 mA (0.1 ms) in axillary and 1.5 mA (0.1 ms) in interscalene position [43, 44]. Furthermore, these studies have not fully taken into account the connective tissue structure of the nerve. The mere apparent contact between needle and nerve is not sufficient to consider that it truly exists. For example, the sciatic nerve within the popliteal fossa, we now know that the sheath (perineurium) which surrounds the nerve plays an important part, and passing through it determines the success of a nerve block [46, 47]. This sheath probably comprises an effective electrical insulator. In our experience, positioning of the needle in contact with the epineurium after crossing the sheath (sensation of a click) enables, in 90% of cases, a response close to 0.5 mA to be obtained (personal data not published). On the contrary, other data suggest that intraneural positioning of the needle is highly probable when a response is present at less than 0.3 mA (0.1 ms).

For some, the unpleasant sensation caused by neurostimulation is an argument to limit its use. While acknowledging this argument, it is not sufficient to permanently prohibit use of this technique. In fact, starting from the principle that it is not the intraneural injection in itself which is harmful, but the intrafascicular injection and fascicular trauma caused by the tip of the needle, an alternative suggestion could be to adjust the intensity of the nerve stimulator to a very low intensity (less than 0.5 mA for 0.1 ms). The occurrence of a motor response or of paraesthesia with such a low quantity of electrical current should suggest a risk of intrafascicular passage of the needle and stop its progression. The benefit is to limit the constraints and discomforts of neurostimulation while maintaining the warning signs. On the contrary, it is essential to use a precise neurostimulator which displays the actual delivered intensity [2, 48, 49].

Ultrasound data have quickly made it possible to consider possible intraneural injections which have gone unnoticed in techniques using neurostimulation alone [50–53]. The possibility of absence of paraesthesia or of pain when the epineurium is crossed by the needle as well as the absence of difficulty in the injection and of pain during the injection during extrafascicular intraneural positioning of the needle calls into question a certain number of dogmas. In this case, under good conditions of echogenicity, it is possible to observe swelling of the nerve and formation of an intraneural halo [1]. Such extrafascicular intraneural injections with limited volumes do not seem to cause neuropathy. The same is not true for intrafascicular injection because the limit crossed in this case is the perineurium, which forms the final barrier and protects the nerve tissue. The systematic search for the disappearance of a sensory motor response when intensity of stimulation is decreased below a certain level presents an additional safety factor [54]. Searing paraesthesia and resistance upon injection of a local anaesthetic are signs of an intrafascicular injection. The most recent neurostimulators display the electrical impedance measured. This measurement is additional information which verifies the correct positioning of the needle. Tsui et al. have shown that elevated impedance before any injection or the increase of this value at a fixed stimulation intensity may suggest intraneural passage [55]. Conversely, the same team showed that the absence of an increase in impedance during injection of dextrose 5% in water solution should suggest an intravascular injection [56]. Of course, it is not necessary to use measurement of impedance each time. But it may be an additional safety element under difficult conditions (poor quality image, deep nerve blocks).

Lastly, the final confirmation of the needle position during injection is achieved by observing the spread of the local anaesthetic in contact with the target. Circumferential diffusion (a "doughnut-like" or "insignia" presentation) (Fig. 3.39) around and along the nerve is conventionally considered to be a measure of efficacy [21]. Dufour et al. confirmed this notion. They studied median nerve block in the elbow by comparing the quality of anaesthesia according to the ultrasound presentation of the spread of the local anaesthetic around the nerve. They showed, first, that the circumferential spread of the LA was in fact associated with better quality anaesthesia and second, that an intraneural injection would not be more effective than a circumferential injection [53]. An ideal needle position should enable this type of spread without swelling of the nerve, eliminating intrafascicular injection. The issue at hand is to determine where the tip of the needle should be located to enable such a result. The beginning of an answer is suggested in an already ageing study by Vloka et al. These authors had shown in cadavers that a single subparaneural injection of the sciatic nerve enabled spread of a dye in a large length of the nerve covering the two components [16]. The pattern of spread of the local anaesthetic can help us to find out if the perineural sheath has been crossed. The liquid tends to spread away from the nerve when we are outside of this structure, while it tends to propagate around the nerve after having crossed this barrier. Anatomical studies are necessary to confirm this visual impression, all the more so since it is sometimes possible to cross several aponeurotic layers. It is useful in this case to reposition the needle for the purpose of obtaining optimal spread of the local anaesthetic around the target. In practice, this involves approaching very close to the nerve; it is advisable to maximise safety precautions in this procedure by maintaining visibility of the needle tip and with the appropriate level of neurostimulation. In addition, it is also preferable to not approach the nerve directly perpendicular to its surface but to approach it tangentially; penetrating the perineurium sheath is sometimes sudden and risk of entering the nerve is real in case of a direct approach. Nevertheless, the importance of the connective tissue in the nerve is an important element in circulation of the local anaesthetic and consequently of its efficacy. A study on cadavers showed that the density of the intra- and perineural connective tissue increased from proximal to more distal areas [57]. In other words, in proximal plexus approaches, the low connective tissue component suggests that the simple crossing of the intermuscular aponeuroses is probably sufficient to penetrate into the perineural space to achieve adequate spread, and to obtain an effective nerve block with low volume of LA; similarly, proximal intraneural puncture may present more risks of fascicular injury than if it is performed more distally from the nerve.

In summary, the findings provided by ultrasound and neurostimulation are different and complementary. At each stage of needle puncture, both are combined to support the procedure. It is illusory, and potentially hazardous, to believe that the view of the tip of the needle and the outer border of the nerve can be constantly maintained. Independently of image quality and of technological progress, the thickness of the ultrasound beam is

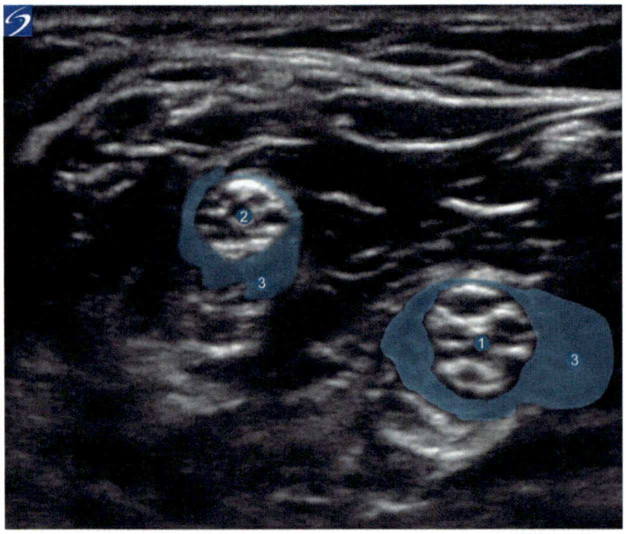

1. tibial n. 2. common fibular n. 3. local anesthetic

Fig. 3.39 "Donut" or "roundel" appearance of the local anaesthetic spread the order of the millimetre. Furthermore, ultrasound in RA requires training that is more or less lengthy depending on the knowledge and ability of each student and the number of procedures that have been achieved. There are two clinical cases of intravascular injection to remind us that the risk of complications persists in spite of use of ultrasound [58, 59]. Hadzic et al. remind us correctly in this regard that "it would be a collective error of judgement to base oneself on this technology (ultrasonography) as an alternative to all existing procedures or as a panacea for all considerations on safety" [60]. In the end, a single principle is to be remembered: "let us treat nerves with great care!"

Algorithm for Practice of Ultrasonography in Combination with Neurostimulation in Conduct of a Perineural Block

The objective of the procedure must be to position the needle in the correct place and in the least traumatic manner possible. As we have stated, all published recommendations emphasise the need to **use techniques for locating structures in combination**. This multimodal approach presents the best benefit/risk ratio in location of structures. Adequate training in the two techniques is the first essential stage [7, 37]. In practice, the procedure is performed on an awake or slightly sedated patient enabling permanent verbal contact [54]. Rules on hygiene proposed in the recommendations must be complied with and all the more so if a catheter is inserted [41]. Before introducing the needle, ultrasound-guided anatomical investigation of the area is essential [61]. Scanning should be wide and systematic for the most precise location possible of anatomical elements, such as nerve and vascular structures. The use of Doppler ultrasound is strongly advised. This locating process will enable to choose the most relevant approach and to plan the trajectory of the needle. Whatever the choice for the approach, in plane or out of plane, the essential item is that, at all times of the procedure, the position of the needle tip should be known and controlled by the operator. In case of improper visualisation of a nerve structure or of the needle, adjustment of the neurostimulator to a stimulation intensity of about 1–1.5 mA for a duration of stimulation of 0.1 ms should be able to serve as a "safeguard". It should be kept in mind that neurostimulation is a process which has very high specificity but low sensitivity [43, 62]. It is always possible to increase the quantity of the current in case of vision of a needle/nerve contact and absence of sensory motor responses. During the procedure, the tip of the needle can be precisely determined by a hydrolocalisation technique [63]. If the image is of good quality with well-identified anatomical structures, the approach under ultrasound should be preferred. Whenever the ultrasound image is of poor quality or when it involves a deep-seated nerve (for example, sciatic nerve by the anterior approach, etc.), neurostimulation will take on more importance and a traditional approach procedure will be chosen, for a short duration and a minimal intensity of stimulation, to locate the nerve. In all cases, finding the minimal intensity of stimulation is essential [54]. The latter provides an indication of the distance between the end of the needle and the closest nerve bundles. If the intensity of stimulation value is considered too low (<0.2 or 0.3 mA for a duration of 0.1 ms), it is always possible to proceed at very low displacement of the needle. On the contrary, in a tangential approach slightly surpassing the nerve in order to obtain circumferential diffusion, the minimal intensity value will necessarily be high and, in this case, the ultrasound image comes into its own.

Complying with rules on safety is essential as, once again, the ultrasound image can give a false sense of security. Easing the pressure exerted by the ultrasound probe makes it possible to visualise the venous structures and to estimate whether or not there is contact between a needle and a vein. It is always essential to conduct an aspiration test before any injection. Since the accuracy of the ultrasound process is approximately in the order of millimetre range, it is essential to position the US beam exactly at the tip of the needle to visualise the hypoechoic spread appearing with its first millilitre [61]. If the minimal intensity for stimulation has been sought, the motor response should disappear starting with the first millilitre of local anaesthetic [64, 65]. A painless injection without resistance remains the criterion for an extrafascicular injection. The appearance of the spread of the solution will confirm whether or not the needle is to be repositioned. Optimum spread, in contact with the nerve and circumferential, will allow a reduction in the volume of local anaesthetic injected provided that the spread also extends along the nerve. Repositioning the needle **should be a considered action, justified by nonoptimal spread of the local anaesthetic** [66]. The aim is to inject the local anaesthetic in the proper place to limit its quantity while administering a sufficient volume. Throughout the procedure, the operator will constantly be evaluating the best benefit/risk ratio for his patient.

Conclusion

The ultimate aim of peripheral RA is to be able to place local anaesthetic in proximity with a nerve structure in order to obtain the highest success rate with the lowest risk of complication. Ultrasound is a complementary method which is added to our arsenal to approach a hypothetical success rate of 100% with 0% complications, but whose use requires a demanding procedure. Against all odds, visualisation of nerve structures can be learned relatively quickly, but the necessary constant control of the tip of the needle is much more complex to master.

It is appropriate to set up a strict procedure for conduct of our perineural blocks under ultrasound, while being aware of the specific requirements for each technique. The algorithm that we are proposing can be simplified based on each one's practice, while keeping in mind that **each item sought corresponds to information lost**.

Mastering the different techniques for locating nerves is desirable in order to combine them to reach the maximum levels of safety and efficacy from this association. Ultrasound is an attractive and very effective method, but which requires rigorous training, in particular in control of the tip of the needle which is an essential (and sometimes overlooked) feature in the safety of the procedure. Adjunctive use of neurostimulation makes it possible to compensate for potential limitations of ultrasound and/or of the operator when identifying nerves and eliciting motor responses. A progressive and rigorous strategy should be standardised precisely in order to be able to adapt to all clinical situations. Whatever the technique for localisation used, "Where is the end of my needle?" should be the question asked constantly throughout the procedure. Knowing the answer to this question is imperative at all times.

General Principles for Conduct of Peripheral Nerve Blocks with a Single Injection, So-Called "Single-Shot" Nerve Blocks, Under Ultrasound Guidance

Rules on Hygiene and on Monitoring

It is necessary to comply with recommended conditions for conventional hygiene (disinfection of the site to be treated, with minimum of two stages, wearing of gloves, mask, and head covering) [8]. Ultrasound guidance requires specific additional rules of hygiene, i.e. in particular use of sterile protective clothing labelled CE whose length, of course, is sufficient to maintain a sterile environment (*see* section "Hygiene and Ultrasound-Guided RA").

Monitoring of the patient should be identical to that of a general anaesthetic [8].

Transcutaneous Location

Use of ultrasound provides a visual control technique for the advancement and positioning of a needle, so the quality of images obtained should be optimum. In this process, the echogenicity characteristics of the patient can play a part, as well as that of the needles, the depth of the nerve block, the angle between the needle and the ultrasound beam; there are many parameters which affect the image (visibility of the needle and of tissue structures) and therefore, potentially, the precision of the procedure. An improperly set ultrasound (gain, focal point, etc.), an inappropriate probe, a poorly positioned ultrasound monitor (reflections, too much light in the environment, etc.), an insufficient quality of gel and poor ergonomic conditions are many parameters increasing the rate of block failure and of block-related complications due to lack of control of the procedure.

When performing an ultrasound-guided nerve block, it's sometimes necessary to forget the usual surface landmarks used in traditional neurostimulation since their utility lies in directing the "blind" needle to its target. However, for certain deep nerve blocks or where the visualisation of the tip of the needle may be difficult, the surface landmarks retain a certain utility: this is the case, for example, of a nerve block of the lumbar plexus by a posterior approach where an inappropriate puncture point and needle angulation may be the cause of potentially serious complications.

This first stage, which above all consists of an ultrasound "status report", aims not only to locate nerve targets, but also to screen for possible anatomical variants. In the surrounding structures, items which present a particular danger are sought (blood vessels, pleura, kidney, etc.) and thus we determine the needle puncture site and ideal trajectory of the needle, enabling minimum iatrogenic exposure. It is of fundamental importance to have good knowledge of anatomy of the area examined in order to read the images on the ultrasound rather than to "go through" them. Emphasis is attached to screening for possible anatomical particularities in the patient which could influence conduct of the anaesthetic procedure. Therefore, it is important to resist the temptation of starting the anaesthesia having examined with ultrasound only the plane of needle puncture and to confirm beforehand the knowledge that was improperly obtained by an overly hasty and limited anatomical study.

The "elevator technique" in this regard consists of a required passage before conducting a true peripheral nerve block. This involves performing an ultrasound scan of an area sometimes greatly exceeding the site of needle puncture upstream and downstream. This examination often turns out to be very informative, enabling a three-dimensional mental reconstruction of the structures visualised by following their position/path throughout translation by the probe. This process also makes it possible to accurately identify all anatomical structures involved. For example, in case of doubt in the axillary area on the exact position of the ulnar and median nerves, by visualising them distally in the arm/forearm and by following them visually during progressive translation of the probe proximally to the axilla, their respective positions can be confirmed. In the same manner, when one has difficulty in specifying the exact position of the interscalene groove and of the nerve roots within it, one can visualise the brachial plexus in the supraclavicular fossa. By moving the probe in a cephalad direction, the cervical nerve roots may be traced proximally to their origin.

Equipment

A relatively broad range of needles exists for performing RA: a cutting bevel with various angles, a Tuohy type needle, a needle with variable diameter and echogenicity enhancements (surface reflectors).

Needles with reflectors are designed to improve their echogenicity characteristics; they are perhaps to be preferred in spite of their cost, which is often higher. They reveal a particular utility in deep nerve blocks (infraclavicular, proximal sciatic) or when the angle of the needle is especially unfavourable for its visibility.

Needle Puncture Sites, Needle Trajectory

The needle puncture *in plane* enables a **longitudinal** vision of the needle; it ensures optimum control of position of the tip. The *out-of-plane* approach, synonymous with vision of the **transversal** section of the body of the needle requires use of indirect techniques to appreciate the position of the bevel of the needle: its principles are to analyse tissue movements induced by the needle and hydrolocalisation. Both require a training phase for the inexperienced practitioner. The out-of-plane approach, even though it often enables to use more "familiar" approaches (the same as in neurostimulation), requires a certain amount of experience in indirectly locating the needle by hydrolocalisation.

The novice in ultrasound guidance, therefore, is often faced with a dilemma: choosing an approach *in plane*, providing safety in his ability to visualise the bevel of the needle, or for an approach *out of plane*, thus less restrictive (no requirement to keep the needle strictly in the ultrasound plane), which seems perhaps simpler but in the end is probably the source of more confusion and requires more experience.

Even if maintaining the needle strictly in the ultrasound plane is sometimes a difficult exercise, this technique is probably the safest. A longitudinal vision of the needle generally requires distancing oneself from the approaches used in neurostimulation alone, but once the obstacle of "novelty" has been crossed, we note that it is easier to avoid the "vulnerable" anatomical structures near to the injection site. Visualisation of the target and of "vulnerable" structures (blood vessels, nerves or other organs) makes it possible to choose the ideal cutaneous needle puncture site with the appropriate needle trajectory for the specific anatomy of the patient. The flexibility of the needle sometimes causes it to bend slightly depending on the constraints of the structures that it crosses, and it is essential to follow progression of the tip of the needle from start of the needle puncture. If the axis of the needle deviates very slightly from the ultrasound plane, the latter does not appear in its entire length, but only in a shorter portion, depending on the angle between the ultrasound plane and the needle: a source of errors in interpreting the position of the tip of the needle and therefore a risk factor for procedure-related incidents (Fig. 3.31).

Use of guides for needle puncture may comprise an aid while enabling the needle to remain more readily in the ultrasound plane. However, the aid provided by current needle guide designs is not always significant, and they are relatively cumbersome. They can offer fixed or variable angulation, with the latter being more appropriate for the practice of ultrasound-guided regional anaesthesia (Figs. 3.40 and 3.41).

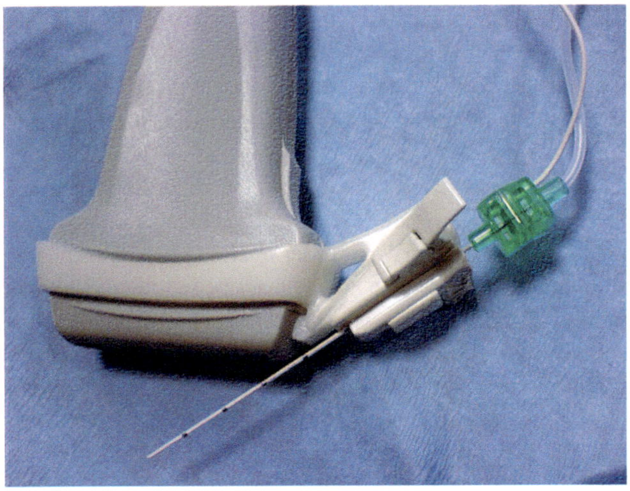

Fig. 3.40 Fixed-angle injection guide

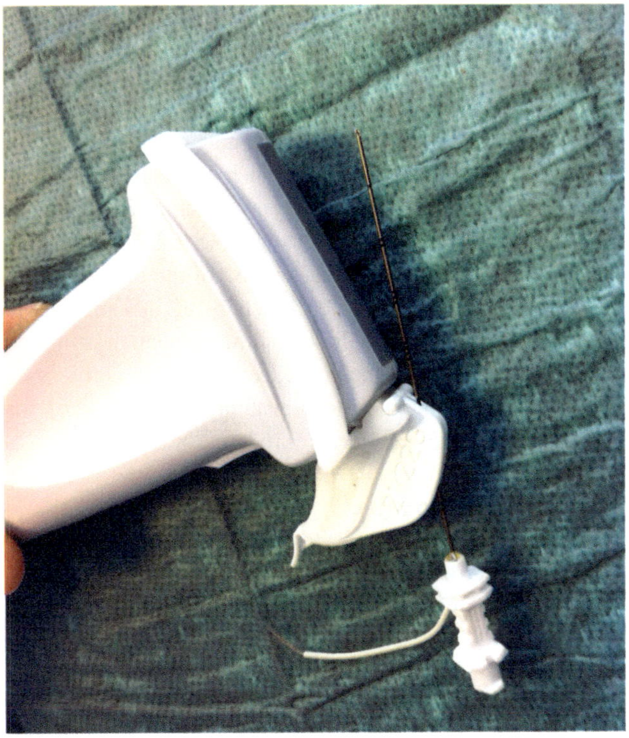

Fig. 3.41 Variable-angle injection guide

Injection Procedures

Approach

In approaching a nerve, a liquid injection creates a space to enable needle tip identification and also to establish an interface which facilitates the visualisation of other structures: this is "**hydrodissection**". Whether it involves a dextrose 5% in water solution (D5W) in order to allow the use of neurostimulation or indeed a local anaesthetic, this technique makes it possible to reinforce the contrast between the different tissues, to visualise the nerves more easily and to establish the potential existence of fasciae or other obstacles to the passage of the local anaesthetic. However, insofar as the vessels are hypoechoic, they are often less visible than when they are surrounded by a local anaesthetic.

The injection of a local anaesthetic results in "teasing out" the different anatomical structures initially stuck together. This makes it possible, if necessary, to redirect the needle and to reposition it in the fluid space created, to optimise diffusion of the local anaesthetic when it is imperfect (Figs. 3.42, 3.43, and 3.44). Injection of air bubbles results in the occurrence of hyperechoic images that can blur the image and alter the precision of the procedure. However, when the position of the tip of the needle is uncertain, injecting a very small volume of air (<1 mL) may sometimes make it possible to localise it. This carries the risk of causing deterioration of the image quality and so may be better achieved by choosing other means to identify the end of the needle (small movements transmitted to the needle, tilting of the probe, activating a specific function on the ultrasound system enabling orientation of the ultrasound beam *(Beam steering)* [67], etc.).

How and When to Inject?

Insofar as possible, the aim is to position the tip of the needle flat against the nerves that you intend to block. As in traditional neurostimulation, the injection should be administered slowly and in a fractionated, painless manner with minimal resistance (injection pressure <10 psi). The patient should be questioned about possible local or radiating pain that can occur at the start of the injection or during it. Aspiration of the syringe should be performed systematically after each movement of the needle or change of the injection site. If the local anaesthetic does not appear on the screen with the injection of the first millilitre, it should immediately be discontinued out of concern for **intravascular** injection (injection into a blood vessel), even if the aspiration test is negative (the latter is of value only when it is positive).

The phenomenon of **excess pressure** during the injection (>20 psi) has several possible causes. It can be due to application of the needle tip against a dense extraneural structure (fascia, aponeurosis, paraneurium, or even the perifascicular epineurium), obstruction of the lumen of the needle, but can also be the result of an intrafascicular intraneural injection [68]. Distensibility of the interfascicular epineurium gener-

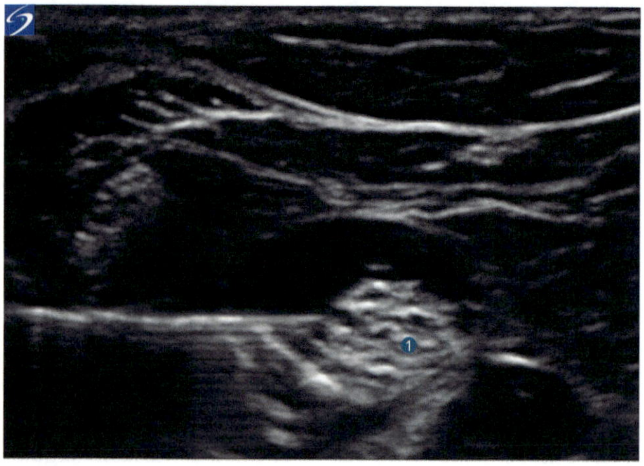

1. tibial n.

Fig. 3.42 First phase of an injection in contact with the tibial nerve facing its superficial hemi-circumference

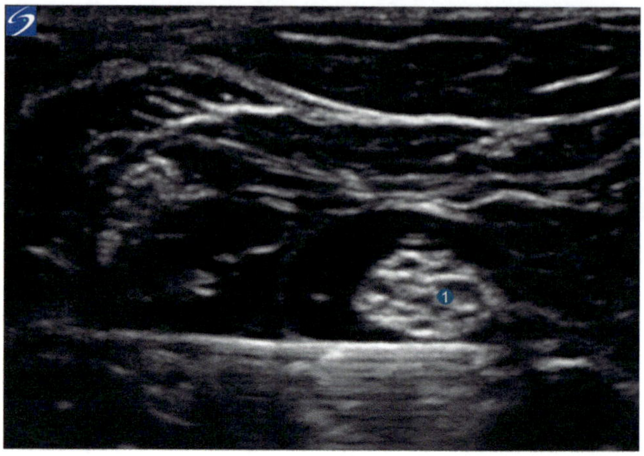

1. tibial n.

Fig. 3.43 Second phase of an injection, this time facing its deep surface after repositioning of the needle

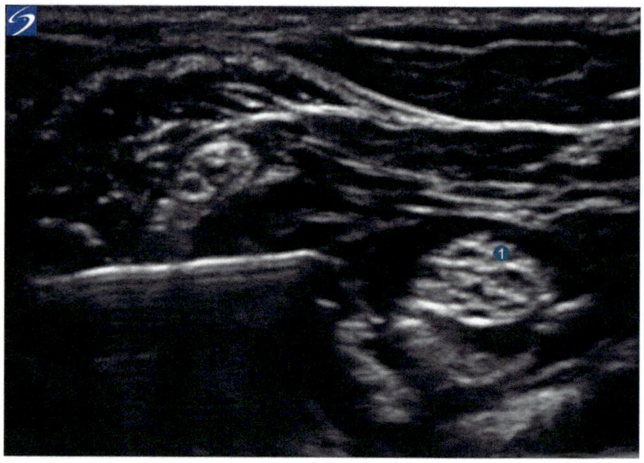

1. tibial n.

Fig. 3.44 Final result of an injection in contact with the tibial nerve, resulting in a circumferential so-called "donut" presentation ("full moon" sign)

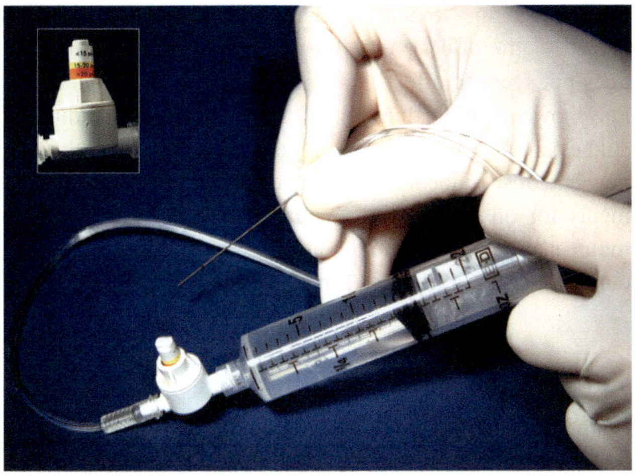

Fig. 3.45 Device for measuring injection pressure ("BSmart", BBraun®). (Iconography: Admir Hadzic)

ally makes the interfascicular intraneural injection easy to perform (<10 psi). Moreover, a high-pressure intraneural injection, even when not intrafascicular, can be the cause of symptomatic nerve lesions [69]. The inability of practitioners to evaluate injection pressure reliably and simply by manual sensitivity has been demonstrated [69, 70]. Although it is not used in routine practice, objective monitoring of injection pressure is possible, available [71] and may be an additional safety component (Fig. 3.45).

In spite of rules of "good practice" in neurostimulation, it has been shown that unintentional **intraneural** injections occurred much more frequently than was thought [51, 52, 72]. The most reliable criteria to detect an intraneural injection are an increase in size of the nerve, possibly with visualisation of spreading of the fascicules, and change in echogenicity [73]. Therefore, any increase in size of the nerve, at any point in the injection (at the very start or even mid-way through), should lead to its immediate discontinuation, even if the patient does not report any specific symptoms. The painless nature of a good number of intraneural injections is to be attributed to distensibility of the interfascicular epineurium in low-pressure low-volume injections [74] and it is not at all synonymous with absence of damage.

Currently, performing deliberate intraneural injection is highly controversial [75]. Some proponents point out the low statistical risk of intraneural injection demonstrated and the gain in onset time that it provides (however, contradicted by a study by Dufour et al. [53]); however, currently the risk-benefit ratio does not support the routine practice of performing intraneural injections, considering the potential lesion-inducing effects [68, 76–80].

Distribution of the local anaesthetic should be done in close contact with the target nerve, ideally achieving circumferential injection around it [82] (the rosette sign or doughnut sign or a "full moon sign" [the author's preferred term]), without interposition of a structure mechanically hindering the spread of the local anaesthetic. This can involve not only the fasciae, aponeurosis, etc., but also membranes combined under the name paraneurium, mesoneurium and adventitia which may hinder contact between the nerve and the local anaesthetic. Frequently the circumferential spread of the local anaesthetic is not seen at the injection site itself, but by moving the ultrasound probe, this process can be observed at a greater or lesser distance from the injection site. This may be simply due to pressure exerted by the probe at the injection site or due to variable "adhesions" along the trajectory of the nerve relating to structures which surround it. However, in addition to circumferential spread, the volume of local anaesthetic in close contact with the nerve and the length of contact (enabling a sufficient number of Ranvier nodes to be "blocked") appear necessary to produce a successful block. The paraneurium is at the centre of many recent communications particularly concerning popliteal fossa block where its presence is especially noteworthy. Generally they conclude that to obtain a nerve block of rapid onset and of good quality, it is desirable to inject the local anaesthetic between the nerve and the paraneurium itself, in order to obtain optimum contact (on the surface and over time) between the actual nerve (perifascicular epineurium) and the anaesthetic [83]. Furthermore, for the popliteal fossa block, it is advisable to place the tip of the needle at the bifurcation between the tibial nerve and the common fibular nerve, within the paraneural sheath. For other nerves and other plexuses, additional studies are necessary to specify the optimum distribution of the local anaesthetic, its volume and the ideal site of injection [74].

> Currently, it is still recommended to avoid needle puncture and not to inject into a nerve [7, 81].

In some **fascial plane blocks**, the circumferential spread of the local anaesthetic is not essential for faster onset of anaesthesia or of better quality. This is the case, for example, in abdominal wall nerve blocks (*transversalis abdominis plan*: TAP), where injection of local anaesthetic is administered between the transversalis abdominis muscles and the internal oblique muscles without the need for location and approach of each small nerve branch [84]. But this process also involves other sites of anaesthesia:

- musculocutaneous nerve block with an intermuscular injection between the coracobrachialis and the biceps brachialis muscles [85]
- obturator nerve block with a proximal or distal intermuscular injection [86–88]

Performing the injection at a distance from the nerves can offer an additional level of safety by limiting the risk of direct needle trauma. This particularly involves an interscalene nerve block for which it has been shown that a voluntarily administered injection between the middle scalene muscle and the brachial plexus offers equivalent nerve block quality to an injection into the sheath of the brachial plexus, while probably limiting the risks inherent in needle-nerve contact [89].

Some recent studies show that an **ultrasound-guided perivascular injection technique**, without seeking to deposit the local anaesthetic selectively in contact with each nerve, is a useful and effective alternative in the axillary [90, 91] and infraclavicular approaches to the brachial plexus [45, 92]. By reducing the need for nerve-needle contact, these approaches do however require **volumes of local anaesthetic that are generally higher** than those routinely used in current ultrasound-guided anaesthesia.

What Volume of Local Anaesthetic to Inject?

The use of ultrasound has enabled improvement in the quality of regional anaesthesia with reduction in time to perform it and the time to onset of anaesthesia [93, 94]. The decrease in volume of local anaesthetic necessary to obtain an effect nerve block comparable to that performed in neurostimulation has been demonstrated several times [95–97]; this phenomenon has particularly been found in axillary [98], femoral [97], sciatic [95, 99] and interscalene [100] nerve blocks. Additionally, in a published study, Riazi et al. demonstrated that limiting the volume of the local anaesthetic may have a favourable impact with respect to the adjacent anatomical structures, limiting the occurrence of phrenic nerve blocks during ultrasound-guided interscalene brachial plexus blocks with 5 mL vs. 20 mL of 5 mg/mL ropivacaine. These major reductions in volume of local anaesthetic injected have been studied, resulting in effective nerve block of mean duration of 160 min in spite of only use of 1 mL of 2% lidocaine in each nerve in the axillary plexus [98, 101].

The majority of studies examining volume effects show that the onset time period is not extended in the "low volume" groups; however, duration of efficacy is reduced in some studies [102–104] but unchanged in others [95].

If (too) low volumes are the cause of a reduction of block duration, is this a volume-effect and/or a dose-effect? Studies comparing high volumes (20 mL vs. 60 mL [105], 20, 28 and 36 mL [106]) had not shown any differences in duration of anaesthetic efficacy. This leads to suggest the possibility of a threshold above which any increase in volume or total dose may no longer have a significant effect on the duration of the block [107].

In practice, ultrasound has made it possible to limit, sometimes significantly, the volume of local anaesthetic injected in peripheral nerve blocks. Its positive impact is the potential limitation of side effects related to the local anaesthetic (systemic toxicity, collateral effects related to local spread of the effect). If a drastic reduction in volume/dose does not seem to affect the time to onset of nerve blocks, it may however limit duration of a sensory block. Additional studies are necessary to determine, by type of block, the volume and dose of local anaesthetic necessary based on requirements for onset time, the quality and duration of sensory/motor nerve block desired, but should also take into account our objective of maximum limitation of potential side effects.

General Principles for Conduct of Continuous Peripheral Nerve Blocks with a Perineural Catheter Under Ultrasound Guidance

Continuous nerve blocks enable better quality analgesia than conventional analgesia with opioids, and this effect exists for the first 72 h after surgery. This fact has been demonstrated for all localisation of perineural catheters. Compared to single-injection nerve blocks, peripheral nerve catheters optimise post-operative analgesia with extended duration, with less use of morphine [108] and, consequently, less frequent adverse events (post-operative nausea and vomiting, pruritus). The benefits, other than decrease in pain at rest and during movement, include reduction in joint inflammation and markers of inflammation, in sleep disorders and in the side effects of morphine, which the patient uses in a lower quantity. Patients are satisfied and their rehabilitation is earlier, reducing duration of hospital stay, accelerating recovery of joint function. Development of post-surgical chronic pain is reduced, thus decreasing cost of management of surgery [109]. Patient satisfaction is improved [110].

Principal Sites of Insertion

- Interscalene
- Supraclavicular
- Infraclavicular
- Axillary
- Distally at the elbow and forearm (in contact with the radial, median or ulnar nerve)
- Lumbar (posterior approach)
- Femoral
- Sciatic by parasacral, subgluteal or popliteal approaches
- Tibial at the ankle

Indications

The catheter is preferably inserted pre-operatively so as to provide benefit from anaesthesia at the outset. The principal indications for peripheral nerve catheter insertion are as follows:

- Post-operative pain predictable for more than 24 h and, consequently, surgery where the pain involved is recognised as intense (procedures involving large joints or the extremities, ligament and tendon surgery, disfiguring procedures such as oncological surgery and surgical reconstruction, procedures involving bone lengthening).
- Surgery for which early post-operative mobilisation is essential.
- Conditions benefiting from prolonged sympathetic nerve block (skin flaps, burns, frostbite, etc.).
- In the case of limb amputation, peripheral nerve catheters may prevent the development of phantom limb pain. Existing phantom limb may also be treated by means of continuous nerve block [111].
- In an emergency, inserting a peripheral nerve catheter improves comfort during transportation and performance of X-rays in patients with fractures (for example, femoral neck fracture, iliofascial or femoral catheter).
- In ambulatory surgery, the routine use of catheters is increasing to enhance the quality of analgesia in the setting of multimodal pain management. The involvement of an urban network is advisable, and rules on use of ropivacaine infusion (order of 26 April 2012) via a peripheral nerve catheter have been well-established [112, 113]: "Conditions stipulated in article L. 5126-4 of the Public Health Code are applicable to medicinal products containing ropivacaine presented as a solution for injection for infusion in a bag at concentration of 2 mg/mL in application of a protocol whose purpose is to accompany and to supervise dispensing and administration of these pharmaceutical products in the setting of post-operative analgesia by perineural catheter in ambulatory practice; available on the ANSM website". This procedure for use sets the roles of each participant (anaesthetist, nurse, etc.) and limits the doses of ropivacaine to 20 mg/h, duration of catheterisation to 72 h and describes the procedure to be followed for continuous analgesia in ambulatory practice.
- In chronic pain, there are many indications for inserting long-duration peripheral nerve catheters:
  complex regional pain syndrome
  shoulder pain
  diabetic neuropathic disorders
  lower limb vascular pain
  cancer pain, etc.

The **rules for prescribing and monitoring** such continuous nerve blocks are currently well-standardised and monitoring procedures make it possible to prevent pressure or compression complications, infections, haematoma, risk of falls or neurological disorders.

Other than the **contraindications** usually identified for RA (infection at needle puncture site, coagulation disorders, patient refusal), the risk of a **compartment syndrome** should be considered. In case of a specific risk inherent in the disorder presented, raised pressure in the **muscular** compartment should be avoided. An increase of 20 mmHg above diastolic blood pressure is an alarm signal. Catheters are to be used with great caution in children under 2 years of age because of the fragility of nerves that are barely covered with myelin.

A specific difficulty exists when using ultrasound guidance to insert peripheral nerve catheters, probably at the origin of anaesthetist communities "lagging" for about 10 years before its use was institutionalised [114]. Yet, analgesic failure in techniques for "blind" placement of perineural catheters is not negligible (around 20–40%), even though its use is an integral part of regional anaesthesia [115] and that the utility of perineural catheters is established in many situations: chronic pain, prevention of neuropathic pain, induction of vasodilator events in the setting of ischaemia or limb re-implantation, but in particular post-operative acute pain generally lasting between 48 and 72 h, possibly even in an outpatient context [116].

The utility of positioning an ultrasound-guided perineural catheter can be justified in the same way as those already mentioned for peripheral nerve block with a single injection (*see* Chap. 2):

- low sensitivity of neurostimulation
- anatomical safety (personalisation of the procedure to the patient's actual anatomy)
- reduction of risk of vascular puncture [117] and systemic toxicity adverse events
- amputees [118], patients with neuropathies
- decrease in doses and/or concentrations of local anaesthetic [119, 120]
- positioning of the catheter in fascial planes not detectable in neurostimulation (TAP block, ilio-inguinal block, iliohypogastric block, rectus sheath block [121, 122])

But the idea can also be proposed that an ultrasound-guided catheter may be of better analgesic efficacy when it is placed along with neurostimulation. This notion has been demonstrated in particular in several studies on femoral [120], popliteal [123] and interscalene [124, 125] approaches. Nevertheless, it remains controversial because of studies that demonstrate no advantage in combining the techniques [126, 127]. Gandhi et al. have shown that under ultrasound guidance, a stimulating femoral catheter did not have better efficacy than that of a non-stimulating catheter [128], and Brull et al. revealed that screening for a quadriceps response (movement of the patella or sartorial response) in placement of a femoral stimulating catheter did not affect the quality of post-operative analgesia [129].

Lastly, and most importantly, it should be noted that contrary to expectations, the time necessary to insert the catheter appears to be reduced by about 30% with ultrasound guidance [117], in interscalene [127], infraclavicular [130], femoral [131] as well as popliteal catheter [132] insertion.

The utility of ultrasound lies in **the placement** and **control of positioning** of the catheter during its entire period of use, enabling examination and potential correction if loss of efficacy occurs. Generally, a catheter is much more visible in a longitudinal section than a transverse one, even if aligning it in the ultrasound beam may sometimes be difficult. However, it is the position of its tip, which is important to us. Following the body of the catheter can bring us to it, but this is inconsistent: "pigtail" models are by definition almost impossible to align in the ultrasound beam, but "rectilinear" models are sometimes also poorly visualised. Only two-thirds of catheters may be located with certainty [117]. Use of indirect means can help in the following:

- **injection of a local anaesthetic** can help in locating the end of the catheter, or the spread in contact with the nerve can confirm the proper positioning of the tip.
- **colour** or **power Doppler** during injection.
- **injection of air:** either a small bubble or a few millilitres of dextrose in water solution previously shaken in the syringe together with a little air [133].
- injection of **radiological contrast medium**.

Rules on Hygiene and Monitoring

It is necessary to comply with "surgical" conditions of hygiene for placement of peripheral nerve catheters (use of gloves and a sterile gown, mask and hat) [8]. Ultrasound guidance requires additional specific rules of hygiene, i.e. in particular, use of a CE-mark sterile sheath whose length should also be sufficient to keep the block environment sterile *(see* section "Hygiene and Ultrasound-Guided RA"*)*. Monitoring should be identical to that of general anaesthesia [8].

Transcutaneous Location

In preparation for insertion of the perineural catheter, the block site is evaluated in the same manner as that for single-injection nerve block. A structured ultrasound-guided examination locates the target nerve(s) and determines its relationship with other adjacent structures. Viewing the area in the short axis is recommended in order to be able to complete safety checks around the nerve and its immediate environment.

Equipment

A relatively wide range of equipment exists in terms of needles and catheters: a short bevel with various angles, a Tuohy tip,

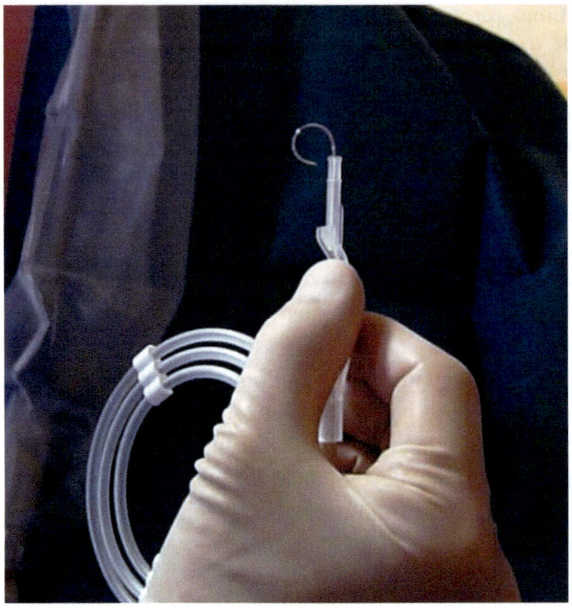

Fig. 3.46 "Pigtail" catheter (Sonolong Curl Sono, Pajunk®)

Fig. 3.47 "Tuohy" type needle with reflectors and a reinforced catheter (Pajunk®). (Iconography: Bertrand Fabre)

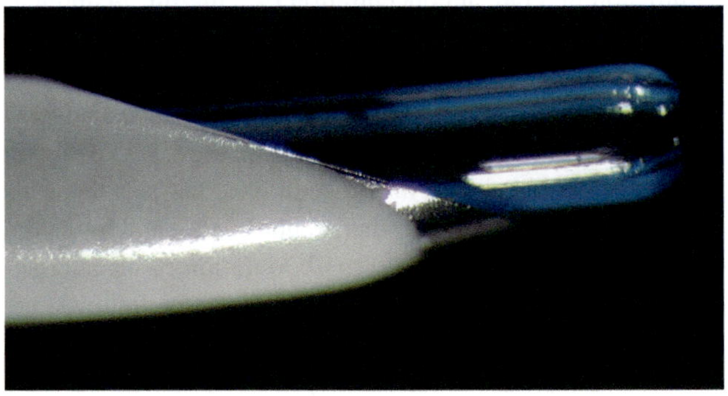

Fig. 3.48 "Standard" bevel insulated needle (Vygon®). (Iconography: Bertrand Fabre)

needle with variable size and echogenicity (surface reflectors), catheters with/without guide-wires/reinforced, a linear profile or in pigtail (Fig. 3.46), single-hole or multi-hole ends, etc.

Needles with reflectors have better echogenicity and, perhaps, are to be preferred in spite of their cost, which is often higher (Fig. 3.47). Insofar as it is desirable to site the catheter in immediate proximity to the nerve, "Tuohy" needle bevels that are less sharp probably carry greater safety. Since catheters with guide-wires/reinforced are appreciably more echogenic, choosing them can facilitate their location (Figs. 3.47 and 3.48).

Needle Puncture Site, Needle Trajectory

The axis of needle puncture is either *in plane* or *out of plane*. Figure 3.49 summarises the three traditionally feasible configurations. Note that the out-of-plane approach with *longitudinal* vision, not represented in this diagram, does not have any specific utility. During insertion of the needle, injection of a D5W solution enables hydrodissection of the anatomical planes, and an opening up of the perineural space where the needle tip is immobilised and the catheter is introduced. Neurostimulation for insertion of perineural catheters remains a useful technique, as already widely detailed for nerve blocks with a single injection. The method and criteria for use are also unchanged, and the D5W solution to perform hydrolocalisation does not compromise its use (contrary to that of a 0.9% NaCl solution or of a local anaesthetic). If the procedure is performed "with two hands" (practitioner only), generally it is necessary to lay aside the ultrasound probe at time of inserting the catheter into the needle. To aid the technique, an assistant (also scrubbed up) can hold the probe during catheter insertion, so that the emergence and final positioning of the catheter can be visualised in "real time".

Nerve in Transverse (Short Axis) View, Needle Puncture Out of Plane (Fig. 3.49b)

This configuration enables the practitioner to use familiar approaches used in neurostimulation, with the exception of the site of introduction for the needle puncture whose position may be changed according to the position of other structures. The introduction of the catheter is performed in the longitudinal axis of the nerve, a theoretical advantage but which sometimes exposes the risk that the end of the catheter could distance itself from the nerve during its advancement, contributing to block failure.

After puncture of the skin, the needle advances with the aid of hydrolocalisation. The tip progressively approaches the target nerve until it appears, using hydrodissection, at the target area for injection. Once the needle is immobilised adjacent to the nerve, an initial bolus dose can potentially be injected through the needle. If not, the catheter is introduced slightly further into the space to stabilise its position. Note: great care must be taken to maintain its proximity to the nerve and not to introduce too much length. Once secured and the tip position confirmed, an initial dose may then be injected.

In some locations, almost the entire length of the catheter introduced can be visualised and controlled with ultrasound.

Nerve in Transverse Section, Needle Puncture in Plane (Fig. 3.49a)

In this approach, so-called "**pigtail**" catheters can have specific utility (Fig. 3.46). This involves a model whose end is deployed on itself in the form of a multiperforated flexible coil (Figs. 3.50, 3.51 and 3.52). The advantage is that the first 3 cm of the catheter coil emerges from the end of the needle, and stays "on site". The end of the catheter, therefore, remains in immediate proximity to the end of the needle, i.e. at the injection site that had been considered as ideal. An unintentional *removal of the first 3 cm of the catheter, i.e.* the length of the coil, does not compromise the proper diffusion of the local anaesthetic.

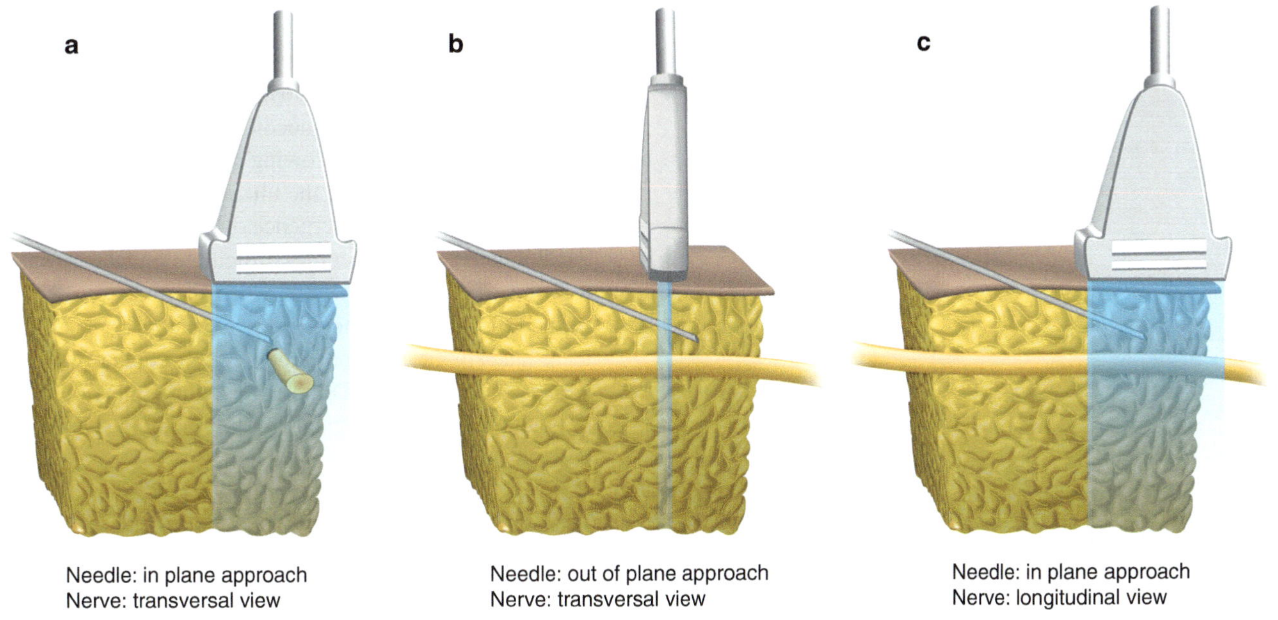

Fig. 3.49 The three principal "needle/nerve" approaches in ultrasound-guided regional anaesthesia

Nerve in Longitudinal Section, Needle Puncture in Plane (Fig. 3.49c)

For practitioners who use this configuration in single-injection nerve block, there is no major change in their technique. After puncture of the skin, insertion of the needle is performed using "hydrodissection" as it approaches the nerve. The tip of the needle is positioned where the end of the catheter is to be placed: that is, in **immediate** contact with the target nerve. Then, the injection of the initial bolus dose of local anaesthetic can be administered through the needle before introducing the catheter, or this can be done through the catheter while observing for appropriate spread.

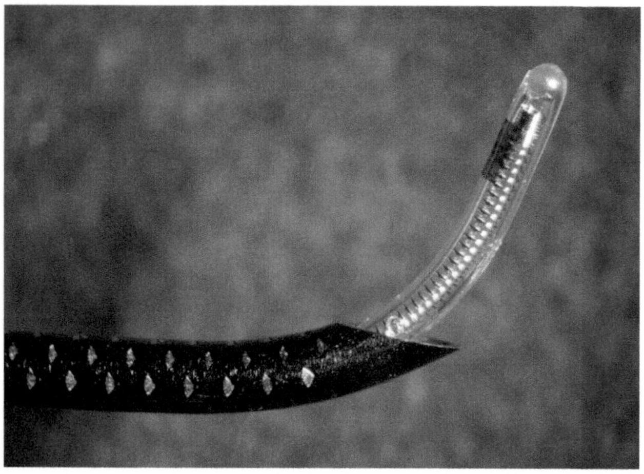

Fig. 3.50 Initial deployment of the loop of a "pigtail" catheter (Sonolong Curl Sono, Pajunk®). (Iconography: Bertrand Fabre)

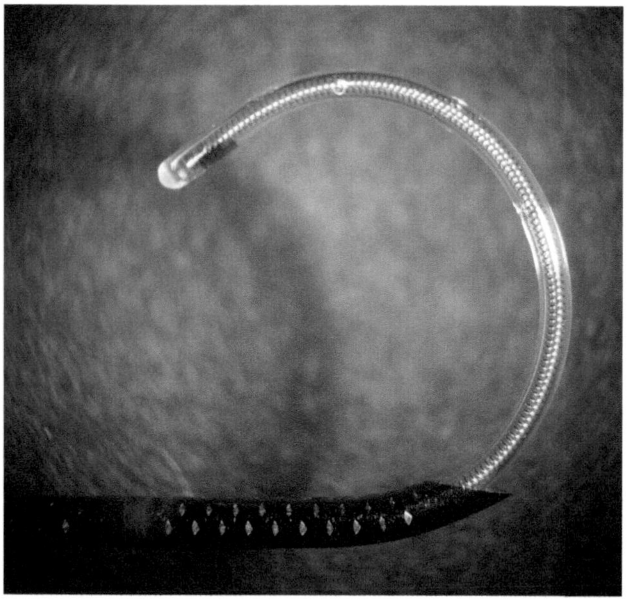

Fig. 3.51 Partial deployment of the loop of a "pigtail" catheter (Sonolong Curl Sono, Pajunk®). (Iconography: Bertrand Fabre)

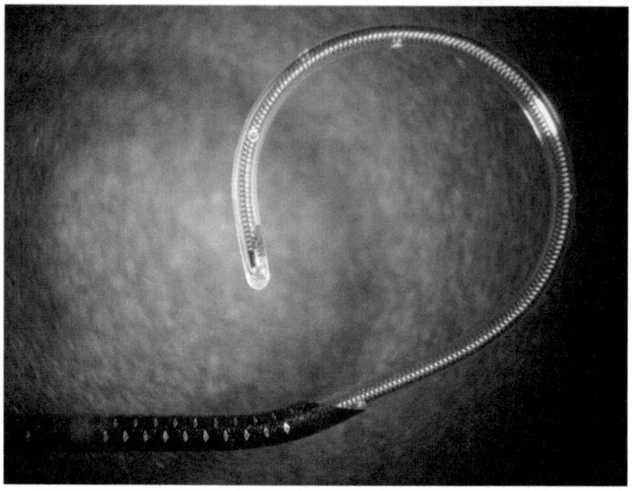

Fig. 3.52 Complete deployment of the loop of a "pigtail catheter" (Sonolong Curl Sono, Pajunk®). (Iconography: Bertrand Fabre)

If initial injection of the local anaesthetic is done through the catheter and not the needle, this enables the efficacy of the catheter to be ensured from the outset (particularly useful in ambulatory practice).

Methods of introduction of the catheter through the needle:

- In the case of a **straight catheter**, it is advanced to the tip of the needle only (sensation of elastic resistance) and the needle is withdrawn **gently** with one hand, maintaining the catheter in place with the other hand. For such "rectilinear" models, it is generally possible to control the position and the end by ultrasound, particularly if it involves a hyperechoic model.
- In case of a **"pigtail" catheter,** immediately before "releasing" the catheter, it is preferable to inject a few millilitres of D5W solution or of local anaesthetic in order to open up space for its deployment. Then the catheter is advanced to the tip of the needle (elastic resistance) and inserted a further 3 cm allowing deployment of the coil. When a third party holds the ultrasound probe, it is possible to see the initial emergence of the catheter, but once the coil has exited, it becomes difficult to find the tip since it is not rectilinear. However, when the injection is performed through the catheter, we see the appearance of the local anaesthetic into the target area in contact with the nerve; colour or power Doppler mode and injection of air or potentially of radiological contrast medium (paravertebral catheter) may also be used.

Verification of the position of the end of the catheter is essential because it guarantees the efficacy of the catheter. It can be performed more naturally with ultrasound guidance either by identifying its position directly or by observing that the injection through the catheter appears in close contact with the nerve.

Although this is still possible, **ultrasound** observation of the proper positioning of the catheter eliminates the need to confirm this by X-ray and thereby reducing patient exposure to radiation.

Specific Aspects in Paediatric Patients

Paediatric RA has undergone major development over the last few years. This method results in quality peri-operative analgesia and enables earlier recovery of consciousness and more rapid resumption of normal activity. However, in order to perform RA in a child, it is essential to understand some specific paediatric aspects.

Pharmacokinetic and Pharmacodynamic Specific Aspects in Children

This mainly concerns infants.

Pharmacokinetics

In young children, fluid compartments are larger than in adults in relative terms. Consequently, the volumes of distribution of hydrophilic compounds, including chiral formulations, are increased.

Local anaesthetics bind to alpha-1-glycoprotein acid proteins (AGP), including orosomucoid, and to albumin. Plasma concentrations of AGP are decreased up to about 12 months of age. This results in an increase in the unbound part of the local anaesthetic which is responsible for its toxicity. This increase in the unbound fraction is only partially compensated by the increase in volume of distribution. Conversely, surgical inflammation results in an increase in plasma concentration of AGP, from around 3 h after surgery, thus increasing the bound part of the local anaesthetic. However, the role of AGP in final toxicity is low. In fact, this particular aspect influences the total concentration, while the concentration of the unbound fraction depends mainly on the intrinsic metabolism of the local anaesthetic in the liver. Clearance of aminoamides occurs mainly in the liver, whose functioning is immature in infants. Therefore, it is decreased. Thus, clearance of ropivacaine is similar to that of adults only in children between 2 and 6 years of age, and that of levobupivacaine starting at 1 year of age. Therefore, a risk of accumulation exists mainly up to about 6 months of age in case of a repeat injection. This means a reduction in dosage by half during continuous administration in this patient population.

The toxic plasma concentration level is not known in children and concentrations are derived from those in adults. The safety margin increases with age and weight, and is higher for single injections compared to continuous administration. It may be difficult to diagnose signs of early LA toxicity in children because most RA is performed under general anaesthesia in this population. Therefore, the first signs of toxicity are mainly cardiac, with general anaesthesia masking the neurological signs. As in adults, **the chiral forms of LA are less cardiotoxic** than the racemic forms, and therefore should be preferred all the more so in a young child.

Pharmacodynamics

Myelination is incomplete at birth and is completed at around age 3–4. The nodes of Ranvier are more spaced out at birth. These two specific aspects, in practice, enable use of low-concentration local anaesthetics for an effective nerve block but at higher volumes.

S enantiomers (ropivacaine and levobupivacaine) produce a lower motor nerve block than bupivacaine with identical efficacy. These two local anaesthetics are therefore preferable, all the more so since children have difficulty tolerating a motor nerve block.

In terms of cardiac status, infants have a high heart rate and cardiac output, which makes them more sensitive to cardiac toxicity of local anaesthetics, hence the importance of using only *S* enantiomers.

In summary (according to RFE ADARPEF/SFAR 2009 recommendations [134]):

- Use low-concentration chiral forms of LA (in spite of limits of authorisation for marketing).
- In children >2 months of age, use a volume of local anaesthetic, which is that much higher compared to weight in a young child.
- In a child <2 years of age, reduce the dosage of local anaesthetic because of high baseline heart rate, which increases vulnerability to cardiac toxicity of local anaesthetics.
- In children <1 year of age, the risk of systemic toxicity is increased. This risk is that much greater before the age of 6 months because of liver immaturity, in particular in case of repeat injections or of continuous administrations.

Practice

RA Under General Anaesthesia: RA in the Awake Patient

In the majority of cases, RA in paediatrics is performed under general anaesthesia, as the result of the psychological and emotional stress of the procedure, the difficulty in understanding the concept of paraesthesia and need sometimes for perfect immobility (deep nerve block, epidural nerve block). This practice is recommended by the experts.

However, RA can also be performed on an awake patient. There are no recommendations regarding the age limit. Clear information for patients, and in particular for the child, must be given. Appropriate premedication, hypno-analgesia and ultrasound-guided RA enable to improve comfort of the child. However, the surgical procedure should be of short duration and the risk/benefit ratio always evaluated.

Equipment

Specific equipment for paediatric RA exists. This should be used and adapted to the child's weight in order to limit complications. Distance between skin and target structures (epidural space or nerve structure) vary according to weight. Therefore, length of the needle should be adjusted to the child's age (abacus). Ultrasonography is therefore useful to verify these distances.

Safety

To prevent complications, certain rules on safety should be complied with

- Compliance with maximum dosages of the different local anaesthetic agents:
  - ropivacaine = 3 mg/kg
  - levobupivacaine = 2.5 mg/kg
- In compliance with dosages for each nerve block for ropivacaine and levobupivacaine [134]:
  - peripheral and nerve block in the trunk: max. = 0.5 mL/kg
  - caudal: max. = 2 mg/kg
  - epidural:
    a single injection: max. = 1.7 mg/kg
    continuous:
    - <1 month = 0.2 mg/kg/h
    - 1–6 months = 0.3 mg/kg/h
    - >6 months = 0.4 mg/kg/h
  - Carrying out a test dose with adrenaline in a central nerve block or a deep peripheral nerve block is always recommended in paediatric patients.
  - Since RA is performed under general anaesthesia, the minimum intensity of stimulation should not be less than 0.7 mA (0.1 ms) because the risk of a nerve lesion is increased without increase in success rate for the block.

RA Under Ultrasound Guidance

Ultrasonography for RA in paediatrics decreases block onset time, increases duration of a sensory nerve block, decreases onset time of a motor block, decreases quantity of local anaesthetic injected and improves the success rate [134].

However, specific paediatric aspects exist.

Equipment

As a result of the nerves being more superficial, a high frequency linear probe should be used, whatever the type of nerve block.

Currently so-called "paediatric" ultrasound probes exist, i.e. of smaller size. These probes are also linear and of high frequency. In order to perform RA under ultrasound with optimum conditions, it is essential to have these probes available, especially in children with low body weight. No recommendations exist on the weight limit for use of these probes, everything depends on size. Probably above a certain weight (at around 20–30 kg), an adult type probe can be used.

Sonoanatomy

The younger the child, the more the structures are superficial and smaller in size; the more the blood vessels are sensitive to pressure of the probe. It should be noted that in the youngest children, even the arteries are compressible.

Up to age of 3–6 months, ossification is not complete. Ultrasound crosses the bone barrier, and the intraspinal structures (dura-mater, spinal cord, etc.) are thus visible.

Rules on safety in adults are also to be applied to children: visualisation of needle tip, injection under visual control to verify proper spread of the local anaesthetic.

Specific Aspects in Paediatrics

Physiologically a child is not a miniature adult, and this is also true anatomically. Therefore, deep nerve blocks in adults are often superficial in children, visibility of structures is often improved and the conduct of certain techniques is simplified, in particular with better spread of the local anaesthetic. The use of "low frequency" probes is no longer necessary, which improves spatial resolution and image precision.

Superficial nerve blocks in adults are even more superficial in children. Consequently, target structures can be too superficial in respect of the technical capacities of the apparatus and probes (≤1 cm). Maximum ultrasound frequencies used in "routine" clinical practice can be insufficient for certain nerve blocks requiring use of higher frequencies (15–18 MHz). Similarly, the size of normal probes can represent an additional difficulty for some already "cramped" approaches in adults, and which are more so in children.

Visualisation of the tip of the needle in relation to the nerve targets should be combined with recommended criteria for neurostimulation, reinforced by an MSI greater than or equal to 0.7 mA (0.1 ms). Although conduct of a peripheral nerve block under general anaesthesia is not advisable in adults, in children it is generally the norm but this practice requires extreme vigilance by the anaesthetist. Visualisation of the needle with ultrasound may offer a greater degree of safety.

Apart from the specific psychological aspect in children which does not always make it possible to perform a peripheral nerve block calmy with an awake child in paediatrics (the stress of the child, of the child's family, difficulty in tolerating pain or the idea of needle puncture, etc.), ultrasound-guided technique often makes it possible to involve the child visually and to improve his or her acceptance by providing a "visual distraction". Moreover, this enables the practitioner to inform the child of possible painful events during conduct of the peripheral nerve block.

Checklist for Conduct of an Ultrasound-Guided Peripheral Nerve Block

Performing an ultrasound-guided peripheral nerve block should be a structured, reproducible process. Use of the checklist principle limits the risk of omissions by systematising the conduct of the procedure in three successive phases (localisation, needle puncture and injection) (Table 3.2).

Table 3.2 Checklist for conduct of an ultrasound-guided peripheral nerve block

Preparation	1. Clean ultrasound system and probes	✓
	2. Nerve stimulator ready and checked	✓
	3. An appropriate probe, equipped with a dedicated sterile cover and single-dose sterile ultrasound gel	✓
	4. Cutaneous aseptic procedure ("4 phase" procedure)	✓
Localisation	5. Ergonomic positioning of the ultrasound system	✓
	6. Targets and adjacent structures visualised ("status report")	✓
	7. "Sensitive" structures to be avoided are identified	✓
	8. Unusual structures or anatomical variations are sought	✓
	9. Optimised angle of the probe (anisotropy)	✓
Needle puncture	10. Appropriate length/calibre of needle for the planned trajectory	✓
	11. "Strategic" needle puncture site to approach target selected	✓
	12. Wiping away excess gel at the needle puncture site	✓
	13. Knowledge at all times of exact position of the needle tip	✓
Injection	14. Safety criteria are complied with: aspiration test, low injection pressure, MSI > 0.3 mA (0.1 ms), no pain on injection, loss of motor response, etc.	✓
	15. Appearance of the local anaesthetic on the screen from the start of the injection with appropriate spread	✓
	16. Redirection of the needle if the injection is not optimal, repeated aspiration tests	✓
	If the needle is redirected in case of multiple targets (for example, axillary nerve block with "multiple ultrasound localisation"), repeat points 14, 15 and 16 as many times	✓

MSI minimal stimulation intensity

Proposal for a Decision-Making Tree for Conduct of an Ultrasound-Guided Peripheral Nerve Block [135] (Fig. 3.53)

New Techniques Under Development

The advent of ultrasonography is undeniably one of the major advances in the field of RA. But changes are emerging which will lead to even more considerable changes in our ultrasound practice. Among these: electromagnetic monitoring of the needle, 3D/4D ultrasound, image fusion, etc.

Electromagnetic Monitoring of the Needle ("GPS")

This is a method of spatial acquisition of the position and direction of the needle in real time. A signal generated by a device integrated in the needle is located in a controlled magnetic field, enabling the system of measurement to determine its position and direction. Parameters calculated are displayed on the ultrasound system screen. The SonixGPS™ system (Ultrasonix, Richmond, BC, Canada) which was recently introduced on the market particularly enables to determine the instantaneous position of the needle and of its movement predicted based on its direction. Among the principal limitations of this system, possible interference gener-

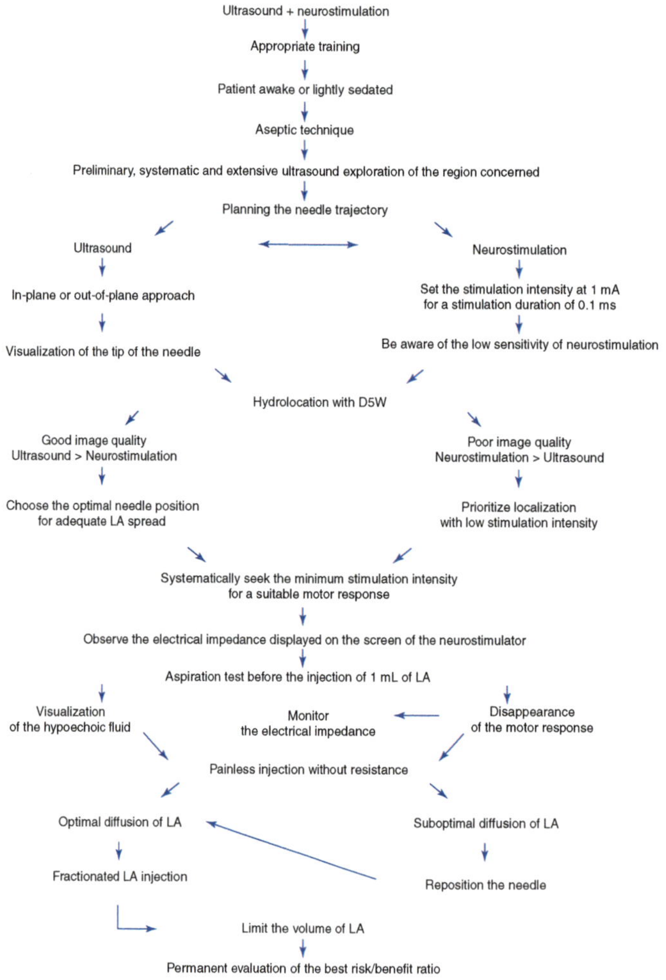

Fig. 3.53 Algorithm for dual-guidance ultrasound imaging associated with neurostimulation during the performance of a perineural block. (From D. Jochum [6] in Chap. 2)

ated by the metal equipment surrounding the electromagnetic field generator is the most inconvenient. The utility of such a device is, in particular, to limit the number of redirections of the needle [136], of facilitating approaches outside of the ultrasound plane [137], of directing the needle towards the deep targets in spite of imperfect visibility of the needle, both in peripheral nerve block [138] and for spinal anaesthesia [139], perispinal anaesthesia [140] or central venous approaches [141].

3D/4D

Three-dimensional ultrasonography consists of representing an instantaneous tissue volume *t*, while so-called four-dimensional ultrasonography adds the time dimension (that is, 3D ultrasound in real time). This technology in RA allows understanding of the anatomy in a new dimension, providing a different view of the nerves and adjacent structures. These technological innovations may provide answers to certain questions in ultrasound-guided RA, and also improve its predictability and safety [84].

3D ultrasound offers the possibility of examining a static volume, which can be "refreshed" at different stages of an anaesthetic procedure, and of studying it by allowing it to "turn", by cutting it in all possible planes, some of which are impossible to obtain with conventional 2D ultrasound. This volumetric approach enables better assessment of the spatial anatomical relations of the area studied, a notion that is found in several studies on the subject [82, 142–145]. Multiplanar visualisation enables to simultaneously obtain a transverse sagittal and coronal vision of the ultrasound volume (Fig. 3.54).

4D ultrasound can enable conduct of ultrasound-guided RA with real-time visualisation of different significant elements (in

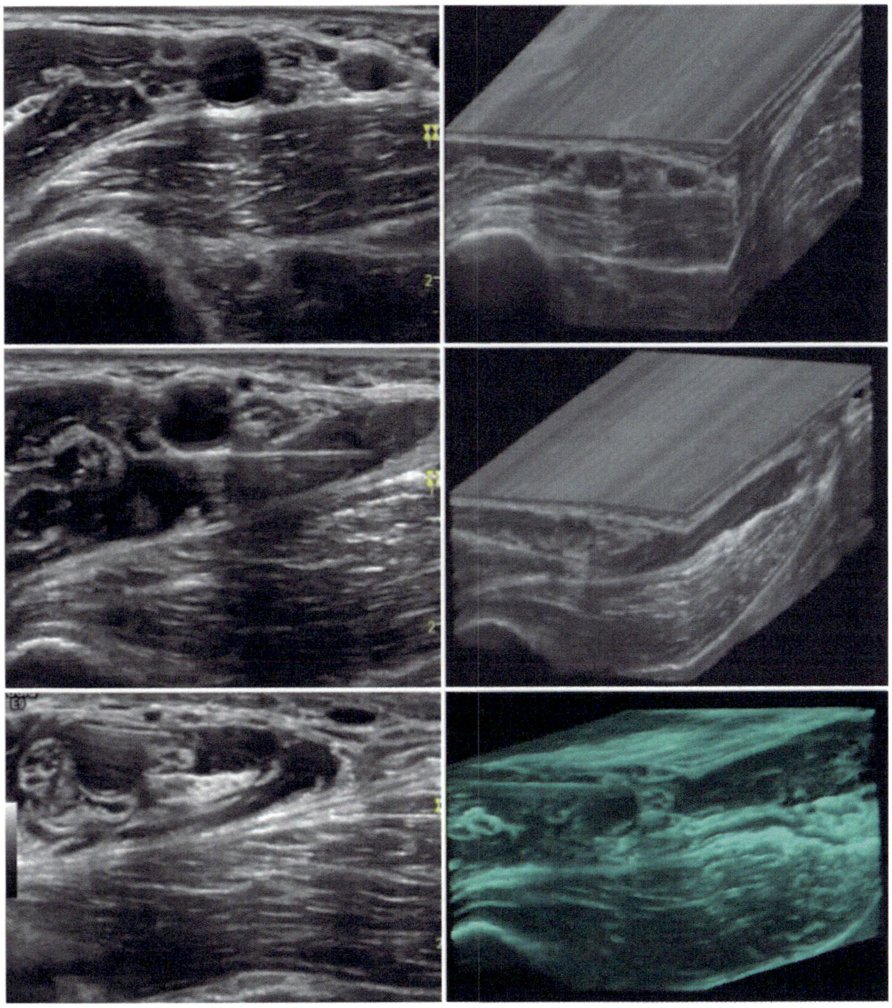

Fig. 3.54 3D reconstruction of different stages in conduct of an ultrasound-guided axillary block. (Iconography: Olivier Choquet)

volume or with multiplanes): anatomical structures, needle, catheter, spread of the local anaesthetic, etc. However, in spite of constant advances, images continue to be insufficiently precise to have real clinical utility, and the ability of ultrasound systems to process images must be considerably increased. A few published articles investigate the scope of four-dimensional ultrasound in regional anaesthesia [146–149].

Fusion Imaging (CT-Scan Ultrasound/ MRI-Ultrasound)

Wing Hong Kwok and Manoj Kumar Karmakar

Today it is possible to perform spinal sonography and obtain high quality ultrasound images of the spine and neuraxis [150, 151]. As a result spinal sonography is gaining popularity in regional anaesthesia and being used increasingly for central neuraxial blocks either as a pre-procedural tool [150, 152] or to guide interventions [153]. However, one of the major limitations of spinal sonography is that the neuraxial structures are enveloped by a bony framework (spine), which reflects ultrasound, allows limited amount of ultrasound energy to enter the spinal canal, and also casts an acoustic shadow [150]. This results in a narrow acoustic window for imaging and one has to rely on insonating the ultrasound beam through the interlaminar or interspinous spaces to visualise the neuraxis. Moreover, in any clinical condition that causes narrowing of the acoustic window, such as in the elderly or patients with osteoarthritis [154], spinal sonography is seriously compromised. Therefore any imaging modality that can complement ultrasound imaging of the neuraxis or facilitate needle guidance at the point-of-care in these difficult conditions is desirable.

Cross-sectional imaging techniques (ultrasound, computed tomography (CT) and magnetic resonance (MR) imaging) provide accurate anatomical information and play an important role in diagnosis and needle guidance (e.g. intrathecal, epidural or facet joint injection). Since each imaging modality functions on a different physical principle, they have their own advantages and limitations as described above for ultrasound imaging during spinal sonography. Also the information obtained from one imaging modality can potentially complement the other. This is the principle behind "fusion imaging", whereby two imaging modalities are combined (fused) together to overcome the limitation of both techniques and improve the overall anatomical information obtained. Since the final image, also referred to as the "third image", is a "hybrid" of the two imaging modalities used, it is also called "hybrid imaging" [155]. In oncology and nuclear medicine [156], anatomical imaging (CT or MRI) is often combined with molecular imaging (single-photon emission computed tomography, SPECT, or positron emission tomography, PET) with improved tumour localisation and clinical decision making [157]. Anatomical imaging modalities can also be fused together such as US/CT image fusion or US/MRI image fusion when real-time ultrasound imaging is fused with previously acquired CT or MRI data. The technique of displaying B-mode ultrasound images with previously acquired CT or MR images in real time has also been referred to as Real-time Virtual Sonography (RVS) [155]. There are published data describing anatomical fusion imaging to assess liver tumours [158], biopsy liver or prostrate lesions [155], and perform percutaneous radiofrequency ablation of liver tumours [159–161]. We believe that anatomical fusion imaging may also be applicable for spinal sonography. Currently, there are no data on fusion imaging in regional anaesthesia and only limited data in interventional pain medicine [162, 163]. This section briefly outlines our preliminary experience with US/CT and US/MRI image fusion for lumbar spinal imaging.

Technical Considerations

The system includes an ultrasound system that is capable of fusion, a transmitter that generates the electromagnetic field, which is placed close to the area scanned, and a magnetic position sensor that is attached to the ultrasound transducer via a bracket (attachment) (Fig. 3.55). The function of the sensor is to sense changes in position; orientation and alignment of the ultrasound transducer during the ultrasound scan and relay this information to the ultrasound system. Once the system is assembled, the CT or MRI dataset is imported (from the hard disc, CD or a compatible portable storage device) to the US system. For the datasets to be compatible with the fusion imaging system, it must be in DICOM (Digital Imaging and Communications in Medicine) format and should be a volume dataset with $\leq$3 mm slice thickness [155]. The ultrasound scan is then performed and displayed on the left half of the display screen while the CT or MRI image is displayed on the right half of the display screen. The next step is "registration" which can be either "image based" or "non-image based" [156]. Image-based registration can be done using an extrinsic method, which utilises a foreign object that is deliberately placed in the image field, and an intrinsic method that involves identification of identical anatomical points (landmarks) in the corresponding images. For registration to be accurate, it is suggested that at least three pairs of anatomical points should be used [156]. In case of a mismatch between the ultrasound and the CT or MR image, further fine adjustments can be made manually or the registration process repeated to correct the mismatch during the examination. While there are published data describing accurate registration and thereby fusion of US/CT images of the liver using the bifurcation of the portal vein or hepatic vein as a registration point there are no comparable data for spinal imaging, which we hope to report in the near future.

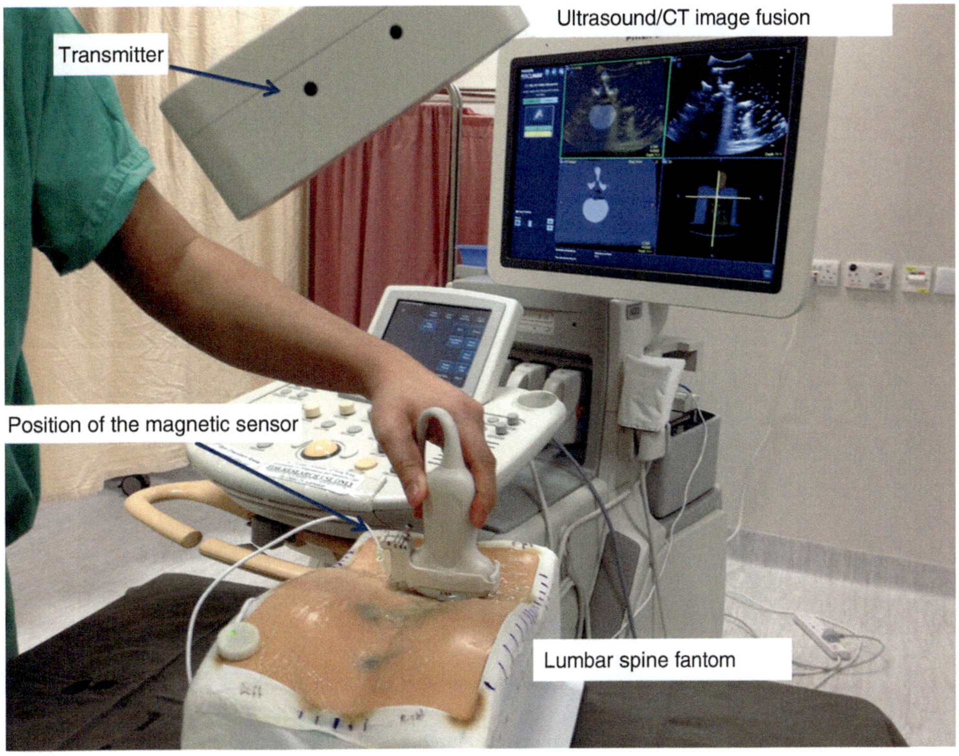

Fig. 3.55 Equipment used to perform a CT-scan ultrasound fusion image on a lumbar spine phantom

Preliminary Experience with US/CT and US/MRI Image Fusion

Lumbar Spine Phantom US/CT Fusion

CT scan of a commercial available lumbar spine phantom (CIRS, model 034 lumbar training phantom, Norfolk, VA, USA) (Fig. 3.55) was first performed. The volume CT data acquired was gapless with a slice thickness of <1 mm. The dataset was obtained from the radiology department server and loaded onto a Philips IU22 ultrasound system (Philips Healthcare, Andover, MA, USA). The magnetic sensor was then attached to the ultrasound transducer (C5-1, 5–1 MHz, curved array transducer equipped with PureWave crystal technology). The images (US and CT) were then registered using common internal anatomical bony landmarks (details to be published), similar to that described above. A quad screen format was used to display the images (Figs. 3.56, 3.57, and 3.58) where the top left image was the fused US/CT image, top right image was the real-time ultrasound image, bottom left image was the CT image and bottom right image was the 3D volume reconstruction of the phantom showing the position and orientation of the ultrasound transducer. Overlay and side by side comparison of the US and CT images was possible which enabled us to correlate the anatomical structures in the CT or MR images with that in the ultrasound image in real time. The amount (%) of overlay or blending could be manually adjusted by the operator for optimal visualisation. Ultrasound imaging in the transverse and sagittal scan plane relevant for ultrasound-guided central neuraxial blocks (spinal and epidural) was then performed and fused with the CT dataset (Figs. 3.56, 3.57, and 3.58).

US/MR Image Fusion of the Lumbar Spine in Volunteers

Young volunteers underwent MR imaging of the lumbar spine in the lateral position (Fig. 3.59a) similar to that adapted during spinal sonography or central neuraxial blocks. The MRI volume dataset was acquired with a slice thickness less than 3 mm. The dataset in DICOM format was transferred to a Logiq E9 (GE Healthcare, USA) and ultrasound imaging was performed using a curved array transducer (C1-6, 1–6 MHz). The basic set-up of the equipment used was similar to that described above and is demonstrated in Fig. 3.59b. The US and MR images were displayed side by side and registration was successfully completed using an image-based method (details to be published) after which the fusion was completed (Fig. 3.60).

Concluding Remarks

We envision with greater availability of high quality anatomical imaging study data (CT and MR) real-time fusion of US and CT, or US and MR images will soon become a reality in the operating room environment in near future. We have demonstrated that fusion imaging relevant for central neuraxial blocks is possible despite some initial difficulties with image registration. Based on our initial experience we don't

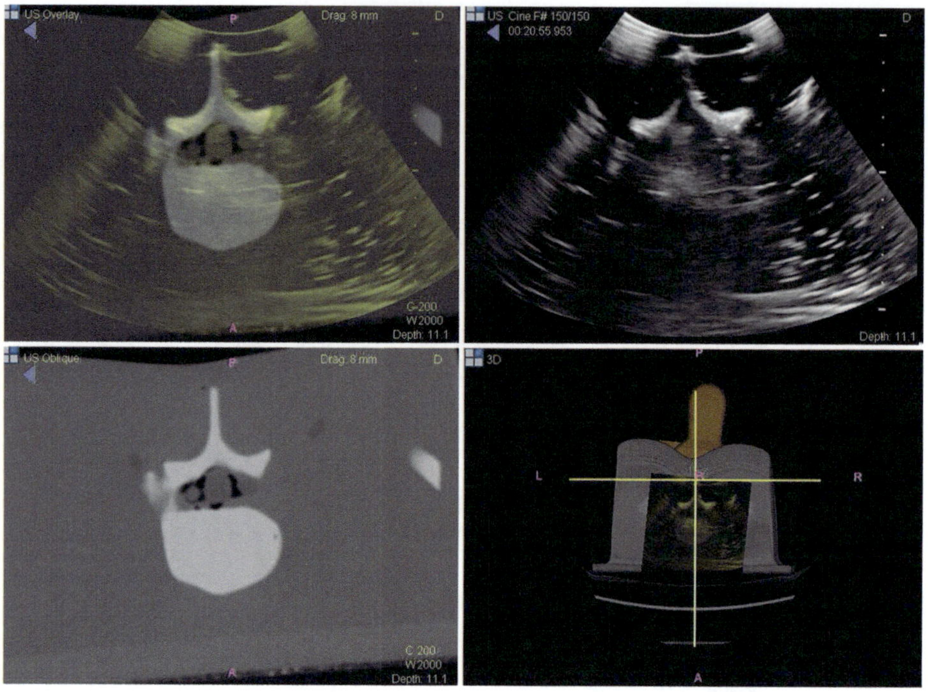

Fig. 3.56 CT-scan ultrasound fusion image, transverse lumbar, at the level of the spinous process of L4

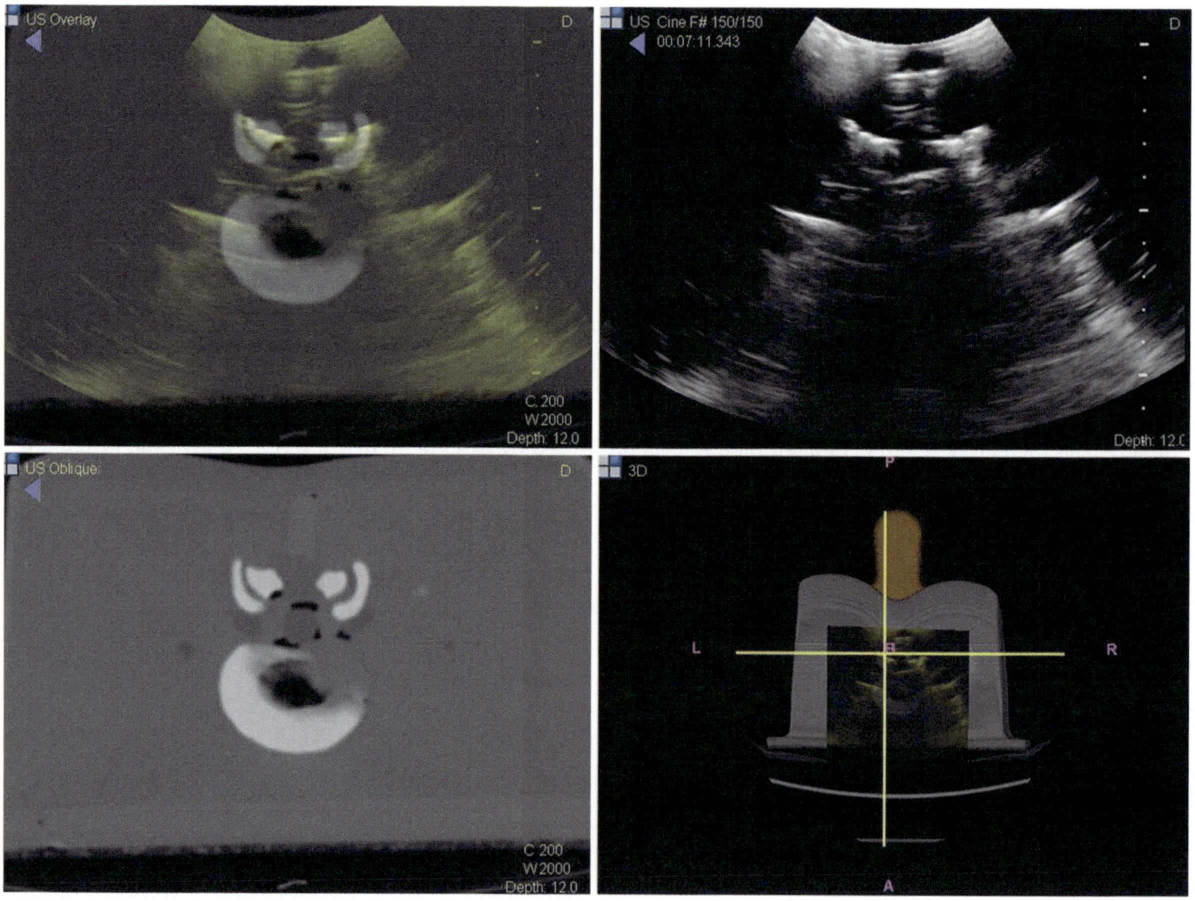

Fig. 3.57 CT-scan ultrasound fusion image, transverse lumbar, at the level of the L4-L5 interspinous space

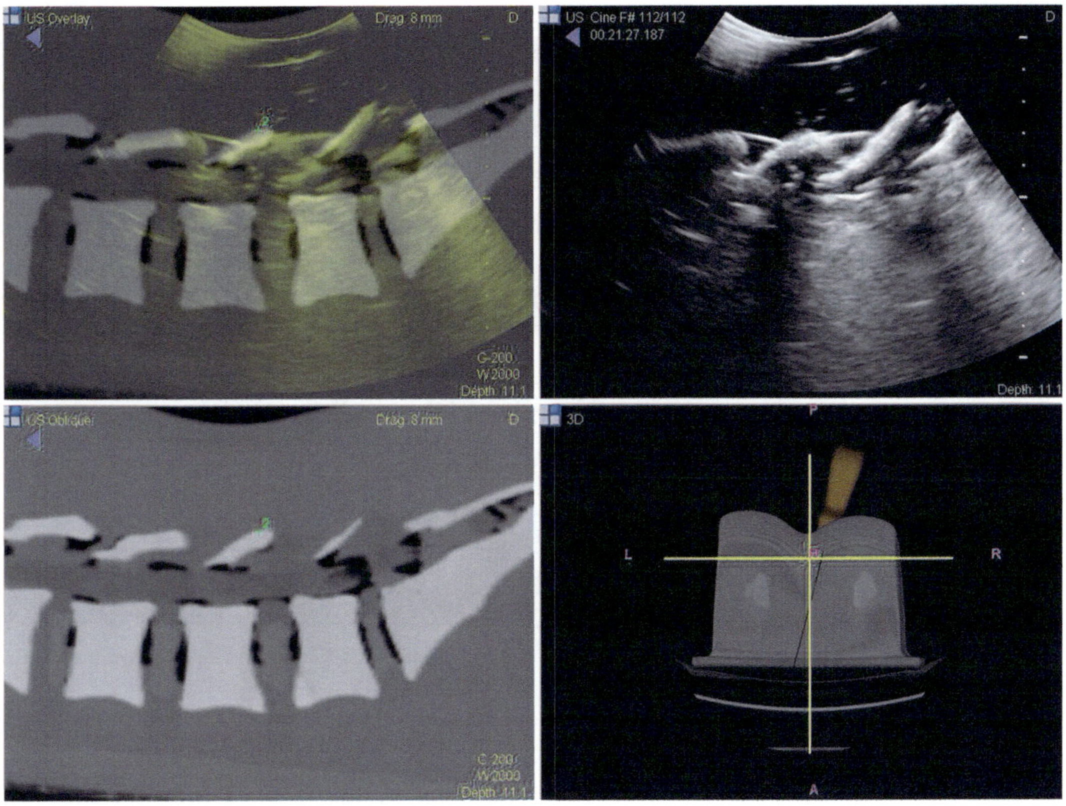

Fig. 3.58 CT-scan ultrasound fusion image on the phantom, showing an oblique paramedian sagittal cut of the lumbar spine, passing through the lamina

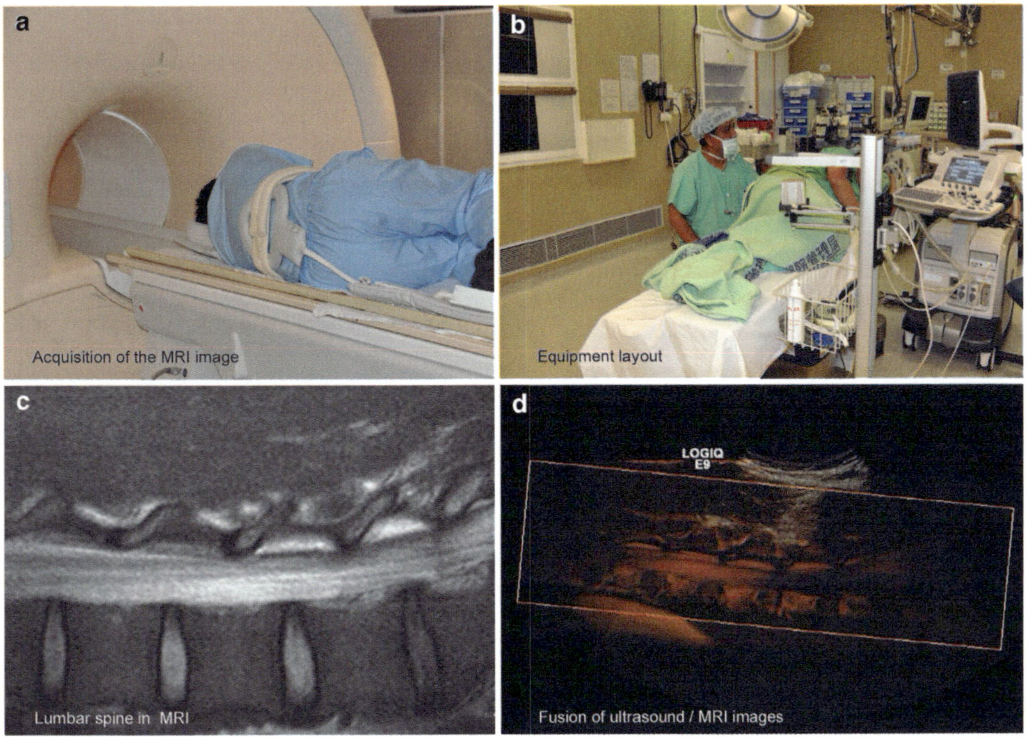

Fig. 3.59 MRI ultrasound fusion image. (**a**) Volunteer subject in lateral recumbent position for conduct of an MRI; (**b**) equipment for obtention of a fusion image; (**c**) MRI scan of the lumbar spine; (**d**) superimposition of the ultrasound image on the MRI scan

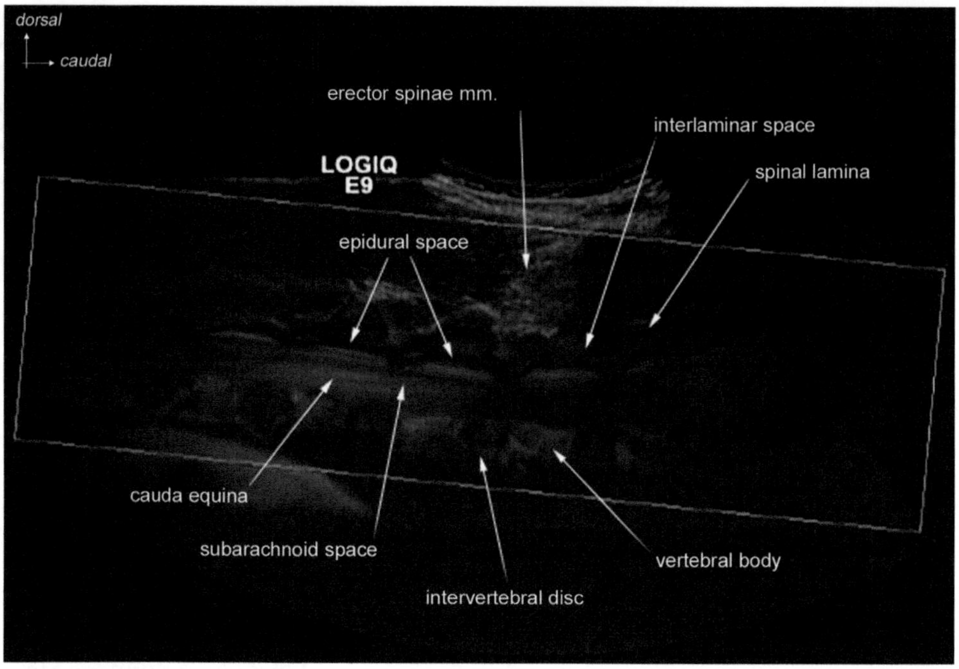

Fig. 3.60 Fusion of ultrasound and MRI images of the lumbar spine, showing neuraxial structures including the outline of the epidural space facing the lamina, normally masked by the acoustic shadow of the lamina in a simple ultrasound image

anticipate fusion anatomical imaging to be used routinely for central neuraxial blocks because ultrasound on its own has proven beneficial in recent years. However, it may play a role in patients that are deemed difficult; e.g. scoliosis, osteoarthritis, operated backs or even considered impossible, etc. Another exciting feature that is available with fusion imaging is volume navigation and GPS needle tracking technology which in principle should improve the accuracy of needle interventions. Only time will tell if fusion imaging will live up to its promise as a guidance tool for central neuraxial blocks.

Acknowledgements Figures 3.55, 3.56, 3.57, 3.58, 3.59 and 3.60 were reproduced with kind permission from www.aic.cuhk.edu.hk/usgraweb. The authors would also like to thank GE Healthcare Hong Kong for providing equipment support to conduct this pilot project.

References

1. Bigeleisen PE. Nerve puncture and apparent intraneural injection during ultrasound-guided axillary block does not invariably result in neurologic injury. Anesthesiology. 2006;105:779–83.
2. Jochum D, Iohom G, Diarra DP, Loughnane F, Dupré LJ, Bouaziz H. An objective assessment of nerve stimulators used for peripheral nerve blockade. Anaesthesia. 2006;61:557–64.
3. Belavy D, Wallis SC. Propylene glycol and glycerol concentration in ultrasound gel. Reg Anesth Pain Med. 2013;38:75–6.
4. Belavy D. Brief reports: regional anesthesia needles can introduce ultrasound gel into tissues. Anesth Analg. 2010;111:811–2.
5. El-Dawlatly A, Kathiry K, Al Rikabi A, Hajjar W, Al Obaid O, Alzahrani T. Ultrasound gel-nerve contact: an experimental animal histologic study. Anesth Analg. 2011;113:657–9.
6. Pintaric TS, Hadzic A, Strbenc M, Podpecan O, Podbregar M, Cvetko E. Inflammatory response after injection of aqueous gel into subarachnoid space in piglets. Reg Anesth Pain Med. 2013;38:100–5.
7. Bouaziz H, Aubrun F, Belbachir AA, et al. Société française d'anesthésie et de réanimation. Échographie en anesthésie locorégionale. Recommandations formalisées d'experts. Ann Fr Anesth Réanim. 2011;30:e33–5.
8. Recommandations pour la pratique clinique: les blocs périphériques des membres chez l'adulte. Paris: SFAR; 2003. http://www.sfar.org/article/155/les-blocs-peripheriques-des-membres-chez-l-adulte.
9. Jochum D. Hygiène relative à l'utilisation des ultrasons. Recommandations formalisées d'experts. Ann Fr Anesth Réanim. 2012;31:e219–21.
10. Hajjar J, Mounier M. Ultrasound equipment in anaesthesia and infection control measures [Letter]. Ann Fr Anesth Réanim. 2009;28:397–8.
11. Kac G, Gueneret M, Rodi A, et al. Evaluation of a new disinfection procedure for ultrasound probes using ultraviolet light. J Hosp Infect. 2007;65:163–8.
12. Bloc S, Mercadal L, Garnier T, et al. Evaluation of a new disinfection method for ultrasound probes used for regional anesthesia: ultraviolet C light. J Ultrasound Med. 2011;30:785–8.
13. Kac G, Podglajen I, Si-Mohamed A, Rodi A, Grataloup C, Meyer G. Evaluation of ultraviolet C for disinfection of endocavitary ultrasound transducers persistently contaminated despite probe covers. Infect Control Hosp Epidemiol. 2010;31:165–70.
14. Gaines de protection à usage unique pour dispositifs médicaux réutilisables: recommandations d'utilisation. Comité technique des infections nosocomiales et des infections liées aux soins. Haut Conseil de la santé publique, 14 décembre 2007. http://www.hcsp.fr/docspdf/avisrapports/hcspr20071214_GainesProtec.pdf.

15. Millesi H, Hausner T, Schmidhammer R, et al. Anatomical structures to provide passive motility of peripheral nerve trunks and fascicles. Acta Neurochir Suppl. 2007;100:133–5.
16. Vloka JD, Hadzic A, Lesser JB, et al. A common epineural sheath for the nerves in the popliteal fossa and its possible implications for sciatic nerve block. Anesth Analg. 1997;84:387–90.
17. Franco CD, Rahman A, Voronov G, et al. Gross anatomy of the brachial plexus sheath in human cadavers. Reg Anesth Pain Med. 2008;33:64–9.
18. Fornage BD. Peripheral nerves of the extremities: imaging with US. Radiology. 1988;167:179–82.
19. Fornage BD. Sonography of peripheral nerves of the extremities. Radiol Med. 1993;85(5 Suppl 1):162–7.
20. Gruber H, Peer S, Kowacs P, Marth R, Bodner G. The ultrasonographic appearance of the femoral nerve and cases of iatrogenic impairment. J Ultrasound Med. 2003;22:163–72.
21. Gray AT. Ultrasound-guided regional anesthesia. Current state of the art. Anesthesiology. 2006;104:368–73.
22. Martinoli C, Bianchi S, Santacroce E, Pugliese F, Graif M, Derchi LE. Brachial plexus sonography: a technique for assessing the root level. Am J Roentgenol. 2002;179:699–702.
23. Silvestri E, Martinoli C, Derchi LE, Bertolotto M, Chiaramondia M, Rosenberg I. Echotexture of peripheral nerves: correlation between US and histologic findings and criteria to differentiate tendons. Radiology. 1995;197:291–6.
24. Créteur V, Bacq C, Widelec J. Échographie des nerfs périphériques. Première partie: membre supérieur. J Radiol. 2004;85:1887–99.
25. Heinemeyer O, Reimers CD. Ultrasound of radial, ulnar, median and sciatic nerves in healthy subjects and patients with hereditary moto rand sensory neuropathies. Ultrasound Med Biol. 1999;25:481–5.
26. Soong J, Schafhalter-Zoppoth I, Gray AT. The importance of transducer angle to ultrasound visibility of the femoral nerve. Reg Anesth Pain Med. 2005;30:505.
27. Perlas A, Chan VW, Simons M. Brachial plexus examination and localization using ultrasound and electrical stimulation: a volunteer study. Anesthesiology. 2003;99:429–35.
28. Selander D, Brattsand R, Lundborg G, et al. Local anesthetics: importance of mode of application, concentration and adrenaline for the appearance of nerve lesions. An experimental study of axonal degeneration and barrier damage after intrafascicular injection or topical application of bupivacaine (Marcain). Acta Anaesthesiol Scand. 1979;23:127–36.
29. Selander D, Dhunér KG, Lundborg G. Peripheral nerve injury due to injection needles used for regional anesthesia. An experimental study of the acute effects of needle bevel trauma. Acta Anaesthesiol Scand. 1977;21:182–8.
30. Sala Blanch X, Ribalta T, Rivas E, Carrera A, Gaspa A, Reina MA, Hadzic A. Structural injury to the human sciatic nerve after intraneural needle insertion. Reg Anesth Pain Med. 2009;34:201–5.
31. Rice AS, McMahon SB. Peripheral nerve injury caused by injection needles used in regional anaesthesia: influence of bevel configuration, studied in a rat model. Br J Anaesth. 1992;69:433–8.
32. Schafhalter-Zoppoth I, McCulloch CE, Gray AT. Ultrasound visibility of needles used for regional nerve block: an *in vitro* study. Reg Anesth Pain Med. 2004;29:480–8.
33. Fornage BD. Preoperative sonographic localization of a migrated transosseous stabilizing wire in the hand. J Ultrasound Med. 1987;6:471–3.
34. Sandhu NS, Capan LM. Ultrasound-guided infraclavicular brachial plexus block. Br J Anesth. 2002;89:254–9.
35. Chan VW, Perlas A, Rawson R, Odukoya O. Ultrasound-guided supraclavicular brachial plexus block. Anesth Analg. 2003;97:1514–7.
36. Bedforth N, Townsley P, Maybin J, Eisenberg E. Single-handed ultrasound-guided regional anaesthesia. Anaesthesia. 2011;66:846.
37. Sites BD, Chan VW, Neal JM, et al. The American Society of Regional Anesthesia and Pain Medicine and the European Society of Regional Anaesthesia and Pain Therapy Joint Committee recommendations for education and training in ultrasound-guided regional anesthesia. Reg Anesth Pain Med. 2009;34:40–6.
38. Chantzi C, Saranteas T, Paraskeuopoulos T, Dimitriou V. Ultrasound and transcutaneous neurostimulator combined technique as a training method for nerve identification in anesthesia residents. Reg Anesth Pain Med. 2007;32:365–6.
39. Tsui BC, Kropelin B. The electrophysiological effect of dextrose 5% in water on single-shot peripheral neurostimulation. Anesth Analg. 2005;100:1837–9.
40. Tsui BC, Kropelin B, Ganapathy S, Finucane B. Dextrose 5% in water: fluid medium for maintaining electrical stimulation of peripheral nerves during stimulating catheter placement. Acta Anaesthesiol Scand. 2005;49:1562–5.
41. Chin KJ, Perlas A, Chan VW, Brull R. Needle visualization in ultrasound-guided regional anesthesia: challenges and solutions. Reg Anesth Pain Med. 2008;33:532–44.
42. Beach ML, Sites BD, Gallagher JD. Use of a nerve stimulator does not improve the efficacy of ultrasound-guided supraclavicular nerve blocks. J Clin Anesth. 2006;18:580–4.
43. Perlas A, Niazi A, McCartney C, Chan V, Xu D, Abbas S. The sensitivity of motor response to neurostimulation and paresthesia for nerve localization as evaluated by ultrasound. Reg Anesth Pain Med. 2006;31:445–50.
44. Sinha SK, Abrams JH, Weller RS. Ultrasound-guided interscalene needle placement produces successful anesthesia regardless of motor stimulation above or below 0.5 mA. Anesth Analg. 2007;105:848–52.
45. Dingemans E, Williams SR, Arcand G, et al. Neurostimulation in ultrasound-guided infraclavicular block: a prospective randomized trial. Anesth Analg. 2007;104:1275–80.
46. Andersen HL, Andersen SL, Tranum Jensen J. Injection inside the paraneural sheath of the sciatic nerve. Direct comparison among ultrasound imaging, macroscopic anatomy, and histologic analysis. Reg Anesth Pain Med. 2012;37:410–4.
47. Missair A, Weisman RS, Suarez MR, Yang R, Gebhard RE. A 3-dimensional ultrasound study of local anesthetic spread during lateral popliteal nerve block. What is the ideal end point for needle tip position. Reg Anesth Pain Med. 2012;37:1–6.
48. Feugeas JL, Choquet O. Comprendre et optimiser la neurostimulation. Congrès national d'anesthésie et de réanimation 2007. Évaluation et traitement de la douleur, vol. 2007. Paris: Elsevier Masson. p. 745–53.
49. Rice AS. Peripheral nerve damage and regional anaesthesia. Br J Anaesth. 1995;75:116.
50. Sala Blanch X, López AM, Carazo J, et al. Intraneural injection during nerve stimulator-guided sciatic nerve block at the popliteal fossa. Br J Anaesth. 2009;102:855–61.
51. Robards C, Hadzic A, Somasundaram L, et al. Intraneural injection with low-current stimulation during popliteal sciatic nerve block. Anesth Analg. 2009;109:673–7.
52. Morau D, Levy F, Bringuier S, et al. Ultrasound-guided evaluation of the local anesthetic spread parameters required for a rapid surgical popliteal sciatic nerve block. Reg Anesth Pain Med. 2010;35:559–64.
53. Dufour E, Cymerman A, Noury G, et al. An ultrasonographic assessment of neurostimulation-guided median nerve block at the elbow: a local anesthetic spread, nerve size, and clinical efficacy study. Anesth Analg. 2010;111:561–7.
54. Jochum D. Quelles recommandations pour les techniques de repérage des blocs plexiques et tronculaires des membres? In La collection de la SFAR. Les blocs périphériques des membres

chez l'adulte. Recommandations pour la pratique clinique. Paris: Elsevier; 2004. p. 69–92.
55. Tsui BC, Pillay JJ, Chu KT, Dillane D. Electrical impedance to distinguish intraneural from extraneural needle placement in porcine nerves during direct exposure and ultrasound guidance. Anesthesiology. 2008;109:479–83.
56. Tsui BCH, Chin JH. Electrical impedance to warn of intravascular needle placement. Reg Anesth Pain Med. 2008;33:A-51.
57. Moayeri N, Gerbrand JG. Difference in quantitative architecture of sciatic nerve may explain differences in potential vulnerability to nerve injury, onset time and minimum effective anesthetic volume. Anesthesiology. 2009;111:128–34.
58. Loubert C, Williams SR, Hélie F, Arcand G. Complication during ultrasound-guided regional block: accidental intravascular injection of local anesthetic. Anesthesiology. 2008;108:759–60.
59. Zetlaoui PJ, Labbe JP, Benhamou D. Ultrasound guidance for axillary plexus block does not prevent intravascular injection. Anesthesiology. 2008;108:761.
60. Hadzic A, Sala Blanch X, Xu D. Ultrasound guidance may reduce but not eliminate complications of peripheral nerve blocks. Anesthesiology. 2008;108:557–8.
61. Brull R, Perlas A, Cheng PH, Chan VW. Minimizing the risk of intravascular injection during ultrasound-guided peripheral nerve blockade [Letter]. Anesthesiology. 2008;109:1142.
62. Sauter AR, Dodgson MS, Stubhaug A, Cvancarova M, Klaastad O. Ultrasound controlled neurostimulation in the elbow region: high currents and short distances needed to obtain motor responses. Acta Anaesthesiol Scand. 2007;51:942–8.
63. Bloc S, Ecoffey C, Dhonneur G. Controlling needle tip progression during ultrasound-guided regional anesthesia using the hydrolocalization technique. Reg Anesth Pain Med. 2008;33:382–3.
64. Dupré LJ. Neurostimulateur en anesthésie locorégionale. Cah Anesthesiol. 1992;40:503–10.
65. Jochum D, Iohom G, Dupré LJ, Bouaziz H. In defense of the peripheral neurostimulation technique [Letter]. Anesth Analg. 2006;103:1038–9.
66. Rosenblatt MA. When is a single-injection nerve block not really a single injection? [Letter]. Anesthesiology. 2008;109:748.
67. Barr RG. Improved needle visualization with electronic beam steering: proof of concept. Ultrasound Q. 2012;28:59–64.
68. Hadzic A, Dilberovic F, Shah S, et al. Combination of intraneural injection and high injection pressure leads to fascicular injury and neurologic deficits in dogs. Reg Anesth Pain Med. 2004;29:417–23.
69. Shah S, Hadzic A, Vloka JD, Cafferty MS, Moucha CS, Santos AC. Neurologic complication after anterior sciatic nerve block. Anesth Analg. 2005;100:1515–7.
70. Claudio R, Hadzic A, Shih H, et al. Injection pressures by anesthesiologists during simulated peripheral nerve block. Reg Anesth Pain Med. 2004;29:201–5.
71. Tsui BC, Knezevich MP, Pillay JJ. Reduced injection pressures using a compressed air injection technique (CAIT): an *in vitro* study. Reg Anesth Pain Med. 2008;33:168–73.
72. Rogriguez J, Taboada M, Blanco M, et al. Intraneural catheterization of the sciatic nerve in humans: a pilot study. Reg Anesth Pain Med. 2008;33:285–90.
73. Moayeri N, Krediet AC, Welleweerd JC, Bleys RL, Groen GJ. Early ultrasonographic detection of low-volume intraneural injection. Br J Anaesth. 2012;109:432–8.
74. Choquet O, Morau D, Biboulet P, Capdevila X. Where should the tip of the needle be located in ultrasound-guided peripheral nerve blocks? Curr Opin Anaesthesiol. 2012;25:596–602.
75. Liu SS, YaDeau JT, Shaw PM, et al. Incidence of unintentional intraneural injection and postoperative neurological complications with ultrasound-guided interscalene and supraclavicular nerve blocks. Anaesthesia. 2011;66:168–74.
76. Neal JM, Wedel DJ. Ultrasound guidance and peripheral nerve injury: is our vision as sharp as we think it is? Reg Anesth Pain Med. 2010;35:335–7.
77. Lupu CM, Kiehl TR, Chan VW, et al. Nerve expansion seen on ultrasound predicts histologic but not functional nerve injury after intraneural injection in pigs. Reg Anesth Pain Med. 2010;35:132–9.
78. Kapur E, Vuckovic I, Dilberovic F, et al. Neurologic and histologic outcome after intraneural injections lidocaine in canine sciatic nerves. Acta Anaesthesiol Scand. 2007;51:101–7.
79. Borgeat A. Regional anesthesia, intraneural injection, and nerve injury: beyond the epineurium. Anesthesiology. 2006;105:647–8.
80. Cohen JM, Gray AT. Functional deficits after intraneural injection during interscalene block. Reg Anesth Pain Med. 2010;35:397–9.
81. Neal JM, Bernards CM, Hadzic A, et al. ASRA practice advisory on neurologic complications in regional anesthesia and pain medicine. Reg Anesth Pain Med. 2008;33:404–15.
82. Brull R, Macfarlane AJ, Parrington SJ, Koshkin A, Chan VW. Is circumferential injection advantageous for ultrasound-guided popliteal sciatic nerve block?: a proof-of-concept study. Reg Anesth Pain Med. 2011;36:266–70.
83. Missair A, Weisman RS, Suarez MR, Yang R, Gebhard RE. A 3-dimensional ultrasound study of local anesthetic spread during lateral popliteal nerve block: what is the ideal end point for needle tip position? Reg Anesth Pain Med. 2012;37:627–32.
84. Carney J, Finnerty O, Rauf J, et al. Studies on the spread of local anaesthetic solution in transversus abdominis plane blocks. Anaesthesia. 2011;66:1023–30.
85. Al-Nasser B, Hubert C, Négre M. Role of local anesthetic spread pattern and electrical stimulation in ultrasound-guided musculocutaneous nerve block. J Clin Anesth. 2010;22:334–9.
86. Sinha SK, Abrams JH, Houle TT, Weller RS. Ultrasound-guided obturator nerve block: an interfascial injection approach without neurostimulation. Reg Anesth Pain Med. 2009;34:261–4.
87. Manassero A, Bossolasco M, Ugues S, et al. Ultrasound-guided obturator nerve block: interfascial injection *versus* a neurostimulation-assisted technique. Reg Anesth Pain Med. 2012;37:67–71.
88. Taha AM. Brief reports: ultrasound-guided obturator nerve block: a proximal interfascial technique. Anesth Analg. 2012;114:236–9.
89. Spence BC, Beach ML, Gallagher JD, Sites BD. Ultrasound-guided interscalene blocks: understanding where to inject the local anaesthetic. Anaesthesia. 2011;66:509–14.
90. Clendenen SR, Riutort K, Ladlie BL, et al. Real-time three-dimensional ultrasound-assisted axillary plexus block defines soft tissue planes. Anesth Analg. 2009;108:1347–50.
91. Imasogie N, Ganapathy S, Singh S, et al. A prospective, randomized, doubleblind comparison of ultrasound-guided axillary brachial plexus blocks using 2 *versus* 4 injections. Anesth Analg. 2010;110:1222–6.
92. Fredrickson MJ, Wolstencroft P, Kejriwal R, et al. Single *versus* triple injection ultrasound-guided infraclavicular block: confirmation of the effectiveness of the single injection technique. Anesth Analg. 2010;111:1325–7.
93. Liu SS, Strodtbeck WM, Richman JM, Wu CL. A comparison of regional *versus* general anesthesia for ambulatory anesthesia: a meta-analysis of randomized controlled trials. Anesth Analg. 2005;101:1634–42.
94. McCartney CJ, Lin L, Shastri U. Evidence basis for the use of ultrasound for upper-extremity blocks. Reg Anesth Pain Med. 2010;35:S10–5.
95. Danelli G, Ghisi D, Fanelli A, et al. The effects of ultrasound guidance and neurostimulation on the minimum effective anesthetic

volume of mepivacaine 1.5% required to block the sciatic nerve using the subgluteal approach. Anesth Analg. 2009;109:1674–8.
96. McNaught A, Shastri U, Carmichael N, et al. Ultrasound reduces the minimum effective local anaesthetic volume compared with peripheral neurostimulation for interscalene block. Br J Anaesth. 2010;106:124–30.
97. Casati A, Baciarello M, Di Cianni S, et al. Effects of ultrasound guidance on the minimum effective anaesthetic volume required to block the femoral nerve. Br J Anaesth. 2007;98:823–7.
98. O'Donnell BD, Iohom G. An estimation of the minimum effective anesthetic volume of 2% lidocaine in ultrasound-guided axillary brachial plexus block. Anesthesiology. 2009;111:25–9.
99. Van Geffen GJ, van den Broek E, Braak GJ, Giele JL, Gielen MJ, Scheffer GJ. A prospective randomised controlled trial of ultrasound guided *versus* neurostimulation guided distal sciatic nerve block at the popliteal fossa. Anaesth Intensive Care. 2009;37:32–7.
100. Riazi S, Carmichael N, Awad I, Holtby RM, McCartney CJ. Effect of local anaesthetic volume (20 *vs* 5 mL) on the efficacy and respiratory consequences of ultrasound-guided interscalene brachial plexus block. Br J Anaesth. 2008;101:549–56.
101. O'Donnell B, Riordan J, Ahmad I, Iohom G. Brief reports: a clinical evaluation of block characteristics using one milliliter 2% lidocaine in ultrasound-guided axillary brachial plexus block. Anesth Analg. 2010;111:808–10.
102. Ponrouch M, Bouic N, Bringuier S, et al. Estimation and pharmacodynamic consequences of the minimum effective anesthetic volumes for median and ulnar nerve blocks: a randomized, double-blind, controlled comparison between ultrasound and neurostimulation guidance. Anesth Analg. 2010;111:1059–64.
103. Fredrickson M, White R, Danesh-Clough T. Low-volume ultrasound-guided nerve block provides inferior postoperative analgesia compared to a higher-volume landmark technique. Reg Anesth Pain Med. 2011;36:393–8.
104. Schoenmakers K, Wegener J, Stienstra R. Effect of local anesthetic volume (15 *vs* 40 mL) on the duration of ultrasound-guided single shot axillary brachial plexus block: a prospective randomized, observer-blinded trial. Reg Anesth Pain Med. 2012;37:242–7.
105. Vester-Andersen T, Husum B, Lindeburg T, Borrits L, Gothgen I. Perivascular axillary block IV: blockade following 40, 50 or 60 mL mepivacaine 1% with adrenaline. Acta Anaesthesiol Scand. 1984;28:99–105.
106. Serradell A, Herrero R, Villanueva JA, Santos JA, Moncho JM, Masdeu J. Comparison of three different volumes of mepivacaine in axillary plexus block using multiple neurostimulation. Br J Anaesth. 2003;9:519–24.
107. McCartney CJ, Patel S. Local anesthetic volume for peripheral nerve blocks: how low can (or should) we go? Reg Anesth Pain Med. 2012;37:239–41.
108. Richman JM, Liu SS, Courpas G, et al. Does continuous peripheral nerve block provide superior pain control to opioids? A meta-analysis. Anesth Analg. 2006;102:248–57.
109. Aguirre J, Del Moral A, Cobo I, Borgeat A, Blumenthal S. The role of continuous peripheral nerve blocks. Anesthesiol Res Pract. 2012;2012:560879.
110. Bingham AE, Fu R, Horn JL, Abrahams MS. Continuous peripheral nerve block compared with single-injection peripheral nerve block: a systematic review and meta-analysis of randomized controlled trials. Reg Anesth Pain Med. 2012;37:583–94.
111. Ilfeld BM, Moeller-Bertram T, Hanling SR, et al. Treating intractable phantom limb pain with ambulatory continuous peripheral nerve blocks: a pilot study. Pain Med. 2013;14:935–42.
112. Recommandations formalisées d'experts 2008. Prise en charge de la douleur post-opératoire chez l'adulte et l'enfant. Ann Fr Anesth Réanim. 2008;17(6):1035–41.
113. Protocole d'utilisation à domicile des spécialités à base de ropivacaïne 2 mg/mL, solution injectable en poche (NAROPEINE et génériques) dans le cadre de l'analgésie post-opératoire par cathéter périnerveux. ANSM; 2012. http://ansm.sante.fr.
114. Swenson JD, Bay NA, Burks RT, et al. Outpatient management of continuous peripheral nerve catheters placed using ultrasound guidance: an experience in 620 patients. Anesth Analg. 2006;103:1436–43.
115. Capdevila X, Barthelet Y, Biboulet P, et al. Effects of perioperative analgesic technique on the surgical outcome and duration of rehabilitation after major knee surgery. Anesthesiology. 1999;91:8–15.
116. Morau D, Biboulet P, Choquet O, Gourari A, Capdevila X. Comment échoguider un cathéter? JEPU. 2012.
117. Minville V, Zetlaoui PJ, Fessenmeyer C, et al. Ultrasound guidance for difficult lateral popliteal catheter insertion in a patient with peripheral vascular disease. Reg Anesth Pain Med. 2004;29:368–70.
118. Tognu A, Borghi B, Gullotta S, White PF. Ultrasound-guided posterior approach to brachial plexus for the treatment of upper phantom limb syndrome. Minerva Anesthesiol. 2011;77:1–2.
119. Fredrickson MJ, Price DJ. Analgesic effectiveness of ropivacaine 0.2% *versus* 0.4% *via* an ultrasound-guided C5-6 root/superior trunk perineural ambulatory catheter. Br J Anaesth. 2009;103:434–9.
120. Aveline C, Le Roux A, Le Hetet H, et al. Postoperative efficacy of femoral nerve catheters sites using ultrasound combined with neurostimulation alone for total knee arthroplasty. Eur J Anaesthesiol. 2010;27:978–84.
121. Gucev G, Yasui GM, Chang TY, Lee J. Bilateral ultrasound-guided continuous ilioinguinal-iliohypogastric block for pain relief after caesarean delivery. Anesth Analg. 2008;106:1220–2.
122. Shido A, Imamachi N, Doi K, Sakura S, Saito Y. Continuous local anesthetic infusion through ultrasound-guided rectus sheath catheters. Can J Anaesth. 2010;57:1046-7.
123. Bendtsen TF, Nielsen TD, Rhode CV, Kibak K, Linde F. Ultrasound guidance improves a continuous popliteal sciatice nerve block when compared with neurostimulation. Reg Anesth Pain Med. 2011;36:181–4.
124. Fredrickson MJ, Ball CM, Dalgleish AJ. Successful continuous interscalene analgesia for ambulatory shoulder surgery in a private practice setting. Reg Anesth Pain Med. 2008;33:122–8.
125. Fredrickson MJ, Ball CM, Dalgleish AJ. A prospective randomized comparison of ultrasound guidance *versus* neurostimulation for interscalene catheter placement. Reg Anesth Pain Med. 2009;34:590–4.
126. Mariano ER, Loland VJ, Sadhu NS, et al. Comparative efficacy of ultrasound-guided and stimulating popliteal-sciatice perineural catheters for postoperative analgesia. Can J Anaesth. 2010;57:919–26.
127. Fredrickson MJ, Ball CM, Dalgleish AJ, et al. A prospective randomized comparison of ultrasound and neurostimulation as needle end points for interscalene catheter placement. Anesth Analg. 2009;108:1695–700.
128. Gandhi K, Lindenmuth D, Hadzic A, Xu D, Maliakal TJ, Gadsen JC. The effect of stimulating *versus* conventional perineural catheters on postoperative analgesia following ultrasound-guided femoral nerve localization. J Clin Anesth. 2011;23:626–31.
129. Brull R, Prasad GA, Gandhi R, Ramlogan R, Khan M, Chan V. Is a patella motor response necessary for continuous femoral nerve blockade performed in conjunction with ultrasound guidance? Anesth Analg. 2011;112:982–6.
130. Mariano ER, Loland VJ, Bellars RH, et al. Ultrasound guidance *versus* electrical stimulation for infraclavicular brachial plexus perineural catheter insertion. J Ultrasound Med. 2009;28:1211–8.
131. Mariano ER, Loland VJ, Sandhu NS, et al. Ultrasound guidance *versus* electrical stimulation for femoral perineural catheter insertion. J Ultrasound Med. 2009;28:1453–60.

132. Mariano ER, Cheng GS, Choy LP, et al. Electrical stimulation *versus* ultrasound guidance for poplitea-sciatic perineural catheter insertion. A randomized controlled trial. Reg Anesth Pain Med. 2009;34:480–5.
133. Dhir S, Ganapathy S. Use of ultrasound guidance and contrast enhancement: a study of continuous infraclavicular brachial plexus approach. Acta Anaesthesiol Scand. 2008;52:338–42.
134. Courrèges P, Dadure C, Ecoffey C, et al. ADARPEF/SFAR. Recommandations formalisées d'experts. Anesthésie loco-régionale en pédiatrie. Association des anesthésistes-réanimateurs pédiatriques d'expression française/Société française d'anesthésie réanimation; 2010. http://www.adarpef.org et www.sfar.org.
135. Jochum D, Bondàr A, Delaunay L, Egan M, Bouaziz H. One size does not fit all: proposed algorithm for ultrasonography in combination with neurostimulation for peripheral nerve blockade. Br J Anaesth. 2009;103:771–3.
136. Tang R, Sawka A, Vaghadia H, Umbarje K. SonixGPS™ needle tracking system for out-of-plane brachial plexus block in human cadavers. Acta Anaesthesiol Scand. 2013;57:398–9.
137. Brinkmann S, Vaghadia H, Sawka A, Tang R. Methodological considerations of ultrasound-guided spinal anesthesia using the Ultrasonix GPS™ needle tracking system. Can J Anaesth. 2013;60:407–8.
138. Umbarje K, Tang R, Randhawa R, Sawka A, Vaghadia H. Out-of-plane brachial plexus block with a novel SonixGPS™ needle tracking system. Anaesthesia. 2013;68:433–4.
139. Wong SW, Niazi AU, Chin KJ, Chan VW. Real-time ultrasound-guided spinal anesthesia using the SonixGPS® needle tracking system: a case report. Can J Anaesth. 2013;60:50–3.
140. Kaur B, Tang R, Vaghadia H, Sawka A. Ultrasound-guided thoracic paravertebral block using the SonixGPS® system in human cadavers. Can J Anaesth. 2013;60:331–2.
141. Kopac DS, Chen J, Tang R, Sawka A, Vaghadia H. Comparison of a novel real-time SonixGPS needle-tracking ultrasound technique with traditional ultrasound for vascular access in a phantom gel model. J Vasc Surg. 2013;58:735–41.
142. Karmakar MK, Li X, Li J, Hadzic A. Volumetric three-dimensional ultrasound imaging of the anatomy relevant for thoracic paravertebral block. Anesth Analg. 2012;115:1246–50.
143. Karmakar M, Li X, Li J, Sala Blanch X, Hadzic A, Gin T. Three-dimensional/four-dimensional volumetric ultrasound imaging of the sciatic nerve. Reg Anesth Pain Med. 2012;37:60–6.
144. Cash CJ, Sardesai AM, Berman LH, et al. Spatial mapping of the brachial plexus using three-dimensional ultrasound. Br J Radiol. 2005;78:1086–94.
145. Choquet O, Capdevila X. Three-dimensional high-resolution ultrasound-guided nerve blocks: a new panoramic vision of local anesthetic spread and perineural catheter tip location. Anesth Analg. 2013;116:1176–81.
146. Clendenen SR, Riutort KT, Feinglass NG, Greengrass RA, Brull SJ. Real-time three-dimensional ultrasound for continuous interscalene brachial plexus blockade. J Anesth. 2009;23:466–8.
147. Feinglass NG, Clendenen SR, Torp KD, Wang RD, Castello R, Greengrass RA. Real-time three-dimensional ultrasound for continuous popliteal blockade: a case report and image description. Anesth Analg. 2007;105:272–4.
148. Clendenen SR, Riutort K, Ladlie BL, Robards C, Franco CD, Greengrass RA. Real-time three-dimensional ultrasound-assisted axillary plexus block defines soft tissue planes. Anesth Analg. 2009;108:1347–50.
149. Foxall GL, Hardman JG, Bedforth NM. Three-dimensional, multiplanar, ultrasound-guided, radial nerve block. Reg Anesth Pain Med. 2007;32:516–21.
150. Chin KJ, Karmakar MK, Peng P. Ultrasonography of the adult thoracic and lumbar spine for central neuraxial blockade. Anesthesiology. 2011;114:1459–85.
151. Karmakar MK, Li X, Kwok WH, Ho AM, Ngan Kee WD. Sonoanatomy relevant for ultrasound-guided central neuraxial blocks via the paramedian approach in the lumbar region. Br J Radiol. 2012;85:e262–9.
152. Chin KJ, Perlas A, Singh M, et al. An ultrasound-assisted approach facilitates spinal anesthesia for total joint arthroplasty. Can J Anaesth. 2009;56:643–50.
153. Karmakar MK, Li X, Ho AM, Kwok WH, Chui PT. Real-time ultrasound-guided paramedian epidural access: evaluation of a novel in-plane technique. Br J Anaesth. 2009;102:845–54.
154. Soh E, Karmakar MK. Assessment of the spine with CT and MRI prior to interspinous/interlaminar spinal procedures: a pictorial review. Br J Radiol. 2013;86:20130066.
155. Sandulescu DL, Dumitrescu D, Rogoveanu I, Saftoiu A. Hybrid ultrasound imaging techniques (fusion imaging). World J Gastroenterol. 2011;17:49–52.
156. Schillaci O, Simonetti G. Fusion imaging in nuclear medicine applications of dual-modality systems in oncology. Cancer Biother Radiopharm. 2004;19:1–10.
157. Schillaci O. Functional-anatomical image fusion in neuroendocrine tumors. Cancer Biother Radiopharm. 2004;19:129–34.
158. Sandulescu L, Saftoiu A, Dumitrescu D, Ciurea T. Real-time contrast-enhanced and real-time virtual sonography in the assessment of benign liver lesions. J Gastrointestin Liver Dis. 2008;17:475–8.
159. Kitada T, Murakami T, Kuzushita N, et al. Effectiveness of real-time virtual sonography-guided radiofrequency ablation treatment for patients with hepatocellular carcinomas. Hepatol Res. 2008;38:565–71.
160. Minami Y, Kudo M, Chung H, et al. Percutaneous radiofrequency ablation of sonographically unidentifiable liver tumors. Feasibility and usefulness of a novel guiding technique with an integrated system of computed tomography and sonographic images. Oncology. 2007;72(Suppl 1):111–6.
161. Clevert DA, Paprottka PM, Helck A, Reiser M, Trumm CG. Image fusion in the management of thermal tumor ablation of the liver. Clin Hemorheol Microcirc. 2012;52:205–16.
162. Galiano K, Obwegeser AA, Bale R, et al. Ultrasound-guided and CT-navigation-assisted periradicular and facet joint injections in the lumbar and cervical spine: a new teaching tool to recognize the sonoanatomic pattern. Reg Anesth Pain Med. 2007;32:254–7.
163. Klauser AS, De ZT, Feuchtner GM, et al. Fusion of real-time US with CT images to guide sacroiliac joint injection *in vitro* and *in vivo*. Radiology. 2010;256:547–53.

General Notions of Anatomy Applied to Regional Anaesthesia

4

Eryk Eisenberg

Contents

Regional Anaesthesia always works – provided you put the right dose of the right drug in the right place (M. Morgan) [1].

The primary objective of this chapter is not to give the reader in-depth information about using local anaesthetics *(the right dose of the right drug)*. Rather, we are going to explore the anatomical elements *(the right place)*: the cornerstone of regional anaesthesia. The study of anatomy is all the more interesting since it has practical applications, and regional anaesthesia is an exercise in applied anatomy (whether performed with neurostimulation or using the technique of ultrasound guidance). The anatomy enables us to develop anaesthetic techniques that are appropriate for the clinical situations that we encounter and to offer better insight in the case of block failure.

The Nervous System

General Organisation

The nervous system consists of the central nervous system (brain and spinal cord) and the peripheral nervous system. The latter is made up of 12 cranial nerve pairs, 31 spinal nerve pairs and the autonomic nervous system (sympathetic and parasympathetic systems).

The spinal nerves are named using the number of the vertebrae below them at the cervical level, except for the pair of spinal nerves which emerge between C7 and T1, which is called the 8th cervical nerve pair (C8). The spinal nerves then take the name of the vertebra above them (the roots of T1 emerge between vertebrae T1 and T2, etc.). Each spinal nerve consists of the union (in the intervertebral foramen) of a dorsal nerve root and of a ventral nerve root. The white and grey communicating branches, containing myelinated and unmyelinated fibres, respectively, connect the ventral branch of the spinal nerve to the homologous ganglion of the laterovertebral sympathetic nerve chain at all thoracic and lumbar levels. The post-ganglionic fibres of the grey communicating branches then provide smooth muscle and exocrine glandular innervation of the trunk and of the limbs via the spinal nerves. The pre- and post-ganglionic fibres which supply internal organs do not join the spinal nerves but are organised in a separate autonomic nervous system (the sympathetic system) which supply the intrathoracic, abdominal and pelvic organs.

E. Eisenberg (✉)
Département d'anesthésie et réanimation, Clinique Pôle Santé République, Clermont-Ferrand, France

E. Eisenberg, E. Gaertner (eds.), *Ultrasound in Peripheral, Neuraxial and Perineuraxial Regional Anaesthesia*,
https://doi.org/10.1007/978-3-031-08804-9_4

Each spinal nerve divides into a ventral branch and a dorsal branch, both comprised of motor, sensory and autonomic fibres. The ventral branches, which are relatively larger, supply the ventral-lateral part of the trunk and limbs, while the thinner dorsal branches are distributed to the skin and muscles of the back.

Each spinal nerve provides the following:

- Sensation of a specific cutaneous segment called the dermatome.
- A segmental motor muscular innervation called the myotome.
- A bony innervation called the osteotome. It is important to note that the topography of the territories of bone innervation does not coincide with the innervation of more superficial structures.

Dermatomes (Figs. 4.1, 4.2, and 4.3)

Myotomes and Osteotomes (Figs. 4.4 and 4.5)

Plexuses

These are complex nerve networks formed by the union of the ventral branches of the spinal nerves. The terminal and collateral branches are relatively constant elements, the anastomoses, on the other hand, are extremely variable.

The four principal plexuses are: the cervical plexus, the brachial plexus, the lumbar plexus and the sacral plexus.

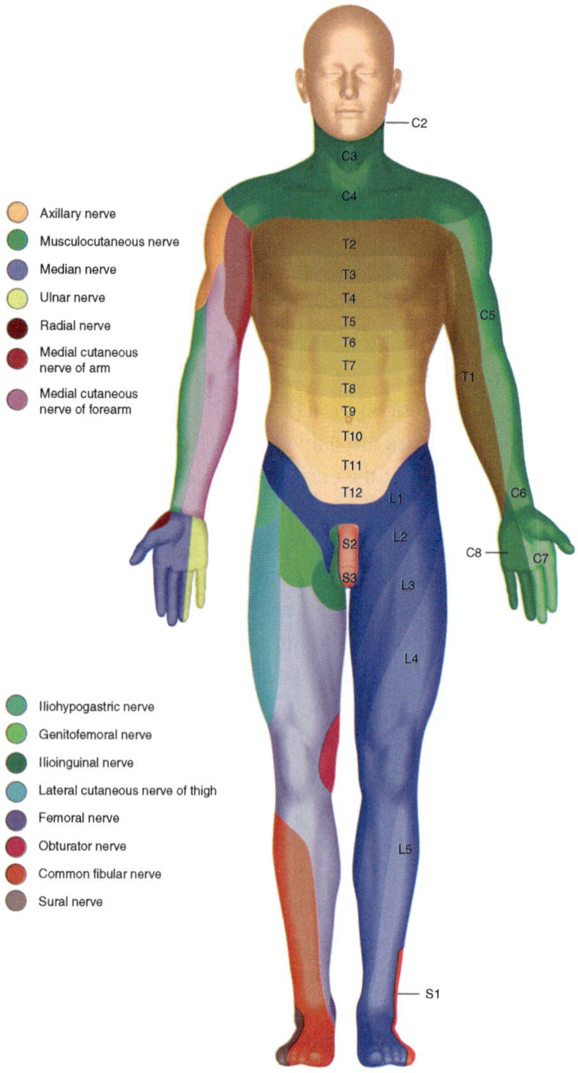

Fig. 4.1 Dermatomes and truncal sensory territories: anterior view

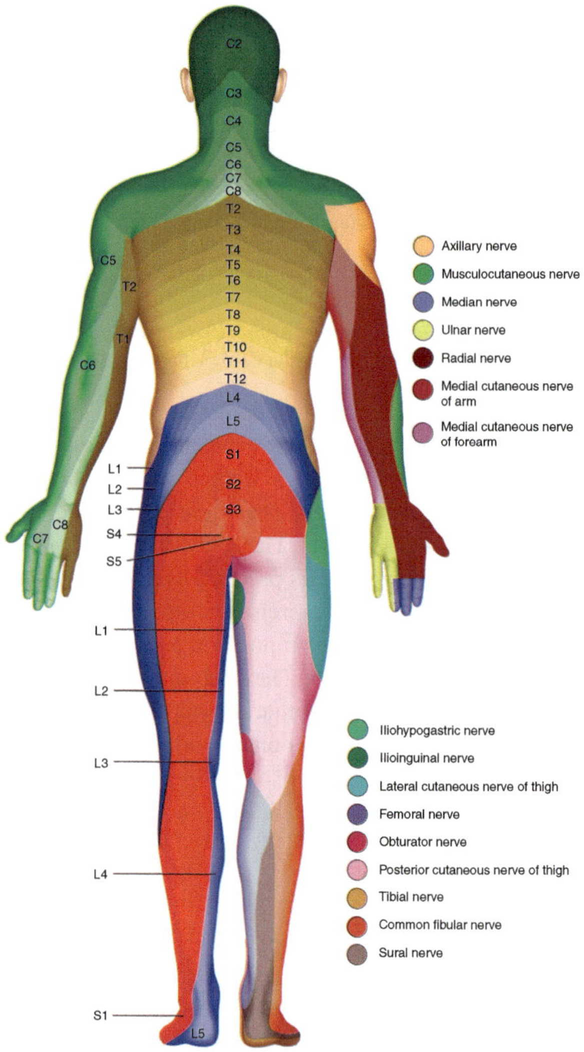

Fig. 4.2 Dermatomes and truncal sensory territories: posterior view

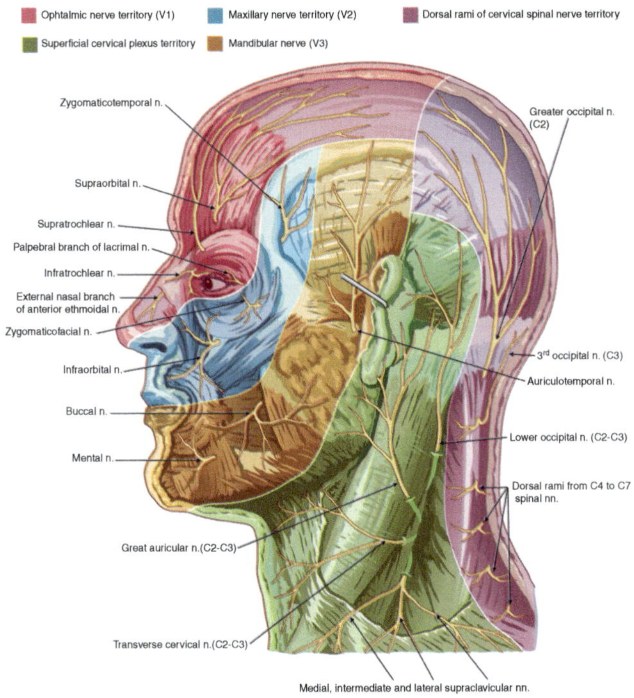

Fig. 4.3 Cutaneous sensory territories of the face and neck

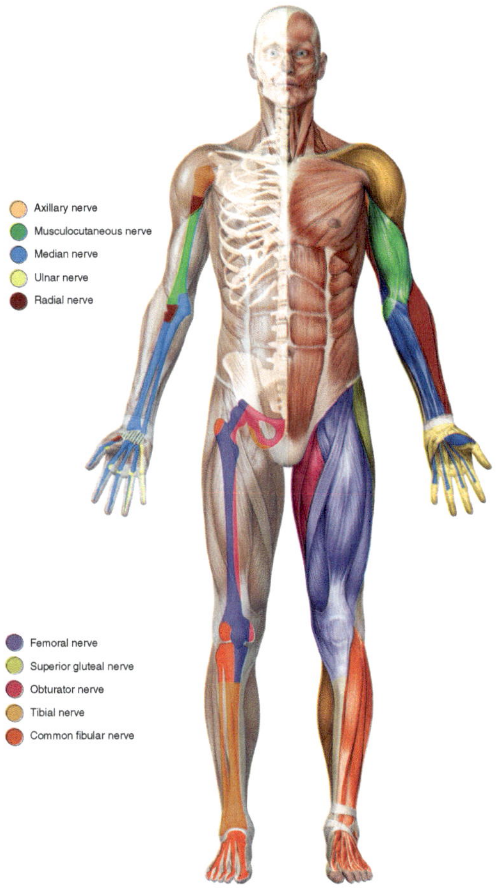

Fig. 4.4 Myotomes and sclerotomes, anterior view

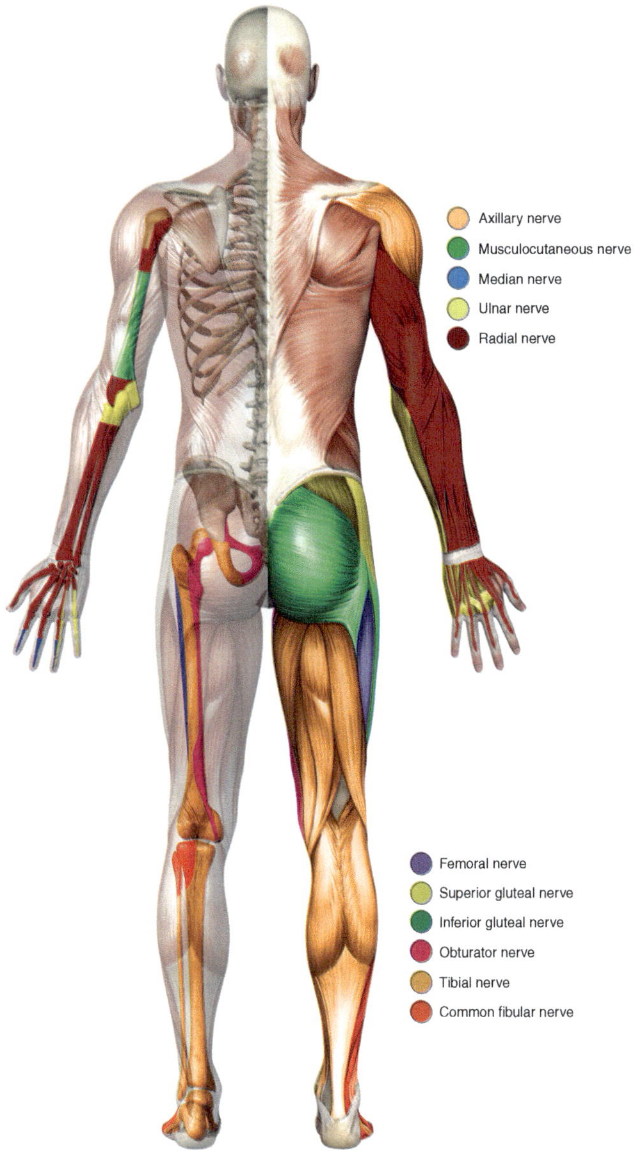

Fig. 4.5 Myotomes and sclerotomes, posterior view

Cervical Plexus (Fig. 4.6)

The cervical plexus consists of the union of the ventral branches of the cervical nerve roots C1 to C5.

It provides:

- Sensory innervation of part of the scalp, of the neck (to the sub-clavicular thoracic area), of the ear and of the cranial area of the shoulder.
- Motor innervation of certain neck muscles:
  sternothyroid, sternohyoid, omohyoid, geniohyoid, thyrohyoid
  levator scapulae, scalenes, sternocleidomastoid, trapezius, platysma muscles
- Innervation of the diaphragm.

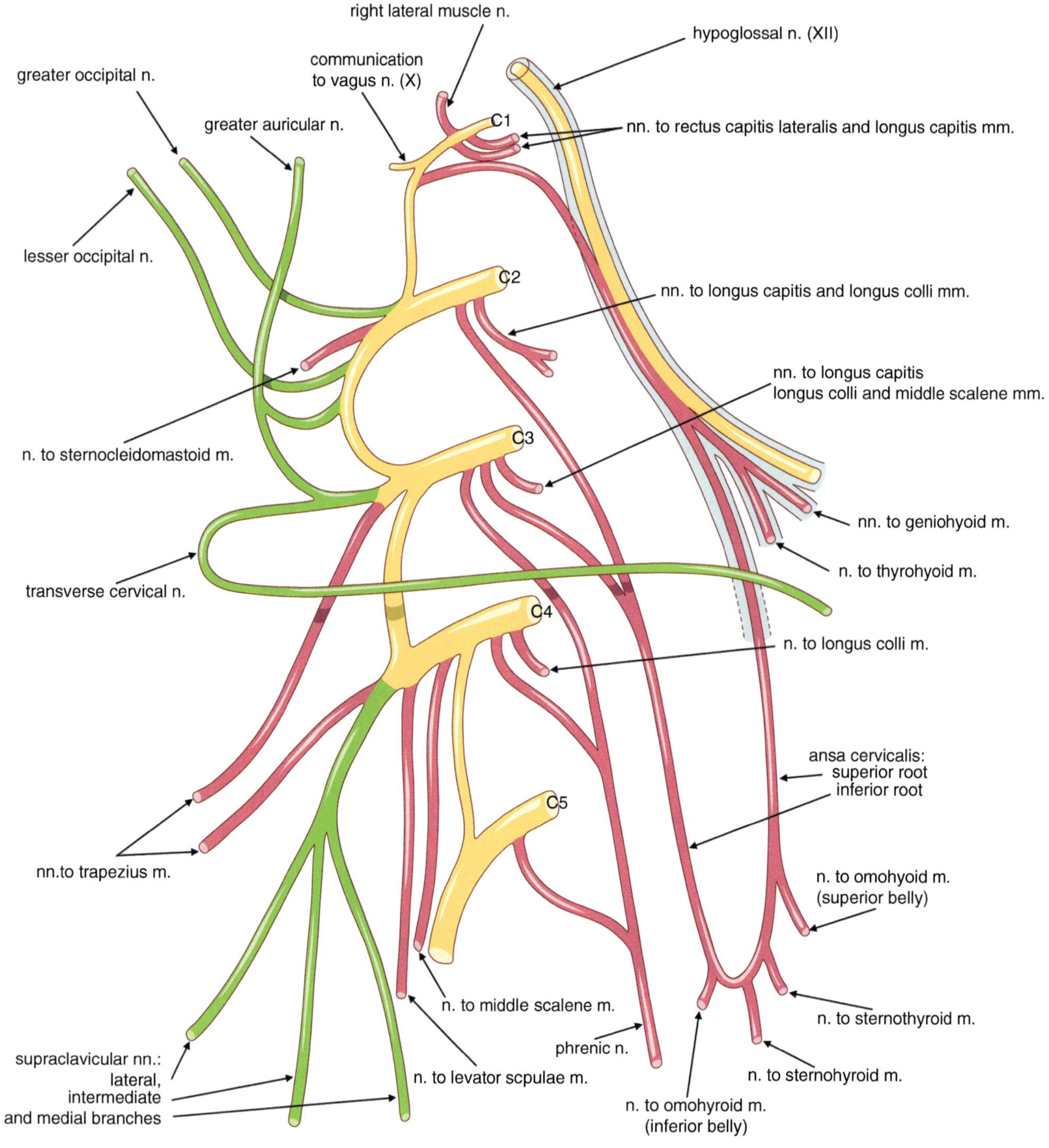

Fig. 4.6 The cervical plexus

Brachial Plexus (Fig. 4.7)

The brachial plexus provides sensory and motor innervation to the shoulder girdle and the entire upper limb (*see below* diagram of motor innervation of the upper limb, Fig. 4.10). It arises from the ventral branches of the spinal nerve C5 to T1. These form the upper trunk (C5-C6), middle trunk (C7) and the lower trunk (C8-T1) which, by the organisation of their anterior and posterior divisions, will comprise the posterior cord, lateral cord and medial cord.

The terminal branches of the brachial plexus end in the lateral border of the pectoralis minor muscle:

- The posterior cord gives rise to:
  the axillary nerve
  the radial nerve

Fig. 4.7 The brachial plexus

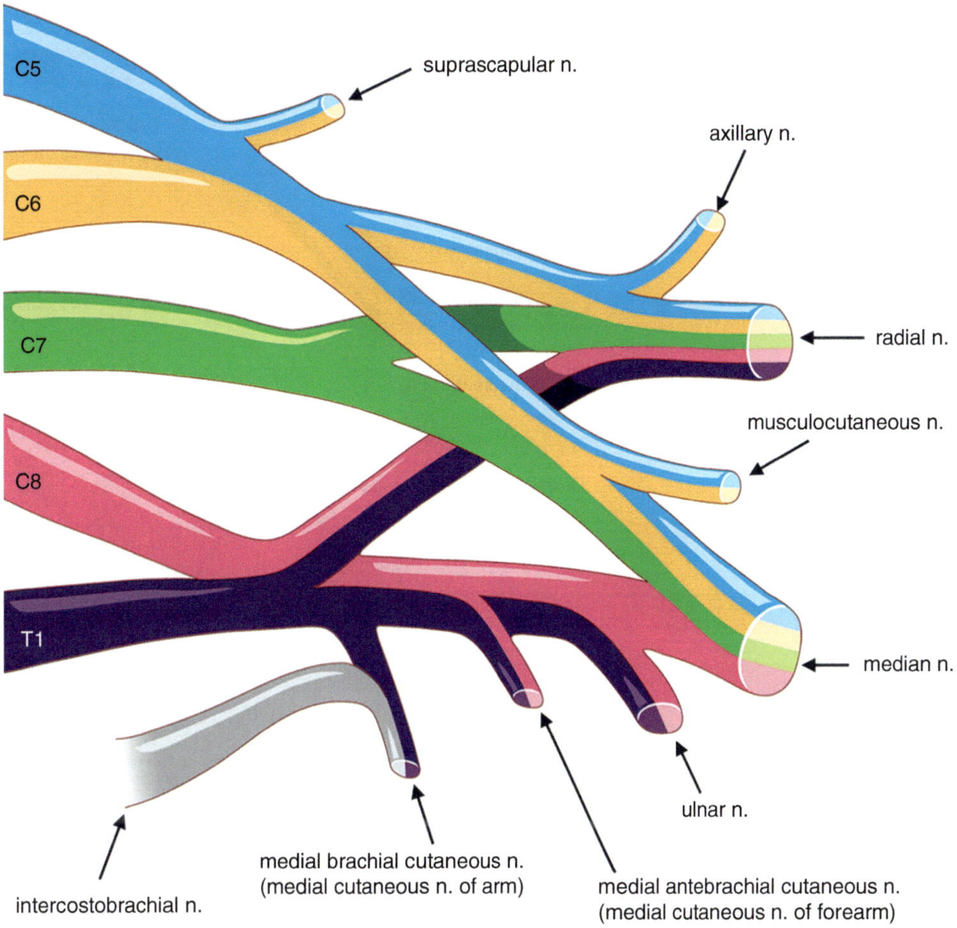

- The lateral cord gives rise to:
  the musculocutaneous nerve
  the lateral root of the median nerve
- The medial cord gives rise to:
  the medial root of the median nerve
  the ulnar nerve
  the medial cutaneous nerve of the forearm
  part of the medial cutaneous nerve of the arm

Collateral branches arise at each level of the plexus, some of which are noteworthy:

- in the ventral branches of the nerve roots:
  dorsal nerve of scapula
  long thoracic nerve
- in the trunks and cords:
  suprascapular nerve
  lateral and medial pectoral nerves (pectoral loop)
  thoracodorsal nerve
  subscapular nerves

Lumbar Plexus (Fig. 4.8)

The lumbar plexus contributes to innervation of the abdominal wall, the external genital organs, and the lower limb (*see below* motor innervation diagram of lower limb, Fig. 4.11). At its origin, it is formed by the union of the ventral branches of the spinal nerves L1, L2, L3 and L4 (a branch of the ventral branch of T12 frequently unites with that of L1). It is placed laterally in relation to the lumbar spine between the two heads of the psoas major muscle, in a cleavable aponeurotic space. The anastomoses and interconnections of these different branches then comprise the terminal branches of the lumbar plexus which are the ilio-hypogastric (T12-L1), ilio-inguinal (L1), genitofemoral (L1-L2), lateral cutaneous nerve of the thigh (L2-L3), femoral (L2-L3-L4) and obturator nerves (L2-L3-L4). Generally, this is as follows [2]:

- The ventral branch of L1 divides into two branches, the common trunk of the ilio-hypogastric and ilio-inguinal nerves, and the superior branch of the genitofemoral nerve.
- The ventral branch of L2 gives rise to the inferior branch of the genitofemoral nerve and three other branches which

Fig. 4.8 The lumbar plexus

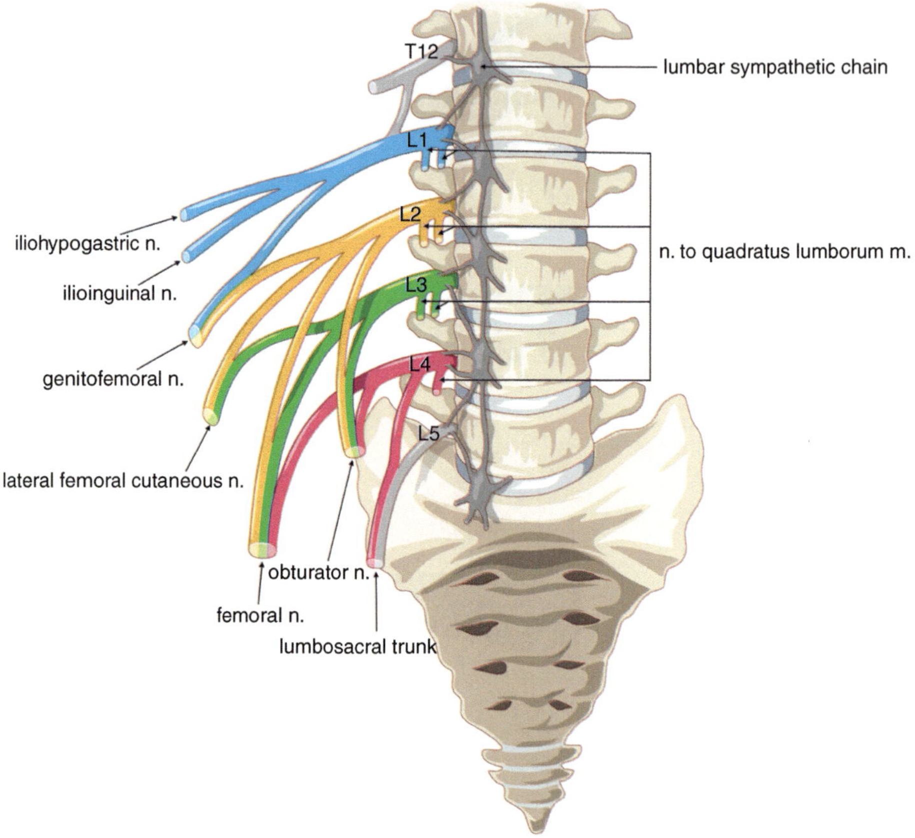

participate respectively in the lateral cutaneous nerve of the thigh, the femoral and obturator nerves.

- The ventral branch of L3 gives rise to three components for the lateral cutaneous nerve of the thigh, the femoral and obturator nerves, respectively.
- The ventral branch of L4 branches into the femoral nerve, the obturator nerve and the lumbo-sacral trunk (L4-L5) which goes to the sacral plexus.

Sacral Plexus (Fig. 4.9)

The sacral plexus contributes to the innervation of the lower limb and of the perineum. It consists of the union of the ventral branches of L5 and the first three sacral spinal nerves (S1 to S3). Two components of fibres that come from the ventral branches of L4 and S4 respectively are added to it. The L4 component joins the ventral branch of L5 to form the lumbosacral trunk which emerges from the medial border of the psoas major muscle. The trunk then crosses the pelvic brim and continues towards the sacro-iliac joint, in the lumbosacral fossa; at this level it joins with the ventral branch of S1.

The ventral branches of the sacral nerve roots emerge from the sacrum via the ventral sacral foramina. The sacral plexus is shaped like a triangle whose base, which is medial, is in relation to the ventral sacral foramina and whose apex, which is lateral, is in relation to the ventrocaudal part of the ischiatic notch [3]. It is directly applied to the ventral aspect of the piriform muscle, in the dorsal part of the pelvic cavity which is covered by the parietal pelvic fascia. It has close anatomical relations with the pelvic organs (ureter, end of the ileum on right side and the sigmoid colon the left side) and the internal iliac and gluteal vessels. The sacral plexus forms connections with the lumbar plexus, the pudendal nerve and the pelvic sympathetic chain.

The sacral plexus, whose sole terminal branch is represented by the sciatic nerve, gives rise, in particular, to the following collateral branches:

Fig. 4.9 The sacral plexus

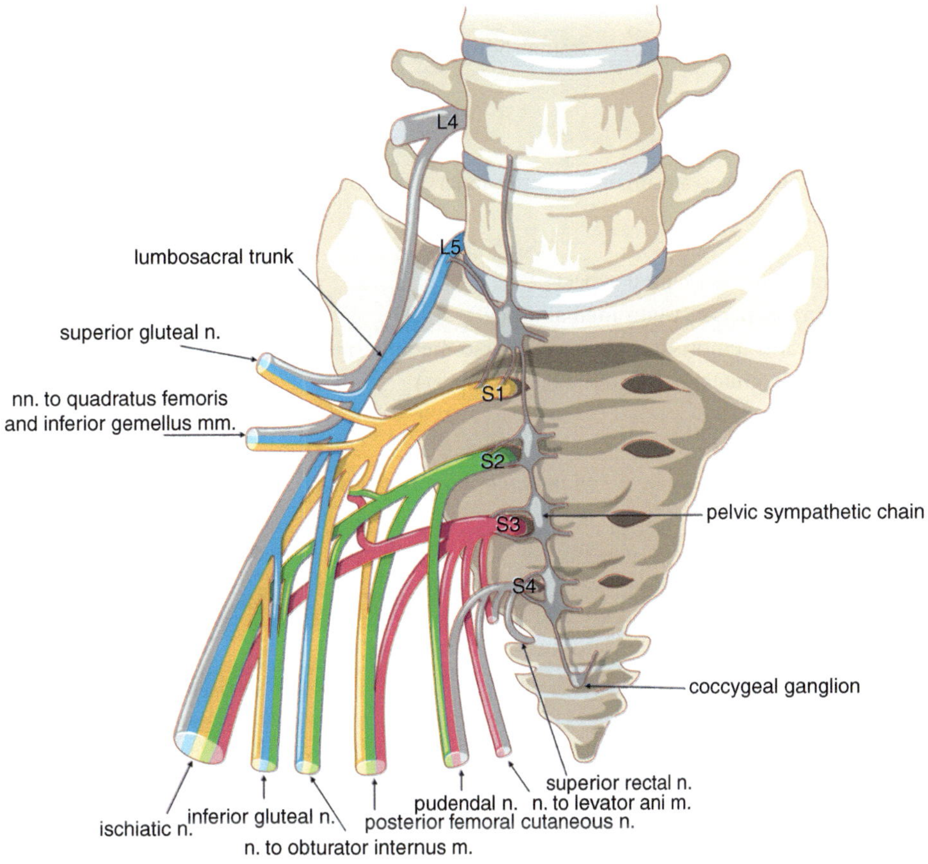

- by its ventral branches:
  nerve to obturator internus (L5-S1-S2)
  nerve to quadratus femoris (L4-L5-S1)
- by its dorsal branches:
  nerve to piriformis (S1-S2)
  superior gluteal nerve (L4-L5-S1)
  inferior gluteal nerve (L5-S1-S2)
  posterior cutaneous nerve of the thigh (S1-S2-S3)

Innervation of the Principal Joints

Shoulder

C5 and C6 are the main nerves involved in innervation of the shoulder joint. This is done through the following nerves:

- subscapular (ventral aspect)
- axillary (ventral and dorsal aspect)
- lateral pectoral (ventral aspect)
- suprascapular (dorsal aspect)

Elbow

The following nerves are involved in innervation of the elbow joint:

- the median nerve (ventral aspect)
- the musculocutaneous nerve (ventral aspect)
- the ulnar nerve (medial and dorsal aspect)
- the radial nerve (dorsal and lateral aspect)

Wrist and Hand

The joints of the wrist and of the hand are innervated primarily by:

- the median nerve
- the ulnar nerve

- the radial nerve
- the musculocutaneous nerve (radial border of the radio-carpal joint)

Hip

The following nerves are involved in the innervation of the hip joint:

- the femoral nerve (ventral aspect)
- the obturator nerve (ventromedial aspect)
- the quadratus femoris muscle nerve (dorso-medial aspect)
- the superior gluteal nerve (dorsolateral aspect)

Knee

The following nerves are involved in innervation of the knee joint:

- the femoral nerve (ventral and medial aspect)
- the sciatic nerve (dorsal and lateral aspect)
- the deep branch of the obturator nerve (dorsal aspect)

Ankle

The following nerves are involved in innervation of the ankle joint:

- the tibial nerve
- the superficial and deep peroneal nerves
- the saphenous nerve (for the skin overlying the medial malleolus)

Motor Innervation of the Upper Limb (from Denis Jochum) (Table 4.1 and Fig. 4.10)

Table 4.1 Myotomes of the upper limb

Muscle name	Root origin	Peripheral innervation
Muscles of the shoulder girdle inserting on the humerus		
Posterior group		
Supra-spinatus muscle	C4-C6	Suprascapular nerve
Infra-spinatus muscle	C4-C6	Suprascapular nerve
Teres minor muscle	C5-C6	Axillary nerve
Deltoid muscle	C4-C6	Axillary nerve
Subscapular muscle	C5-C7	Subscapular nerve
Teres major muscle	C6-C7	Teres major nerve
Latissimus dorsi muscle	C6-C8	Thoraco-dorsal nerve
Anterior group		
Coraco-brachialis muscle	C6-C7	Musculocutaneous nerve
Pectoralis minor muscle	C7-T1	Inferior pectoralis nerve
Pectoralis major muscle	C5-T1	Pectoralis nerve
Muscles from the trunk inserting on the shoulder girdle		
Posterior group		
Rhomboid major muscle	C4-C5	Dorsal nerve of scapula
Rhomboid minor muscle	C4-C5	Dorsal nerve of scapula
Levator scapulae muscle	C4-C5	Dorsal nerve of scapula
Serratus anterior muscle	C5-C7	Long thoracic nerve
Anterior group		
Subclavian muscle	C5-C6	Subclavian nerve
Omohyoid muscle	C1-C3	Cervical loop
Muscle of the head inserting on the shoulder girdle		
Trapezius muscle	X1 and C2-C4	Spinalis nerve and trapezius
Sternocleidomastoid muscle	X1 and C2-C3	Spinalis nerve and cervical branches
Muscles of the arm		
Anterior group		
Brachialis muscle	C5-C6	Musculocutaneous and radial nerve
Biceps brachialis muscle	C5-C6	Musculocutaneous nerve
Posterior group		
Triceps brachialis muscle (long, lateral and medial heads)	C6-C8	Radial nerve
Anconeus muscle	C7-C8	Radial nerve
Muscles of the forearm		
Anterior muscles (superficial plane)		
Pronator teres muscle	C6-C7	Median nerve
Flexor digitorum superficialis	C7-T1	Median nerve
Flexor carpi radialis muscle	C6-C7	Median nerve
Palmaris longus muscle	C7-C8	Median nerve
Flexor carpi ulnaris muscle	C7-C8	Ulnar nerve
Anterior muscles (deep plane)		
Pronator quadratus muscle	C8-T1	Median nerve
Flexor digitorum profundus	C7-T1	Median and ulnar nerve
Flexor pollicis longus muscle	C7-C8	Median nerve
Radial muscles		
Extensor carpi radialis brevis muscle	C6-C7	Radial nerve
Extensor carpi radialis longus muscle	C6-C7	Radial nerve

Table 4.1 (continued)

Muscle name	Root origin	Peripheral innervation
Brachioradialis muscle	C5-C6	Radial nerve
Posterior muscles (superficial plane)		
Extensor digitorum muscle	C6-C8	Radial nerve
Extensor digiti minimi muscle	C6-C8	Radial nerve
Extensor carpi ulnaris muscle	C7-C8	Radial nerve
Posterior muscles (deep plane)		
Supinator muscle	C5-C6	Radial nerve
Abductor pollicis longus muscle	C7-C8	Radial nerve
Extensor pollicis brevis muscle	C7-C8	Radial nerve
Extensor pollicis longus muscle	C7-C8	Radial nerve
Extensor indicis muscle	C6-C8	Radial nerve
Short muscles of the hand		
Muscles of the metacarpal		
Palmar interosseous muscle	CS-T1	Ulnar nerve
Dorsal interosseous muscle	CS-T1	Ulnar nerve
Lumbrical muscles	C8-T1	Median and ulnar nerve
Muscles of the thenar eminence		
Abductor brevis pollicis muscle	C6-C7	Median nerve
Flexor pollicis brevis muscle	C6-C7 and C8-T1	Median and ulnar nerve
Adductor pollicis muscle	C8-T1	Ulnar nerve
Opponens pollicis muscle	C6-C7	Median nerve
Muscles of the hypothenar eminence		
Abductor digiti minimi muscle	C8-T1	Ulnar nerve
Flexor digiti minimi brevis muscle	C8-T1	Ulnar nerve
Opponens digiti minimi muscle	C8-T1	Ulnar nerve

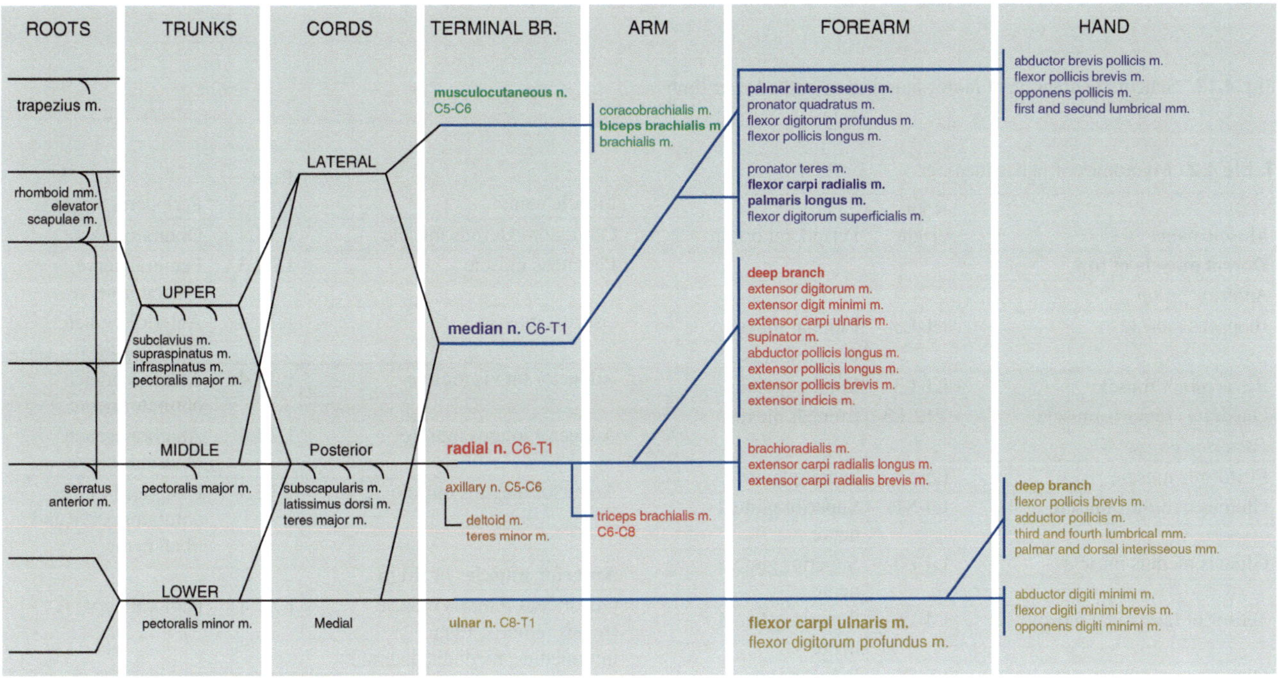

Fig. 4.10 Schematic diagram of motor innervation of the upper limb

Motor Innervation of Lower Limb (from Denis Jochum) (Fig. 4.11 and Table 4.2)

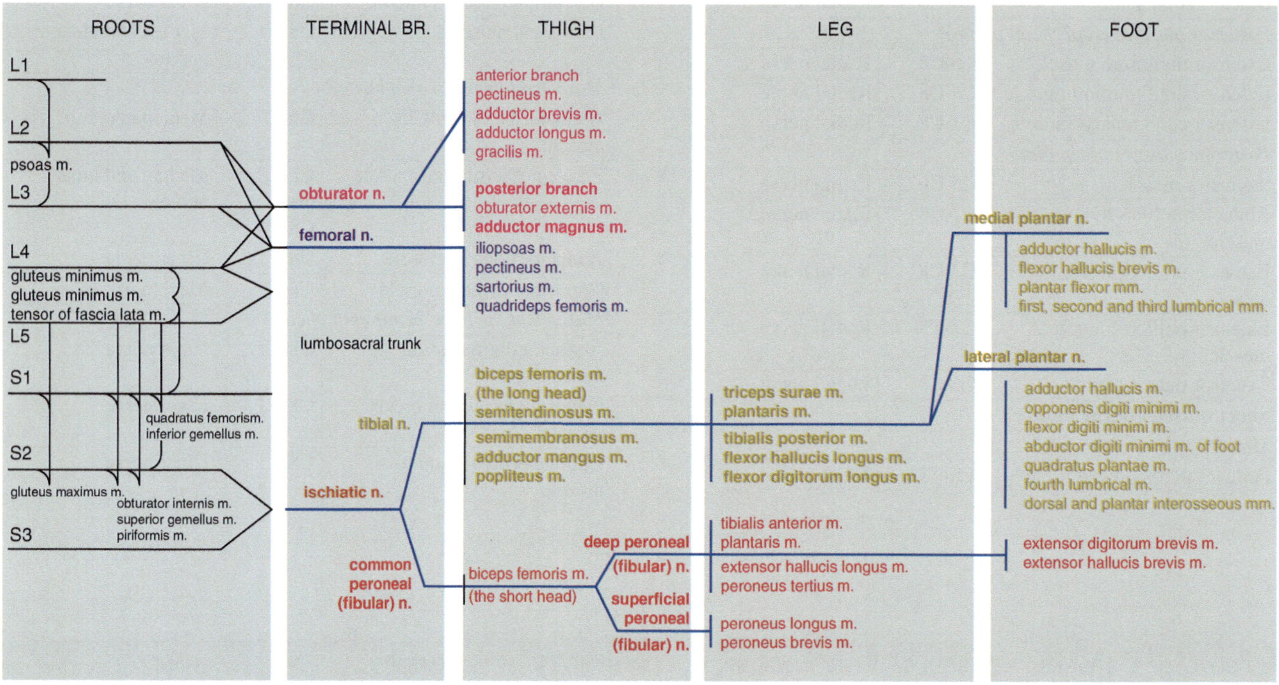

Fig. 4.11 Schematic diagram of motor innervation of the lower limb

Table 4.2 Myotomes of pelvic member

Muscle name	Root origin	Peripheral origin
Dorsal muscle of hip		
Anterior group		
Iliopsoas muscle	L1-L3	Lumbar plexus
		Femoral nerve
Psoas minor muscle	L1-L3	Lumbar plexus
Quadratus femoris muscle	T12-L3	Lumbar plexus
Posterior group		
Piriformis muscle	L5-S2	Sacral plexus
Gluteus minimus muscle	L4-S1	Superior gluteal nerve
Gluteus medius muscle	L4-L5	Superior gluteal nerve
Tensor of fascia lata muscle	L4-L5	Superior gluteal nerve
Gluteus maximus muscle	L5-S2	Inferior gluteal nerve
Ventral muscles of hip and adductors of the thigh		
Obturator internus muscle	L5-S2	Sacral plexus
		Inferior gluteal nerve
Gemellus muscle	L5-S2	Sacral plexus
		Inferior gluteal nerve
Quadratus femoris muscle	L5-S2	Sacral plexus
		Inferior gluteal nerve
Obturator externus muscle	L3-L4	Obturator nerve
Pectineus muscle	L2-L3	Femoral nerve, obturator nerve
Gracilis muscle	L2-L4	Anterior branch obturator nerve
Adductor brevis muscle	L2-L4	Anterior branch obturator nerve
Adductor longus muscle	L2-L4	Anterior branch obturator nerve
Adductor magnus muscle	L3-L5	Posterior branch obturator nerve and tibial nerve
Anterior muscles of thigh		
Quadriceps femoris muscle (rectus femoris, vastus intermedius, medialis, lateralis)	L2-L4	Femoral nerve
Sartorius muscle	L2-L3	Femoral nerve
Posterior muscles of thigh		
Biceps femoris muscle	L5-S2	Tibial and fibular nerve
Semitendinosus muscle	L5-S2	Tibial nerve
Semimembranosus muscle	L5-S2	Tibial nerve
Popliteus muscle	L4-S1	Tibial nerve
Anterior muscles of leg		
Group of extensors		
Tibialis anterior muscle	L4-L5	Deep fibular nerve

Muscle name	Root origin	Peripheral origin
Extensor digitorum longus of foot	L4-S1	Deep fibular nerve
Extensor digitorum brevis muscle	L5-S1	Deep fibular nerve
Extensor hallucis longus	L4-S1	Deep fibular nerve
Lateral fibular muscle group		
Peroneus longus muscle	L5-S1	Superficial fibular nerve
Peroneus brevis muscle	L5-S1	Superficial fibular nerve
Posterior muscles of leg		
Superficial plane		
Triceps surae muscle (soleus, gastrocnemius)	L5-S2	Tibial nerve
Plantaris muscle	L5-S2	Tibial nerve
Deep plane		
Tibialis posterior muscle	L4-L5	Tibial nerve
Flexor hallucis longus	L5-S2	Tibial nerve
Flexor digitorum longus	L5-S2	Tibial nerve
Short muscles of the foot		
Dorsal muscles		
Extensor digitorum brevis	L5-S1	Deep fibular nerve
Extensor hallucis brevis	L5-S1	Deep fibular nerve

Muscle name	Root origin	Peripheral origin
Plantar muscles		
Abductor hallucis	L5-S1	Medial plantar nerve
Flexor hallucis brevis	L5-S1	Medial plantar nerve
Adductor hallucis	S1-S2	Lateral plantar nerve
Opponens digiti minimi	S1-S2	Lateral plantar nerve
Flexor digiti minimi	S1-S2	Lateral plantar nerve
Abductor digiti minimi of foot	S1-S2	Lateral plantar nerve
Lumbrical muscles of the foot	L5-S2	Plantar nerve
Quadratus plantae muscle	S1-S2	Lateral plantar nerve
Dorsal and plantar interosseous muscles	S1-S2	Lateral plantar nerve

References

1. Denny NM, Harrop-Griffiths W. Editorial. Location, location, location! Ultrasound imaging in regional. Br J Anaesth. 2005;94:1–3.
2. Kamina P. Précis d'anatomie clinique. Tome 1. Paris: Maloine; 2003.
3. Rouvière H, Delmas A. Anatomie humaine, descriptive, topographique et fonctionnelle. Tome III. Paris: Masson; 1984.

Brachial Plexus Blocks

5

Eryk Eisenberg, Elisabeth Gaertner, and Philippe Clavert

Contents

Anatomy

The brachial plexus (BP) is entirely responsible for innervation of the upper limb (Figs. 5.1, 5.2, 5.3, and 5.4). It has the shape of an hourglass whose middle part, narrowed, is crossed by the clavicle which determines a supraclavicular, cervical part and an infraclavicular, axillary part [1] (Figs. 5.5, 5.6, 5.7, 5.8, and 5.9). It is formed at its origin by the ventral branches of the cervical nerve roots of C5 to C8 and of the thoracic nerve root T1 (an anastomosis of C4 and/or T2 can occur). When passing between the anterior and middle scalene muscles, these branches develop interconnections and form three trunks:

- *The upper trunk*, formed by the union of ventral branches C5 and C6.
- *The middle trunk*, comprised of the ventral branch C7.
- *The inferior trunk,* formed by union of the ventral branches C8 and T1. Behind and above the clavicle, each trunk divides into two branches (one anterior and the other posterior) which give rise to the following cords:
- *The posterior cord,* resulting from the union of the three dorsal branches.
- *The lateral cord*, resulting from the union of the ventral branches of the upper and middle trunks.
- *The medial cord* comprised of the ventral branch of the inferior trunk.

Several variations of this configuration of interconnections exist, in particular, as the result of the variable distributions between the ventral branch of the middle trunk (C7)

The original version of the chapter has been revised. A correction to this chapter can be found at https://doi.org/10.1007/978-3-031-08804-9_13

Supplementary Information The online version contains supplementary material available at https://doi.org/10.1007/978-3-031-08804-9_5.

E. Eisenberg (✉)
Département d'anesthésie et réanimation, Clinique Pôle Santé République, Clermont-Ferrand, France

E. Gaertner
Département d'anesthésie et réanimation chirurgicale, Hôpitaux Civils de Colmar, Colmar, France

P. Clavert
Service de chirurgie du membre supérieur, Centre de Chirurgie Orthopédique et de la Main (CCOM), Illkirch, France

E. Eisenberg, E. Gaertner (eds.), *Ultrasound in Peripheral, Neuraxial and Perineuraxial Regional Anaesthesia*,
https://doi.org/10.1007/978-3-031-08804-9_5

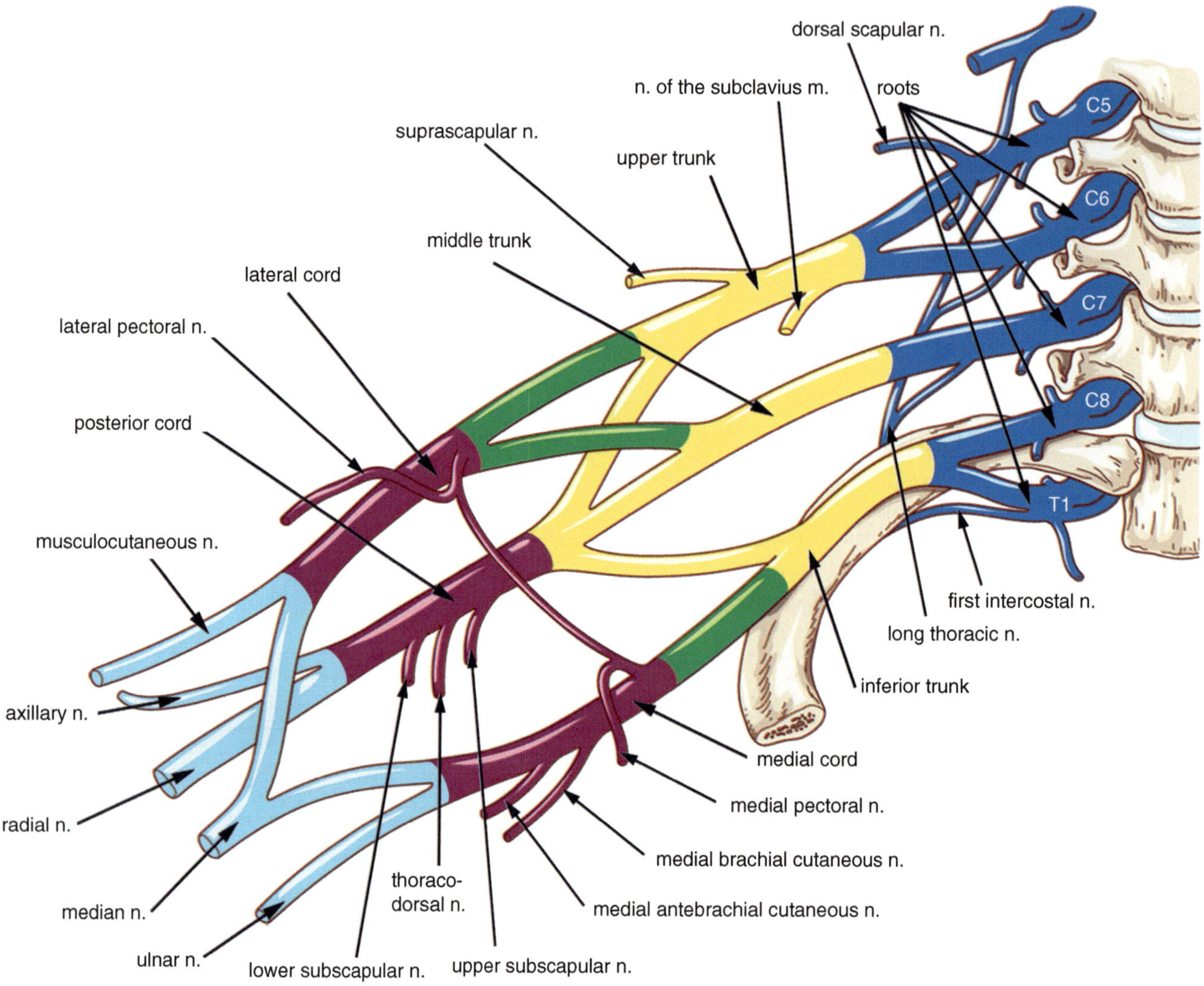

Fig. 5.1 Schematic diagram of brachial plexus

and the lateral or medial cord [2]. These variations have minimal clinical implications relating to the ulnar nerves (C8-T1) and axillary nerve (C5-C6), although they may affect the nerve root origin of some nerve fibres contained in the terminal branches of the brachial plexus [3]. For example, the musculocutaneous nerve (usually C5-C6) may contain fibres from the nerve root of C7. The brachial plexus gives rise to terminal branches in the lateral border of the pectoralis minor muscle:

- The posterior cord gives rise to the axillary (C5-C6) and radial (C5-T1) nerves.
- The lateral cord gives rise to the musculocutaneous nerves (C5-C6-C7) and the lateral root of the medial nerve (C5-C6-C7).
- The medial cord gives rise to the medial root of the median nerve (C6-C7), as well as the ulnar nerve (C8-T1), the medial cutaneous nerve in the arm and the medial cutaneous nerve in the forearm.

Among the Collateral Branches

The Dorsal Scapular Nerve

The dorsal scapular nerve is a collateral branch of the brachial plexus, usually arising from the C5 root, just before the upper trunk formation; a small quota of fibre of C3 and C4 roots may associate with it. It is a motor nerve emerging at the supraclavicular part of the plexus. Once the dorsal scapular nerve pierces the middle scalene muscle, it continues downwards and backwards, deep to levator scapulae then rhomboid muscles. It provides motor innervation to the levator scapulae muscle and the minor and major rhomboid muscles. The nerve is accompanied by either the dorsal scapular

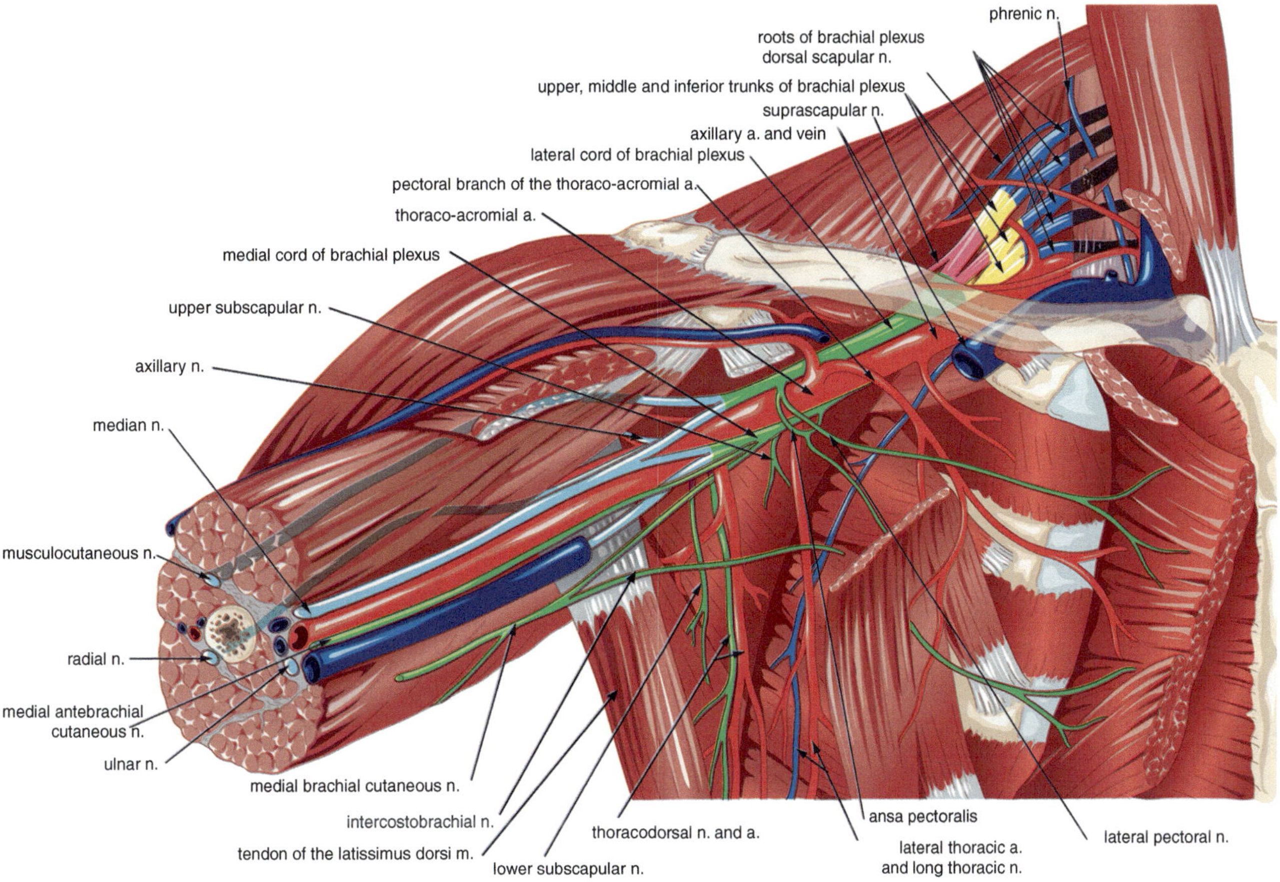

Fig. 5.2 Brachial plexus. (From Netter FH. *Atlas of human anatomy*. 3rd ed. Paris: Masson, 2007)

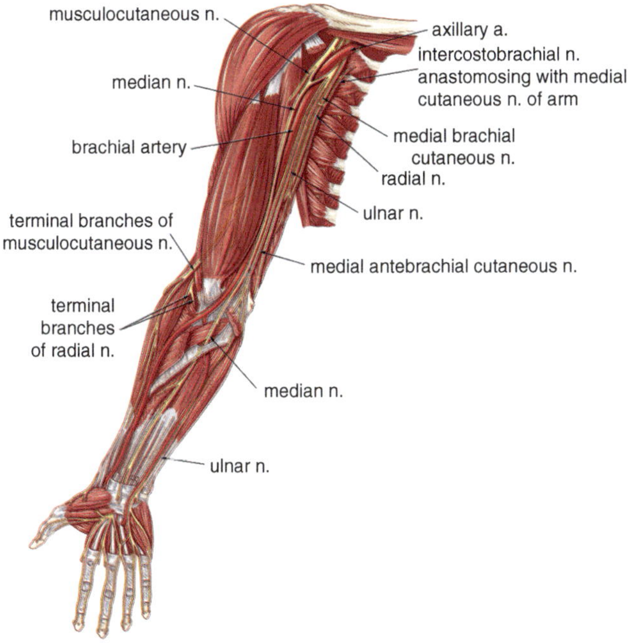

Fig. 5.3 Ventral aspect of upper limb. Terminal branches of brachial plexus

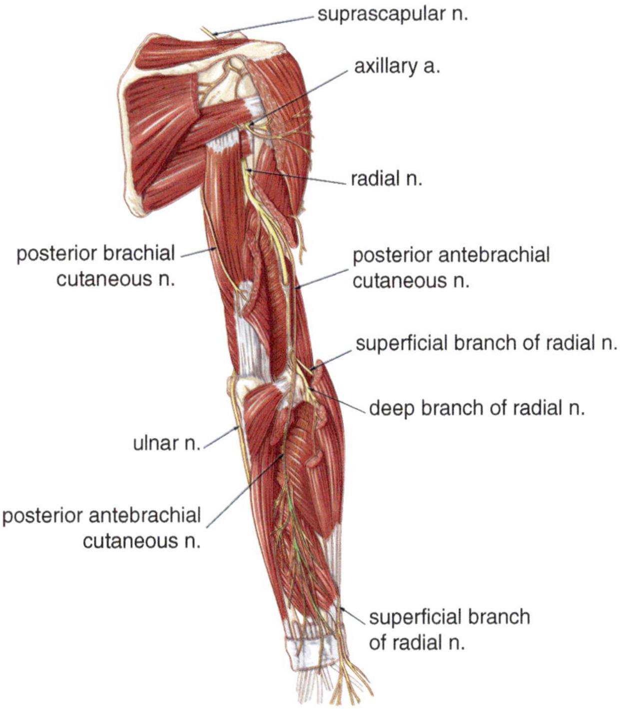

Fig. 5.4 Dorsal aspect of upper limb. Terminal branches of brachial plexus

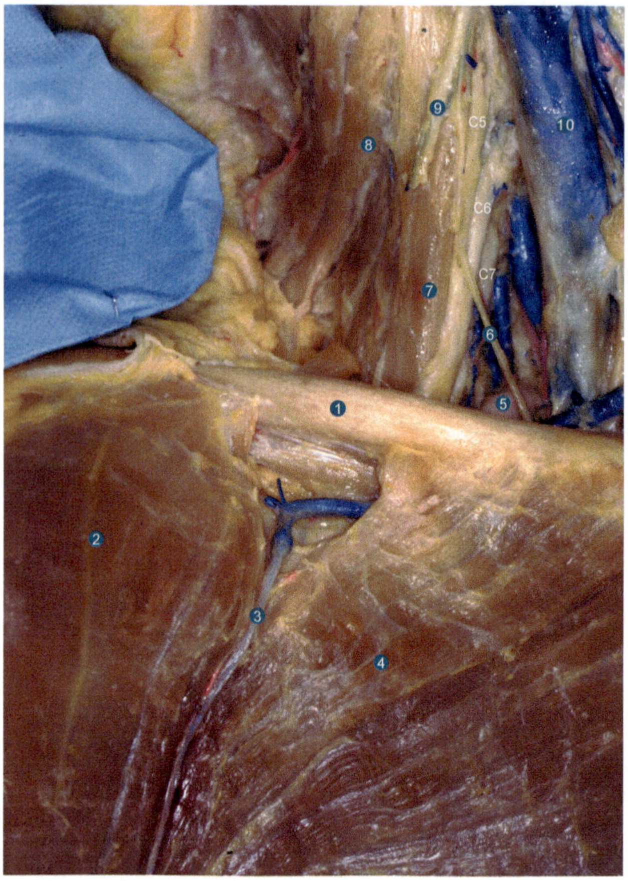

1. clavicle 2. deltoid m. 3. cephalic vein 4. pectoralis major m.
6. phrenic n. (anterior scalene m. removed) 7. middle scalene m.
8. levator scapular m. 9. suprascapular n.

Fig. 5.5 Cervical, supra- and infraclavicular areas: superficial view. (Dissection: Bertrand Fabre)

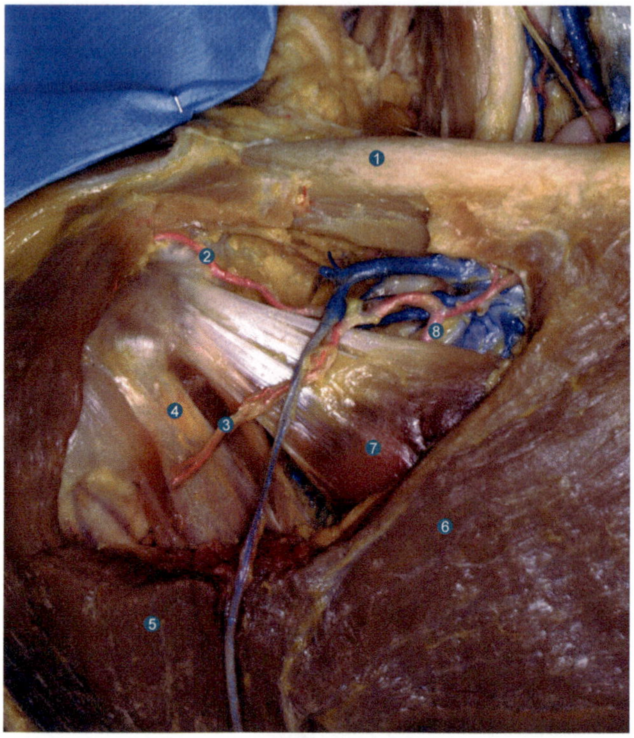

1. clavicle 2. acromial branch of the thoracoacromial a.
3. deltoid branch of the thoracoacromial a.
4. coracobrachialis m. 5. deltoid m. 6. pectoralis major m.
7. pectoralis minor m. 8. thoracoacromial a.

Fig. 5.6 Cervical, supra- and infraclavicular areas: resection of a portion of the pectoralis major muscle. (Dissection: Bertrand Fabre)

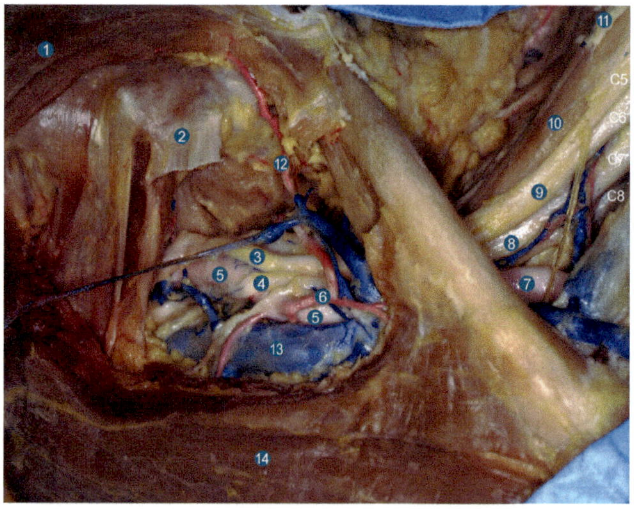

1. Deltoid m. 2. tendon of the pectoralis minor m.
3. lateral cord 4. medial cord
5. axillary a. 6. thoracoacromial a.
7. subclavian a. 8. middle trunk
9. upper trunk 10. middle scalene m.
11. suprascapular n. (removed)
12. acromial branch of the thoracoacromial a.
13. axillary v. 14. pectoralis major m.

Fig. 5.7 Cervical, supra- and infraclavicular areas: brachial plexus, principal vascular structures, relations with the clavicle. (Dissection: Bertrand Fabre)

artery (branching off the third part of the subclavian artery), or the deep branch of the cervical transverse artery (branching off the thyrocervical trunk) when the dorsal scapular artery is absent.

The Long Thoracic Nerve

The long thoracic nerve, also called the "Charles Bell nerve", is a nerve that is essentially a motor nerve, responsible exclusively for innervation of the serratus anterior muscle. It arises from the anterior branches of roots of C5 and C6, of C7 in over two-thirds of cases, and of C4 and C8 very rarely. It is described as having two portions:

- **Cervical pathway:** Branches from C5 to C6 most often pass through the middle scalene muscle, and then joins with the branch that arises from C7 which passes between the middle scalene muscle and the anterior scalene muscle, behind the vascular bundle. The nerve then leaves the remainder of the axillary neurovascular bundle opposite

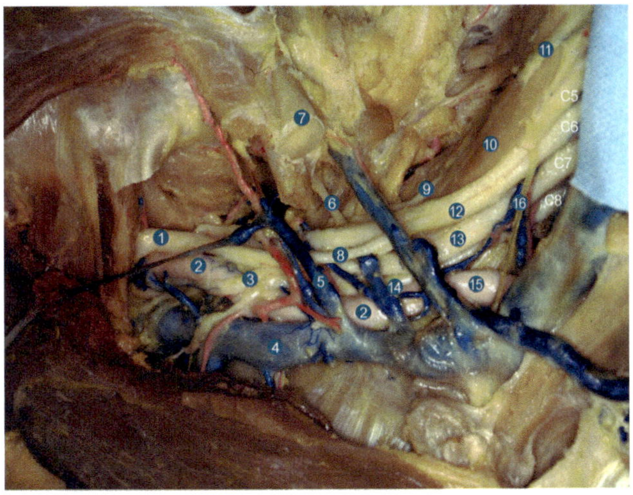

1. lateral cord
2. axillary a.
3. medial cord
4. axillary v.
5. cephalic v.
6. n. to subclavius m.
7. clavicle (removed)
8. anterior division of upper trunk
9. dorsal scapular n.
10. middle scalene m.
11. suprascapular n. (removed)
12. upper trunk
13. middle trunk
14. inferior trunk
15. subclavian a.

Fig. 5.8 Cervical, supra- and infraclavicular areas: brachial plexus, principal vascular structures, resection of the clavicle. (Dissection: Bertrand Fabre)

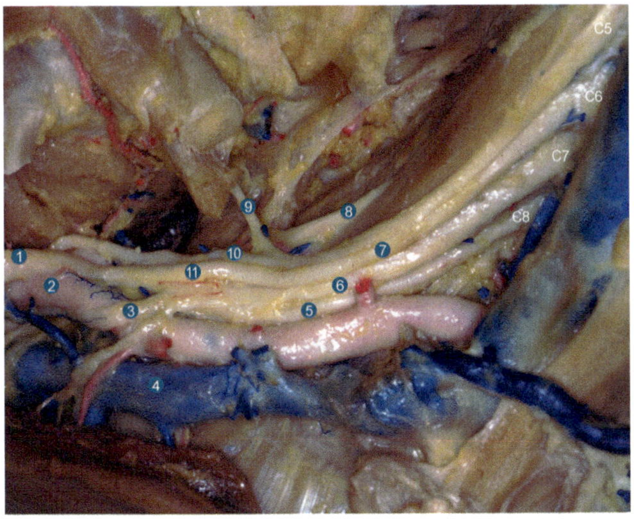

1. lateral cord
2. axillary a.
3. medial cord
4. axillary v.
5. inferior trunk
6. middle trunk
7. upper trunk
8. dorsal scapular n.
9. n. to subclavius m.
10. posterior division of upper trunk
11. anterior division of upper trunk

Fig. 5.9 Brachial plexus in the interscalene and supraclavicular area. (Dissection: Bertrand Fabre)

the second rib. Between its exit from the scalene muscle and the posterior angle of the second rib, it then descends obliquely from front to back creating an angle of about 30° with the horizontal plane in the sheath of the brachial plexus. By reflecting on the second rib, the nerve then entirely changes direction and becomes vertical.

- **Thoracic trajectory:** The thoracic part of the trajectory of the long thoracic nerve is relatively constant. It emerges from the axillary fossa under the pectoral muscles, opposite the fourth or fifth digitation. In the projection, the long thoracic nerve is always behind the midaxillary line, that is, posterior to the cutaneous branches which emerge from the intercostal bundle. In the distal part of its trajectory, it is joined by the branch of the anterior division of the thoracodorsal artery. This artery passes in front of the nerve at the point that it penetrates under the fascia of the serratus anterior muscle. The nerve divides into three terminal branches in 96% of cases for the middle and lower segments of the serratus anterior muscle. This division lies between 3 and 5 cm in front of the apex of the scapula.

The length of the nerve, after the union of the proximal branches, varies from 24 to 30 cm. The diameter of the nerve varies from 2.2 to 1.4 mm between its proximal portion and its ending.

Pectoral Nerves Loop (Ansa Pectoralis), Medial and Lateral Pectoral Nerves

The loop of the pectoral nerves is an anterior collateral branch of the brachial plexus resulting from branches from the lateral and medial cords: the lateral pectoral arises from the lateral cord, and the medial pectoral nerve arises from the medial cord. The ansa pectoralis finds its origin in the roots C5, C6, C7, C8 and T1. It is located close to the thoracoacromial artery, after the departure of the branch which goes to the latissimus dorsi muscle.

The pectoralis major muscle is innervated by the lateral pectoral nerve and the medial pectoral nerve resulting from the loop of the pectoral nerves (branches resulting from the C5-C6-C7). Each muscle head (clavicular, sternal and abdominal) has its own neurovascular pedicle.

The pectoralis minor muscle is innervated by a single branch of the pectoral loop (branches resulting from the C7-C8-T1 nerves).

Thoracodorsal Nerve

The thoracodorsal nerve arises from the posterior bundle of the brachial plexus (from C6-C7-C8 roots). The posterior bundle gives rise to the thoracodorsal nerve, the subscapular nerves, the axillary nerve and lastly the radial nerve. The thoracodorsal nerve emerges just downstream from the emergence of the axillary nerve.

This nerve passes inferiorly along the posterior wall of the axillary fossa and penetrates the medial aspect of the latissimus dorsi, where it becomes tendinous. It also innervates the teres major muscle. The end part of the nerve lies in front of the scapular lymphatic chain and the subscapular artery.

Terminal Branches of the Brachial Plexus

Axillary Nerve

The axillary nerve is a mixed nerve. Its roots of origin are C5 and C6; its fibres then follow the upper trunk and then the posterior cord. It passes behind the axillary artery and joins the interior border of the subscapular muscle. It passes under the capsule of the scapulohumeral joint and through the quadrangular space. It connects dorsally to the surgical head of the humerus. It gives rise to branches for the subscapular muscle, the ventral and dorsal aspects of the capsule, the teres minor muscle, the long head of the triceps brachialis muscle, one branch for each of the deltoid muscle, and it innervates the skin in the upper/outer arm. It can also be damaged along its trajectory around the humerus in case of a fracture of the proximal end of the humerus or of gleno-humeral dislocation. Damage to it results in complete paralysis of the abduction of the arm (while leaving intact triggering of its movement related to contraction of the supraspinous muscle).

Radial Nerve

This is a mixed nerve and the nerve of extension (elbow, wrist, fingers). It is formed of fibres which emerge from roots C5 to T1. As the terminal branch of the posterior cord in the brachial plexus, it leaves the axillary fossa, passing under the teres major, and gives rise to its first collaterals. It continues caudally, dorsally and laterally, penetrates into the posterior compartment of the arm via the triangular space to reach the spiral groove of the humerus. The nerve goes around the humerus to lie at the lateral aspect of the arm. In its brachial trajectory, it is in contact with the bone along 8–10 cm in length. It crosses the lateral intermuscular septum at 10–12 cm above the elbow joint interline to pass into the anterior compartment of the arm, where it leads into the lateral bicipital groove. Up until this level, it gives rise to the following collateral branches:

- Posterior cutaneous nerve of the arm.
- Nerve of the long head of the triceps muscle.
- Upper nerve of medial head of triceps muscle.
- Lower nerve of medial head of triceps muscle (or nerve of medial head and of the anconeus).
- Nerve of lateral head of triceps muscle, inferior lateral cutaneous nerve of the arm.
- Nerve of the brachioradialis muscle.
- Nerve of the extensor carpi radialis longus muscle.
- Posterior cutaneous nerve of the forearm; at the emergence from the radial groove, this sensory branch detaches and, after emerging in the subcutaneous plane in the lateral border of the tendon of the triceps muscle, continues between the lateral epicondyle and the olecranon process, before it branches to provide the sensation of the posterior aspect of the forearm.

At the lateral epicondyle, the radial nerve divides into its two terminal branches:

- The superficial branch, solely sensory, descends vertically against the deep aspect of the brachioradialis muscle, accompanied by the radial artery, to arrive, in the lower third of the forearm, in the posterior antebrachial area by superficially crossing the tendon of the brachioradialis muscle. It provides sensation to the lateral aspect of the back of the hand located outside of a line passing through the axis of the third finger, except for the middle and distal phalanges of the second and third fingers.
- The deep branch, mainly motor, passing between the two heads of the supinator muscle, and then emerging in the dorsal compartment of the forearm between the two muscle planes where it divides into its terminal branches. Other than the posterior antebrachial interosseous nerve, which descends up to the dorsal aspect of the carpus that it innervates, it gives rise to motor branches of the following muscles:
  extensor carpi radialis brevis
  supinator
  extensor digitorum
  extensor digiti minimi
  extensor carpi ulnaris
  abductor pollicis longus
  extensor of pollicis longus
  extensor indicis
  extensor pollicis brevis

Characteristic signs of damage to the radial nerve, depending on level of the lesion, are paralysis of extension of the forearm, the wrist, the fingers, with anaesthesia reaching all or part of the posterior aspect of the arm, the forearm and the area of the back of the hand located outside of a line passing through the axis of the third finger (apart from the middle and distal phalanges of this same digit and of the index finger).

Musculocutaneous Nerve

The musculocutaneous nerve is a mixed nerve. Comprised of fibres from C5 and C6, it arises from the lateral cord of the brachial plexus in the axillary region. It passes through the coracobrachialis muscle and continues caudally and laterally. It travels between the biceps brachialis and the brachialis muscle. It emerges at the inferior part of the arm in the lateral bicipital groove, between the brachioradialis muscle and the biceps brachii muscle. It passes through the brachial facia, becomes subcutaneous, thus comprising the lateral cutaneous nerve of forearm, and gives rise to its two terminal branches:

- Ventral, which supplies skin sensation to the lateral part of the ventral aspect of the forearm and the lateral half of the elbow fold
- Dorsal, which is responsible for dorsal and lateral forearm skin sensation

In its brachial trajectory, the musculocutaneous nerve gives rise to the:

- Diaphyseal nerve of the humerus
- Generally two branches of the coracobrachialis muscle
- The vascular bundle for the brachial artery
- The nerve of the biceps muscle (one branch for each head)
- The nerve of the brachialis muscle
- An articular branch for the elbow joint

Signs of nerve lesion (or anaesthesia) in the musculocutaneous nerve are: pronation of the forearm, absent biceps contraction, an impaired ability to flex the elbow. Flexion is partially compensated by the brachioradialis and brachialis muscles, but no opposite flexion of the elbow is possible.

Median Nerve

The median nerve is a mixed nerve. It consists of fibres from C5 to C8. It is the nerve which enables flexion of the fingers and the wrist, and pronation of the forearm. It is formed by the lateral and medial cords, and arises within the axillary fossa. It descends into the brachial canal in the medial aspect of the arm, in contact with the brachial artery, describing an *S* around it (it passes from the lateral border to the ventral aspect and then to the medial border of the artery). It passes anterior to the elbow, an area where it is very exposed to injury, under the aponeurotic arch of the biceps muscle. At the elbow, it continues in the medial bicipital groove between the biceps muscle on the outside and the humeral head of the pronator teres muscle medially. At this level, it gives rise to the articular nerves for the ventral and medial aspect of the elbow and the nerve of the branching of the brachial artery. It then slides under the fibrous arch of the flexor digitorum superficialis muscle and descends in the ventral compartment of the forearm between the superficial flexor tendons which go to the second and third fingers. A few centimetres above the wrist, it separates laterally from the tendons of this muscle and becomes superficial; it then passes along the inner aspect of the tendon of the flexor carpi radialis muscle. It then enters into the carpal tunnel in front of the tendon of the flexor digitorum superficialis muscle and, at its emergence in the tunnel, it divides into a motor branch for the thenar eminence (abductor pollicis brevis, opponens pollicis and flexor pollicis brevis muscles) and the two lateral lumbrical muscles, and a sensory branch which gives rise to the common palmar digital nerves 1, 2 and 3. The motor and sensory nerves can anastomose with the ulnar nerve.

The median nerve gives rise to the following collateral branches:

- Nerve to the brachial artery
- Articular branch to the elbow
- The nerve of the medial epicondylar muscles (flexor carpi radialis, palmaris longus, flexor digitorum superficialis, ulnar head of the pronator teres)
- The anterior antebrachial interosseous nerve (flexor pollicis longus, the two lateral heads of the flexor digitorum profundus, the pronator teres muscle and the wrist joints)
- The palmar branch of the median nerve

From a sensory perspective, the median nerve supplies the lateral part of the palm of the hand (except for the lateral part of the thenar eminence where the radial nerve participates) and the palmar aspect of fingers 1–3, as well as the lateral half of the fourth finger. On the dorsal aspect, it provides sensation for the proximal and middle phalanges of fingers 2 and 3 and the lateral half of phalanges 2 and 3 of the 4th finger.

The median nerve forms connections in the arm with the musculocutaneous nerve, and in the forearm with the ulnar nerve and radial nerve (influencing innervation of the thenar eminence/thumb). The existence of these anastomoses and articular branches explains the need to block the median nerve more often than its cutaneous distribution would suggest.

It should be noted that it is the nerve which presents the largest number of anatomical variations, difficult to present and summarise in this description.

Ulnar Nerve

The ulnar nerve is a mixed nerve. It arises from C8 to T1, emerges from the axillary fossa and ends in the hand. The terminal branch of the medial cord of the brachial plexus descends first between the brachial artery and vein. In the axillary area, it is deep, located at the medial aspect of the artery. In the middle part of the arm, it penetrates the medial intermuscular septum, thus going from the anterior brachial compartment to the posterior brachial compartment, in contact posteriorly with the medial head of the triceps muscle. In the dorsal medial aspect of the elbow, it passes superficially into the medial epicondyle-olecranon groove. Like the median nerve, it does not give rise to any collaterals in the arm, but provides an articular branch for the elbow. It enters the forearm by passing between the two heads of the flexor carpi ulnaris muscle, to which it gives a muscular branch. In the upper two-thirds of the forearm, it passes on the medial aspect and then the ventral aspect of the flexor digitorum profundus, covered by the flexor carpi ulnaris in the lateral aspect of which it positions itself. The ulnar nerve is joined at the mid-portion of the forearm by the ulnar artery, which lies laterally. It then descends into the ventral compartment of the forearm up to the wrist, where after penetrating the antebrachial fascia, it continues into the ulnar tunnel accompanied by the ulnar artery.

Its collateral branches are:

- Articular branches for the posterior aspect of the elbow
- Muscular branches for the flexor carpi ulnaris muscle and the medial half of the flexor digitorum profundus muscle
- The dorsal branch of the ulnar nerve which arises in the lower one-third of the forearm and after reaching the posterior aspect of the wrist, gives rise to the medial and lateral dorsal digital nerves of the fifth and fourth digits, and the medial dorsal digital nerve of the third digit
- The palmar branch which innervates the skin of the hypothenar eminence

Its terminal branches separate at the end of the ulnar canal and are represented by:

- A mixed superficial branch, innervating the palmar muscles and which provides complete palmar sensitivity to the fifth digit and medial sensitivity to the fourth digit
- A deep branch, in contact with the skeleton, which divides into several branches to innervate the eight interosseous muscles, the third and fourth lumbricals and a few muscles of the thenar eminence (adductor pollicis and deep head of flexor pollicis brevis) where there is a possible anastomosis with the medial nerve

Its motor function is not essential in the forearm, where it innervates few muscles. It is essential in the hand, where it innervates the largest part of the muscles, apart from the first and second lumbricals, opponens pollicis, the superficial head of flexor pollicis brevis and the abductor pollicis brevis muscles, which are innervated by the median nerve.

In terms of sensory function, the ulnar nerve, by its hypothenar branch, innervates the palmar aspect of the fifth finger and the palmar medial course of the fourth digit. By its dorsal branch, it innervates the dorsal aspect of the fifth finger, the dorsal aspect of the first phalanx and of the medial half of the second and third phalanges of the fourth digit, as well as the dorsal aspect of the medial half of the first phalanx of the third digit.

Medial Cutaneous Nerve of the Arm

This is a small sensory nerve and which innervates the axillary fossa and overlaps a little on the medial aspect of the arm and the chest wall. The second collateral branch of the medial cord of the brachial plexus arises from T1. It forms an anastomosis usually with the intercosto-brachial nerve, which does not belong to the brachial plexus and which arises from the lateral perforating branches of the first two intercostal nerves (T1-T2), sometimes also from the third one (T3).

Medial Cutaneous Nerve of the Forearm

The third collateral branch of the medial cord of the brachial plexus is a sensory nerve which arises from C8-T1. In the medial part of the arm, it continues superficially beside the basilic vein. Starting from the upper third of the arm it gives rise to the sensory perforating branches, crosses the aponeurosis in the lower one-third and becomes superficial at the elbow. It provides sensory innervation of the medial half of the lower two-thirds of the arm and the medial half of the forearm.

Although the medial cutaneous nerve of the forearm generally is anaesthetised during conduct of a brachial plexus nerve block in the axillary fossa or in the humeral tunnel, the subcutaneous portion of the medial cutaneous nerves in the arm and the medial cutaneous nerve of the forearm make them accessible to subcutaneous injection.

Brachial Plexus Nerve Blocks

Anaesthesia of the nerves of the brachial plexus can be planned in the plexus area strictly speaking (interscalene, supraclavicular and infraclavicular nerve block) (Fig. 5.9), where it can be considered that a single injection may effectively block the entire upper limb. This is not exactly the case, since both in the interscalene and supraclavicular level, there is a significant incidence of block failure in the C8-T1 distribution [4]. This is less common in the infraclavicular approach which generally enables anaesthesia of all areas of the distal half of the arm up to the end of the fingers [5]. Approaching the plexus more distally, because of the divergence of the terminal branches of the plexus, in order to anaesthetise an equivalent area, it is necessary to administer separate, targeted nerve blocks. This is the basis of multistimulation techniques with the infraclavicular and axillary approaches [6, 7].

In deciding which type of nerve block is to be performed (which depends on the indication and the desired extent of anaesthesia), it is desirable to consider those for which discomfort for the patient and iatrogenic harm are the lowest. Therefore, insofar as possible, the practitioner will strive to limit the number of injections necessary and will choose the approaches which present the fewest risks.

For all brachial plexus nerve blocks, in practice it is possible to perform either a single block or a continuous block with placement of a perineural catheter. The expected result should take into account the anatomical specific aspects of each site. If a catheter for analgesia is inserted during an axillary block, depending on the precise place where it is located, its efficacy may be manifested only in the area of the radial and/or ulnar nerve, or indeed the median nerve, or even the musculocutaneous nerve. And if in the initial phase of the block, an injection of a local anaesthetic is administered to all of the nerves, initial overall efficacy will be followed subsequently by more selective anaesthesia or analgesia, depending on the precise site of distribution of the low flow of local anaesthetic through the catheter. This process should be thoroughly understood in order to guide positioning of the tip of the catheter so that the latter is placed in proximity to the nerve(s) for which prolonged anaesthesia is desired after regression of the initial block. During injection through a catheter, spread of the local anaesthetic occurs via the openings of the catheters; some catheters have a single perforation (end), others multiple perforations (along the last centimetres). Therefore, spread of the anaesthetic can take place over a certain length of the catheter starting from its end. This is all the more so when retrograde spread of the local anaesthetic occurs (called backflow) along the course of the catheter. This should be used and exploited advantageously in the strategy for placement of the catheter. For example, also in the axillary area, to achieve the most effective block of the brachial plexus while using a multiperforated catheter, it may be possible to insert it in contact with the different target nerves in order to guide the backflow of the local anaesthetic.

Interscalene Block (Fig. 5.10)

Indications

An interscalene block (ISB) is recommended for anaesthesia and post-operative analgesia in shoulder surgery [8]. In fact, the analgesic potency that it provides is the most appropriate to this type of surgery, with high levels of satisfaction and excellent tolerability by patients. Alternatives to this block in the presence of a contraindication are, by order of efficacy on post-operative pain, a suprascapular block (with or without an axillary nerve block), a subacromial injection, and lastly by patient-controlled analgesia (PCA) with intravenous morphine [9, 10]. The volumes of local anaesthetics used have been greatly decreased as a result of ultrasound-guided technique [11, 12].

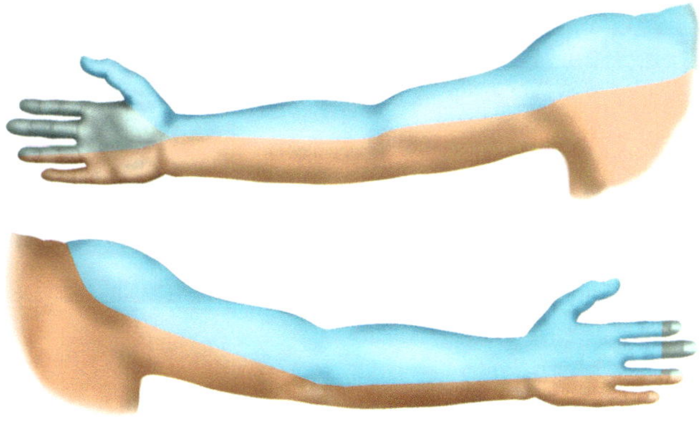

Fig. 5.10 Area of coverage of an interscalene block

Although the time necessary for conduct of an ISB is slightly longer than that of general anaesthesia, the time for installation and emergence from the OR are shortened, as well as duration of stay in the post-operative recovery room and hospital stay in ambulatory arthroscopic surgery. This block is not indicated for distal surgery of the upper limb, even by deep injection of the local anaesthetic, in the lower roots of the brachial plexus. Ilfeld's team demonstrated that only 15% of patients obtained a surgical block of the hand and the forearm 30 min after conduct of an ISB [13].

Conduct of an interscalene block is not a contraindication to ambulatory surgery [14, 15]. This block is used essentially in arthroscopies, for which low doses of local anaesthetics are sufficient to improve post-operative analgesia (10 mL of 0.5% ropivacaine according to Krone et al. [15]).

It is possible to perform an interscalene block with a single injection or to insert a perineural catheter, ideally with use of a PCA pump (patient-controlled analgesia) connected to the catheter, which improves the quality and increases the duration of post-operative analgesia [16–19]. This is the case for procedures where predictable "VAS" (visual analogue scale) scores are greater than 3 and where duration of post-operative pain exceeds 24 h, i.e. repair of the rotator cuff, arthrolysis, a shoulder block, shoulder arthroplasty (prosthesis), fracture of the humeral head and tumours. Although it was frequently said that shoulder arthroscopy was not painful, advances in surgical techniques have led to the indication with use of a catheter in the majority of therapeutic arthroscopies. In fact, currently many repair procedures for shoulder surgery are performed with arthroscopy: rotator cuff rupture, Bankart repair, bone blocks, acromioplasties, arthrolysis and tendonplasty. A catheter can also be inserted for certain physical therapy procedures or for cancer-related pain [20].

Continuous interscalene block (interscalene catheter) can also be part of an analgesic strategy enabling ambulatory surgery of the shoulder, including complicated surgery [21–24].

Type of probe: linear, 5–10 MHz or 6–13 MHz.
Axis of probe: transverse (Fig. 5.11a, b).
Configuration: nerves in the small axis, needle in the plane.
Studied depth: 2–4 cm.
Neurostimulation: enables further identification of the nerves visualised and can limit, by determination of an MIS > 0.3 mA (0.1 ms), the risk of accidental intrafascicular injection. Furthermore, it can supplement locating nerves when conditions of visibility are unfavourable.
Needle: 50–80 mm isolated, 22G.
Utility of Doppler ultrasound: cardiac, internal jugular, vertebral artery and vein that can be encoun-

tered at the tip of the needle, sometimes blood vessels within the interscalene area (transverse cervical artery which may be encountered inconsistently, ascending cervical artery).

Approach, Ultrasound Anatomy

Two principal approaches can be described for an interscalene block with a single injection or for insertion of a perineural catheter: posterolateral (in plane) and in the axis of the interscalene groove (out of plane).

Posterolateral Approach (In Plane) (Fig. 5.12)

This is the preferred approach when performing interscalene brachial plexus block. Although not routinely adopted when using the traditional landmark/neurostimulation technique, when using ultrasound it offers the optimal view of the underlying anatomy, the needle approach to components of the plexus and for visualising the spread of local anaesthetic.

Nerve Localisation

The head is turned to the opposite side to relax the interscalene portion of the sternocleidomastoid muscle. Finding the optimum injection site is achieved by moving the probe along the interscalene groove to indentify, in a single view in the ultrasound plane, the nerve roots C5 and C6 (or their branches of ventral and dorsal division) and C7 root (middle trunk) which lies at the centre of the brachial plexus (Figs. 5.13, 5.14, and 5.15). Accurate identification of the nerve roots is based on presence and shape of the cervical transverse processes. Transverse processes of C3-C6 have an anterior and posterior tubercle (Figs. 5.16, 5.17, 5.18, and 5.19). However, the C7 transverse process has only a single posterior tubercle (Fig. 5.20) and thus is readily differentiated from the others. This enables identification of the cervical nerve roots [25]. When it is possible to differentiate C8 more in depth, the upper and middle trunks divided into the ventral and dorsal branches (as well as structures corresponding to the collateral branches of the plexus) are seen on

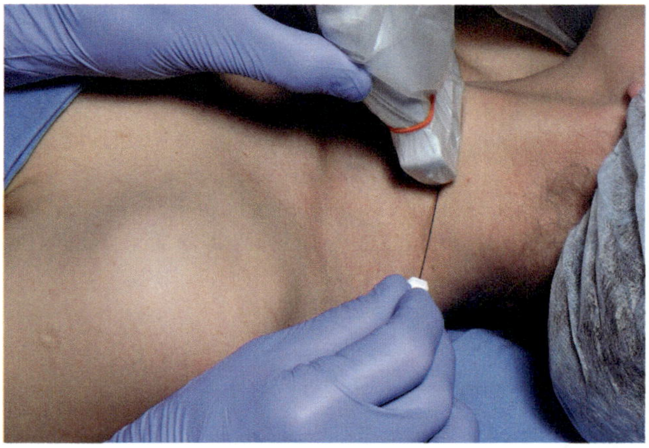

Fig. 5.12 Ultrasound-guided interscalene block. Needle insertion in-plane

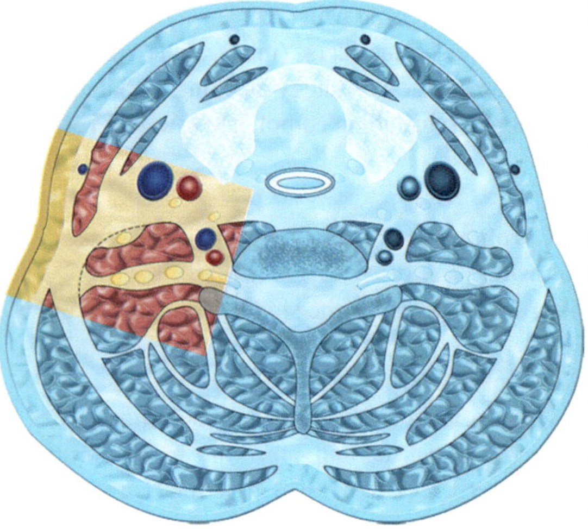

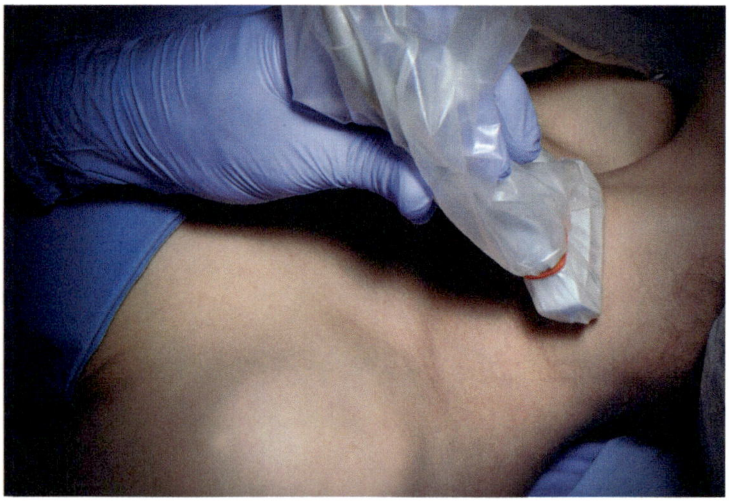

Fig. 5.11 Cervical transverse section at the level of C6 with materialisation of the ultrasound beam. Position of the probe

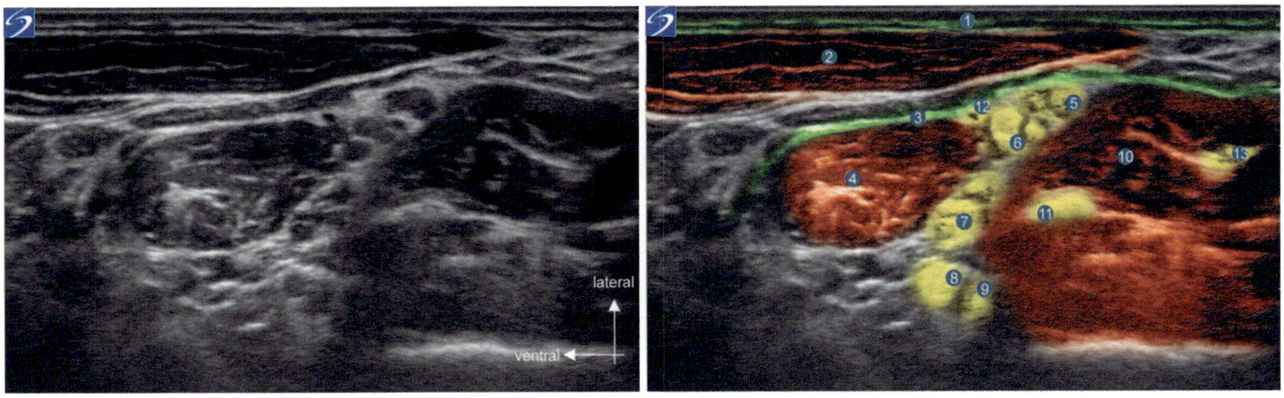

1. superficial layer of cervical fascia
2. sternocleidomastoi m.
3. C5 n. root
6. C6 n. root
7. C7 n. root
8. C8 n. root
9. T1 n. root
10. middle scalene m.
11. long thoracic n.
12. phrenic n.
13. dorsal scapular n.

Fig. 5.13 Low transverse ultrasound section of the interscalene area

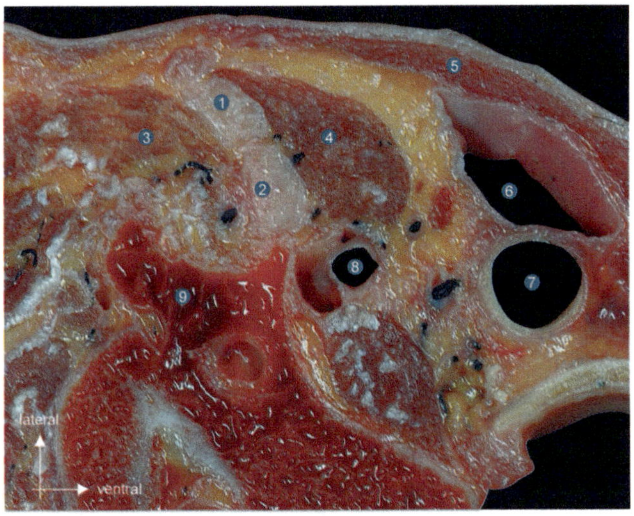

1. C5 and C6 n. roots
2. C7 n. root
3. middle scalene m.
4. anterior scalene m.
5. sternocleidomastoid m.
6. internal jugular v.
7. common carotid a.
8. vertebral a.
9. transverse process of C7 vertebra

Fig. 5.14 Transverse section passing through transverse process of C7. (Iconography: Admir Hadzic)

superficial level of the interscalene space (Fig. 5.13). Under good conditions, the phrenic nerve can be identified lying on the lateral aspect and then the ventral aspect of the anterior scalene muscle (Figs. 5.13 and 5.21). The suprascapular nerve can be seen to lie on the lateral aspect of the middle scalene muscle which separates from the brachial plexus, by continuing dorsally (Fig. 5.22). The dorsal nerve of the scapula can also be found along with the long thoracic nerve passing through or crossing the body of the middle scalene muscle (Figs. 5.9, 5.13 and 5.23).

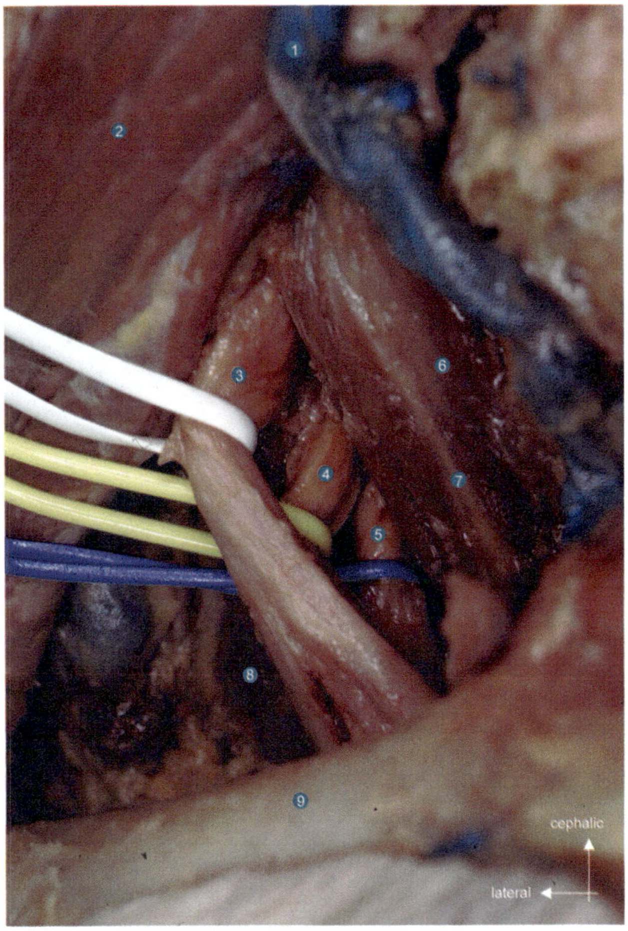

1. external jugular v.
2. sternocleidomastoid m.
3. C5 n. root
4. C6 n. root
5. C7 n. root
6. anterior scalene m.
7. phrenic n.
8. middle scalene m.
9. clavicle

Fig. 5.15 Brachial plexus at the interscalene level. (Dissection: Bertrand Fabre)

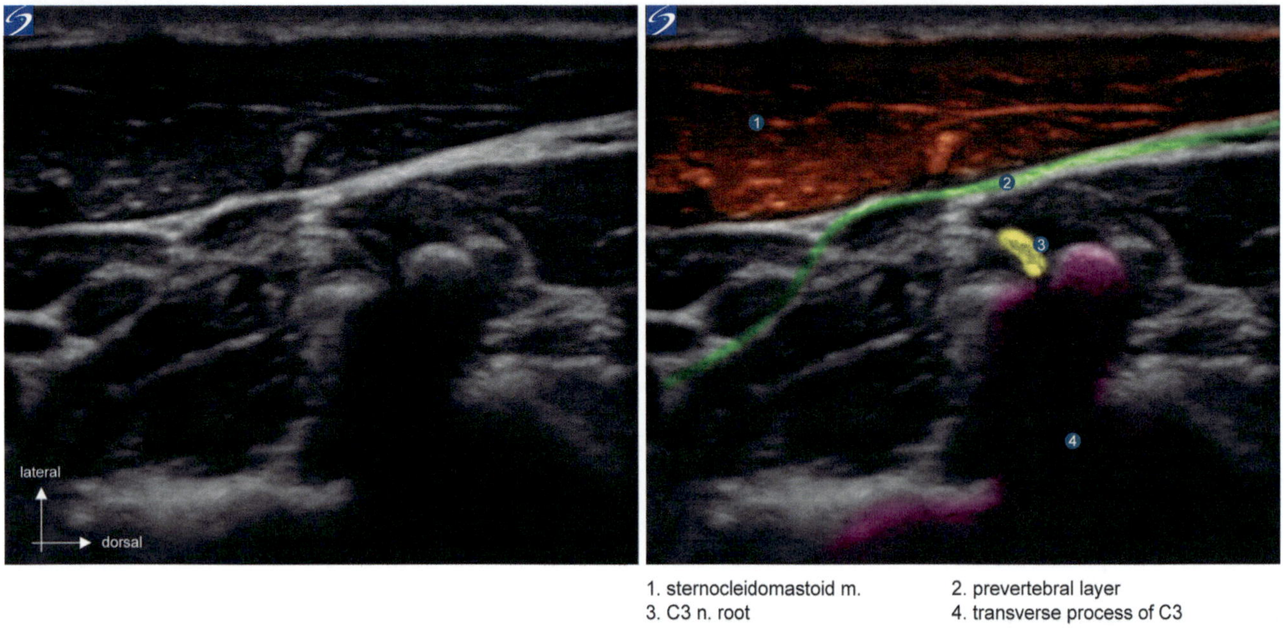

Fig. 5.16 Transverse ultrasound section of neck at the level of the transverse process of C3

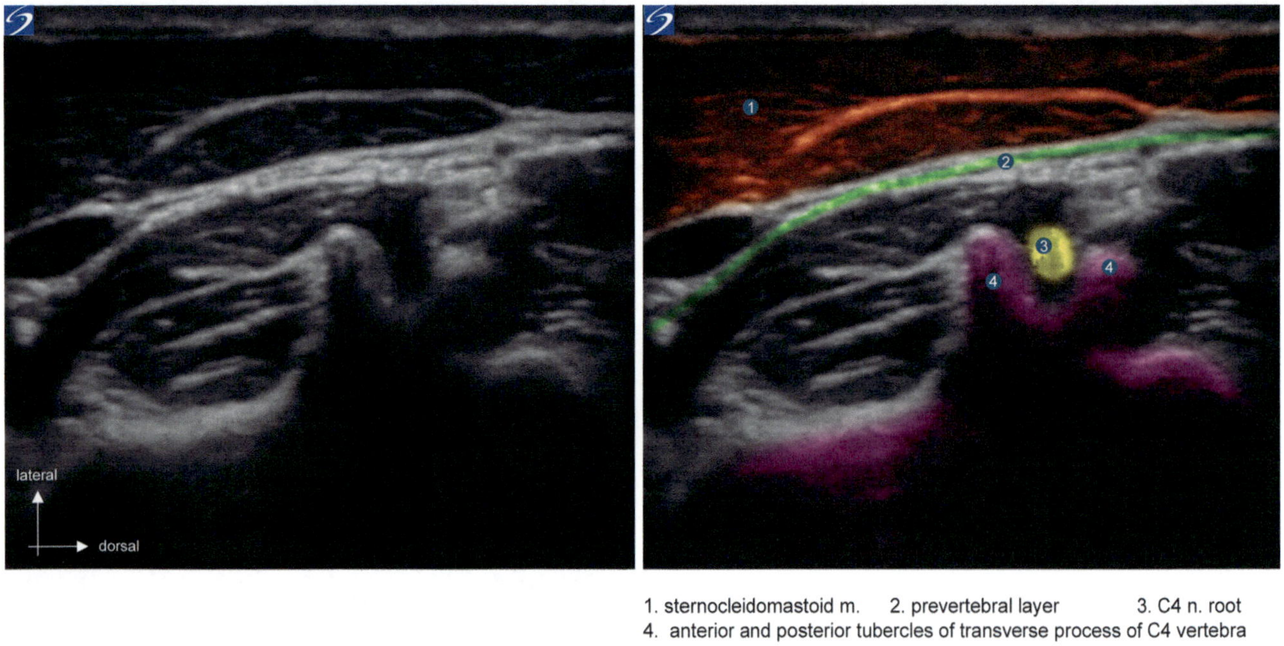

Fig. 5.17 Transverse ultrasound section of neck at the level of the transverse process of C4

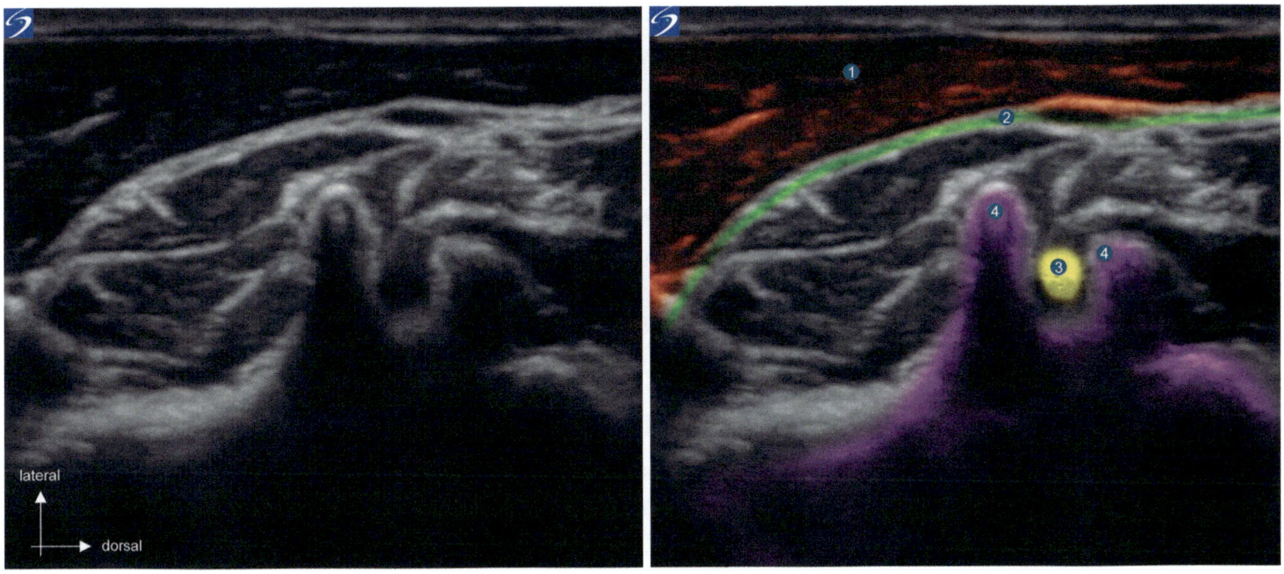

1. sternocleidomastoid m. 2. prevertebral layer 3. C5 n. root
4. anterior and posterior tubercles of transverse process of C5 vertebra

Fig. 5.18 Transverse ultrasound section of neck at the level of the transverse process of C5

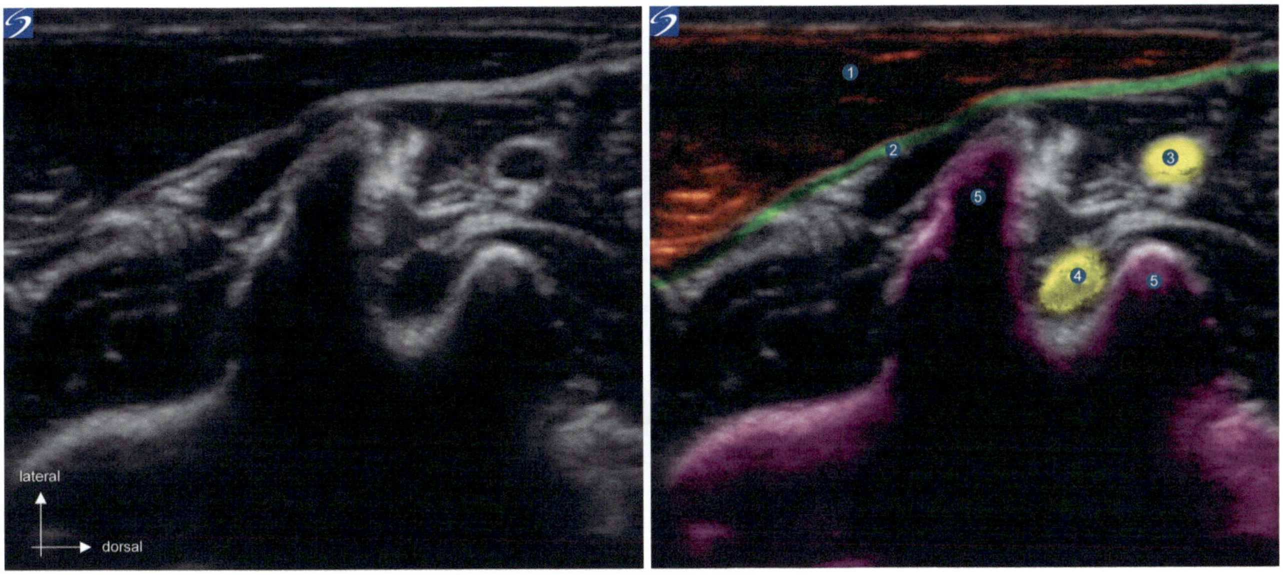

1. sternocleidomastoid m. 2. prevertebral layer
3. C5 n. root 4. C6 n. root
5. anterior and posterior tubercles of transverse process of C6 vertebra

Fig. 5.19 Transverse ultrasound section of neck at the level of the transverse process of C6

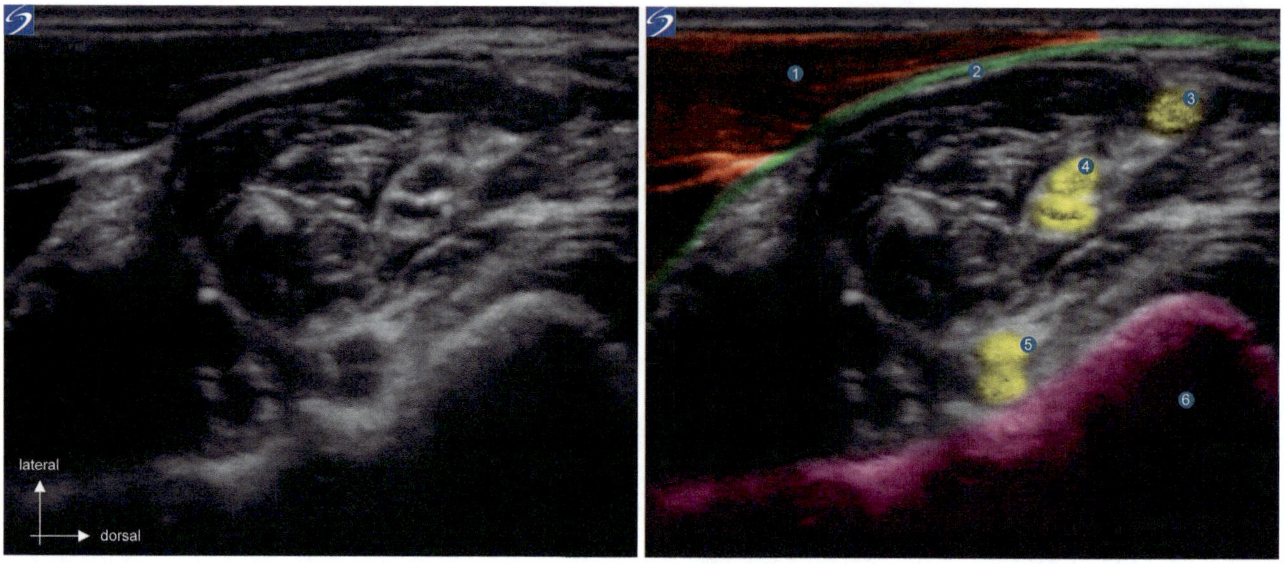

1. sternocleidomastoid m.
2. prevertebral layer
3. C5 n. root
4. C6 n. root
5. C7 n. root
6. posterior tubercles of transverse process of C7 vertebra

Fig. 5.20 Transverse ultrasound section of neck at the level of the transverse process of C7

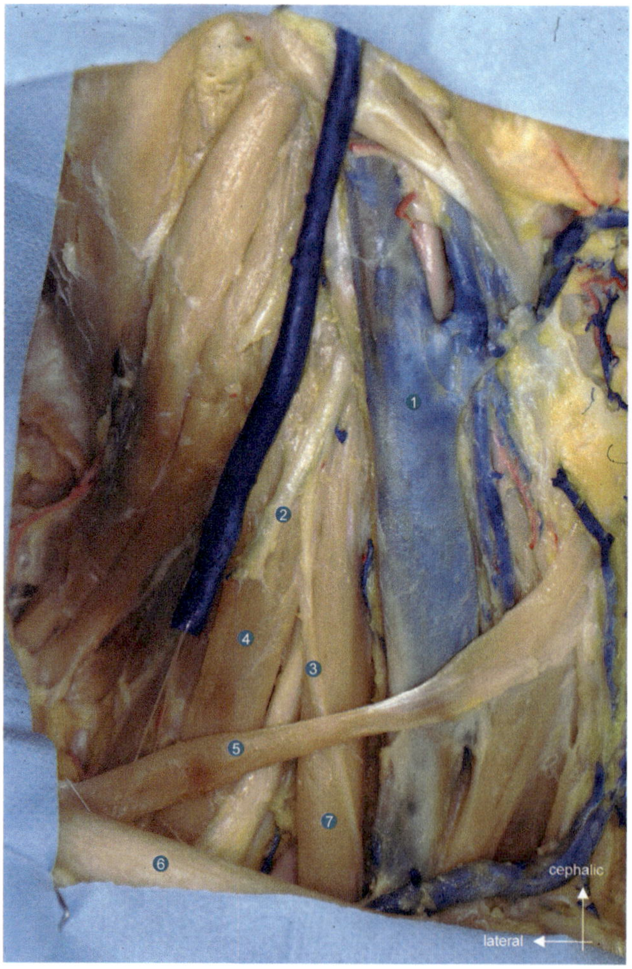

1. internal jugular v.
2. suprascapular n.
3. phrenic n.
4. middle scalene m.
5. omohyoid m.
6. clavicle
7. anterior scalene m.

Fig. 5.21 Lateral aspect of the neck, sternocleidomastoid muscle. (Dissection: Bertrand Fabre)

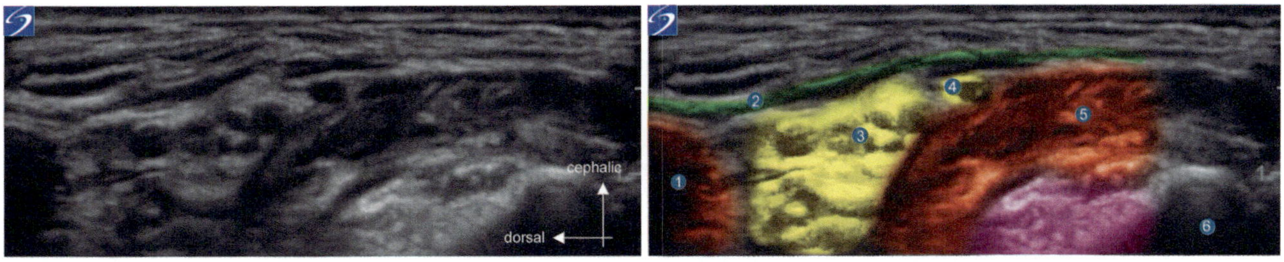

1. anterior scalene m.
2. prevertebral layer
3. branchial plexus (upper and middle trunk, collateral branches)
4. suprascapular n.
5. middle scalene m.
6. first rib

Fig. 5.22 Position of the suprascapular nerve on the surface of the middle scalene muscle, coming off the rest of the brachial plexus

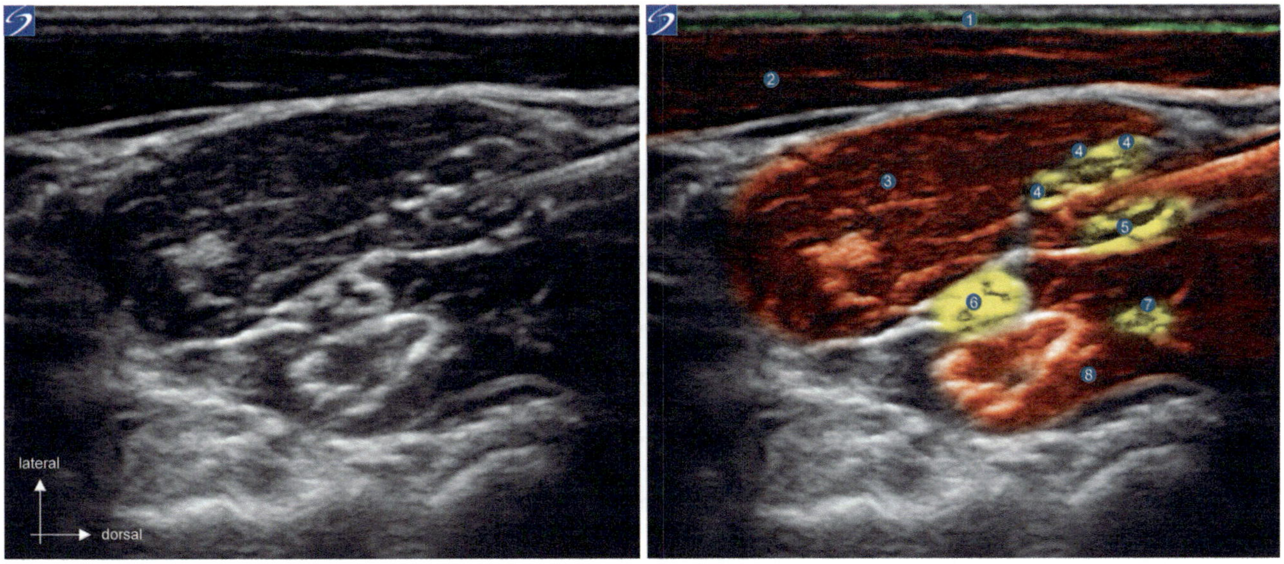

1. superficial layer of cervical fascia
2. sternocleidomastoid m.
3. anterior scalene m.
4. C5 and C6 n. divisions
5. dorsal scapular n.
6. C7 n. root
7. long thoracic n.
8. middle scalene m.

Fig. 5.23 Transverse ultrasound section of neck. Scapulodorsal and long thoracic nerves

The vertebral artery is an essential vascular element to positively identify during conduct of an ISB in order to avoid accidentally injuring it. In its ascending pathway, traditionally it crosses the transverse processes starting from C6 (but variations are possible in which it crosses only the transverse process starting from C5 or even C4) (Figs. 5.24, 5.25, 5.26, 5.27, 5.28, 5.29, 5.30, and 5.31).

In ultrasound-guided localization, it is possible to discover variations of the paths of the roots of the brachial plexus [26]: Fig. 5.32 shows the roots C5 and C6 (already divided) which have crossed the anterior scalene muscle to join its lateral aspect. They are very clearly separated and distant from the root of C7 which is well-placed in the interscalene pathway. Harry et al. [27] have described several types of variations in the pathway of nerve roots C5 and C6 after their emergence from the intervertebral foramina (Fig. 5.33).

Injection

The needle is inserted after infiltration of the skin with LA, at the posterior end of the probe (Fig. 5.12). It should be inserted and guided in the ultrasound plane. It should remain visible along its entire length, in a ventromedial direction in order to cross the axis of the interscalene contents. By avoiding an injection point that is too dorsal, after crossing the prevertebral lamina, the intramuscular course through the

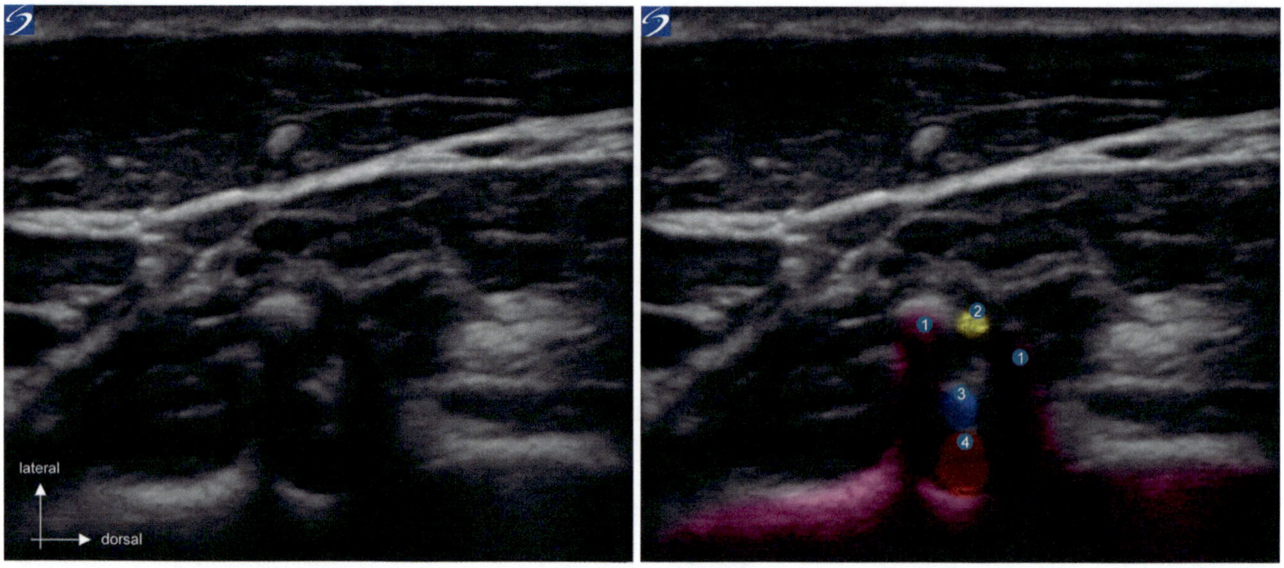

Fig. 5.24 Transverse ultrasound section passing through the transverse process of C3

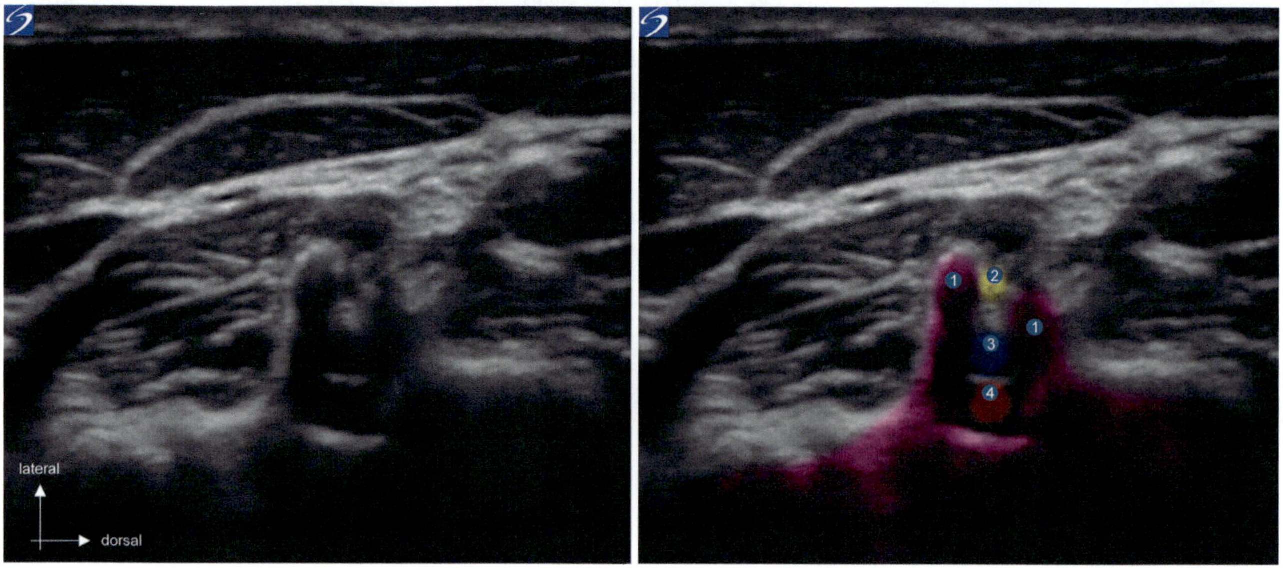

Fig. 5.25 Transverse ultrasound section passing through the transverse process of C4

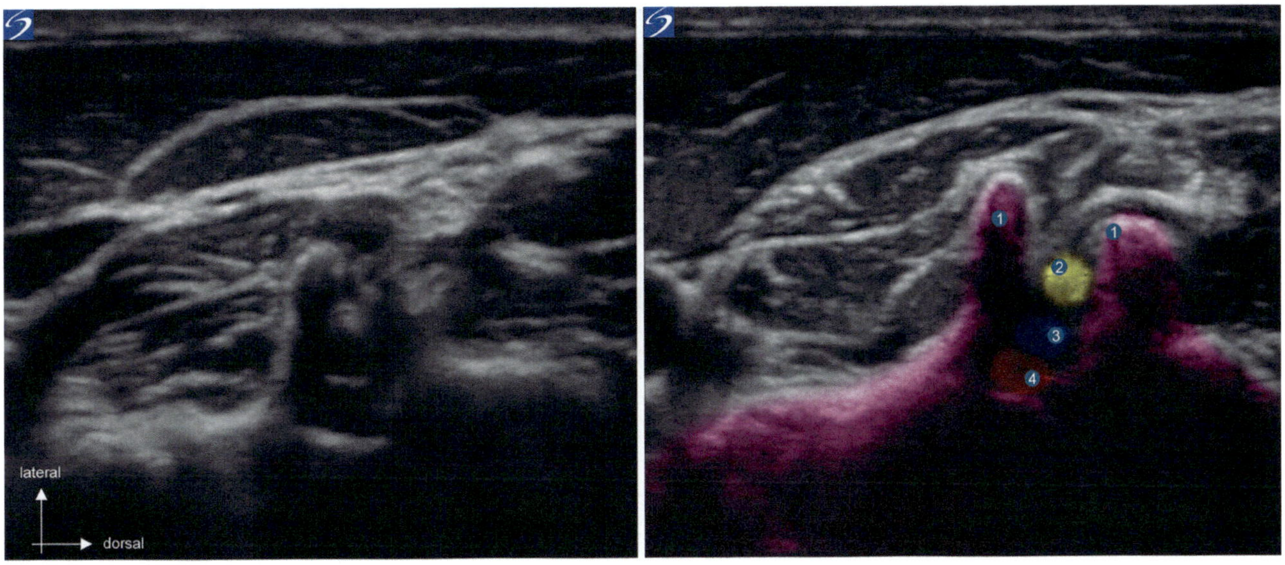

1. anterior and posterior tubercles of transverse process of C5 vertebra
2. C5 n. root
3. vertebral v.
4. vertebral a.

Fig. 5.26 Transverse ultrasound section passing through the transverse process of C5

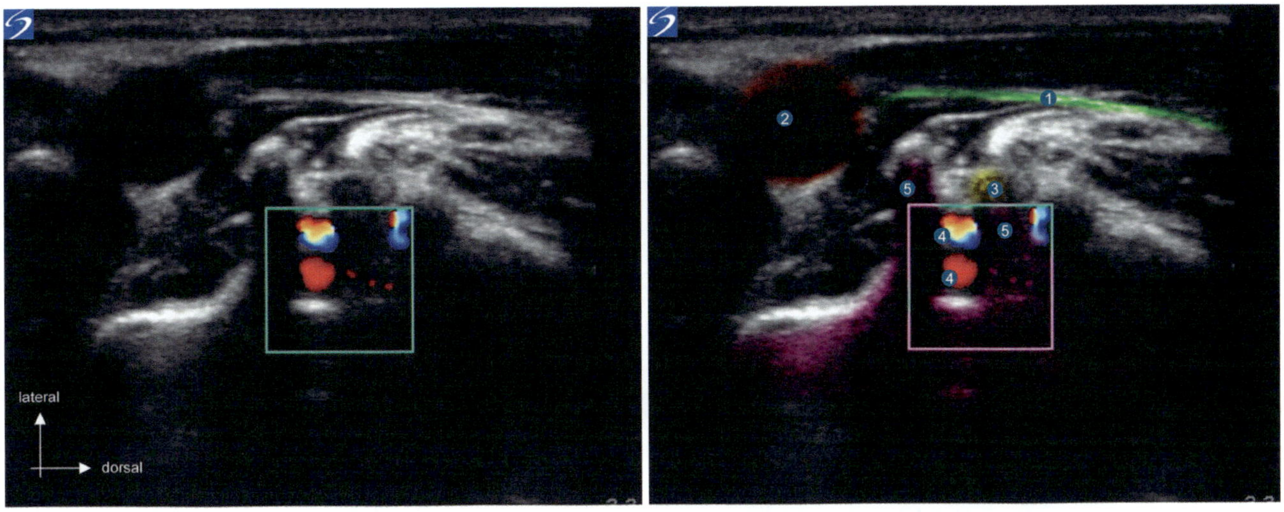

1. prevertebral layer
2. common carotid a.
3. C5 n. root
4. vertebral vessels
5. anterior and posterior tubercles of transverse process of C5 vertebra

Fig. 5.27 Transverse ultrasound section passing through the transverse process of C4: Doppler signal in vertebral vessels

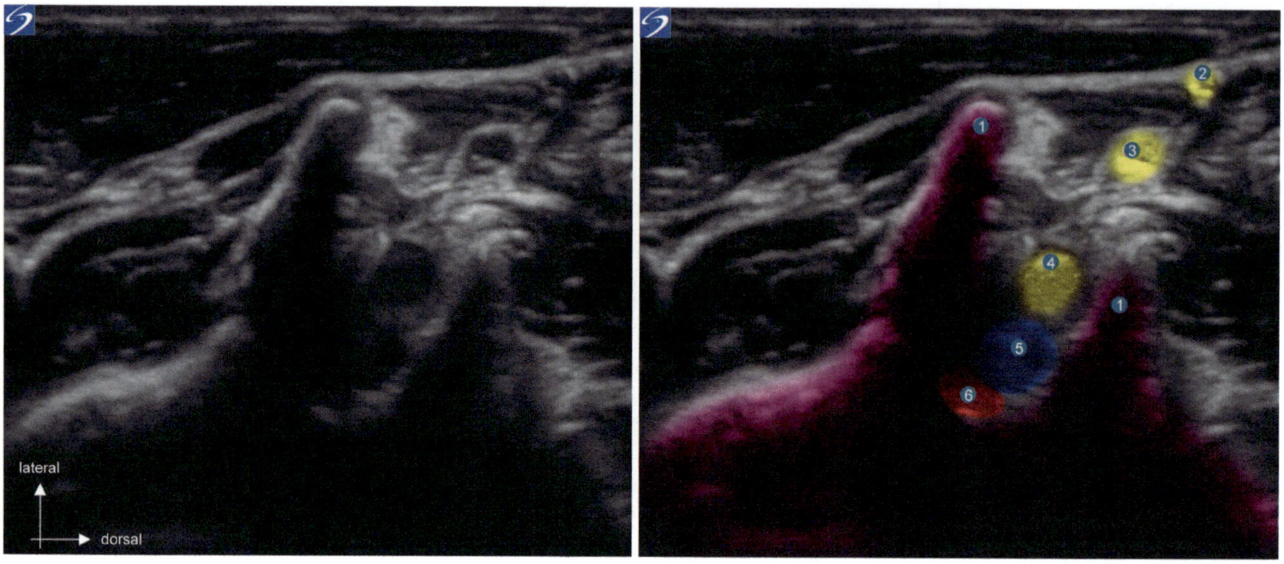

1. anterior and posterior tubercles of transverse process of C5 vertebra
2. phrenic n. 3. C5 n. root 4. C6 n. root
5. vertebral v. 6. vertebral a.

Fig. 5.28 Transverse ultrasound section passing through the transverse process of C6

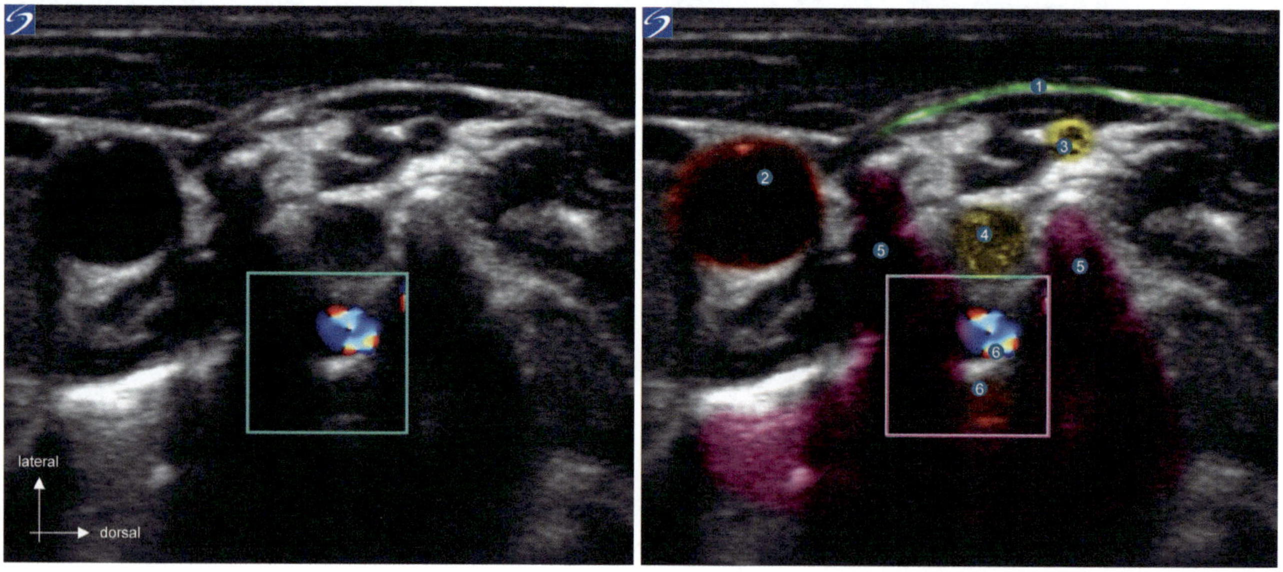

1. prevertebral layer 2. common carotid a.
3. C5 n. root 4. C6 n. root
5. anterior and posterior tubercles of transverse process of C6 vertebra
6. vertebral vessels

Fig. 5.29 Transverse ultrasound section passing through the transverse process of C6: Doppler signal in vertebral vessels

middle scalene muscle is limited. Thus, it is possible to progress almost parallel to the dorsal plane of the plexus, without penetrating the middle scalene muscle, while remaining in front of its anterior aponeurosis. In the final approach, it may be possible to inject a few millilitres of a D5W solution in order to perform "hydrodissection", thus facilitating ultrasound differentiation of the structures. This process also makes it possible to determine if the tip of the needle has in fact crossed the anterior aponeurosis of the middle scalene muscle. If this is not the case, we see fluid injected collect in

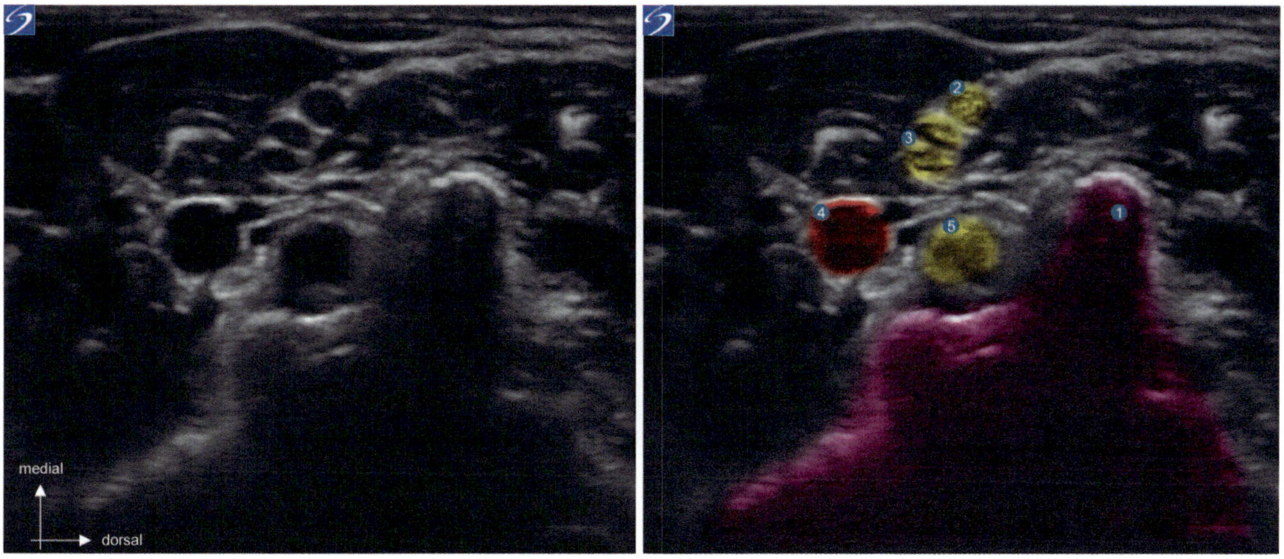

1. anterior and posterior tubercles of transverse process of C6 vertebra
2. C5 n. root 3. C6 n. root 4. vertebral a. 5. C7 n. root

Fig. 5.30 Transverse ultrasound section passing through the transverse process of C7

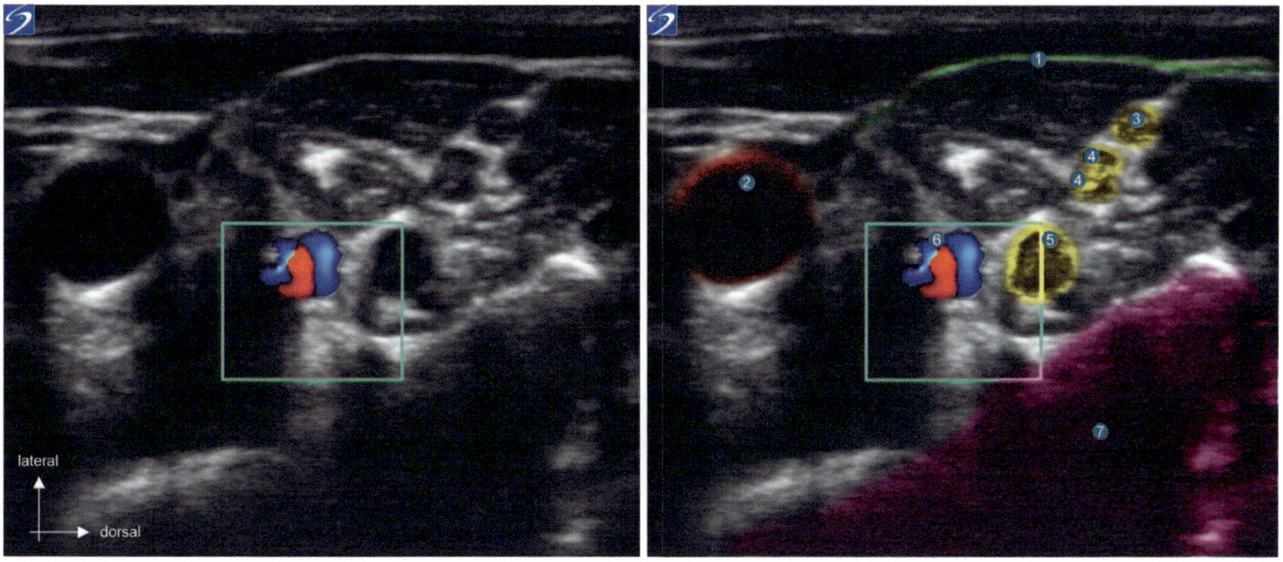

1. prevertebral layer 2. common carotid a.
3. C5 n. root 4. C6 n. root 5. C7 n. root
6. vertebral vessels 7. transverse process of C7 vertebra

Fig. 5.31 Transverse ultrasound section passing through the transverse process of C7: Doppler signal in vertebral vessels

this muscle and not in the interscalene plane. It is preferable to position the tip of the needle between roots of C6 and C7 (in proximity to the deep plane of C6) (Figs. 5.34 and 5.35). This posterolateral approach moreover offers the benefit in limiting the risk of direct traumatic injury to the phrenic nerve (sectioning by the needle). In fact, cases of persistent diaphragmatic paralysis after interscalene nerve block exist whose cause remains unclear [28–30]: toxicity of local anaesthetics, intraneural injection, direct injury to the nerve, etc. Since the phrenic nerve is not always visible with ultrasound (and thus not always "avoidable"), it is recommended to not perform ISB by approaching the plexus anteriorly while crossing the anterior scalene muscle (injection at the anterior end of the probe).

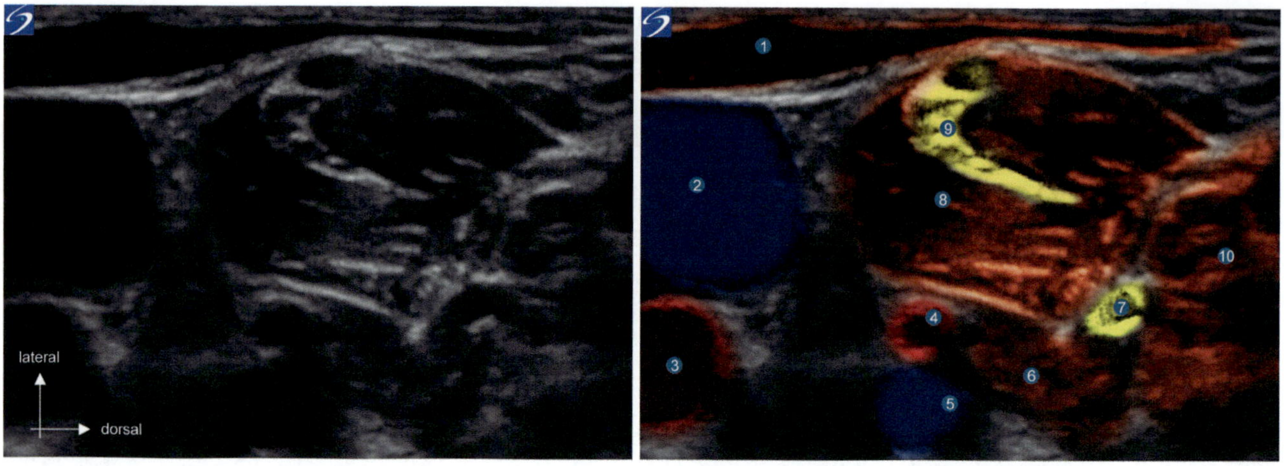

1. sternocleidomastoid m. 2. internal jugular v.
3. common carotid a. 4. vertebral a.
5. vertebral v. 6. longus colli m.
7. C7 root 8. anterior scalene m.
9. C5 and C6 n. roots 10. middle scalene m.

Fig. 5.32 Interscalene block. Variation in position of nerve roots: C5 and C6 (already divided) crossing the anterior scalene muscle, while C7 is located in the interscalene pathway

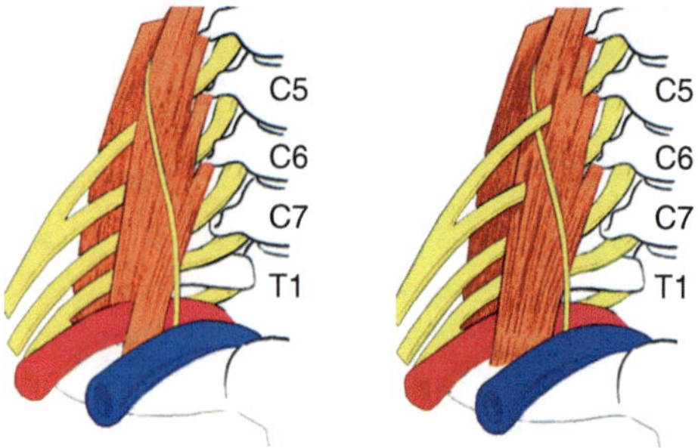

C5 and C6 n. roots through all the anterior scalene m.	15%
C5 n. root alone piercing the anterior scalene m.	13%
C5 and C6 n. roots piercing separately the anterior scalene m.	6%
C5 n. root in front of the anterior scalene m.	3%

Fig. 5.33 Variations in course of nerve roots C5 and C6

Interscalene Catheter by Posterolateral Approach (in the Ultrasound Plane) (Figs. 5.36, 5.37, 5.38, 5.39, 5.40, 5.41, and 5.42)

For insertion of an interscalene catheter, the injection procedure is absolutely identical to a single shot block. The objective is to position the **tip of the needle** at the place where we wish to position the **end of the catheter**. The needle trajectory should be planned to minimise tissue trauma, if possible not crossing the middle scalene muscle but without taking the risk of direct injury to the branches of the brachial plexus. Figures 5.43, 5.44, 5.45, 5.46, and 5.47 show several steps in insertion of an interscalene catheter with an injection trajectory which leaves the middle scalene muscle intact, with the tip of the needle placed in the superficial plane of C7, enabling the catheter to be placed at this level and then withdrawn slightly in order to be positioned in the deep plane of C6; as it can be seen in Fig. 5.47, the spread of the local anaesthetic is optimum around C5, C6 and partly in C7.

Injection

Direct visualisation of spread of the local anaesthetic injected enables possible repositioning of the needle in case of atypical or non-optimal spread. During the injection, the interscalene pathway "fills up" progressively with an increase in contrast of the adjacent structures (scalene muscles on both sides, nerves, aponeurosis, etc.) (Fig. 5.48). Depending on context, a volume of between 10 and 20 mL is injected.

Short/Transverse Axis Approach (Needle Out of Plane)

This approach which maintains the needle in the traditional pathway in the "anatomical" axis of the interscalene groove, in particular appears useful for insertion of a perineural catheter since it facilitates its insertion. The level of injection is

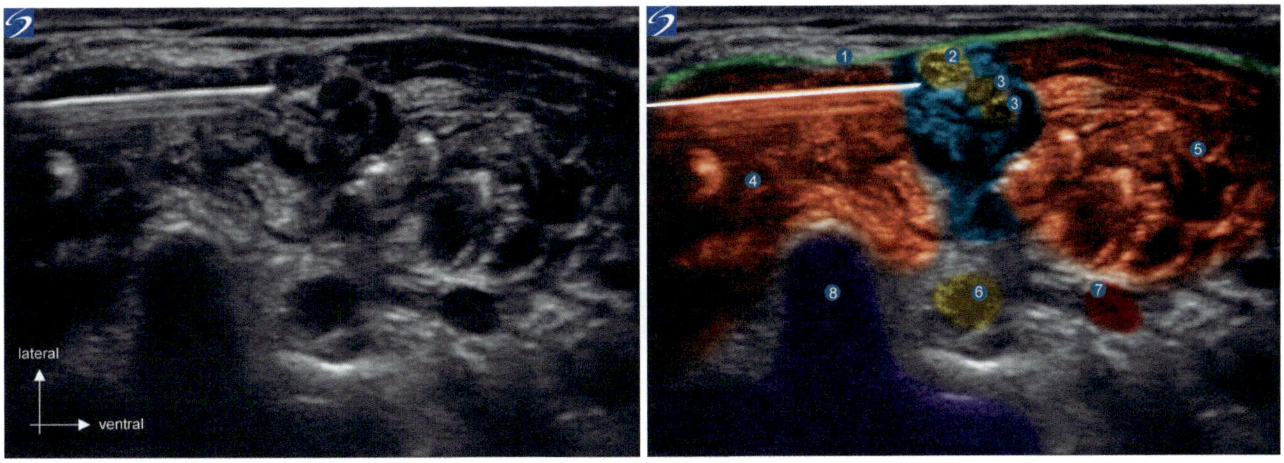

1. prevertebral layer
2. C5 n. root
3. C6 n. root
4. middle scalene m.
5. anterior scalene m.
6. C7 n. root
7. vertebral a.
8. transverse process of C7 vertebra

Fig. 5.34 Interscalene block. Tip of needle deep to C5

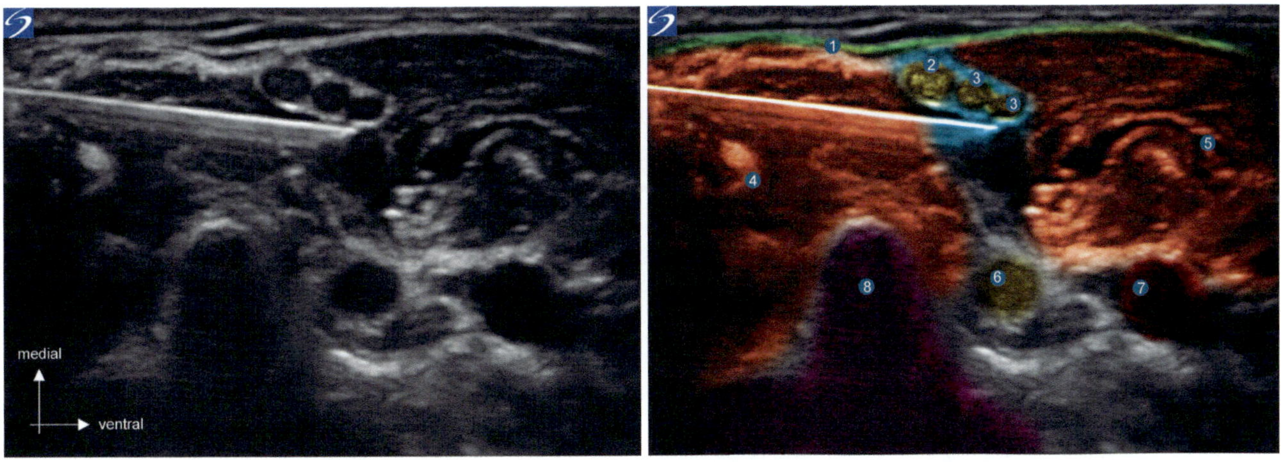

1. prevertebral layer
2. C5 n. root
3. C6 n. root
4. middle scalene m.
5. anterior scalene m.
6. C7 n. root
7. vertebral a.
8. transverse process of C7 vertebra

Fig. 5.35 Interscalene block. Tip of needle deep to C6

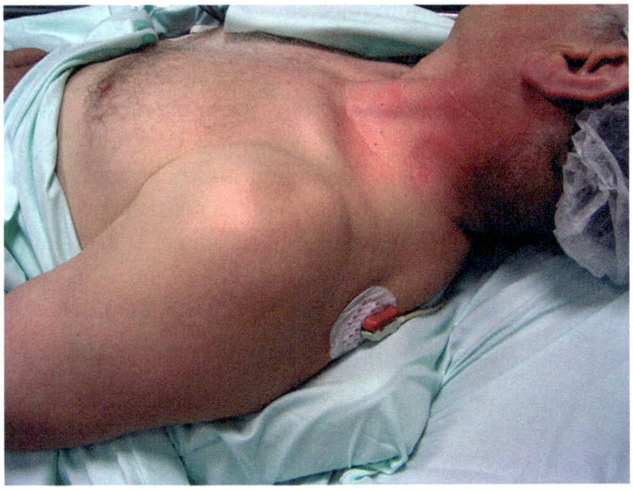

Fig. 5.36 Position of patient for placement of an ultrasound-guided interscalene perineural catheter

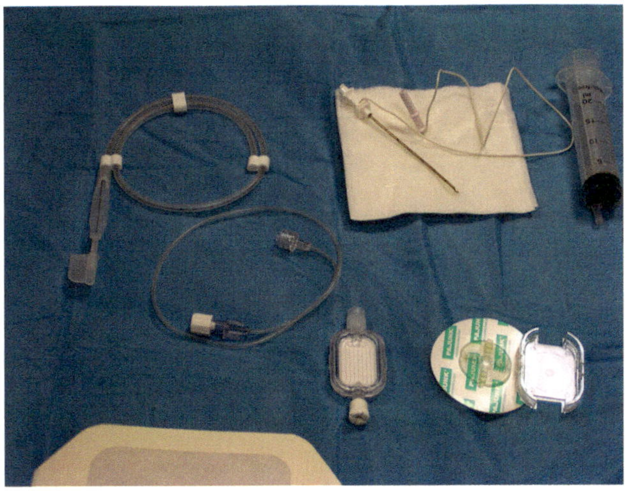

Fig. 5.37 Interscalene catheter: equipment

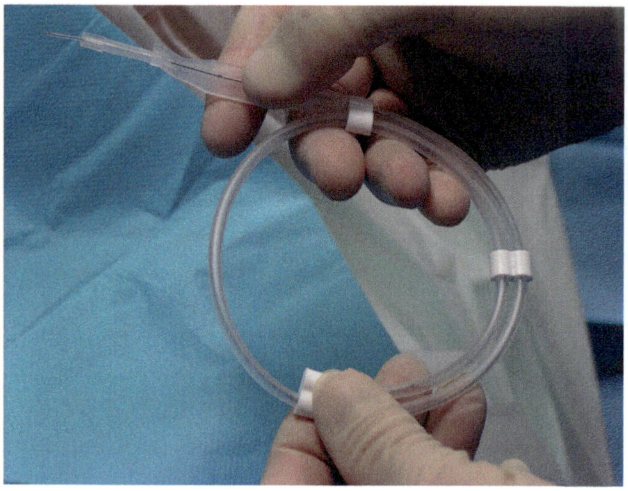

Fig. 5.38 Catheter in its introducer sheath

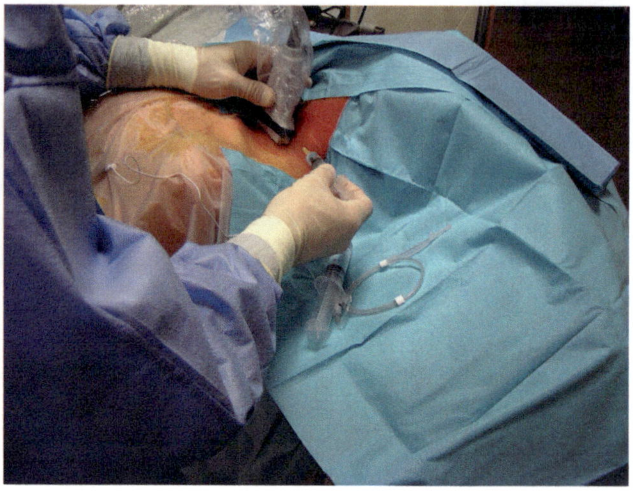

Fig. 5.40 Positioning for placement of an interscalene catheter. Injection in-plane. Initial cutaneous local anaesthesia

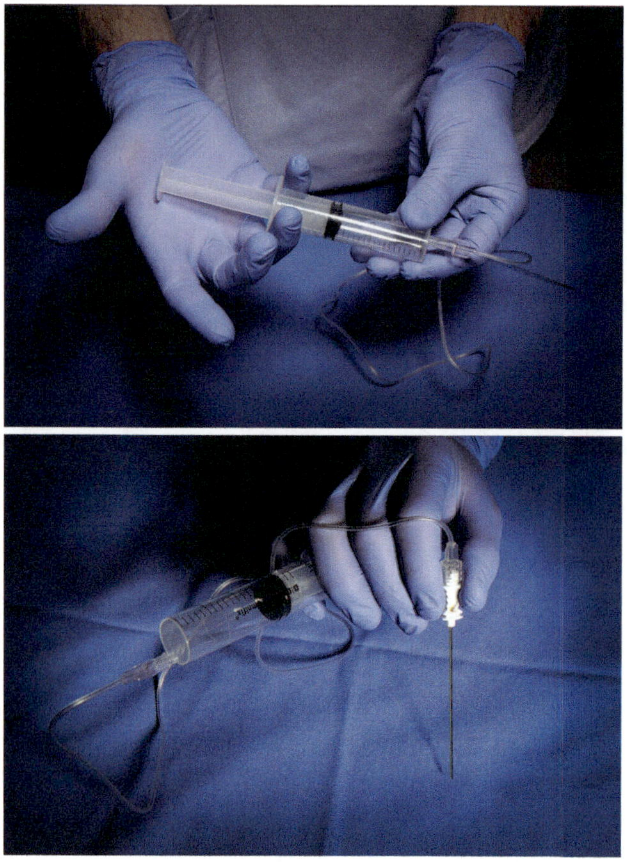

Fig. 5.39 Method of holding syringe and needle enabling needle puncture and simultaneous injection with one hand, without alternately releasing them

determined by a preliminary ultrasound scan and should not be dependent on surface landmarks alone. It is necessary to visualise the brachial plexus in a transverse section (short axis). In fact, if the probe is rotated and placed in the long axis, the structures are seen longitudinally and recognition of structures and positioning of the needle become very difficult.

This "out of plane" approach involves a transverse view and therefore only a pinpoint image of the needle, making the procedure more delicate to perform. In this configuration, control of the tip of the needle requires a cautious approach, good needle and probe control and experience with the procedure. The technique of combining neurostimulation with ultrasound reveals its advantage in especially delicate situations such as this one. The technique of hydrolocalisation is crucial during this approach. This consists of injecting successively small volumes of liquid as the needle advances, enabling the position of the needle tip to either be seen directly, or to know the plane in which it is located. The liquid injected is ideally a D5W solution in order to avoid both local and systemic toxicity (reabsorption or accidental intravascular injection of the local anaesthetic), and also to permit the simultaneous use of neurostimulation.

Whether for a single injection or insertion of a catheter, the objective is to position the tip of the needle in proximity to C5 or C6. It is imperative to avoid injury to the phrenic nerve during insertion of the needle. For a continuous nerve block, once the tip of the needle is in place, the catheter is

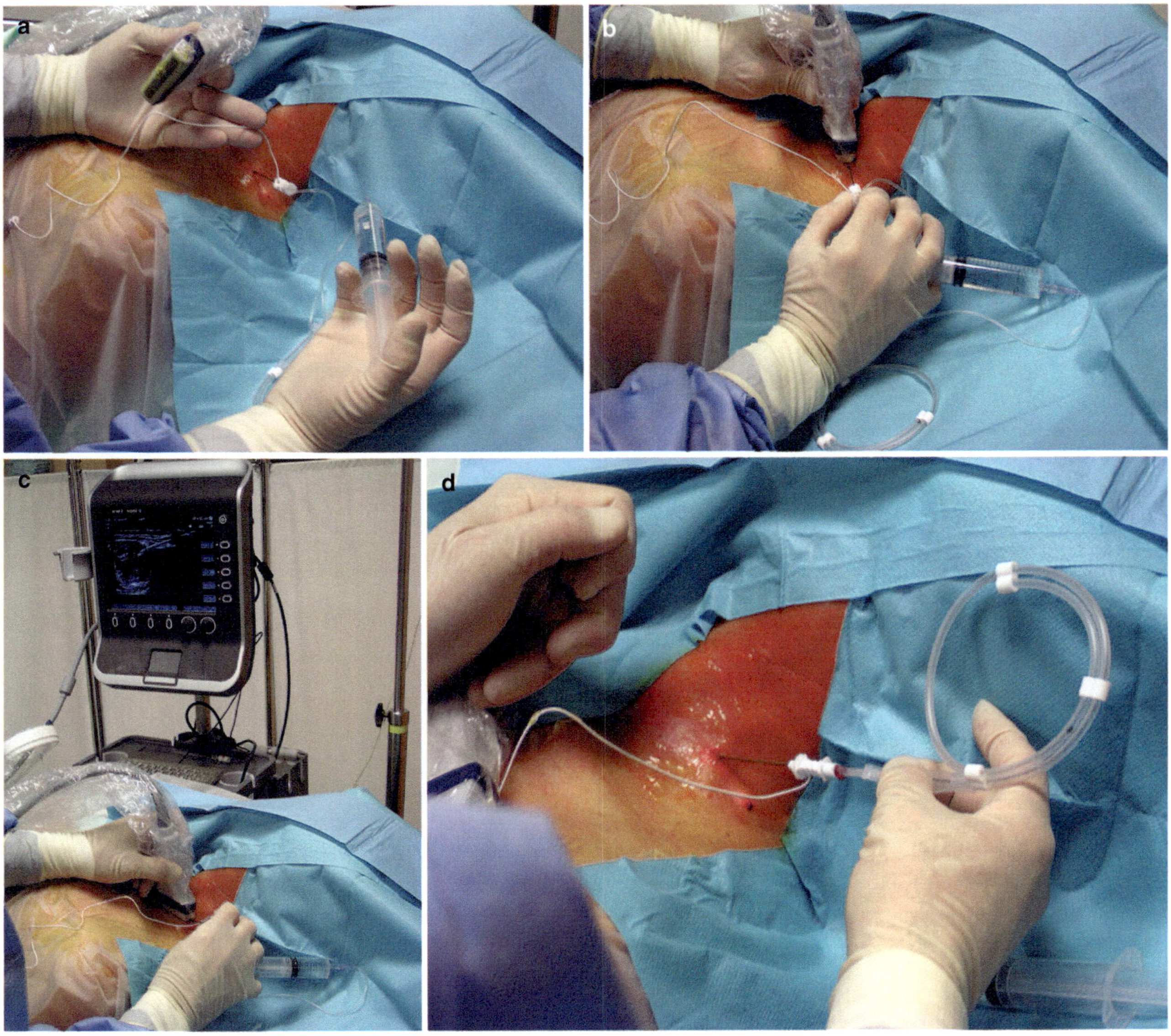

Fig. 5.41 Procedure for placement of an interscalene catheter. (**a**) Hold on the syringe. (**b**) Simultaneous needle insertion and injection with one hand. (**c**) Visualisation of needle position and spread of the local anaesthetic. (**d**) Introduction of catheter through the needle

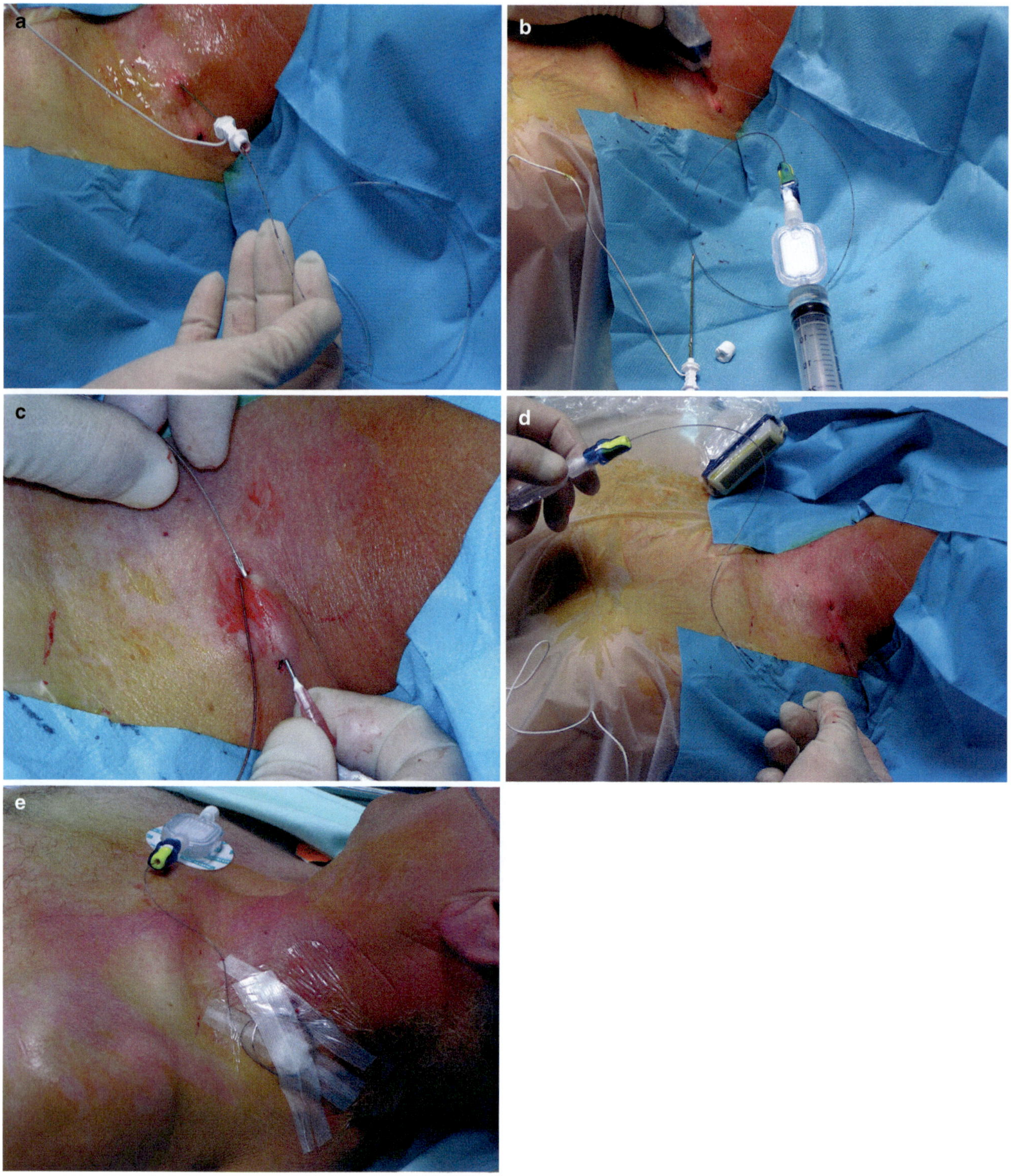

Fig. 5.42 Removal of the catheter introducer sheath and then of the needle (**a**). Verification of proper spread of local anaesthetic through the catheter (**b**). Tunnelling of the catheter (**c**). Fixation and dressing (**d**, **e**)

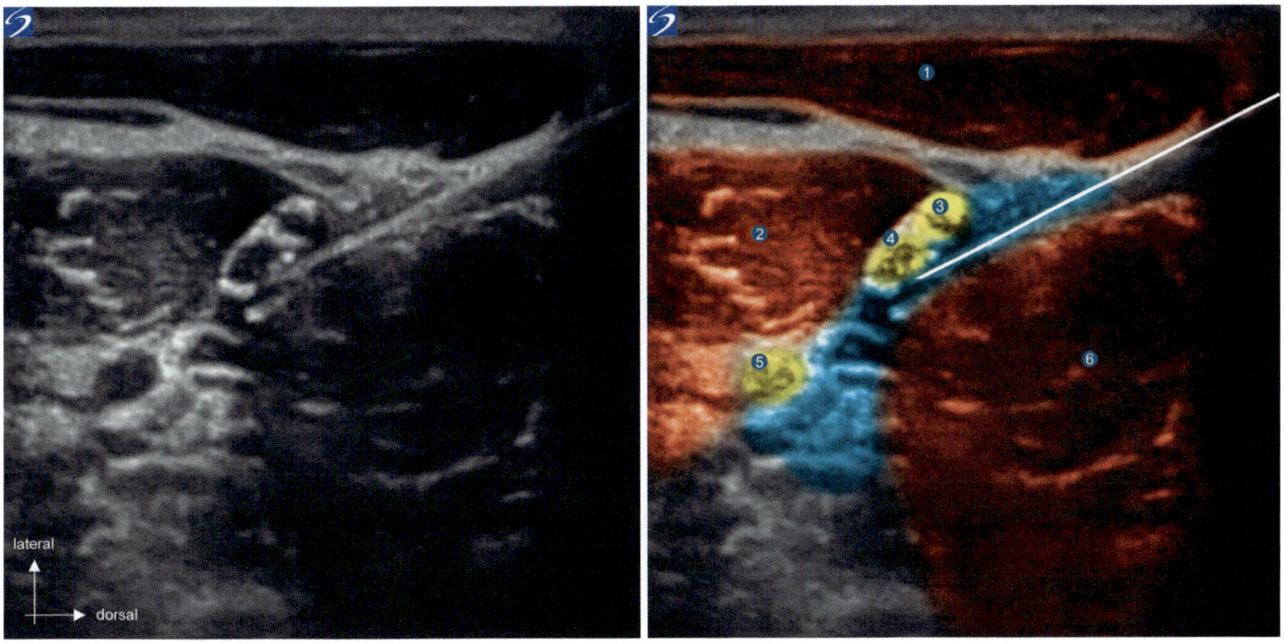

1. sternocleidomastoid m.
2. anterior scalene m.
3. C5 n. root
4. C6 n. root
5. C7 n. root
6. middle scalene m.

Fig. 5.43 Placement of an interscalene catheter. Injection trajectory not involving the middle scalene muscle

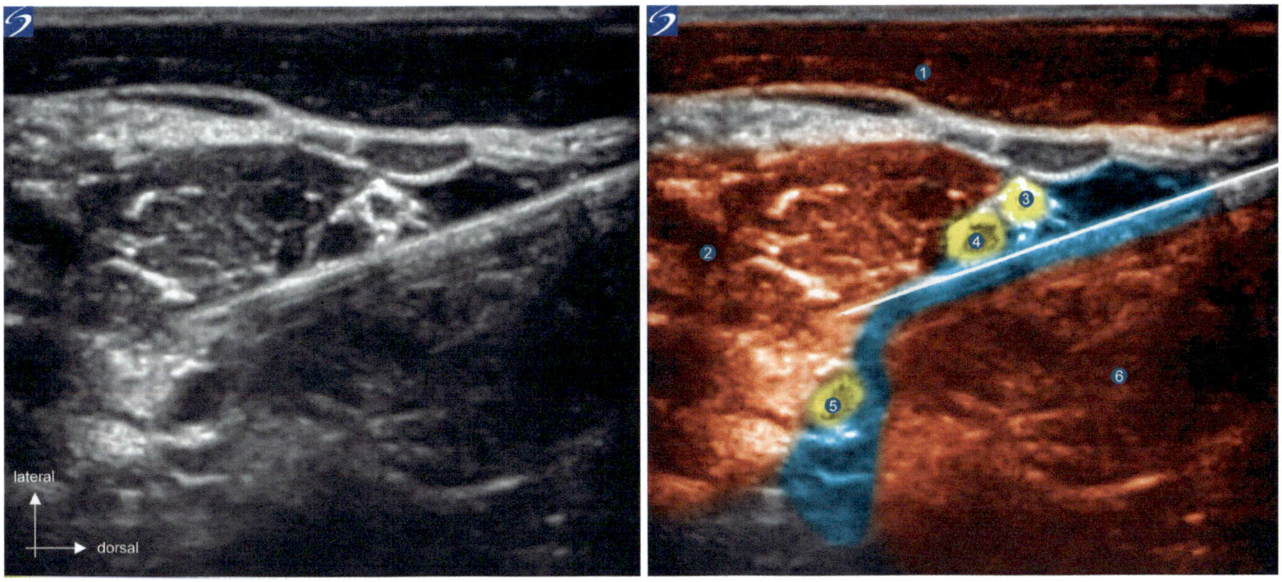

1. sternocleidomastoid m.
2. anterior scalene m.
3. C5 n. root
4. C6 n. root
5. C7 n. root
6. middle scalene m.

Fig. 5.44 Placement of an interscalene catheter. Bevel of needle placed superficial to the C7 nerve root

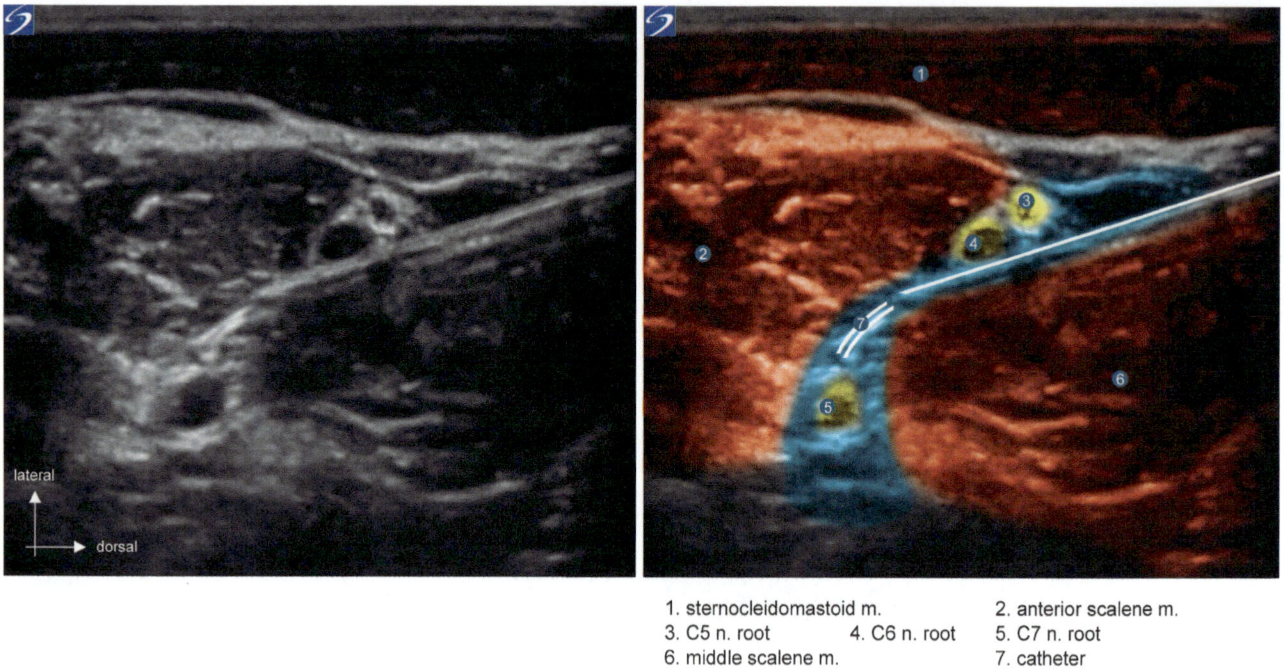

Fig. 5.45 Interscalene catheter "dropped" superficial to the C7 nerve root

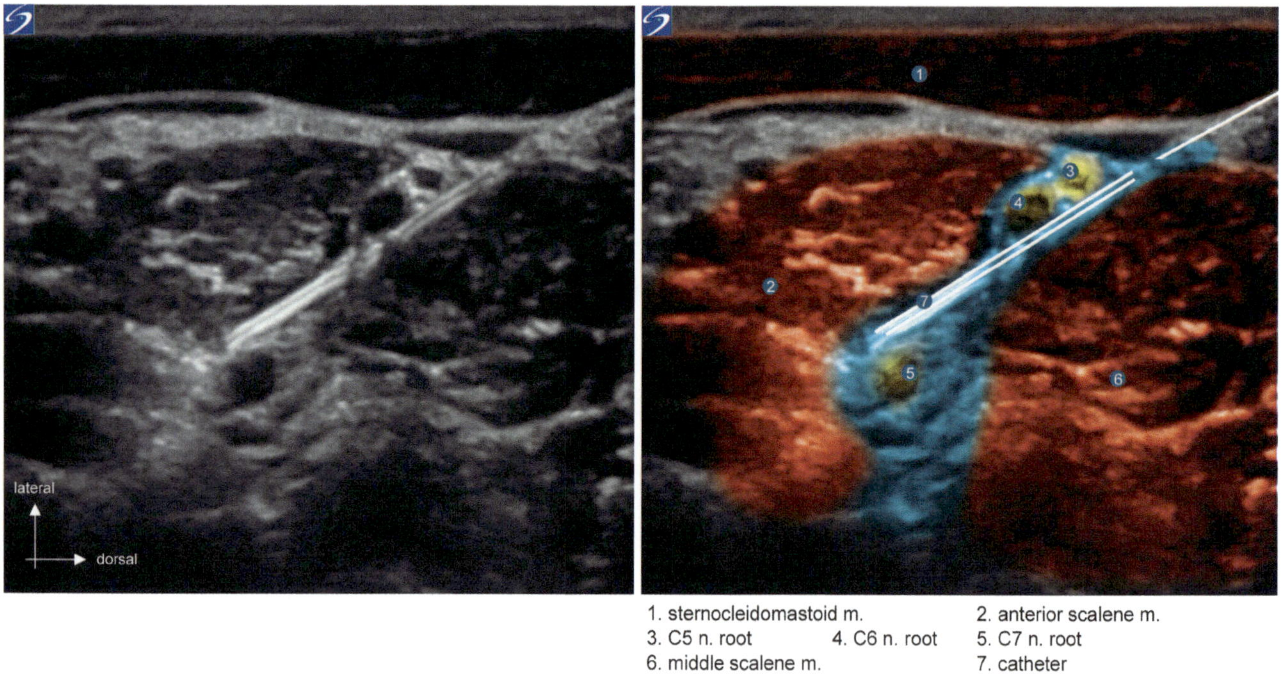

Fig. 5.46 Interscalene catheter positioned superficial to the C7 nerve root after removal of needle

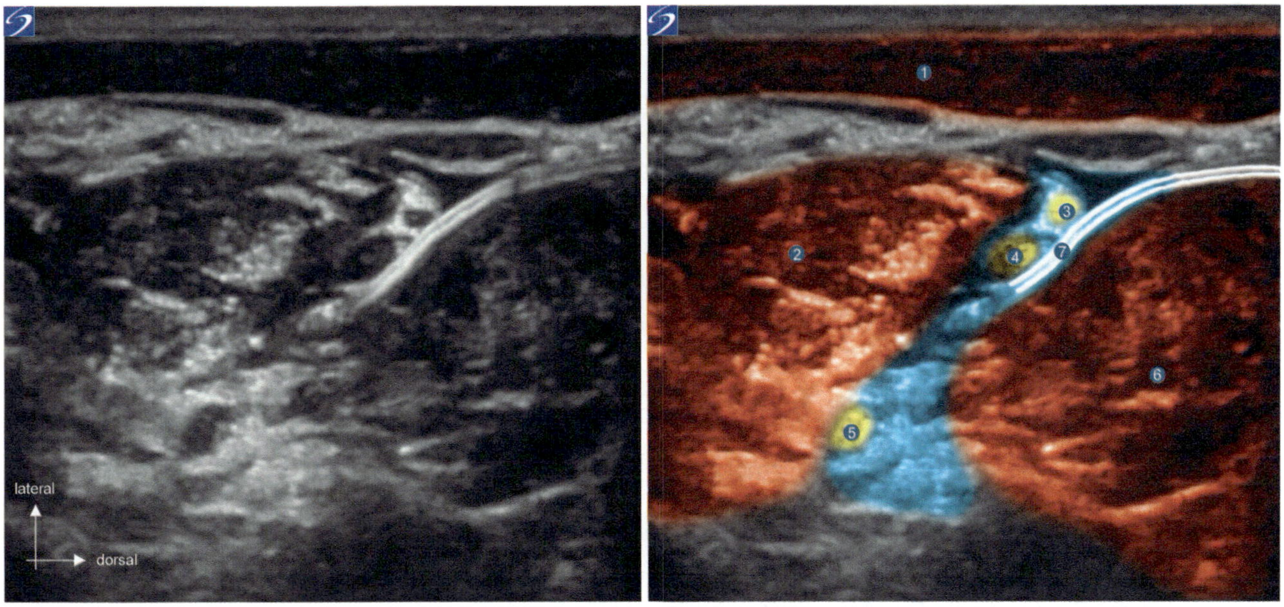

Fig. 5.47 Interscalene catheter positioned deep to the C6 nerve root. Optimal distribution of local anaesthetic around nerve roots C5, C6 and partly around C7

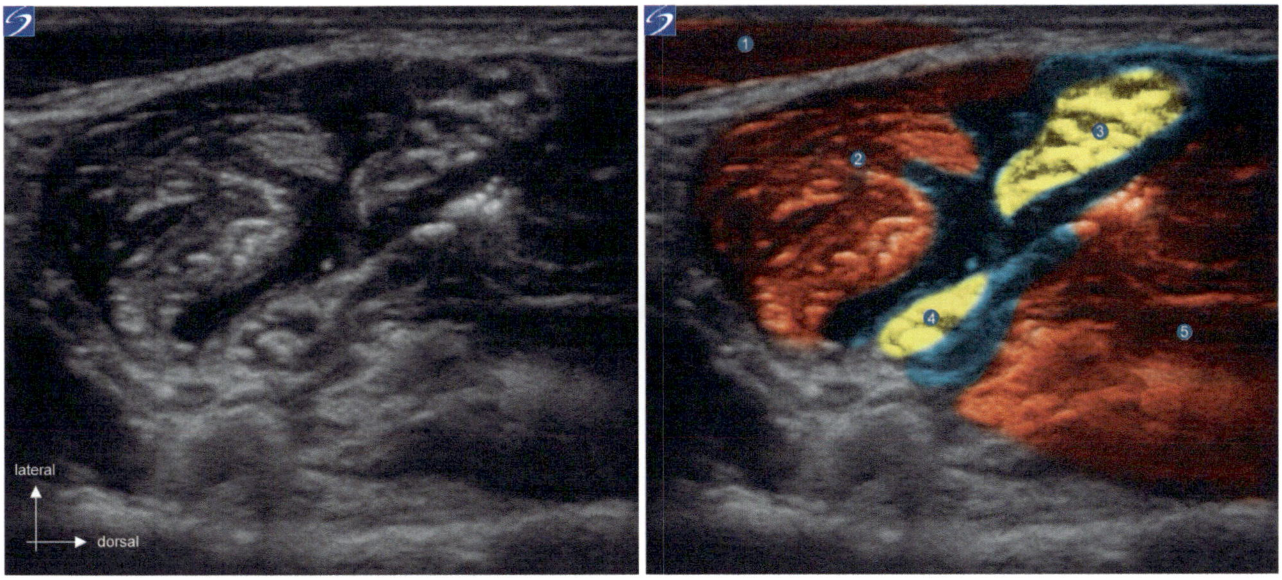

Fig. 5.48 Spread of the local anaesthetic in the interscalene area surrounding the superior (C5 + C6) and middle (C7) trunks

inserted into the needle in the same manner as an approach with neurostimulation only; in other words, it is advanced a few centimetres further in order to stabilise it.

Paediatrics

The shallow depth and small size of the anatomical structures should encourage use of caution. It is necessary to use a high frequency probe and to insist on optimising the ultrasound adjustment settings.

Supraclavicular Block (Fig. 5.49)

Traditional indications for supraclavicular block (SCB) are surgery on the upper end of the humerus with distal incision and elbow surgery. The supraclavicular approach does not require mobilisation of the upper limb, which makes it one of the techniques of choice in trauma cases, in the same category as infraclavicular block. It can be relatively easy in an obese subject where it can be a second-line choice for shoulder arthroscopy (the success rate is lower in an obese subject than in the non-obese patient, but the complication rates are identical [31]). It can replace an axillary or humeral BP approach for surgery of the hand and forearm in patients with a history of ipsilateral lymph node dissection [8]. Extension of anaesthesia occurs more distally than with an interscalene block, which makes supraclavicular block useful in surgery of the arm and elbow, whenever the incisions are made on the posterior aspect of the arm (for example, in fixation of humeral head or neck fracture) [32]. SCB is not indicated for hand surgery because blockade of the ulnar nerve is not constant [33]. Extension of analgesia to the axillary nerve makes it preferable to infraclavicular block for surgery requiring prolonged abduction of the arm (arthroscopy of the elbow), which is uncomfortable for the patient.

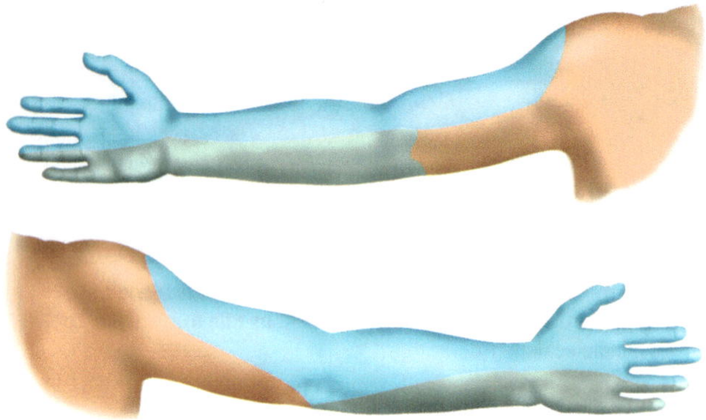

Fig. 5.49 Area of coverage of a supraclavicular block

It is in the supraclavicular level that the nerve structures of the brachial plexus (lateral, posterior and medial cords) are closest to each other. Induction of anaesthesia is obtained preferably in the area of the axillary, radial and musculocutaneous nerves. Due to the deeper position of the medial cord, median nerve block and in particular ulnar nerve block occurs later and regularly incomplete. With neurostimulation only, occurrence of a motor response in the area of the ulnar nerve is a warning sign because it corresponds to stimulation of the medial cord which is very close to the dome of the pleura. In case of mediocre or poor visibility, ultrasound guidance allows the needle to be directed superficial to the cords of the plexus in order to avoid the pleura [9].

In light of the proximity of the pleura and of the brachial plexus, this nerve block, with traditional neurostimulation, would carry the risk of pneumothorax. Many techniques have been described in order to decrease its incidence.

The incidence of pneumothorax ranges from 0.36% with Kulenkampff's traditional technique [34] to 6% [35]. The risk was higher in thin, slender subjects, and more frequent in the right lung rather than the left [6]. Since signs of pneumothorax are often minimal and of late onset, supraclavicular block was contraindicated in ambulatory surgery [36]. To avoid injury to the lung tissue, the proposed techniques involve assessment of the depth of the plexus and of the pleura, the length of the needle, the landmarks at the injection site and the direction of injection, and other aids in locating the target [37]. The most recent technique is ultrasound, which readily enables to recognise and avoid "susceptible" structures near the plexus, namely blood vessels, the pleura and the lung.

Estimating the depth of the brachial plexus varies depending on technique. The relative complexity of these methods of assessment [38] makes them unsuitable for everyday use.

The majority of historical techniques for supraclavicular approach to the brachial plexus use the subclavian artery as an essential landmark [39–43]. However, its location sometimes is clinically difficult to determine but greatly facilitated by use of Doppler method or with ultrasound. Easier to use surface landmarks have been proposed: middle of the clavicle [44], the scalene muscle area [45], the external jugular vein and trapezius muscle [42], and the heads of the sternocleidomastoid muscle [46].

Traditionally, the direction of injection was caudal, slightly medial and posterior [41], and is towards the inner aspect of the first rib towards the dome of the pleura. According to Winnie [43], the direction of the needle, in particular, should not be posterior nor medial but should remain parallel to the lateral border of the scalene muscles. Since these muscles insert on the first rib, the needle will necessarily pass on the outside. In order to avoid the pleura, the axis of injection can be lateral [42–45], introducing the notion of tangential approach to the plexus.

In these approaches, for a better definition, the axis of injection passes through a fixed point such as the subclavian artery [45], the earlobe [42] or the middle of the clavicle [44]. A totally different concept of injection has been proposed with the technique of a "plumb line" [46] where the plexus is approached in a strict anteroposterior sagittal plane, above the clavicle, outside of the insertion of the sternocleidomastoid muscle. This approach does not protect from pneumothorax any more than the other techniques.

All these techniques which attempt to avoid hazardous structures are no longer used since the advent of ultrasound. Visualisation of the pleura enables to remain at a distance from it, under reservation of seeing the tip of the needle. Use of ultrasound in fact does not completely eliminate the risk of pneumothorax [47].

Type of probe: linear, 6–13 MHz.
Axis of probe: transversal/short axis (Fig. 5.50).
Configuration: nerves in the short axis, needle in plane.
Study depth: 2–3 cm.
Neurostimulation: enables further identification of the nerves visualised and can limit, by determination of an MIS >0.3 mA (0.1 ms), the risk of accidental intrafascicular injection. Furthermore, it can supplement locating nerves when conditions of visibility are unfavourable.
The high prevalence of anaesthetic failure in the territories C8-T1 during supraclavicular nerve block is related to the anatomical isolation (aponeurotic interpositions, independent sheath) of C8-T1 components from the remainder of the plexus at this level. Neurostimulation can be useful in positioning of the injection in contact with these nerves in order to anaesthetise them correctly and thus restore supraclavicular nerve block as being reproducible, effective on the entire brachial plexus.
Needle: 50–80 mm isolated, 22G.
Utility of Doppler: subclavian vessels, transverse cervical vessels, dorsal artery of the scapula, costocervical trunk.

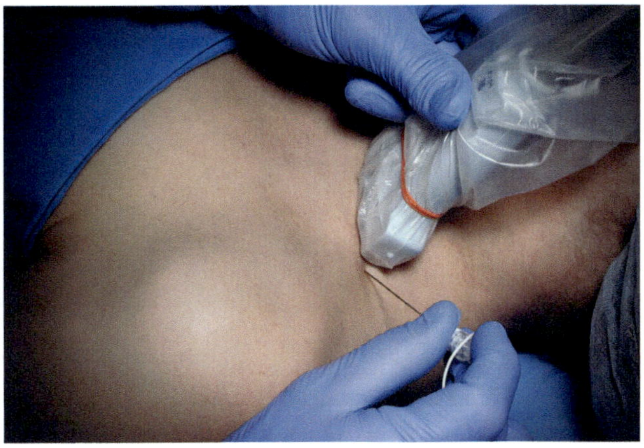

Fig. 5.50 Supraclavicular block. Position of the probe parallel to the clavicle, injection in-plane, lateral approach

Echoanatomy
The brachial plexus, at this level, is cephalic and lateral in relation to the subclavian artery, which generally is in close contact with the first rib and the pleura (Figs. 5.51 and 5.52). Very frequently, the plexus is crossed by the dorsal artery of the scapula (which should be sought), a branch of the subclavian artery or of the thyrocervical trunk, whose diameter can be relatively large, and which generally continues between the trunks of the plexus (Figs. 5.53 and 5.54). Generally, the latter is "tied" by the transverse artery of the neck, also a branch of the thyrocervical trunk in cephalic position. It is sometimes also possible to locate the costocervical trunk which arises from the posterior aspect of the subclavian artery and which can be in close contact with the brachial plexus in its medial part. These arteries, most of the time together with the veins, can be located with Doppler ultrasound to avoid accidentally injuring them (Fig. 5.55). Note that cases of accidental injection of a blood vessel have been reported in spite of an in-plane injection [48, 49]. In the brachial plexus, it can be observed that the nerve roots of C8-T1, which comprise the lower trunk, are located more in proximity to the subclavian artery, generally in the deep layer, against the first rib.

Pathways of Approach

As with the interscalene block, the supraclavicular block of the brachial plexus should be considered differently to the technique with neurostimulation. It is possible primarily to differentiate the lateral approach and the medial approach. Both involve the trajectory of the needle in the direction both of blood vessels and of the lung. Since the latter structures

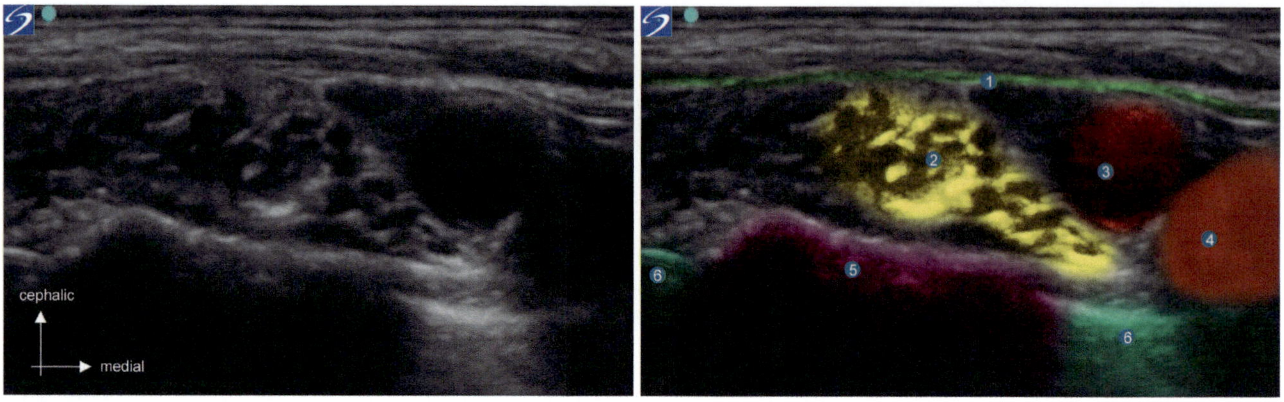

Fig. 5.51 Ultrasound section at the supraclavicular level

can be readily identified with ultrasound, they are avoided by maintaining careful strict visual control of the needle.

For each of the two approaches, the probe is positioned in the supraclavicular fossa, parallel to and immediately above the clavicle. Use of a small probe is comfortable and sometimes necessary in patients with a small supraclavicular fossa (children). The medial approach is often more useful in patients with a deep supraclavicular fossa or a protruding trapezius muscle which hinders the trajectory of the needle from the lateral end of the probe.

Lateral Approach

Nerve Localisation

The first phase consists of confirming the exact nature of the structures visualised by scanning between the interscalene and supraclavicular regions. Thus, it is possible to see, during descending scanning, that the roots C5, C6 and C7 progressively become superficial in the interscalene approach, comprise the upper and middle trunks, form anterior and posterior divisions and give rise to their first collaterals and are joined by C8 and T1, lastly, to regroup in a cranial lateral position in relation to the subclavian artery (Fig. 5.51). This "elevator" technique enables to define the outline of the plexus.

Needle Insertion

The needle is inserted, after local anaesthesia infiltration of the skin, at the lateral end of the probe (Fig. 5.50). It should be inserted and guided in the ultrasound plane to remain visible constantly along its length. In the event that vascular structures have been visualised in the plexus, they are avoided by directing the ultrasound plane and thus the pathway of the needle, either to no longer see them and consequently to not cross them, or on the contrary, to visualise them well and thus be able to control them.

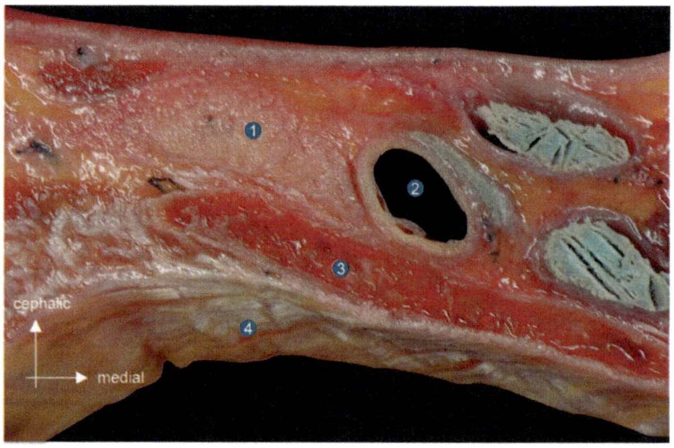

Fig. 5.52 Anatomical relations of the brachial plexus with blood vessels and pleura in the supraclavicular area. (Iconography: Admir Hadzic)

Injection

The elements/components of the plexus at this level are all stuck together and it is almost impossible to differentiate a small nerve from a large bundle. Therefore, it is entirely possible, by advancing the needle, in this "mass" of nerve structures, to involuntarily perform an intraneural puncture followed by intraneural injection, even though it is possible for

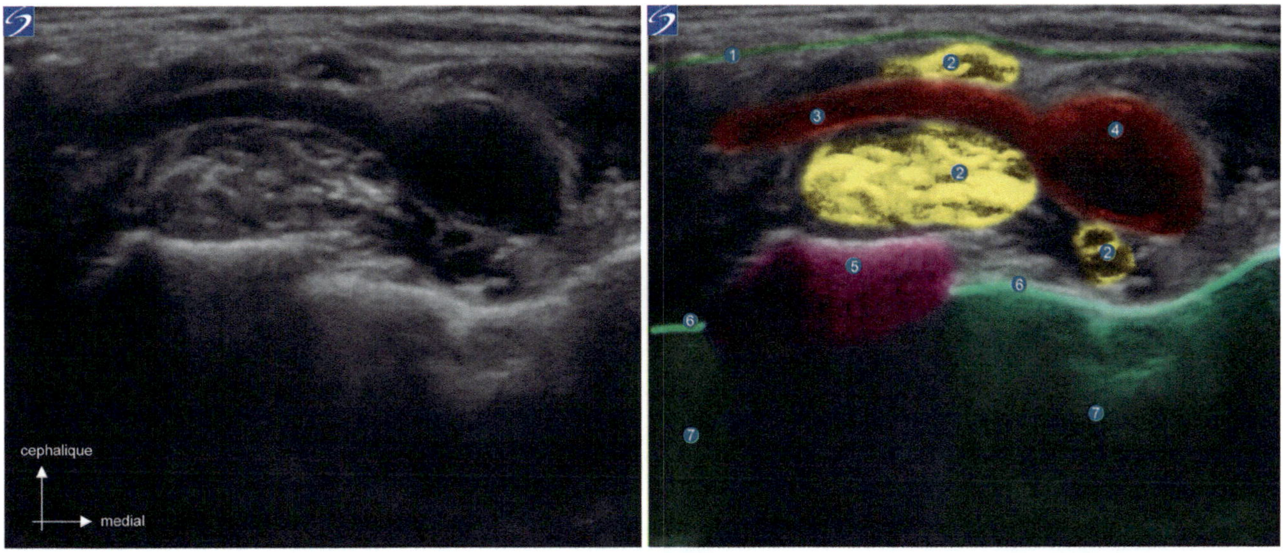

1. superficial layer of cervical fascia 2. brachial plexus
3. dorsal scapular a. 4. subclavian a.
5. first rib 6. pleura 7. lung

Fig. 5.53 Dorsal scapular artery crossing the brachial plexus at the supraclavicular level

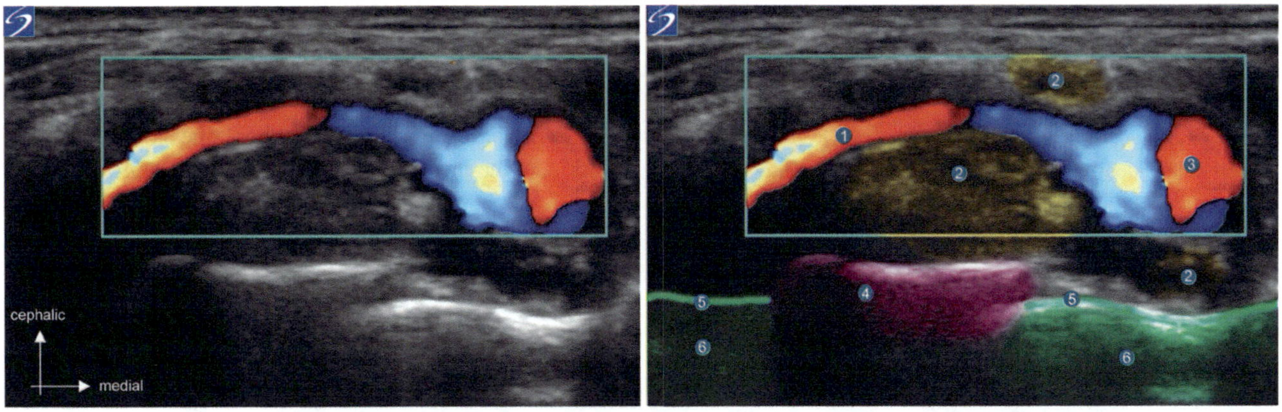

1. dorsal scapular a. 2. brachial plexus
3. subclavian a. 4. first rib
5. pleura 6. lung

Fig. 5.54 Dorsal scapular artery at the supraclavicular level. Confirmation by Doppler ultrasound

the clinician to think that the needle is located between the nerves. Consequently, **the author strongly suggests not** to place the needle in the middle of the plexus and to inject right into it in this manner, contrary to what may sometimes be suggested by other commentators. Thus, the safest approach consists of starting the injection on the surface of the plexus (Fig. 5.56) and possibly, depending on spread of the local anaesthetic, to redirect it in order to optimise its distribution in contact with the nerve branches. If the objective is to perform anaesthesia on the entire brachial plexus (including C8-T1), the needle should be cautiously redirected towards the roots C8 and T1 which lie in the deep layer of the plexus, generally in the confines of the superficial plane of the first rib and of the subclavian artery (Fig. 5.57). This procedure is done by avoiding the plexus at its lateral edge, taking care to avoid any risk to the adjacent pleura and lung. It is often possible to make this procedure safe by positioning the probe in a manner so that the plexus is placed on the first rib and so that the latter serves as a "barrier" to the pleura and to the lung (Fig. 5.58).

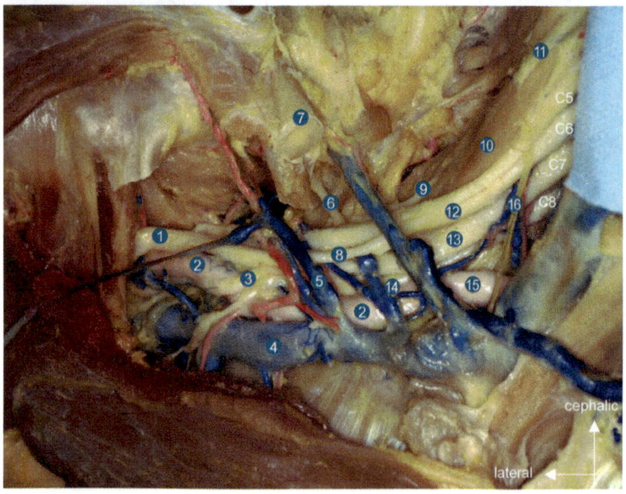

1. lateral cord
2. axillary a.
3. medial cord
4. axillary v.
5. cephalic a.
6. n. to subclavius m.
7. clavicle (removed)
8. anterior division of upper trunk
9. dorsal scapular n.
10. middle scalene m.
11. suprascapular n. (removed)
12. upper trunk
13. middle trunk
14. inferior trunk
15. subclavian a.
16. phrenic n.

Fig. 5.55 Dissection of the brachial plexus in the supraclavicular level. Anatomical relations with blood vessels. (Dissection: Bertrand Fabre)

Medial Approach

The probe is positioned in the supraclavicular fossa, generally slightly more laterally than the lateral approach. This approach can be useful when the patient has a trapezius muscle that is especially protruding and/or when the supraclavicular fossa is deeper.

Nerve Localisation

Just as with the lateral approach, it is necessary first to determine the exact position of the plexus and of its general outline. This step makes it possible to position the probe in "strategically": the puncture point should enable the trajectory of the needle to lie between the subclavian artery and the plexus, which will facilitate injection of the local anaesthetic more in contact from C8 to T1, and thus anaesthetise the corresponding areas. This technique is similar to supraclavicular block by inter-sternocleidomastoid approach although, with ultrasound guidance, the injection point generally is lateral compared to the clavicular head of the sternocleidomastoid muscle.

Needle Insertion

The needle is inserted after local anaesthesia infiltration of the skin, at the medial end of the probe (Fig. 5.59). It should

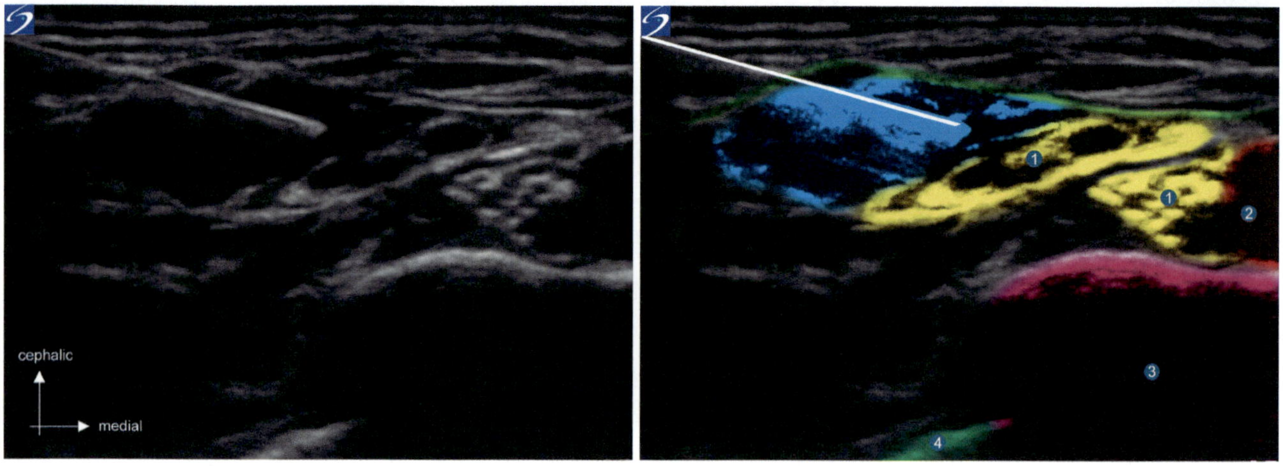

1. branchial plexus
2. subclavian a.
3. first rib
4. pleura and lung

Fig. 5.56 Supraclavicular block. Injection of local anaesthetic superficial to the plexus

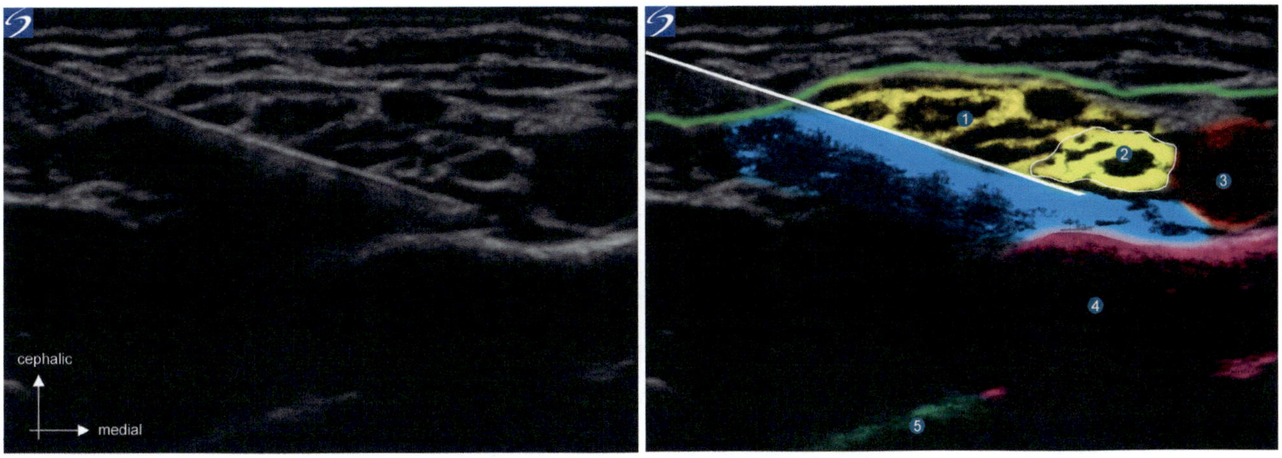

1. area of upper and middle trunks (C5-C6-C7)
2. area of inferior trunk (C8-T1)
3. subclavian a.
4. first rib
5. pleura and lung

Fig. 5.57 Supraclavicular block. Needle in contact with nerve roots C8-T1

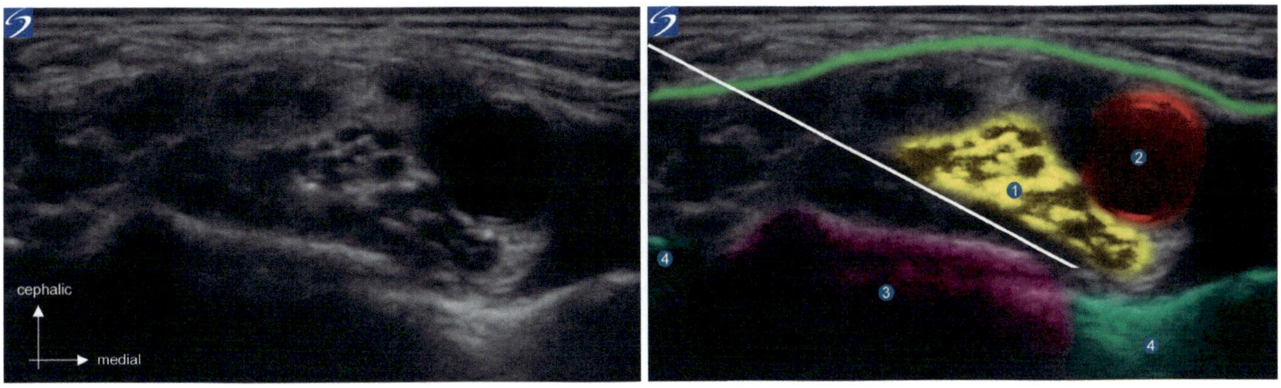

1. brachial plexus
2. subclavian a.
3. the first rib doesn't protect the pleura from needle progression
4. pleura and lung

Fig. 5.58 Supraclavicular block. Needle in contact with nerve roots C8-T1. Safe procedure positioning the probe such that the plexus lies on the first rib acting as a "shield" for the pleura and lung

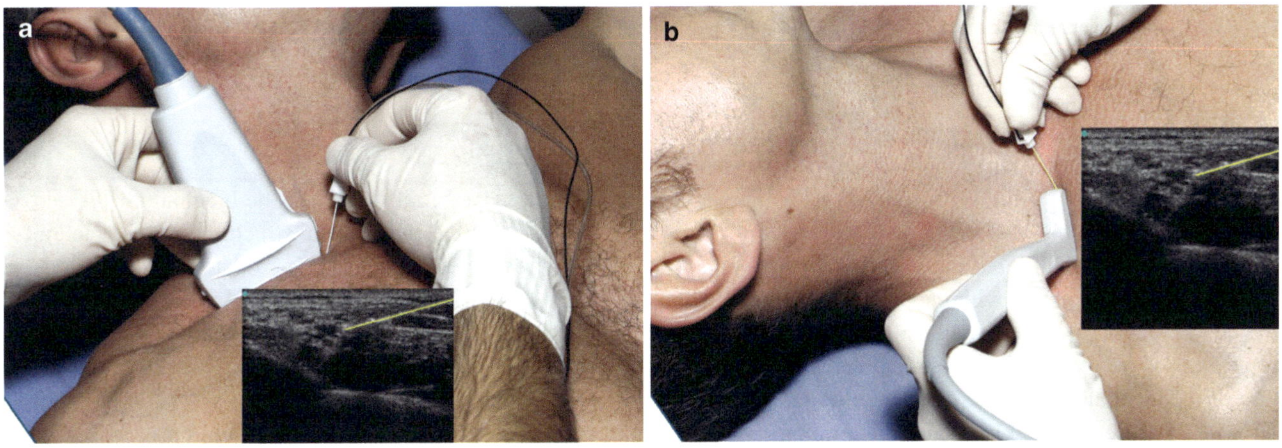

Fig. 5.59 Supraclavicular block: medial approach with two different probes (as this is a demonstration, the probe has not been covered with a dedicated cover that would be used during the actual performance of the block)

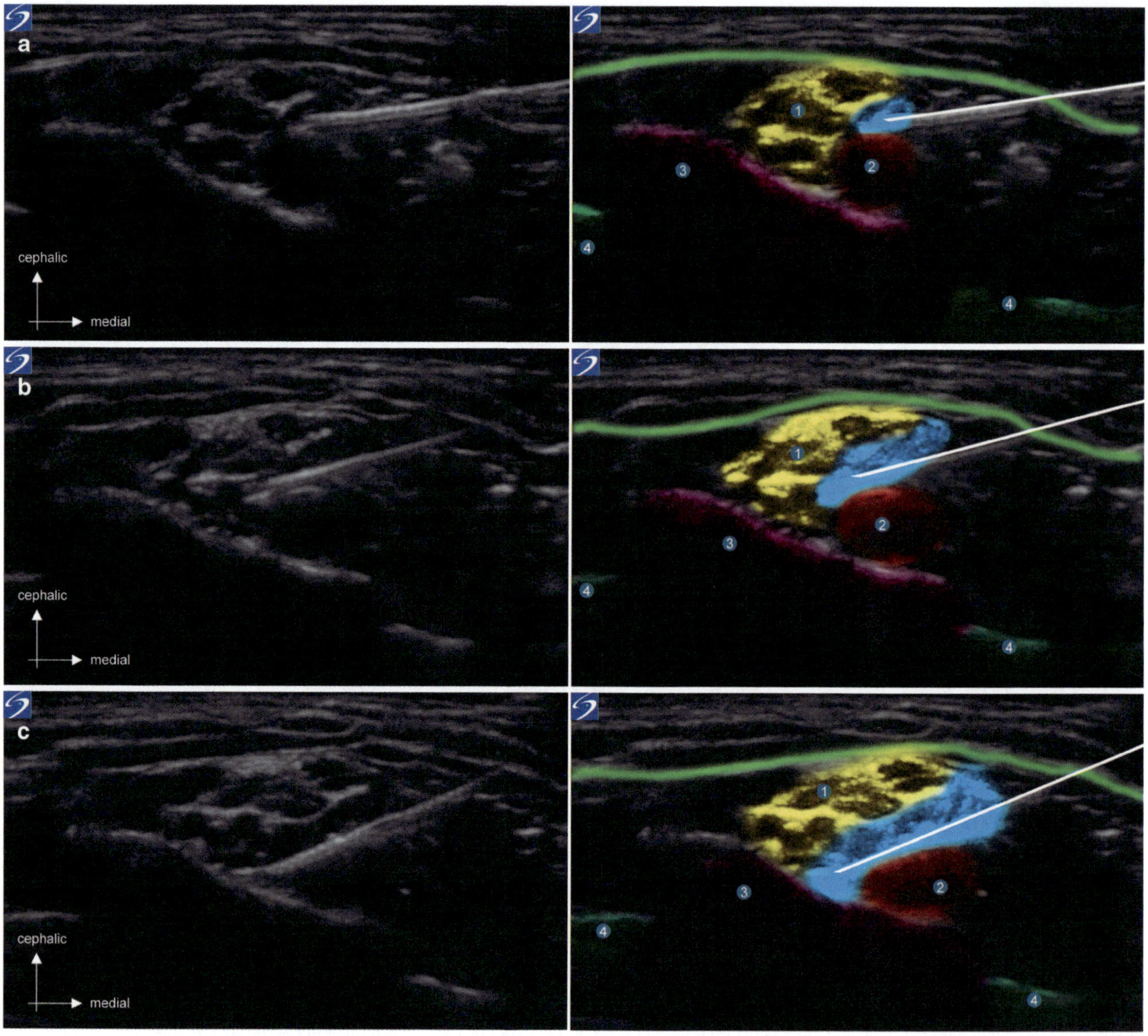

1. brachial plexus
2. subclavian a.
3. first rib
4. pleura and lung

Fig. 5.60 Supraclavicular block: medial approach. Delicate progression of the needle between medial border of the plexus and subclavian artery

be inserted and guided in the ultrasound plane in order to remain visible along its entire length (Fig. 5.60).

Injection

The injection is administered in contact with the elements/components of the plexus, with the possibility of redirecting the needle and the spread of the local anaesthetic during the procedure, if necessary. This approach can facilitate injection of the local anaesthetic in contact with the components C8 and T1 (inferior trunk) by carefully sliding the needle between the subclavian artery and the medial limit of the plexus, and by avoiding crossing the remainder of the plexus to reach them (Fig. 5.60c).

Paediatrics

The close proximity between the brachial plexus and the "susceptible" anatomical structures, i.e. the pleura, subclavian vessels and their collateral branches, is such so that this block is rarely used in paediatric practice with neurostimulation alone. The visual control dimension provided by ultrasound can make it possible to enhance the place given to this approach in the arsenal of peripheral nerve blocks in children.

Infraclavicular Block (Fig. 5.61)

Indications

Indications for infraclavicular block (ICB) are upper limb surgeries, from the proximal third of the humerus to the hand. Cases of trauma to the upper limb is the indication of choice because of the advantage of not moving the arm for performing the block, whatever the technique used.

There are many traditional pathways of approach. Their classification will depend on their direction in relation to the plexus, that is, their position in the upper part of the thorax:

- The perpendicular pathway (Kilka [50, 51] and subcoracoid approach [52]) or tangential to the brachial plexus (Raj [53], Sims [54] and Borgeat [55] pathways)
- The very medial pathways (Kilka) or extra-thoracic (subcoracoid)

In the medical literature, the risk of pneumothorax is low and these approaches are indicated for insertion of an indwelling catheter. The area of insertion is hairless, relatively immobile and clean, with very low risk of infection. Transmuscular injection, crossing the pectoralis major muscle, seems to be the psychological limit of subclavicular approaches. These blocks are deep and consequently are contraindicated in the presence of coagulation disorders because of difficulty in compressing a blood vessel which has been injured by direct needle puncture.

Infraclavicular nerve block is indicated primarily for trauma surgery of the arm, elbow or forearm. For insertion of catheters, a tangential technique will be chosen. With neurostimulation only, multi-stimulation makes it possible to decrease the total volume of local anaesthetic, to increase the success rate and accelerate the time to onset of the nerve block [6]. Injection of a local anaesthetic around the different cords of the plexus has this advantage. The use of ultrasound is an essential too; to guide the injection around the artery, on the three cords of the brachial plexus. In fact, by observing spontaneous spread of the local anaesthetic, we note that the latter is not distributed systematically in contact with the three components, including when the injection is administered deep to the axillary artery in immediate proximity to the posterior cord. However, it is here that is advisable to inject a local anaesthetic.

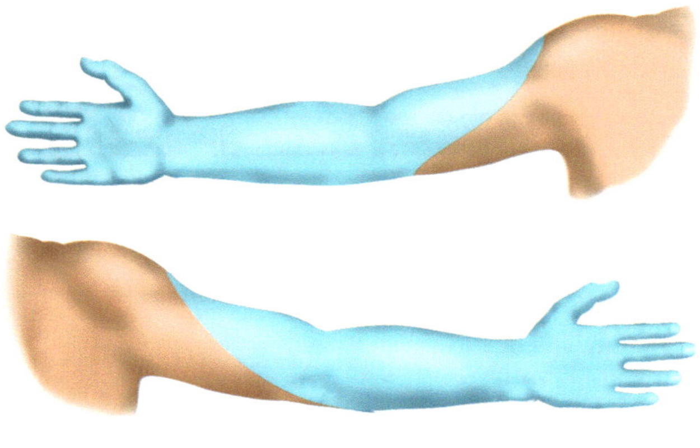

Fig. 5.61 Area of coverage of the infraclavicular block

The indication for conduct of an infraclavicular block on the side of an arteriovenous fistula should be weighed conscientiously in light of the impossibility of exerting pressure in case of a haematoma. In the event that another anaesthetic technique is not feasible, it is absolutely necessary to use rigorous ultrasound-guided technique.

Type of probe: linear, 5–10 or 6–13 MHz.
Axis of probe: parasagittal (Fig. 5.62).
Configuration: nerves in the short axis, needle in plane.
Studied depth: 2–6 cm.
Neurostimulation: enables further identification of the nerves visualised and can limit, by determination of an MIS >0.3 mA (0.1 ms), the risk of accidental intrafascicular injection. Furthermore, it can supplement locating nerves when conditions of visibility are unfavourable.
Needle: 80 mm isolated, 22 G.
Utility of Doppler ultrasound: primarily axillary vessels, thoracoacromial, upper and lateral thoracic or cephalic vein.

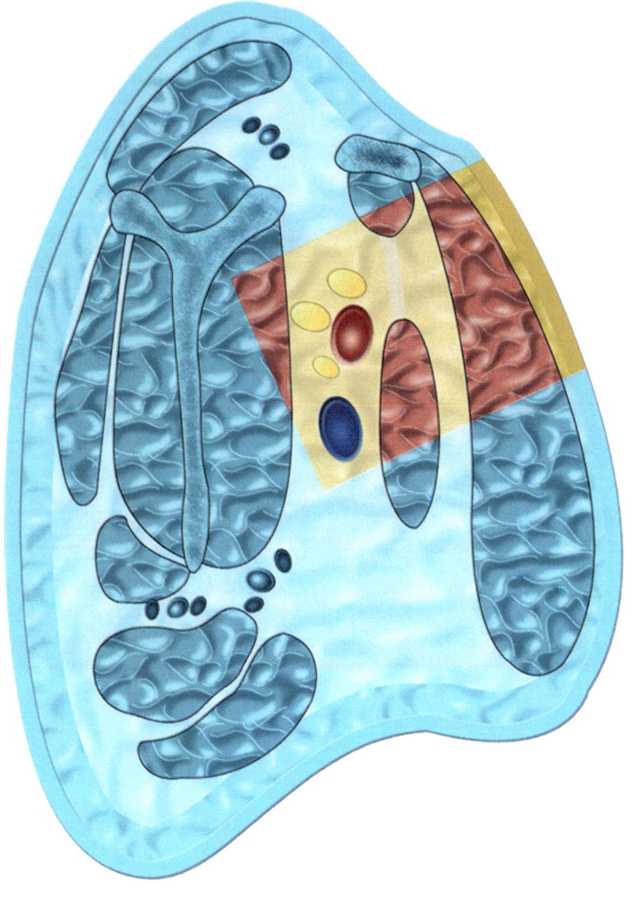

Fig. 5.62 Parasagittal section of the axillary fossa with materialisation of the ultrasound beam

Echoanatomy

As it enters the axillary region, the brachial plexus is in lateral position in relation to the blood vessels. All of this neurovascular bundle located deep to the pectoralis major and pectoralis minor muscles, continues in an oblique inferior pathway, running laterally and dorsally, and continues tangentially with the rib cage and then distances itself from it to enter the medial aspect of the arm. At the start of their journey through the axillary region, the lateral, medial and posterior cords are initially stuck together (Figs. 5.63 and 5.64), but they then rearrange around the blood vessels to position themselves laterally, medially and posteriorly in relation to the artery (Fig. 5.65). The terminal branches of the brachial plexus then arise opposite the shoulder joint. Major variations in position of the bundles around the blood vessels may possibly exist. Moayeri et al. [56] have showed that the relations of the nerve structures with respect to the axillary artery are changed as their "distal end" is approached, but that they also vary in patients in a given sectional plane (Fig. 5.66).

An infraclavicular block is qualified as deep also because of the difficulty regularly encountered in correctly visualising the different bundles. Perlas et al. identified them precisely in only 27% of cases [57].

Approach

The principal technique is an approach in the ultrasound plane, by cranial-caudal direction. The probe is positioned according to a parasagittal infraclavicular axis, its cranial end is placed just medially to the coracoid process. The ideal cutaneous puncture site lies between the clavicle and the cranial end of the probe, and once the cutaneous plane has been crossed, it makes it possible to direct it in contact with the three cords by simple redirection during the injection.

Parasagittal Probe, Needle in Plane

Nerve Localisation

Deep to the pectoralis major and minor muscles, the lateral, medial and posterior cords are organised around the axillary artery. By ultrasound scanning, it is necessary to determine their relative positions in the different levels of the infraclavicular area. Consideration must be given to the deep position of the plexus, the inability to apply pressure in the case of inadvertent vascular puncture, the existence and position of the thoracoacromial blood vessels and of the cephalic vein in particular. These are all important parameters in the choice of the injection site for an infraclavicular block (Fig. 5.67). The probe is gen-

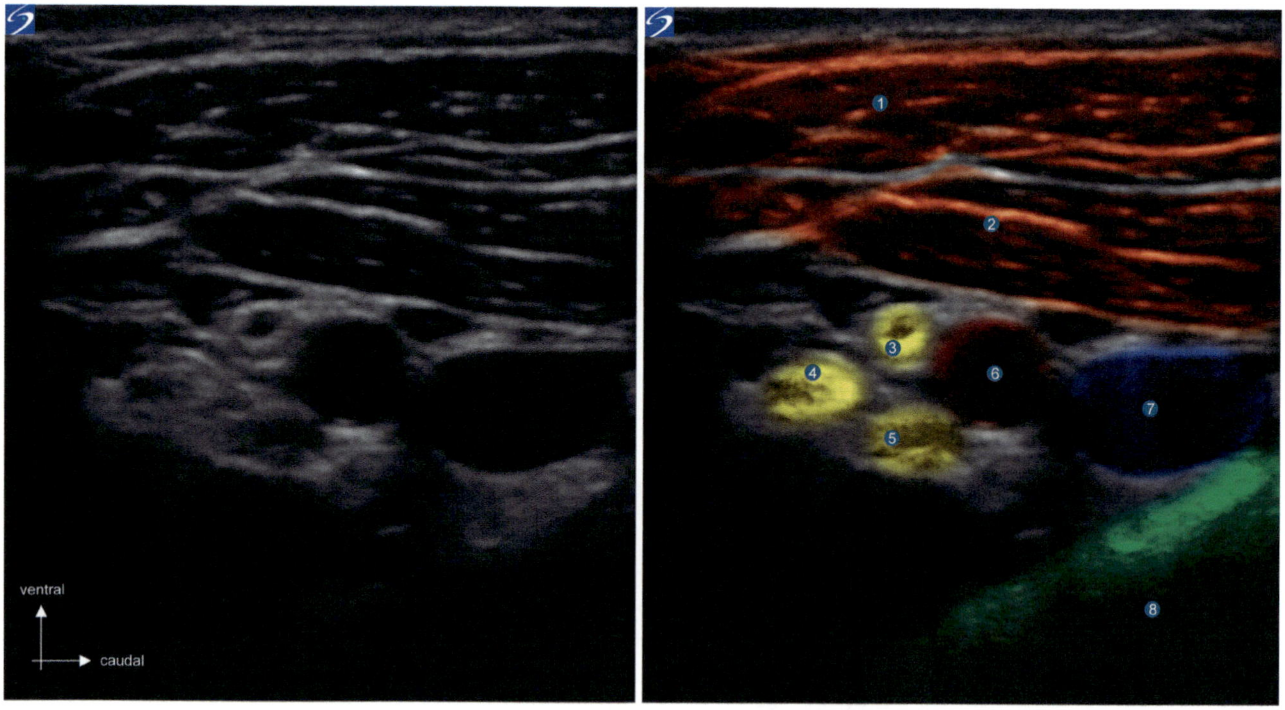

1. pectoralis major m.
2. pectoralis minor m.
3. lateral cord
4. posterior cord
5. medial cord
6. axillary a.
7. axillary v.
8. lung

Fig. 5.63 Sonoanatomical section at the infraclavicular level

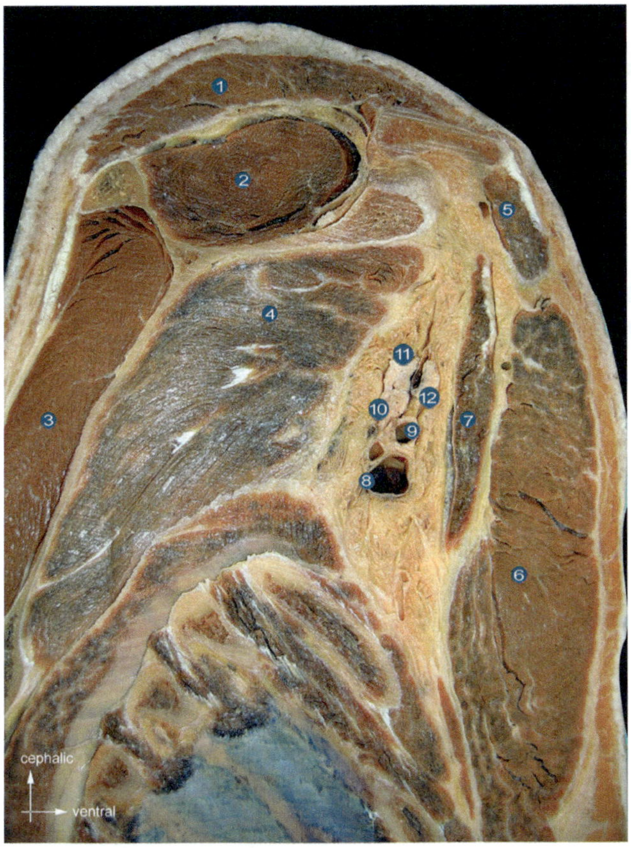

1. deltoid m.
2. supraspinatus m.
3. infraspinatus m.
4. subscapularis m.
5. subclavius m.
6. pectoralis major m.
7. pectoralis minor m.
8. axillary v.
9. axillary a.
10. medial cord
11. posterior cord
12. lateral cord

Fig. 5.64 Parasagittal section of the brachial plexus, axillary fossa

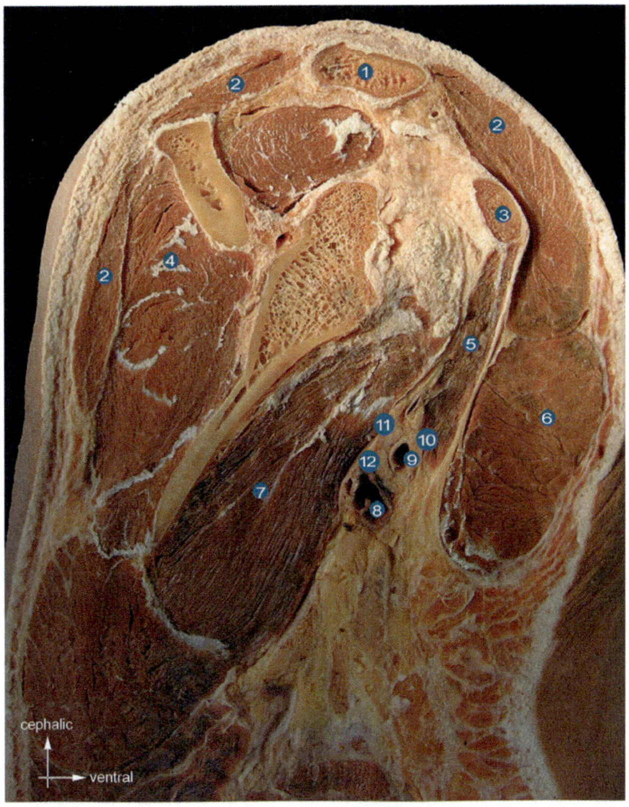

1. clavicle
2. deltoid m.
3. coracoid process
4. infraspinatus m.
5. pectoralis minor m.
6. pectoralis major m.
7. subscapularis m.
8. axillary a.
9. axillary v.
10. lateral cord
11. posterior cord
12. medial cord

Fig. 5.65 Parasagittal section of the brachial plexus cutting through the coracoid process

erally positioned just medially to the coracoid process (Figs. 5.62 and 5.68). Its axis is not systematically strictly parasagittal and may sometimes be directed more or less medially depending on if it is desired "to cut" the neurovascular axis perpendicularly. The pectoralis muscles, the axillary artery and vein, their efferent and afferent vessels, and the nerve structures are visualised (the cords and collateral branches of the brachial plexus) and, more caudally, the lung, pleura (Fig. 5.69), as well as most of the time, the second rib.

Needle Insertion

The ideal cutaneous puncture point, at the cephalic end of the probe, should enable the three cords of the plexus to be reached by means of successive needle redirections. After local anaesthesia infiltration of the skin, the needle

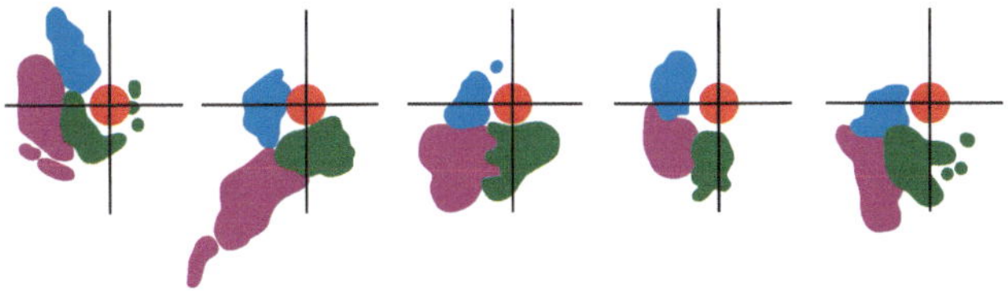

• lateral cord • posterior cord • medial cord

Fig. 5.66 Distribution of the cords of the brachial plexus around the axillary artery. (From [56])

crosses successively the pectoralis major and the pectoralis minor muscles. It is only once the deep aponeurosis of the pectoralis minor muscle has been crossed that neurostimulation is started, thus making it possible to avoid patient discomfort related to stimulating the pectoral muscles. The high angulation of the needle necessary to reach

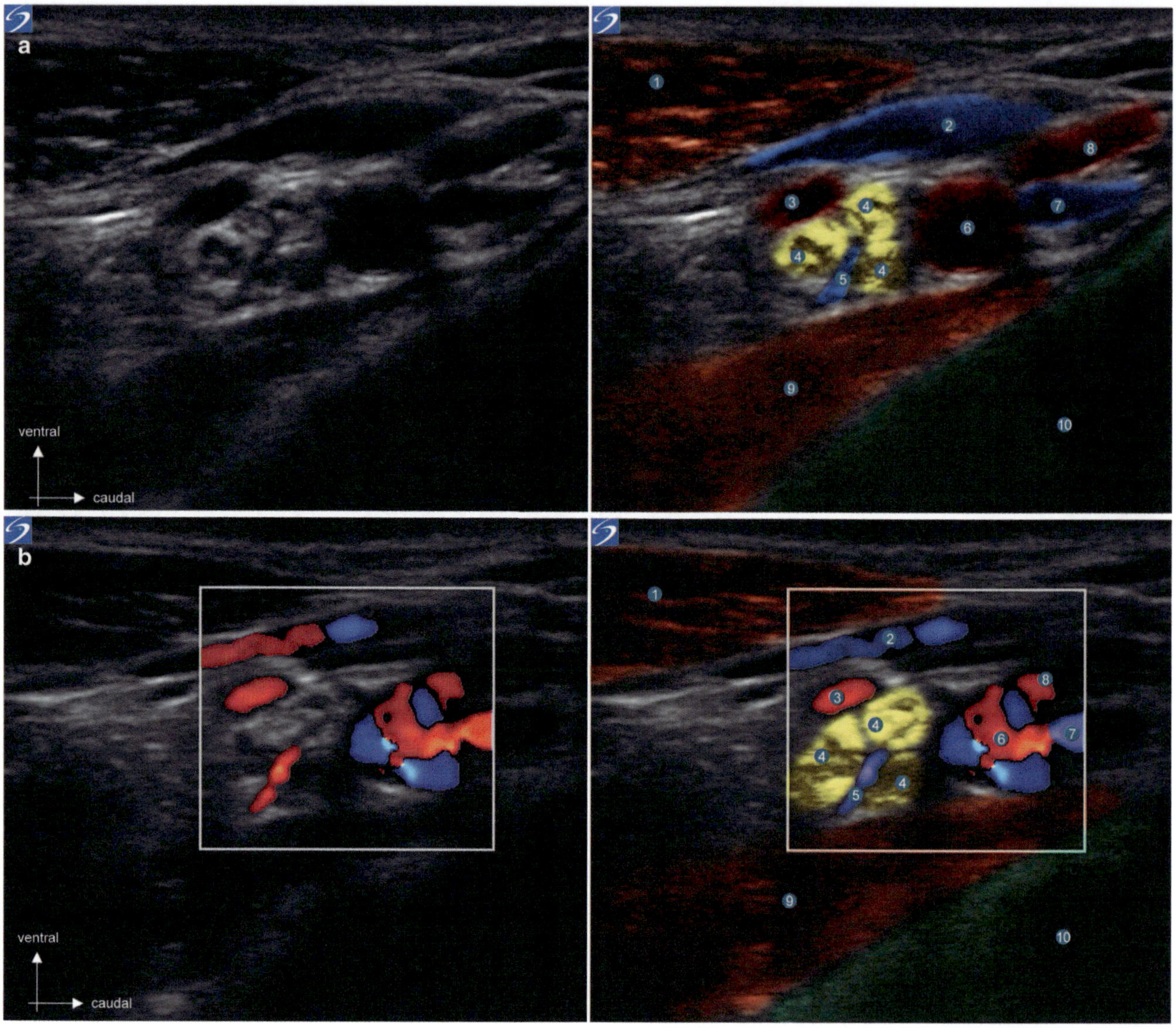

1. pectoralis minor m.
2. cephalic v.
3. d ltoid branch of the thoracoacromial a.
4. cords of brachial plexus
5. vein between the cords of brachial plexus
6. axillary a.
7. axillary v.
8. pectoral branch of the thoracoacromial a.
9. serratus m.
10. lung

Fig. 5.67 B mode (**a**) ultrasound and colour Doppler (**b**) ultrasound section of the infraclavicular region showing the many vascular structures that can be found in it

Fig. 5.68 Ultrasound-guided infraclavicular block: position of the probe. Needle inserted in-plane

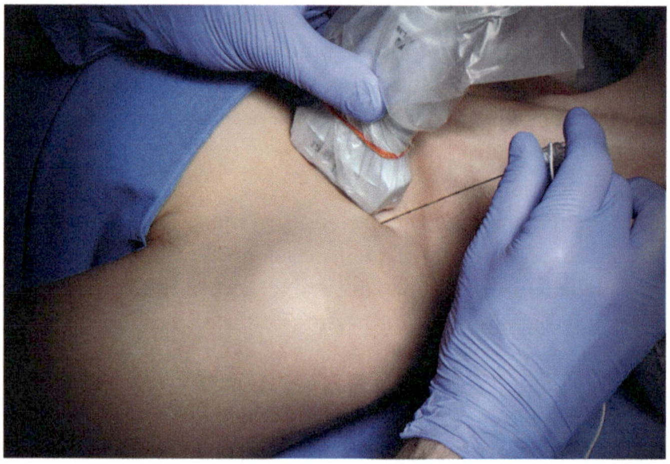

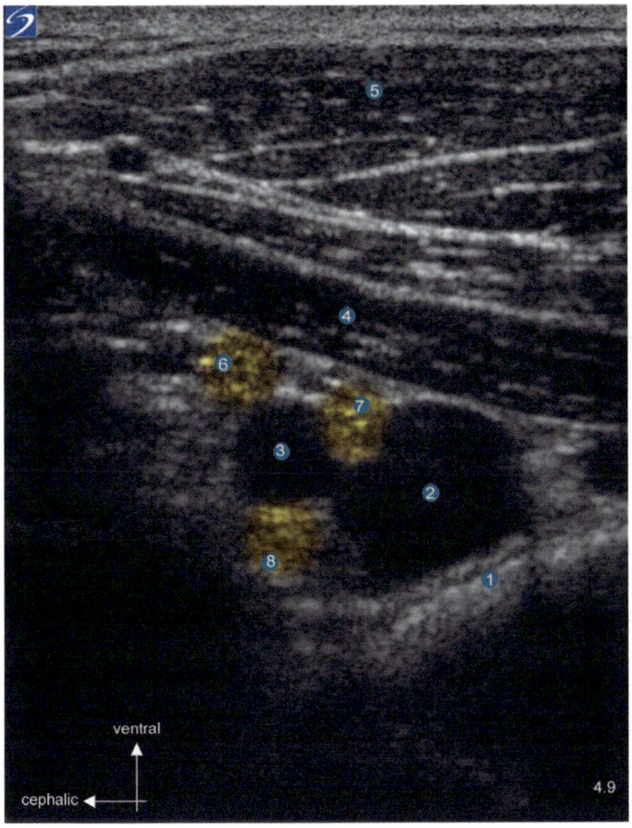

1. pleura 2. axillary v. 3. axillary a.
4. pectoralis minor m. 5. pectoralis major m. 6. lateral cord
7. medial cord 8. posterior cord

Fig. 5.69 Parasagittal infraclavicular ultrasound section of the brachial plexus

the target area encourages to use a so-called "hyperechoic" needle for this block. Additionally, neurostimulation is a useful safety aid during conduct of an ICB, especially considering the frequently poor visibility of the cords of the brachial plexus. Ideally, the objective is to position the tip of the needle in the dorsal aspect of the axillary artery, in proximity to the posterior cord. Generally, this leads to good local anaesthetic spread around the artery, in contact at least with the posterior and medial cords, but often of the three components and a complete block of the brachial plexus [58, 59]. By going from the superolateral border of the axillary artery to reach this position, the tip of the needle will first encounter the lateral cord. It is necessary to circumvent it dorsally, visually, and/or with the aid of neurostimulation ("median" and/or "musculocutaneous" responses). At this level, it is also possible to perform hydrodissection by injecting a few millilitres of a D5W solution which, by reinforcing the contrast between the different tissues, enables better identification of the lateral cord and facilitates its avoidance. It is necessary to be careful to avoid accidental pleural, pulmonary or vascular injury, by always keeping in mind the position of the tip of the needle and the structures to be avoided.

Injection

Once the needle visually is in place, obtaining "radial" responses to neurostimulation confirms the proximity of the posterior cord, and thus the correct position of the needle tip. However, it is also possible to observe other types of response (median or ulnar) if the medial cord is stimulated, often still

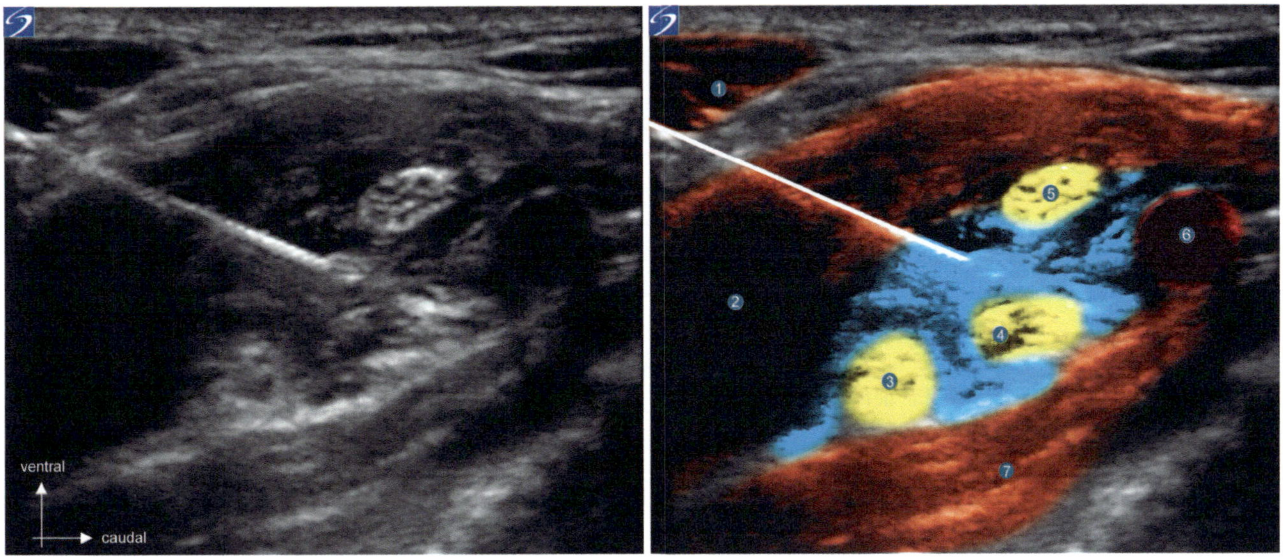

1. pectoralis major m. 2. pectoralis minor m.
3. posterior cord 4. medial cord
5. lateral cord 6. axillary a. 7. serratus m.

Fig. 5.70 Infraclavicular block. Spread of the local anaesthetic in contact with the cords

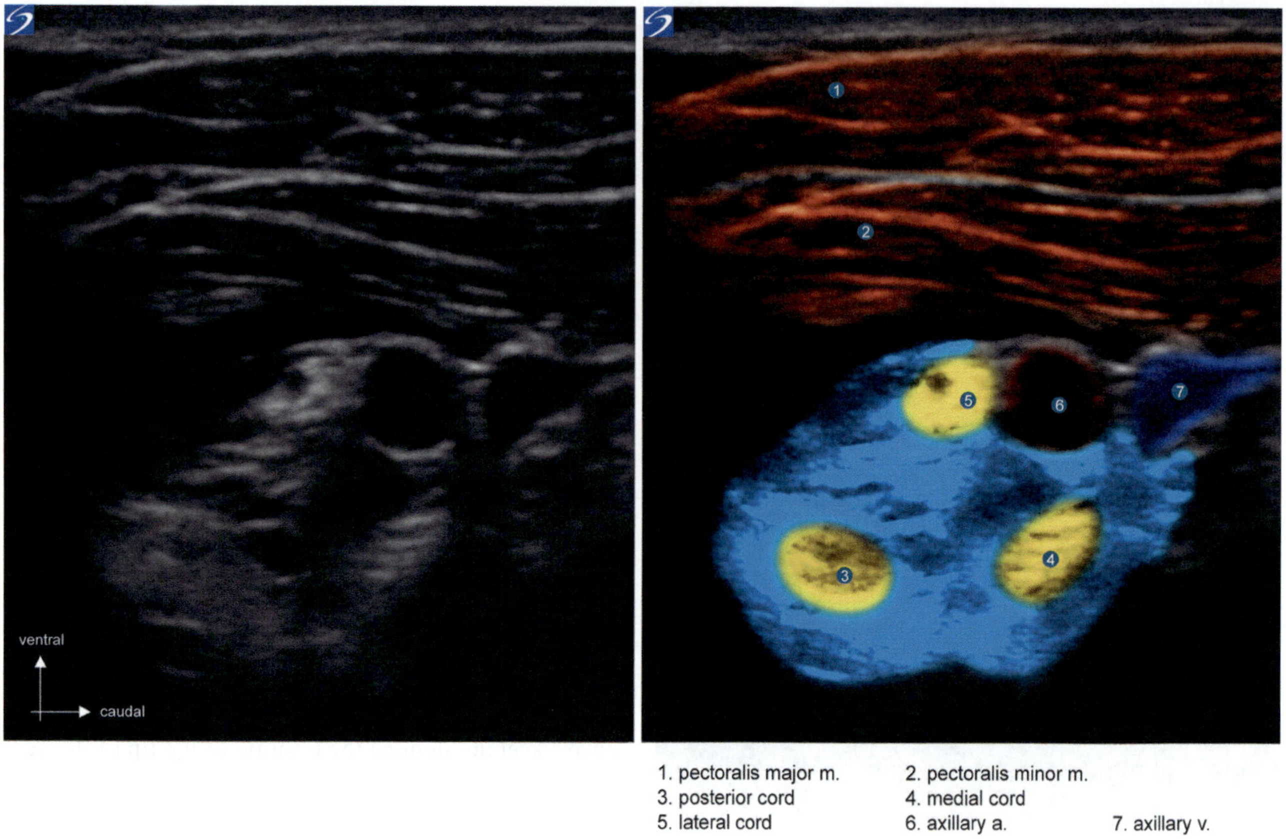

Fig. 5.71 Posterior surrounding of the axillary artery by the local anaesthetic injected with spread in contact with the three cords

placed behind the axillary artery (Fig. 5.70). The appearance of spread of the local anaesthetic is predictive of the efficacy of the block. Therefore, a main diffusion close to the posterior aspect of the axillary artery is synonymous with high probability of spread of the anaesthetic to the three cords (Fig. 5.71). In case of posterior "partitioning" of the local anaesthetic, after removing the needle, it is also possible to inject around the lateral cord and then to position oneself in front between the axillary artery and vein, and this time to inject in close to the medial cord. By going to the anterior aspect of the axillary artery, the lateral pectoral nerve and/or anastomotic branch with the medial pectoral nerve (pectoral muscles loop) can be found. In adults, a maximum volume of local anaesthetic of 20 mL is injected.

Paediatrics

In paediatrics, the infraclavicular nerve block is little used. In fact, although the target structures often are less deep than in adults (Fig. 5.72), the risks that it carries (pneumothorax, vascular injection) do not enable it to be used in routine practice. The axillary approach in principle should be preferred to it [60].

Axillary Nerve Block (Fig. 5.73)

Indications

The axillary block is indicated for hand surgery up to the elbow. With neurostimulation only, it is effective in almost 90% of cases. A single injection performed at the lower border of the deltoid muscle makes it possible to anaesthetise all of the terminal branches of the brachial plexus at this level as they form around the axillary artery. The injection can also selectively target the musculocutaneous nerve, which often takes a divergent course at this level and the smaller cutaneous nerves which supply sensation to the arm and forearm which may be involved in surgery. Compared to the supraclavicular and infraclavicular techniques, this approach eliminates the risk of pneumothorax and of phrenic nerve paresis. The trans-arterial technique no longer is recommended because of its low success rate and the high risk of haematoma compared to neurostimulation (professional consensus). As the result of the rarity of complications with the axillary approach, it remains the block technique of choice for distal surgery. The only drawback in trauma cases is the need to position of the patient's arm in 90° abduction. In this situation it is preferable to perform the infraclavicular approach [61].

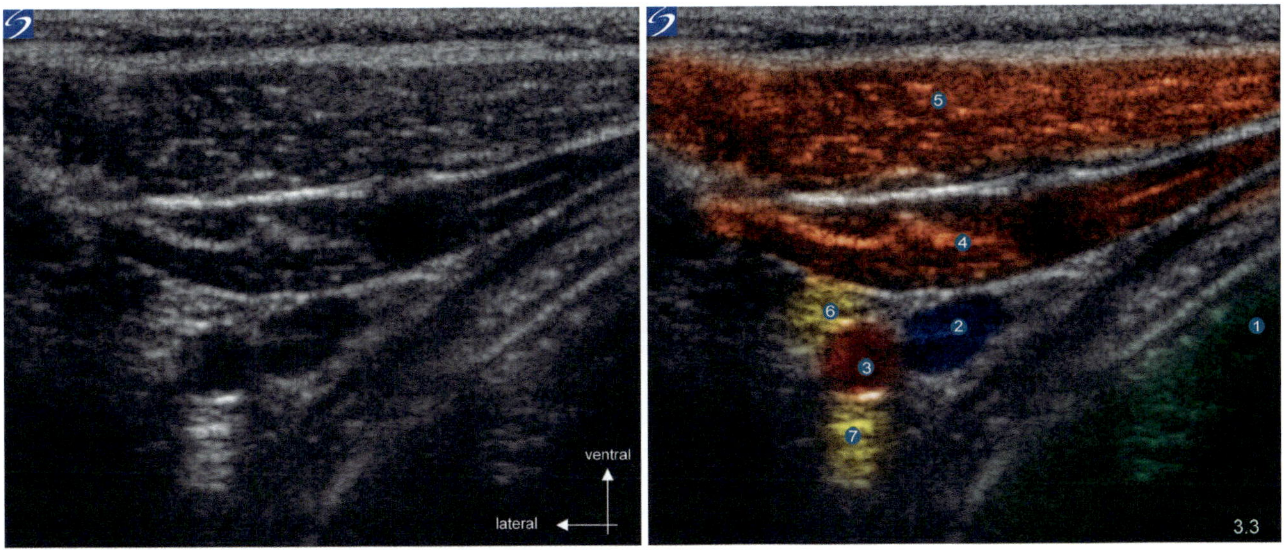

Fig. 5.72 Infraclavicular parasagittal ultrasound section in the brachial plexus in a 3-year-old child

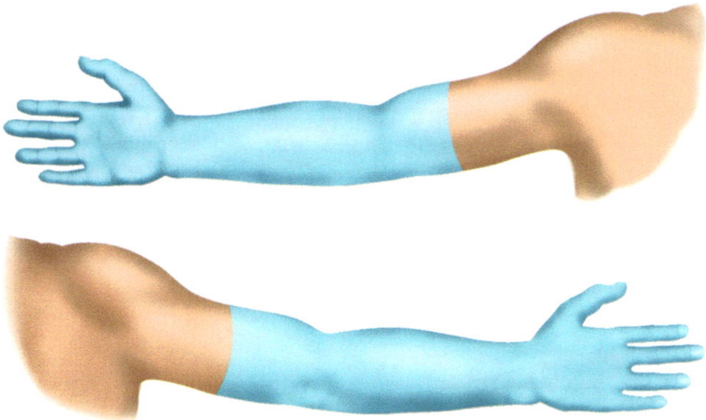

Fig. 5.73 Area of coverage of the axillary block

It is also possible to site perineural catheters at this level.

Type of probe: linear, 5–10 or 6–13 MHz.
Axis of probe: transversal/short axis (Fig. 5.74).
Configuration: nerves in the short axis, needle in or out of the plane.
Depth studied: from 1 to 5 cm.
Neurostimulation: enables further identification of the nerves visualised and can limit, by determination of an MIS >0.3 mA (0.1 ms), the risk of accidental intrafascicular injection. Furthermore, it can supplement locating nerves when conditions of visibility are unfavourable.
Needle: 80 mm isolated, 22 G.
Utility of Doppler ultrasound: the brachial artery and veins, basilic vein.

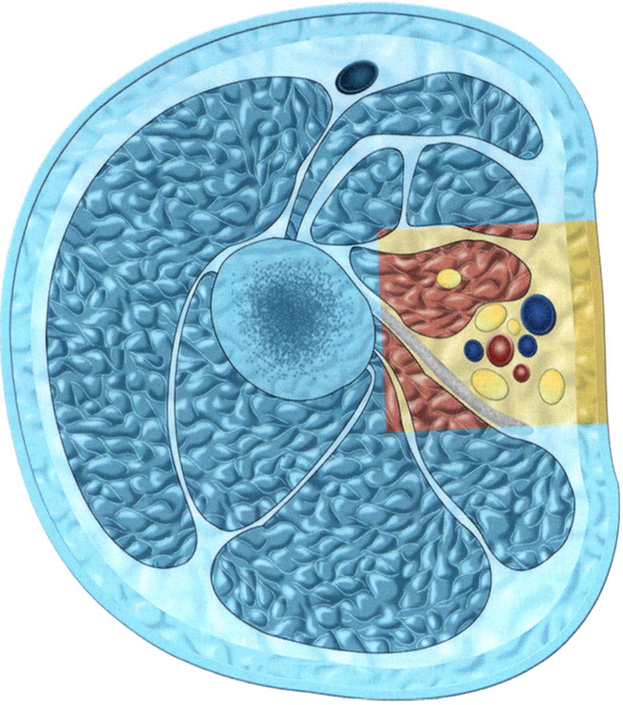

Fig. 5.74 Section of the proximal aspect of the arm with materialisation of the ultrasound beam for axillary block

Echoanatomy and Approach

In order to obtain a precise and comprehensive view of the neurovascular structures, the probe is positioned transversely (Fig. 5.75). The ultrasound study in the upper brachial and axillary area is facilitated by the shallow depth of these structures in the majority of patients. Moreover, the predominant muscle constitution of the arm makes contrast of inter-tissue echogenicity favourable in locating and following the blood

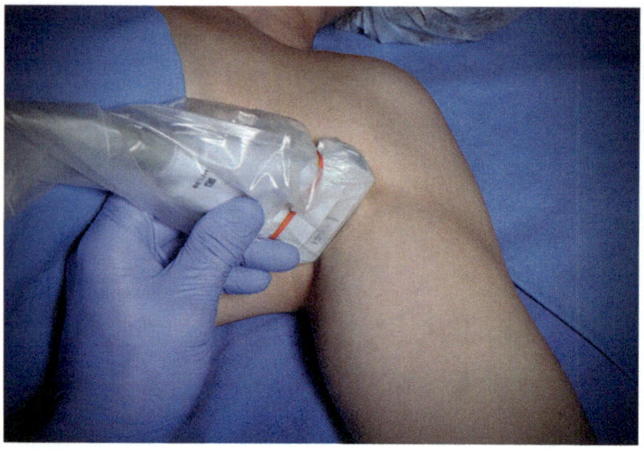

Fig. 5.75 Ultrasound-guided axillary block. Position of probe

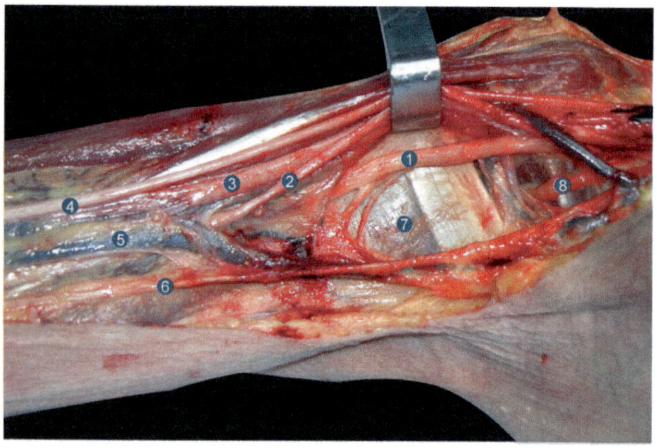

1. radial n. 2. ulnar n. 3. brachial a.
4. medial n. 5. basilic v.
6. medial antebrachial cutaneous n.
7. latissimus dorsi tendon 8. axillary n.

Fig. 5.76 Dissection of proximal end of the arm. Relations of the brachial plexus nerves with the tendon of the latissimus dorsi muscle. (Dissection: Bertrand Fabre)

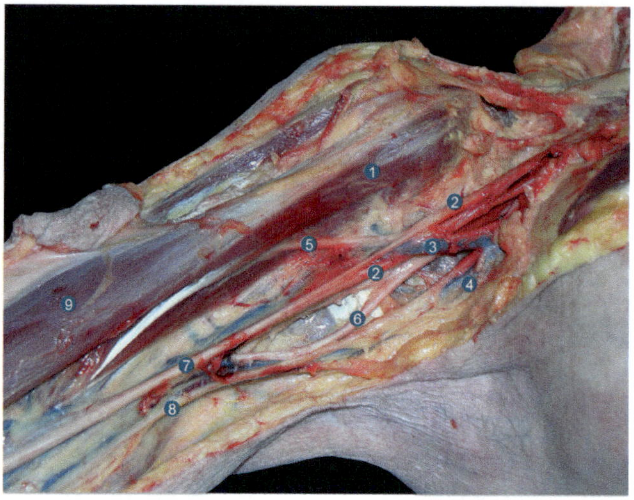

1. coracobrachialis m.
2. terminal branches of lateral and medial cord composing median n.
3. brachial v. 4. basilic v. 5. musculocutaneous n.
6. ulnar n. 7. median n. 8. medial antebrachial cutaneous n.
9. biceps brachii m.

Fig. 5.77 Dissection of the brachial plexus in the proximal aspect of the arm. (Dissection: Bertrand Fabre)

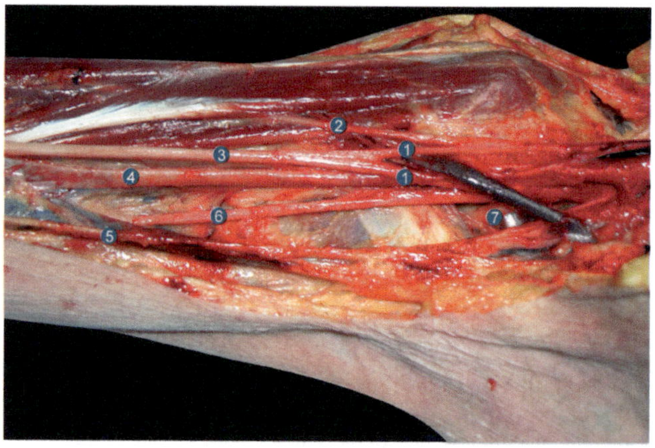

1. terminal branches of lateral and medial cord composing median n.
2. musculocutaneous n. 3. median n.
4. brachial a. 5. medial antebrachial cutaneous n.
6. ulnar n. 7. axillary n.

Fig. 5.78 Dissection at the proximal aspect of the arm. Relations of the brachial plexus nerves with the tendon of the latissimus dorsi muscle. (Dissection: Bertrand Fabre)

vessels and nerves. With ultrasound of the axillary fossa, with the patient's arm in abduction and probe perpendicular to the neurovascular axis (transverse), the terminal branches of the brachial plexus and the axillary blood vessels can be viewed together in a relatively compact neurovascular "bundle" (apart from the axillary nerve, which is no longer visible (Fig. 5.76) and the musculocutaneous nerve which often takes a divergent course at this level) (Figs. 5.77 and 5.78). However, in some cases the musculocutaneous nerve is immediately adjacent to the median nerve at this level. This observation can be compared to the cases of axillary block with pure neurostimulation during which the musculocutaneous nerve "cannot be found", and yet after injection of the local anaesthetic following a median nerve stimulus, it is anaesthetised by spread of the LA due to this close relationship. In many cases, precise identification of the nerve structures requires an ultrasound "status report": the distal and proximal translation of the probe (so-called "elevator" technique) is an excellent means to confirm identification of the anatomical elements. In fact, if ultrasound examination of a nerve is performed distally where its identity can be confirmed, it is then possible to track it proximally to determine its position in an area where its identification was less clear.

Just as for the more proximal approaches of the brachial plexus, the axillary approach enables anaesthesia of the terminal branches of the brachial plexus with a single cutaneous injection and a minimum of repositioning of the needle and a minimum of injections (apart from the subcutaneous injection for the medial cutaneous nerve in the arm).

In light of the size, position and orientation of the probe, the point of injection lies next to the anterior axillary fold (the groove which separates the biceps from the caudal border of the pectoralis major muscle). Therefore, this is a "proximal brachial approach", equivalent in terms of spread of the local anaesthetic to an "axillary" block with neurostimulator. For this reason, technically, the term *brachial* artery/vein should be used and not *axillary* artery/vein.

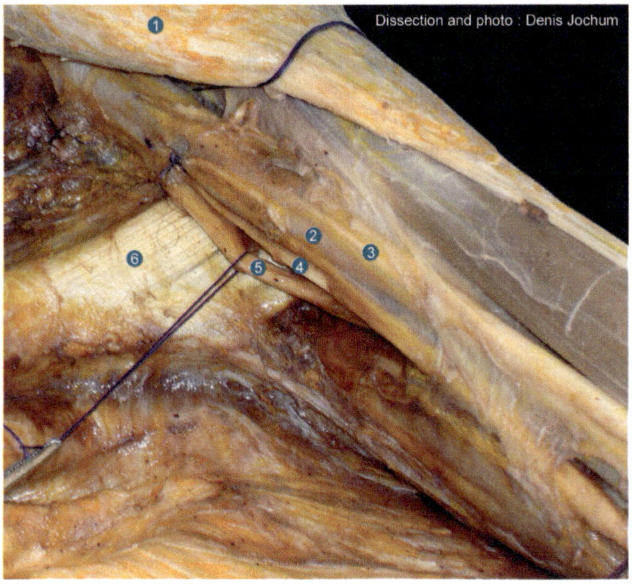

1. deltoid m. 2. brachial a. 3. median n.
4. ulnar n. 5. radial n. 6. latissimus dorsi tendon

Fig. 5.79 Position of the brachial plexus nerves overlying the tendon of the latissimus dorsi muscle

Nerve Localisation

To safely and successfully perform an axillary block is necessary to visualise and positively identify several structures: muscles, vessels (brachial artery and veins, basilic vein) and, of course, nerves. It is **essential** to locate the tendon of the latissimus dorsi muscle and its humeral insertion. Anterior to this tendon are combined all the terminal branches of the brachial plexus (apart from the axillary nerve) (Figs. 5.76, 5.77, 5.78 and 5.79). At the inferior border of this tendon the terminal branches of the brachial plexus begin their divergent pathways, particularly the radial nerve which quickly passes posteriorly into the radial groove of the humerus. If the clinician especially needs to target the radial nerve this position is ideal to perform the block.

By carefully positioning the probe, it is usually possible to visualise all the nerves to be anaesthetised on the same screen (Fig. 5.80). Therefore, from a single cutaneous injection site this enables an approach to each nerve with simple redirections of the needle.

Anatomical variations [62] are commonly detected during this initial scanning phase (Fig. 5.81) [63].

They can concern all the nerves in the brachial plexus. Note, for example in Fig. 5.82, that the median nerve is located at "6 o'clock" to the brachial artery. Figure 5.83 shows the median

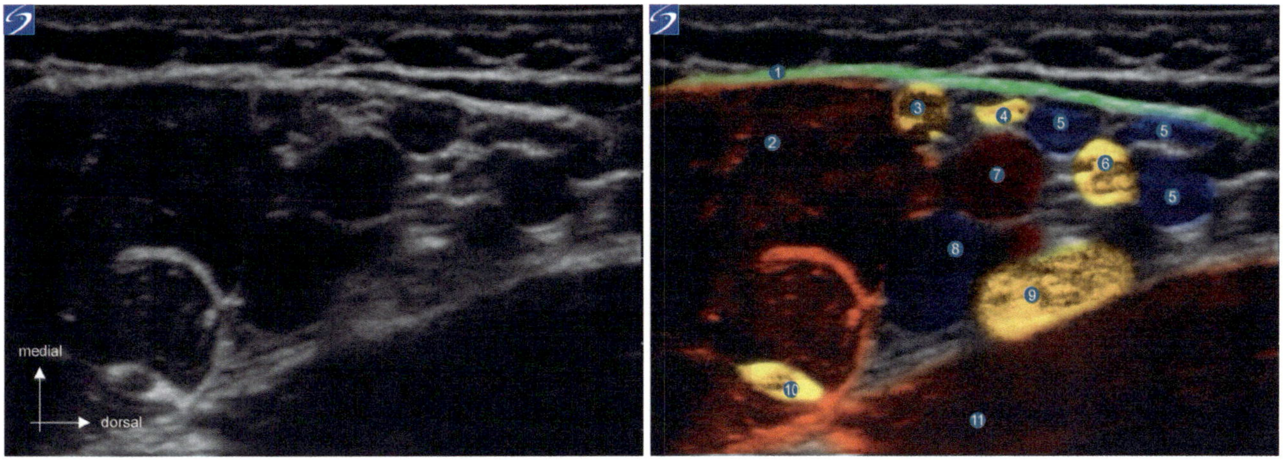

1. brachial fascia 2. biceps brachii and coracobrachialis mm.
3. median n. 4. medial antebrachial cutaneous n.
5. vv. 6. ulnar n.
7. brachial a. 8. brachial v.
9. radial n. 10. musculocutaneous n.
11. latissimus dorsi m.

Fig. 5.80 Ultrasound-guided axillary block: overall view of neurovascular structures before performance of block

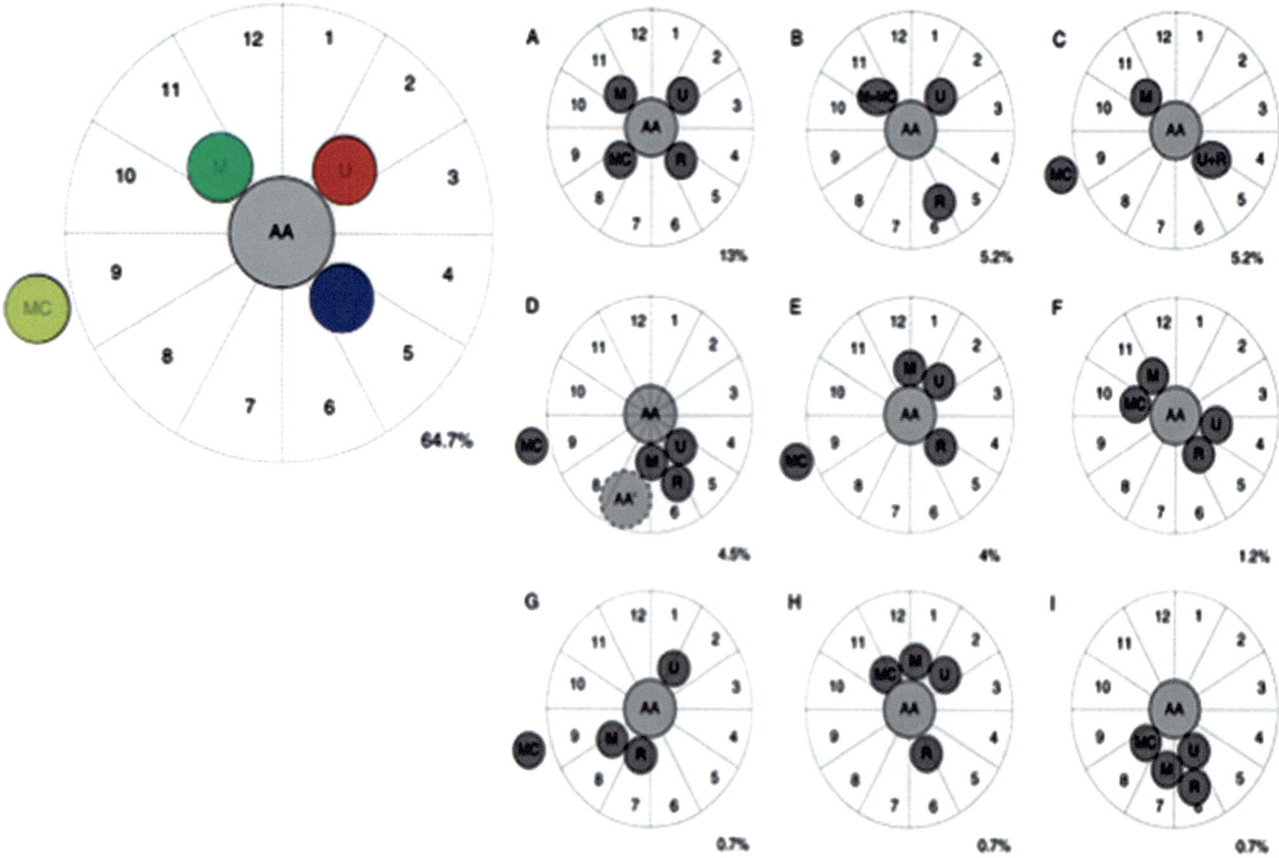

Fig. 5.81 Distribution of terminal branches of brachial plexus. (From [63])

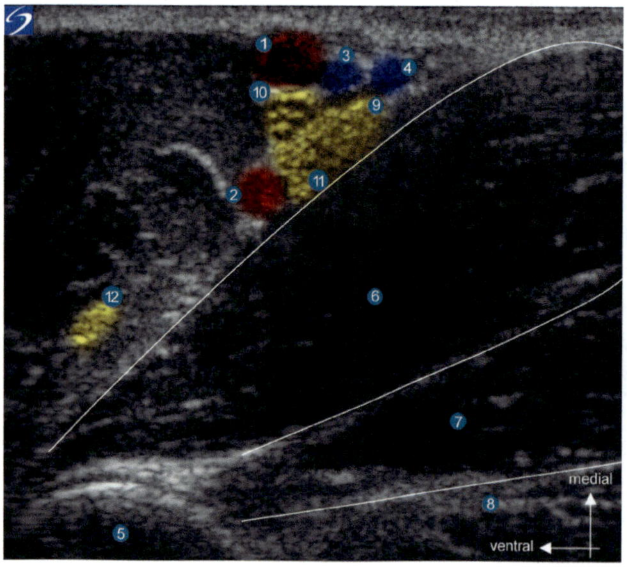

1. brachial a.
2. deep brachial (profunda brachii) a.
3. brachial v.
4. basilic v.
5. humerus
6. latissimus dorsi tendon
7. teres major tendon
8. long head of triceps brachii m.
9. ulnar n.
10. median n.
11. radial n.
12. musculocutaneous n.

Fig. 5.82 "Axillary" transverse ultrasound section showing median nerve deep to the brachial artery

nerve located at the posterior border of the brachial artery and Fig. 5.84 represents the median nerve at the proximal brachial level consisting of two separate components which then unite more distally to form the single median nerve.

The position of the musculocutaneous nerve is also variable [64]. The latter is often found distant from the remainder of the neurovascular bundle between the biceps brachialis and coracobrachialis muscles. This can often require repositioning of the probe during the block but rarely a second cutaneous injection. Conversely, it is sometimes often found in immediate proximity to the median nerve with which it can give the impression of a "merging" [65].

Injection

Once the probe has been positioned, there are two methods to perform this block.

Needle in Plane (Fig. 5.85)

With the probe placed in the anterior axillary fold, the needle is inserted after local anaesthesia infiltration of the skin at the anterior end of the probe. The injection point is relatively similar to that of the technique of the infraclavicular block described by Dalens in children [66]. In order to maintain optimum visibility of all neurovascular structures throughout the procedure, it is important to start by injecting the local

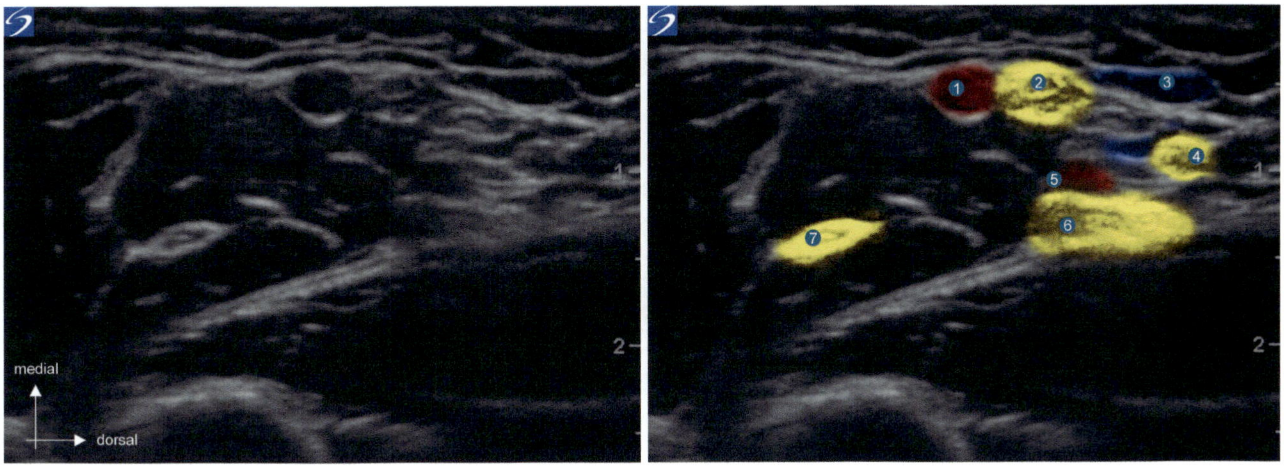

1. brachial a. 2. median n. 3. basilica v.
4. ulnar n. 5. deep brachial (profunda brachii) a.
6. radial n. 7. musculocutaneous n.

Fig. 5.83 "Axillary" transverse ultrasound section showing median nerve dorsal to the brachial artery

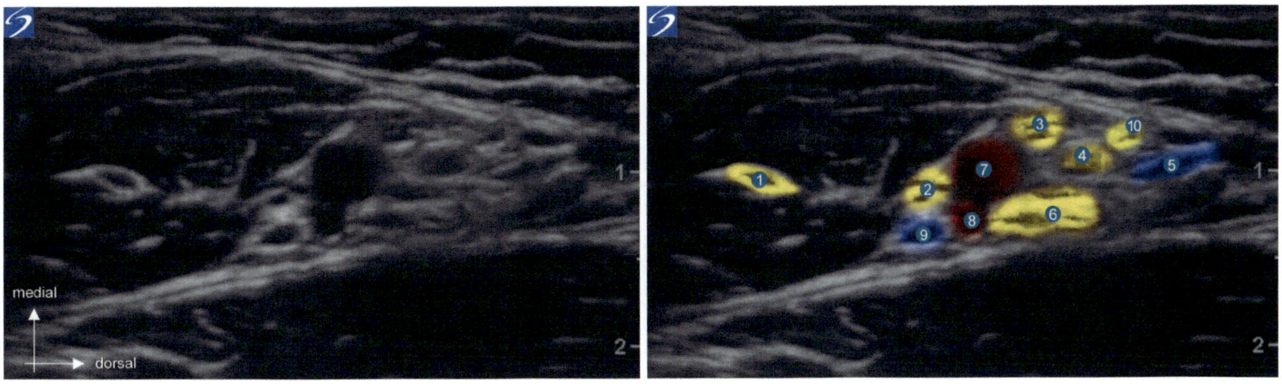

1. musculocutaneous n. 2. lateral median n. 3. medial median n.
4. ulnar n. 5. basilic v. 6. radial n.
7. brachial a. 8. deep brachial (profunda brachii) a.
9. brachial v. 10. medial antebrachial cutaneous n.

Fig. 5.84 "Axillary" transverse ultrasound section showing median nerve still comprised of two components which will merge more distally

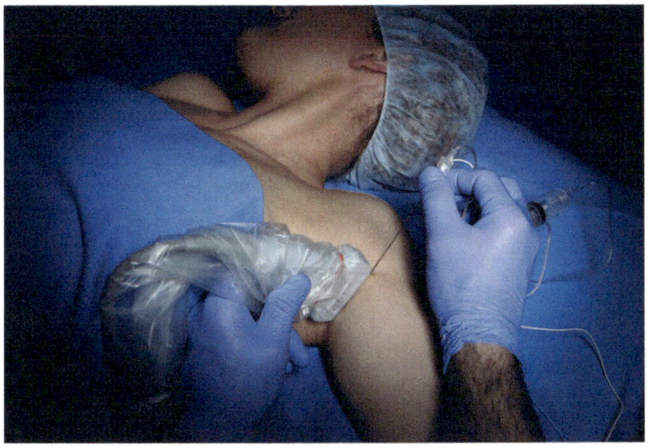

Fig. 5.85 Ultrasound-guided axillary block. Position of probe. Needle inserted in-plane

anaesthetic around the nerves that are furthest from the cutaneous injection site, i.e. the deepest ones, and then to block successively the more superficial nerves by withdrawing and/or redirecting the needle. In the case of inadvertent injection of a small amount of air or simply as the result of sonoanatomical degradation subsequent to the injection of the local anaesthetic, needle visibility may deteriorate and continuation of the block is then made more difficult and hazardous.

After crossing the aponeurosis of the biceps brachialis muscle, the needle is directed towards the radial nerve (Fig. 5.86a, b) which lies behind and deep to the brachial artery, often in the area of its acoustic enhancement. In this situation the combined use of US with neurostimulation is useful, by confirming that the tip of the needle is close to the nerve. Identification of the radial nerve can be aided by the noting the presence of the deep brachial artery, which

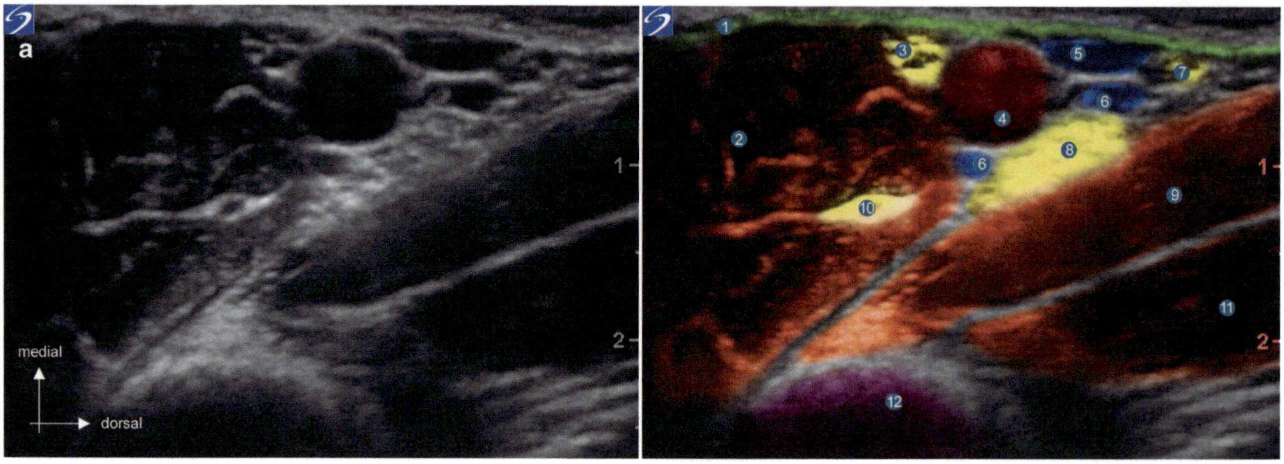

1. brachial fascia
2. biceps brachii and coracobrachialis mm.
3. median n.
4. brachial a.
5. basilic v.
6. brachial v.s
7. ulnar n.
8. radial n.
9. latissimus dorsi m.
10. musculocutaneous n.
11. teres major m.
12. humerus

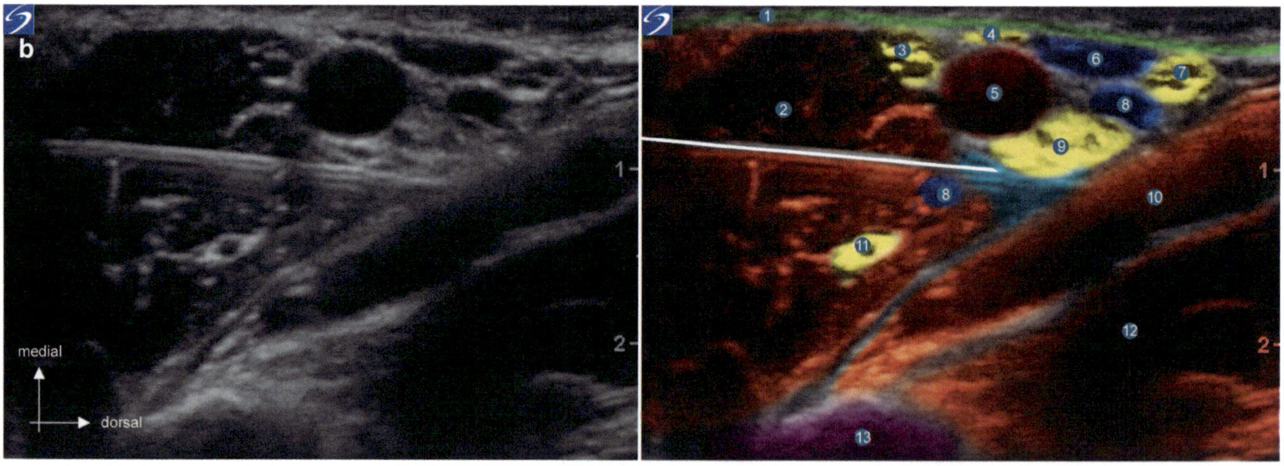

1. brachial fascia
2. biceps brachii and coracobrachialis mm.
3. median n.
4. medial antebrachial cutaneous n.
5. brachial a.
6. basilic v.
7. ulnar n.
8. brachial v.s
9. radial n.
10. latissimus dorsi m.
11. musculocutaneous n.
12. teres major and triceps brachii mm.
13. humerus

Fig. 5.86 Ultrasound-guided axillary block performed using an in-plane approach. (**a**) Transverse sonoanatomical section in axillary area. (**b**) Axillary block. Tip of needle in contact with radial nerve. (**c**) Axillary block: initial injection of local anaesthetic in contact with radial nerve. (**d**) Axillary block. Spread of local anaesthetic around radial nerve. (**e**) Axillary block. Needle in contact with median nerve, progressing towards the ulnar nerve, with aid of hydrodissection. (**f**) Axillary block. Needle in contact with medial cutaneous nerve of forearm, progressing towards the ulnar nerve with aid of hydrodissection. (**g**) Axillary block. Needle in contact with the ulnar nerve. (**h**) Axillary block. Injection of local anaesthetic in contact with ulnar nerve. (**i**) Axillary block. Spread of the local anaesthetic around the ulnar nerve. (**j**) Axillary block. Needle in contact with the median nerve, start of injection of local anaesthetic. (**k**) Axillary block. Injection and spread of local anaesthetic around the median nerve. (**l**) Axillary block. Injection of local anaesthetic in contact with the musculocutaneous nerve after sharply redirecting the needle

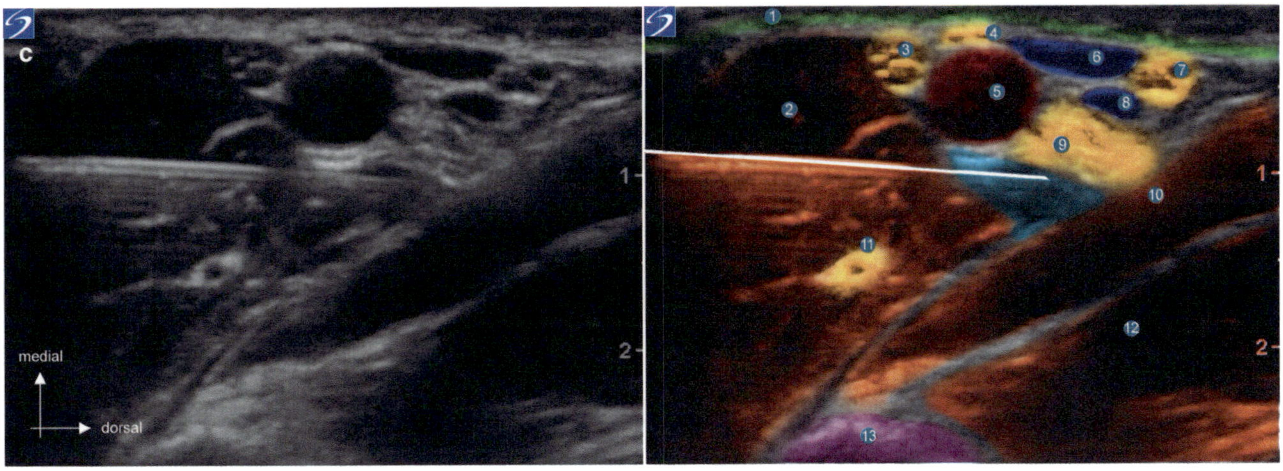

1. brachial fascia
2. biceps brachii and coracobrachialis mm.
3. median n.
4. medial antebrachial cutaneous n.
5. brachial a.
6. basilic v.
7. ulnar n.
8. brachial v.
9. radial n.
10. latissimus dorsi m.
11. musculocutaneous n.
12. teres major and triceps brachii mm.
13. humerus

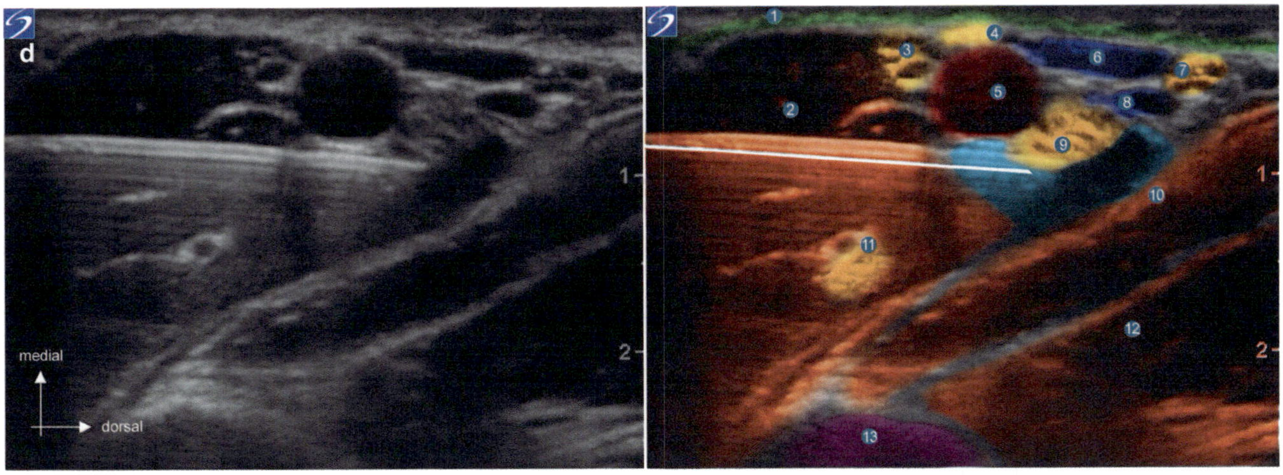

1. brachial fascia
2. biceps brachii and coracobrachialis mm.
3. median n.
4. medial antebrachial cutaneous n.
5. brachial a.
6. basilic v.
7. ulnar n.
8. brachial v.
9. radial n.
10. latissimus dorsi m.
11. musculocutaneous n.
12. teres major and triceps brachii mm.
13. humerus

Fig. 5.86 (continued)

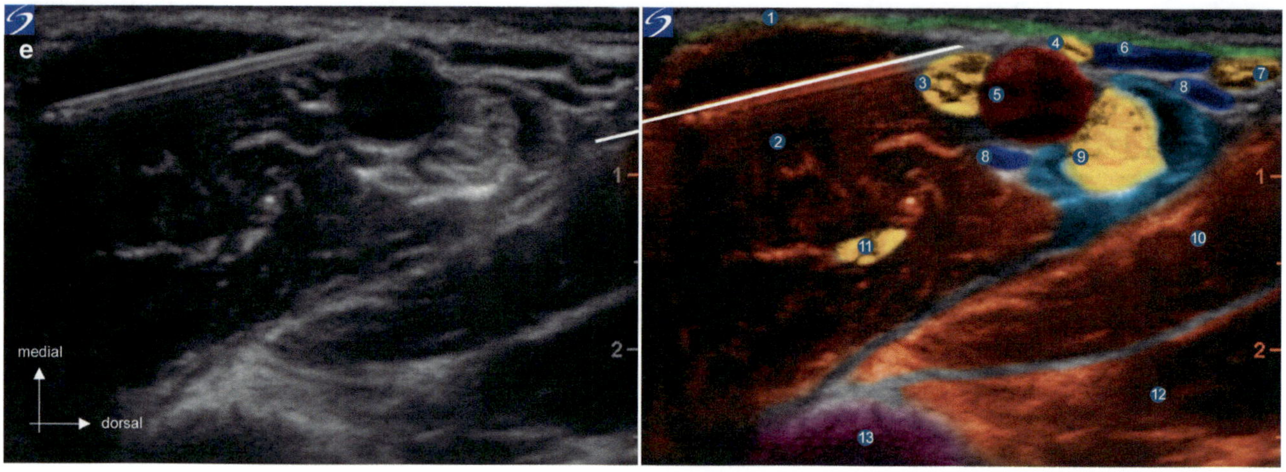

1. brachial fascia
2. biceps brachii and coracobrachialis mm.
3. median n.
4. medial antebrachial cutaneous n.
5. brachial a.
6. basilic v.
7. ulnar n.
8. brachial v.s
9. radial n.
10. latissimus dorsi m.
11. musculocutaneous n.
12. teres major and triceps brachii mm.
13. humerus

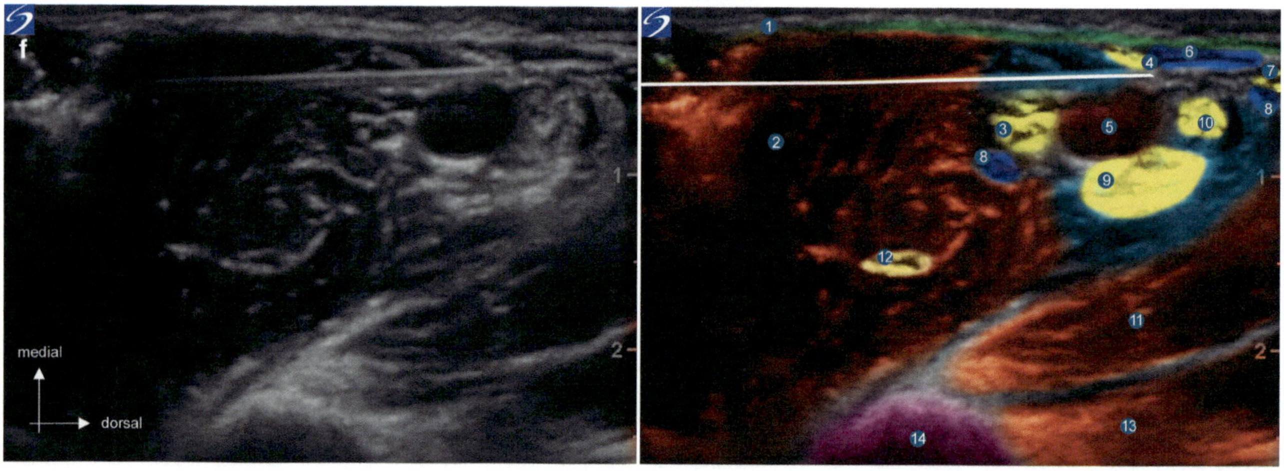

1. brachial fascia
2. biceps brachii and coracobrachialis mm.
3. median n.
4. medial antebrachial cutaneous n.
5. brachial a.
6. basilic v.
7. ulnar n.
8. brachial v.s
9. radial n.
10. motor branch of the long head of the triceps brachii m.
11. latissimus dorsi m.
12. musculocutaneous n.
13. teres major m.
14. humerus

Fig. 5.86 (continued)

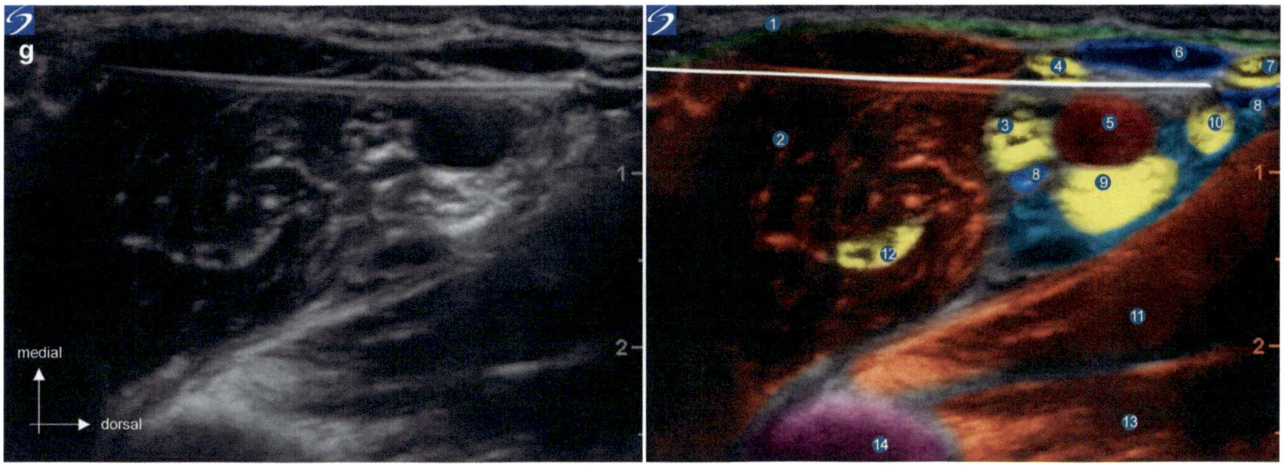

1. brachial fascia
2. biceps brachii and coracobrachialis mm.
3. median n.
4. medial antebrachial cutaneous n.
5. brachial a.
6. basilic v.
7. ulnar n.
8. brachial v.s
9. radial n.
10. motor branch of the long head of the triceps brachii m.
11. latissimus dorsi m.
12. musculocutaneous n.
13. teres major m.
14. humerus

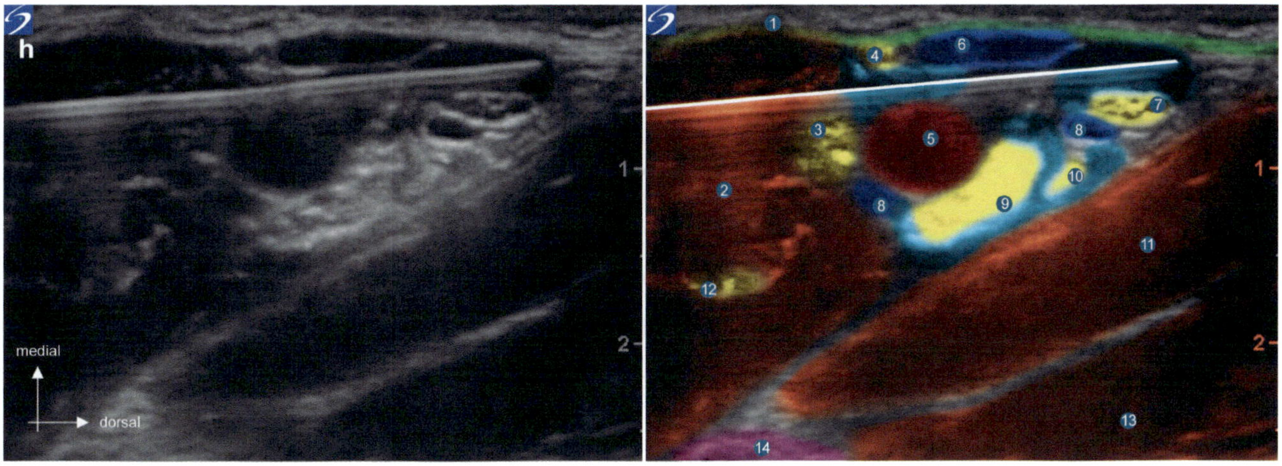

1. brachial fascia
2. biceps brachii and coracobrachialis mm.
3. median n.
4. medial antebrachial cutaneous n.
5. brachial a.
6. basilic v.
7. ulnar n.
8. brachial v.s
9. radial n.
10. motor branch of the long head of the triceps brachii m.
11. latissimus dorsi m.
12. musculocutaneous n.
13. teres major m.
14. humerus

Fig. 5.86 (continued)

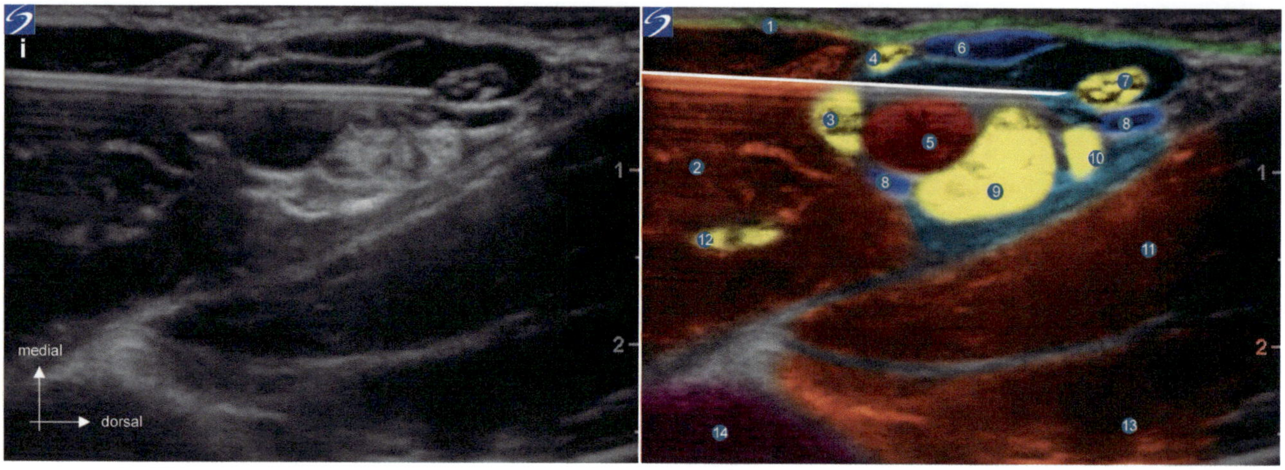

1. brachial fascia
2. biceps brachii and coracobrachialis mm.
3. median n.
4. medial antebrachial cutaneous n.
5. brachial a.
6. basilic v.
7. ulnar n.
8. brachial v.s
9. radial n.
10. motor branch of the long head of the triceps brachii m.
11. latissimus dorsi m.
12. musculocutaneous n.
13. teres major m.
14. humerus

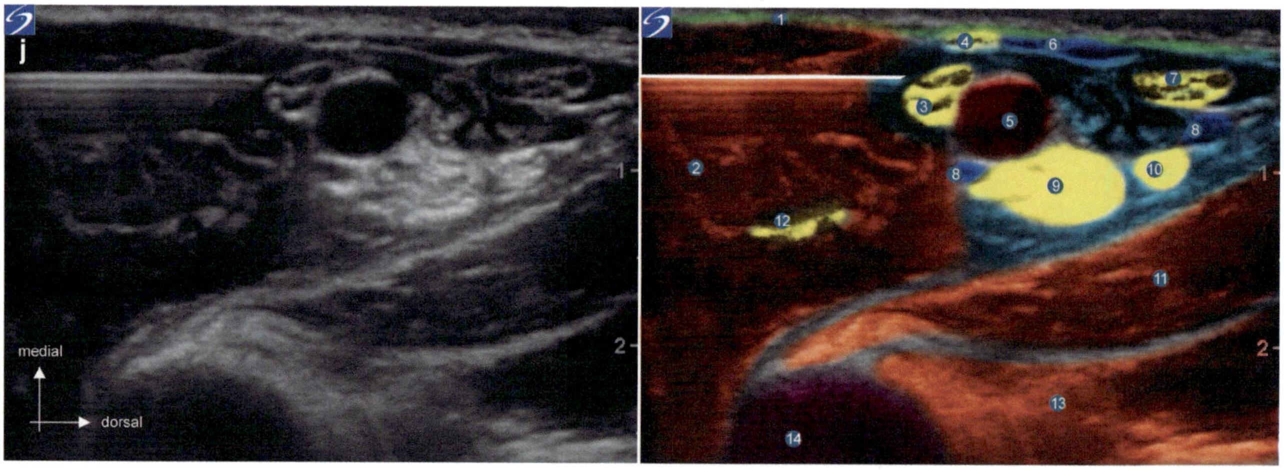

1. brachial fascia
2. biceps brachii and coracobrachialis mm.
3. median n.
4. medial antebrachial cutaneous n.
5. brachial a.
6. basilic v.
7. ulnar n.
8. brachial v.s
9. radial n.
10. motor branch of the long head of the triceps brachii m.
11. latissimus dorsi m.
12. musculocutaneous n.
13. teres major m.
14. humerus

Fig. 5.86 (continued)

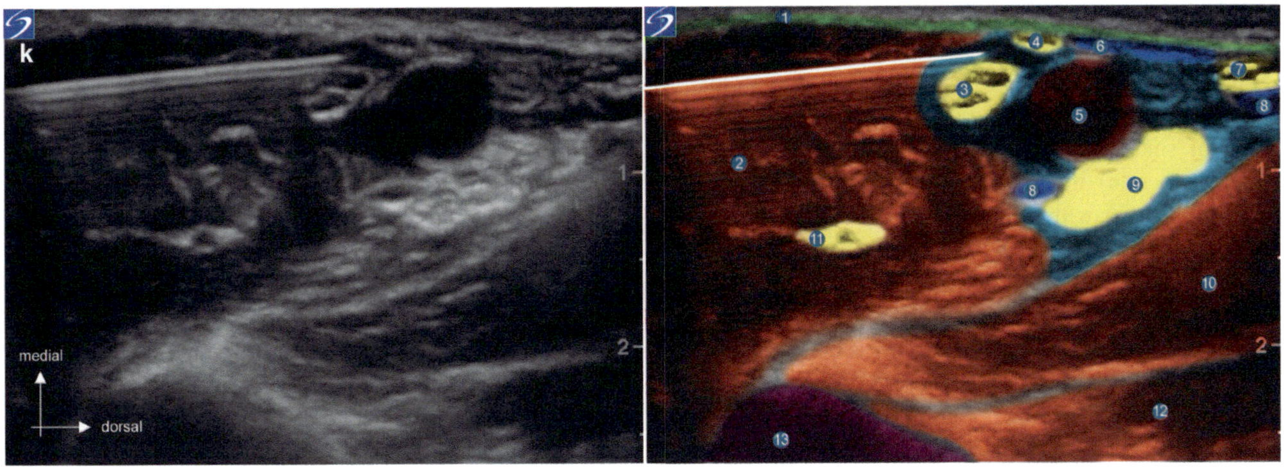

1. brachial fascia
2. biceps brachii and coracobrachialis mm.
3. median n.
4. medial antebrachial cutaneous n.
5. brachial a.
6. basilic v.
7. ulnar n.
8. brachial v.s
9. radial n.
10. motor branch of the long head of the triceps brachii m.
11. latissimus dorsi m.
12. musculocutaneous n.
13. teres major m.
14. humerus

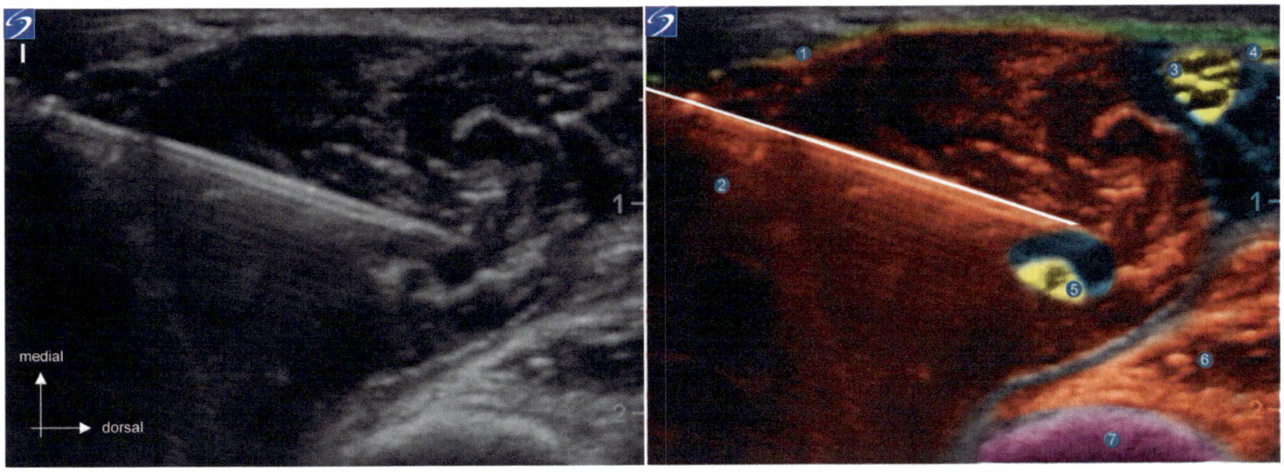

1. brachial fascia
2. biceps brachii and coracobrachialis mm.
3. median n.
4. medial antebrachial cutaneous n.
5. musculocutaneous n.
6. latissimus dorsi and teres major mm.
7. humerus

Fig. 5.86 (continued)

accompanies the nerve as it enters the radial groove of the humerus. By performing small proximal-distal movements of the probe, this artery can be visualised and thus define the position of the radial nerve. By injecting next to the radial nerve (Fig. 5.86b–d), the clinician looks to see if local anaesthetic spreads towards the ulnar nerve, which is often the case. Selective injection next to the ulnar nerve would then no longer be necessary [67–69]. If, however, the local anaesthetic does not spread towards the ulnar nerve, the needle is repositioned in order to block it next (Fig. 5.86e–i). After withdrawing and redirecting the needle, the local anaesthetic is then injected next to the median nerve (Fig. 5.86j, k). If the musculocutaneous nerve has not been covered by the local anaesthetic intended for the median nerve, the block is completed by withdrawing the needle almost to the subcutaneous level and redirecting it towards this nerve (Fig. 5.86l). At the end of the procedure, it should be verified that all the nerves have been adequately covered by the local anaesthetic. Considering the common anatomical variations that exist in the relative positions of each nerve and in the relative positions, number and size of the brachial blood vessels, the actual needle-nerve pathway to follow for each nerve is necessarily variable. Often it is necessary to pass the needle between the brachial artery and veins, or between the brachial artery and the median nerve. These adjustments are generally simple to perform but should be done attentively to avoid inadvertent vascular or nerve puncture. During needle insertion, it may be helpful to inject of small volumes (0.5–2 mL) of a D5W solution (hydrodissection) to create space for the needle and to improve contrast within the tissues (for example, to reach the ulnar nerve [Fig. 5.86e–g]). Note that it is necessary to perform medial subcutaneous injection only if it is desired to anaesthetise the area supplied by the medial cutaneous nerve **of the arm**. In fact, the medial cutaneous nerve **of the forearm**, which at this level is close to the basilic vein, is usually blocked during injection of the median nerve and/or ulnar nerve.

Needle Out of Plane

This is a technique which is rarely performed. In the author's opinion, it is useful only for a slightly shorter trajectory of the needle to reach the target nerves. However, this creates a greater risk of not identifying the true position of the needle tip possibly leading to inadvertent vascular/neural injection. In this configuration, constant pressure applied by the needle (and/or of the probe) results in compression of the venous network. Therefore, the potential for accidental venous puncture is greater and is accompanied by the risk of intravenous injection of LA or the rapid reabsorption of LA into the "injured" blood vessels. It should not to be used by inexperienced staff.

The probe is always positioned in the anterior axillary fold in order to obtain an image close to that of Fig. 5.86a. This time, the needle is inserted perpendicularly to the probe (Fig. 5.87), as in an axillary block with neurostimulation only. The practitioner should postively identify the position of the tip of the needle by scrupulously observing tissue movement. When combined with hydrolocalization these movements are a valuable indicator of position of the needle tip. As for the "in plane" approach previously described, it is necessary to

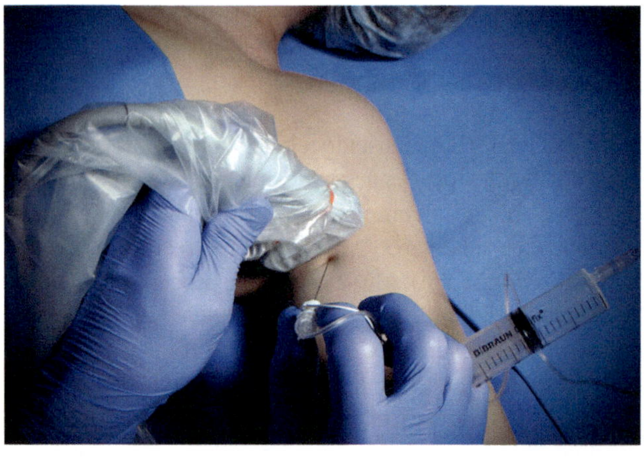

Fig. 5.87 Axillary block. Injection using an out-of-plane approach

start by blocking the deepest nerves. The local anaesthetic is injected successively next to the radial nerve, the ulnar nerve and median nerve, and lastly the musculocutaneous nerve.

Insertion of a Perineural Catheter

As has been already stated, if we insert a perineural catheter during an axillary block, depending on where the tip is finally positioned its effect may be more pronounced in the sensory territory of one nerve or another, e.g. in the radial and/or ulnar nerve area, or median nerve area or even the musculocutaneous nerve area. If when establishing the block, LA injection is performed around all the nerves, the initial overall efficacy will subsequently give rise to more selective anaesthesia/analgesia, depending on the precise site of distribution of low flow of the LA through the catheter. This process should guide positioning of the tip of the catheter so that it can be placed near to the nerve(s) which will produce the best quality of analgesia continuing beyond the regression of the initial block.

Paediatrics

The depth of the neurovascular bundle is that much shallower in the younger child. This requires appropriate equipment if one wishes to obtain precise images of anatomical structures, their position and the injection procedure. Therefore, higher frequency probes (15 MHz) should be used with appropriate adjustment of the focal point to optimise the quality of the scan.

Patients with an Arteriovenous Fistula (AVF)

The axillary block is a routine procedure in the case of AVF surgery in a patient with renal impairment [70], and also in the setting of nonvascular surgery (orthopaedics and trauma). When scanning the axilla in a limb which has an AVF, there is a large increase in size of the brachial, axillary and basilic veins (Fig. 5.88a). The nerves of the brachial plexus are found entangled between the blood vessels and there is a high risk of intravascular injection. To avoid this, ultrasound guidance should be the technique of choice in such fragile patients. The images in Fig. 5.88a–g summarise a safer approach to the axillary block in this situation. This involves an axillary block with a transversely applied probe, with the needle inserted and visualised in-plane (the out-of-plane technique is strongly not recommended).

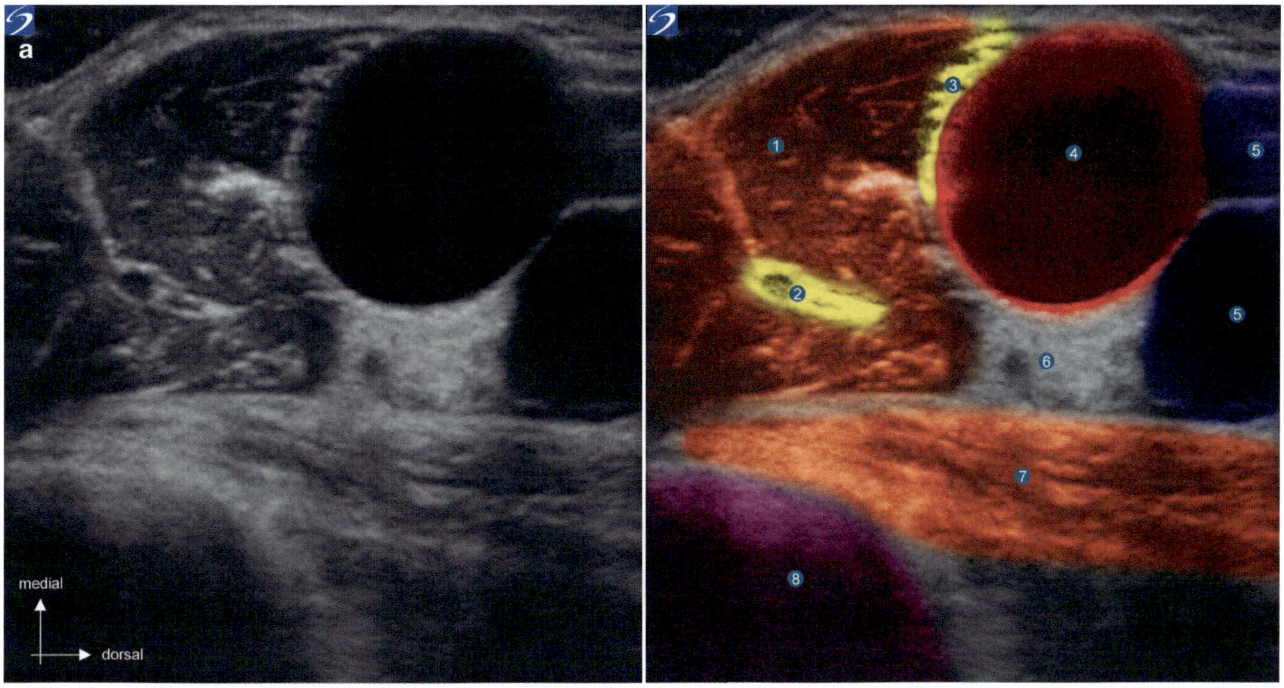

1. biceps brachii and coracobrachialis mm.
2. musculocutaneous n.
3. median n.
4. brachial a.
5. arterialized vv.
6. posterior reinforcement
7. latissimus dorsi m.
8. humerus

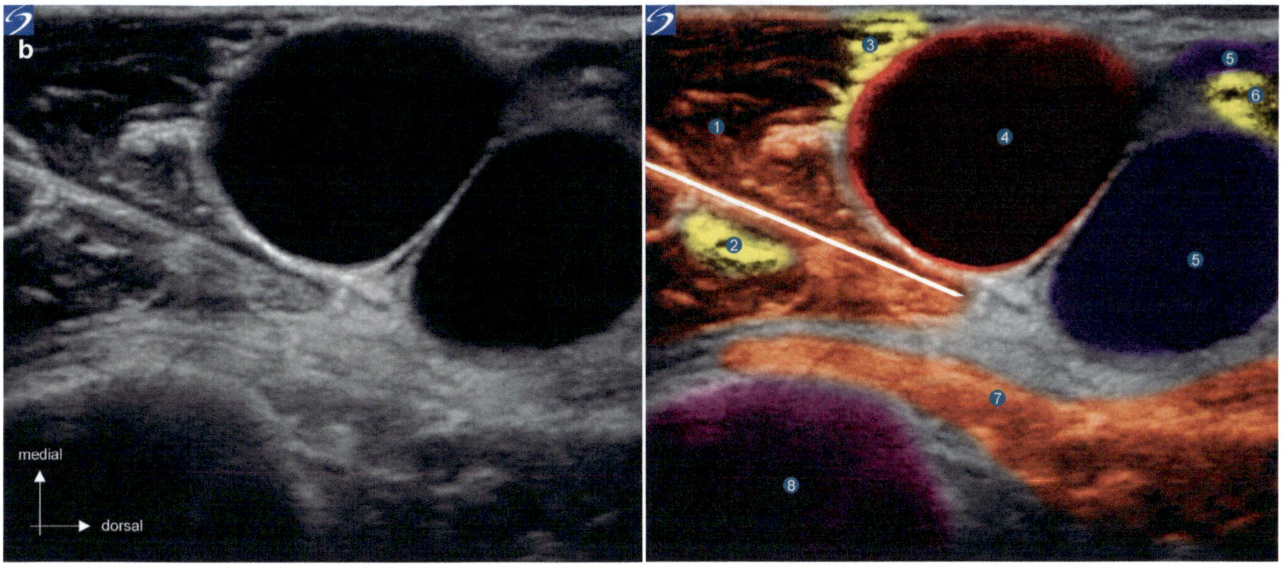

1. biceps brachii and coracobrachialis mm.
2. musculocutaneous n.
3. median n.
4. brachial a.
5. arterialized vv.
6. ulnar n.
7. latissimus dorsi m.
8. humerus

Fig. 5.88 Axillary block in a patient with an arteriovenous fistula. (**a**) Transverse section in axillary level. (**b**) Initial approach of needle towards the plane deep to the blood vessels. (**c**) Injection of local anaesthetic in contact with radial nerve. (**d**) Redirection of the needle in a plane superficial to blood vessels aiming towards the ulnar nerve. (**e**) Approach and injection of local anaesthetic in contact with the ulnar nerve with the aid of hydrodissection. (**f**) Withdrawal of the needle with, in passing, an injection of local anaesthetic in contact with the median nerve. (**g**) After redirection of the needle, the local anaesthetic is deposited in contact with the musculocutaneous nerve

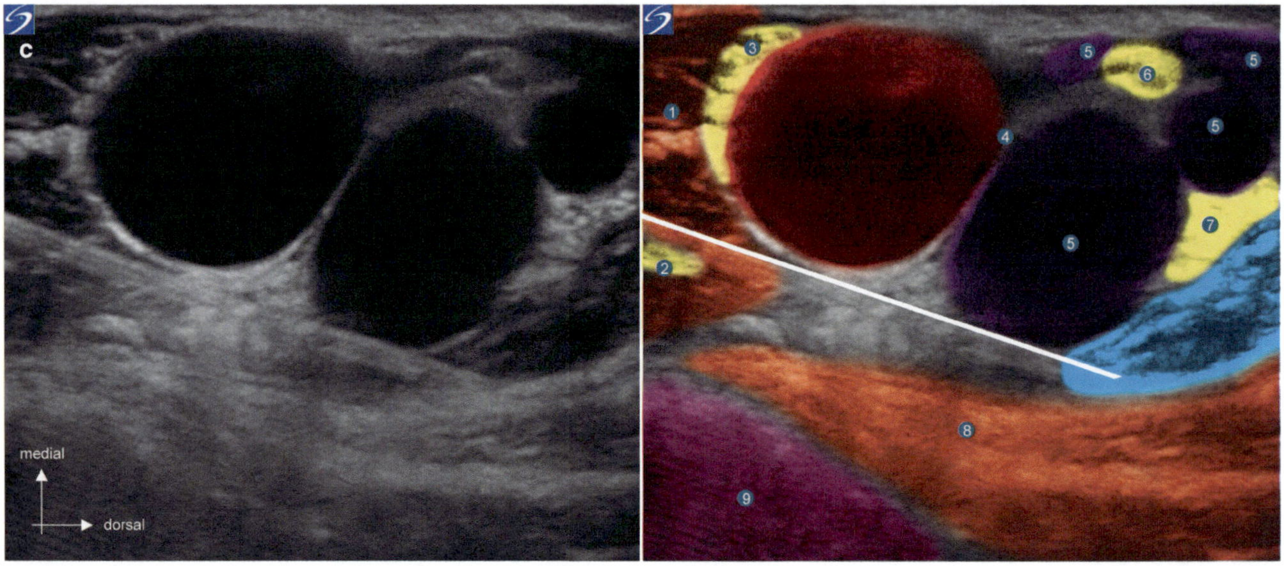

1. biceps brachii and coracobrachialis mm. 2. musculocutaneous n.
4. brachial a. 5. arterialized vv. 6. ulnar n.
7. radial n. 8. latissimus dorsi m. 9. humerus

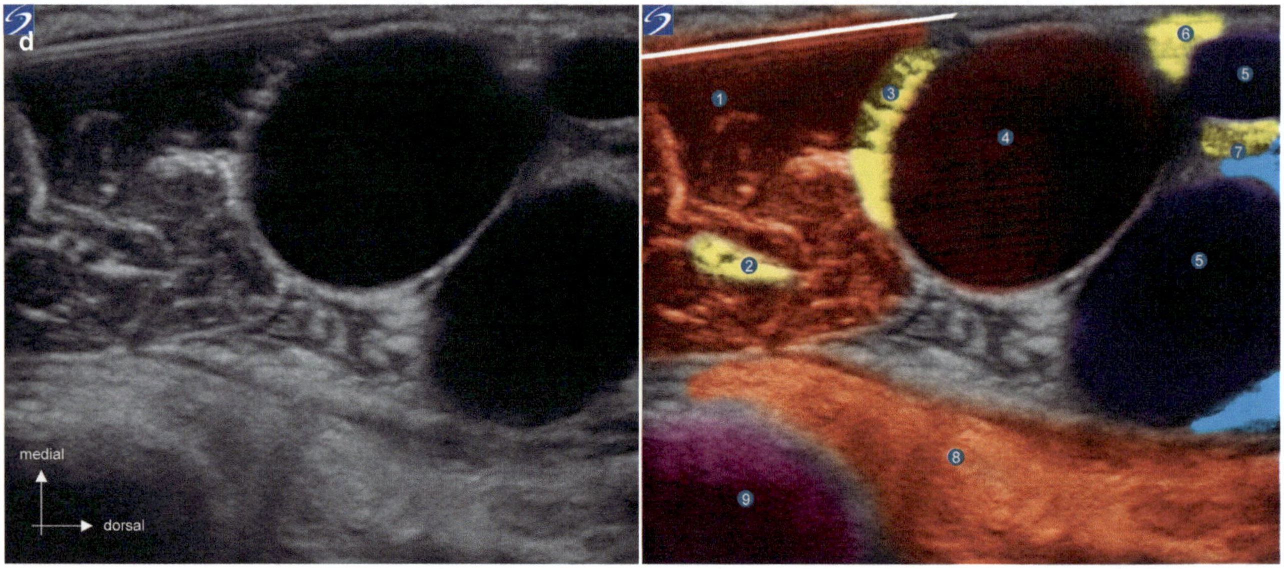

1. biceps brachii and coracobrachialis mm. 2. musculocutaneous n.
3. median n. 4. brachial a. 5. arterialized vv.
6. ulnar n. 7. radial n. 8. latissimus dorsi m.
9. humerus

Fig. 5.88 (continued)

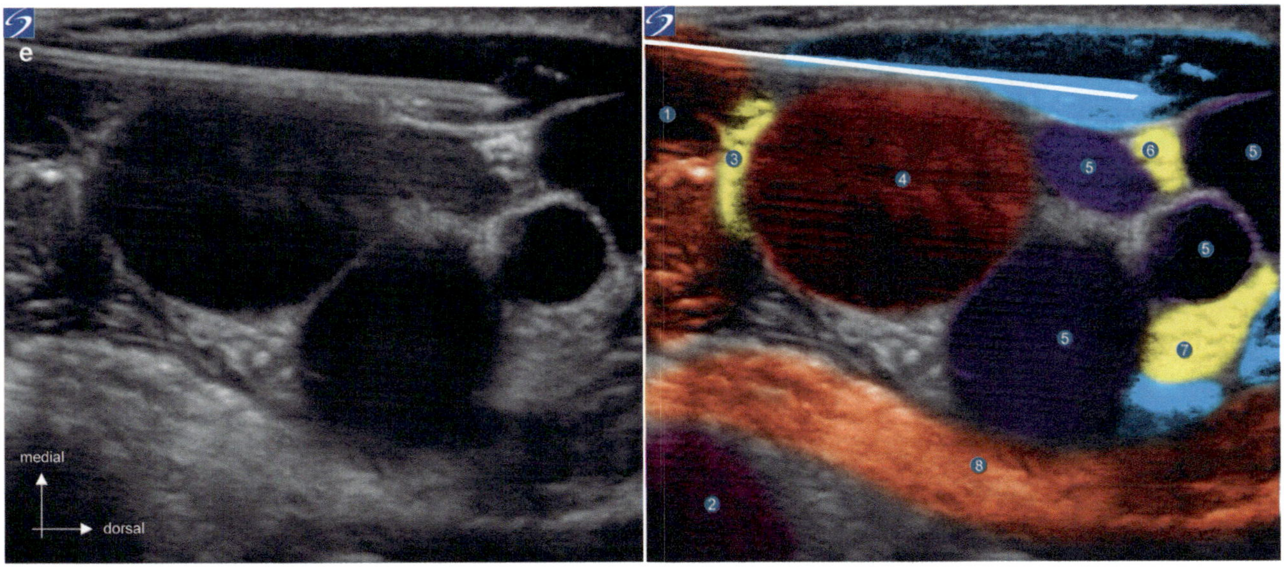

1. biceps brachii and coracobrachialis mm. 2. humerus
3. median n. 4. brachial a. 5. arterialized vv.
6. ulnar n. 7. radial n. 8. latissimus dorsi m.

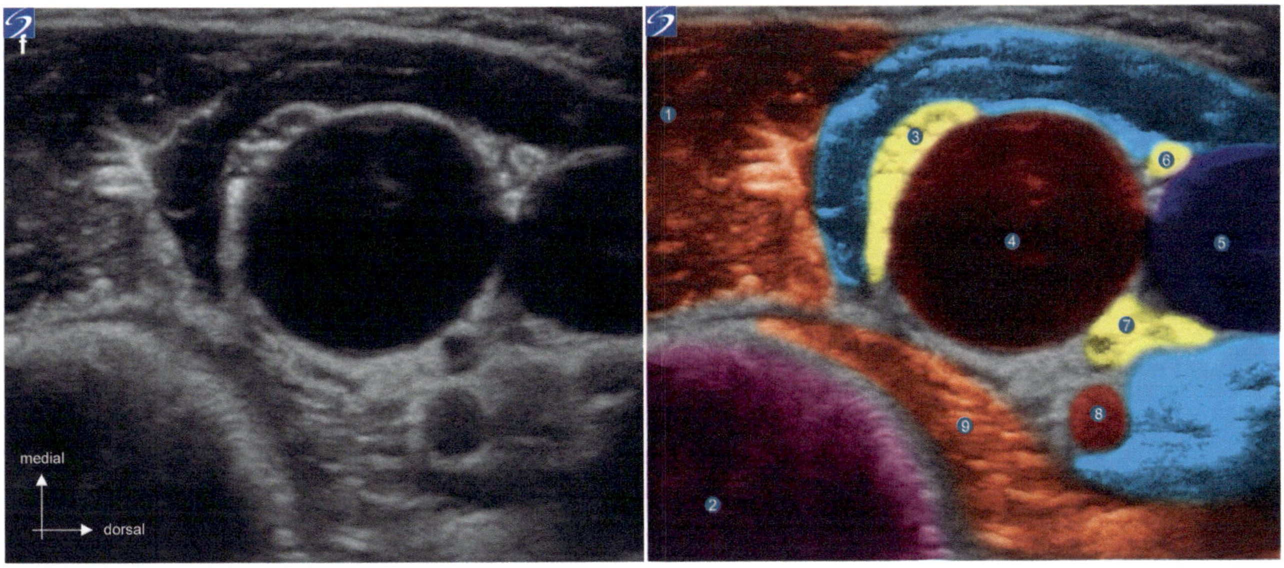

1. biceps brachii and coracobrachialis mm. 2. humerus
3. median n. 4. brachial a. 5. arterialized vv.
6. ulnar n. 7. radial n.
8. deep brachial (profunda brachii) a. 9. latissimus dorsi m.

Fig. 5.88 (continued)

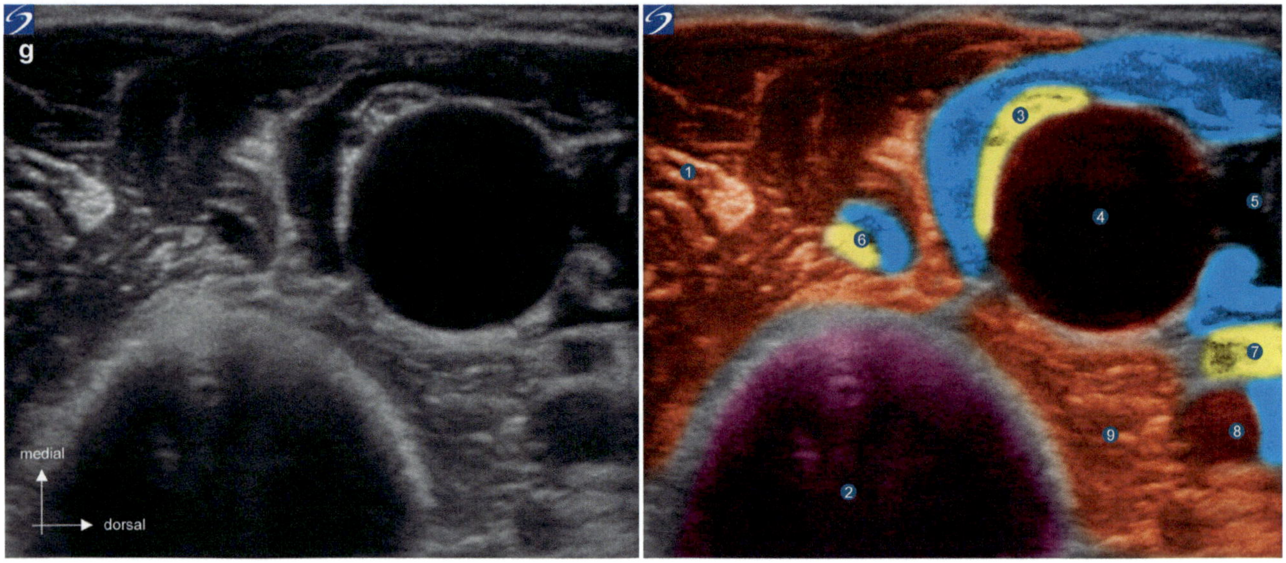

1. biceps brachii and coracobrachialis mm.
2. humerus
3. median n.
4. brachial a.
5. arterialized vv.
6. musculocutaneous m.
7. radial n.
8. deep brachial (profunda brachii) a.
9. triceps brachii m.

Fig. 5.88 (continued)

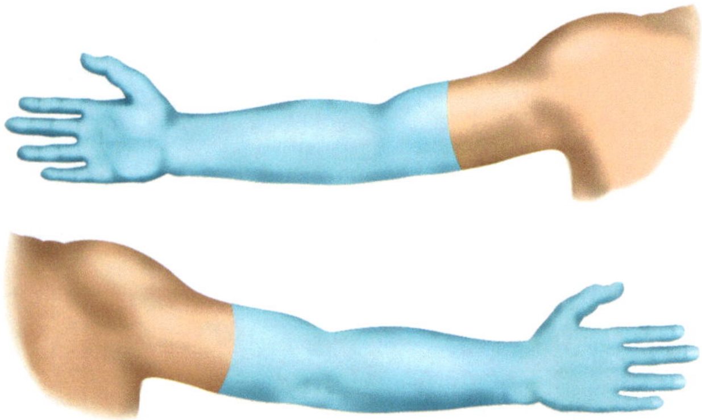

Fig. 5.89 Area of coverage of the block in the brachial canal

Brachial Canal Block (Fig. 5.89)

Indications

The nerve block in the brachial canal (humeral canal) is routinely used for surgery of the upper limb (professional consensus). Good results are obtained by anaesthesiologists whether experienced or not, and for Carles et al. [70], the failure rate is less than 5%. The order of injection varies according to authors: median, radial, ulnar, and then musculocutaneous nerve according to Gaertner et al. [71, 72] while Güntz [73] recommends starting with the radial nerve, the deepest.

It allows differential anaesthesia to be performed on the 4 main nerves of the upper limb. With neurostimulation alone, there may be concerns in stimulating nerves that are situated close together and possibly injuring a nerve already anaesthetised which has become non-stimulable with a traditional moderate electrical current (1.5 mA for 0.1 ms). Visual control enabled by using ultrasound can offer some reassurance on this point. This block enables surgery of the hand up to the elbow, sometimes with need to block the branches of the medial cutaneous nerve in the arm by subcutaneous infiltration (and also the posterior cutaneous nerve in the arm for elbow surgery). Indications for brachial canal block are minor surgery from the elbow down and hand surgery. As for an axillary block, the position of the patient's arm in abduction may be a drawback for injured patients.

Compared to the previously described axillary block, the utility of ultrasound-guided brachial canal approach is essentially the possibility of performing selective or differential blocks, with the nerves being further from each other than in the axillary level. A perineural catheter also can be inserted in contact with the nerve trunk for which prolonged action is desired (anaesthetic or analgesic).

Type of probe: linear, 5–10 or 6–13 MHz.
Axis of probe: transversal (Fig. 5.90).
Configuration: nerves in short axis, needle in or out of the plane.
Studied depth: 1–5 cm.
Neurostimulation: enables further identification of the nerves visualised and can limit, by determination of an MIS >0.3 mA (0.1 ms), the risk of accidental intrafascicular injection. Furthermore, it can supplement locating nerves when conditions of visibility are unfavourable.
Needle: 50–80 mm isolated, 22 G.
Utility of Doppler ultrasound: brachial artery and veins, basilic vein.

Echoanatomy

As in the (proximal) axillary block, US scanning in the humeral canal generally benefits from the shallow depth of the neurovascular structures in the majority of patients. This is less true for the radial nerve, which has already started its descent into the radial groove of the humerus. It is much deeper and sometimes difficult to visualise in the mid-humeral position. However, apart from the axillary and medial cutaneous nerve of the arm, the humeral canal approach allows anaesthesia of all major terminal branches of the brachial plexus from a single cutaneous injection site.

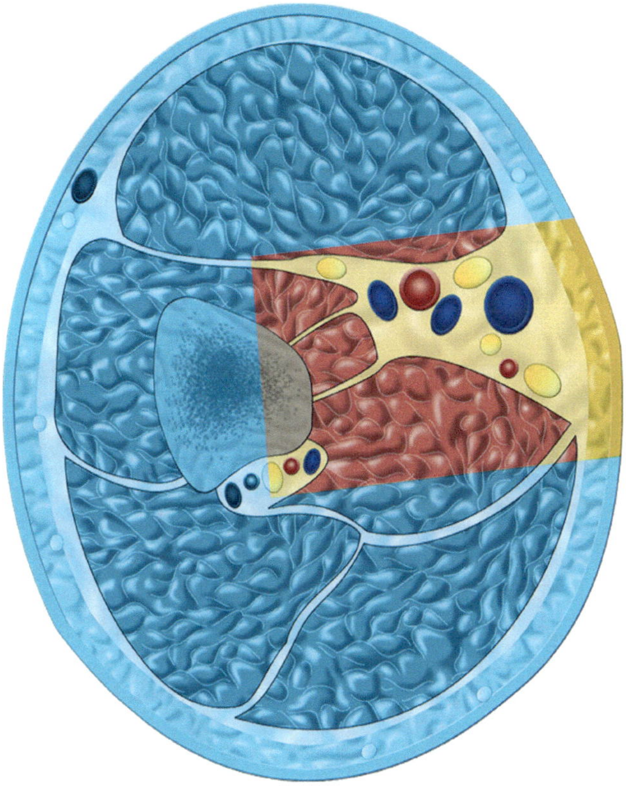

Fig. 5.90 Section of arm in the brachial canal with materialisation of the ultrasound bundle

Localization

The probe is positioned transversely. The important structures to visualise are vascular (brachial artery and veins, basilic vein), muscular and bone (humerus). Although it is not constant, it is often possible to visualise on the same screen all the terminal branches of the brachial plexus at this level. When this is the case, it is possible with a single injection site, to approach each nerve with simple axial redirection of the needle. However, the musculocutaneous nerve may already be laterally distant from the other nerve elements, and divided into its different branches (motor branches and terminal sensory branch). Occasionally, the musculocutaneous nerve is in immediate proximity to the neurovascular bundle (Fig. 5.91). The radial nerve, which is deep, may already be behind the acoustic shadow of the humerus. This can be an obstacle to a single injection, thus requiring an additional injection at the posterior end of the probe or a separate, more distal block of the radial nerve performed at the elbow. The medial cutaneous nerve in the forearm, which follows the basilic vein, often is relatively well-visualised (Fig. 5.92). Sometimes, the median nerve may be seen already posterior to the brachial artery (Fig. 5.93).

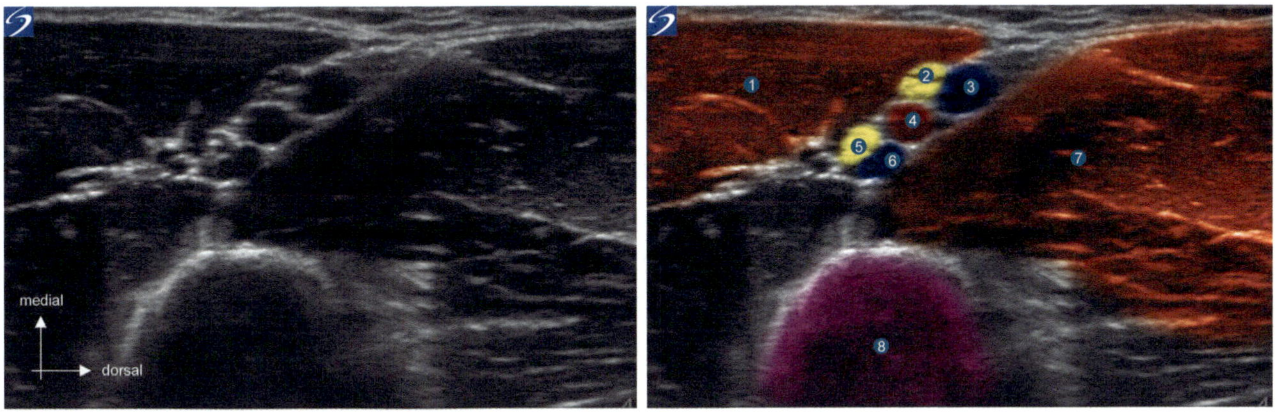

1. biceps brachii m. 2. median n. 3. basilic v.
4. brachial a. 5. musculocutaneous n. 6. brachial v.
7. triceps brachii m. 8. humerus

Fig. 5.91 Brachial canal block. Musculocutaneous nerve in proximity to the neurovascular bundle

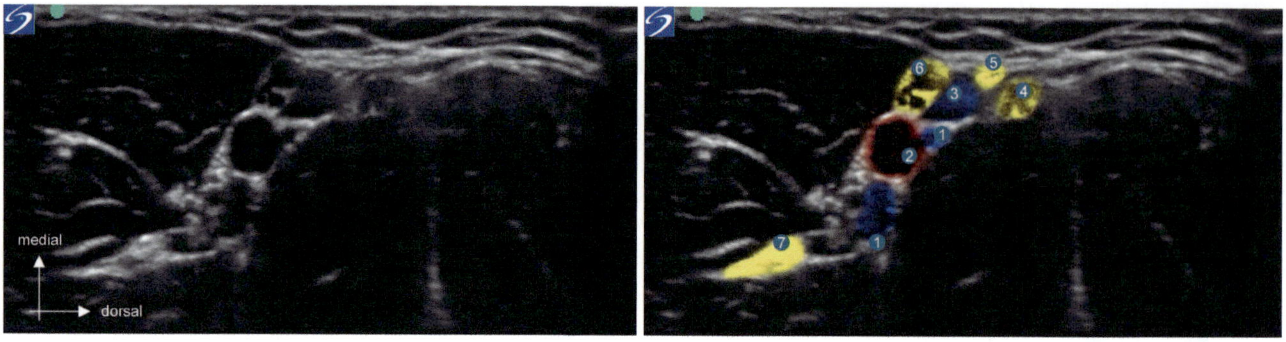

1. brachial vv. 2. brachial a. 3. basilica v.
4. ulnar n. 5. medial antebrachial cutaneous n.
6. median n.

Fig. 5.92 Brachial canal block. The medial cutaneous nerve of the forearm is clearly visible

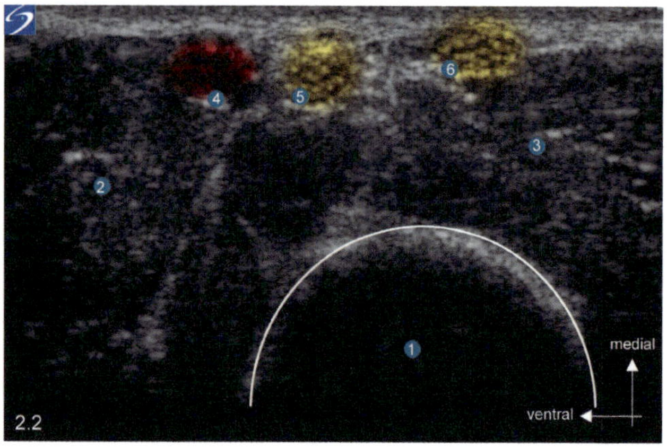

1. humerus 2. biceps brachii m.
3. triceps brachii m. 4. brachial a.
5. median n. 6. ulnar n.

Fig. 5.93 Transverse ultrasound section in the brachial canal displaying the median nerve posterior to the artery while the musculocutaneous and radial nerves no longer appear in the field

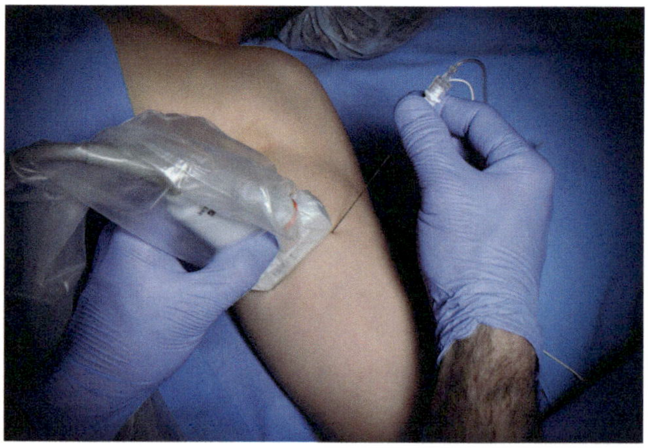

Fig. 5.94 Block in the brachial canal. Needle inserted in-plane

Injection

As with performing a block at the axillary level, the needle can be visualised in-plane with it being inserted either at the anterior end (Fig. 5.94) or posterior end of the probe, or placed out-of-plane, by inserting it laterally to the probe. Once the skin has been punctured the process of needle redirection is the same as that for the axillary approach. However, due to the major nerves diverging at this level, when using an in-plane approach it may be necessary to make a second posterior injection to place the needle more closely to the radial nerve (Fig. 5.95). As in the axillary approach, it is recommended to start by injecting the local anaesthetic around the deepest nerves (radial and musculocutaneous), in order to not be hindered by injection artefacts, as may be the case when starting the block by injecting the LA around most superficial nerves.

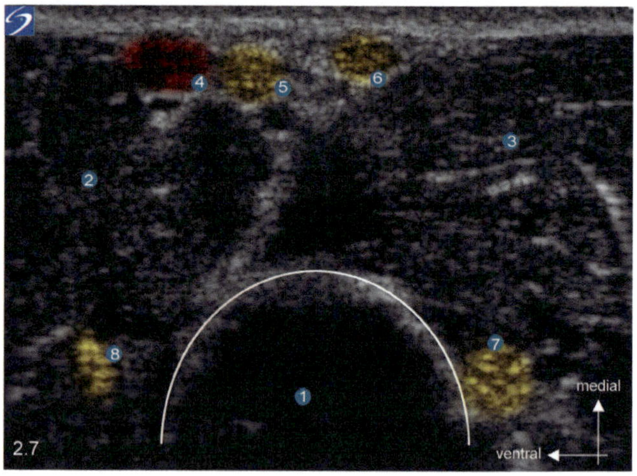

1. humerus 2. biceps brachii m. 3. triceps brachii m.
5. median n. 6. ulnar n. 7. radial n.
8. musculocutaneous n.

Fig. 5.95 Transverse ultrasound section in the brachial canal: needle insertion near the posterior end of the probe, path of needle (in yellow) to reach radial nerve

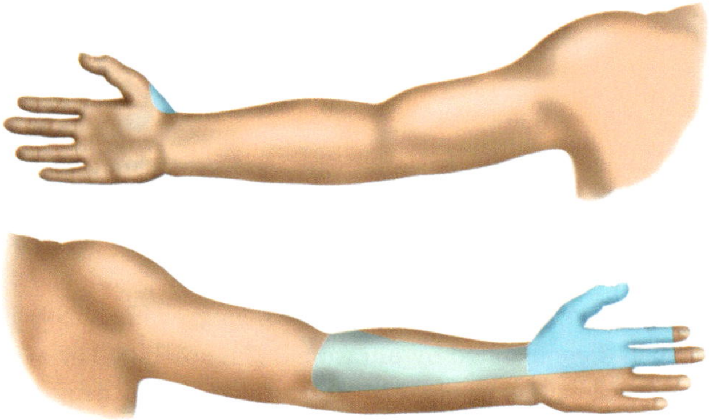

Fig. 5.96 Area of coverage of the truncal blocks of the radial nerve. The lighter blue area shows the innervation by the posterior cutaneous nerve of the forearm

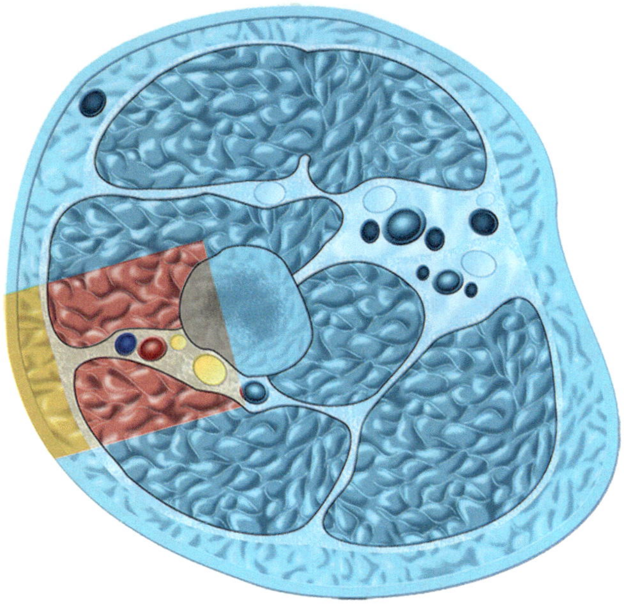

Fig. 5.97 Section of arm at the distal end of the radial groove of the humerus with materialisation of ultrasound beam

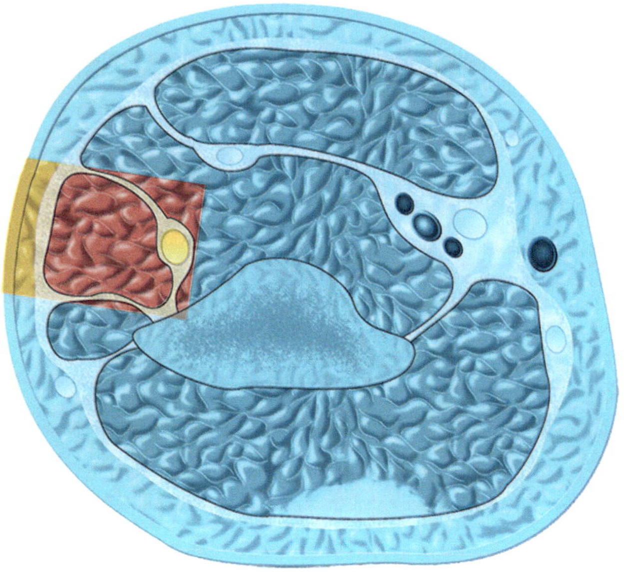

Fig. 5.98 Section in ¼ distal portion of the arm with materialisation of ultrasound beam for performance of a truncal block of the radial nerve

Distal Blocks of the Radial Nerve (Fig. 5.96)

Indications

This block can be performed for distal surgery solely in the radial nerve territory or to supplement a proximal brachial plexus block. It is also possible to insert a radial perineural catheter to prolong post-operative analgesia in this area.

Type of probe: linear, 5–10 or 6–13 MHz.
Axis of probe: transversal (Figs. 5.97 and 5.98).
Configuration: nerve in the short axis, needle in or out of plane.
Depth studied: 2–4 cm.
Neurostimulation: enables further identification of the nerves visualised and can limit, by determination of an MIS >0.3 mA (0.1 ms), the risk of accidental intrafascicular injection. Furthermore, it can supplement locating nerves when conditions of visibility are unfavourable.
Needle: 50 or 80 mm isolated, 22 G.
Utility of Doppler ultrasound: deep brachial artery and veins.

Echoanatomy

At the emergence from the radial groove of the humerus, the neurovascular bundle distances itself from the diaphysis by continuing in the intermuscular space which makes its visibility especially favourable with ultrasound. It consists of the posterior cutaneous nerve of the forearm (PCFA), the terminal branch of the radial nerve and the deep vessels of the arm. Separation of the PCFA often is easy to demonstrate (Figs. 5.99 and 5.100). This is not the case when a more distal approach (from 2 to 3 cm above the flexion point of the elbow) where the PCFA risks not being blocked by the local anaesthetic due to insufficient cephalic spread.

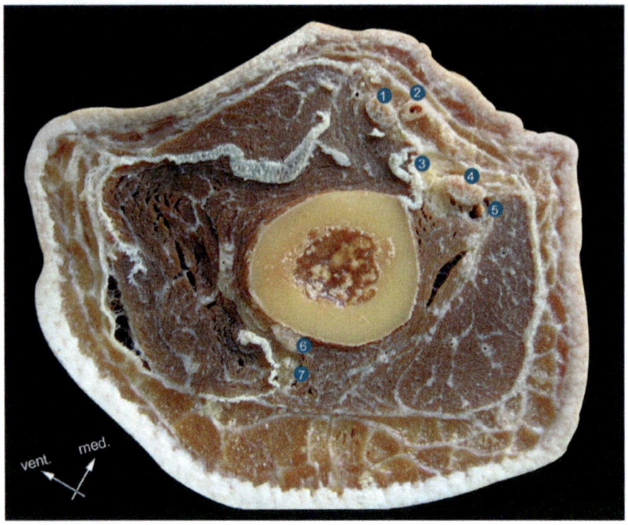

1. median n.
2. brachial a.
3. medial intermuscular septum
4. ulnar n.
5. superior ulnar collateral a.
6. radial n.
7. posterior antebrachial cutaneous n.

Fig. 5.99 Anatomical section in the distal end of radial groove of the humerus

Localisation and Injection

The probe is positioned transversely to the path of the radial nerve.

1. **If injection of local anaesthetic is performed at the emergence of the radial groove**, anaesthesia is obtained in the entire territory of the radial nerve below the elbow. The patient's upper limb and the probe are positioned as in Fig. 5.101. The underlaying anatomy (Fig. 5.99) is well demonstrated by ultrasound (Fig. 5.100). The needle is inserted in-plane at the anterior end of the probe. It passes between the brachialis muscle and the lateral head of the triceps brachialis muscle up to the nerve bifurcation at the emergence from the radial groove (Fig. 5.102) where 5–7 mL of local anaesthetic are sufficient to ensure anaesthesia of both components of the nerve (radial and PCFA). (By turning the ultrasound probe in the long axis of the radial nerve, a longitudinal section is obtained where the bifurcation of the radial nerve and of its PCFA branch are seen (Fig. 5.103). This image is relatively difficult to stabilise and is shown for interest only.)

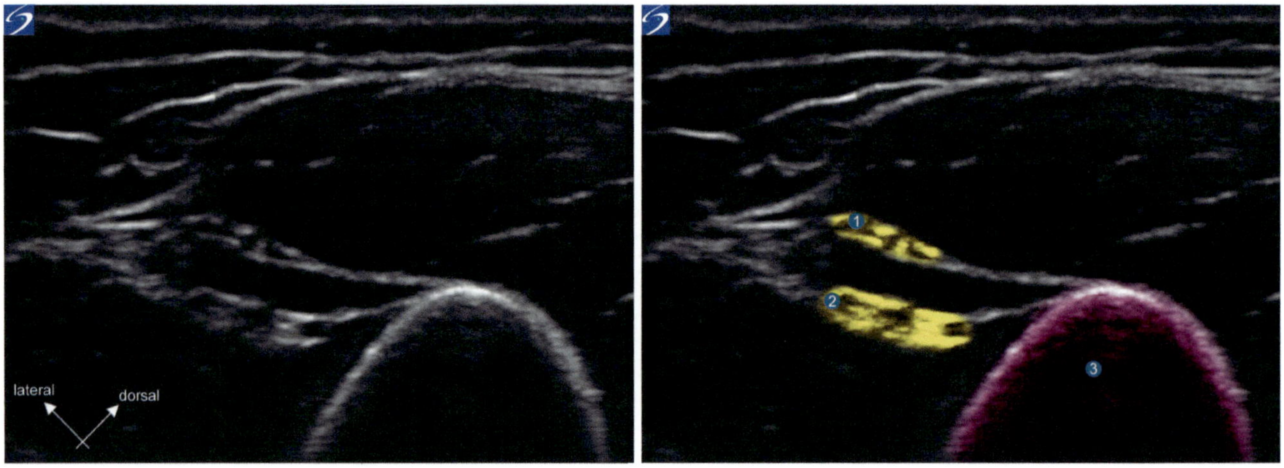

1. posterior antebrachial cutaneous n.
2. radial n.
3. humerus

Fig. 5.100 Truncal block of the radial nerve. Echoanatomy at distal emergence of radial groove of the humerus: separation of the posterior cutaneous nerve of the forearm

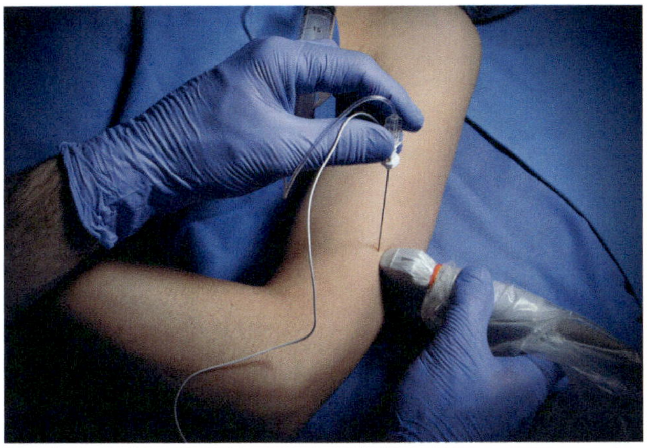

Fig. 5.101 Truncal block of radial nerve at level of its emergence from the radial groove of humerus

2. **If it is chosen to inject the local anaesthetic above the point of flexion of the elbow (the traditional technique with neurostimulation)**, the probe should be placed as in Fig. 5.104. The anaesthetised area then is, at least, that of the superficial and deep branches of the radial nerve that are visualised with ultrasound (Fig. 5.105), which correlates well with the underlying anatomy (Fig. 5.106).

Anaesthesia in the area of the PCFA (Fig. 5.96) is inconsistent with this approach. In fact, the latter is already located at a distance, following in the subcutaneous tissue of the lateral epicondyle-olecranon groove. It is necessary to rely on sufficient cephalic spread of the local anaesthetic to reach it at the level of bifurcation, which

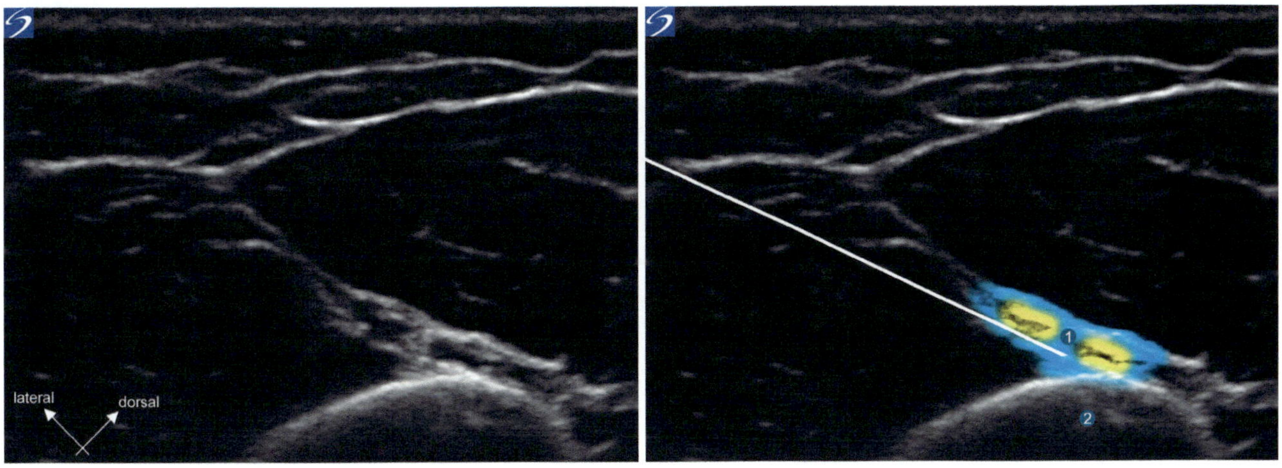

1. radial n. 2. humerus

Fig. 5.102 Truncal block of radial nerve at level of its emergence from the radial groove of humerus: transverse section

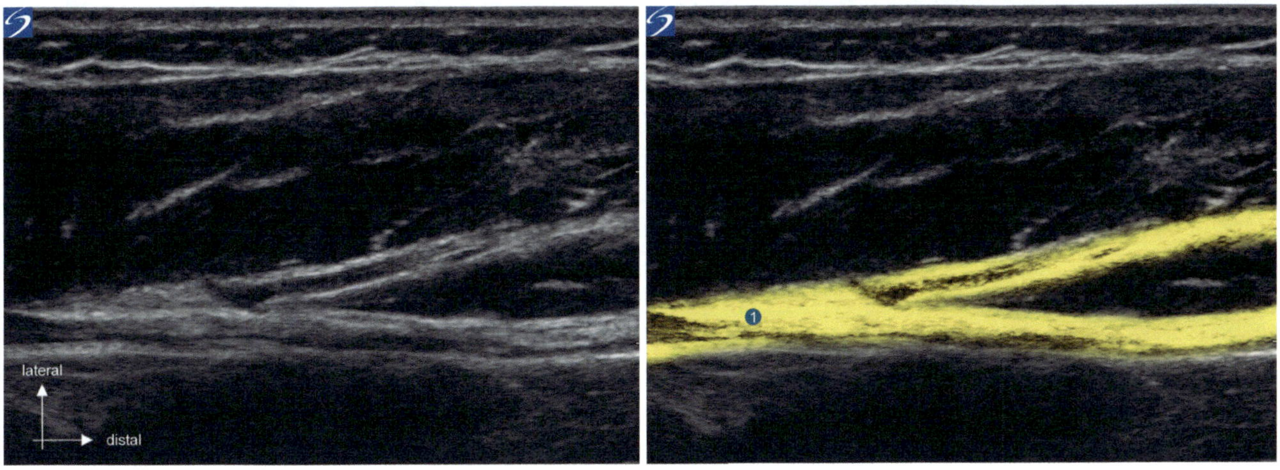

1. radial n.

Fig. 5.103 Truncal block of radial nerve at level of its emergence from the radial groove of humerus: longitudinal section showing the posterior cutaneous nerve of the forearm separating from the radial nerve

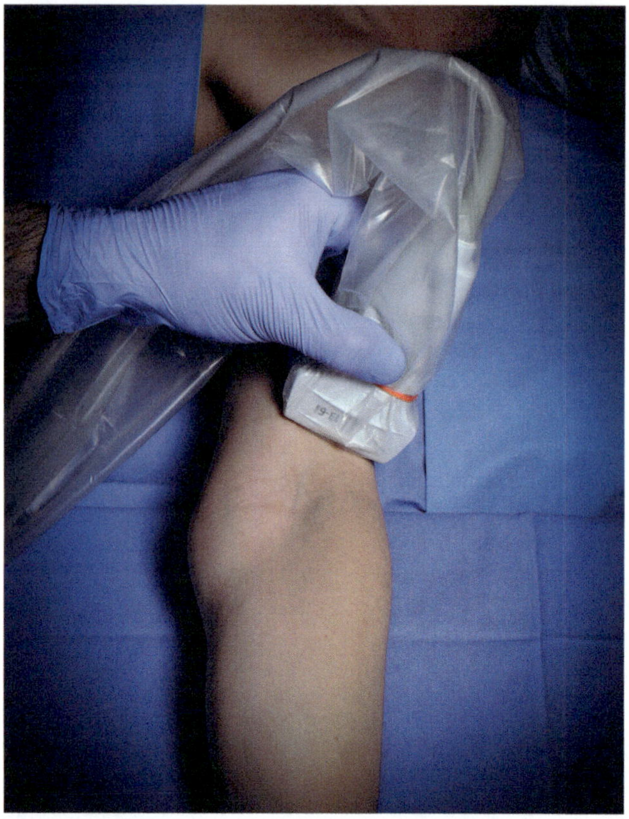

Fig. 5.104 Truncal block of radial nerve at the elbow: position of the ultrasound probe

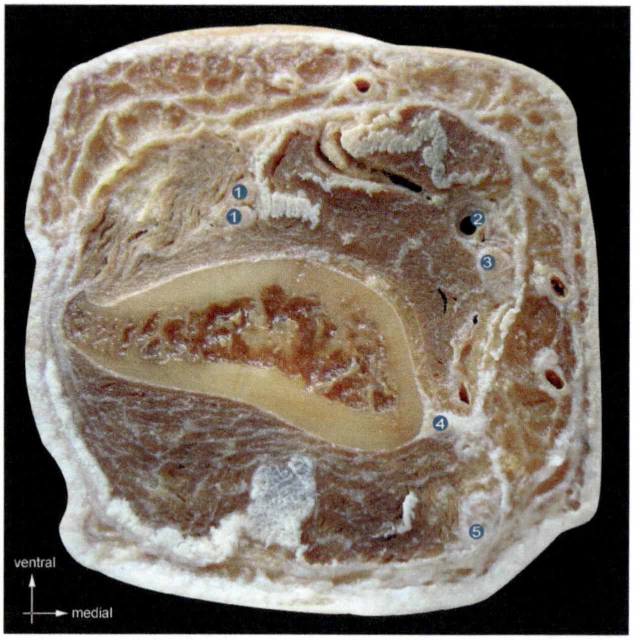

1. radial n. (superficial and deep branches)
2. brachial a.
3. median n.
4. medial intermuscular septum
5. ulnar n.

Fig. 5.106 Anatomical section in ¼ distal portion of the arm

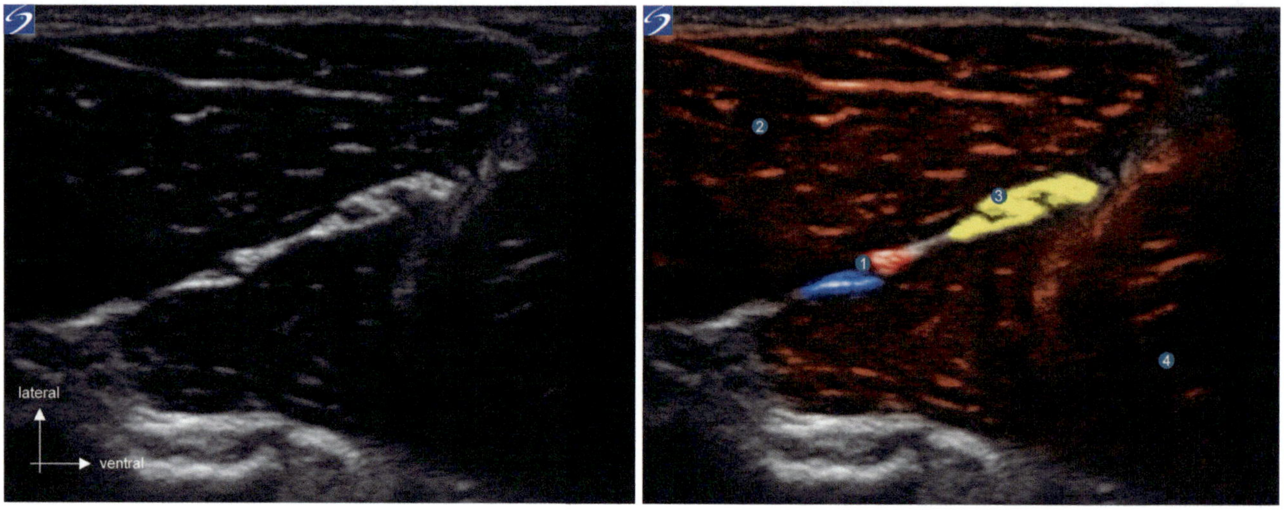

1. radial recurrent a. and v.
2. brachioradialis m.
3. superficial and deep branches of radial n.
4. brachialis m.

Fig. 5.105 Truncal block of radial nerve at the elbow: transverse section visualising superficial and deep branches of radial nerve

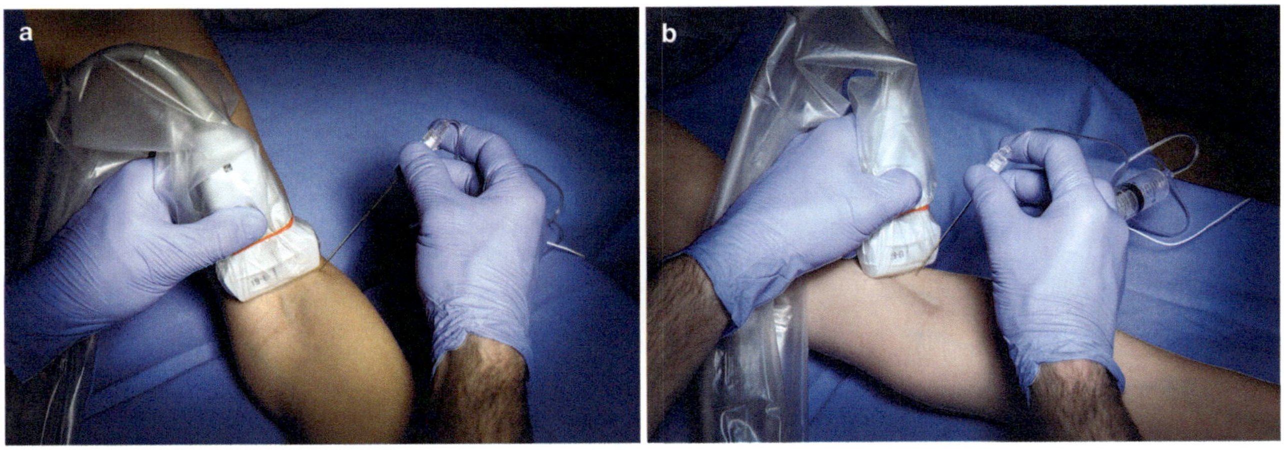

Fig. 5.107 Truncal block of radial nerve at the elbow. (**a**) In-plane. (**b**) Out-of-plane

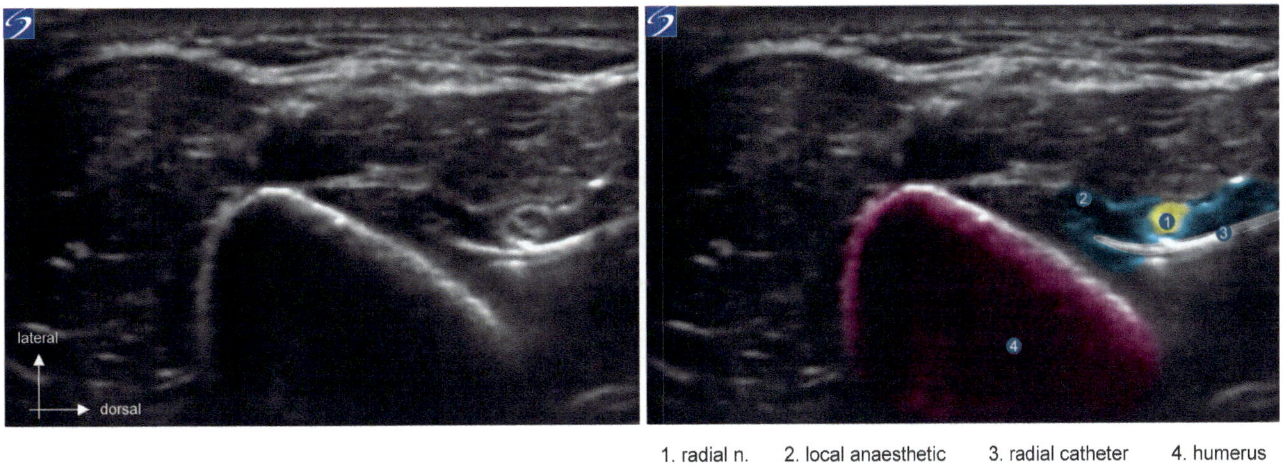

Fig. 5.108 Continuous block of radial nerve: catheter placed in-plane at the emergence from the radial groove of humerus. (Iconography: Nicolas Dufeu)

may not occur with a conventional volume of LA. A block can be performed either in-plane (Fig. 5.107a), or out-of-plane with the aid of the hydrolocalization process (Fig. 5.107b). Five to 7 mL are sufficient to ensure effective anaesthesia.

Figure 5.108 shows a catheter positioned in-plane at the emergence of the radial groove of the humerus.

Distal Blocks of the Median Nerve (Fig. 5.109)

Indications

This block can be performed for distal surgery solely in the median nerve territory or to supplement a proximal brachial plexus block.

Type of probe: linear, 5–10 or 6–13 MHz.
Axis of the probe: transversal (Fig. 5.110).
Configuration: nerve in short axis, needle in or out of plane.
Studied depth: 1–4 cm.
Neurostimulation: enables further identification of the nerves visualised and can limit, by determination of an MIS >0.3 mA (0.1 ms), the risk of accidental intrafascicular injection. Furthermore, it can supplement locating nerves when conditions of visibility are unfavourable.
Needle: 50 or 80 mm isolated, 22 G.
Utility of Doppler ultrasound: brachial artery and veins, superior ulnar collateral artery.

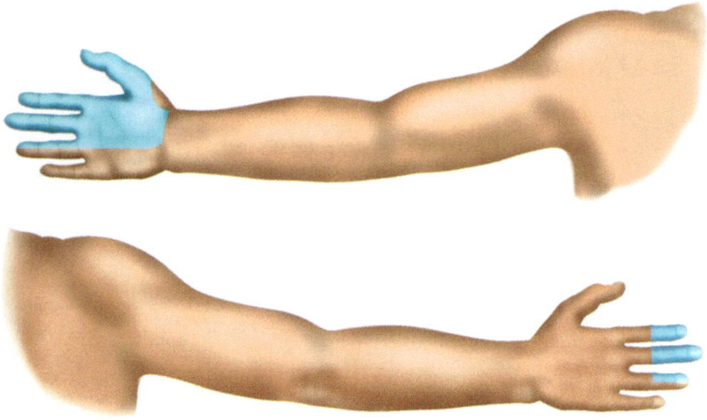

Fig. 5.109 Area of coverage of a truncal block of the median nerve

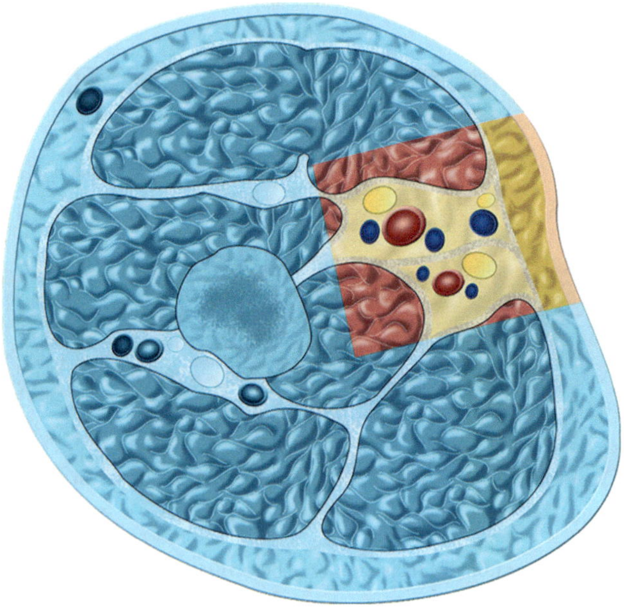

Fig. 5.110 Section of the arm at the middle third-distal third junction of the arm with materialisation of ultrasound beam

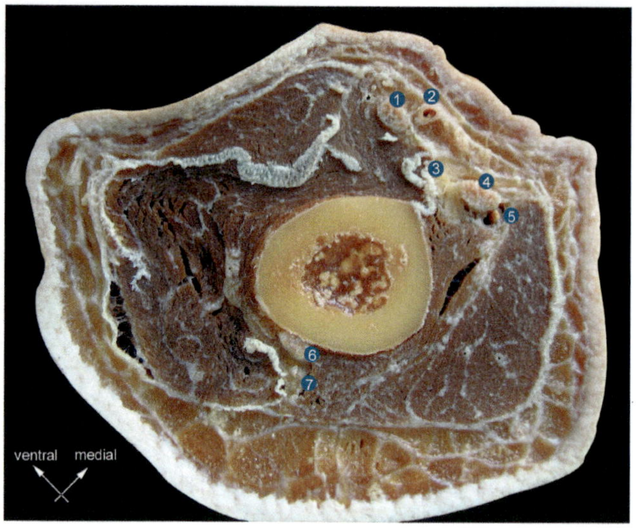

1. median n. 2. brachial a. 3. medial septum
4. ulnar n. 5. superior ulnar collateral a.
6. radial n. 7. posterior antebrachial cutaneous n.

Fig. 5.111 Anatomical section at level of the distal end of the radial groove of humerus

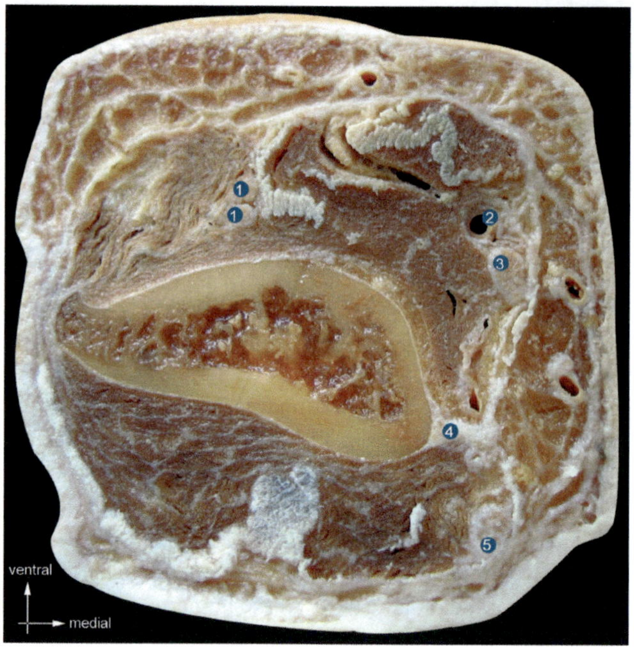

1. radial n. (superficial and deep branches)
2. brachial a. 3. median n.
4. medial intermuscular septum 5. ulnar n.

Fig. 5.112 Anatomical section at the distal quarter of arm

Echoanatomy

At the distal end of the arm, the median nerve is located in immediate proximity to the brachial artery and veins (Figs. 5.111 and 5.112), in front of the medial brachial intermuscular septum. Ultrasound makes it possible to easily visualise these neurovascular structures (Fig. 5.113). At this level, a simple and superficial approach makes it possible to perform, if necessary, a combined block of the median and ulnar nerves with a single injection site with a single redirection of the needle, or by a separate approach when they are relatively distant from each other.

At the end of its path in the anterior compartment of the forearm, the median nerve becomes superficial and a few centimetres above the wrist it is positioned deep to the tendons of the flexor carpi radialis muscle laterally, and the palmar longus muscle medially (Figs. 5.114 and 5.115). Lying superficial to the deep flexors, the long flexor of the thumb and the pronator teres muscle, the median nerve is differentiated from the adjacent tendons by its anisotropic characteristics and its mobility during flexion of the fingers. Identification of the median nerve is facilitated by the so-called “elevator” technique, which consists of following it from a more proximal level where its identification is more certain.

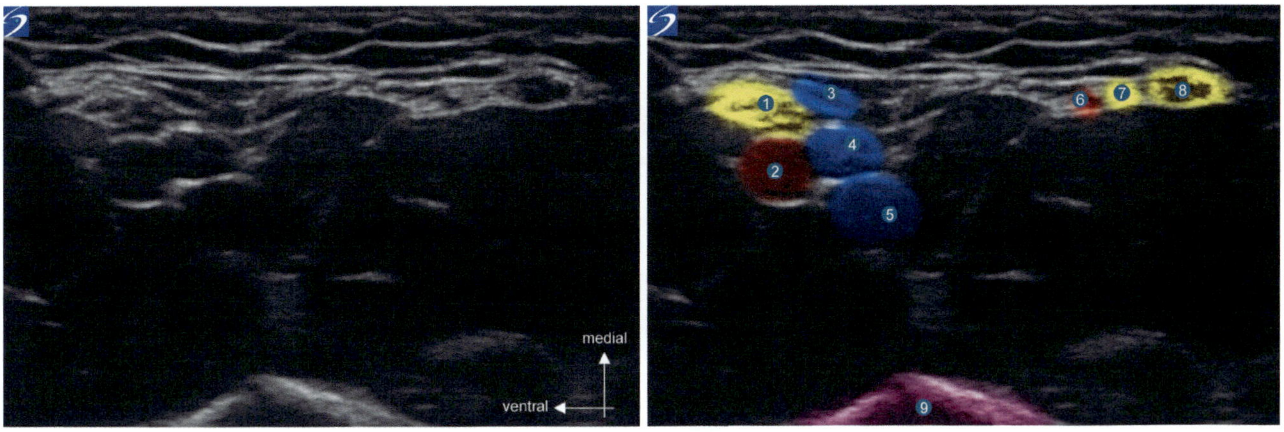

1. median n. 2. brachial a. 3. basilic v.
4. brachial v. 5. brachial v. 6. superior ulnar collateral a.
7. medial antebrachial cutaneous n. 8. ulnar n. 9. humerus

Fig. 5.113 Transverse ultrasound section at middle third-distal third junction of the arm

Fig. 5.114 Distal anatomical section of the forearm cutting through the radioulnar joint

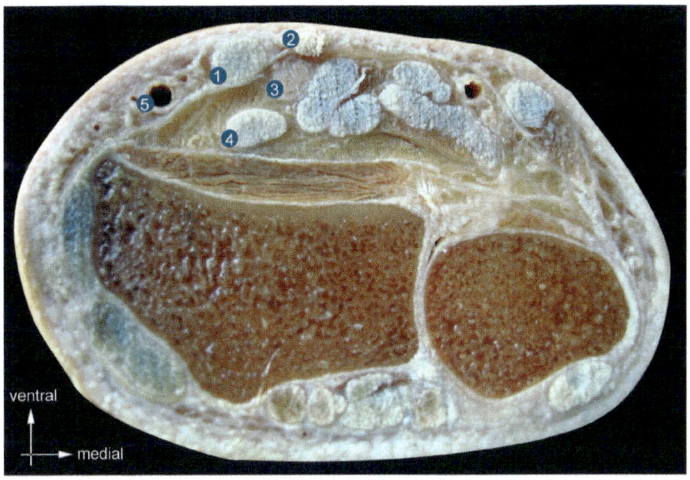

1. flexor carpi radialis tendon 2. palmaris longus tendon
3. median n. 4. flexor pollicis longus tendon
5. radial a.

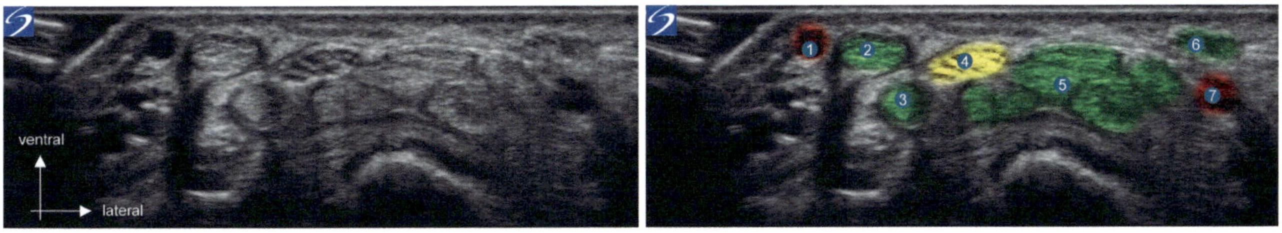

1. radial a. 2. flexor carpi radialis tendon
3. flexor pollicis longus tendon 4. median n.
5. flexor digitorum profundus tendon 6. flexor carpi ulnaris tendon
7. ulnar a.

Fig. 5.115 Transverse ultrasound section: median nerve at the wrist

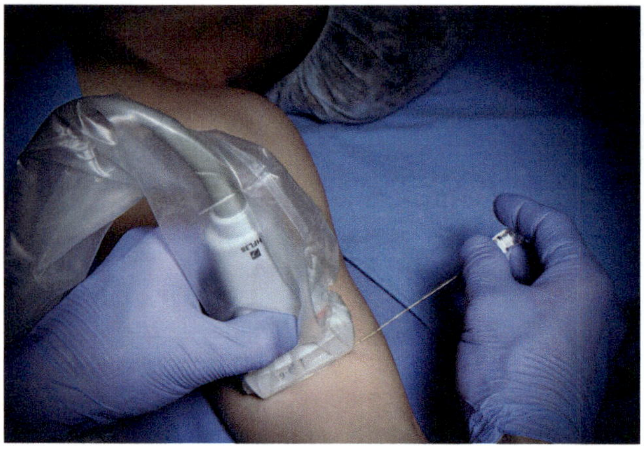

Fig. 5.116 Truncal block of the median nerve at the middle third-distal third junction of the arm. Needle inserted in-plane

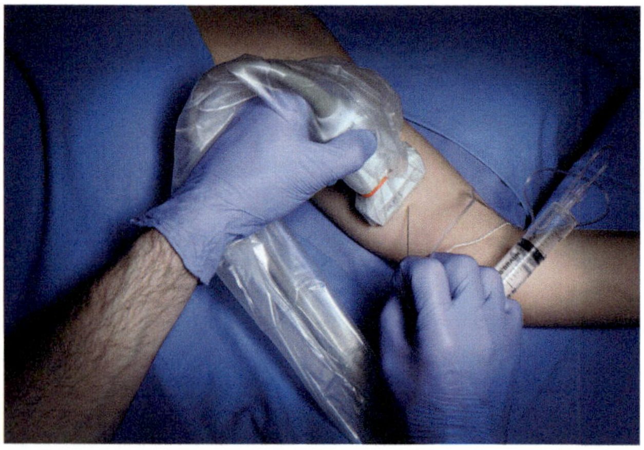

Fig. 5.117 Truncal block of median nerve at the middle third-distal third junction of arm. Needle inserted out-of-plane

Localisation and Injection

Above the Elbow Joint

The probe is positioned transversely, in the medial aspect of the inferior end of the arm, at a level where visualisation of the target structures is optimum (Fig. 5.116). The neurovascular structures are seen in their short axis (transversal section) (Fig. 5.110). The needle is introduced in-plane at the anterior extremity of the probe (Fig. 5.116). However, the nerve can be approached out-of-plane using hydrolocalisation to visualise the needle tip (Fig. 5.117). Note that the nerve lies in a very shallow position and this approach could risk accidental nerve puncture if the needle tip is not quickly visualised. At this level, a simple redirection of the needle also makes it possible to perform anaesthesia of the ulnar nerve.

Forearm

The probe is placed transversely between 5 and 8 cm above the point of flexion of the wrist according to the quality of visualisation of the median nerve (Figs. 5.118 and 5.119). This position ensures a block of the palmar cutaneous branch

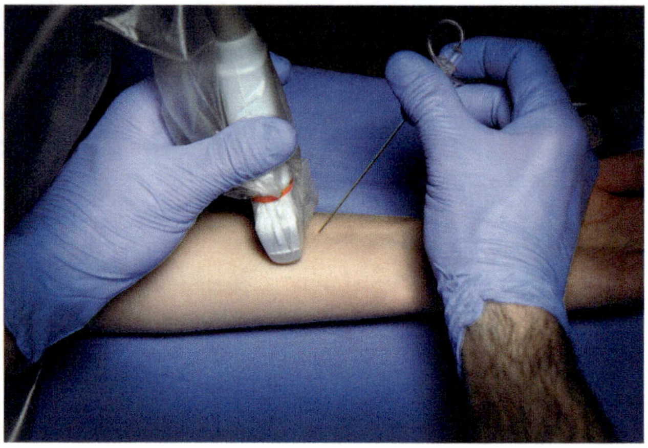

Fig. 5.118 Truncal block of median nerve at the forearm. Needle inserted out-of-plane

of the median nerve prior to its emergence. The existence of numerous muscles and tendons in the anterior compartment of the forearm at this distal site promotes the use of an out-of-plane approach, combined with a hydrolocalisation technique to identify the needle tip during its insertion. The needle is advanced between the tendons of the palmar longus and flexor carpi radialis muscles and placed next to the nerve. However, an in-plane approach may be preferred by targeting the median nerve from the lateral side (the motor nerves of the thenar branch which are located on the lateral side of the nerve would be more accessible to neurostimulation). Five to 7 mL of local anaesthetic are sufficient to obtain anaesthesia.

Distal Blocks of the Ulnar Nerve (Fig. 5.120)

Indications

This block can be performed for distal surgery solely in the ulnar nerve territory or to supplement a proximal brachial plexus block.

Type of probe: linear, 5–10 or 6–13 MHz.
Axis of probe: transversal (Fig. 5.121).
Configuration: nerve in short axis, needle in or out of plane.
Depth studied: 1–4 cm.
Neurostimulation: enables further identification of the nerves visualised and can limit, by determination of an MIS >0.3 mA (0.1 ms), the risk of accidental intrafascicular injection. Furthermore, it can supplement locating nerves when conditions of visibility are unfavourable.
Needle: 50 or 80 mm isolated, 22 G.
Utility of Doppler ultrasound: brachial artery and veins and upper ulnar collateral artery for approach in the arm; ulnar artery for approach in the lower third of the forearm.

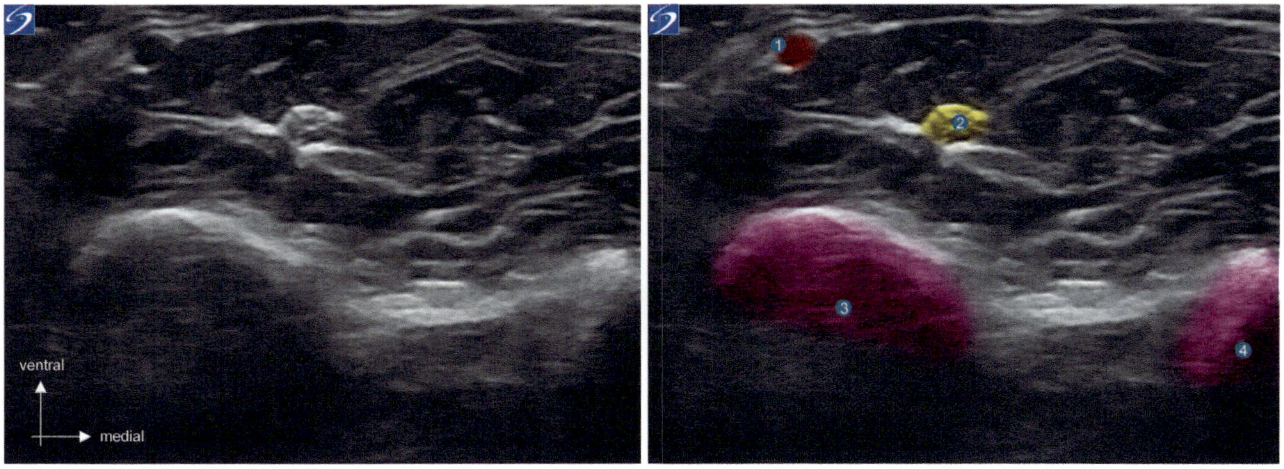

Fig. 5.119 Transverse ultrasound section in the distal third of the forearm

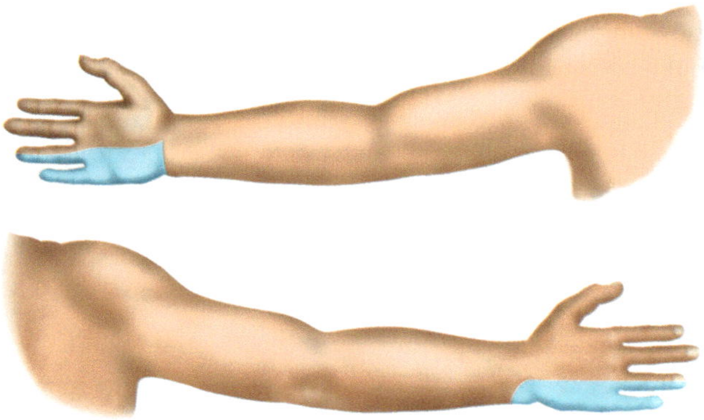

Fig. 5.120 Area of coverage of a truncal block of the ulnar nerve

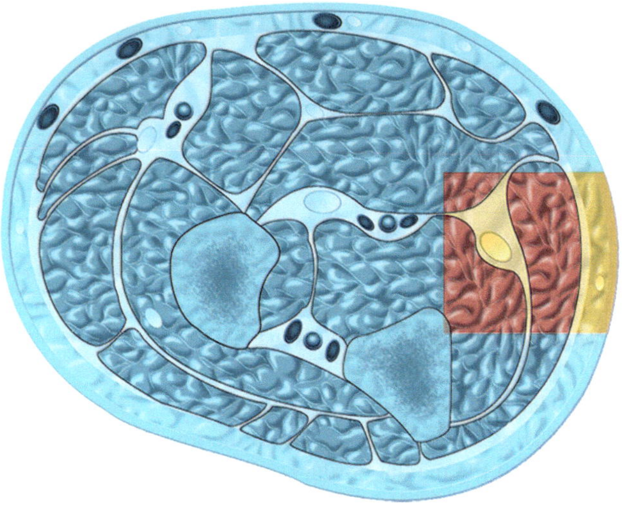

Fig. 5.121 Materialisation of ultrasound beam for block of ulnar nerve in the proximal aspect of the forearm

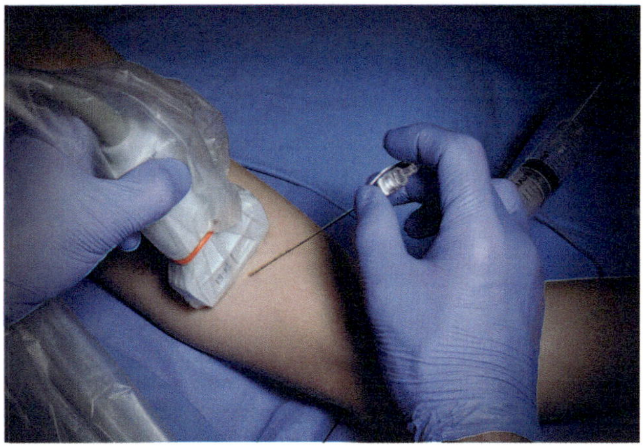

Fig. 5.122 Ulnar truncal block in the distal third of the arm. Needle inserted out-of-plane

Echoanatomy and Localization

In the distal third of the arm, the ulnar nerve is easy to locate. It is located immediately deep to the brachialis fascia, in immediate proximity to the upper ulnar collateral artery, behind the medial brachial intermuscular septum. Ultrasound makes it possible to easily locate these neurovascular structures (Fig. 5.113). At this level, a simple and superficial approach makes it possible to perform, if necessary, a combined block of the median and ulnar nerves via a single injection site with a single redirection of the needle (Figs. 5.116 and 5.122). Depending on their relative positions, a separate block of both nerves may be required.

At the elbow, nerve blocks performed in the olecranon groove should be avoided because of its lack of compliance and the potential for risk of traumatic/compressive damage to the ulnar nerve. An approach in the proximal forearm is

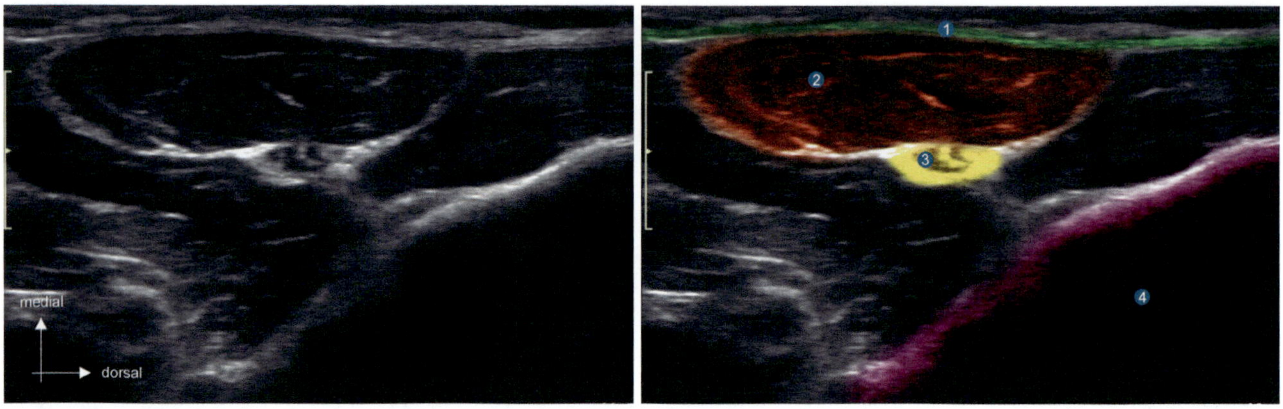

Fig. 5.123 Transverse ultrasound section of ulnar nerve in the distal third of the forearm

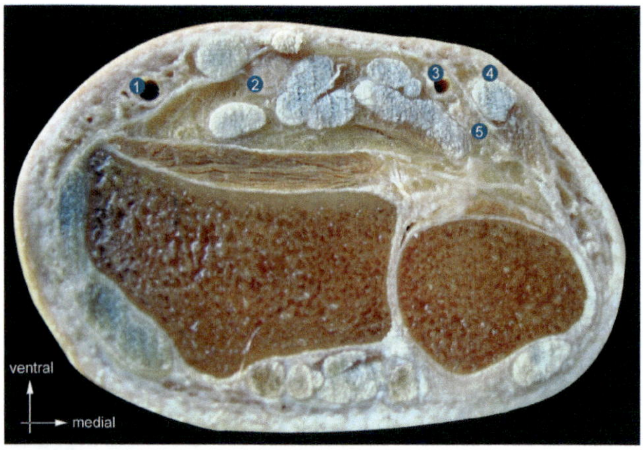

Fig. 5.124 Distal anatomical section of the forearm cutting through radioulnar joint

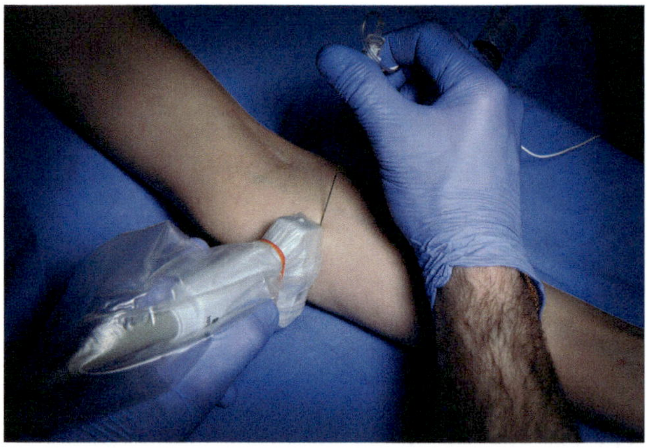

Fig. 5.125 Ulnar block at the proximal aspect of the forearm. Needle inserted in-plane

preferred after emergence of the nerve from the olecranon groove. In this region the ulnar nerve is easy to locate and approach, located deep to flexor carpi ulnaris muscle (FCU) (Fig. 5.123).

In the distal forearm, the ulnar nerve is easy to locate laterally to the tendon of the FCU (Fig. 5.124), accompanied by the ulnar artery from the middle third of the forearm.

Injection

Ulnar Nerve Block in the Distal Arm

The probe is positioned transversely on the medial aspect of the distal arm, as with block of the median nerve. For an in-plane approach, the needle is inserted at one end of the probe, and advanced until adjacent to the ulnar nerve (Fig. 5.116). However, the nerve can also be approached out-of-plane, with the aid of hydrolocalisation to identify the needle tip during its insertion (Fig. 5.122).

Ulnar Nerve Block in the Proximal Forearm

With the probe positioned tranversely in relation to the nerve axis, the ulnar nerve is located deep to FCU and can be approached in or out of plane (Figs. 5.125 and 5.126).

Ulnar Nerve Block in the Distal Forearm

The probe is placed transversely to the longitudinal axis of the nerve, at the medial border of the distal forearm. Due to the tendon of the FCU muscle lying ventrally (with the ulna lying dorsally), the needle is more easily inserted out-of-plane (Fig. 5.127). The nerve is cautiously approached by performing hydrolocalisation, taking care to not puncture it or the adjacent ulnar blood vessels (Fig. 5.128).

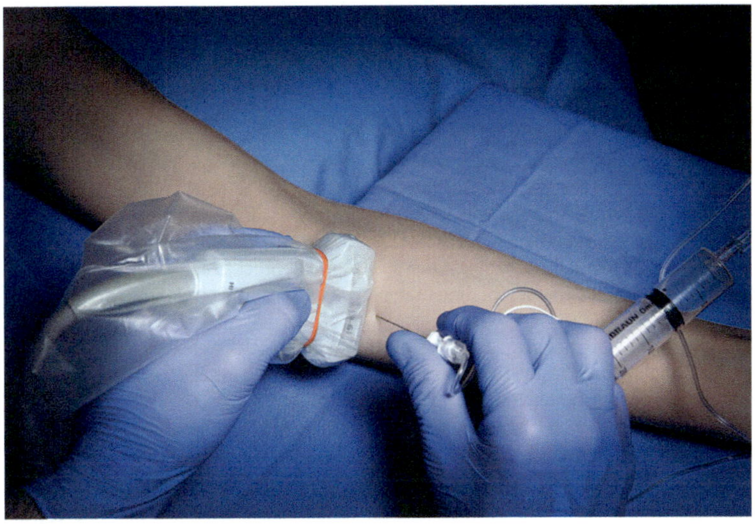

Fig. 5.126 Ulnar block at the proximal aspect of the forearm. Needle inserted out-of-plane

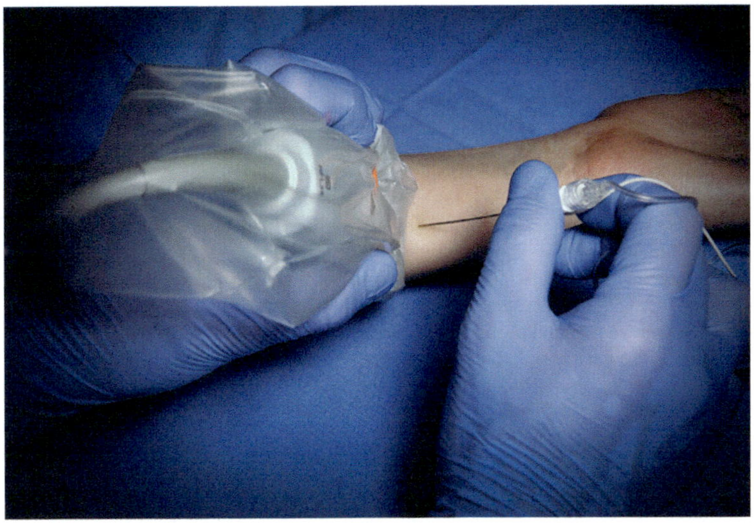

Fig. 5.127 Ulnar block at the distal third of the forearm. Needle inserted out-of-plane

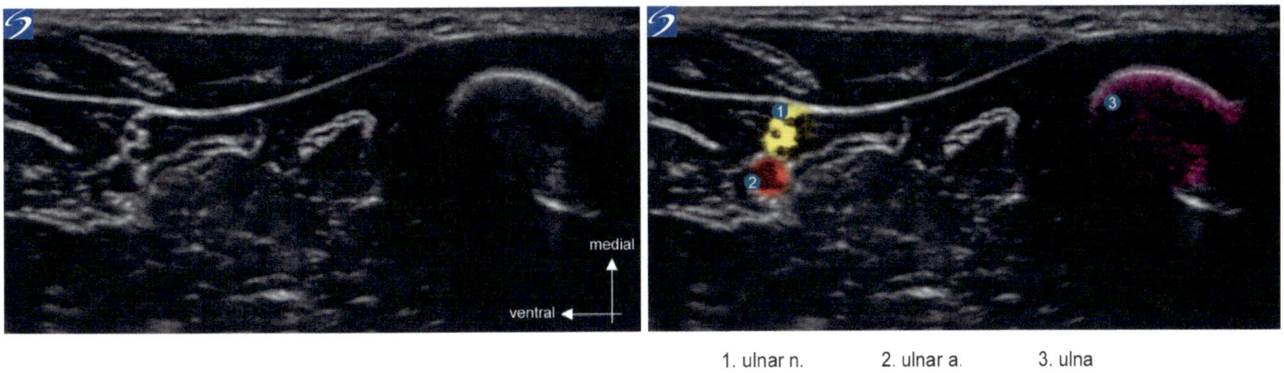

Fig. 5.128 Sonoanatomy. Ulnar nerve at the distal aspect of the forearm

References

1. Kamina P. Précis d'anatomie clinique. Tome 1. Paris: Maloine; 2003.
2. Seddon HJ. Surgical disorders of the peripheral nerves. Édimbourg: Churchill Livingstone; 1972.
3. Borene SC, Edwards JN, Boezaart AP. At the cords, the pinkie towards: interpreting infraclavicular motor response to neurostimulation. Reg Anesth Pain Med. 2004;29:125–9.
4. Lanz E, Theiss D, Jankovic D. The extent of blockade following various techniques of brachial plexus block. Anesth Analg. 1983;62:55–8.
5. Gaertner E, Choquet O, Macaire P, Zetlaoui PJ. Anesthésie régionale, anesthésie tronculaire et plexique de l'adulte. Rueil-Malmaison: Arnette; 2001.
6. Gaertner E, Estebe JP, Zamfir A, Cuby C, Macaire P. Infraclavicular plexus block: multiple injection versus single injection. Reg Anesth Pain Med. 2002;27:590–4.
7. Lavoie J, Martin R, Tetrault JP, Cote DJ, Colas MJ. Axillary plexus block using a peripheral nerve stimulator: single or multiple injections. Can J Anaesth. 1992;39:583–6.
8. Gaertner E, Bouaziz H. Ultrasound-guided interscalene block. Ann Fr Anesth Réanim. 2012;31:e213–8.
9. Nadeau MJ, Lévesque S, Dion N. Ultrasound-guided regional anesthesia for upper limb surgery. Can J Anaesth. 2013;60:304–20.
10. Meier G, Bauereis C, Maurer H. The modified technique of continuous suprascapular nerve block. A safe technique in the treatment of shoulder pain. Anaesthesist. 2002;51:747–53.
11. Vandepitte C, Gautier P, Xu D, Salviz EA, et al. Effective volume of ropivacaine 0.75% through a catheter required for interscalene brachial plexus blockade. Anesthesiology. 2013;118:863–7.
12. Falcão LF, Perez MV, de Castro I, et al. Minimum effective volume of 0.5% bupivacaine with epinephrine in ultrasound-guided interscalene brachial plexus block. Br J Anaesth. 2013;110:450–5.
13. Madison SJ, Humsi J, Loland VJ, et al. Ultrasound-guided root/trunk (interscalene) block for hand and forearm anesthesia. Reg Anesth Pain Med. 2013;38:226–32.
14. Liu SS, Gordon MA, Shaw PM. A prospective clinical registry of ultrasound-guided regional anesthesia for ambulatory shoulder surgery. Anesth Analg. 2010;111:617–23.
15. Krone SC, Chan VW, Regan J, et al. Analgesic effects of low-dose ropivacaïne for interscalene brachial plexus block for outpatient shoulder surgery: a dose-finding study. Reg Anesth Pain Med. 2001;26:439–43.
16. Fredrickson MJ, Abeysekera A, Price DJ, et al. Patient-initiated mandatory boluses for ambulatory continuous interscalene analgesia: an effective strategy for optimizing analgesia and minimizing side-effects. Br J Anaesth. 2011;106:239–45.
17. Borgeat A, Schappi B, Biasca N, Gerber C. Patient-controlled analgesia after major surgery: patient controlled interscalene analgesia vs patient-controlled analgesia. Anesthesiology. 1997;87:1343–7.
18. Borgeat A, Tewes E, Biasca N, Gerber C. Patient-controlled interscalene analgesia with ropivacaine after major surgery: PCIA vs PCA. Br J Anaesth. 1998;81:603–5.
19. Singelyn FJ, Seguy S, Gouverneur JM. Interscalene brachial plexus analgesia after open shoulder surgery: continuous versus patient-controlled infusion. Anesth Analg. 1999;89:1216–20.
20. Sato S, Yamashita S, Iwai M, Mizuyama K, Satsumae ET. Continuous interscalene block for cancer pain. Reg Anesth Pain Med. 1994;19:385–6.
21. Rawal N, Axelsson K, Hylander J, et al. Post-operative patient-controlled local anesthetic administration at home. Anesth Analg. 1998;86:86–9.
22. Macaire P, Gaertner E, Capdevila X. Continuous post-operative regional analgesia at home. Minerva Anestesiol. 2001;67:109–16.
23. Nielsen KC, Greengrass RA, Pietrobon R, Klein SM, Steele SM. Continuous interscalene brachial plexus blockade provides good analgesia at home after major shoulder surgery. Report of four cases. Can J Anaesth. 2003;50:57–61.
24. Ilfeld BM, Enneking FK. A portable mechanical pump providing over four days of patient-controlled analgesia by perineural infusion at home. Reg Anesth Pain Med. 2002;27:100–4.
25. Martinoli C, Bianchi S, Santacroce E, Pugliese F, Graif M, Derchi LE. Brachial plexus sonography: a technique for assessing the root level. Am J Roentgenol. 2002;179:699–702.
26. Cash CJ, Sardesai AM, Berman LH, et al. Spatial mapping of the brachial plexus using three-dimensional ultrasound. Br J Radiol. 2005;78:1086–94.
27. Harry WG, Bennett JD, Guha SC. Scalene muscles and the brachial plexus: anatomical variations and their clinical significance. Clin Anat. 1997;10:250–2.
28. Ediale KR, Myung CR, Neuman GG. Prolonged hemidiaphragmatic paralysis following interscalene brachial plexus block. J Clin Anesth. 2004;16:573–5.
29. Bashein G, Robertson HT, Kennedy WF Jr. Persistent phrenic nerve paralysis following interscalene brachial plexus block. Anesthesiology. 1985;63:102–4.
30. Robaux S, Bouaziz H, Boisseau N, Raucoules-Aime M, Laxenaire MC. Persistent phrenic nerve paralysis following interscalene brachial plexus block. Anesthesiology. 2001;95:1519–21.
31. Franco CD, Gloss FJ, Voronov G, Tyler SG, Stojilkovic LS. Supraclavicular block in obese population: an analysis of 2020 blocks. Anesth Analg. 2006;102:1252–4.
32. Kjoscielnak-Nielsen ZJ, Frederiksen BS, Rasmussen H, et al. A comparison of ultrasound-guided supraclavicular and infraclavicular blocks for upper extremity surgery. Acta Anaesthesiol Scand. 2009;53:620–6.
33. Fredrickson MJ, Patel A, Young S, Chinchanwala S. Speed of onset of « corner pocket supraclavicular » and infraclavicular ultrasound guided brachial plexus block: a randomised observer-blinded comparison. Anesthesia. 2009;64:738–44.
34. Moore DC. Complications induced by block techniques. In: Anesthésiques locaux en anesthésie et réanimation. Paris: Arnette; 1973. p. 397–412.
35. Brand L, Papper EM. A comparison of supraclavicular and axillary techniques for brachial plexus block. Anesthesiology. 1961;22:226–9.
36. SFAR. Recommandations pour la pratique clinique. Les blocs nerveux périphériques chez l'adulte. Paris: SFAR; 2003. www.sfar.org.
37. Dupré LJ. Le bloc du plexus brachial par voie sus-claviculaire. www.i-alrf.asso.fr, techniques.
38. Cornish PB, Greenfield LJ, O'Reilly M, Allan L. Indirect versus direct measurement of brachial plexus depth. Anesth Analg. 1999;88:1113–6.
39. Kulenkampff D, Persky A. Brachial plexus anesthesia: its indications, technique and dangers. Ann Surg. 1928;87:883–91.
40. Pauchet V, Sourdat P, Labat G, de Butler d'Ormont R. Anesthésie du plexus brachial. In: Pauchet V, editor. L'anesthésie régionale. Paris: Doin; 1927. p. 146–68.
41. Moore DC. Supraclavicular approach for block of the brachial plexus. In: Moore DC, editor. Regional block: a handbook for use in the clinical practice of medicine and surgery. 4th ed. Springfield, IL: CC Thomas Publisher; 1965. p. 221–42.
42. Dupré LJ, Danel V, Legrand JJ, Stieglitz P. Surface landmarks for supraclavicular block of the brachial plexus. Anesth Analg. 1982;61:28–31.
43. Winnie AP, Collins VJ. The subclavian perivascular technique of brachial plexus anesthesia. Anesthesiology. 1964;25:353–63.
44. Pham-Dang C, Gunst JP, Gouin F, et al. A novel supraclavicular approach to brachial plexus block. Anesth Analg. 1997;85:111–6.

45. Vongvises P, Panijayanond T. A parascalene technique of brachial plexus anesthesia. Anesth Analg. 1979;58:267–73.
46. Brown DL, Cahill DR, Bridenbaugh LD. Supraclavicular nerve block: anatomic analysis of a method to prevent pneumothorax. Anesth Analg. 1993;76:530.
47. Gupta K, Bhandari S, Singhal D, et al. Pneumothorax following ultrasound guided supraclavicular brachial plexus block. J Anaesthesiol Clin Pharmacol. 2012;28:543–4.
48. Sandhu NS, Capan LM. Ultrasound-guided infraclavicular brachial plexus block. Br J Anaesth. 2002;89:254–9.
49. Chan VW, Perlas A, Rawson R, Odukoya O. Ultrasound-guided supraclavicular brachial plexus block. Anesth Analg. 2003;97:1514–7.
50. Kilka HG, Geiger P, Mehrkens HH. Infraclavicular vertical brachial plexus blockade. A new method for anesthesia of the upper extremity. An anatomical and clinical study. Anaesthesist. 1995;44:339–44.
51. Neuburger M, Kaiser H, Ass B, Franke C, Maurer H. Vertical infraclavicular blockade of the brachial plexus (VIP). A modified method to verify the puncture point under consideration of the risk of pneumothorax. Anaesthesist. 2003;52:619–24.
52. Whiffler K. Coracoid block: a safe and easy technique. Br J Anaesth. 1981;53:845–8.
53. Raj PP, Montgomery SJ, Nettles D, Jenkins MT. Infraclavicular brachial plexus block: a new approach. Anesth Analg. 1973;52:897.
54. Sims JK. A modification of landmarks for infraclavicular approach to brachial plexus block. Anesth Analg. 1977;56:554–6.
55. Borgeat A, Ekatodramis G, Dumont C. An evaluation of the infraclavicular block via a modified approach of the Raj technique. Anesth Analg. 2001;93:436–41.
56. Moayeri N, Renes S, van Geffen GJ, Groen GJ. Vertical infraclavicular brachial plexus block: needle redirection after elicitation of elbow flexion. Reg Anesth Pain Med. 2009;34:236–41.
57. Perlas A, Chan VW, Simons M. Brachial plexus examination and localization using ultrasound and electrical stimulation. Anesthesiology. 2003;99:429–35.
58. Bloc S, Garnier T, Komly B, et al. Single-stimulation, low-volume infraclavicular plexus block: influence of the evoked distal motor response on success rate. Reg Anesth Pain Med. 2006;31:433–7.
59. Bloc S, Garnier T, Komly B, et al. Spread of injectate associated with radial or median nerve type motor response during infraclavicular brachial plexus block: an ultrasound evaluation. Reg Anesth Pain Med. 2007;32:130–5.
60. Dalens B. Anesthésie locorégionale de la naissance à l'âge adulte. Paris: Pradel; 1993.
61. Swerdlow M. Complications of local anesthetic neural blockade. In: Cousins MJ, Bridenbaugh P, editors. Neural blockade in clinical anesthesia. Philadelphia: Lippincott; 1980. p. 526–42.
62. Retzl G, Kapral S, Greher M, Mauritz W. Ultrasonographic findings of the axillary part of the brachial plexus. Anesth Analg. 2001;92:1271–5.
63. Chriblockhe JL, Berthier F, Boillot A, et al. Assessment of topographic brachial plexus nerves variations at the axilla using ultrasonography. Br J Anaesth. 2009;103:606–12.
64. Remerand F, Laulan J, Couvret C, et al. Is the musculocutaneous nerve really in the coracobrachialis muscle when performing an axillary block? An ultrasound study. Anesth Analg. 2010;110: 1729–34.
65. Testut L. Les anomalies musculaires chez l'homme expliquées par l'anatomie comparée. Leur importance en anthropologie. Paris: Masson; 1884. p. 411.
66. Dalens BJ. Infraclavicular brachial plexus blocks. In: Dalens B, editor. Pediatric regional anesthesia. Boca Raton: CRC Press; 1990. p. 241–55.
67. Coventry DM, Barker KF, Thomson M. Comparison of two neurostimulation techniques for axillary brachial plexus blockade. Br J Anaesth. 2001;86:80–3.
68. Sia S, Lepri A, Ponzecchi P. Axillary brachial plexus block using peripheral nerve stimulator: a comparison between double- and triple-injection techniques. Reg Anesth Pain Med. 2001;26:499–503.
69. Adnan T, Elif A, Ayse A, Gülnaz A. Clonidine as an adjuvant for lidocaine in axillary brachial plexus block in patients with chronic renal failure. Acta Anesthesiol Scand. 2005;49:563–8.
70. Carles M, Pulcini A, Macchi P, Duflos P, Raucoules-Aime M, Grimaud D. An evaluation of the brachial plexus block at the humeral canal using a neurostimulator (1417 patients): the efficacy, safety, and predictive criteria of failure. Anesth Analg. 2001;92:194–8.
71. Gaertner E, Kern O, Mahoudeau G, Freys G, Golfetto T, Calon B. Block of the brachial plexus branches by the humeral route. A prospective study in 503 ambulatory patients. Proposal of a nerve-blocking sequence. Acta Anaesthesiol Scand. 1999;43: 609–13.
72. Gaertner E. Pourquoi je fais un bloc au canal huméral. Prat Anesth-Reanim. 2001;5:277–9.
73. Güntz E, Herman P, Delbos A, Sosnowski M. The radial nerve should be blocked before the ulnar nerve during a brachial plexus block at the humeral canal. Can J Anaesth. 2004;51:354–7.

Lumbar Plexus Blocks

6

Eryk Eisenberg, Elisabeth Gaertner, and Philippe Clavert

Contents

The original version of the chapter has been revised. A correction to this chapter can be found at https://doi.org/10.1007/978-3-031-08804-9_13

Supplementary Information The online version contains supplementary material available at https://doi.org/10.1007/978-3-031-08804-9_6.

E. Eisenberg (✉)
Département d'anesthésie et réanimation, Clinique Pôle Santé République, Clermont-Ferrand, France

E. Gaertner
Département d'anesthésie et réanimation chirurgicale, Hôpitaux Civils de Colmar, Colmar, France

P. Clavert
Service de chirurgie du membre supérieur, Centre de Chirurgie Orthopédique et de la Main (CCOM), Illkirch, France

E. Eisenberg, E. Gaertner (eds.), *Ultrasound in Peripheral, Neuraxial and Perineuraxial Regional Anaesthesia*,
https://doi.org/10.1007/978-3-031-08804-9_6

Anatomy

The lumbar plexus (LP) (Fig. 6.1a) participates in innervation of the abdominal wall, the external genitalia and the lower limb (Fig. 6.1b). It consists at origin of the union of the ventral branches of the spinal nerves L1, L2, L3 and L4 (a branch of the ventral branch of T12 frequently joins it with L1). It is located between the two heads of the psoas major muscle laterally in relation to the lumbar vertebral column, in a cleaved, aponeurotic space, demarcated dorsally by muscle fibres which insert on the transverse process (TP) and ventrally by fibres which insert on the lateral aspects of the vertebral bodies and on the corresponding discs. In this space, the ascending lumbar vein also passes. Anastomoses and interconnections of these different branches eventually form the terminal branches of the lumbar plexus (Fig. 6.2) which are the **iliohypogastric** (L1), **ilioinguinal** (L1), **genitofemoral** (L1-L2), **lateral cutaneous of the thigh** (L2-L3), **femoral** (L2-L3-L4) and **obturator nerves** (L2-L3-L4) as follows [1]:

- The anterior ramus of L1 divides into two branches, the iliohypogastric and ilioinguinal common nerve trunk, and the upper branch of the genitofemoral nerve.
- The anterior ramus of L2 gives rise to the inferior branch of the genitofemoral nerve and the other three branches which participate in the lateral cutaneous nerve, femoral and obturator nerve of the thigh, respectively.
- The anterior ramus of L3 gives rise to three components for the cutaneous lateral nerve of the thigh, the femoral and obturator nerve, respectively.
- The anterior ramus of L4 participates in the femoral nerve, the obturator nerve, and in the lumbosacral trunk which goes to the sacral plexus.

a

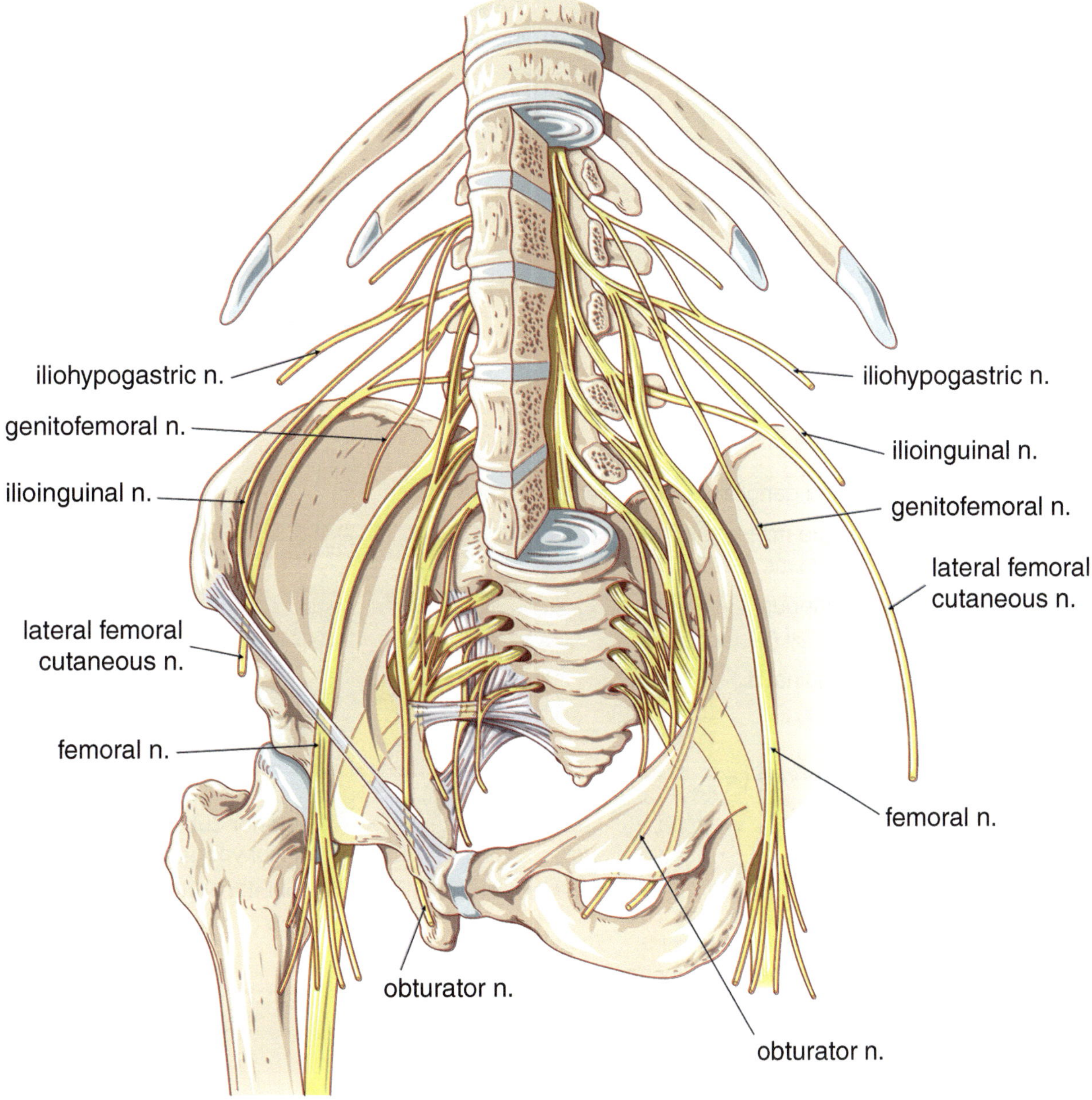

Fig. 6.1 (**a**) Diagram for overall organisation of the lumbar, sacral and pudendal nerve plexus (from Netter FH. *Atlas of human anatomy*. third ed. Paris: Masson, 2007). (**b**) Nerves of the abdominal wall (from Netter FH. *Atlas of human anatomy*. third ed. Paris: Masson, 2007)

b

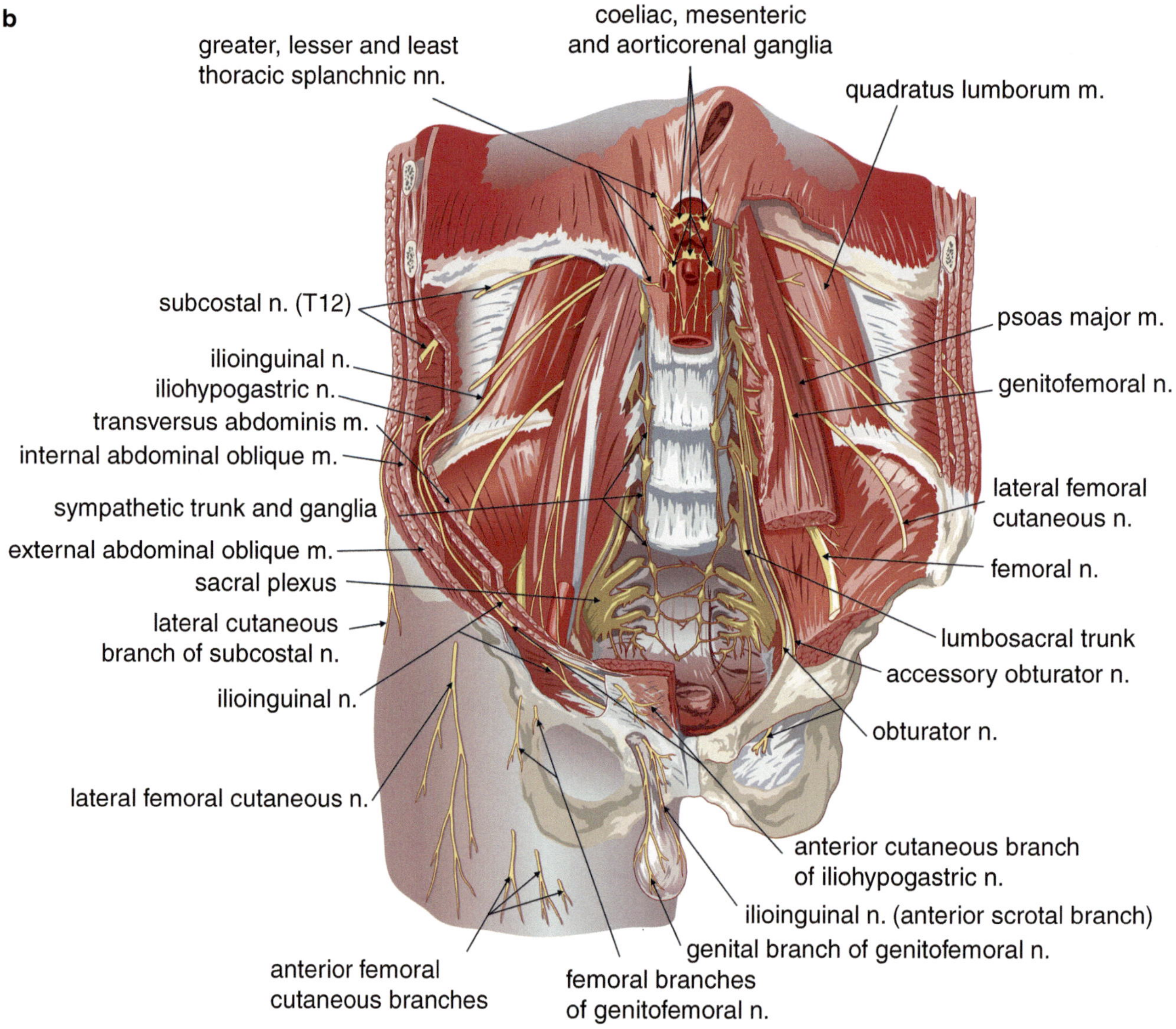

Fig. 6.1 (continued)

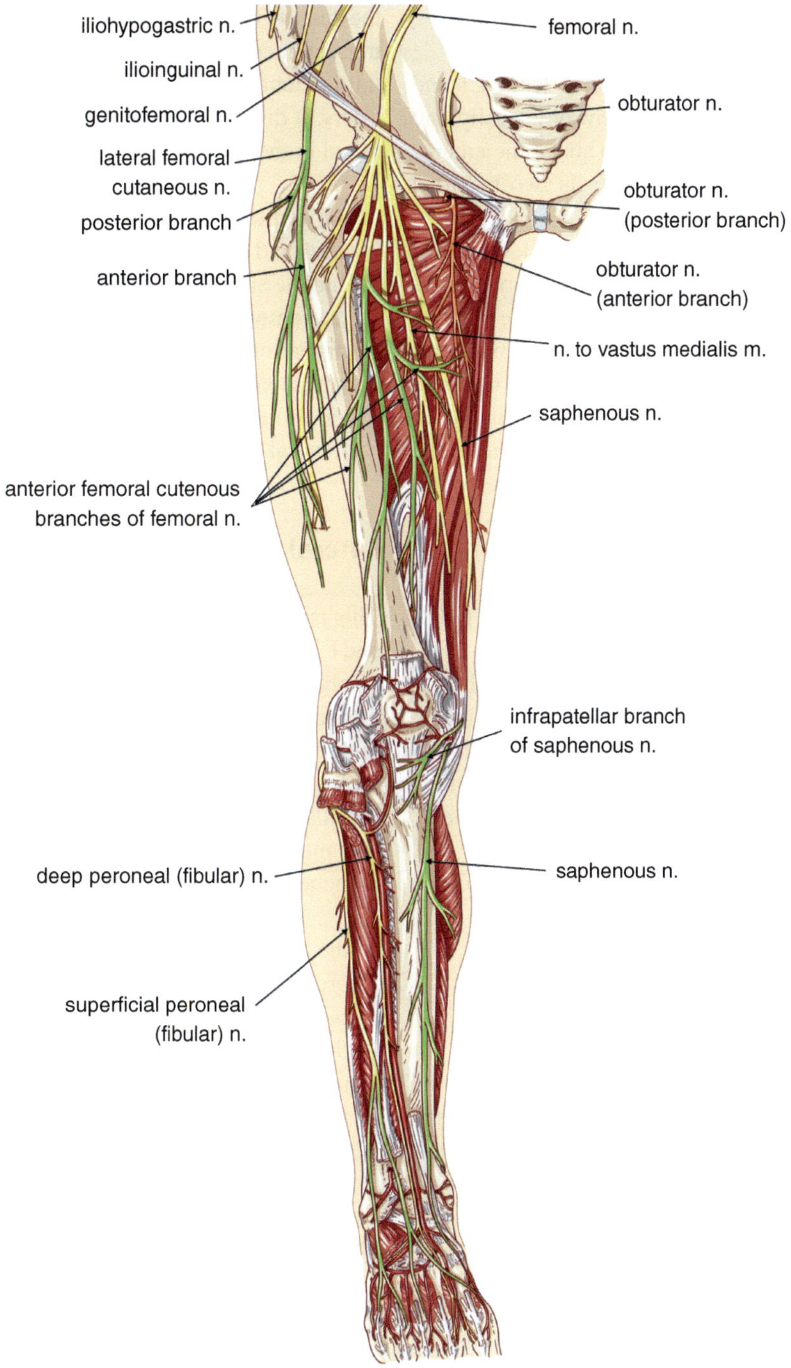

Fig. 6.2 Anterior view of the lower limb with depiction of nerve elements which branch off mainly from the lumbar plexus

Iliohypogastric Nerve (IH)

This is a **mixed** nerve which receives fibres from L1 and T12 and which follows a curvilinear pathway laterally and caudally, within the abdominal wall up to above the antero-superior iliac spine where it divides into its two terminal branches.

Successively, after its emergence behind the lateral border of the psoas major muscle, it crosses the ventral aspect of the fascia and the quadratus lumborum muscle passing under the 12th intercostal nerve and behind the renal compartment, and then continues above the iliac crest between transverses abdominus and internal oblique muscles to which it gives rise to its collateral motor branches. The terminal branches of the iliohypogastric nerve are represented by:

- the **lateral** cutaneous branch which, after crossing the external oblique muscle above the iliac crest, innervates the skin of the anterior gluteal area.
- the **anterior** sensory-motor cutaneous branch, which continues along the cephalad border of the inguinal ligament, gives rise to the muscular branches for the pyramidalis and rectus abdominis muscles, and cutaneous brunches for the inguinal fold, the pubis, the scrotum or the labia majora.

This nerve anastomoses with the ilioinguinal nerve and the subcostal nerve (12th intercostal nerve).

Ilioinguinal Nerve (II)

It is a **sensory** nerve which also receives fibres from L1 and T12. It has the same pathway and the same relations as the iliohypogastric nerve below which it lies. By its cutaneous branches, it provides sensation to the supramedial area of the thigh, as well as the scrotum or the labia majora.

Genitofemoral Nerve (GF)

It is a **mixed** nerve which originates from L1 and L2. It continues caudally and ventrally across the psoas major muscle from which it emerges opposite L3. While descending under the fascia iliaca and the ventral aspect of the psoas major muscle, it crosses the ureter dorsally and the gonadal vessels until it continues along the lateral border of the external iliac artery. It divides above the inguinal ligament into its two terminal branches:

- the **genital** branch, medially, which follows the spermatic cord and innervates the cremaster muscle, the scrotum and the skin of the pubis.
- the **femoral** branch, laterally, which follows the external iliac artery, passes under the inguinal ligament in the femoral sheath and then penetrates into the fascia lata to provide innervation to the skin of the proximal part of the femoral triangle.

Lateral Cutaneous Nerve of the Thigh (LCT)

This is a **sensory** nerve whose origins are at the level of L2 and L3. After crossing the psoas major muscle, from which it emerges laterally at the iliac crest, it continues its oblique pathway caudally and laterally, covered by the peritoneum, on the surface of the iliac muscle. It passes under or across the inguinal ligament, about 1 cm inside of the anterosuperior iliac spine. It then crosses the ventral aspect of the sartorius muscle where, usually, it divides into its two terminal branches:

- the **dorsal** branch, which, after crossing the fascia lata, provides cutaneous innervation of the supero-lateral area of the thigh.
- the **ventral** branch, which descends vertically and, after crossing the fascia lata a little lower, will innervate the anterolateral cutaneous area of the thigh up to the knee.

Obturator Nerve

It is a **mixed** nerve, a terminal branch of the lumbar plexus, whose origins are at L2, L3 and L4. Taking a vertical pathway, it emerges at the medial border of the psoas major muscle, it then crosses behind the bifurcation of the common iliac vessels and penetrates into the pelvic cavity laterally to the lumbosacral trunk, ventrally in relation to the sacroiliac joint. It then continues laterally, ventrally and caudally, along the internal obturator muscle, and emerges from the pelvis via the obturator canal where it gives rise to its collateral branches:

- the **articular** branches for the ventral aspect of the hip joint.
- the nerve of the **external obturator** muscle.

  It then divides into two terminal branches:

  the **anterior** branch which lies between pectineus and adductor longus muscles ventrally, and the obturator externus and adductor brevis muscles dorsally. It innervates the adductor longus and brevis, gracilis, often pectineus, and inconsistently [2] gives rise to a cutaneous branch to the medial thigh.

  the **dorsal** branch which descends between the adductor brevis and longus muscles, provides their motor innervation and also articular branches for the hip joint. It then follows the femoral artery and popliteal artery and innervates the dorsal aspect of the knee joint and popliteal artery.

The obturator nerve anastomoses with the saphenous nerve comprising the sub-sartorial plexus, and with the accessory obturator nerve, when present.

Accessory Obturator Nerve

This **mixed** small nerve, which is found in about 12% of cases, should be considered as an erratic branch of the femoral nerve and/or of the obturator nerve. It originates from L3 and L4, travels along with and cranially in relation to the obturator nerve, and then deviates from it to pass ventrally into the pubis, medially to the iliopectinate eminence and ends dorsally in relation to pectineus. It gives rise to an articular branch to the hip joint, and its terminal branches are represented by the following:

- **cutaneous** branches for the proximal part of the femoral triangle,
- a **muscular** branch for pectineus,
- a **vascular** branch for the femoral artery,

Femoral Nerve

This is a **mixed** nerve which originates from the anterior rami of L2, L3 and L4. It emerges laterally from psoas major along the groove formed by this muscle and iliacus, where it is covered by the fascia iliaca. It descends vertically, passing in front of and then along the outer aspect of psoas major, it passes under the inguinal ligament laterally in relation to the artery from which it is separated by the iliopectinate arch and enters the thigh. Its **collateral** branches which arise cranially in relation to the inguinal ligament are:

- **muscular** branches for iliacus, psoas major and pectineus,
- a **vascular** branch for the femoral artery.

Its terminal branches are muscular and cutaneous.

Muscular Branches

These are:

- nerves to **sartorius muscle**,
- nerves to **quadriceps femoris muscle**:

  nerve to rectus femoris muscle,

  nerve to vastus lateralis, which lies deep to sartorius,

  nerve to vastus medialis, which descends into the adductor canal, laterally to the blood vessels and saphenous nerve, and gives rise to an articular branch for the knee,

  nerve to vastus intermedius,
- nerves to the medial group of muscles, which cross anterior to the femoral vessels to innervate pectineus and adductor longus muscles, the hip joint and the profunda femoris artery.

Cutaneous Branches

These are:

- the **anterior cutaneous nerves of the thigh**.
- the **saphenous nerve**. The longest and largest terminal branch of the femoral nerve, it is first located lateral to the

femoral vessels, running with the nerve to vastus medialis. Then, it crosses the femoral artery and continues along its medial border to the caudal part of the adductor canal, where it passes deep to sartorius and its aponeurotic sheath, generally accompanied by the descending geniculate artery. Descending deep to sartorius, it is in proximity to the knee joint interspace where it again pierces the aponeurosis to become subcutaneous. Then, while continuing its descent it is adjacent to the long saphenous vein. At this level it has already given rise to collateral branches which are **vascular** (for the femoral artery) and **cutaneous** (infra-patellar branch and medial cutaneous branches of the leg). Its ventral and dorsal terminal branches spread out on the back of the foot, particularly along on its medial border in a variable manner.

– the **accessory saphenous nerve** which follows the femoral artery medially and gives rise to a superficial branch which accompanies the long saphenous vein and a deep branch, which after travelling adjacent to the femoral artery, enters the sub-sartorial fascia and participates in the plexus of the same name to innervate the anteromedial aspect of the knee.

The femoral nerve forms **anastomoses** with the genito-femoral nerve in the upper part of the thigh (ventral cutaneous nerves of the thigh) and with the anterior branch of the obturator nerve which forms, along with the saphenous nerve, the sub-sartorial plexus.

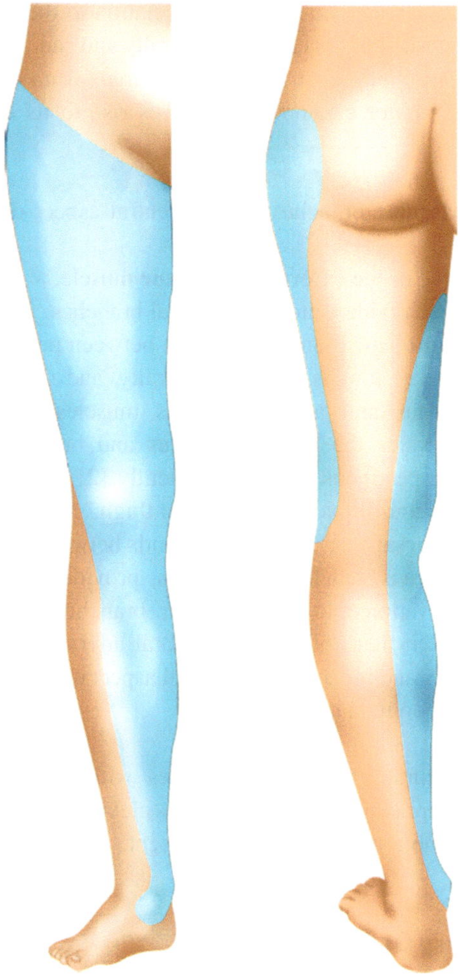

Fig. 6.3 Area of coverage of the lumbar plexus block by posterior approach

Lumbar Plexus Blocks

Several techniques exist to perform anaesthesia of the lumbar plexus. The posterior approaches which have been described [3–7] generally enable anaesthesia of all the aforementioned terminal branches with a single injection, because it involves true plexus block. However, in an analysis of the literature, serious adverse events of posterior approach to the LP are not uncommon, even if performed by experienced anaesthesiologists [8]. The risks inherent in these approaches are related to the proximity of the spinal canal and the potential for peri-spinal spread. Whenever the aim is to anaesthetise the nerves of the lumbar plexus by a more distal approach, usually it is necessary to perform separate distal blocks. In fact, in a so-called “3-in-1” or “ilio-fascial” block, extension of anaesthesia to all of the femoral, obturator and lateral cutaneous nerves of the thigh is inconsistent [9], which can be readily understood when topographic divergence of these terminal branches is observed.

Lumbar Plexus Block by Posterior Approach (Fig. 6.3)

Indications

A posterior lumbar plexus block (PLPB) is indicated for complicated surgery of the knee and of the hip (anaesthesia for femoral neck and peri-trochanteric fractures, among other fractures [10]). A subcutaneous injection in the apex of the surgical approach supplements cutaneous anaesthesia in the area of the cluneal and subcostal nerves. In combination with sciatic nerve block, it enables surgery of the lower limb, with a 12 to 22% failure rate [11–14]. It also enables early post-operative analgesia after hip arthroplasty [15]. Compared to opioids after total hip replacement surgery, it decreases pain for 4–8 hours post-operatively. This effect is prolonged by placement of a catheter. Compared to other regional techniques for post-operative analgesia of the hip, a continuous lumbar block offers analgesia equivalent to epidural anaesthesia and is more effective than a continuous femoral block. For post-operative analgesia of the knee, this technique is reserved for complicated surgery (total knee replacement). As for the hip, it offers analgesia equivalent to epidural anaesthesia; it is as effective as a femoral block in the areas of the femoral nerve and the lateral cutaneous nerve of the thigh but extends more regularly to the obturator nerve [11, 15]. Complications of lumbar plexus block by posterior approach are of concern and have made anaesthesiologists reticent to perform them. Epidural extension is the most common adverse event. It can be due to spread of the local anaesthetic into the epidural nerve sleeves [16], or to direct needle puncture or to an aberrant pathway of the catheter in the epidural space [17]. Its incidence varies depending on technique: 16% according to Parkinson [6], 88 to 90% for Chayen [5], and 1 to 9% for Piffaut [18]. Usually it has no clinical impact but justifies weighing the indication in a patient who is in a precarious haemodynamic condition. It should be noted that a death has been reported by the SOS LRA group of an ASA4 patient with epidural spread of a lumbar block [19]. Spinal puncture is a major adverse event. It can be diagnosed at time of injection by repeated aspiration techniques, a test dose, a slow and fractionated injection and radiological control of the catheter. These preventive measures are recommended to avoid total spinal anaesthesia which may have dramatic consequences [20]. Since this is a deep block and direct vascular pressure cannot be applied, it is not recommended to perform it in a patient with acquired or innate coagulation disorders. Retroperitoneal haematomas have been reported in the literature [20]. Intravascular injection [21, 22], which is not specific to this block, should be sought systematically, all the more so since the aspiration test is useful only if it is positive. Slow injection of local anaesthetics is the best guarantee of prevention of initial toxic adverse events. Systemic toxic events are also promoted by excessive doses in combined nerve block which require a higher total dosage of local anaesthetic.

The utility of the posterior approach of the lumbar plexus is related to better LA coverage of the obturator nerve which is inconsistently blocked by classical anterior approaches. The technique for PLPB currently recommended with neurostimulation differs from Winnie's traditional approach by more cephalic and medial injection. Search for contact of the needle with the bony landmark of the transverse process of L4, before electrostimulation of a branch of the plexus, is a "safeguard" enabling better prediction of the depth of the plexus. The localisation of L4 transverse process can be done visually under ultrasound control [23, 24]. This may limit the risk of an injection that is too deep (peritoneal or renal puncture) or of an injection made too medially (total spinal anaesthesia, epidural anaesthesia). Figure 6.4 is a decision-making tree for the injection procedure, in which progression 1 corresponds to the ideal injection. Ultrasound enables us either to obtain direct guidance to the transverse process (which limits "wandering" of progressions 2 and 4), or immediately a progression 3 type. The correct position of the needle is confirmed by neurostimulation and obtaining the appropriate motor response, i.e. quadricipital muscle twitches (ascension of the patella).

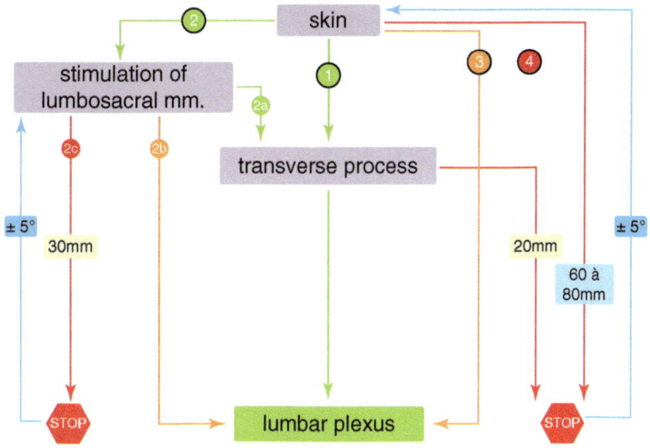

Fig. 6.4 Decision-making algorithm for performance of a lumbar plexus block by posterior approach [23]

What ultrasound provides:

- **visualisation of the transverse processes**: Their location is determined, thus avoiding screening for them "blindly" as a "safeguard". Their depth can be calculated and that of the lumbar plexus when it is sufficiently visible. This enables an estimate of the distance from the transverse process (bone contact sought with the needle in a block performed with neurostimulation only) to the lumbar plexus in the psoas muscle (Fig. 6.5);
- Under good conditions of echogenicity, the **position of the lumbar plexus within the psoas muscle** can be visualised (the psoas compartment) (Figs. 6.6, 6.7, 6.8, 6.9 and 6.10). The intramuscular space in which the plexus continues is clearly visible in thin adults and in children, and provides a relatively specific image (Fig. 6.5). In this case, echoguidance of the needle up to contact of the plexus is possible [25, 26], the principal difficulties remaining are optimum position of the probe, visibility of the target and of the needle tip, which one must seek to place strictly in the ultrasound plane along its entire pathway.

Figures 6.11 and 6.12 show a lumbar plexus root on an ultrasound section, paramedian, sagittal and transversal, respectively.

Conduct of a lumbar plexus block by ultrasound-guided posterior approach nevertheless runs into difficulty with increasing depth: it is not always possible to obtain a good quality image. Therefore, since ultrasound does not enable a very precise visual control of the relations between the tip of the needle and the nerves, it is neurostimulation which confirms the correct position of the needle (if it is not extraneural, at least extrafascicular [SMI > 0.3 mA; 0.1 mS]).

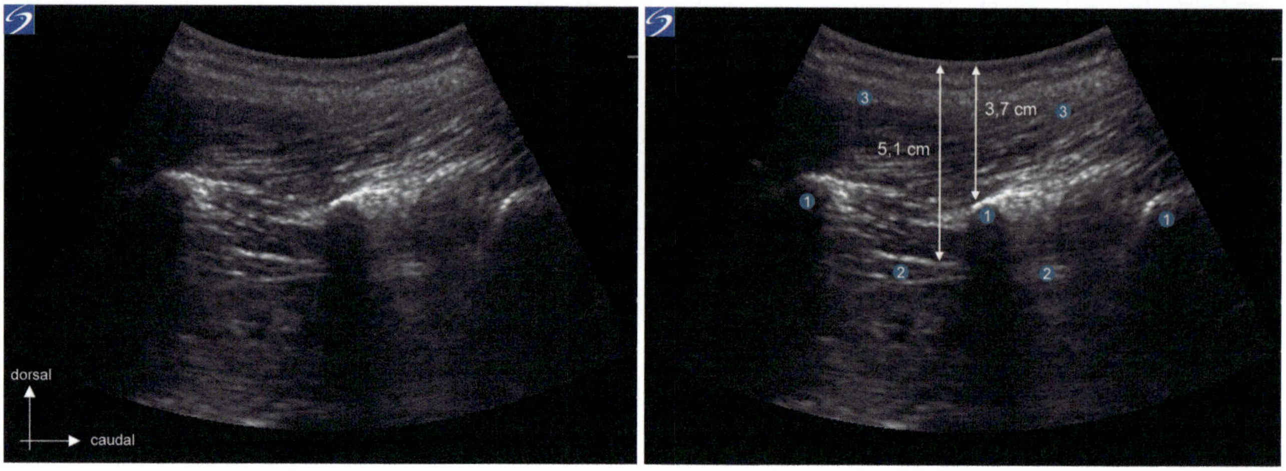

1. transverse process 2. lumbar plexus 3. erector spinae mm.

Fig. 6.5 Paramedian sagittal ultrasound plane. Visualisation of the lumbar plexus for performance of a lumbar plexus block by posterior approach

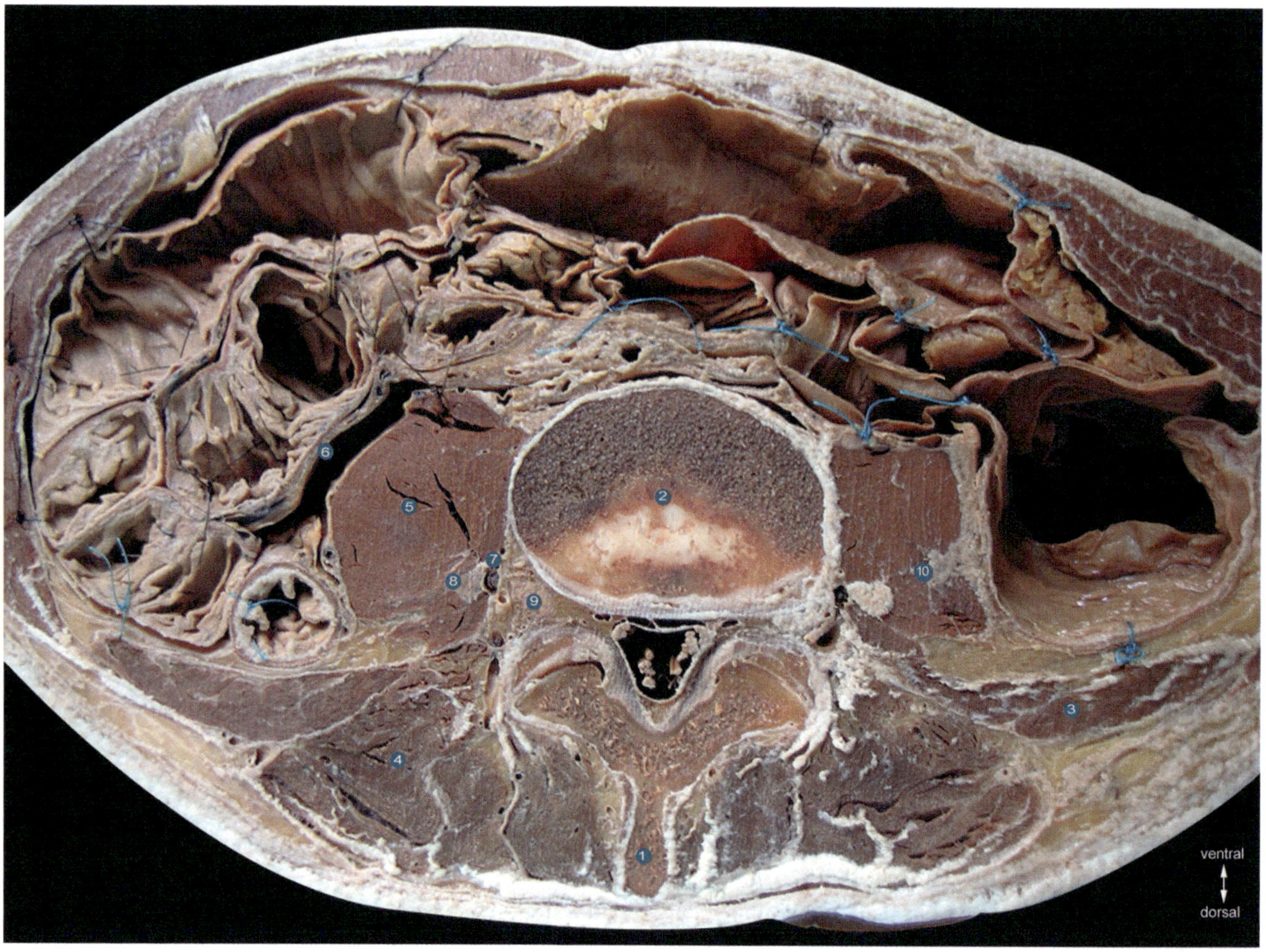

1. spinous process (L5)
2. intervertebral disk (L4-5)
3. quadratus lumborum m.
4. erector spinae m.
5. psoas major m.
6. peritoneum
7. ascending lumbar v.
8. lumbar plexus (femoral and obturator nn.)
9. L4 n. root
10. ilioinguinal and iliohypogastric nn.

Fig. 6.6 Transverse anatomical section cutting through the body of L4

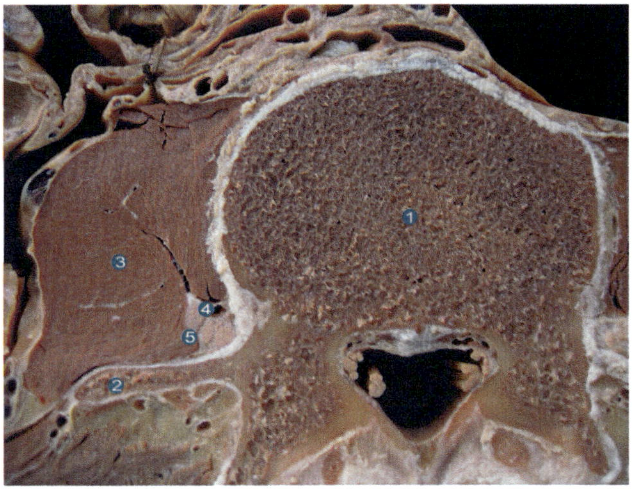

1. vertebral body
2. costiform process
3. psoas major m.
4. ascending lumbar v.
5. lumbar plexus

Fig. 6.7 Anatomical section cutting through the body of L4 and its transverse process

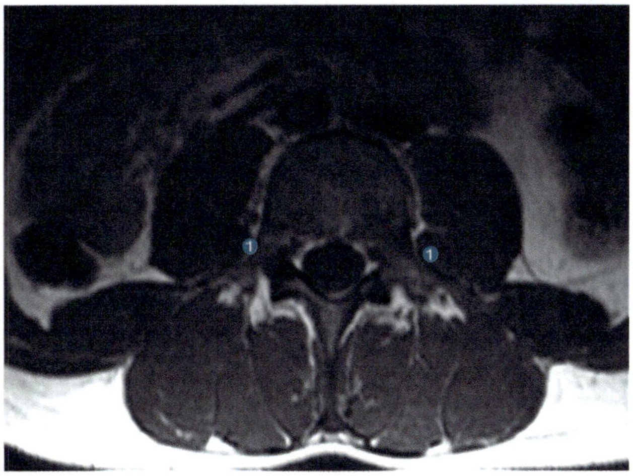

1. lumbar plexus

Fig. 6.8 Lumbar MRI-scan: transverse section revealing the plane of cleavage corresponding to the "psoas compartments"

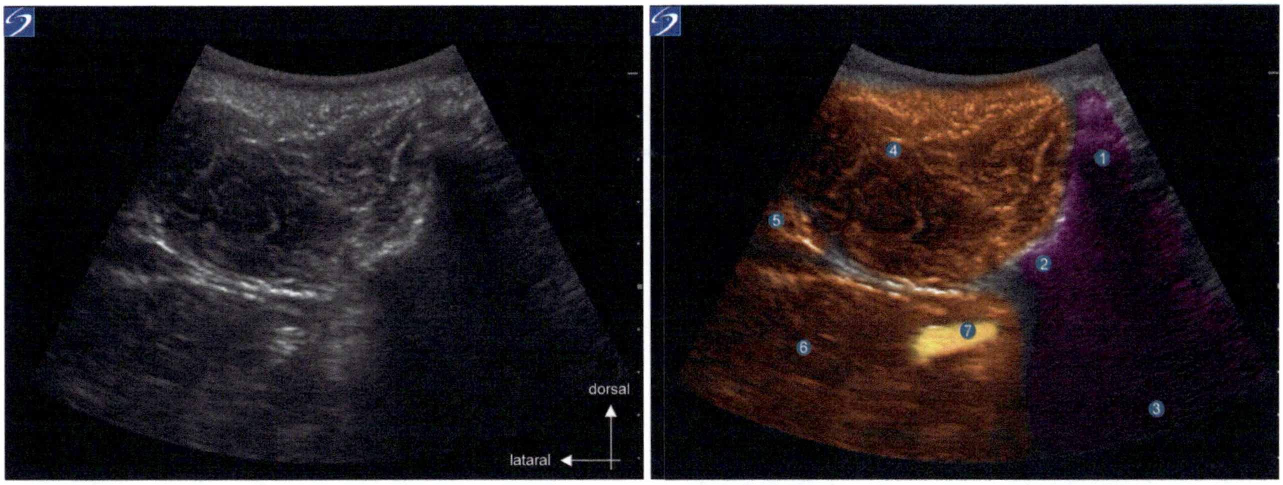

1. spinous process
2. vertebral lamina
3. vertebral body
4. erector spinae mm.
5. quadratus lumborum m.
6. psoas major m.
7. lumbar plexus

Fig. 6.9 Lumbar plexus: transverse ultrasound section

The needle should be long enough for insertion in the ultrasound plane, which often lengthens its pathway slightly. If the procedure can be performed with a "micro-convex" probe, whose smaller size limits the shift of the cutaneous injection point, the angle of the needle with respect to the micro-convex probe however is less favourable for its good visibility than with an "abdominal" curved probe. The procedure can be performed with transverse or paramedial sagittal ultrasound section.

Type of probe: "low frequency" curvilinear, 2 to 5 MHz.
Depth studied: 6 to 9 cm in "average" adults.
Position of the probe: It is necessary to have a low frequency probe that can be placed laterally to the lumbar spine in either paramedial sagittal position (Fig. 6.15) or transverse position (Fig. 6.16).
Neurostimulation: enables further identification of the nerves visualised and can limit, by determination of an MIS > 0.3 mA (0.1 ms), the risk of accidental intrafascicular injection. Furthermore, it can supplement locating nerves when conditions of visibility are unfavourable.
Needle: isolated, 22 G; and 100 mm.

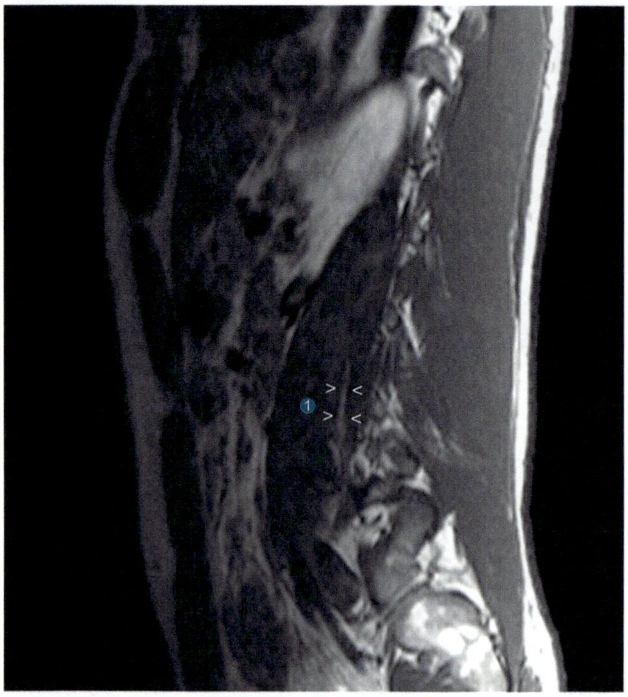

1. psoas compartment

Fig. 6.10 Lumbar MRI-scan: parasagittal section revealing the plane of cleavage corresponding to the "psoas compartment"

Echoanatomy and Location

The transverse ultrasound plane (Figs. 6.9, 6.13 and 6.14) and paramedian sagittal plane (Fig. 6.5) can be used to study the sonoanatomy of the area.

Lumbar spine: the spinous processes are located in the midline, more laterally the lamina, and then the transverse processes.

Muscle masses and adjacent organs: the erector spinae muscles, the quadratus lumborum and psoas major muscles, the kidney and the inferior vena cava.

Lumbar plexus: within the psoas major, approximately at the junction of the anterior two-thirds and posterior one-third of this muscle, immediately lateral to the posterior part of the vertebral body.

Conduct of the Block

With the patient lying on one side, in "foetal position". The block can be performed by visualising the lumbar plexus in a transverse or sagittal section. Since one of the serious complications of a lumbar plexus block is epidural spread of the anaesthetic, or even the occurrence of spinal anaesthesia, it is

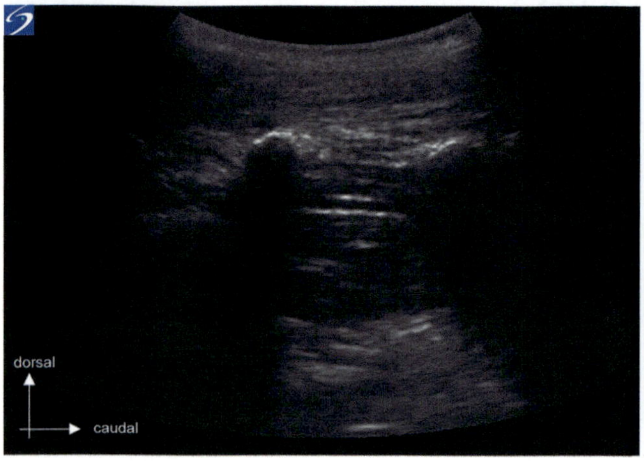

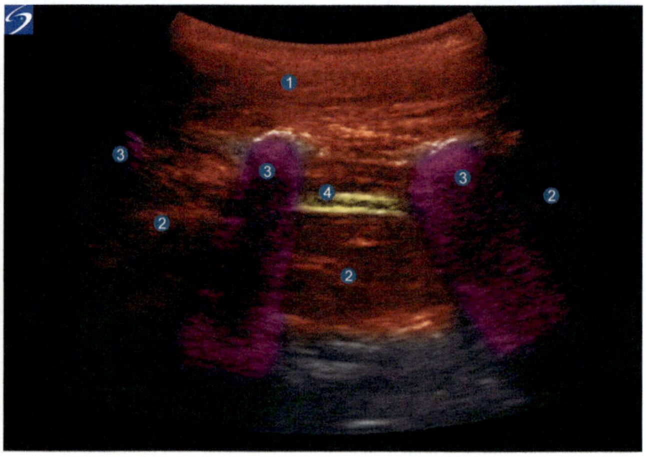

1. erector spinae mm. 2. psoas major m.
3. lumbar transverse processes : "trident sign"
4. lumbar plexus in psoas major m.

Fig. 6.11 Paramedian sagittal ultrasound section of a nerve root of the lumbar plexus

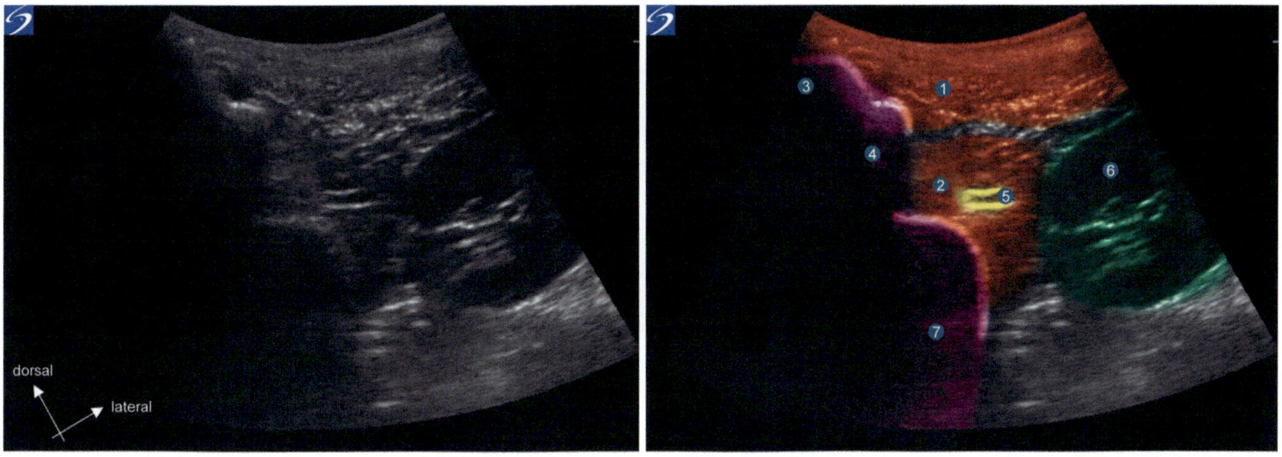

1. erector spinae mm.
2. psoas major m.
3. spinous process
4. vertebral lamina
5. lumbar plexus in psoas major m.
6. kidney
7. vertebral body

Fig. 6.12 Transverse ultrasound section of a nerve root of the lumbar plexus (same root as Fig. 6.11)

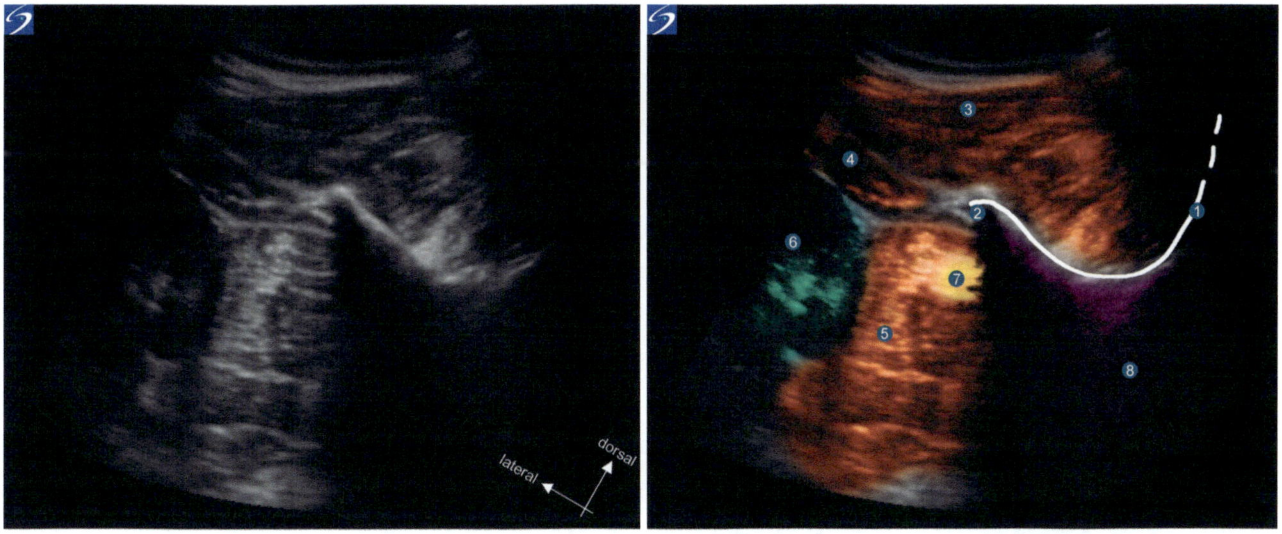

1. spinous process
2. costiform process
3. erector spinae mm.
4. quadratus lumborum m.
5. psoas major m.
6. kidney
7. part of lumbar plexus
8. acoustic shadow of the vertebra

Fig. 6.13 Transverse ultrasound section of the lumbar plexus at the level of a transverse process. The latter may be an obstacle to visualisation of nerve structures and therefore to a block

recommended to perform this block with the patient lying on the side to be anaesthetised; this perhaps may limit medial spread (arguably this postion would encourage the injected LA to move laterally under the influence of gravity).

This block can be planned overall in two manners:

- either the position or the depth of the transverse processes is determined by ultrasound (and even to guide the needle up to contact with bone), and then the injection is completed with the aid of neurostimulation only, with reference to the measured depth of the TP. In this case, the **paramedian sagittal** ultrasound section plane preferably will be used;
- or ultrasound-guided advancement of the needle can be performed in real time from the skin up to the lumbar plexus itself. In a second case, the block can be performed by using a **paramedian sagittal** ultrasound section, or a **transverse** section.

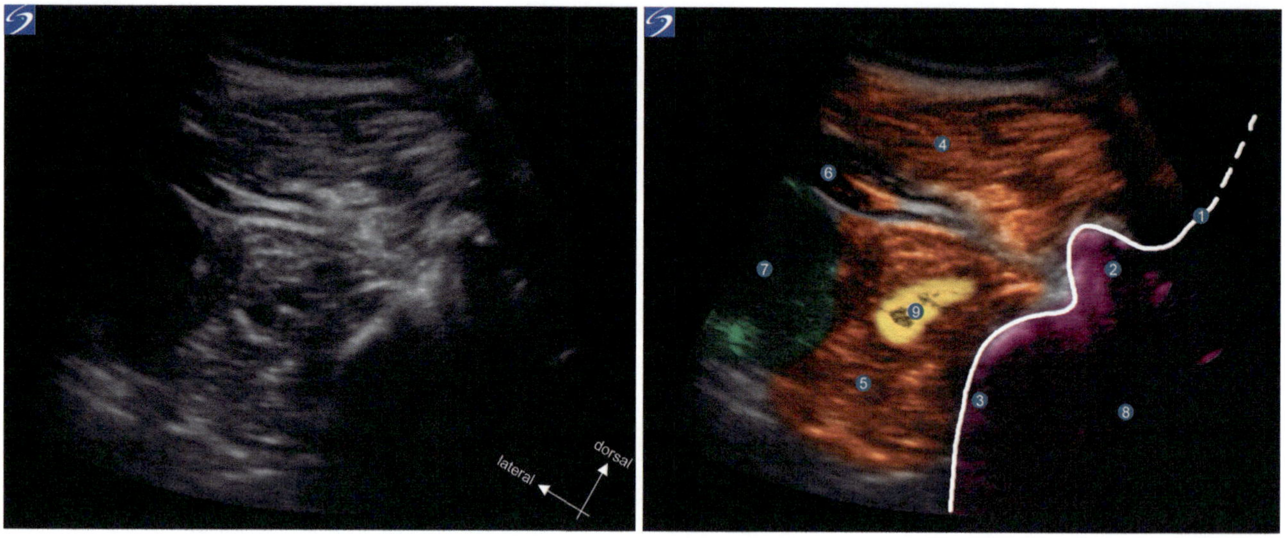

1. spinous process
2. vertebral lamina
3. vertebral body
4. erector spinae mm.
5. psoas major m.
6. quadratus lumborum m.
7. kidney
8. acoustic shadow of the vertebra
9. part of lumbar plexus

Fig. 6.14 Transverse ultrasound section of the lumbar plexus between the two transverse processes enabling to reveal more easily the position of the nerve structures (*see* Fig. 6.13)

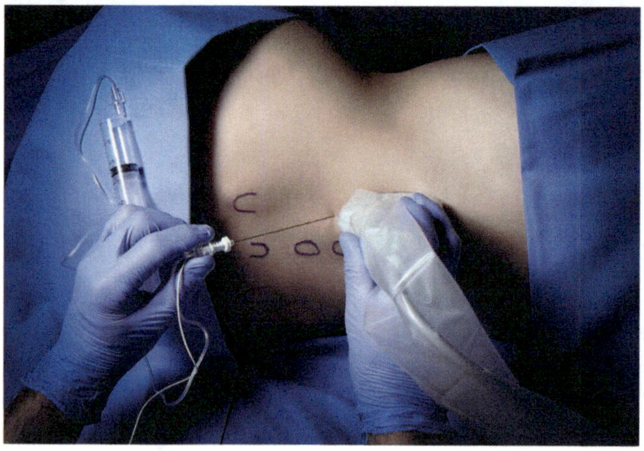

Fig. 6.15 Lumbar plexus block by posterior approach. Probe in paramedian sagittal position

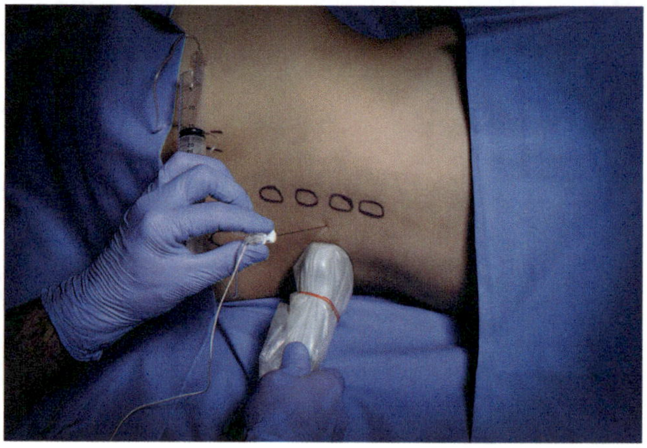

Fig. 6.16 Lumbar block by posterior approach. Probe in transverse position, patient lying down on the side to be anaesthetised perhaps to limit epidural spread of the local anaesthetic

Block performed using a paramedian sagittal ultrasound section (Fig. 6.15):

- a curved linear probe (2–5 MHz) in the paramedian sagittal plane, 4 to 5 cm from the spinous processes.
- needle ≥100 mm, inserted at the caudal end of the probe in direction of the transverse process of L4 that is crossed in the lateral one-third of its cranial border. The needle progresses by crossing the dorsal plane of the transverse processes between L3 and L4, and a femoral motor response is obtained generally at less than 2 cm beyond the transverse process. A response from the obturator component (adduction of the thigh) signifies a position that is too medial, that of a lumbosacral component (sciatic type) evidences a position that is too medial and caudal; these responses require the needle to be re-positioned/–directed.

Block performed by using a transverse ultrasound section (Fig. 6.16):

- At the level of the transverse processes of L3 and L4, the needle is inserted between the TPs under ultrasound guidance until contact with the lumbar plexus, whose position is visualised.

- The curvilinear probe (2–5 MHz) is positioned in a transverse plane (axial), laterally to the point of the cutaneous injection. The tip of the needle is inserted in immediate proximity to the structures identified as containing the components of the lumbar plexus. The correct positioning ideally is confirmed by neurostimulation which induces a femoral type motor response.

A volume of 10 to 20 mL of local anaesthetic can be injected when the needle is in place. The precision of guidance can enable to limit volumes of the local anaesthetic, the key element enabling to limit risk of spread into the spine.

In Practice

It is sometimes difficult to perform real-time ultrasound-guided insertion of the needle up to contact with the lumbar plexus. Sonoanatomy of this area in fact is complex, and quality of echogenicity is variable in particular due to the depth of the lumbar plexus. However, it is almost always possible to use ultrasound to locate the lumbar transverse processes L3 and L4, to measure their depth and to use these landmarks for distance to safely perform the block under traditional neurostimulation without having to look for physical contact between the needle and the transverse process. Sometimes this block is reserved for experienced or supervised practitioners.

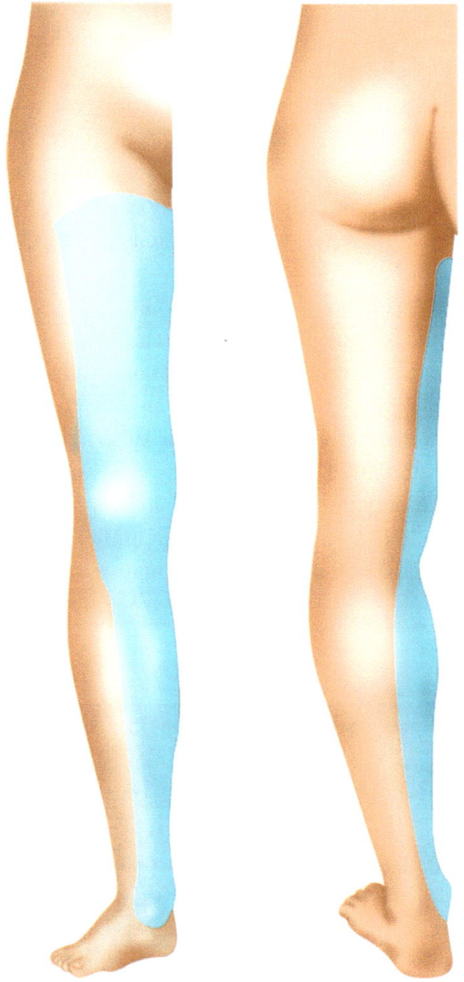

Fig. 6.17 Area of coverage of a femoral nerve block

Femoral Nerve Block (Fig. 6.17)

Indications

A femoral nerve block is the most commonly performed nerve block of the lower limb. Easily located, the femoral nerve however lies at a variable depth and ultrasound can accelerate the procedure for experienced practitioners. Used alone, it ensures analgesia for surgery on the surface of the anterior aspect of the thigh (wounds, collection of skin grafts). In combination with sciatic and obturator nerve blocks, it enables the performance of all types of surgery of the lower limb starting with the middle one-third of the femur. It is also indicated for emergency analgesia in femoral fractures, including prior to arrival in the hospital [27]. The association of femoral, sciatic and obturator blocks is becoming more widely applied in specialist centres. The femoral continuous nerve block (with catheter) is the method of choice and the best validated for analgesia of all types of major surgery of the knee [28–31]. It is more effective than intravenous analgesia with patient-controlled anaesthesia (PCA) with morphine in terms of analgesia, with fewer side effects. It is as effective as epidural anaesthesia in terms of analgesia, also with fewer side effects. Maintenance of the catheter is best done with PCRA (patient-controlled regional anaesthesia). Complications of this block are rare; proximity of the femoral blood vessels creates the potential for inadvertent vascular puncture, but ultrasound makes it possible to avoid this drawback.

Type of probe: "high frequency" linear, 5 to 10 MHz or 6 to 13 MHz.
Axis of probe: transversal, slightly oblique caudally and medially, the probe is placed in the skin crease of the thigh (Fig. 6.18).
Configuration: nerve in its short axis, needle in or out of plane.

Depth studied: up to about 4 to 5 cm depending on the patient's body build (body weight).
Neurostimulation: enables further identification of the nerves visualised and can limit, by determination of an MIS > 0.3 mA (0.1 ms), the risk of accidental intrafascicular injection. Furthermore, it can supplement locating nerves when conditions of visibility are unfavourable.
Needle: 50 to 80 mm isolated, 22 G.
Utility of Doppler ultrasound: Enables visualisation of the femoral arteries and veins, the long saphenous vein with its arch and the profunda femoris artery. Less frequently, the superficial and/or deep circumflex iliac artery may be seen in front of the femoral nerve in the lateral and cephalic direction, the lateral and medial circumflex arteries of the thigh and the external pudendal vessels.

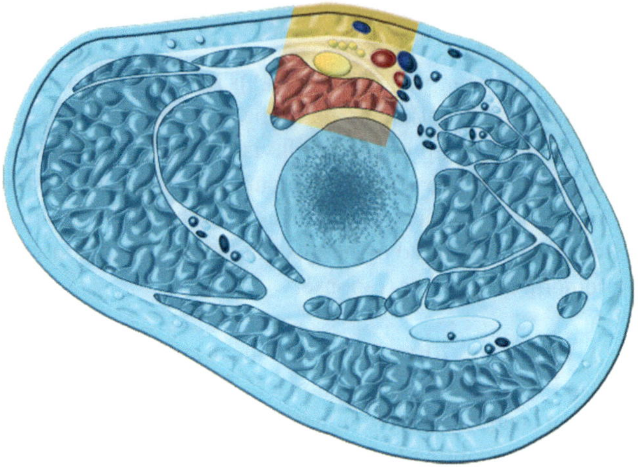

Fig. 6.18 Section at the base of the thigh with materialisation of ultrasound beam for performance of a femoral nerve block

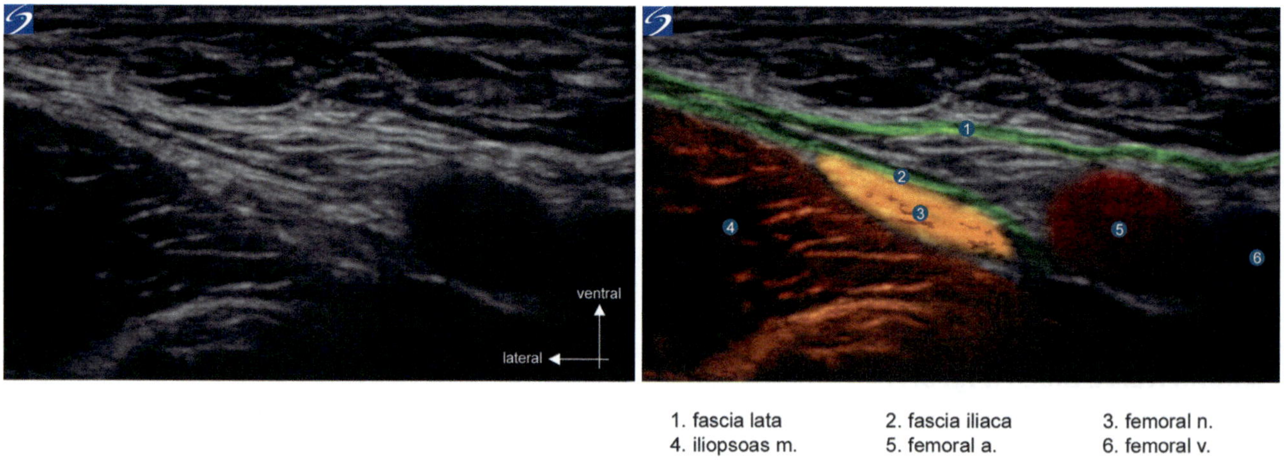

Fig. 6.19 Transverse ultrasound section at the level of the inguinal fold

Echoanatomy and Location of the Nerve

By positioning the probe in the skin crease of the groin, the vascular structures are easily visualised and, in particular, the femoral artery which is used as an internal landmark (Fig. 6.19). During this procedure, one may sometimes observe the existence of atheromatous plaques in the lumen of the artery (Figs. 6.20 and 6.21). The nerve is located immediately lateral to the artery, on the ventral aspect of the iliopsoas muscle, deep to the fascia iliaca (Figs. 6.22, 6.23, 6.24 and 6.25). It lies within a triangle created medially by the lateral wall of the femoral artery, anteriorly by the fascia lata and posteriorly by the ventral aponeurosis of the iliopsoas muscle. Seen in the "short axis", the femoral nerve presents at this level a flattened oval shape. Generally, it has already branched and its different components are often difficult to differentiate, providing it most of the time with an "alveolar" appearance. By slowly angling the ultrasound probe it is possible to alter the echogenicity of the iliopsoas muscle, femoral artery and femoral nerve. Using this effect of "anisotropy" the visibility of the femoral nerve can be optimised. However the anatomy is visualised, it is important to keep in mind that at this level the femoral nerve is always deep to the fascia iliaca. Only the smaller cutaneous

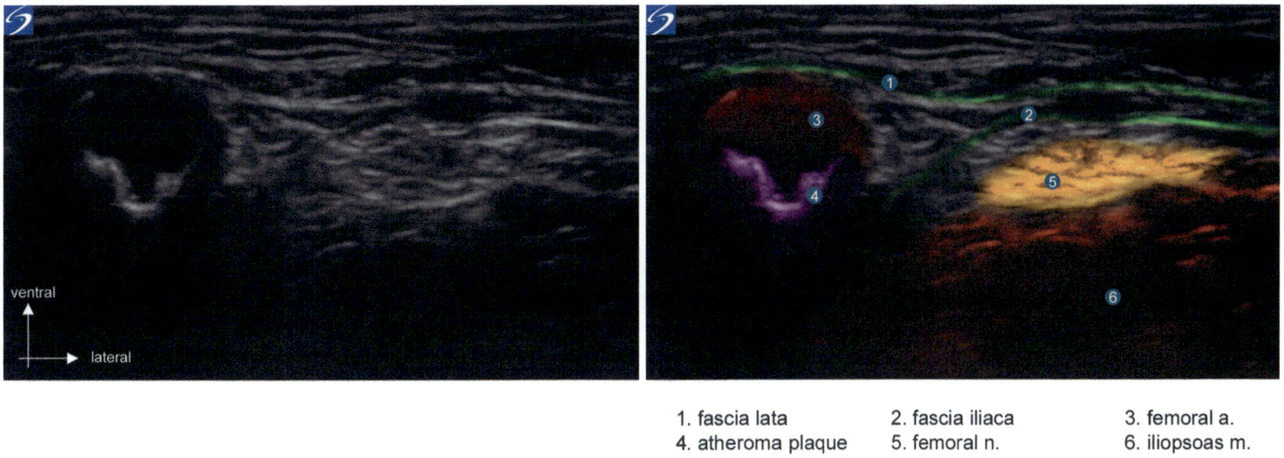

Fig. 6.20 Atheroma plaque in the lumen of the femoral artery

Fig. 6.21 Atheroma plaque in the lumen of the femoral artery. Doppler ultrasound

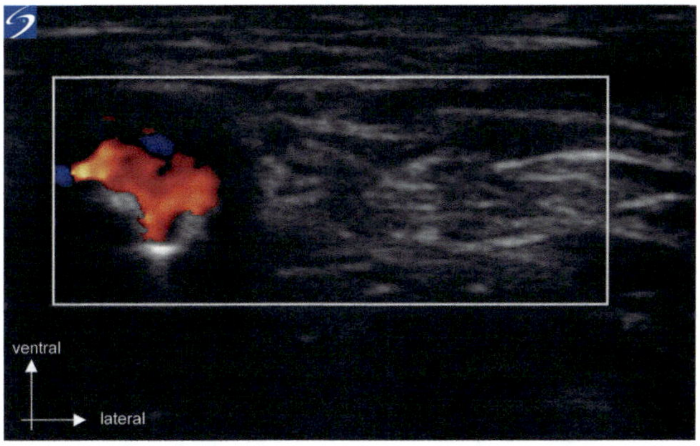

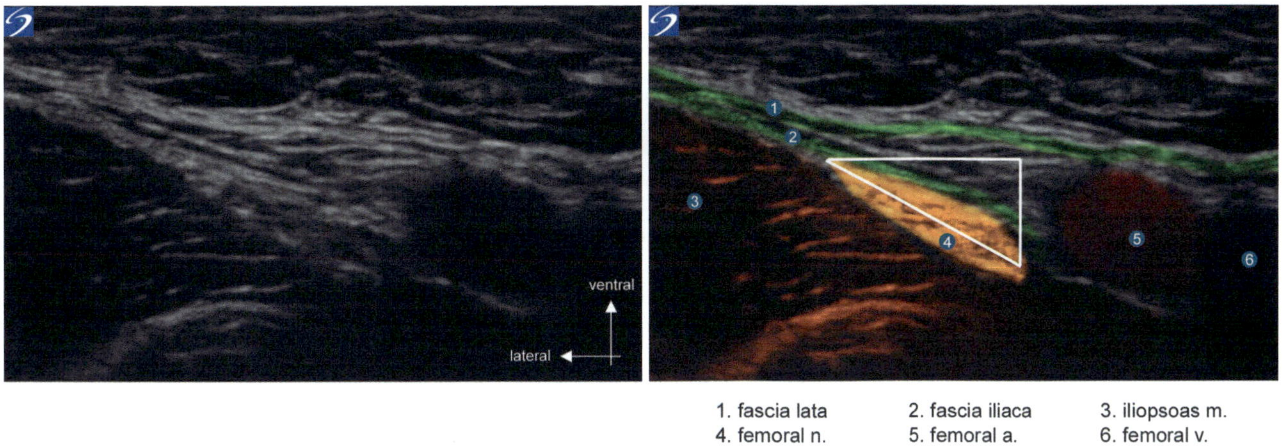

Fig. 6.22 Transverse sonoanatomical section at the level of the inguinal fold

branches which supply sensation to the anterior aspect of the thigh have become superficial and may lie between the fascia lata and fascia iliaca, laterally to the femoral artery.

Sometimes, the femoral nerve is located more laterally to the femoral artery, again under the fascia iliaca (Fig. 6.26), in particular in the most cephalic part of the inguinal area.

Sometimes, a circumflex iliac artery is observed which, arising from the femoral artery, crosses in front of the femoral nerve to continue cranially and laterally in the direction of the iliac crest (Fig. 6.27).

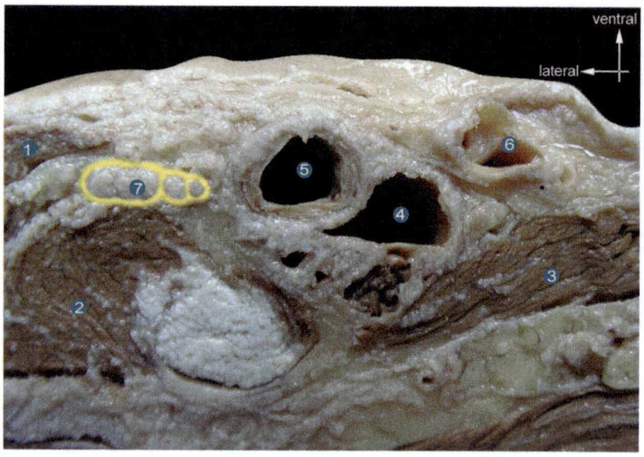

1. sartorius m.
2. iliopsias m. and tendon
3. pectineus m.
4. femoral v.
5. femoral a.
6. greater saphenous v.
7. branches of femoral n.

Fig. 6.23 Anatomical section at the level of the inguinal fold

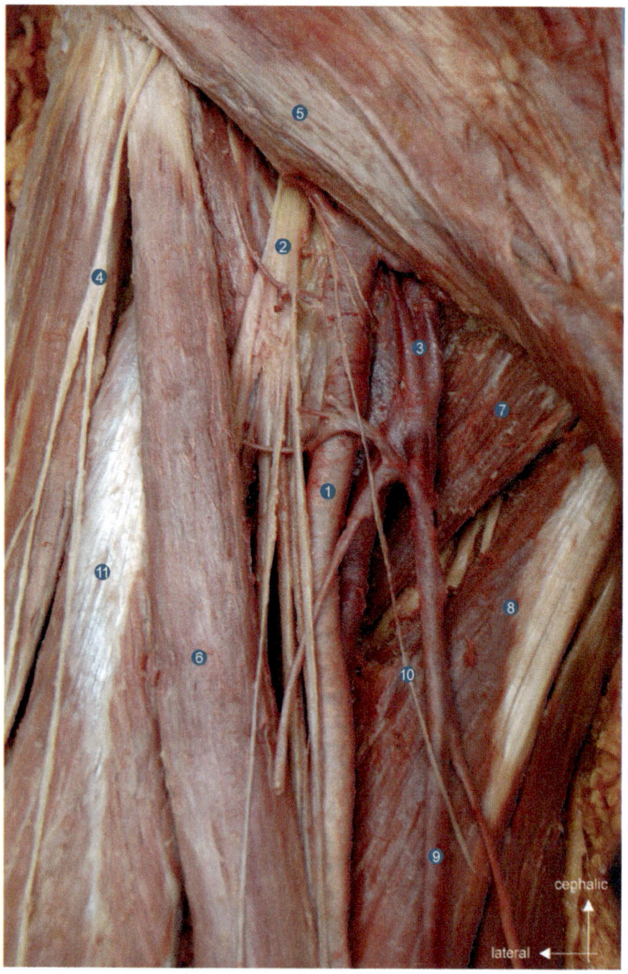

1. femoral a.
2. femoral n.
3. femoral v.
4. lateral femoral cutaneous n.
5. inguinal ligament
6. sartirius m.
7. pectineus m.
8. adductor longus m.
9. greater saphenous v.
10. femoral branches of genitofemoral n.
11. rectus femoris m.

Fig. 6.24 Anatomy of the base of the thigh. Anterior view. Iconography: Admir Hadzic

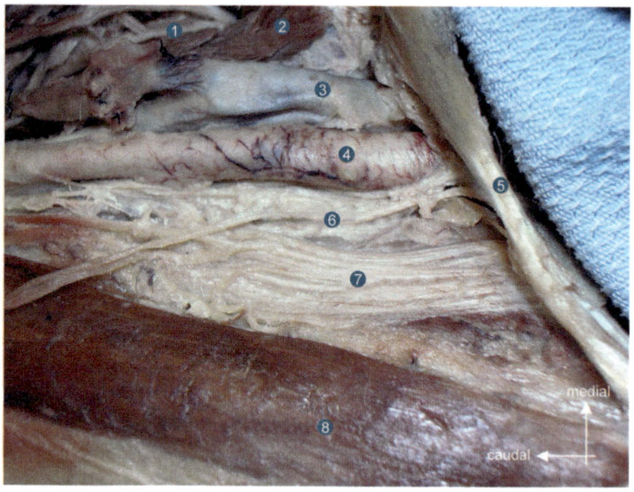

1. obturator n.
2. pectineus m.
3. femoral v.
4. femoral a.
5. inguinal ligament
6. genitofemoral n.
7. femoral n.
8. sartorius m.

Fig. 6.25 Anatomy of left inguinal region. Iconography: Admir Hadzic

Utility of Doppler Ultrasound

Generally, the femoral artery and vein, the long saphenous vein with its arch, and the profunda femoris artery are visualised. Less commonly, the superficial and/or deep iliac circumflex artery can be seen passing in front of the femoral nerve going in a lateral and cephalic direction, and the lateral and medial circumflex arteries of the thigh and the external pudendal blood vessels.

Approach

The ideal position for the probe is the skin crease of the groin, where the femoral nerve is at its most superficial. It can be approached in a traditional axial manner, as in the block technique performed with neurostimulation (out of plane), or laterally by choosing a skin puncture site at the lateral end of the probe (in plane) (Fig. 6.28).

Axial Approach The main advantage may be of not changing the usual puncture site with relation to the technique in neurostimulation. The disadvantage of this approach is the position of the needle "out of plane" and the difficulty of visualising the needle tip with ultrasound.

Lateral Approach (Fig. 6.28) The puncture site lies at the lateral end of the probe, which enables an "in plane" needle trajectory, with excellent visualisation of the needle tip.

If, in the axial approach, the puncture is in cephalic direction, perpendicular to the ultrasound plane, in both approaches, the nerve is seen on transverse section, which makes it more useful for location with control of its position and spread of the local anaesthetic.

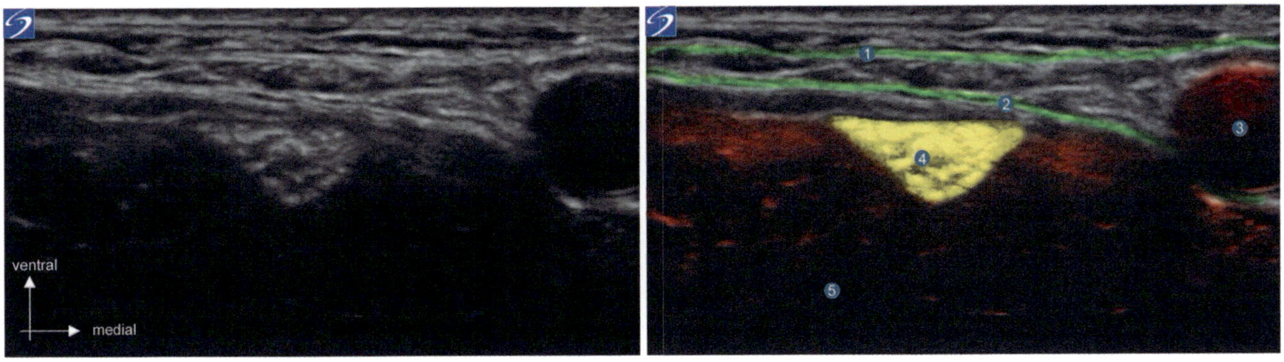

1. fascia lata 2. fascia iliaca 3. femoral a.
4. femoral n. 5. iliopsoas m.

Fig. 6.26 Transverse ultrasound section in the cephalic area of the inguinal region

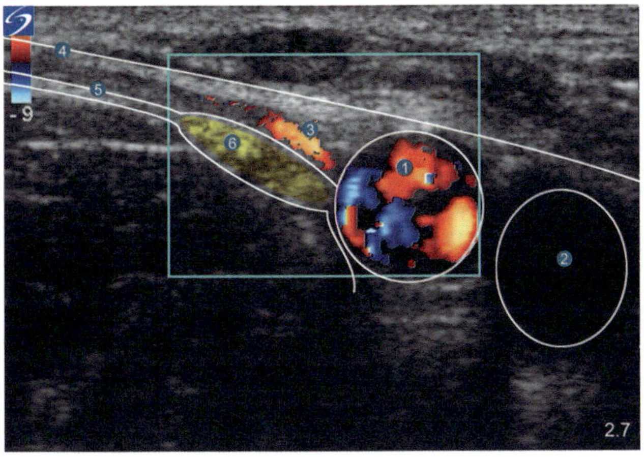

1. femoral a. 2. femoral v. 3. circumflex iliac a.
4. fascia lata 5. iliac fascia 6. femoral n.

Fig. 6.27 Transverse section of the inguinal fold. The iliac circumflex artery pre-crosses the femoral nerve

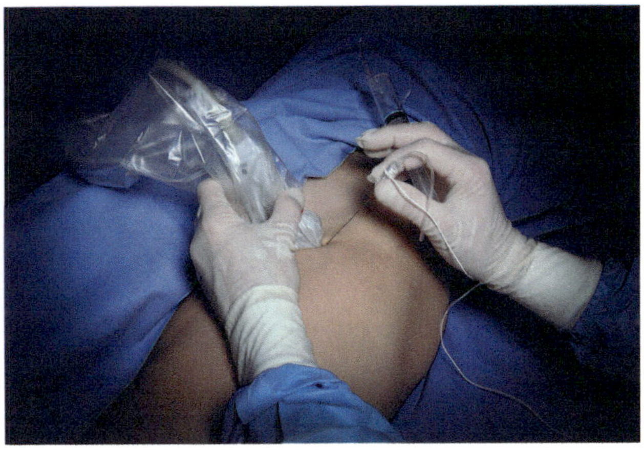

Fig. 6.28 Femoral block: needle insertion in-plane

Injection

In the **axial** (out of plane) approach, the needle is inserted 1 or 2 cm distally from the probe and is advanced obliquely in the cephalic direction. The needle tip can be difficult to see in transverse section. To be more sure of its position, it is necessary to visualise the transmitted tissue movements and to utilise hydrolocalisation as the needle tip advances. As for all "out of plane" approaches, it is necessary to maintain the very tip of the needle constantly within the plane of the ultrasound beam. Ideally, the needle should approach the nerve towards its lateral border, and not directly towards it (to minimise direct injury with the needle). Although neurostimulation is useful, it is the spread of the local anaesthetic/D5W solution visualised during injection which confirms the correct position of the tip of the needle, allowing repositioning if the spread is not optimal. We remind the reader that spread of the local anaesthetic/D5W solution used for hydrolocalisation must always be observed deep to the fascia iliaca.

In a **lateral** approach, the needle is visualised longitudinally in the ultrasound plane. The tip of the needle should be positioned at the lateral edge of the femoral nerve, and LA spread seen in the plane between the fascia iliaca and the ventral aponeurosis of the iliopsoas muscle.

For a nerve block with a **single injection**, the injection of local anaesthetic should be initiated at the lateral border of the nerve. Initially a virtual space, the placement of more local anaesthetic here encourages spread, generally developing in the plane over the femoral nerve. This pattern of spread leaves the nerve in position lying on the ventral aponeurosis of the iliopsoas muscle (Figs. 6.29 and 6.30). Occasionally, spread of the local anaesthetic occurs spontaneously between the femoral nerve and the ventral aponeurosis of the iliopsoas muscle, lifting the nerve off the muscle. Nevertheless, both patterns of spread provide an excellent nerve block, and it is not necessary to look absolutely for circumferential spread (Fig. 6.31).

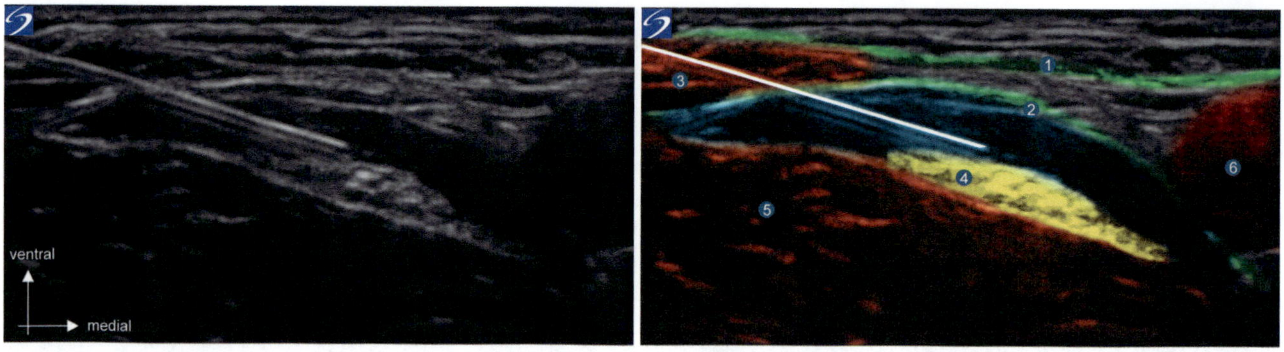

Fig. 6.29 Ultrasound-guided femoral block: lateral in-plane needle insertion

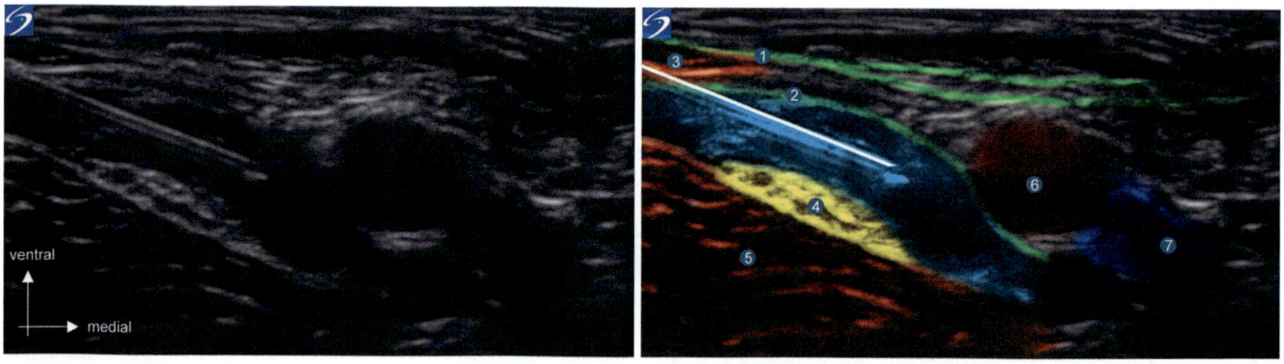

Fig. 6.30 Ultrasound-guided femoral block: lateral in-plane needle insertion

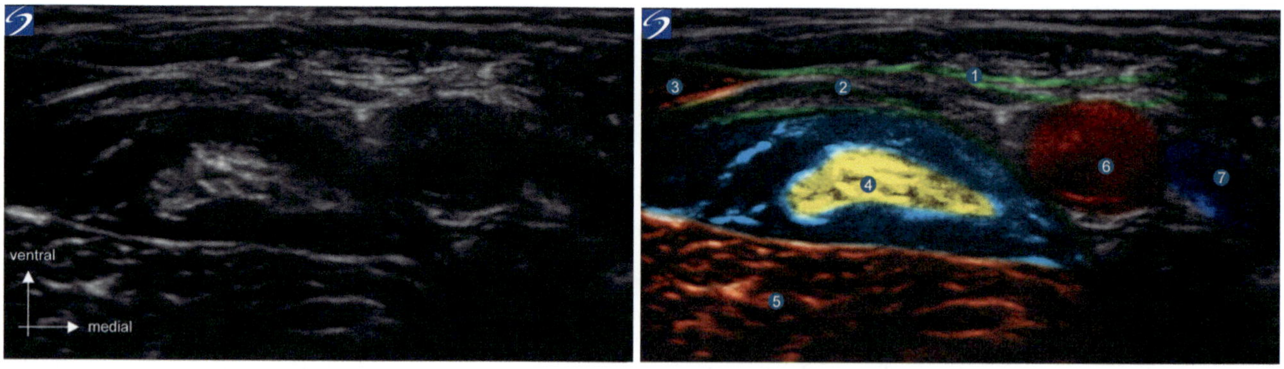

Fig. 6.31 Femoral block: circumferential spread of local anaesthetic

For placement of a **perineural catheter**, the needle is inserted at a greater distance from the lateral border of the femoral nerve allowing a shallow approach to the target plane between the fascia iliaca and the aponeurosis of the iliopsoas. When this space is evidenced by injection (Fig. 6.32), the needle is advanced progressively by "opening up" of the plane by hydrodissection. Generally, the D5W solution is distributed between the ventral aspect of the femoral nerve and the fascia iliaca. The needle then is slowly advanced with caution into this space (Fig. 6.33), as it further opens up with the injectate until the medial border of the nerve is reached (adjacent to the femoral artery) (Fig. 6.34).

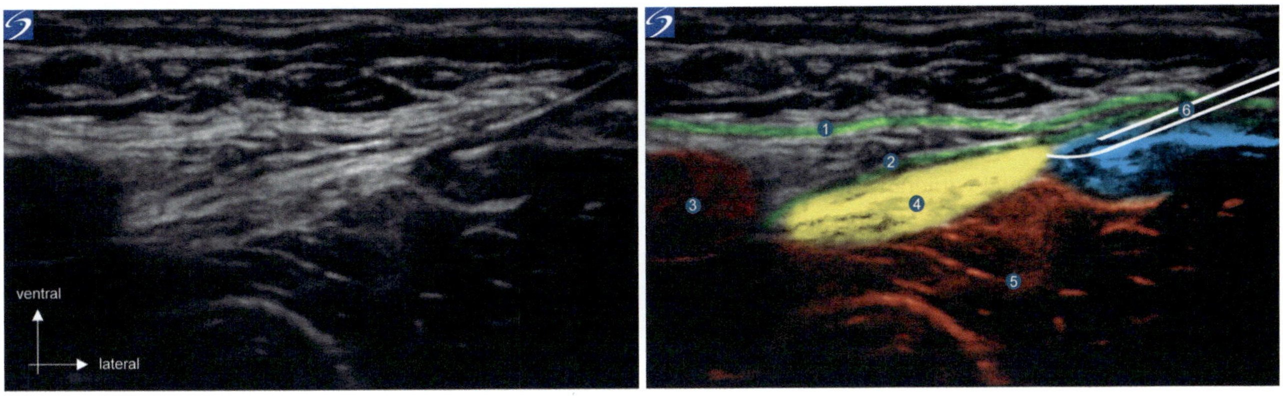

1. fascia lata 2. fascia iliaca 3. femoral a.
4. femoral n. 5. iliopsoas m. 6. Tuohy needle

Fig. 6.32 Placement of an ultrasound-guided femoral catheter: hydrodissection

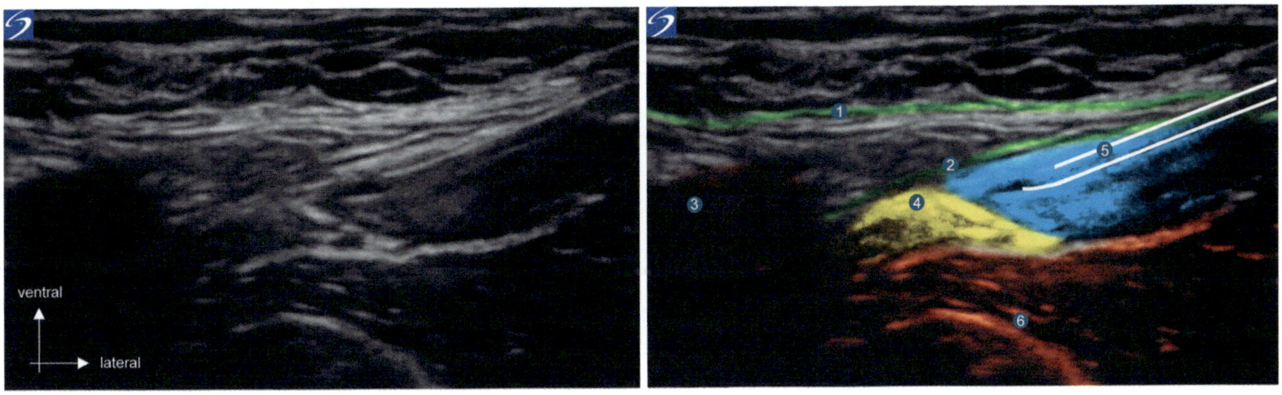

1. fascia lata 2. fascia iliaca 3. femoral a.
4. femoral n. 5. Tuohy needle 6. iliopsoas m.

Fig. 6.33 Placement of a femoral catheter: "opening" of space between fascia iliaca and aponeurosis of iliopsoas muscle thanks to continuation of hydrodissection through a Tuohy needle

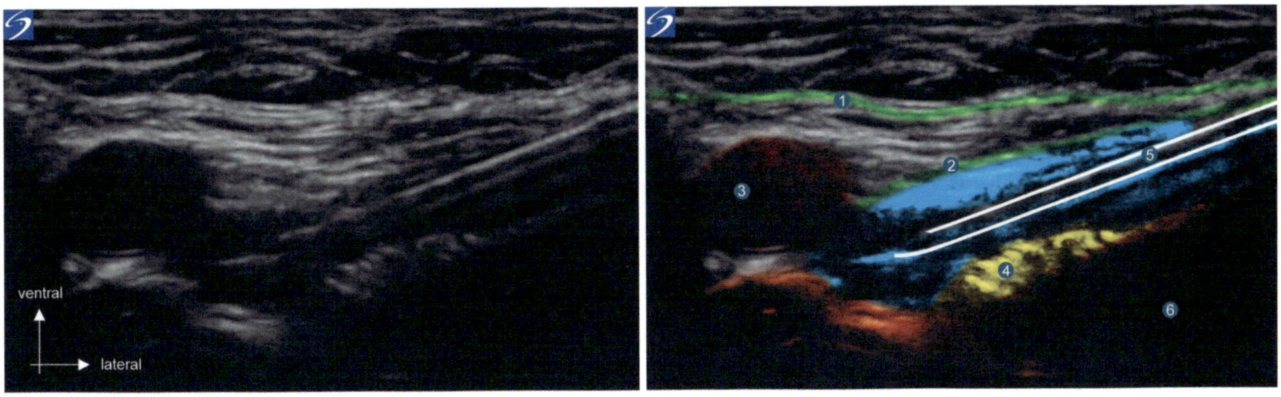

1. fascia lata 2. fascia iliaca 3. femoral a.
4. femoral n. 5. Tuohy needle 6. iliopsoas m.

Fig. 6.34 Placement of an ultrasound-guided femoral catheter: result of hydrodissection through the Tuohy needle, the tip of the Tuohy needle is eventually positioned by the medial border of the femoral nerve

The catheter then is introduced into the needle up to its end only, which will place it across the femoral nerve, under the fascia iliaca (Fig. 6.35). With this insertion technique, the catheter crosses the fascia iliaca in a more lateral position in relation to the nerve. This provides the catheter with a larger margin for accidental withdrawal while enabling its end to remain under the fascia iliaca, and thus in the proper plane for the effective spread of LA (Fig. 6.36). The option of using a pigtail catheter (Sonolong Curl Sono, Pajunk) provides for a more secure positioning of the catheter. In fact, the 3 cm of the loop deployed provides an additional margin of safety in case of accidental partial withdrawal.

When performing the block, some cutaneous sensory branches of the anterior aspect of the thigh (anterior cutaneous nerves of the thigh) may sometimes be found in front of the fascia iliaca. Therefore, it can be wise to inject a few millilitres of local anaesthetic, immediately above of the femoral artery anterior to plane of the fascia iliaca, in order to anaesthetise them. Their presence, which is often difficult to ascertain with ultrasound in light of their size, may be confirmed possibly by sensory neurostimulation.

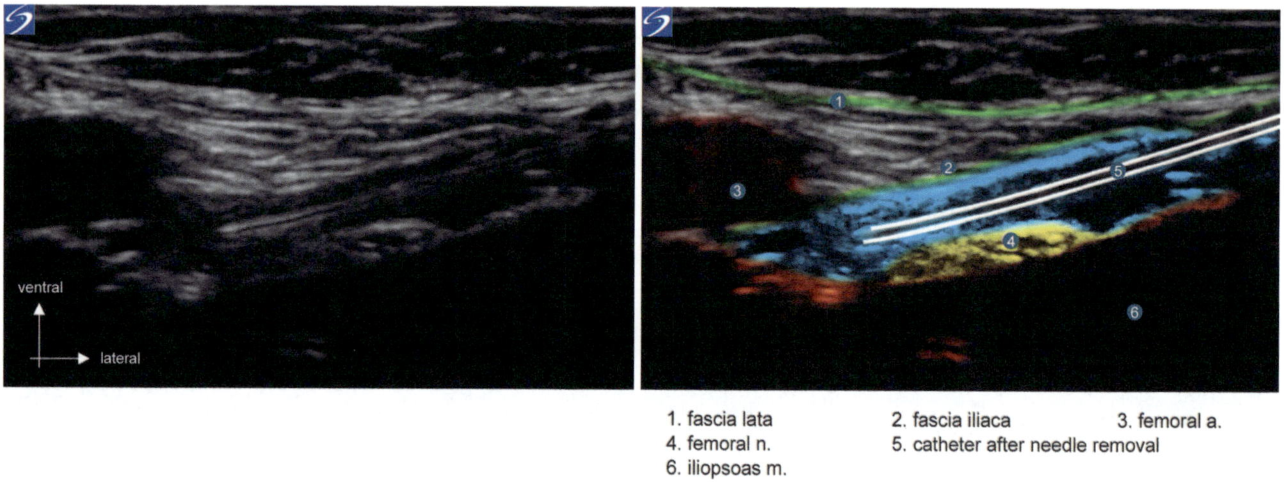

Fig. 6.35 Femoral catheter in place under the fascia iliaca, in the correct plane of spread of the local anaesthetic, with its tip located by the medial border of the nerve

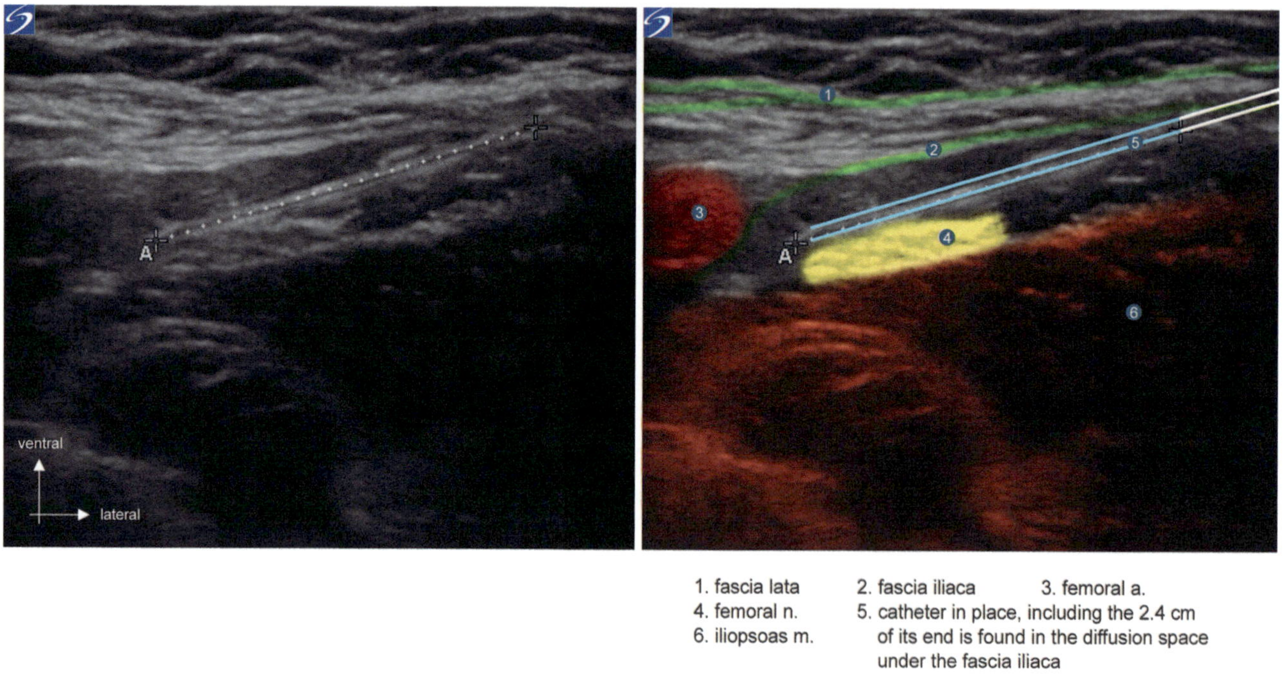

Fig. 6.36 Ultrasound-guided femoral catheter in place, with the distal 2.4 cm located in the diffusion space under the fascia iliaca

Paediatrics

Use of a high frequency probe (≥ 13 MHz) is strongly recommended in light of the shallow depth of the femoral nerve in children.

As in adult cases, ultrasound guidance can probably improve the success of the block by optimising the position of the needle and injection of the local anaesthetic (Fig. 6.37).

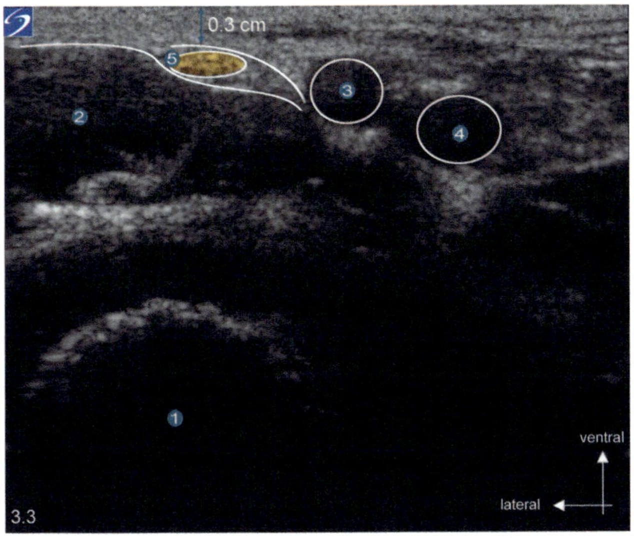

1. femur
2. iliopsoas m.
3. femoral a.
4. femoral v.
5. femoral n.

Fig. 6.37 Transverse section at the level of the inguinal fold in a 3-year-old child

Lateral Cutaneous Nerve Block of the Thigh (Fig. 6.38)

Indications

The lateral cutaneous nerve of the thigh (LCNT) block provides anaesthesia to the anterolateral aspect of the thigh and knee, for the ventral branch, and the supero-lateral part of the thigh for the dorsal branch. Since a pure femoral block does not anaesthetise the skin on the lateral side of the patellar tendon, blocking the LCNT provides supplementary anaesthesia for insertion of the lateral ports during arthroscopy of the knee. This block is indicated for superficial surgery of the lateral aspect of the thigh (e.g. collection of a skin graft [32]). It can also supplement a "3-in-1 block" to extend anaesthesia to the lateral aspect of the upper of the thigh. The LCNT block is also used to treat meralgia paraesthetica.

Type of probe: linear, 6 to 13 MHz.
Axis of probe: transversal (Fig. 6.39).
Configuration: nerve in short axis, needle in plane.
Depth studied: 0.5 to 3 cm depending on thickness of the adipose fold.
Neurostimulation: It may be useful to elicit paraesthesiae in the area of the lateral cutaneous nerve of the thigh when ultrasound identification of the nerve is unclear.
Needle: 50 or 80 mm isolated, 22 G.

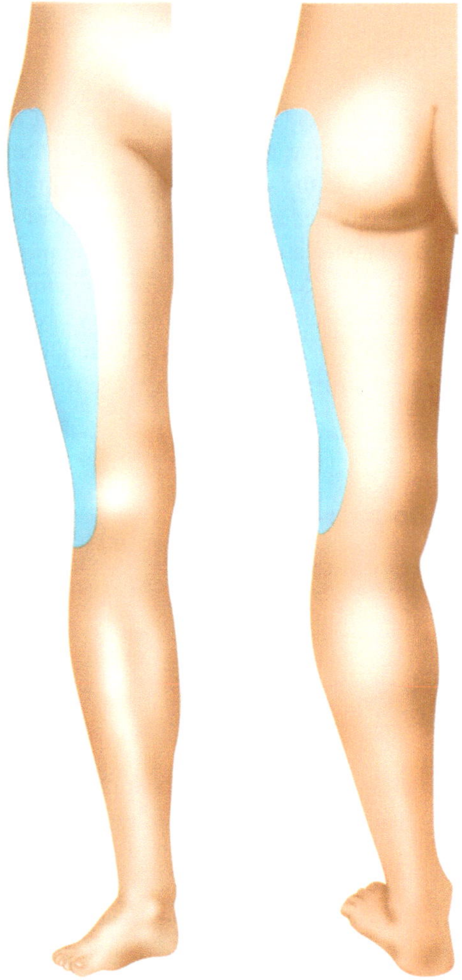

Fig. 6.38 Area of coverage of the lateral cutaneous nerve block in the thigh

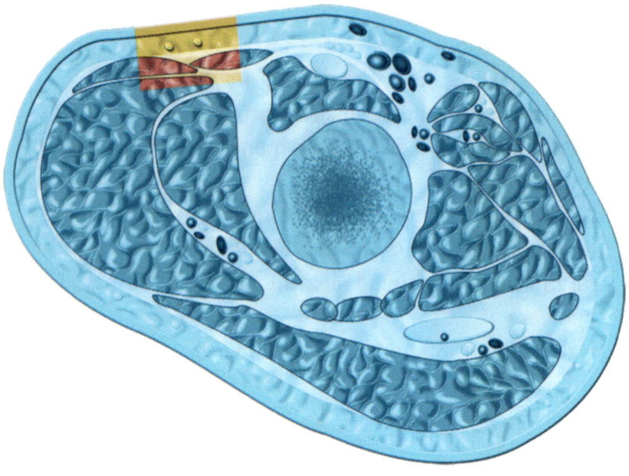

Fig. 6.39 Transverse section at the level of the base of the thigh with materialisation of the ultrasound beam for lateral femoral cutaneous nerve block

Echoanatomy

The injection site for an ultrasound-guided LCNT block lies between the skin and the aponeurosis of sartorius and tensor fascia lata muscles in the supero-lateral aspect of thigh. The US probe is placed medially in the skin crease of the groin and, after the femoral artery is identified, the probe is slid slowly laterally. In succession, the following structures are seen in transverse section: the femoral nerve lying on the ventral aspect of the iliopsoas muscle, laterally, sartorius lying superficial to iliopsoas, and lastly tensor fascia lata (Fig. 6.24).

Utility of Doppler Ultrasound

Enables confirmation of the presence/absence of a superficial or deep iliac circumflex artery.

Localisation

The lateral cutaneous nerve of the thigh passes through (or deep to) the inguinal ligament medial to the anterosuperior iliac spine (ASIS) (Fig. 6.40). It then moves laterally to lie superficially between sartorius and tensor fascia lata muscles (Fig. 6.41), and divides into its ventral and dorsal terminal branches. Due to its small size, it is not always identifiable with ultrasound.

Approach

Anaesthesia of the lateral cutaneous nerve of the thigh is performed often in combination with a femoral nerve block. Since the latter is usually performed by lateral approach, the same point of injection can be used to achieve a block of the lateral cutaneous nerve of the thigh. It is necessary to reposition the probe lateral to the initial injection point, in the inguinal skin crease, and to direct the needle laterally (Fig. 6.42). Under visual control, in the axis of the ultrasound plane, the needle is advanced between the fascia lata and the ventral aponeurosis of the sartorius muscle just beyond its lateral border. On approach to the lateral cutaneous nerve of the thigh, neurostimulation triggers paraesthesiae of the vento-lateral aspect of the thigh, aiding identification of the nerve if this has not been confirmed visually.

Injection

In the inguinal skin crease, the LCNT is usually located deep to the fascia lata, at the junction between the sartorius and the tensor fascia lata muscles (Fig. 6.41). Sometimes the individual branches are separated by fascial components, which appear as hyper-echoic strands. The injection is performed in contact with the nerves when they are identifiable (Fig. 6.43) or by more "empirical" injection if they are not visualised; then injection is performed preferably by withdrawing after having inserted the needle along the entire distance to be infiltrated. With ultrasound, spread of the local anaesthetic is very easily seen. In adults, a volume of 4 to 7 mL is sufficient.

Paediatrics
This block is useful for surgery with a lateral thigh incision, in combination with a femoral nerve block. Of course, it is not indicated if the child has undergone a lumbar plexus block by posterior approach or a "3-in-1" block.

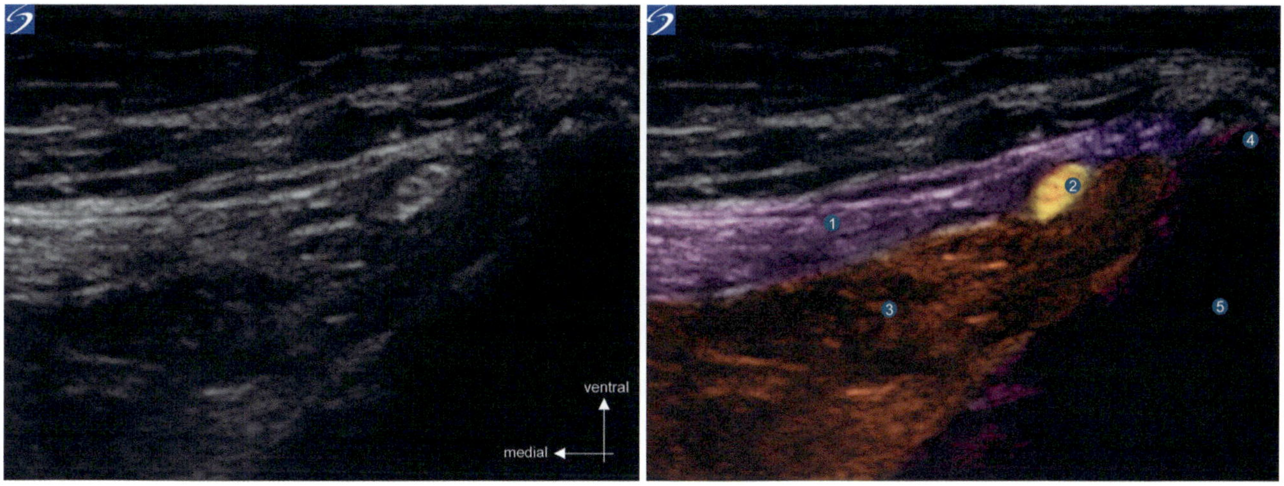

1. inguinal ligament
2. lateral femoral cutaneous n.
3. iliacus m.
4. anterosuperior iliac spine
5. acoustic shadow of the ilium

Fig. 6.40 Transverse section of the lateral femoral cutaneous nerve at the level of the anterior-superior iliac spine

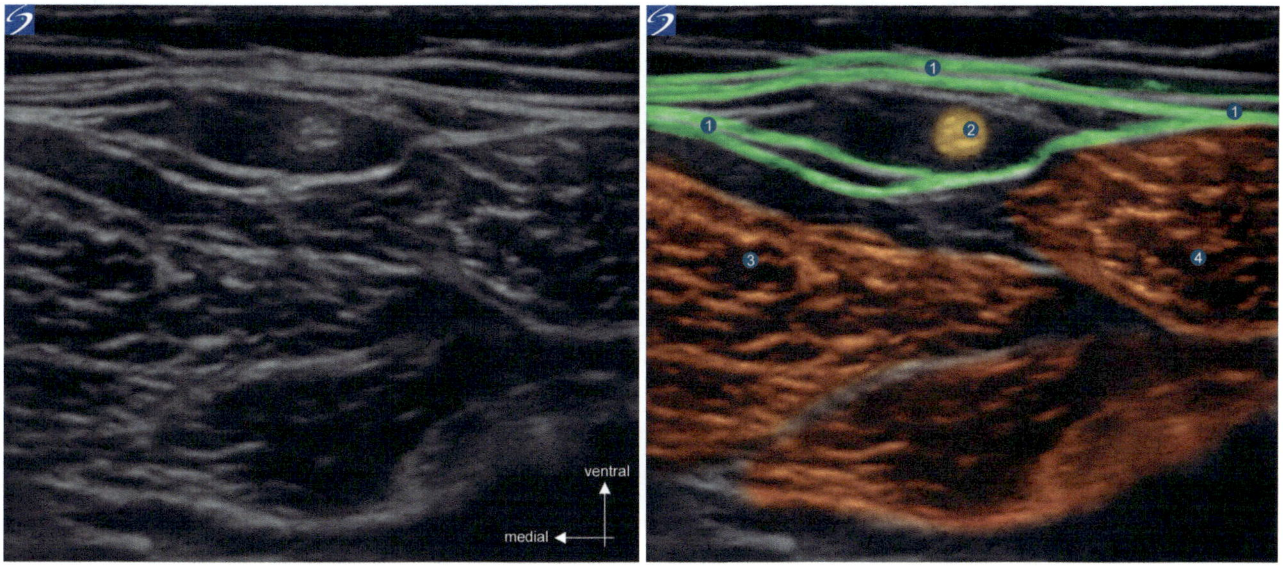

1. layers of fascia lata
2. lateral femoral cutaneous n.
3. sartorius m.
4. tensor fasciae latae m.

Fig. 6.41 Lateral femoral cutaneous nerve block: transverse section in the space between the sartorius and tensor fascia lata muscles

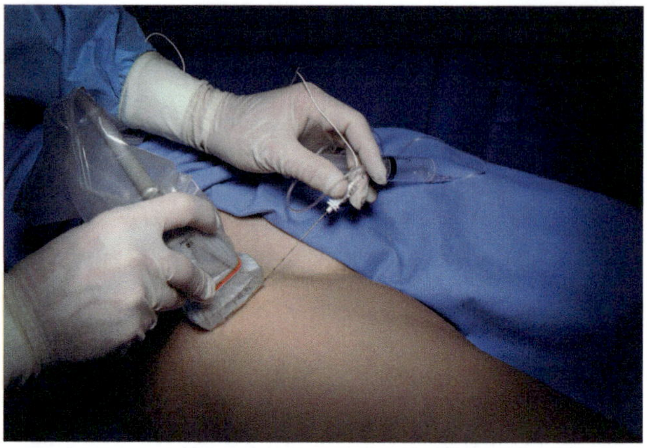

Fig. 6.42 Lateral femoral cutaneous nerve block with an in-plane approach, using the same needle entry point as for the femoral nerve block that was just performed

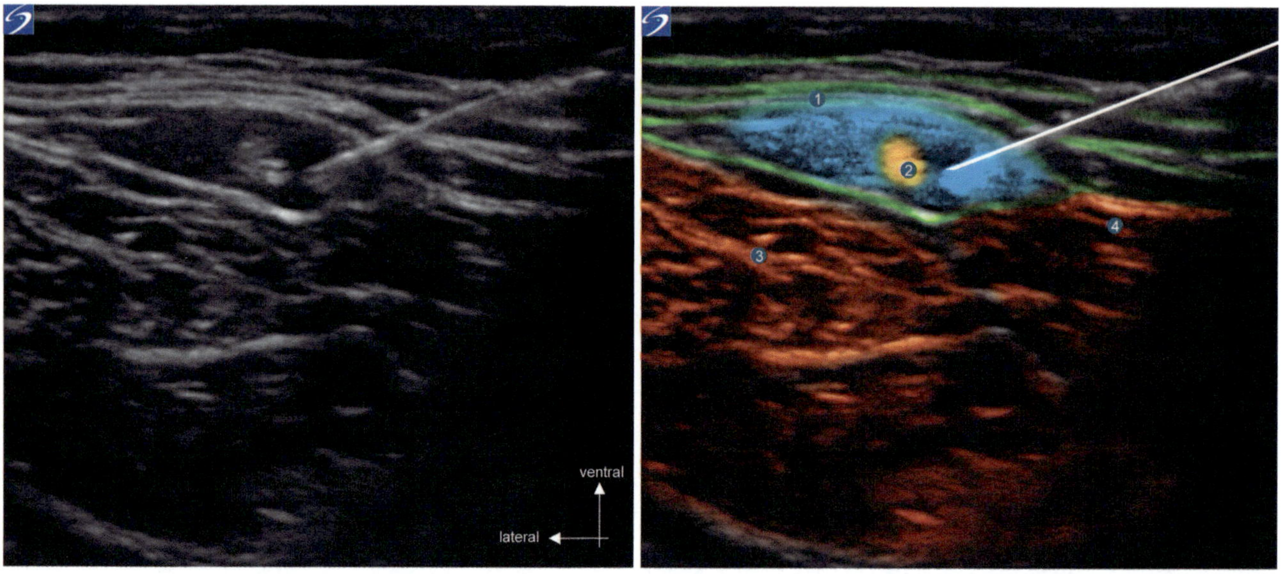

Fig. 6.43 Lateral femoral cutaneous nerve block. Needle insertion in-plane

Saphenous Nerve Block (Fig. 6.44)

Indications

The saphenous nerve is purely sensory. It provides innervation of the ventromedial aspect of the leg extending to the medial malleolus, but it can sometimes supply the skin of the medial aspect of the foot up to the head of the first metatarsal. A saphenous nerve block is suitable for anaesthesia in foot, ankle and tibia surgery and possibly, although inconsistently, for surgery on the great toe.

The saphenous nerve can be blocked above the knee (perifemoral, sub-sartorial and trans-sartorial location [33–35]), at the knee (medial femoral condyle with or without neurostimulation [36, 37]); a catheter can be inserted at mid-thigh level to prolong post-operative analgesia after a total knee replacement (TKR) in combination with injection of local anaesthetic into the tissues [38]. A saphenous nerve block can also be performed below the knee (posterior to the insertion of the muscles of the "pes anserinus") [39, 40], or with a "perivenous" injection [41]. Lastly, a subcutaneous injection can be administered just above the medial malleolus [39, 40]. A subcutaneous injection, whatever its level, can be painful. Its efficacy can be unpredictable, particularly in the upper one-third of the leg, perhaps because of the uncertainty about the exact depth to which

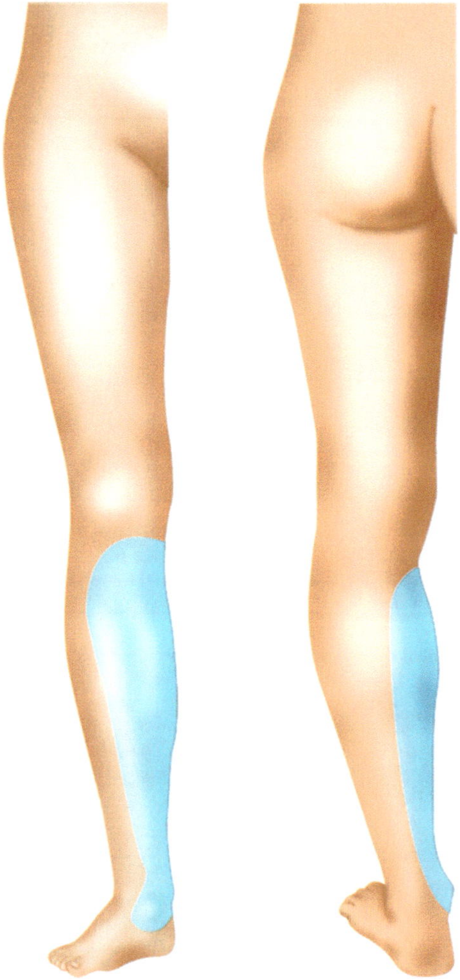

Fig. 6.44 Area of coverage of a saphenous nerve block

the local anaesthetic should be injected, which is affected by the thickness of the patient's adipose tissue. In our experience, a local anaesthetic injection near to the long saphenous vein under ultrasound guidance a few centimetres below the knee joint (medial tibial tuberosity) [42] produces unpredictable results. This is most likely related to the "dispersal" of the subcutaneous branches of the saphenous nerve, and also because of variations in the position of the long saphenous vein at this level.

The multiplicity of described approaches to anaesthetise the saphenous nerve probably relates to the inconsistency of results of these techniques. A trans-sartorial block, described in the first edition of this work, produced a good success rate [43], although injection of the local anaesthetic is given within the fascial plane without direct visualisation of the nerve. Nevertheless, in our experience, the greatest efficacy is obtained with injection of a few millilitres of local anaesthetic in direct contact with the saphenous nerve, found at the deep to the sartorius muscle, at the junction of the middle and distal one-third of the thigh. This approach causes less pain and can be easily performed with ultrasound with reproducible results due to the consistency of the visualised anatomy.

> **Type of probe:** linear, 5 to 10 MHz or 6 to 13 MHz.
> **Axis and position of the probe:** transversal, at junction of the middle one-third and distal one-third of the thigh (Fig. 6.45).
> **Depth studied:** up to 6 cm depending on patient build.
> **Neurostimulation:** It may be useful to elicit paraesthesiae in the distribution of the saphenous nerve when ultrasound identification of the nerve is unclear (Fig. 6.47). By determination of MIS > 0.3 mA (0.1 ms), the risk of accidental intrafascicular injection is reduced.
> **Needle:** 50 to 80 mm isolated, 22 G.

Echoanatomy

The block is performed deep to the sartorius muscle in the mid to distal third of the thigh. The saphenous nerve lies in close proximity to the superficial femoral artery, which is used as an anatomical landmark to aid nerve identification. At this level the nerve and vessels run within a fascial plane, demarcated ventrally by the vastus medialis, dorsally by the adductor magnus and semimembranosus, and medially by the sartorius and gracilis. These muscle masses are of variable size depending on patient build but generally have a homogenous echogenicity and clear aponeurotic boundaries, demarcating an inter-muscular space within which it is easy to identify the neurovascular structures. Sometimes the motor branch of the vastus medialis is identified in near to the artery (Figs. 6.46 and 6.47).

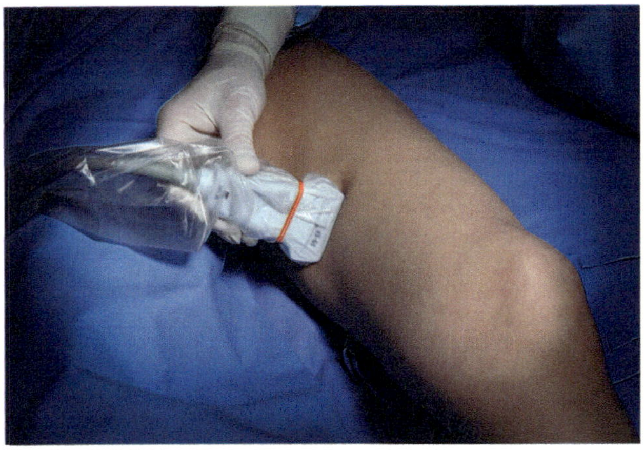

Fig. 6.45 Saphenous nerve block: position of the probe

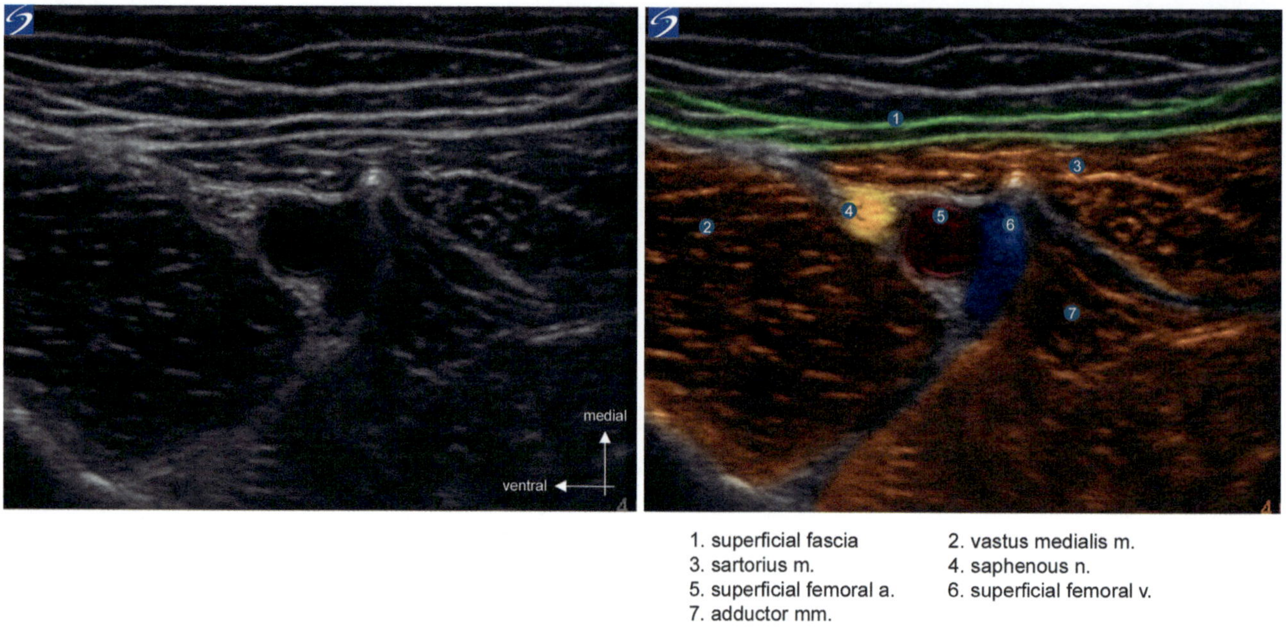

Fig. 6.46 Saphenous nerve block. Transverse ultrasound section at the middle third-distal third junction of the thigh

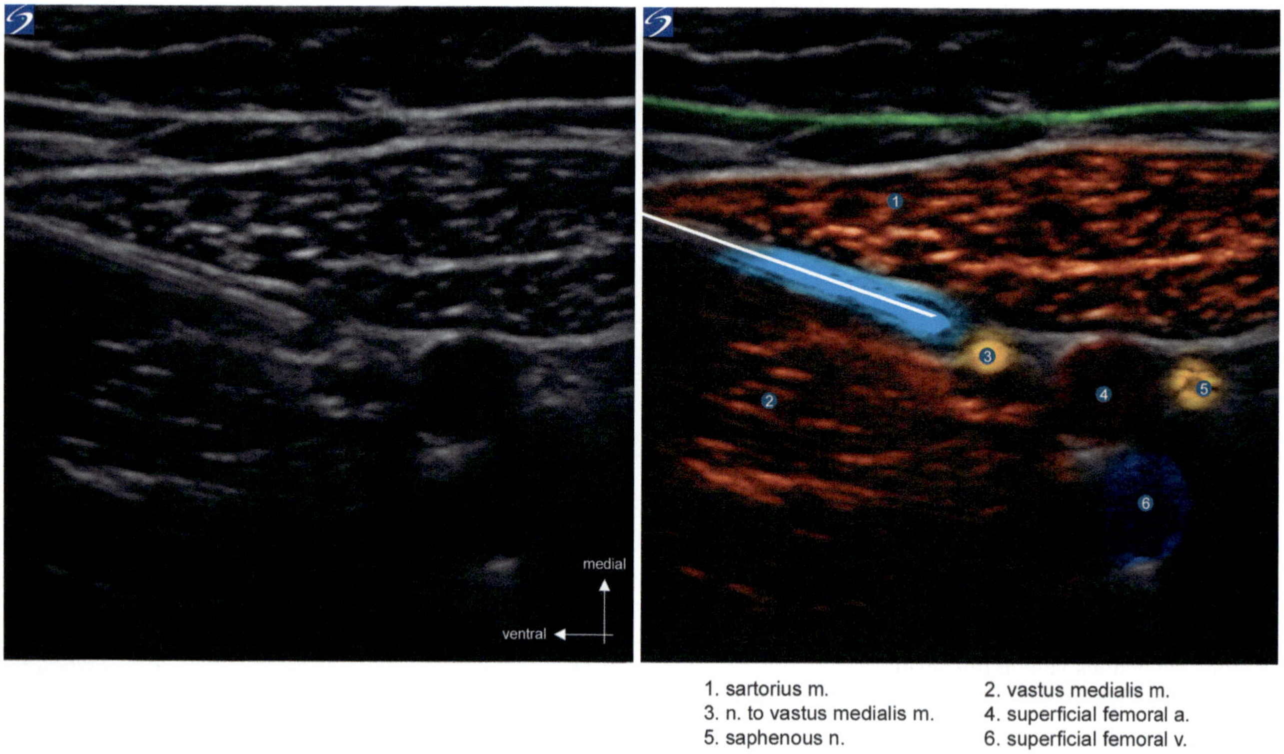

Fig. 6.47 Saphenous nerve block: course of the needle between the sartorius and vastus medialis muscles. Identification of the nerve to the vastus medialis (3) and of the saphenous nerve (5) was done post hoc, using neurostimulation, evidence of its usefulness

Utility of Doppler Ultrasound

It enables localisation of the artery and of the superficial femoral vein.

Approach and Localisation

The probe is placed transversely at the medial aspect of the thigh, opposite the sartorius muscle, 10 to 15 cm above the upper border of the patella (Fig. 6.45).

Injection

After anaesthesia of the skin, the needle is advanced in the ultrasound plane, at the anterior end of the probe (Fig. 6.48). The needle is inserted between vastus medialis and sartorius muscles. The needle is advanced using hydrodissection into the inter-muscular space until it lies adjacent to the superficial femoral artery (Fig. 6.47). Hydrodissection performed with a D5W solution, enables use of neurostimulation to confirm the presence of nerve structures near to the superficial femoral artery (differentiating, for example, the motor nerve of the vastus medialis from the sensory saphenous nerve) (Fig. 6.47).

Once the needle has been correctly positioned, the block may be performed by injection of 5–7 mL of local anaesthetic. If the nerve is not clearly identified, it is possible to use a larger volume of local anaesthetic to "fill" the perivascular space to promote its effect. Repositioning of the needle may also improve the spread of the local anaesthetic and enhance the block.

> **Paediatrics**
>
> It may be necessary to use a higher frequency probe (15 MHz) to perform this block when the depth of injection is particularly shallow. As in adults, this block can be used for analgesia of the medial malleolus and the anteromedial aspect of the leg without performing a complete femoral nerve block, so avoiding its associated motor effects.

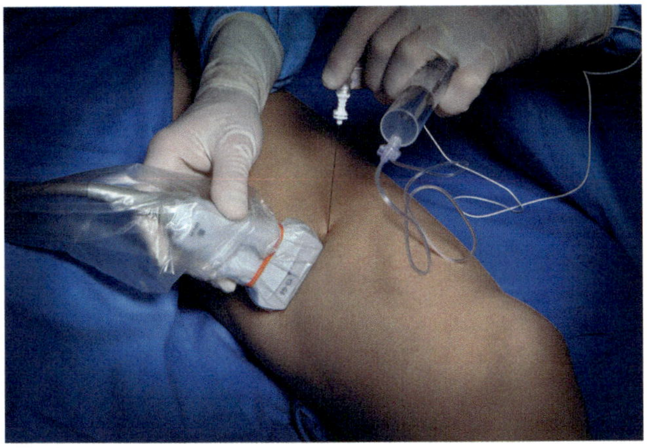

Fig. 6.48 Saphenous nerve block. Position of probe, needle inserted in-plane

Obturator Nerve Block (Fig. 6.49)

Indications

An obturator nerve block can be performed as a supplement to a femoral-sciatic "two-nerve block", when anaesthesia is insufficient in the obturator territory. The test of a functioning obturator nerve block is motor—a patient already blocked in the sciatic and femoral nerve territories should not be able to perform adduction of the thigh. For surgery of the knee, some practitioners prefer a "three-nerve block" at the outset. The latter is more effective and results in improved postoperative analgesia [44]. An obturator block can be used in cases of chronic pain and in the treatment of adductor spasticity [45].

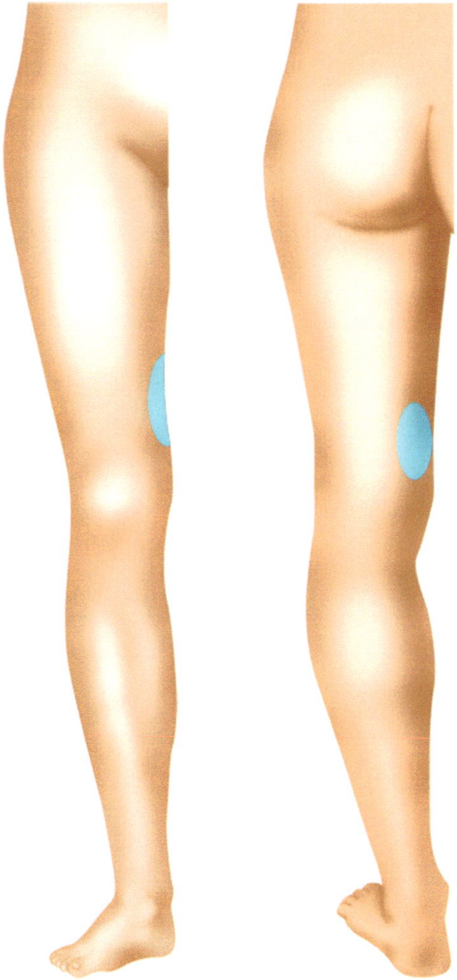

Fig. 6.49 Area of coverage of the obturator nerve (inconstant cutaneous area)

Type of probe: linear, 5 to 10 MHz or 6 to 13 MHz. Relatively wide vision of the area of the same ultrasound image is desirable (38 mm probe).
Axis and position of the probe: transversal, in the skin crease of the thigh, medial to the femoral vessels (Fig. 6.50).
Configuration: nerve in short axis, needle in-plane.
Depth studied: variable, generally up to maximum 7 cm depending on the patient build.
Neurostimulation: enables further identification of the nerves visualised and can limit, by determination of an MIS > 0.3 mA (0.1 ms), the risk of accidental intrafascicular injection. Furthermore, it can supplement locating nerves when conditions of visibility are unfavourable. In case of an approach very proximal to the nerve, it also makes it possible to differentiate the motor responses induced by stimulation of the anterior branch (adductor longus, adductor brevis, gracilis and pectineus) from those due to stimulation of the posterior branch (adductor magnus).
Needle: 80 mm isolated, 22 G.
Utility of Doppler ultrasound: obturator vessels accompany the anterior and posterior branches of the nerve.

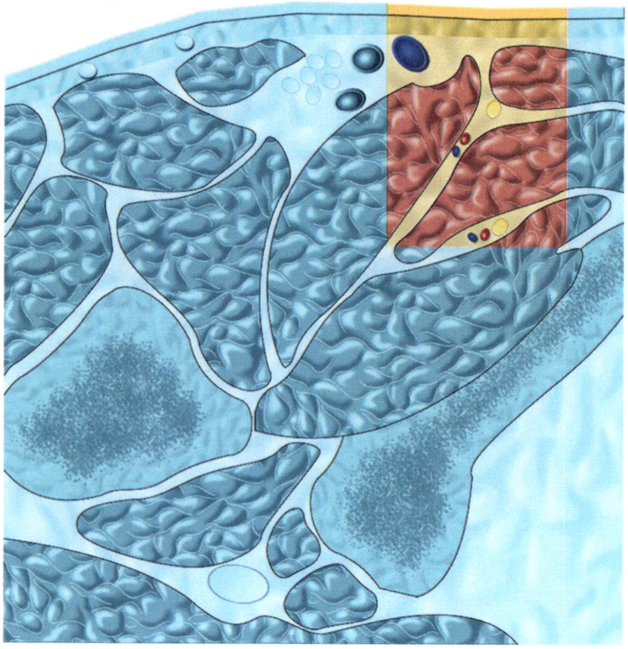

Fig. 6.50 Transverse section of the base of the thigh with materialisation of ultrasound beam for obturator nerve block

Echoanatomy and Location

As it emerges from the obturator foramen, the obturator nerve (which may already be divided into its two terminals, anterior and posterior branches) lies deep to the pectineus and anterior to obturator externus (Figs. 6.51 and 6.52). More distally, its branches progressively diverge (Figs. 6.52, 6.53 and 6.54): the anterior branch continues between pectineus and adductor brevis, and the posterior branch between the adductor brevis and adductor magnus. Their inter-muscular positions make them relatively easily identifiable, and accurate localisation is aided by slight movements of the probe. Hypo- and hyper-echoic heterogenous fusiform images are sought in between the muscles and associated aponeurotic layers. The branches of the obturator nerve are small, but can be seen with ultrasound with a probe of frequency greater than or equal to 7.5 MHz.

Utility of Doppler Ultrasound

Aids identification of the femoral vessels (generally easy), the arch of the long saphenous vein, and the superficial and deep external pudendal vessels. Frequently, it is possible to differentiate the obturator vessels and their branches which follow the anterior and posterior branches of the obturator nerve (Fig. 6.55).

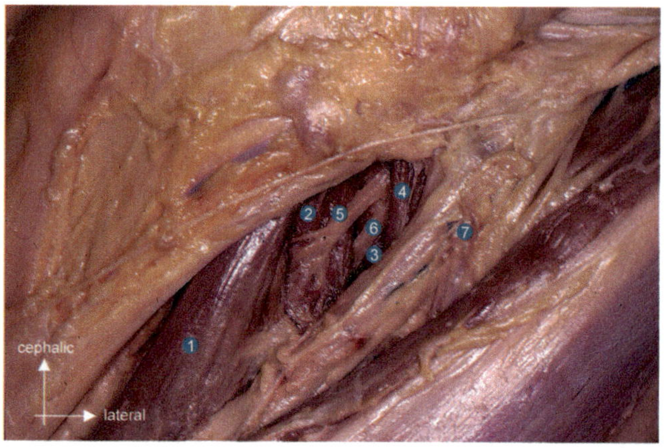

1. adductor longus m.
2. adductor brevis m.
3. adductor magnus m.
4. pectineus m.
5. obturator n. (anterior branch)
6. obturator n. (posterior branch)
7. femoral a., v. and n.

Fig. 6.51 Anterior view of the two branches of the obturator nerve. Iconography: Admir Hadzic

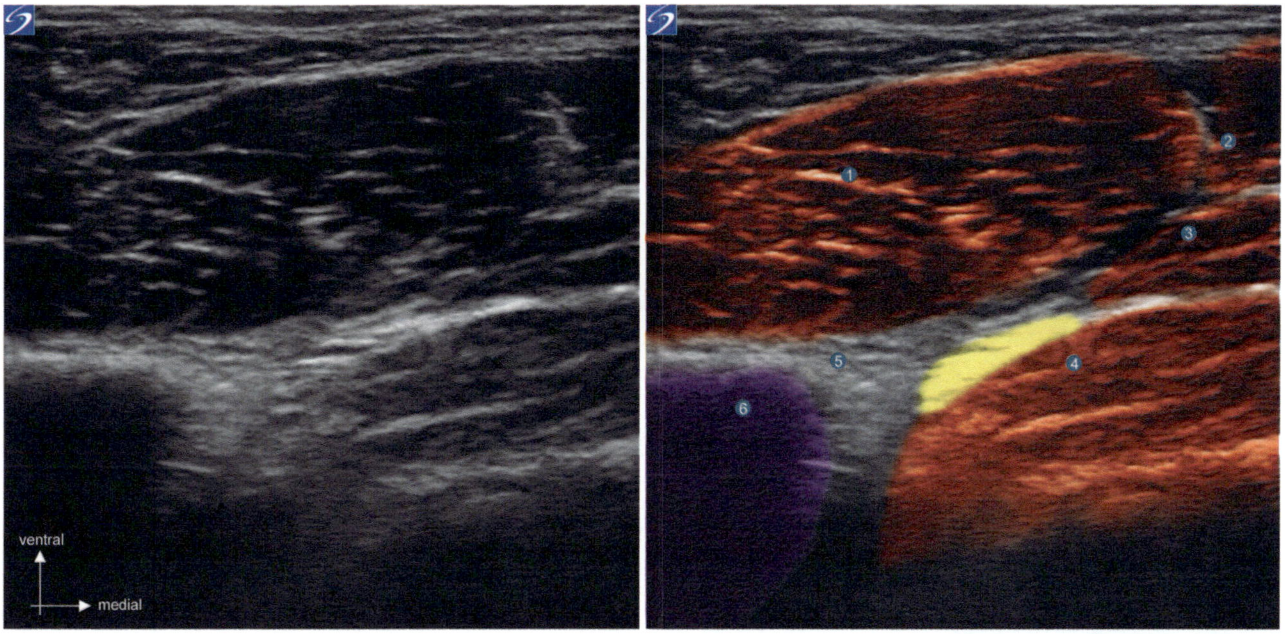

1. pectineus m.
2. adductor longus m.
3. adductor brevis m.
4. obturator externus m.
5. obturator n.
6. iliopubic eminence

Fig. 6.52 Transverse ultrasound section of the obturator nerve, between the pectineus and obturator externus muscles, immediately after its emergence from the obturator foramen

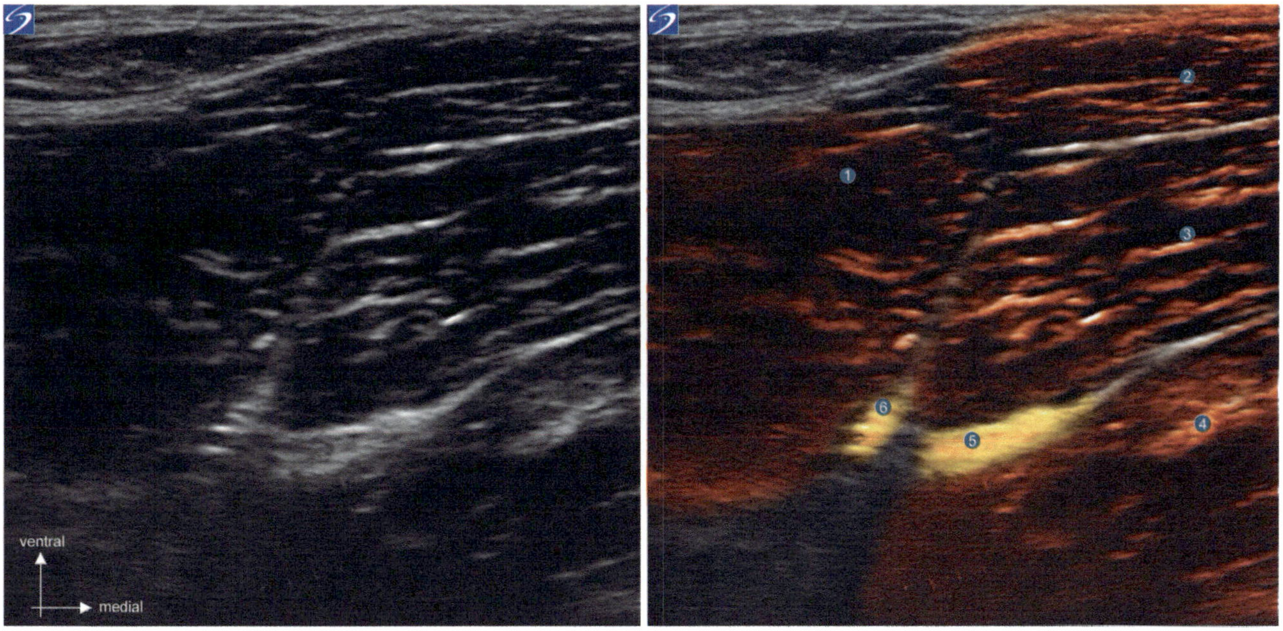

1. pectineus m.
2. adductor longus m.
3. adductor brevis m.
4. adductor magnus m.
5. obturator n. (posterior branch)
6. obturator n. (anterior branch)

Fig. 6.53 Transverse ultrasound section of the obturator nerve at the base of the thigh. These two branches get progressively further away from each other as they course on either side of the adductor brevis muscle

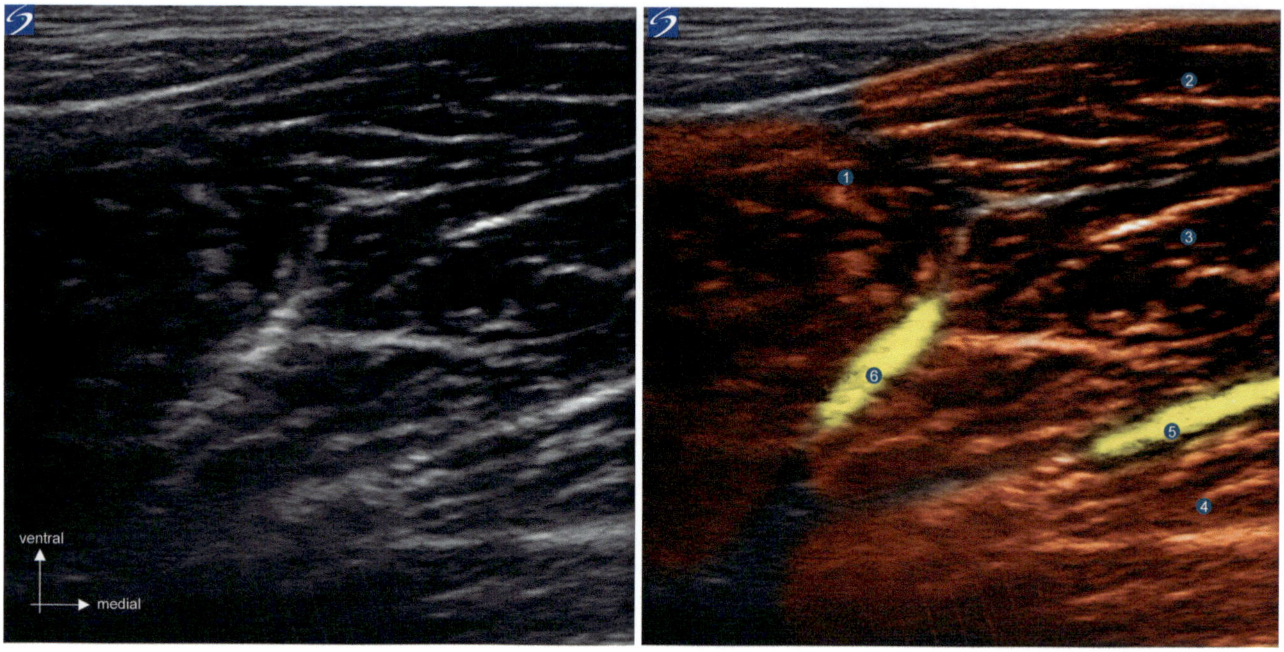

1. pectineus m.
2. adductor longus m.
3. adductor brevis m.
4. adductor magnus m.
5. obturator n. (posterior branch)
6. obturator n. (anterior branch)

Fig. 6.54 Transverse ultrasound section of the obturator nerve. The two branches of division of the obturator nerve are now located on either side of the adductor brevis muscle

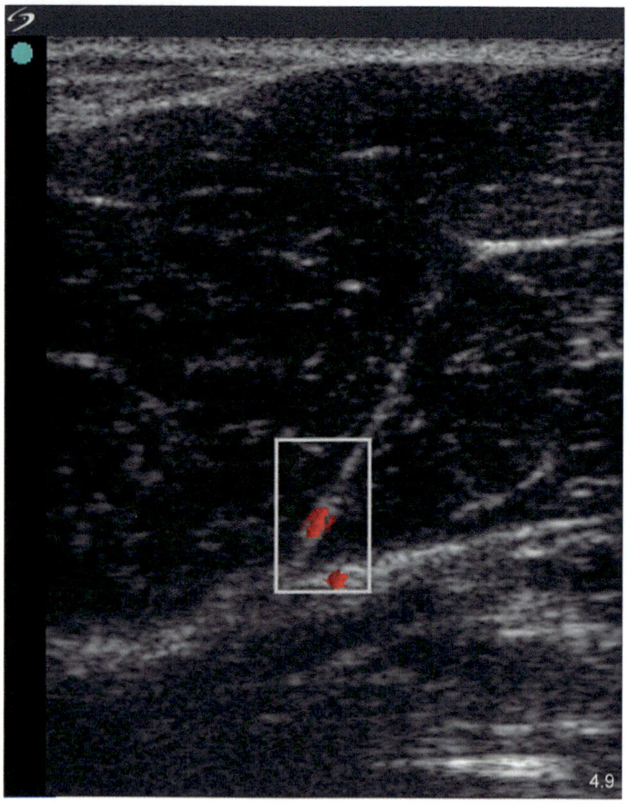

Fig. 6.55 Obturator blood vessels accompanying the anterior and posterior branches of the obturator nerve

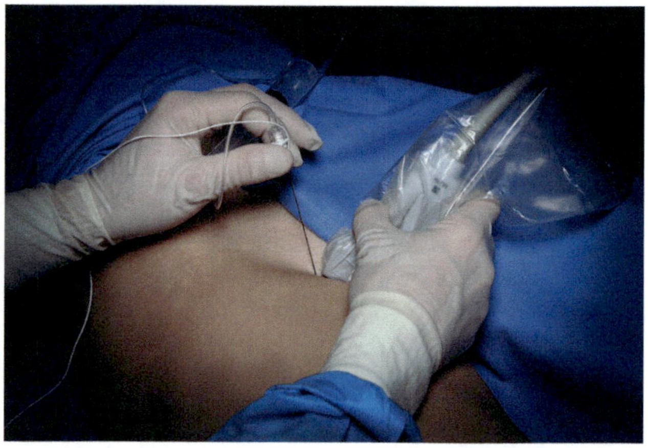

Fig. 6.56 Obturator nerve block: needle inserted in-plane

Approach and Injection

Although it is possible to perform an out of plane approach, an in-plane technique is simple, safe (Fig. 6.56), and offers easier identification of the position of the needle tip. Whatever the option chosen, tissue movements related to the advancement of the needle should be closely observed and it should be possible to precisely locate the position of the tip. In case of doubt, injection and observation of the spread of a few millilitres of a D5W solution

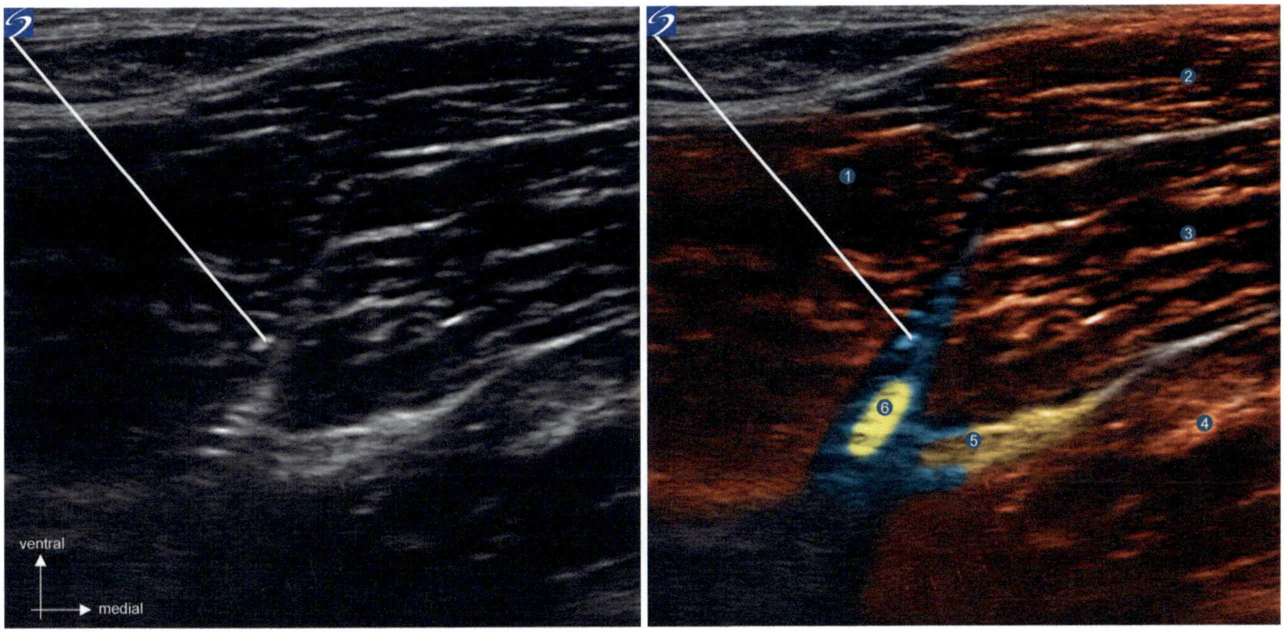

1. pectineus m.
2. adductor longus m.
3. adductor brevis m.
4. adductor magnus m.
5. obturator n. (posterior branch)
6. obturator n. (anterior branch)

Fig. 6.57 Obturator nerve block. Injection of the local anaesthetic that "dissects" the space between the pectineus and adductor brevis muscles

aids in determining the exact position of the needle tip (hydrolocalisation).

The injection site lies medially in the inguinal skin crease, between the lateral border of the ultrasound probe, and the femoral vein/long saphenous vein lying laterally. After piercing the skin, the needle is directed medially into the fascial plane between pectineus and the adductor brevis, towards the anterior branch of the obturator nerve (Fig. 6.57). It is possible to angle the probe laterally in order to obtain a better visualisation of the needle, by creating a more perpendicular relationship between needle and ultrasound beam.

Once pectineus muscle has been crossed, neurostimulation can confirm the proximity of the obturator nerve and allows for adjustment of the needle position. Since the common obturator nerve is accessible only very "cranially", the needle generally approaches the terminal nerves downstream of its bifurcation.

Injection

It is possible to administer the injection beside each component (anterior and posterior branches) of the obturator nerve, but injecting only in proximity to the anterior branch seems to achieve equivalent effects. If the needle is in the correct inter-muscular position, the local anaesthetic spreads between both pectineus and adductor brevis (Fig. 6.57). Visualisation of the obturator nerve surrounded by the local anaesthetic injected is not consistent, because it depends on conditions of echogenicity. If it is observed that the injection is not within the ***inter*-muscular** space, then the tip of the needle is most probably ***intra*muscular** (either within the pectineus or adductor brevis). When the anterior branch of the obturator nerve is approached at short distance beyond the obturator nerve bifurcation, it does not seem necessary to administer a second injection in contact with its posterior branch, since the local anaesthetic spreads cranially towards it within the inter-muscular space. Generally, a volume of 7–10 mL of LA is injected for clinical effect.

References

1. Kamina P. Précis d'anatomie clinique. Tome 1. Paris: Maloine; 2003.
2. Bouaziz H, Vial F, Jochum D, et al. An evaluation of the cutaneous distribution after obturator nerve block. Anesth Analg. 2002;94:445–9.
3. Zetlaoui PJ. Les blocs du plexus lombaire. Cah Anesthesiol. 1994;42:771–80.
4. Winnie AP. Regional anesthesia. Surg Clin North Am. 1975;54:861–92.
5. Chayen D, Nathan H, Chayen M. The psoas compartment block. Anesthesiology. 1976;45:95–9.
6. Parkinson S, Mueller JB, Little WL, Bailey S. Extend of blockade with various approaches to the lumbar plexus. Anesth Analg. 1989;65:185–7.

7. Hanna MH, Peat SJ, D'Costa F. Lumbar plexus block: an anatomical study. Anesthesia. 1993;48:675–8.
8. Macaire P, Gaertner E, Choquet O. Le bloc du plexus lombaire est-il dangereux? Épidémiologie des accidents et complications des techniques de bloc du plexus lombaire. Comment éviter l'échec? Comment prévenir les complications? In: Évaluation et traitement de la douleur. Paris: Elsevier SAS/SFAR; 2002. p. 37–50.
9. Capdevila X, Biboulet P, Bouregba M, Barthelet Y, Rubenovitch J, d'Athis F. Comparison of the three-in-one and fascia iliaca compartment blocks in adults: clinical and radiographic analysis. Anesth Analg. 1998;86:1039–44.
10. Eyrolle L, Zetlaoui P, Belbachir A, Conseiller C. Regional anesthesia for femoral neck fracture surgery: comparison of lumbar plexus block and spinal anesthesia. Br J Anaesth. 1998;78:A383.
11. Touray ST, de Leeuw MA, Zuurmond WW, Perez RS. Psoas compartment block for lower extremity surgery: a meta-analysis. Br J Anaesth. 2008;101:750–60.
12. Farny J, Drolet P, Girard M. Posterior approach to the lumbar plexus combined with a sciatic nerve block using lidocaine. Can J Anaesth. 1994;41:486–91.
13. de Leeuw MA, Zuurmond WW, Perez RS. The psoas compartment block for hip surgery: the past, present, and future. Anesthesiol Res Pract. 2011;2011:159541.
14. Ilfeld BM, Loland VJ, Mariano ER. Prepuncture ultrasound imaging to predict transverse process and lumbar plexus depth for psoas compartment block and perineural catheter insertion: a prospective, observational study. Anesth Analg. 2010;1(110):1725–8.
15. Siddiqui ZI, Cepeda MS, Denman W, et al. Continuous lumbar plexus block provides improved analgesia with fewer side effects compared with systemic opioids after hip arthroplasty: a randomized controlled trial. Reg Anesth Pain Med. 2007;32:393–8.
16. Bromage P. Connections between extradural and Para spinal tissue spaces. In: Epidural analgesia. Philadelphie: Sanders; 1978. p. 31–2.
17. De Biasi P, Lupescu R, Burgun G, Lascurain P, Gaertner E. Continuous lumbar plexus block: use of radiography to determine catheter tip location. Reg Anesth Pain Med. 2003;28:135–9.
18. Piffaut V, Bruelle P, VanRoy C, et al. Bloc du plexus lombaire par voie postérieure. Ann Fr Anesth Réanim. 1996;15:R041.
19. Auroy Y, Benhamou D, Bargues L, et al. Major complications of regional anesthesia in France: the SOS regional anesthesia hotline service. Anesthesiology. 2002;97:1274–80.
20. Dauri M, Faria S, Celidonio L, et al. Retroperitoneal haematoma in a patient with continuous psoas compartment block and enoxaparin administration for total knee replacement. Br J Anaesth. 2009;103:309–10.
21. Pham-Dang C, Beaumont S, Floch H, Bodin J, Winer A, Pinaud M. Acute toxic accident following lumbar plexus block with bupivacaine. Ann Fr Anesth Réanim. 2000;19:356–9.
22. Huet O, Eyrolle LJ, Mazoit JX, Ozier YM. Cardiac arrest after injection of ropivacaine for posterior lumbar plexus blockade. Anesthesiology. 2003;99:1451–3.
23. Marhofer P, Harrop-Griffiths W, Willschke H, et al. Fifteen years of ultrasound guidance in regional anesthesia: part 2-recent developments in block techniques. Br J Anaesth. 2010;104:673–83.
24. Capdevila X, Macaire P, Dadure C, et al. Continuous psoas compartment block for post-operative analgesia after total hip arthroplasty: new landmarks, technical guidelines and clinical evaluation. Anesth Analg. 2002;94:1606–13.
25. Kirchmair L, Entner T, Kapral S, Mitterschiffthaler G. Ultrasound guidance for the psoas compartment block: an imaging study. Anesth Analg. 2002;94:706–10.
26. Kirchmair L, Enna B, Mitterschiffthaler G, et al. Lumbar plexus in children, a sonographic study and its relevance to pediatric regional anesthesia. Anesthesiology. 2004;101:445–50.
27. Gros T, Viel E, Ripart J. Prehospital analgesia with femoral nerve block following lower extremity injury. A 107 cases survey. Ann Fr Anesth Réanim. 2012;31:846–9.
28. Paul JE, Arya A, Hurlburt L. Femoral nerve block improves analgesia outcomes after total knee arthroplasty: a meta-analysis of randomized controlled trials. Anesthesiology. 2010;113:1144–62.
29. Salinas FV, Liu SS, Mulroy MF. The effect of single-injection femoral nerve block versus continuous femoral nerve block after total knee arthroplasty on hospital length of stay and long-term functional recovery within an established clinical trajectory. Anesth Analg. 2006;102:1234–9.
30. Sakai N, Inoue T, Kunugiza Y, Tomita T, et al. Continuous femoral versus epidural block for attainment of 120° knee flexion after total knee arthroplasty: a randomized controlled trial. J Arthroplast. 2013;28:807–14.
31. Fowler SJ, Symons J, Sabato S, et al. Epidural analgesia compared with peripheral nerve blockade after major knee surgery: a systematic review and meta-analysis of randomized trials. Br J Anaesth. 2008;100:154–64.
32. Shteynberg A, Riina LH, Glickman LT, et al. Ultrasound guided lateral femoral cutaneous nerve (LFCN) block: safe and simple anesthesia for harvesting skin grafts. Burns. 2013;39:146–9.
33. Bouaziz H, Benhamou D, Narchi P. A new approach for saphenous nerve block (letter). Reg Anesth. 1996;21:490.
34. Bouaziz H, Narchi P, Zetlaoui PJ, Paqueron X, Benhamou D. Lateral approach to the sciatic nerve at the popliteal fossa combined with saphenous nerve block. Tech Reg Anesth Pain Manag. 1999;3:19–22.
35. Van der Wal M, Lang SA, Yip RW. Transsartorial approach for saphenous nerve block. Can J Anaesth. 1993;40:542–6.
36. Bonica JJ. Regional analgesia with local anesthetics. In: The management of pain. Philadelphie: Lea Febiger; 1990. p. 1883–966.
37. Katz J. Saphenous nerve. In: Atlas of regional anesthesia. Norwalk: Appleton Lange; 1994. p. 164–7.
38. Comfort VK, Lang SA, Yip RW. Saphenous nerve anesthesia: a nerve stimulator technique. Can J Anaesth. 1996;43:852–7.
39. Bridenbaugh PO. The lower extremity: somatic blockade, neural blockade. In: Cousins MJ, Bridenbaugh PO, editors. Clinical anesthesia and management of pain. New York: JB Lippincott; 1988. p. 417–41.
40. Rogers JN, Ramamurthy S. Lower extremity blocks. In: Brown DL, editor. Regional anesthesia and analgesia. Philadelphie: WB Saunders; 1996. p. 279–91.
41. De Mey JC, Deruyck LJ, Cammu G, De Baerdemaeker LE, Mortier EP. A paravenous approach for the saphenous nerve block. Reg Anesth Pain Med. 2001;26:504–6.
42. Gray AT, Collins AB. Ultrasound guided saphenous nerve block. Reg Anesth Pain Med. 2003;28:148.
43. Benzon HT, Sharma S, Calimaran A. Comparison of the different approaches to saphenous nerve block. Anesthesiology. 2005;102:633–8.
44. Choquet O, Capdevila X, Bennourine K, Feugeas JL, Bringuier-Branchereau S, Manelli JC. A new inguinal approach for the obturator nerve block: anatomical and randomized clinical studies. Anesthesiology. 2005;103:1238–45.
45. Yavuz F, Yasar E, Ali Taskaynatan M, et al. Nerve block of articular branches of the obturator and femoral nerves for the treatment of hip joint pain. Musculoskelet Rehabil. 2013;26:79–83.

Sacral Plexus Blocks

7

Eryk Eisenberg, Elisabeth Gaertner, and Philippe Clavert

Contents

E. Eisenberg (✉)
Département d'anesthésie et réanimation, Clinique Pôle Santé République, Clermont-Ferrand, France

E. Gaertner
Département d'anesthésie et réanimation chirurgicale, Hôpitaux Civils de Colmar, Colmar, France

P. Clavert
Service de chirurgie du membre supérieur, Centre de Chirurgie Orthopédique et de la Main (CCOM), Illkirch, France

E. Eisenberg, E. Gaertner (eds.), *Ultrasound in Peripheral, Neuraxial and Perineuraxial Regional Anaesthesia*,
https://doi.org/10.1007/978-3-031-08804-9_7

Anatomy

The sacral plexus is formed by the union of the ventral branches of L5 and of the first three spinal sacral nerves (S1 to S3) (Figs. 7.1, 7.2 and 7.3). Components of fibres arising from the ventral branches of L4 and from S4 also contribute to this plexus. The L4 element joins the ventral branch of L5 to form the lumbosacral trunk which emerges at the medial border of the psoas major muscle. This trunk then crosses the pelvic inlet and continues across the sacral iliac joint, in the lumbosacral fossa, in which it joins with the ventral branch of S1.

The ventral branches of the sacral roots, whose size decreases from upper to lower, emerge from the sacrum via the ventral sacral foramina. They continue laterally, caudally, and ventrally. The first one crosses the cranial border of the piriformis obliquely, the second continues ventrally in relation to this muscle, and the third continues along its caudal border. Thus, the sacral plexus has the shape of a triangle for which the base is in relation to the ventral sacral foramina medially, and the apex at the ventral caudal part of the sciatic foramen laterally [1]. It is directly applied to the ventral aspect of the piriformis in the dorsal part of the pelvic cavity, covered by the pelvic parietal fascia. It has close anatomical relations with the pelvic internal organs (the ureter, the end of the ileum on right side and the sigmoid colon on left side) and the gluteal and internal iliac blood vessels. The superior gluteal artery passes between the lumbosacral trunk and S1, the lateral sacral artery passes ventrally in relation to S1, the lower gluteal artery passes between S2 and S3, and the internal pudendal artery passes below the plexus. The sacral plexus anastomoses with the lumbar plexus, the pudendal plexus, and the pelvic sympathetic lymph node chain.

The sacral plexus, whose single terminal branch is represented by the sciatic nerve, gives rise to the following collateral branches.

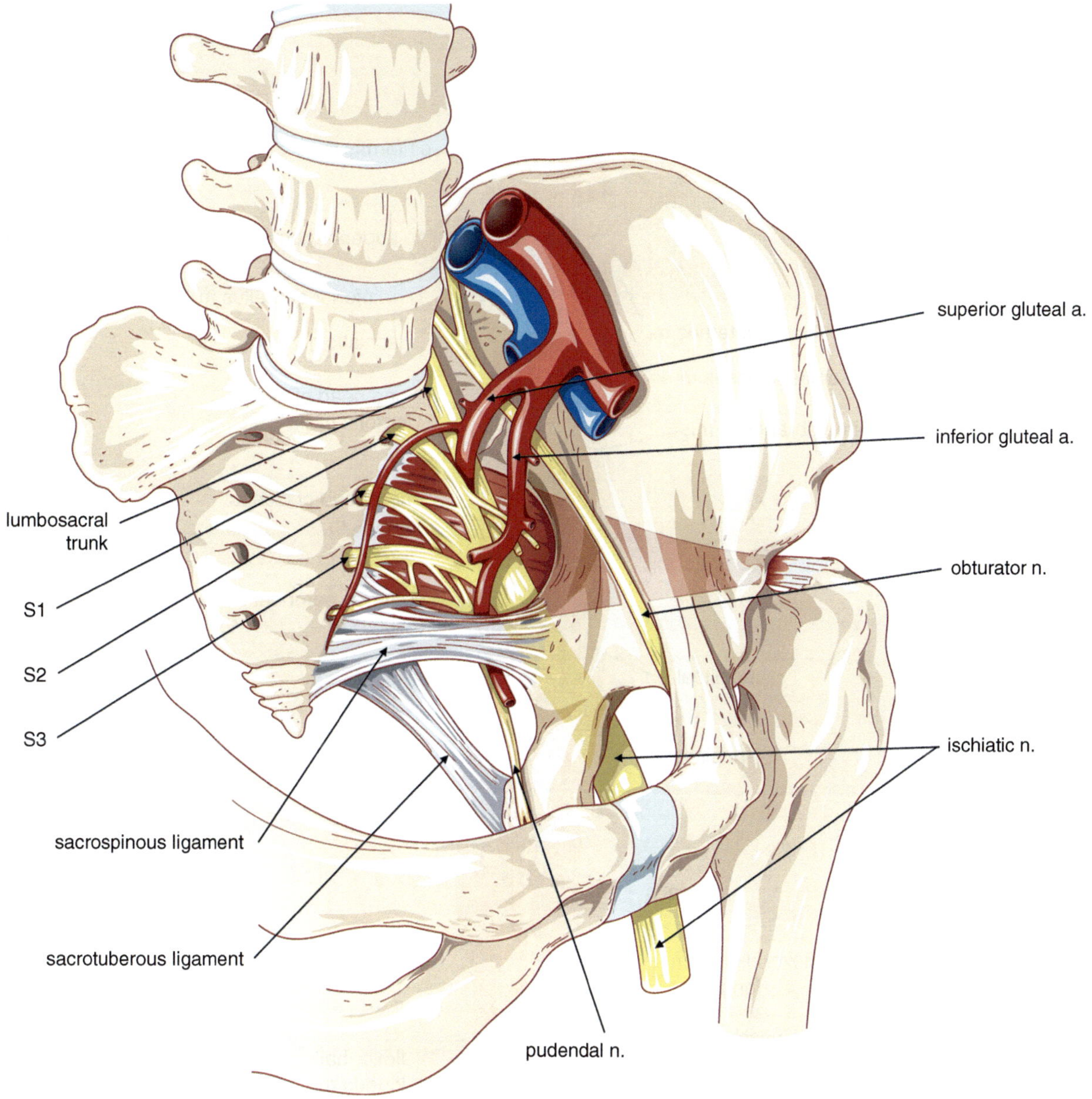

Fig. 7.1 Schematic diagram of the overall organisation of sacral plexus

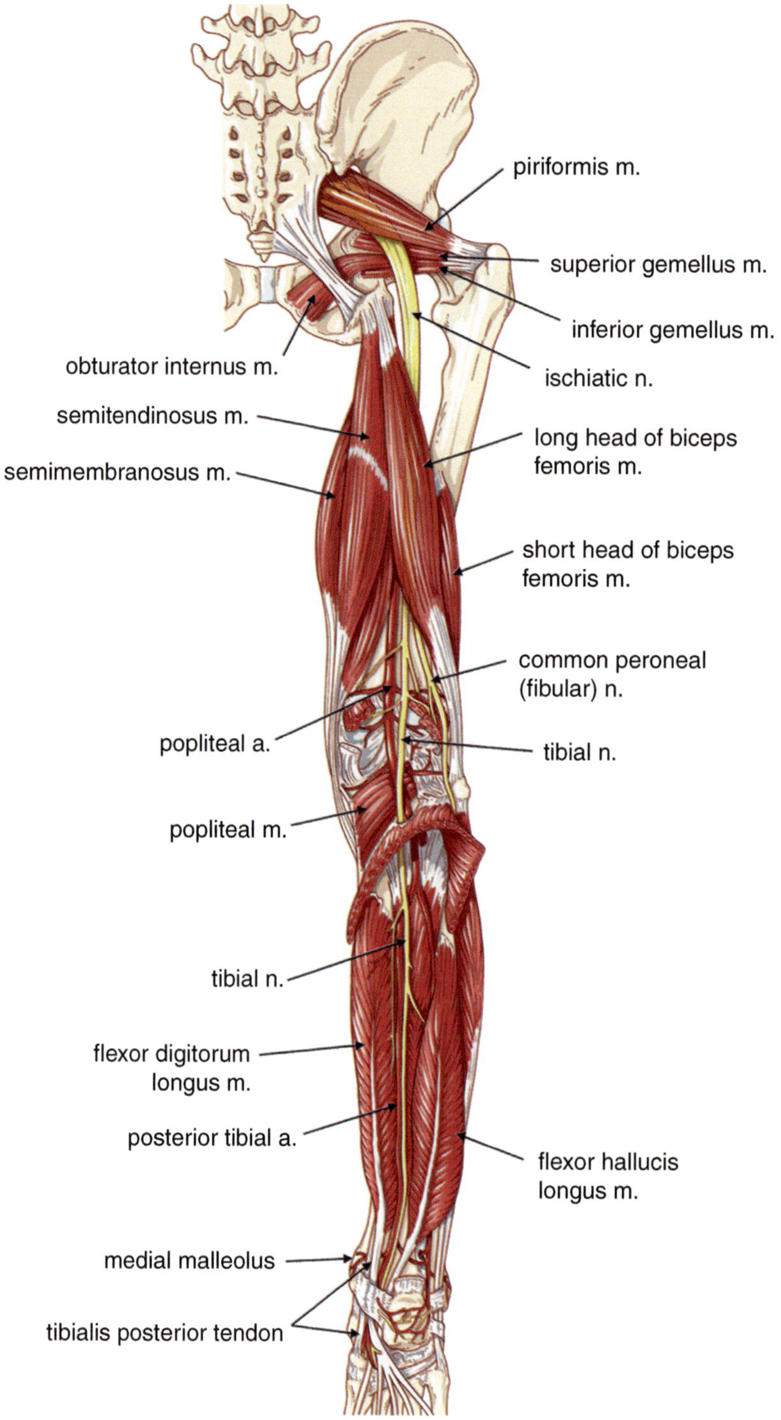

Fig. 7.2 Posterior view of the lower limb with depiction of nerve structures emerging from the sacral plexus

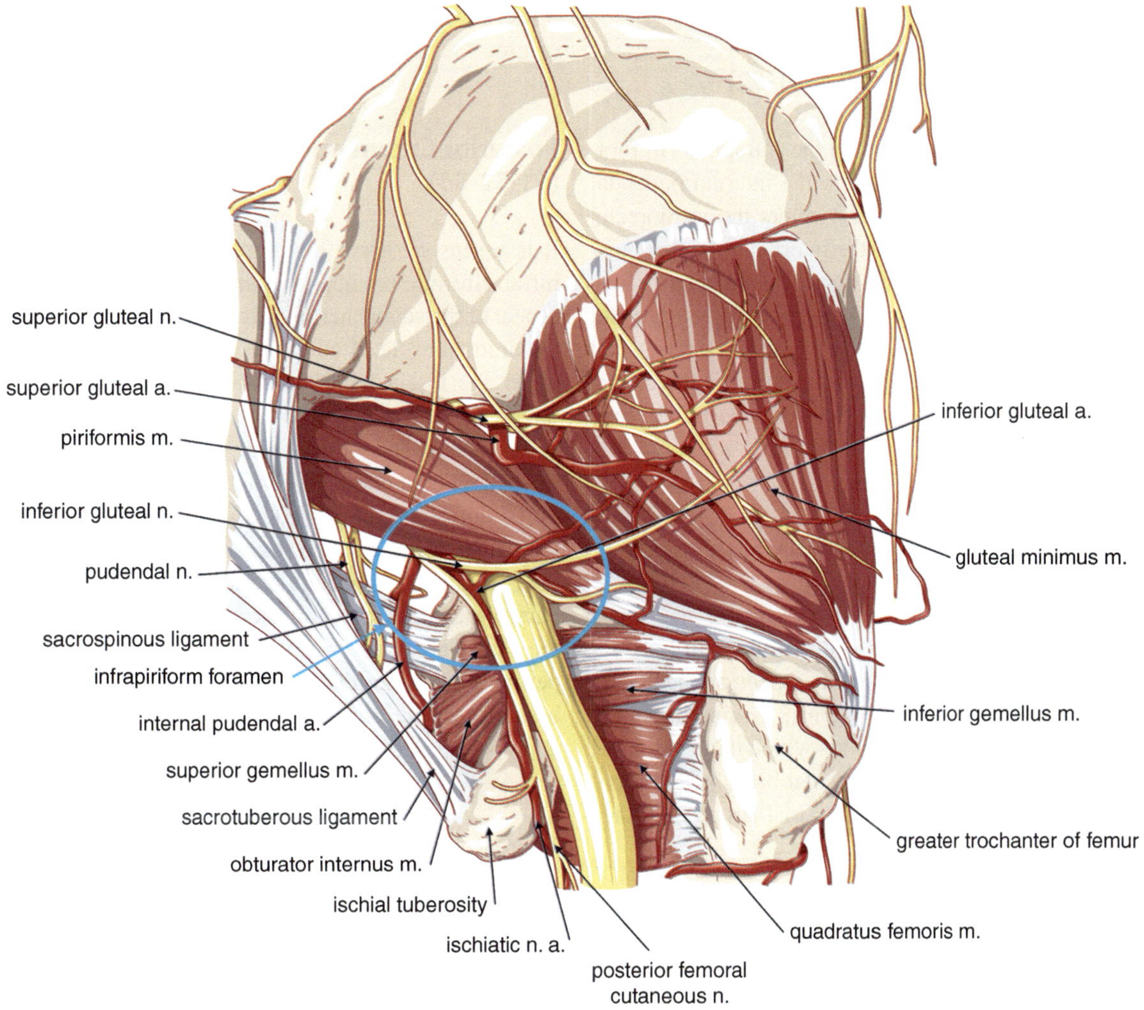

Fig. 7.3 Diagram of the sacral plexus in a posterolateral view. Visualisation of the infrapiriform foramen

Ventral Branches

Nerve to Obturator Internus (L5-S1-S2)

It emerges from the pelvic cavity via the infrapiriform foramen, continues laterally in relation to the internal pudendal vessels in the gluteal area, and then enters the ischiorectal fossa beside the medial aspect of the internal obturator muscle that it innervates. It sometimes gives a muscular branch to the superior gemellus muscle when the latter does not arise independently from the ventral aspect of the sacral plexus. In the latter case, it descends in front of the sciatic nerve and enters into the muscle by its superficial, deep aspect, or by its upper border.

Nerve to Quadratus Femoris (L4-L5-S1)

It follows the pathway of the internal obturator nerve and emerges laterally from the pelvis via the infrapiriform foramen. It then continues on the ventral aspect of the sciatic nerve, the gemelli and obturator internis, terminating at the ventral aspect of the quadratus femoris which it innervates. It gives rise during its passage to a muscular branch which supplies the inferior gemellus as well as a small articular branch to the hip joint.

Dorsal Branches

Nerve to Piriformis (S1-S2)

It terminates on the ventral aspect of the muscle, in its intrapelvic portion.

Superior Gluteal Nerve (L4-L5-S1)

It emerges from the pelvis through the suprapiriform foramen, together with the superior gluteal vessels. Continuing laterally, it divides in the buttock into cranial and caudal branches, respectively, together with the cranial and caudal branches of the superior gluteal artery. Passing between and innervating both gluteus minimus and gluteus medius (cranial and caudal branches), it terminates in tensor fascia lata (supplying it via the inferior branch).

Inferior Gluteal Nerve (L5-S1-S2)

It emerges from the pelvis via the infrapiriform foramen and is distributed to the gluteus maximus muscle that it innervates via its deep aspect. Inferior gluteal nerve and posterior cutaneous nerve of the thigh often have the same origin.

Posterior Cutaneous Nerve of the Thigh (S1-S2-S3)

It is a sensory nerve that emerges from the pelvis via the infrapiriform foramen, passing along the posteromedial aspect of the sciatic nerve. After being covered by the gluteus maximus muscle, it continues its vertical, distal path lying under the fascia lata. It crosses over the long head of the biceps femoris muscle and, accompanied by the short saphenous vein, crosses the popliteal fascia dorsal to the knee joint. It ends in cutaneous branches down to the middle part of the dorsal aspect of the leg where anastomoses exist with the medial sural cutaneous nerve. During its course, it provides a gluteal cutaneous branch for the caudal and lateral part of the buttock, the inferior cluneal nerves for the skin of the sacral area, a peroneal branch to supply the superficial perineum, scrotum/labia majora, as well as cutaneous branches to the dorsal aspect of the thigh, popliteal fossa, and upper part of the leg.

Terminal Branch

The terminal branch of the sacral plexus is composed of the sciatic nerve which arises from L4-L5-S1-S2 and S3 and is the longest and largest nerve in the body. It is a mixed nerve which, at its origin, measures 5 mm in thickness and 10–15 mm in width. Traditionally, it descends vertically into the dorsal area of the thigh up to the popliteal area where it gives rise to its two terminal branches: the tibial nerve and the common fibular nerve. This description is subject to many variations, primarily concerning the level of division of the nerve.

It emerges from the pelvis through the infrapiriform foramen, usually between piriformis (above), the superior gemellus (below), the inferior gluteal vessels and the internal pudendal vessels lying medially, the inferior gluteal nerve and posterior cutaneous nerve of the thigh lying dorsally. However, in 15–30% of cases [2–4], variations exist regarding relations of the nerve to piriformis (Fig. 7.4). In the 1937 study concerning 1510 sciatic nerves [2]:

- In 0.13% of cases, the entire sciatic nerve passes through piriformis.
- In 0.86% of cases, the fibular component passes cephalad to piriformis.
- In 11% of cases, the fibular component passes through piriformis.
- In 88% of cases, the entire sciatic nerve emerges “traditionally” below the caudal border of piriformis.

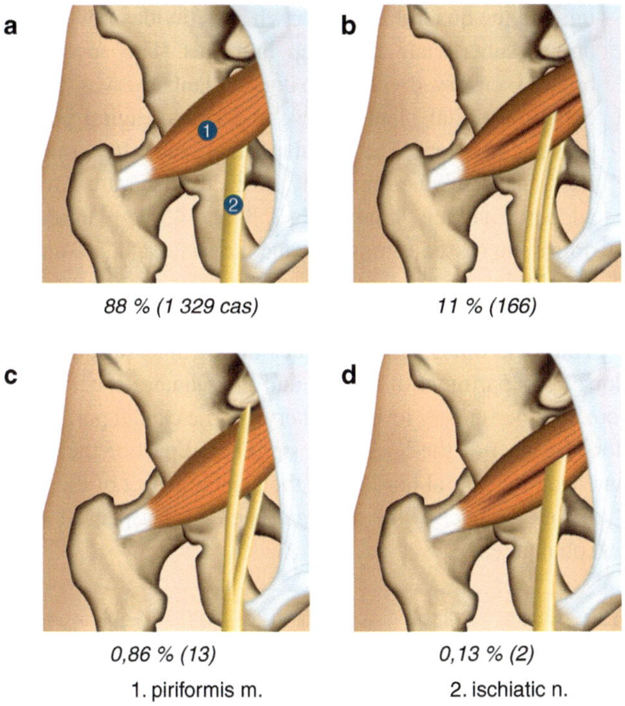

Fig. 7.4 Anatomical variations in relations of the sciatic nerve with the piriformis muscle (from [2])

Pokorny et al. [5] found the same variations in 91 cadavers in proportions of 2.2%, 4.4%, and 14.3%, respectively, with a "traditional" infrapiriform emergence of the complete sciatic nerve in 79.1% of cases. Other variations have been described. Cases exist where a fibular component passes into the cephalic border and a tibial component across the piriform muscle (from 0 to 1.5% of cases) as well as cases where the complete sciatic nerve emerges from the cephalic border of the piriform muscle (from 0 to 2.9% of cases).

In the gluteal area, the sciatic nerve runs vertically between:

- Obturator internus and gemelli muscles, and then quadratus femoris ventrally.
- Gluteus maximus muscle dorsally.
- The ischial tuberosity medially.
- The greater trochanter laterally.

Other relations are: the inferior gluteal, obturator and internal pudendal vessels, the pudendal and internal obturator nerves lie medially; the posterior cutaneous nerve of the thigh and the gluteal branch of the inferior gluteal artery lie dorsally.

In the posterior femoral area, the nerve is accompanied by the artery of the sciatic nerve, a branch of the inferior gluteal artery. The nerve is separated successively from the femoral shaft by adductor longus, and then the short head of biceps femoris. Dorsally, it is initially covered by gluteus maximus, then it is crossed by the long head of biceps femoris, which continues distally on the lateral side of the nerve. The sciatic nerve then lies beneath the aponeurotic layer which joins the sheath of biceps femoris to that of semitendinosus. When these muscles diverge to demarcate the upper boundaries of the popliteal fossa, the sciatic nerve divides into its two terminal branches.

The collateral branches of the sciatic nerve consist of:

- Articular branches:
  To the hip, supplying the dorsal aspect.
  To the knee, arising high up in the thigh (from the ventral aspect of the sciatic nerve or from the nerve to the short head of the biceps femoris) and innervating the posterolateral aspect of the knee joint.
- Muscular branches arising from the sciatic nerve along its pathway in the dorsal femoral area:
  The upper and lower nerves to semitendinosus.
  The nerve to semimembranosus.
  The nerve to the long head of the biceps femoris.
  The nerve to the short head of the biceps femoris.
- The nerve to the hamstring part of adductor magnus.

The terminal branches of the sciatic nerve generally arise in the popliteal fossa and are represented by the tibial nerve and the common fibular nerve.

Tibial Nerve

It is a mixed nerve that arises from L4-L5-S1-S2 and S3. In the thigh, it passes vertically downwards (as the medial component of the sciatic nerve), and then, in the leg, it passes obliquely/medially to the medial sub-malleolar area to divide into the medial and lateral plantar nerves.

In the popliteal fossa, it is located under the fascia of the same name, dorsally and laterally in relation to the popliteal vein, which itself is located dorsally and laterally with respect to the artery. Its relations are:

- Cranially: Lies medial to semimembranosus and semitendinosus, and lies lateral to biceps femoris.
- Caudally: Lies medial to the medial head of gastrocnemius, and lateral to the lateral head of gastrocnemius and plantaris.

At this level, its collateral branches arise:

- The dorsal branch of the knee.
- The medial and lateral nerves to gastrocnemius.
- The upper nerve to soleus.
- The nerve to plantaris.
- The nerve to popliteus.

- The crural interosseous nerve.
- The medial sural cutaneous nerve. Although it may be sometimes possible to find its origin in a cephalic position in relation to the apex of the popliteal fossa, usually it arises towards the middle or the caudal part of this area. It continues into the groove separating the two heads of gastrocnemius dorsally up to its caudal part where it penetrates the crural facia, together with the short saphenous vein. By anastomosing with the fibular communicating branch, it gives rise to the sural nerve. The latter follows the lateral border of the calcaneal tendon, circumscribes the lateral malleolus to give rise to the lateral calcaneal branches and the lateral dorsal cutaneous nerve of the foot, which ends in the lateral dorsal digital nerve of V.

In the dorsal area of the leg, the tibial nerve relations are:

- Lying ventrally, tibialis posterior, and flexor digitorum longus.
- Lying dorsally, in its cranial two-thirds, soleus and, in its caudal one-third, the skin via the crural fascia.
- Lying laterally, flexor hallucis longus.
- Lying medially, flexor digitorum longus.

At this level, the following collateral branches arise:

- The inferior nerve to the soleus.
- The nerve to tibialis posterior.
- The nerve to flexor digitorum longus.
- The nerve to flexor hallucis longus.
- Articular and vascular branches.
- The medial calcanean branch which innervates the skin of the heel.

The terminal branches of the tibial nerve are the medial and lateral plantar nerves which arise in the medial submalleolar area:

- The medial plantar nerve follows the lateral border of the medial plantar artery and, in the plantar aspect of the foot, under the abductor hallucis and flexor digitorum brevis (which it innervates). Lying on the quadratus plantae and the tendon flexor digitorum longus, it divides into its medial and lateral terminal branches. The latter provide sensation to the medial two-thirds of the plantar aspect of the foot and motor function to flexor hallucis brevis and the first lumbrical. It can also innervate the second lumbrical muscle. Its collateral branches include the joint branches for joints in the tarsal and tarso-metatarsal joints.
- The lateral plantar nerve runs ventrally and laterally, reaching the medial border of the lateral plantar artery. Its collateral branches are muscular for the quadratus plantae, flexor digiti minimi and abductor digiti minimi, and vascular structures. It lies between the flexor digiti minimi and the quadratus plantae where it divides into its terminal, superficial, and deep branches. The superficial branch, sensory, divides into digital plantar nerves (fourth common digital plantar nerve and lateral digital plantar nerve of fifth toe). The deep branch, mixed, after continuing transversely and medially, provides innervation to the adductor hallucis, lumbricals 2, 3, and 4, the plantar and dorsal interosseous muscles, and the tarsal and tarsometatarsal joints.

Thus, the tibial nerve provides motor innervation to the dorsal compartment of the leg and the plantar aspect of the foot. On the sensory level, it innervates the caudal part of the dorsal aspect of the leg, the posterolateral part of the ankle and heel, the lateral border and the plantar aspect of the foot, the palmar aspect of the toes, and the dorsal aspect of their last phalange.

Common Fibular Nerve

It is the terminal branch of the lateral component of the sciatic nerve. Smaller than the tibial nerve, the common fibular nerve contains fibres that emerge from the nerve roots of L4, L5, S1, and S2. It arises in the upper part of the popliteal fossa, and then continues caudally and laterally, closely related to biceps femoris. It passes under the fleshy muscular fibres and then under the tendon of biceps femoris up to its insertion on the head of the fibula. During its pathway, it passes successively on the lateral head of gastrocnemius and on soleus, which separates it from the head of the fibula. The common fibular nerve circumscribes the neck of the fibula, directly in contact with the periosteum, crossing the lateral intermuscular septum via an osteofibrous orifice. It then divides into two planes at the insertion of peroneus longus. At this level, it gives rise to its two terminal branches: the superficial fibular nerve and the deep fibular nerve.

During its course, it gives rise to:

- Articular branches, for the lateral aspect of the knee and for the proximal tibiofibular joint.
- Muscular branches, two in number, for tibialis anterior. They arise in the lateral compartment, cross the anterior intermuscular septum and innervate the cranial part of tibialis anterior.
- Cutaneous branches:

  The lateral sural nerve, which innervates the anterolateral part of the knee and the cranial two-thirds of the lateral aspect of the leg. It overlaps more on the ventral aspect than on the dorsal aspect of the latter.

  The fibular communicating branch, which arises in the popliteal fossa and descends to the dorsal aspect of the leg with the small saphenous vein. It crosses the superficial aponeurosis at mid-height of the leg. Most often, it anastomoses with the medial sural nerve to form the common sural nerve. Sometimes, it remains individualised and ends in the lower part of the lateral aspect of the leg and the lateral malleolar area.

Superficial Peroneal Nerve

Its course is vertical. It quickly diverges from the deep peroneal nerve, continues beside the fibula, covered by peroneus longus, and then enters peroneus brevis located dorsally in relation to the nerve and the peroneus longus, which covers it. In the middle part of the leg, it emerges ventrally in relation to peroneus longus, in contact with the ventral interosseous membrane of the leg, whose dorsal aspect it then follows, up to the caudal one-third of the leg where it penetrates the aponeurosis of the sheath of the leg. It then gives rise to its two terminal branches: the medial dorsal cutaneous nerve and the intermediate dorsal cutaneous nerve. During its course, it innervates the following:

- Peroneus longus by two or three branches.
- Peroneus brevis.
- Peroneus tertius.

The medial and intermediate dorsal cutaneous nerves branch out on the dorsum of the foot and innervate the cranial part of the foot medially in relation to a line passing through the middle of the fourth toe, and sometimes by the fifth toe (except the lateral border), the distal phalanx of the corresponding toes and the first inter-digital membrane.

Deep Peroneal Nerve

It continues ventrally and caudally, between the two planes of insertion of peroneus longus. It then divides the caudal insertion of the peroneus longus into two. It then pierces the anterior inter-muscular septum to join extensor digitorum longus, extensor hallucis longus, and the interosseous membrane. It thus joins the vascular bundle for which it will be a companion; first located laterally in relation to the vessels, it continues medially in relation to the middle of the leg. In the instep of the foot, it gives rise to its two terminal branches: a lateral branch and a medial branch.

During its course, it innervates the muscles of the anterior compartment of the leg:

- Tibialis anterior.
- Extensor hallucis longus.
- Extensor digitorum longus.

Variably, it can give rise to a branch to peroneus tertius.

Thus, it is responsible for movements of flexion of the foot and extension of the toes.

Concerning the terminal branches of the deep peroneal nerve:

- The lateral branch is directed ventrally and laterally under extensor digitorum brevis and extensor hallucis brevis. It supplies these muscles and bones and joints of the back of the foot up to the metatarsophalangeal joints.
- The medial branch is directed ventrally and medially, between the tendons of extensor hallucis longus located medially and the dorsal blood vessels of the foot located laterally. It innervates the bones and joints in the adjacent area, and ends by becoming superficial in the first interdigital space where it innervates the aspects opposite the first and second toes, except for the distal phalanx.

Blocks

Sciatic Nerve Block by Infrapiriform Parasacral (IPPS) and Transpiriform Parasacral (TPPS) Approach (Fig. 7.5)

Indications

A parasacral block is a true nerve plexus block, including all the different branches of the sacral plexus and, thus, all collaterals of the sciatic nerve (posterior cutaneous nerve of the thigh and the gluteal nerves). A study has shown that a single stimulus, a response of the tibial or peroneal component, was necessary and sufficient to perform a sciatic nerve block by parasacral approach [6]. In fact, this approach enables spread of the local anaesthetic in contact with the nerve structures before they separate the medial (tibial nerve), lateral (common peroneal nerve) components, as well as a good number of collateral branches. A few teams insert catheters by this approach [7, 8]. Its utility is in surgery of the popliteal fossa area, the knee, and the lower limb with a tourniquet making placement of a popliteal catheter inappropriate. Performed alone, a parasacral block is the rarest indication for difficult and prolonged surgery of the foot or involving the area of the tibial and common peroneal nerves below the knee. An approach in the popliteal area might be preferable for such surgery below the knee.

In combination with a lumbar plexus block or a femoral block, a parasacral block enables all types of surgery or analgesia of the lower limb starting with the middle one-third of the femur [9].

Several teams have examined parasacral block or have recently proposed techniques for localisation and sacral plexus block in the parasacral area [10–14].

The IPPS and TPPS ultrasound-guided approaches already described in the first edition of this work (2007) comprise an approach which, under visual control, enables positioning of the tip of the needle precisely in contact with the sciatic nerve and of controlling spread of the local anaesthetic around this target; they use a pathway for injection which distances itself from the pelvic cavity and its vessels. This is not the case when a parasacral block is performed

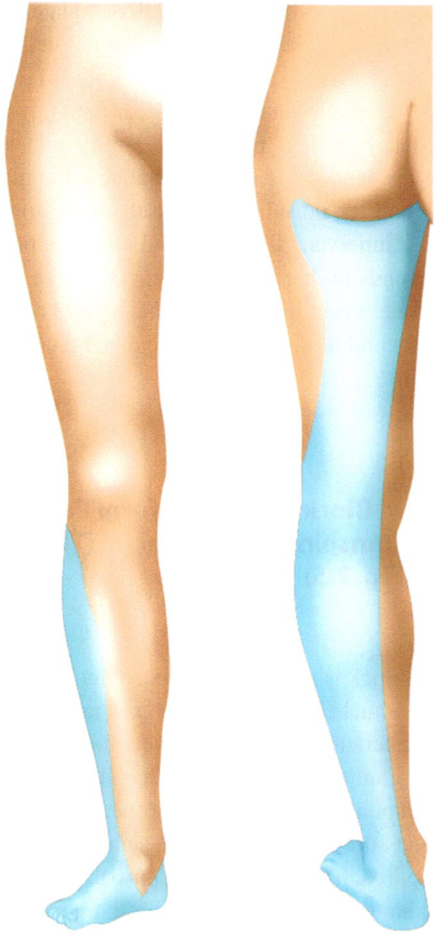

Fig. 7.5 Area of extension of the parasacral block

according to traditional approach where the ultrasound system enables the location of the piriform muscle or its crossing with the needle, but during which generally the sacral roots cannot be identified and the method of distribution of the local anaesthetic in their contact. The IPPS and TPPS approaches enable a true plexus block in over 80% of cases or when ascending spread of the local anaesthetic, insufficient, spares only the superior gluteal nerve.

Type of probe: "low frequency" convex in adults (2 to 5 MHz), linear "high frequency" (5–10 or 6–13 MHz) possible in children or in very thin adults. A curved probe enables a wide field of study without having to mobilise its position.
Configuration: Nerve in long axis, needle in plane.
Axis of the probe: Oblique, in the initial axis of emergence of the sciatic nerve which one seeks to visualise longitudinally (Figs. 7.6 and 7.7).
Studied depth: Generally 10 cm at most. Adjusted according to patient build.

Neurostimulation: Enables further identification of the nerves visualised and can limit, by determination of an MIS > 0.3 mA (0.1 ms), the risk of accidental intrafascicular injection. Furthermore, it can supplement locating nerves when conditions of visibility are unfavourable.
Needle: 100 mm isolated, 21 G.
Utility of Doppler ultrasound: Inferior gluteal (+++) (Fig. 7.8), internal pudendal (+), superior gluteal arteries.

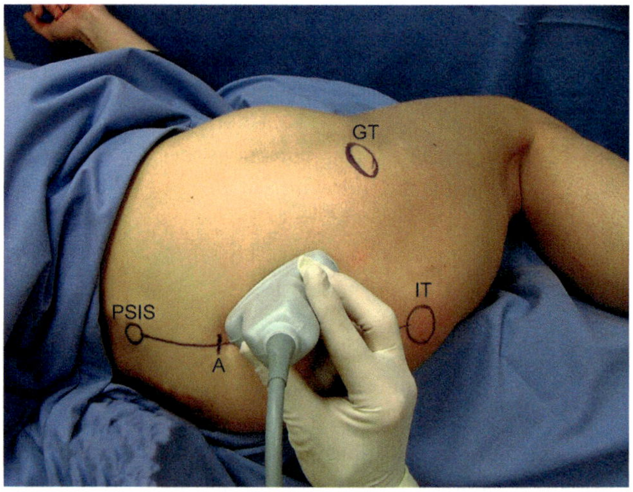

Fig. 7.6 Sciatic nerve block by infra-piriformis parasacral approach (IPPS) and trans-piriformis parasacral approach (TPPS): position of patient, surface landmarks, overall position of probe
GT: greater trochanter; PSIS: posterosuperior iliac spine; IT: ischial tuberosity; A: theoretical needle puncture site

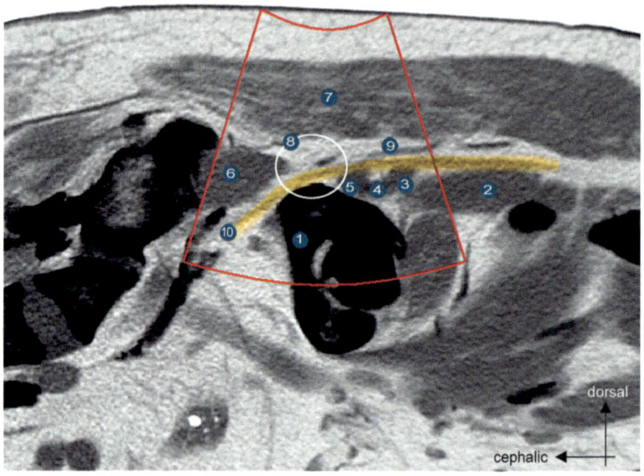

1. ischium
2. quadratus femoris m.
3. inferior gemellus m.
4. obturator internus tendon
5. superior gemellus m.
6. piriformis m.
7. gluteus maximus m.
8. infrapiriform foramen
9. inferior gluteal a.
10. ischiatic n.

Fig. 7.7 Sciatic nerve and its relations along its course in the infrapiriform foramen. Reconstructed CT scan image. Materialisation of the ultrasound beam

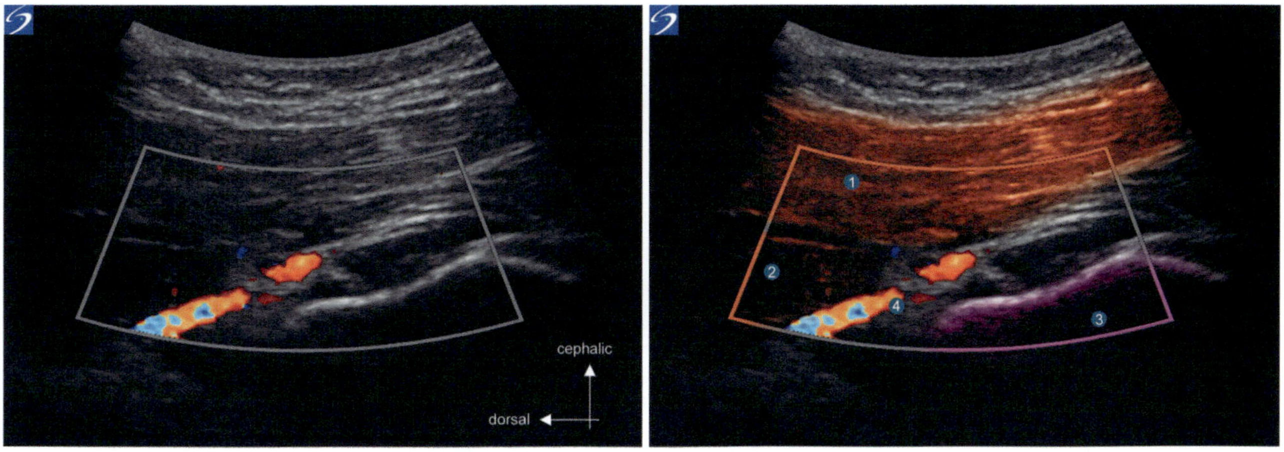

1. gluteus maximus m. 2. piriformis m.
3. ilium 4. inferior gluteal a.

Fig. 7.8 Sciatic nerve block by IPPS and TPPS approaches: visualisation of the inferior gluteal artery using colour Doppler ultrasound

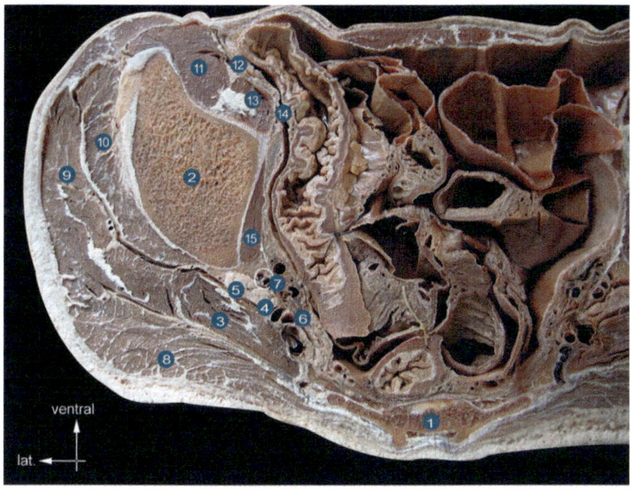

1. sacrum 2. ischium 3. piriformis m.
4. pudendal n. 5. ischiatic n. 6. inferior gluteal a.
7. internal pudendal a. 8. gluteus maximus m. 9. gluteus medius m.
10. gluteus minimus m. 11. iliacus m. 12. femoral n.
13. psoas major m. 14. obturator n.
15. obturator internus m.

Fig. 7.9 Transverse anatomical section cutting through the sacral plexus near the level of the ischial spine

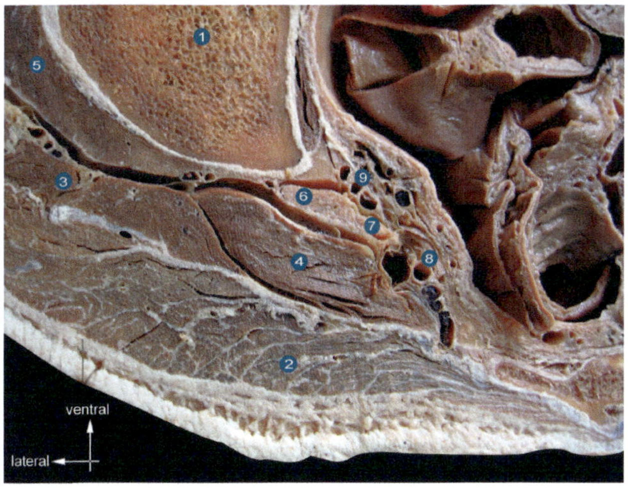

1. ischium 2. gluteus maximus m.
3. gluteus medius m. 4. piriformis m.
5. gluteus minimus m. 6. ischiatic n.
7. pudendal n. 8. inferior gluteal a.
9. internal pudendal a.

Fig. 7.10 Transverse anatomical section centred on the sciatic nerve and its relations at the ventral aspect of the piriformis muscle

Echoanatomy

The gluteal area is located deep and is anatomically complex (Figs. 7.9, 7.10, 7.11, 7.12, 7.13 and 7.14). Depending on position of the probe, the images will be very different and good anatomical knowledge is essential. Echogenicity is variable, and depends on patient age, size, the relative proportions of fat and muscle tissue, and other parameters including hydration.

As always, three principal tissue components are found: fatty, muscular, and bony. Although the latter varies little according to subjects, this is not the case with the other two. The ultrasound and anaesthetic procedure is performed in Sims position with the hip flexed ideally at 90° (Fig. 7.15). The utility of this position lies in the fact that it makes it possible to place the tissues under tension and to compress the fat and muscles, so that it decreases the depth of the "target" zone represented by the sciatic nerve in its proximal position. By carefully observing the image from superficial to deep, it is important to identify structures in sequence: from the superficial subcutaneous fatty tissue to the plane beneath gluteus maximus, which is immediately adjacent and which itself covers the other structures (muscular, nervous, vascular, and bony). The sciatic nerve is not always easy to see at first glance, so it is necessary to locate

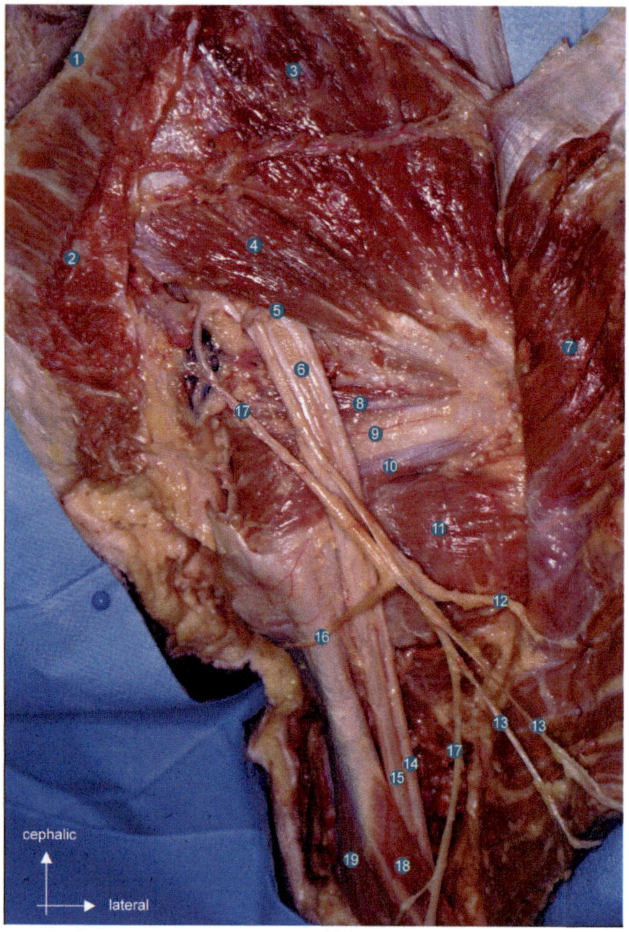

1. iliac crest
2. gluteus maximus m. (removed)
3. gluteus medius m.
4. piriformis m.
5. infrapiriform foramen
6. ischiatic n.
7. gluteus maximus m. (removed)
8. superior gemellus m.
9. obturator internus m.
10. inferior gemellus m.
11. quadratus femoris m.
12. motor n. of gluteus maximus m.
13. inferior clunial nn.
14. ischiatic n. (fibular/peroneal division)
15. ischiatic n. (tibial division)
16. perineal branch of posterior femoral cutaneous n.
17. posterior femoral cutaneous n.
18. long head of biceps femoris m.
19. semitendinosus m.

Fig. 7.11 Gluteal area, gluteus maximus muscle reflected. Dissection Bertrand Fabre

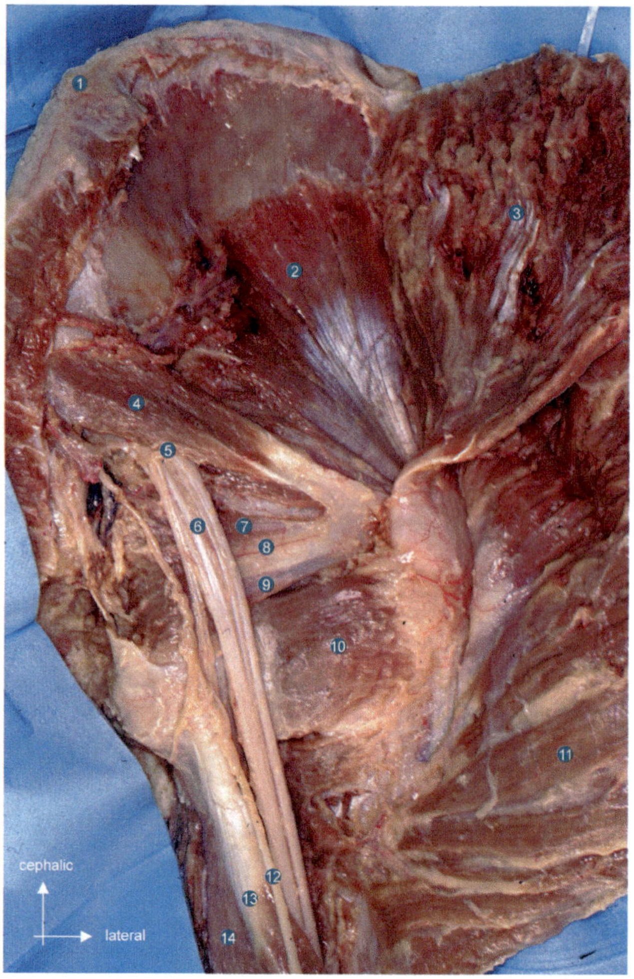

1. iliac crest
2. gluteus minimus m.
3. gluteus medius m. (removed)
4. piriformis m.
5. infrapiriform foramen
6. ischiatic n.
7. superior gemellus m.
8. internal obturator n.
9. inferior gemellus m.
10. quadratus femoris m.
11. gluteus maximus m. (removed)
12. posterior femoral cutaneous n.
13. long head of biceps femoris m.
14. semitendinosus m.

Fig. 7.12 Gluteal area, gluteus maximus muscle reflected. Dissection Bertrand Fabre

the structures which surround it, namely, gluteus maximus, with piriformis beneath along with the superior and inferior gemelli accompanying the tendon of obturator internus. The sciatic nerve emerges from the deep (ventral) aspect of piriformis via the infrapiriform foramen, appearing as a "cord" whose echogenicity generally increases toward the distal end, clearly showing anisotropic change with gentle tilting of the probe. The muscles are distinguished from each other not only by their aponeuroses, often visible, but also by their echogenicity or even dissimilar echostructures inherent in the different directions of their fibres. This is the case when this nerve block is performed, with piriformis often appearing as more hypoechoic in comparison to gluteus maximus.

Approach and Location

To aid in optimum positioning of the probe, four surface markers are used to aid in this procedure (Fig. 7.6): the apex

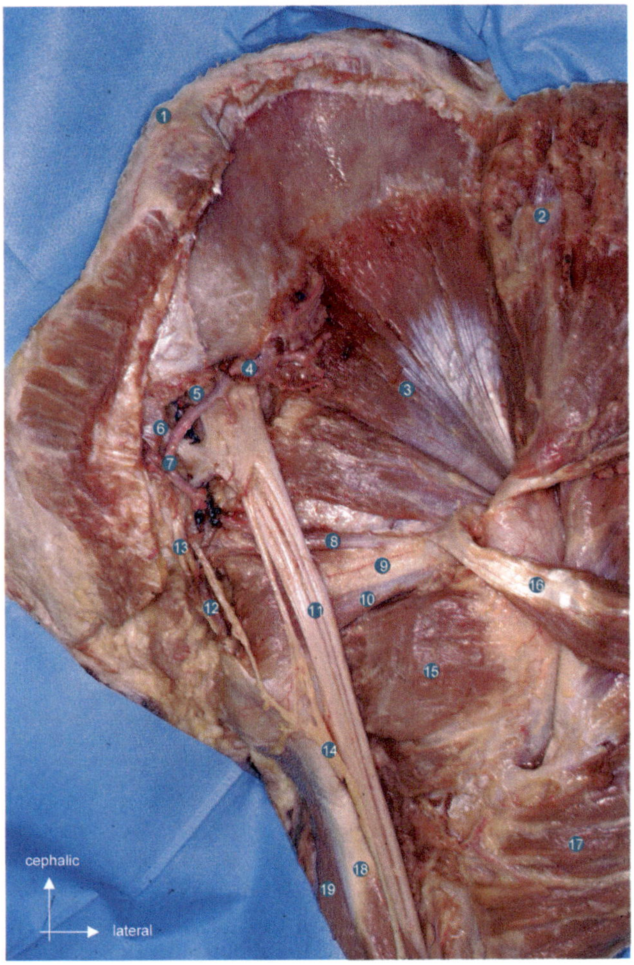

1. iliac crest
2. gluteus medius m. (removed)
3. gluteus minimus m.
4. superior gluteal a.
5. S2 n. root
6. S3 n. root
7. inferior gluteal a.
8. superior gemellus m.
9. obturator internus m.
10. inferior gemellus m.
11. ischiatic n.
12. internal pudendal a.
13. pudendal n.
14. posterior femoral cutaneous n.
15. quadratus femoris m.
16. piriformis m. (removed)
17. gluteus maximus m. (removed)
18. long head of biceps femoris m.
19. semitendinosus m.

Fig. 7.13 Gluteal area, piriformis muscle reflected. Dissection Bertrand Fabre

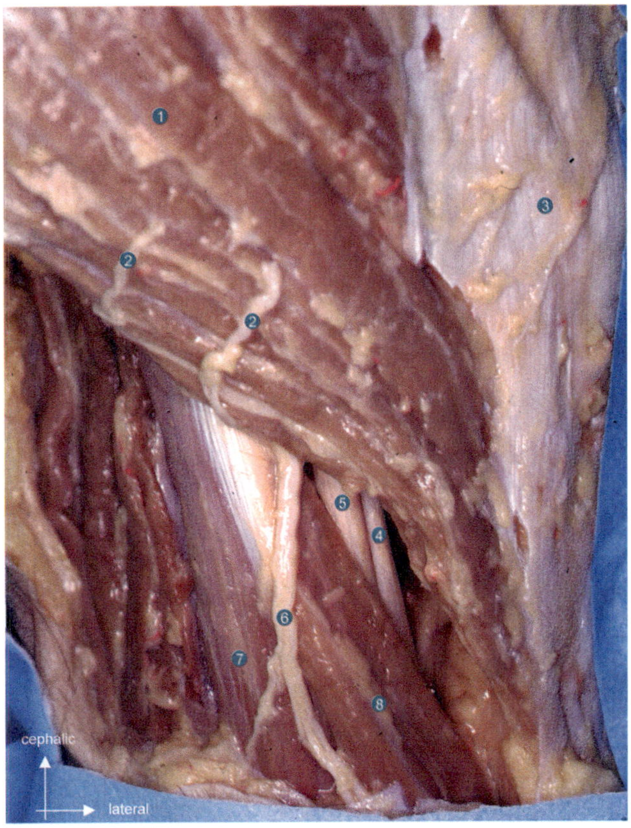

1. gluteus maximus m.
2. inferior clunial nn.
3. fascia lata
4. ischiatic n. (fibular/peroneal division)
5. ischiatic n. (tibial division)
6. posterior femoral cutaneous n.
7. semitendinosus m.
8. long head of biceps femoris m.

Fig. 7.14 Subgluteal area. Dissection Bertrand Fabre

of the greater trochanter (GT), the posterosuperior iliac spine (PSIS), the apex of the ischial tuberosity (TI), and the theoretical point of injection of the parasacral block with neurostimulation, described by Mansour [15], 6 cm distal to the PSIS, that we will call A. The probe is first positioned on the line bisecting the angle GT-A-TI, with its cephalic end in immediate proximity to point A. It is from this initial configuration then that readjustments of the probe can be undertaken intended to correctly visualise the sciatic nerve in

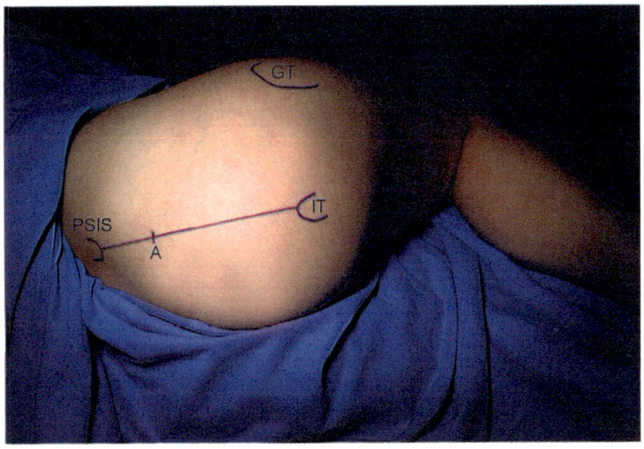

Fig. 7.15 Position for performance of an infra- or trans-piriformis parasacral block

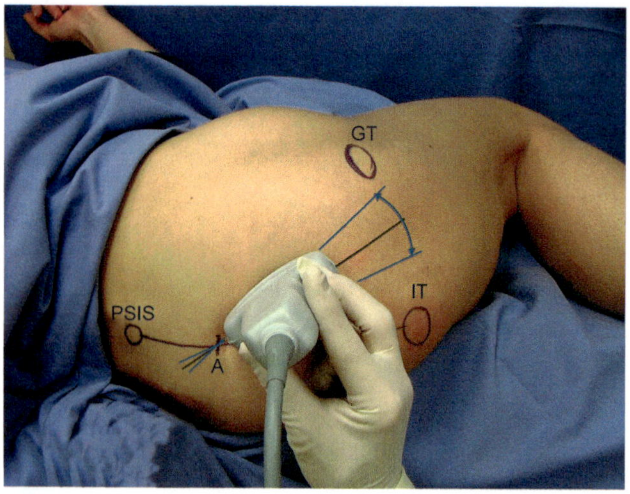

Fig. 7.16 IPPS and TPPS sciatic nerve block: movement of probe rotating around point "A" used as an axis
GT: greater trochanter; PSIS: posterosuperior iliac spine; IT: ischial tuberosity; A: theoretical point of injection

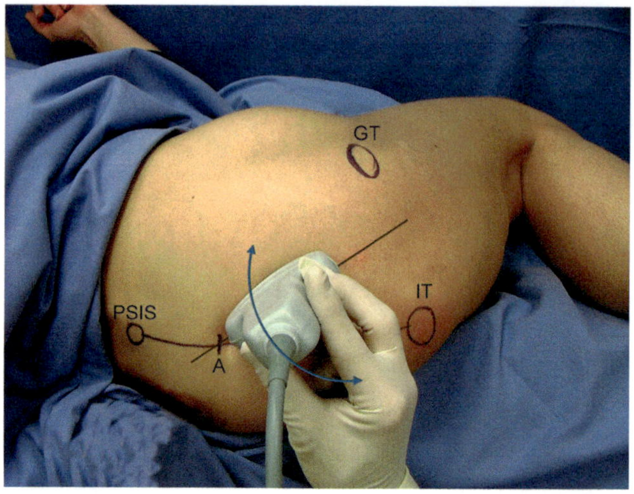

Fig. 7.18 IPPS and TPPS sciatic nerve block: tilting of the probe
GT: greater trochanter; PSIS: posterosuperior iliac spine; IT: ischial tuberosity; A: theoretical point of injection

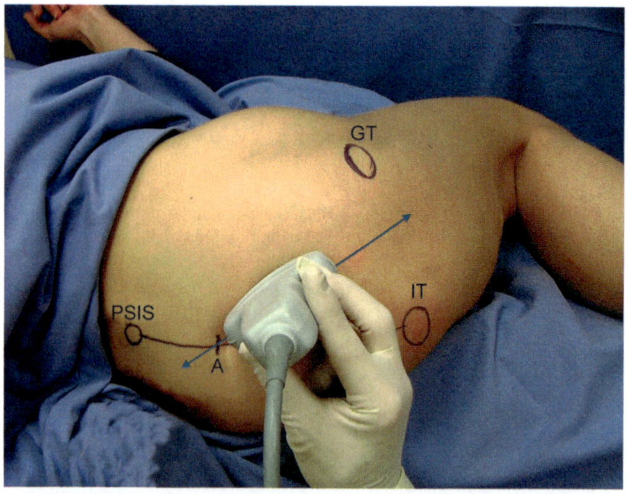

Fig. 7.17 IPPS and TPPS sciatic nerve block: sliding of the probe on its axis
GT: greater trochanter; PSIS: posterosuperior iliac spine; IT: ischial tuberosity; A: theoretical point of injection

approach of the needle. In order to obtain the ideal position of the probe, it is possible to move it in several ways:

- Rotation, around point A enables to vary its oblique direction downward and outward (Fig. 7.16).
- Axial translation and longitudinal inclination of the probe enable to position the lower border of piriformis in the middle of the probe (and therefore of the image) and to use the effect of anisotropy to optimise visualisation of the sciatic nerve (Fig. 7.17).
- Eventually, the lateral tilting of the probe enables to refine visualisation of the infrapiriform foramen and emergence of the sciatic nerve (Fig. 7.18).

These movements are performed in a relatively simultaneous manner until an "optimum" image is obtained (Fig. 7.19).

The sciatic nerve continues laterally to the ischial spine. The latter comprises the caudal end of the greater sciatic notch, separating it from the lesser sciatic notch. The "craniodorsal" apex of the ischial tuberosity is relatively easy to visualise and thus is a good deep bony marker (Fig. 7.20). Although it is visible in the ultrasound field, generally it is sufficient to turn the probe laterally around point A (from 10 to 20°) to be in a position favourable to visualisation of the sciatic nerve. During this rotation (Fig. 7.21), we see appear successively the following:

- The inferior gluteal artery (Fig. 7.8).
- The craniocaudal succession of the superior gemellus, obturator internus (of a more tendinous nature at this level here, and thus more echoic), and the inferior gemellus which comprise the ventral plane on which the sciatic nerve continues which lies on the flat bony surface located laterally to the ischial spine (Fig. 7.19).

In this area located on the inner aspect of the ischial tuberosity, it is possible to visualise the internal pudendal vessels which accompany the pudendal nerve.

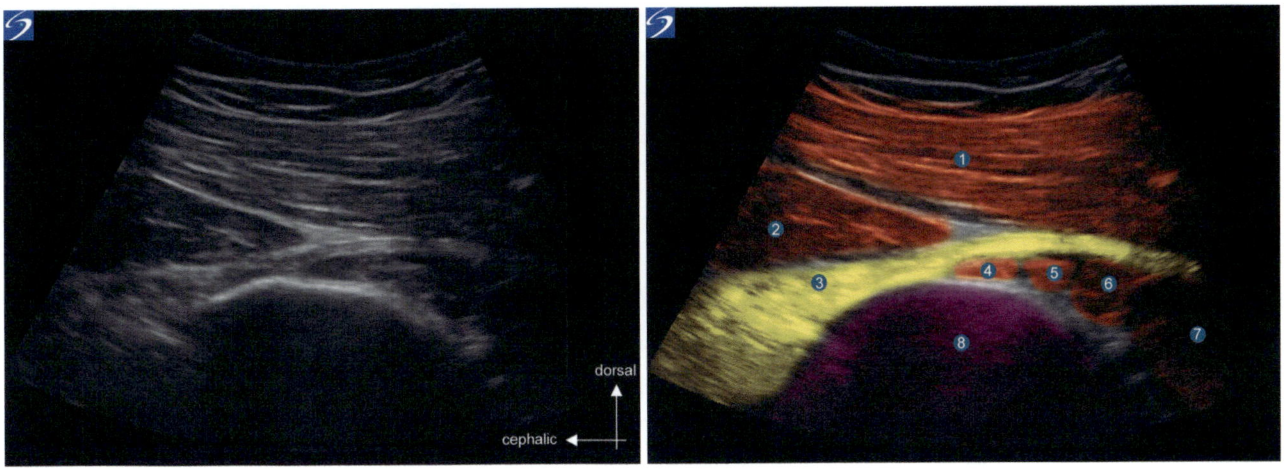

1. gluteus maximus m. 2. piriformis m. 3. ischiatic n.
5. obturator internus m. 6. inferior gemellus m.
7. quadratus femoris m. 8. ilium

Fig. 7.19 IPPS and TPPS sciatic nerve block. Ideal ultrasound image (corresponds to ultrasound plane no. 3 in fig. 7.21)

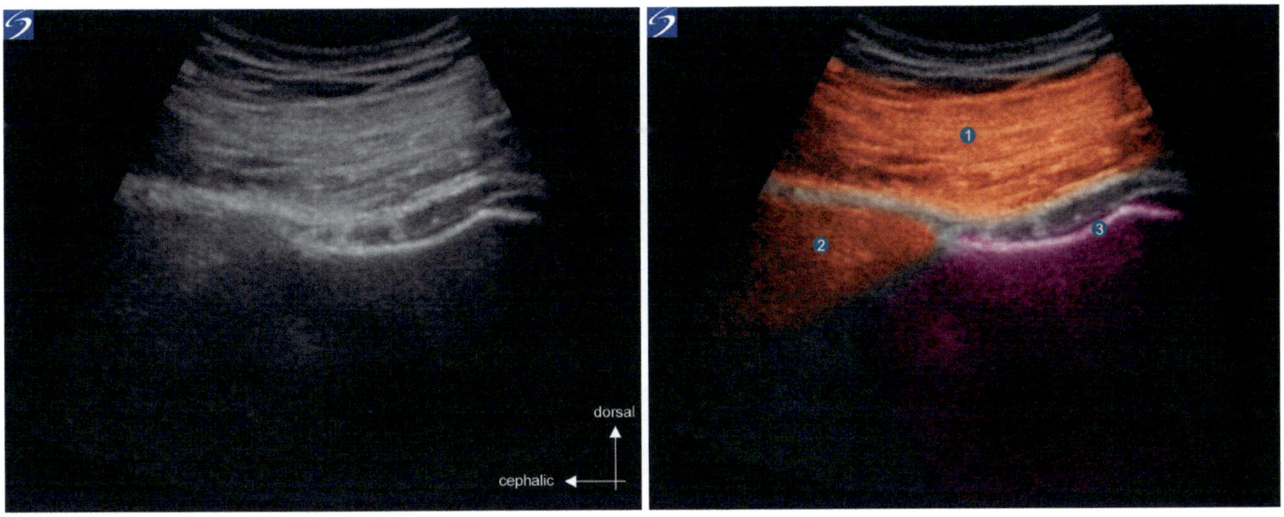

1. gluteus maximus m. 2. piriformis m.
3. "craniodorsal" apex of ischial tuberosity

Fig. 7.20 Sciatic nerve block by IPPS and TPPS approach. Ultrasound section visualising the "craniodorsal" apex of the ischial tuberosity (corresponds to ultrasound plane no. 1 in fig. 7.21)

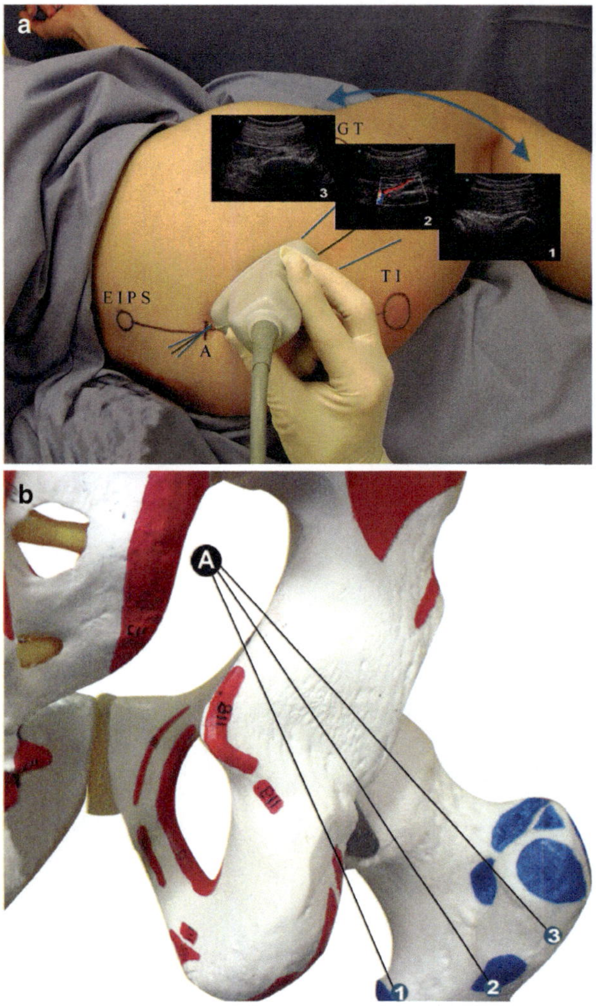

Fig. 7.21 Sciatic nerve block by IPPS and TPPS approach: succession of different characteristic images depending on the alignment of the probe (a); in b, materialisation on a skeleton of the ultrasound planes 1, 2, and 3 noted in a

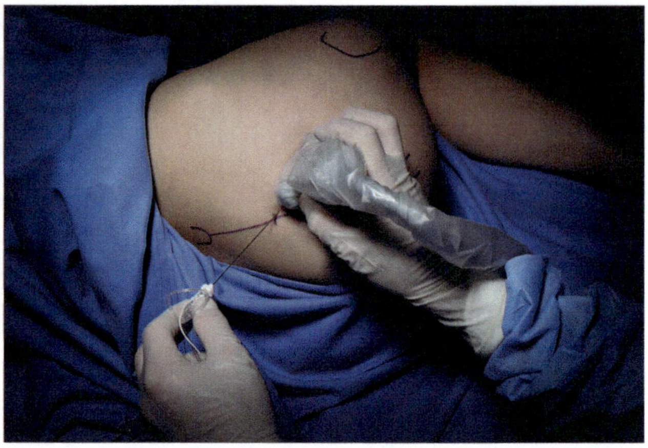

Fig. 7.22 Sciatic nerve block by IPPS and TPPS approach: position of probe and of needle

Pathway

The point of puncture at the cephalic end of the probe enables advancement of the needle in the ultrasound plane (Fig. 7.22). It is inserted in an oblique direction laterally and caudally up to the target nerve structure. The end point can be the sciatic nerve in the infrapiriform foramen (infrapiriform parasacral block, IPPS) (Fig. 7.23), but it is also possible to cross the caudal part of piriformis to position the tip of the needle immediately in front of the latter, between its deep aponeurosis and the bony outline (corresponding to the lower part of the greater sciatic notch), against the sciatic nerve which is being comprised (transpiriform parasacral block, TPPS) (Fig. 7.24).

However, the latter is not always clearly visible. The coupling of ultrasound-guided technique with neurostimulation for this block is recommended in situations where the images obtained are sub-optimal. Eliciting a tibial or fibular motor response enables confirmation of the correct needle position in close proximity to the nerve. A minimal intensity of stimulation between 0.3 and 0.5 mA (0.1 ms), which, although not absolutely confirming the extraneural position of the needle, is a surrogate end-point for its extra-fascicular position. As always, the occurrence of mechanical paraesthesia, of continuing pain when approaching the nerve or at start of the injection should immediately lead to cessation of the injection and to careful repositioning of the needle.

This approach to the sciatic nerve, during which advancement of the needle is not made in the direction of the pelvis, makes it possible to avoid a peritoneal, digestive tract, or iliac vascular needle puncture that can occur accidentally in case of an overly medially directed injection during conduct of a parasacral block with traditional neurostimulation. Angulation of the needle during an IPPS or TPPS sciatic nerve block, furthermore, is entirely favourable for the insertion of a perineural catheter (Fig. 7.25).

Injection

When combing the US-guided technique with neurostimulation it is helpful to initiate the injection with a 5% dextrose solution. This allows for the position of the needle tip to be more easily visualised whilst allowing for the possibility to reposition it whilst still generating useful motor/sensory responses. It is also possible to visualise fluid spread which is not optimal, e.g. too superficial (dorsal), or deep to piriformis and the sciatic nerve, in spite of proper neurostimulation criteria (Fig. 7.26).

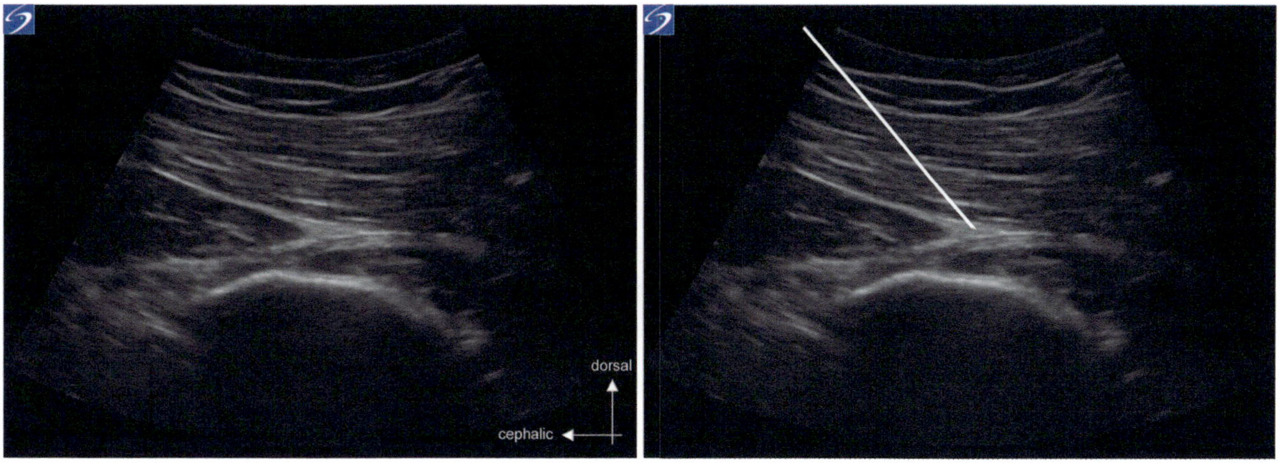

Fig. 7.23 Infrapiriformis parasacral block. The extremity of the needle is located in contact with the sciatic nerve in the infrapiriformis foramen

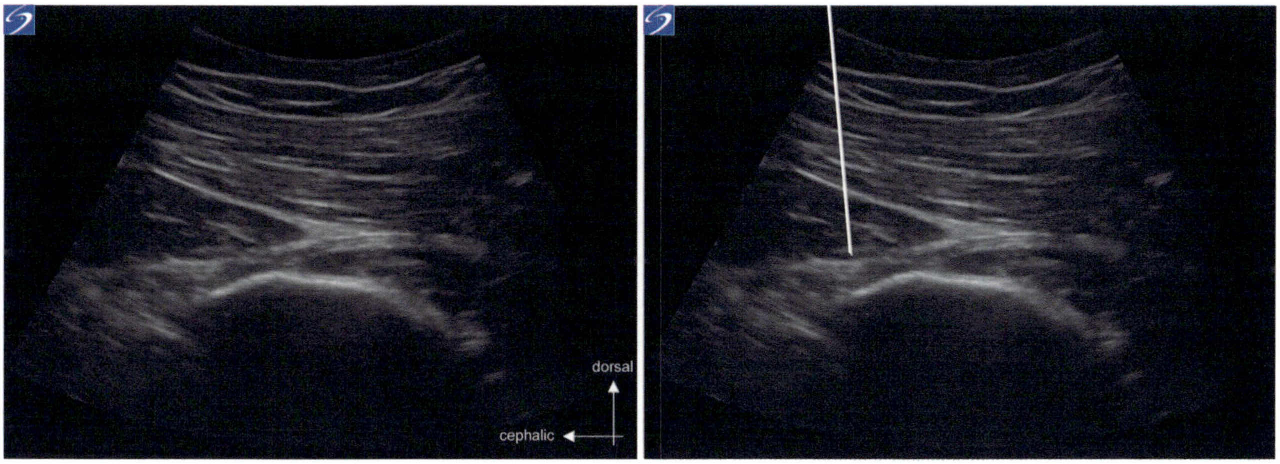

Fig. 7.24 Transpiriformis parasacral block. The needle crosses the caudal part of the piriformis muscle

Fig. 7.25 Sciatic catheter inserted by an infrapiriformis parasacral approach. Philippe Gautier iconography

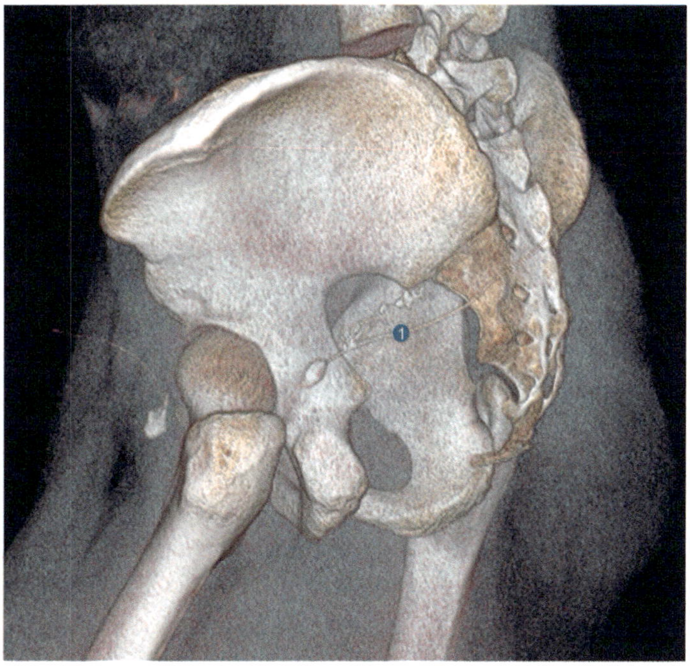

1. ischiatic catheter

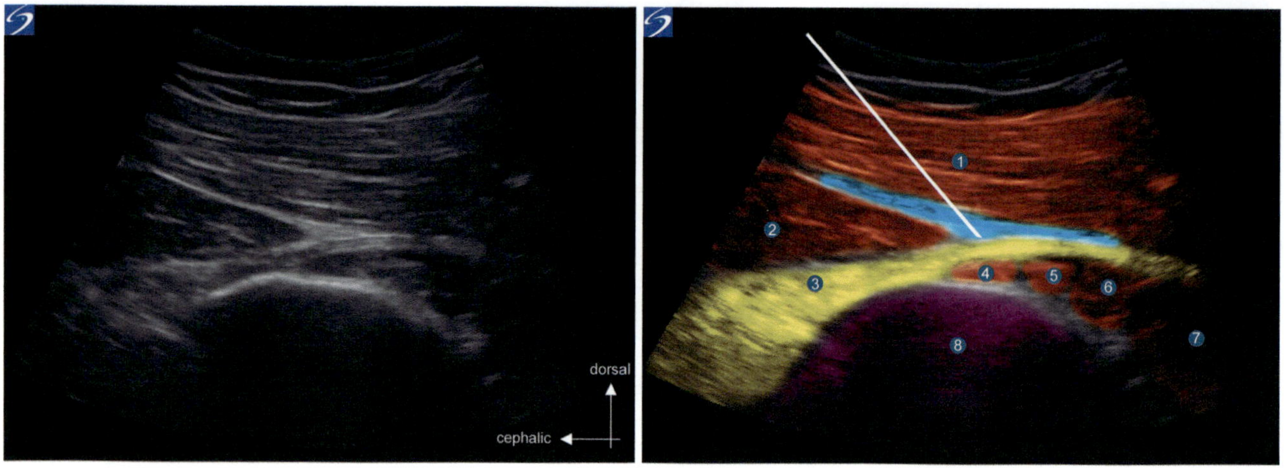

1. gluteus maximus m. 2. piriformis m. 3. ischiatic n.
4. superior gemellus m. 5. obturator internus m.
6. inferior gemellus m. 7. quadratus femoris m. 8. ilium

Fig. 7.26 Too superficial spread of local anaesthetic

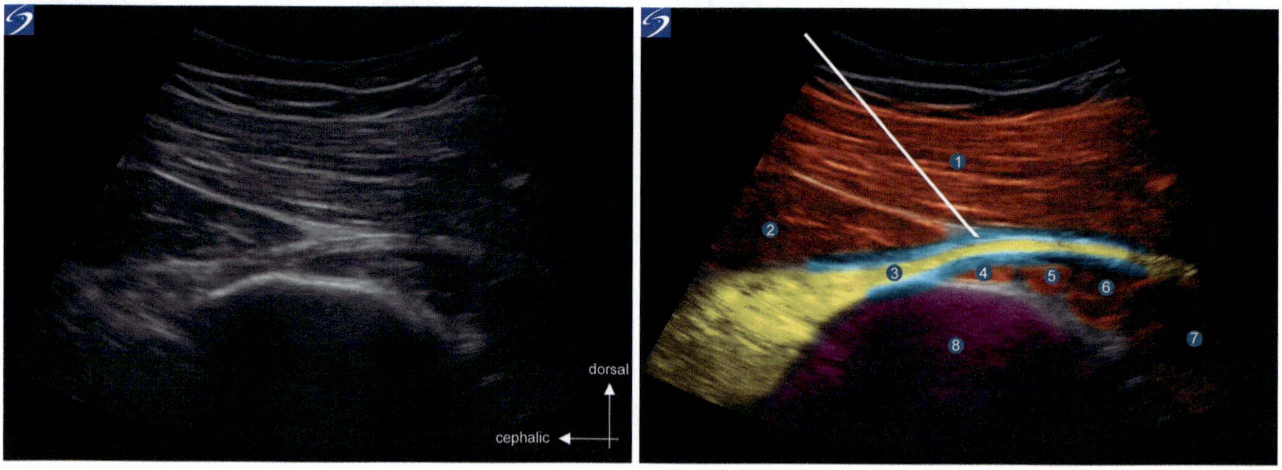

1. gluteus maximus m. 2. piriformis m. 3. ischiatic n.
4. superior gemellus m. 5. obturator internus m.
6. inferior gemellus m. 7. quadratus femoris m. 8. ilium

Fig. 7.27 Centrifugal and centripetal spread of local anaesthetic along sciatic nerve

Centrifugal spread (in a distal direction along the path of the sciatic nerve) and centripetal spread (in a proximal direction along the sciatic nerve) (Fig. 7.27) in a plane deep to piriformis, are both considered as positive predictive criteria for the effectiveness of the nerve block. Proximal spread of the local anaesthetic to the sacral roots may also achieve a "truly plexic" effect of anaesthesia (Fig. 7.27). Extensive distal spread of local anaesthestic along the sciatic nerve is often observed (Fig. 7.28).

Figs. 7.29, 7.30 and 7.31 show the excellent spread of the local anaesthetic in a parasacral nerve block performed under ultrasound guidance according to the infrapiriform technique (IPPS) with insertion of a perineural catheter.

Paediatrics

In children, the images are of the same pattern as in adults. The more shallow depth in which the infrapiriform foramen lies promotes better visibility of the nerve structures and a more confident appreciation of the relationship between the needle and the nerve. However, the shorter distances between the target nerves of the pelvis and the associated blood vessels require increased caution in respect of needle positioning, as well as with the volumes of the local anaesthetic used. The benefit/risk ratio should be carefully weighed.

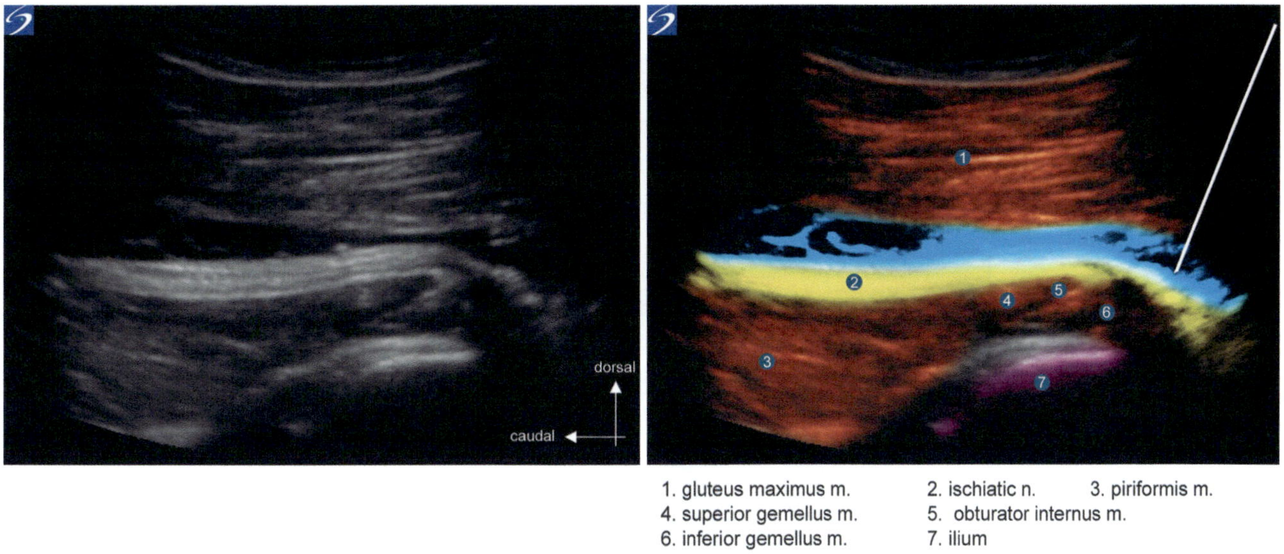

Fig. 7.28 Extensive spread of local anaesthetic distally along sciatic nerve

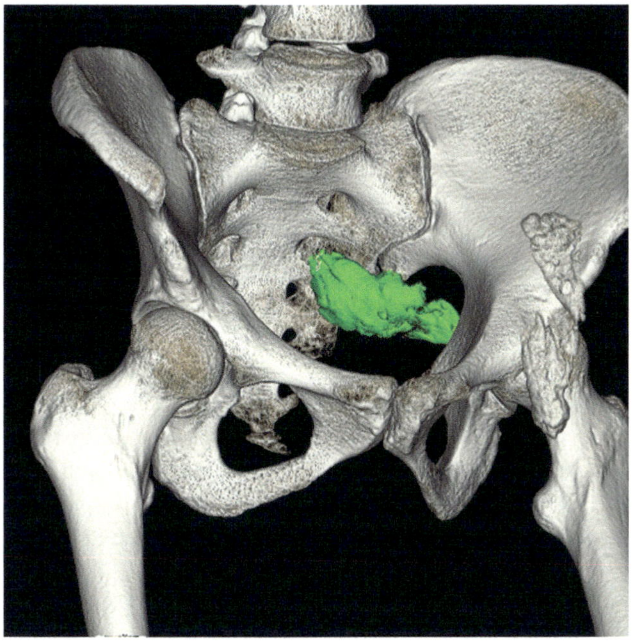

Fig. 7.29 3D CT-scan reconstitution of spread of local anaesthetic in an infrapiriformis parasacral block, anterior view. Philippe Gautier iconography

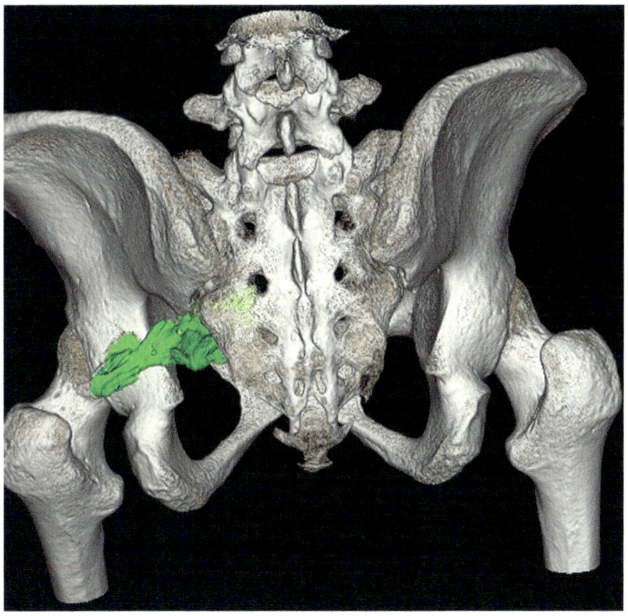

Fig. 7.30 3D CT scan reconstitution of spread of local anaesthetic in an infrapiriformis parasacral block, posterior view. Philippe Gautier iconography

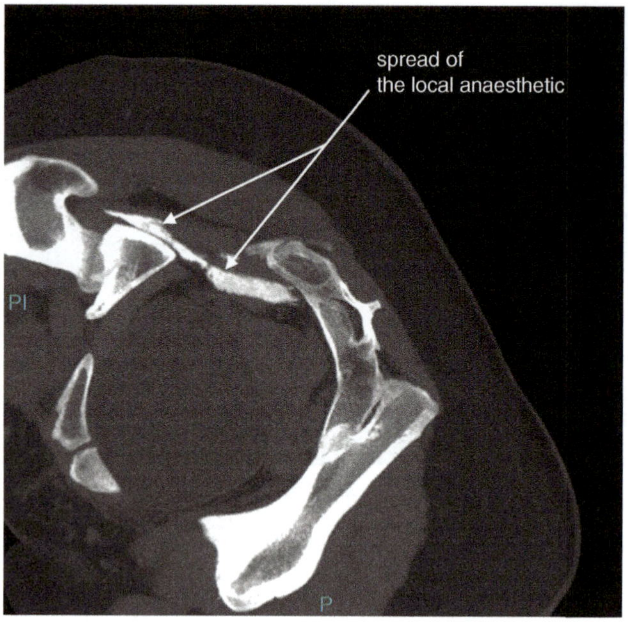

Fig. 7.31 Sciatic nerve block by IPPS approach: CT scan section evidencing spread of the local anaesthetic. Philippe Gautier iconography

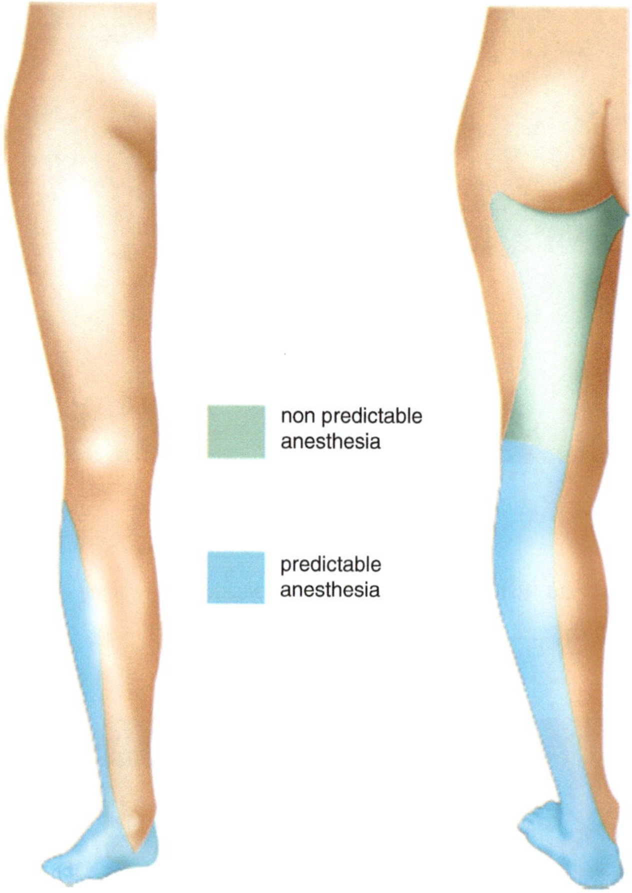

Fig. 7.32 Area of extension of a subgluteal ischiatic block

Sciatic Nerve Block in the Buttock

Ultrasound of the sciatic nerve in the gluteal area (via the trans-gluteal approach) is not particularly useful. Often, the sciatic nerve is situated at a depth that often makes it less easily visible than in the subgluteal and parasacral areas.

Sciatic Nerve Block by Subgluteal Approach (Fig. 7.32)

Indications

In the "blind" technique, the subgluteal approach to the sciatic nerve was first described by Raj [16], with an uncomfortable position for the patient and a non-negligible risk of injury to the nerve. It then was revised by Di Benedetto [17], with the patient positioned on his side with a lower risk of needle stick injury to a nerve because of the more tangential direction of the needle (70° versus 90° with the Raj technique). The surface landmarks are simple, a line is traced between the greater trochanter and the ischial tuberosity. The site of injection lies 4 cm caudally on the midpoint of this line. This block is indicated, as with other proximal approaches to the sciatic nerve when combined with femoral block, for lower limb surgery starting from the middle third of the femur. It has the disadvantage of randomly blocking the posterior cutaneous nerve of the thigh. Insertion of a perineural catheter is possible, but this site does not always enable a perfect aseptic procedure for the injection site. It may be preferable to insert a parasacral or popliteal perineural catheter.

Type of probe: 2 to 5 MHz curved in adults (5 to 10, or 6 to 13 MHz linear probe is possible in children or in very thin adults).
Axis of the probe: Initially transverse, in a line joining the apex of the greater trochanter to the ischial tuberosity (Figs. 7.38 and 7.39).
Configuration: Initially nerve in short axis, rotating the probe to the longitudinal view, i.e. nerve in long axis.
Depth studied: 5 to 8 cm generally, adjusted according to patient build.
Neurostimulation: Enables further identification of the nerves visualised and can limit, by determination of an MIS >0.3 mA (0.1 ms), the risk of accidental intrafascicular injection. Furthermore, it can supplement locating nerves when conditions of visibility are unfavourable.
Needle: 100 mm isolated, 21 G.

Echoanatomy and Location

Apart from the subcutaneous adipose tissue of variable extent and the heterogenous ultrasound presentation, this area is comprised primarily of structures with homogenous echostructure, with clear interfaces. The following are found in it:

- A bony component represented by the ischial tuberosity medially and the greater trochanter laterally.
- A muscular component consists of the quadratus femoris muscle (ventrally, extending between the lateral border of the ischial tuberosity and the intertrochanteric crest of the femur) and the gluteus maximus muscle (the most superficial) covering the entire osteomuscular structure dorsally. The sciatic nerve, the oval structure in transverse section, hyperechoic flattened in the frontal plane, continues between these four components: the gluteus maximus muscle dorsally and the quadratus femoris ventrally, the greater trochanter laterally and the ischial tuberosity medially (which it is generally slightly closer than to the greater trochanter) (Figs. 7.33, 7.34 and 7.35). However, its good visibility is not constant and it then is possible to use the structures in the surrounding area in an attempt to position it on the image (Fig. 7.36).

If the ultrasound section is slightly more distal, passing under the ischial tuberosity, the latter no longer appears and the sciatic nerve then is in relation medially to the hamstring muscles (semimembranosus, semitendinosus, and long head of biceps femoris). It is often possible to differentiate in the medial border of the sciatic nerve and in the same interaponeurotic space, a satellite structure corresponding to the posterior cutaneous nerve of the thigh. It is sometimes also possible to note the Doppler signal of the artery of the sciatic nerve and/or of the femoral branch of the inferior gluteal artery.

Approach with Ultrasound Guidance

The patient is placed in the lateral recumbent position (Sims position), lying on the non-operative side, with the "upper" hip flexed (Fig. 7.37). The probe is positioned transversely, in the middle of a line connecting the greater trochanter to the apex of the ischial tuberosity [18], the sciatic nerve is visualised in transverse section. We have the choice between an approach in plane, inserting the needle at the lateral end of the probe (Fig. 7.38), or indeed out of plane (Fig. 7.39).

Needle Puncture and Injection

After anaesthesia of the skin, advancement of the needle in the direction of the sciatic nerve through the gluteus maximus muscle is performed without neurostimulation in order to not cause unnecessary and painful contractions. If the anaesthetist has chosen the in-plane approach, advancement of the needle occurs up until it reaches the immediate proximity of the sciatic nerve. The needle in the majority of cases is well-visible considering the homogenous echostructure of the gluteus maximus muscle. As is recommended, the nerve should be approached tangentially in order to avoid any direct injury by the tip of the needle. Using this approach the needle tip is placed at the lateral border of the nerve, as it lies in the subgluteal space [19], whereupon the injection is started.

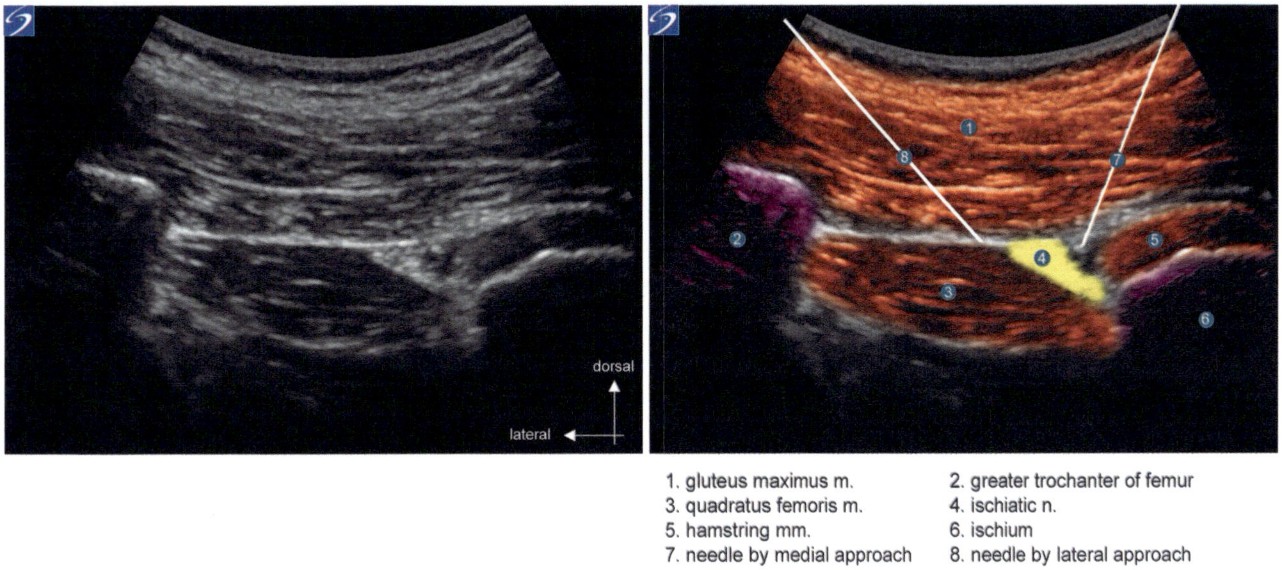

Fig. 7.33 Sciatic nerve block by subgluteal approach. Transverse ultrasound section. Possible trajectory of the needle in the ultrasound plane

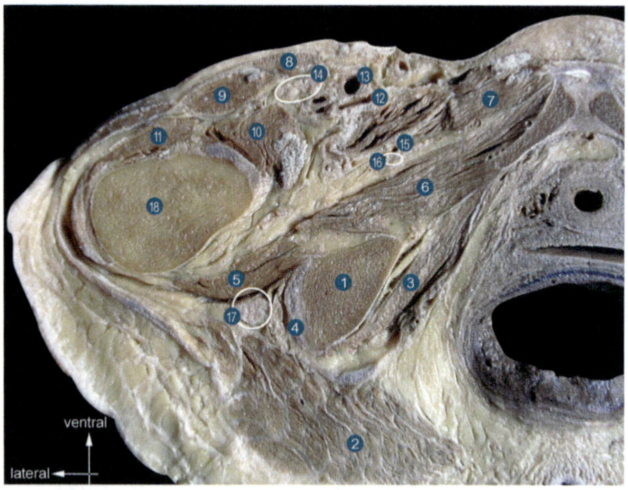

1. ischial tuberosity
2. gluteus maximus m.
3. obturator internus m.
4 . hamstring mm. tendon
5. quadratus femoris m.
6. obturator externus m.
7. adductor longus m.
8. sartorius m.
9. rectus femoris m.
10. vastus lateralis m.
11. tensor of fascia lata m.
12. femoral v.
13. superficial femoral a.
14. branches of femoral n.
15. obturator a.
16. obturator n.
17. ischiatic n.
18. greater trochanter of femur

Fig. 7.34 Axial anatomical section cutting through the pubic symphysis, the ischial tuberosity, and the greater trochanter

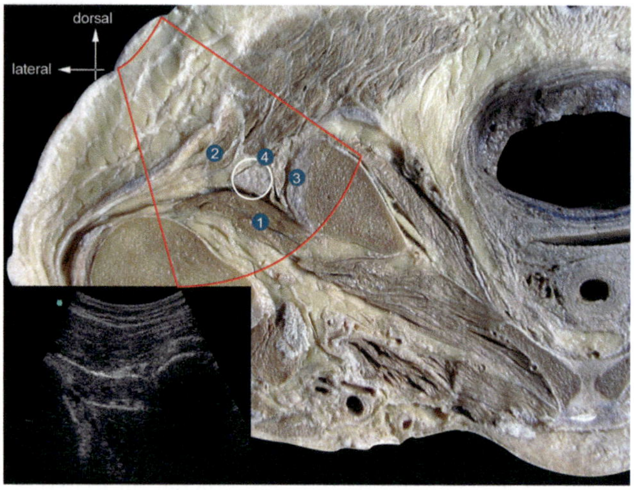

1. quadratus femoris m.
2. gluteus maximus m
3. hamstring mm. tendon
4. ischiatic n.

Fig. 7.35 Subgluteal block of the sciatic nerve: sonoanatomical correlation

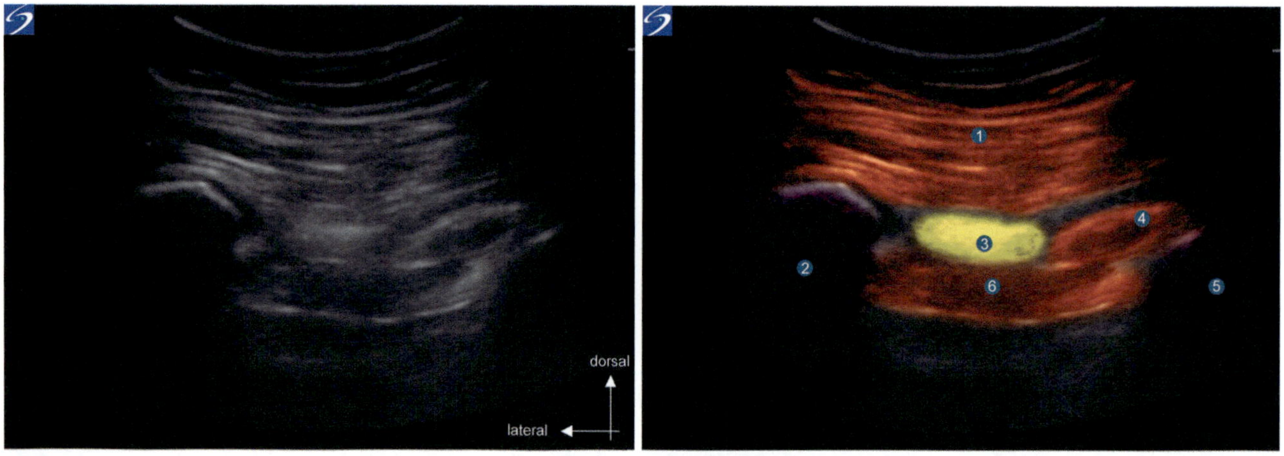

1. gluteus maximus m.
2. greater trochanter of femur
3. ischiatic n. area
4. hamstring mm.
5. ischium
6. quadratus femoris m.

Fig. 7.36 Ultrasound section at subgluteal level not enabling to precisely localise position of sciatic nerve. Contribution of the adjacent anatomical elements

However, if an out-of-plane approach has been chosen, care must be taken to correctly identify the position of the tip of the needle (hydrolocalisation with a 5% dextrose solution and/or tissue movements caused by the needle, followed by introducing the bevel of the needle into the ultrasound plane). In the immediate vicinity of the nerve during an out-of-plane injection, it is sometimes possible to rotate the probe through 90° in order to achieve the long axis view, enabling visualisation of both the nerve and the needle (Fig. 7.40). This will enable more precise positioning of the needle tip in relation to the nerve, a procedure which is also aided and refined by neurostimulation. Attentive visual control of the needle has the objective of ensuring that the local anaesthetic is in direct contact with the nerve (Fig. 7.41) and without interposition

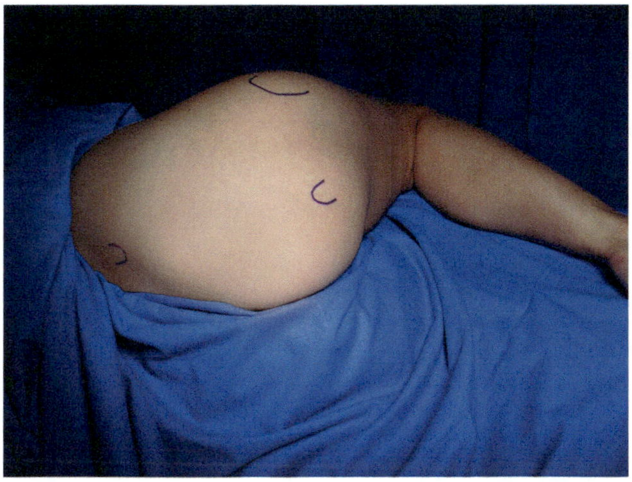

Fig. 7.37 Position of patient to undergo ultrasound-guided sciatic nerve block by subgluteal approach

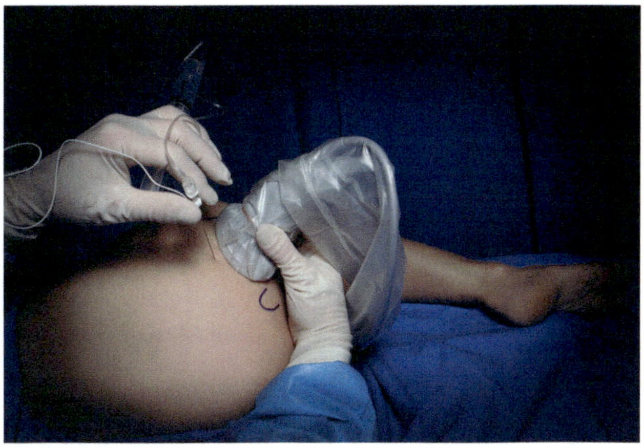

Fig. 7.39 Sciatic nerve block by subgluteal approach: probe in transverse position, needle inserted out of plane

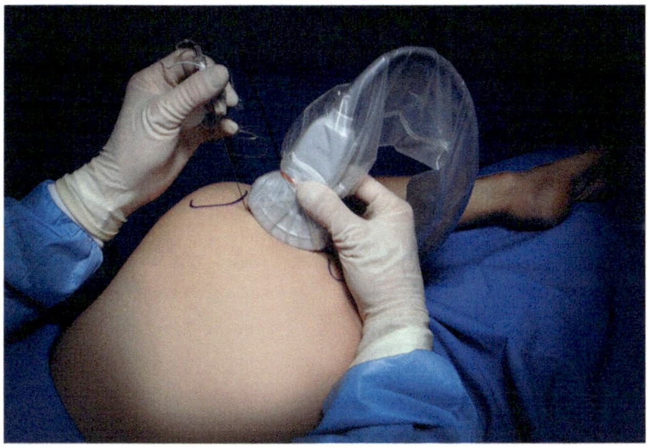

Fig. 7.38 Sciatic nerve block by subgluteal approach: probe in transverse position, needle inserted in plane

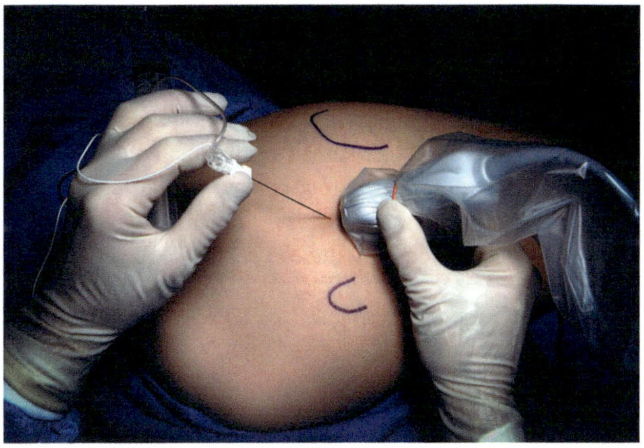

Fig. 7.40 Position of the probe for longitudinal imaging of sciatic nerve

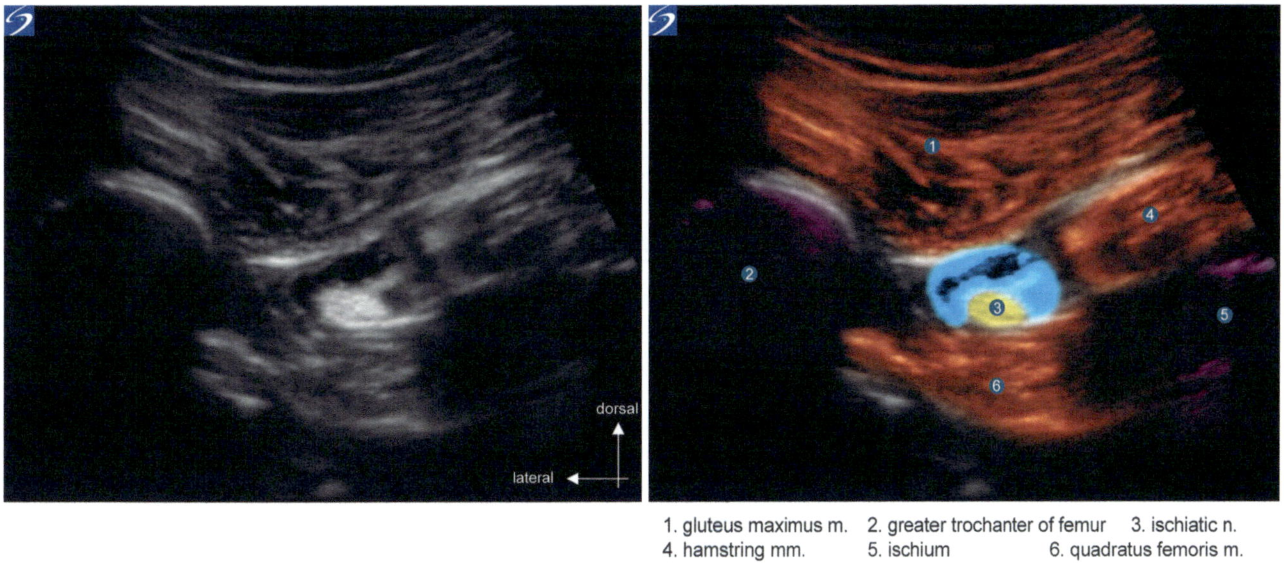

Fig. 7.41 Sciatic nerve block in subgluteal area. Spread of local anaesthetic in immediate proximity to nerve

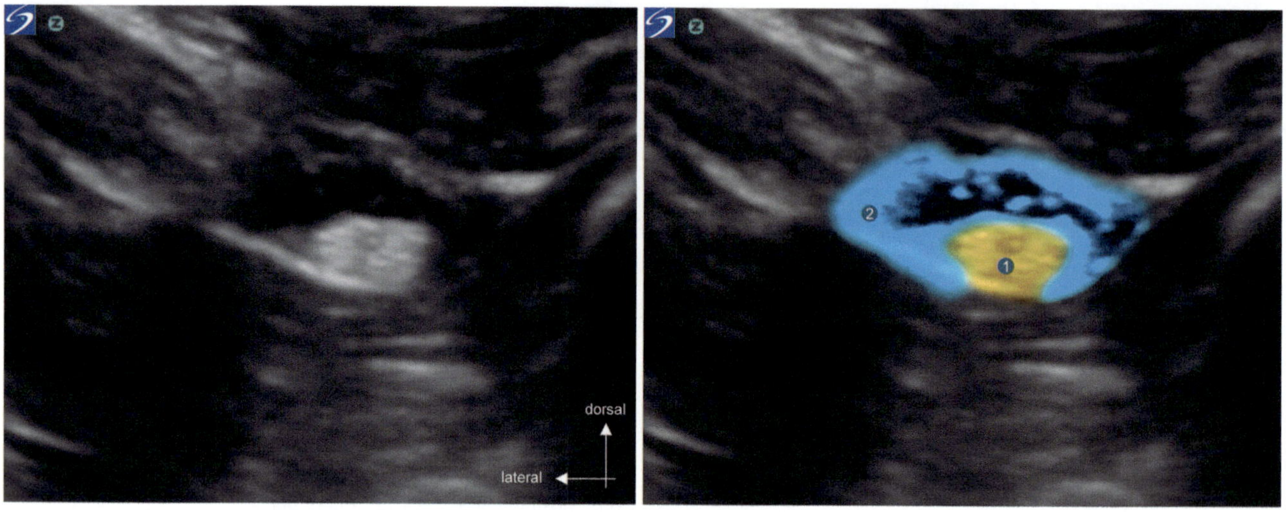

Fig. 7.42 Sciatic nerve block in subgluteal area. Local anaesthetic well-positioned comes into close contact with outline of nerve, with no tissue interposition

of the aponeurosis (Fig. 7.42). If this is the case, cautious movement of the needle during the injection process may be necessary. Although ideal spread of the local anaesthetic is probably circumferential, direct contact of the local anaesthetic with the sciatic nerve in a portion at least greater than half of its circumference produces a nerve block with perfect efficacy.

> **Paediatrics**
> This nerve block can usually be performed with ease in children as it presents very favourable characteristics of depth and echogenicity.

Sciatic Nerve Block in the Popliteal Space (Fig. 7.43)

Indications

Sciatic nerve block in the popliteal fossa is reserved primarily for distal surgery of the leg, foot, and ankle; it can be combined with a saphenous nerve block performed at a level to obtain anaesthesia of the medial aspect of the leg and of the foot. (However, if a thigh tourniquet is necessary for surgery, a femoral block should be performed in order to ensure optimal analgesia). Two approaches are possible, the posterior and lateral approaches. An injection of both components of the sciatic nerve (tibial and common peroneal) is indicated [20]. The primary risk is injection of the artery or of the popliteal vein, in particular for the approach close to the flexion

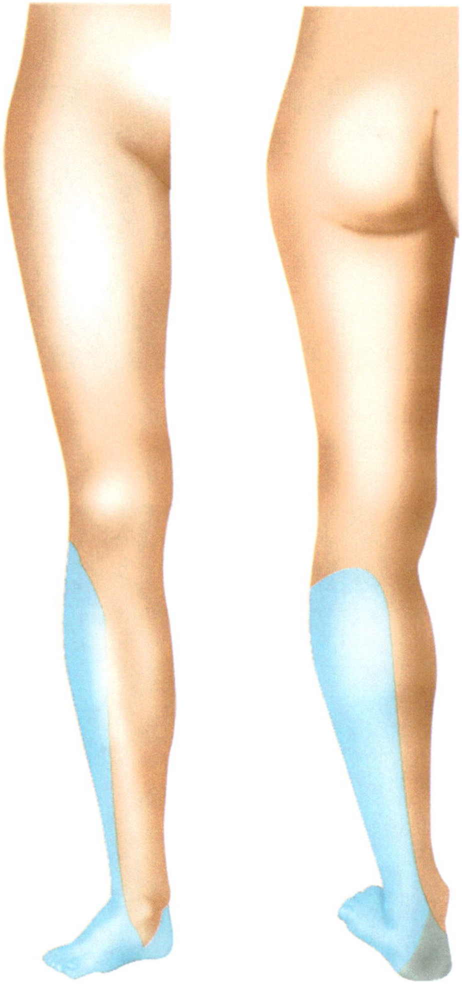

Fig. 7.43 Area of extension of a sciatic nerve block by the popliteal approach

fold of the knee. Ultrasound reduces this risk by allowing visualisation of spread of the local anaesthetic. With double stimulation, Vloka and Hadzic [21] showed that the two approaches have the same efficacy. The indication for one or the other depends in particular on the team, and the ability to move the patient into prone or lateral position (on an injured patient). The popliteal approaches have a longer time of onset than the proximal approaches. The popliteal approach enables insertion of a catheter for post-operative analgesia and by posterior or lateral approach.

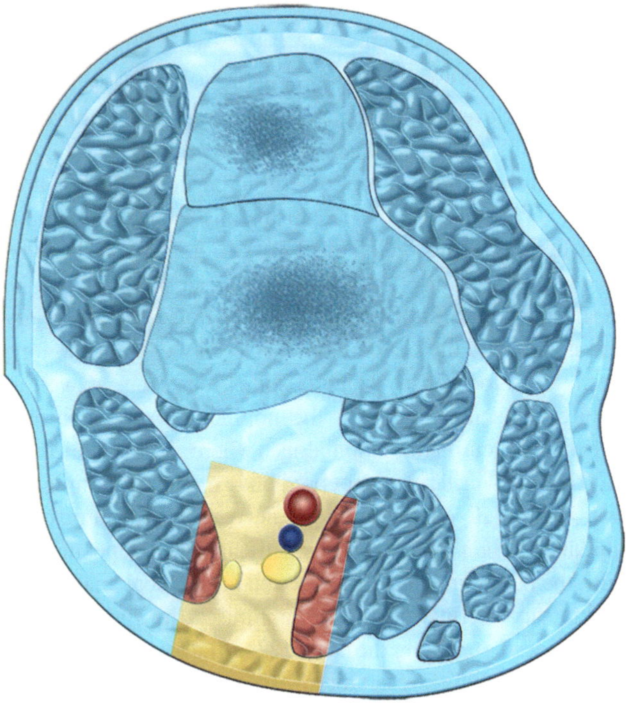

Fig. 7.44 Transverse section of the popliteal region with materialisation of ultrasound beam

Type of probe: Linear, 5 to 10 MHz or 6 to 13 MHz.
Axis of the probe: Transversal (Fig. 7.44).
Configuration: Nerve in short axis, needle in plane.
Studied depth: 3 to 5 cm depending on patient build.
Neurostimulation: Enables further identification of the nerves visualised and can limit, by determination of an MIS > 0.3 mA (0.1 ms), the risk of accidental intrafascicular injection. Furthermore, it can supplement locating nerves when conditions of visibility are unfavourable.
Needle: 50 to 80 mm, 22 G.
Utility of Doppler ultrasound: Aid in locating the popliteal vessels which are very close to the tibial nerve component (Fig. 7.45).

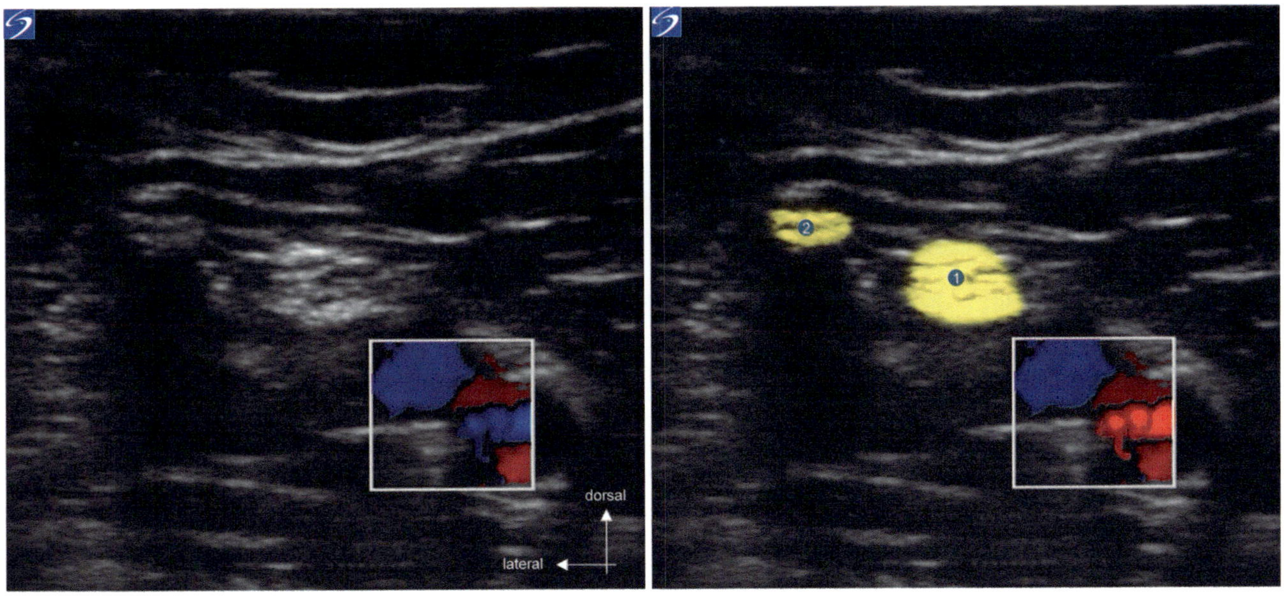

1. common peroneal (fibular) n.
2. tibial n.
3. popliteal v.
4. popliteal a.

Fig. 7.45 Sciatic nerve block in the popliteal fossa: transverse ultrasound section showing proximity of vessels with Doppler signal

Histology

In the popliteal fossa, the sciatic nerve is surrounded by an especially developed perineural structure. The connective tissue forms a perineural sheath which promotes sliding of nerves in their environment, useful in particular in this area which is subject to high mechanical constraints (high amplitude flexion-extension of the knee) (Fig. 7.46). However, these concentric envelopes create a barrier to spread of the local anaesthetic if it has been deposited outside its boundary and not around the neural target. This may explain the long latency of block onset which is sometimes seen, as well as failures of sciatic block in the popliteal fossa performed either with neurostimulation alone (where the usual electrophysiological criteria are not predictive of an optimum injection in intimate contact with the nerve), or with ultrasound guidance (when the local anaesthetic has been deposited out with the perineural sheath). Injection of the local anaesthetic within this "perineural sheath" optimises the spread of the local anaesthetic along the nerve [22–25]. However, it cannot be ruled out that a "reservoir effect" may be at the origin of an enhanced toxic risk by prolongation of contact time and the increase in surface contact between nerve tissue and the local anaesthetic. This very likely encourages the practitioner to use less concentrated local anaesthetics and to decrease the volumes of local anaesthetic used (a trend that is entirely evident under ultrasound control).

Echoanatomy

The popliteal fossa is comprised primarily of intermuscular tissue whose echogenicity generally is very heterogenous. A search for the sciatic nerve with its two components should be performed in the area demarcated posteriorly by the skin, laterally by the biceps femoris muscle, medially by the semimembranosus muscle, and the tendon of the semitendinosus muscle, and, lying deep, the popliteal vessels. Ultrasound guidance allows for visualisation of the level of the sciatic nerve bifurcation and promotes the positive identification of each of the components of the sciatic nerve. The tibial and common peroneal nerves which are stuck together or separated according to level (Figs. 7.47, 7.48 and 7.49) are in close relation in depth with the popliteal vein, which itself is placed dorsally to the popliteal artery.

It is sometimes possible to see the distal branch of the posterior cutaneous nerve of the thigh which is located more superficially in the popliteal fossa (Fig. 7.50), before it pierces the superficial aponeurosis and is distributed to the skin of the craniodorsal part of the calf muscle. Similarly, it is often possible to visualise the lateral sural cutaneous nerve detached from the common peroneal nerve (Figs. 7.51 and 7.52) and, sometimes, to visualise the medial sural cutaneous nerve (Fig. 7.53). In order to differentiate the latter from the posterior cutaneous nerve of the thigh, trace the structure(s) caudally. The medial sural cutaneous nerve continues in the groove separating the gastrocnemius muscles dorsally, enters the crural fascia and descends posterior to the lateral malleolus. If the structure is the terminal branch of the posterior cutaneous nerve of the thigh it quickly pierces the superficial fascia and becomes lost in the subcutaneous tissue in the proximal part of the calf. In a relatively distal popliteal approach, a gap in the area of anaesthesia may be noted in the sural area when its emergence is very proximal in the popliteal fossa.

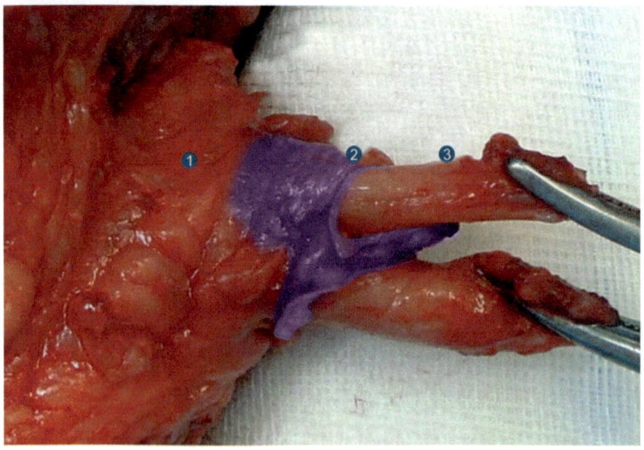

Fig. 7.46 Dissection of common peroneal and tibial nerves demonstrating the importance of the perineural tissue surrounding them. Dissection: Denis Jochum

Approaches with Ultrasound Guidance

It is essential to visualise the tibial and the common peroneal nerves in transverse section (whether they are stuck together or not). The longitudinal view of the needle could allow a better precision for the conduct of this block for which the onset time can be prolonged in case of imprecise distribution of the local anaesthetic. With an in-plane approach, the probe is placed transversely to the neurovascular axis, and the block is performed by inserting the needle at the lateral aspect of the probe (Figs. 7.54 and 7.55). The patient is placed in the supine or prone position according to habits of the team or other requirements (e.g. an injured patient in pain). In order to create space for correct probe position in the popliteal fossa with the patient in supine position, it is necessary to elevate the calf and the foot, placing them on cushions (Fig. 7.56).

If the decision is made to perform an out-of-plane approach, it is necessary to place the patient in a prone or lateral position.

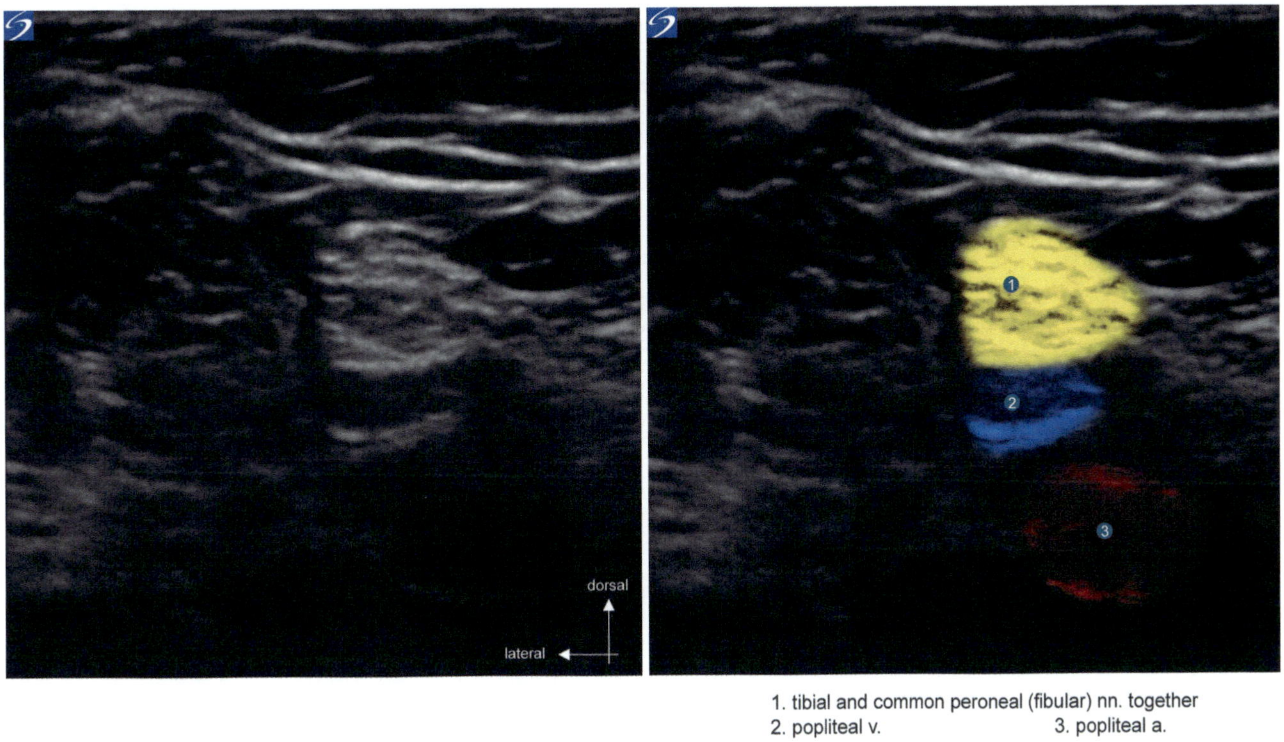

Fig. 7.47 Sonoanatomy of the popliteal fossa: sciatic nerve upstream of its bifurcation, in contact with the popliteal vein

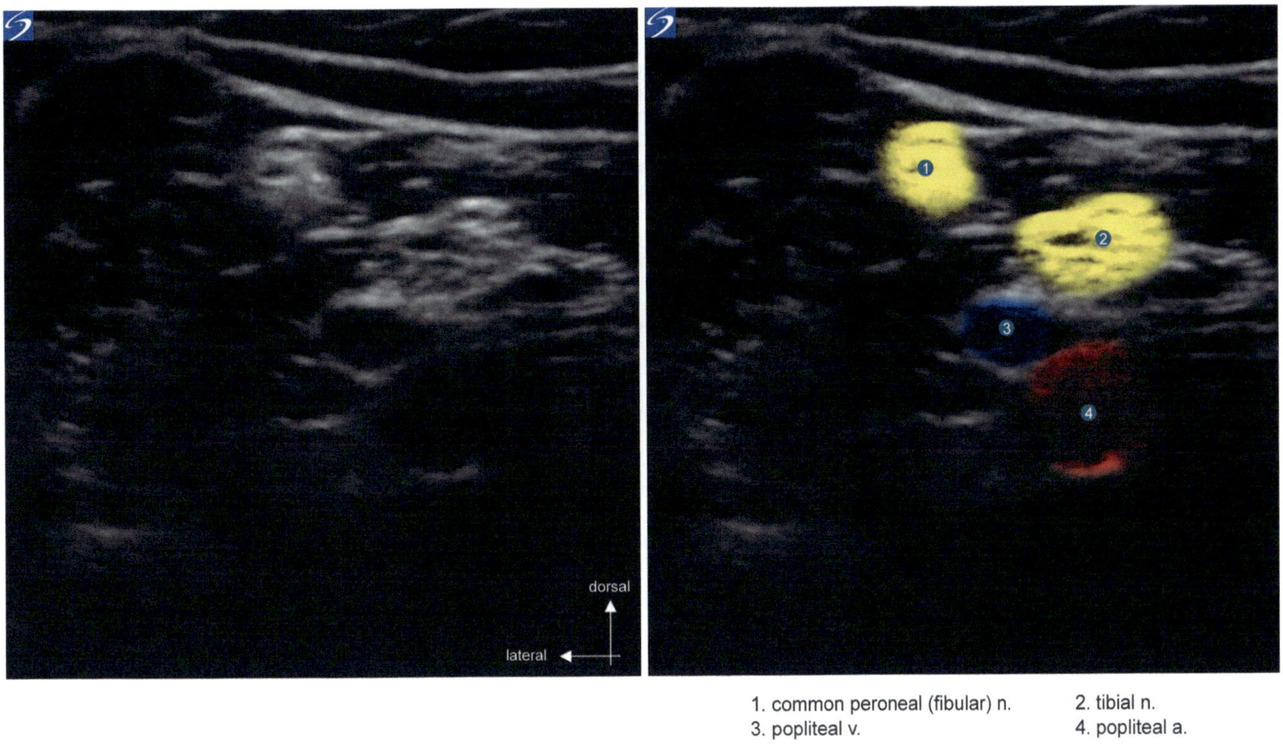

Fig. 7.48 Sonoanatomy of the popliteal fossa: sciatic nerve seen immediately after its bifurcation

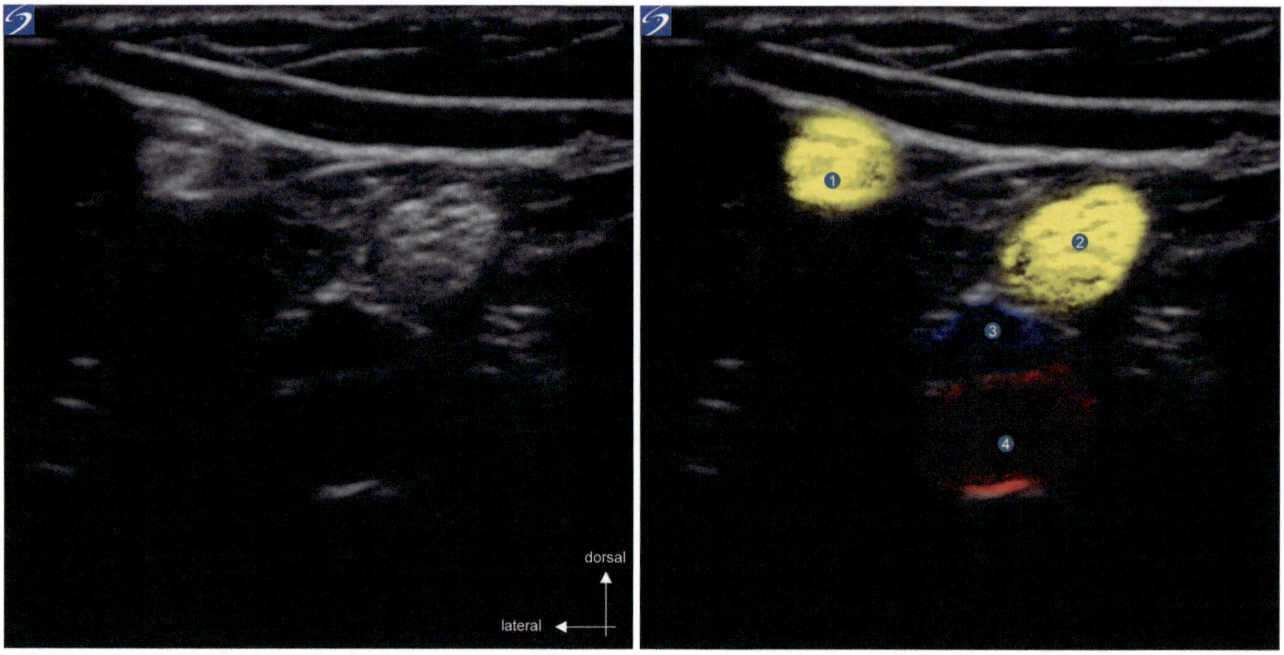

Fig. 7.49 Sonoanatomy of the popliteal fossa: progressive divergence of the two branches of bifurcation. The common peroneal nerve moves laterally, while the tibial nerve remains closer to the midline of the leg, in proximity to the popliteal vein

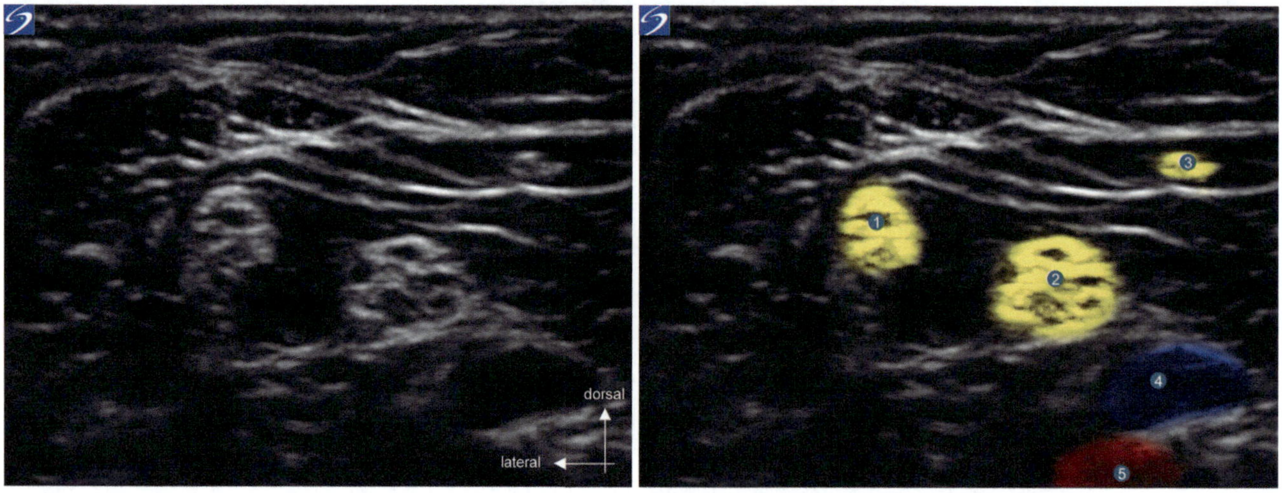

Fig. 7.50 Sonoanatomy of the popliteal fossa: visualisation of the distal branch of the posterior cutaneous nerve of thigh

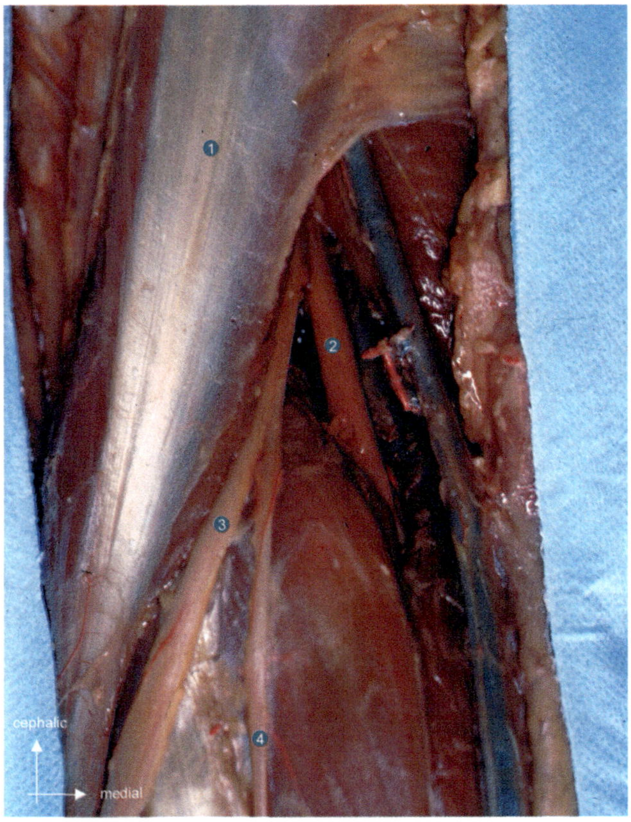

1. biceps femoris m. 2. tibial n.
3. common peroneal (fibular) n. 4. lateral sural cutaneous n.

Fig. 7.51 Popliteal fossa: the lateral sural cutaneous nerve detaches from the common peroneal nerve. Dissection: Bertrand Fabre

Location

The fatty nature of the tissue surrounding the tibial and common peroneal nerves in the popliteal fossa (Figs. 7.57 and 7.58) makes their localisation highly dependent on inclination of the probe (anisotropy). This manoeuvre makes these structures "appear" literally in an environment whose very heterogenous echogenicity varies little (Figs. 7.59, 7.60, 7.61 and 7.62). During movement of the probe performed in locating the nerves, the perineurium is not easily identifiable. It is more so during approach and injection, when the needle can move it and the local anaesthetic separate it from the nerve strictly speaking. The perineurium, in fact, is a structure separate from the epineurium, and sub-perineural injection is not synonymous with sub-epineural injection, confusion found in initial descriptions of this type of ultrasound-guided nerve block.

Approach

The lateral approach (Fig. 7.54) allows for a more optimum angulation between probe and needle which improves the visibility of the needle. The dorsolateral approach (Fig. 7.55), in the majority of cases, requires a needle angulation which is less favourable to its visibility, even though it has the obvious advantage of a shorter pathway for the injection and so may be less painful. The block can be performed upstream of sciatic nerve bifurcation or further downstream depending on individual nerve visibility and the level to which the gross

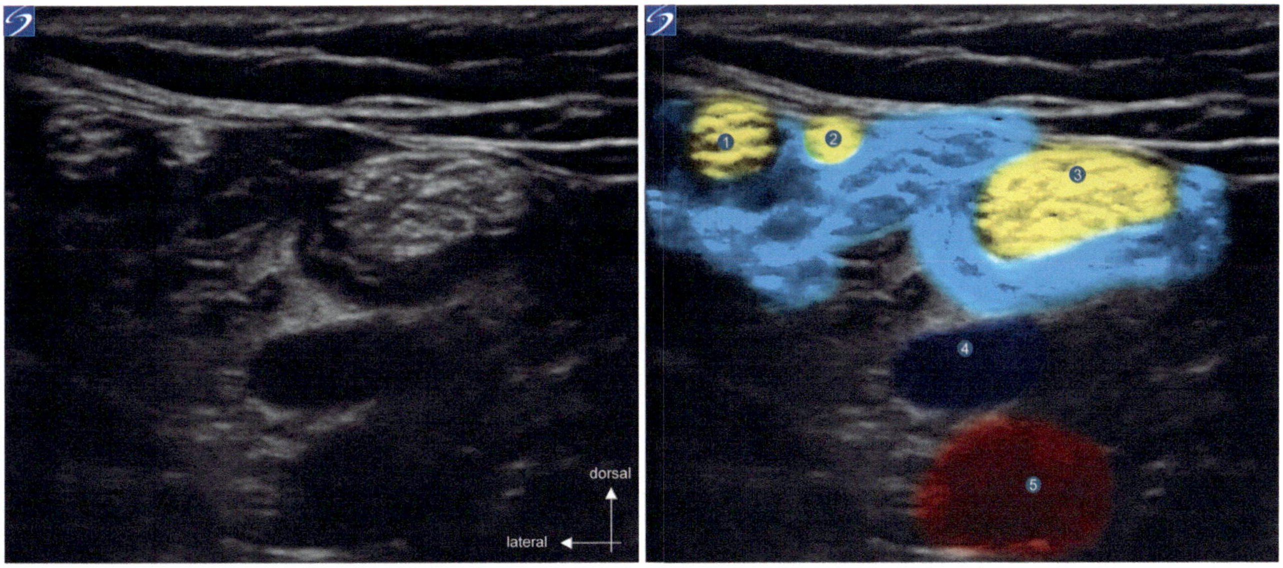

1. common peroneal (fibular) n. 2. lateral sural cutaneous n.
3. tibial n. 4. popliteal v. 5. popliteal a.

Fig. 7.52 Sonoanatomy of popliteal fossa: separation of lateral sural cutaneous nerve

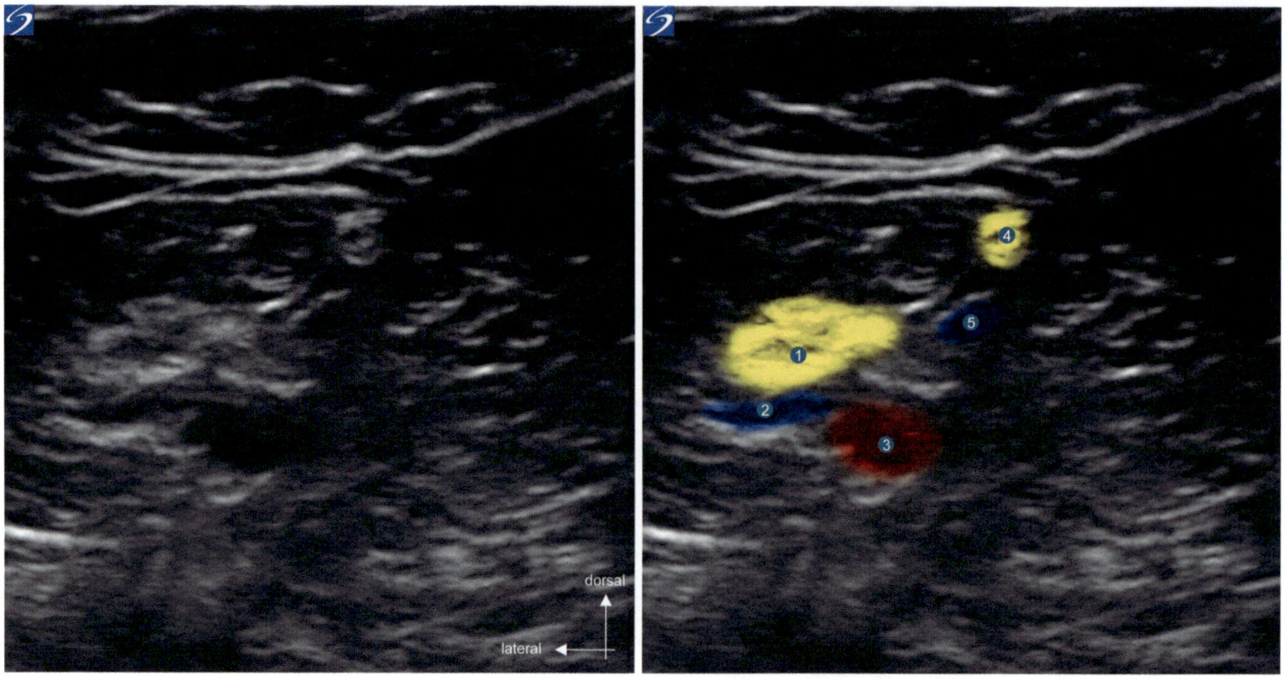

1. tibial n. 2. popliteal v. 3. popliteal a.
4. lateral sural cutaneous n. 5. small saphenous v.

Fig. 7.53 Sonoanatomy of popliteal fossa: medial sural cutaneous nerve splitting off from the tibial nerve

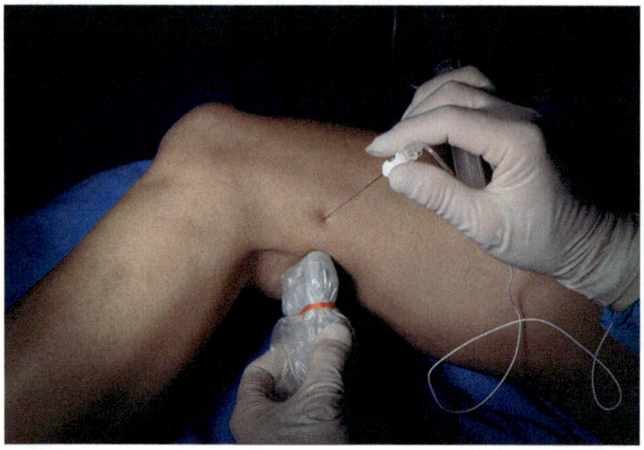

Fig. 7.54 Sciatic popliteal block by a lateral approach

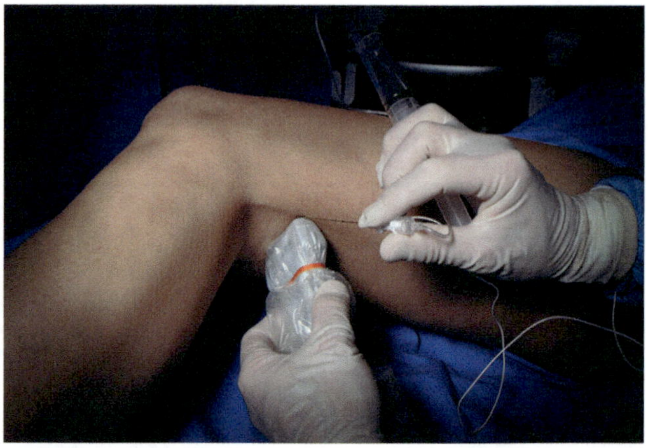

Fig. 7.55 Sciatic popliteal block by a dorsolateral approach

separation of the two components occurs. Generally, the physician is guided in this choice by the constraints of depth and visibility of the nerves. Better visibility is often found in the approaches slightly or frankly downstream of the bifurcation, in particular, due to the more superficial position of the nerves. Although studies on the involvement of the perineurium in sciatic popliteal nerve blocks do not provide a single answer to the question of the ideal site for conduct of the nerve block (at the sciatic nerve bifurcation, upstream or downstream of it?), nevertheless they show a decrease in the time to onset when injection of the local anaesthetic is done under the perineural sheath in intimate contact with the nerve structure [24, 25].

Injection

Injection upstream or at the level of the sciatic bifurcation in principle enables a single injection for the two nerve components. On the contrary, an approach downstream of the bifurcation most of the time involves injection of the local anaesthetic in contact with each of the two nerves. However,

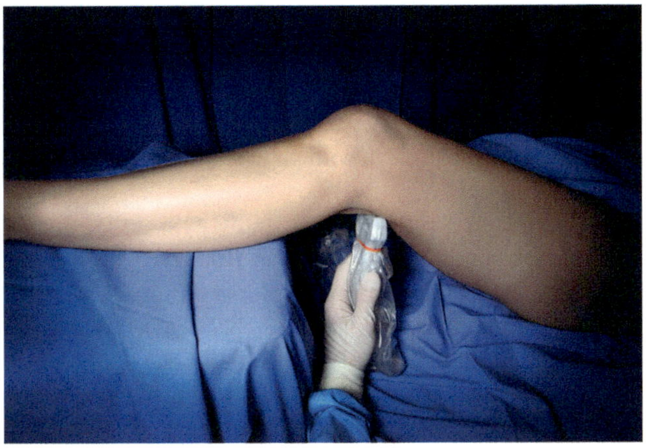

Fig. 7.56 Sciatic popliteal block by a lateral approach: leg elevated, placed on cushions to free up the space behind the popliteal fossa

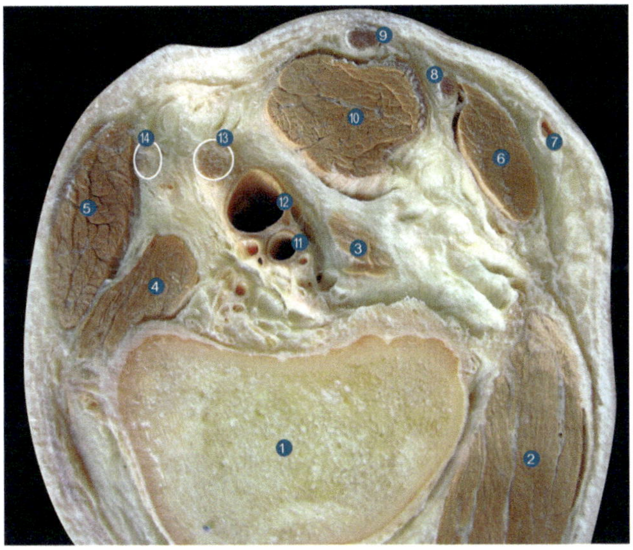

1. femur
2. vastus lateralis m.
3. gastrocnemius m. (medial head)
4. gastrocnemius m. (lateral head)
5. biceps femoris m.
6. sartorius m.
7. greater saphenous v.
8. gracilis m.
9. semitendinosus m.
10. semimembranosus m.
11. popliteal a.
12. popliteal v.
13. tibial n.
14. common peroneal (fibular) n

post.
lat.

Fig. 7.57 Transverse anatomical section in the popliteal area

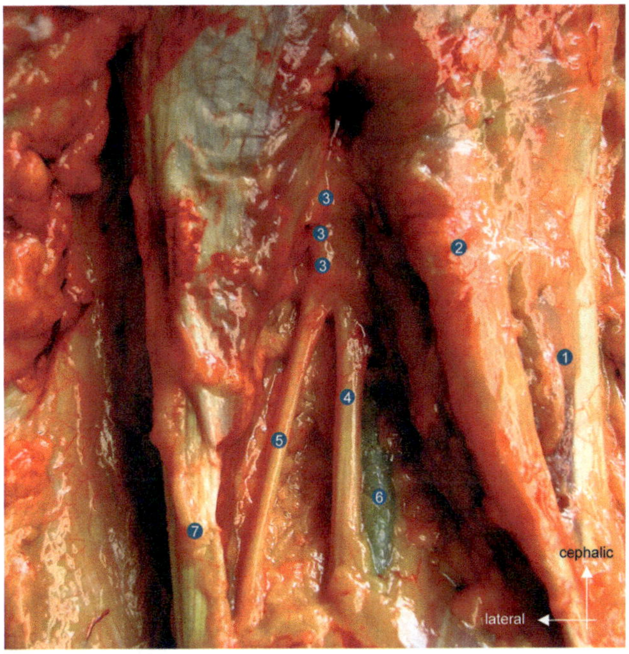

1. semimembranosus m.
2. semitendinosus m.
3. paraneurium
4. tibial n.
5. common peroneal (fibular) n.
6. popliteal v.
7. biceps femoris m.

Fig. 7.58 Popliteal fossa. The nerves and vessels are located in the intermuscular connective space. Dissection: Bertrand Fabre

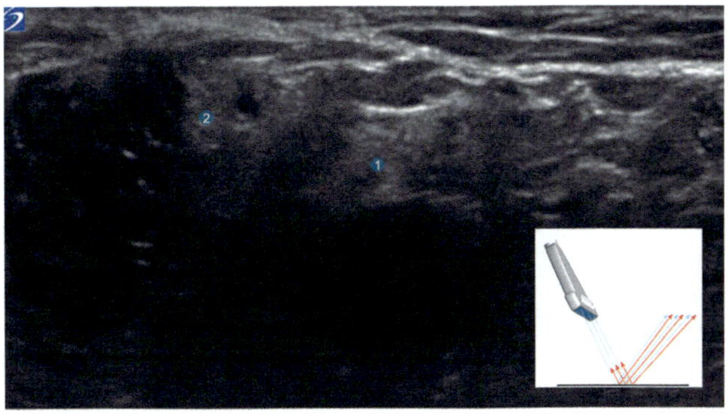

1. tibial n.
2. common peroneal (fibular) n.

Fig. 7.59 Tilting of probe poorly showing nerve structures

the appearance of the local anaesthetic around the common peroneal nerve is observed regularly during injection in contact with the tibial nerve; this phenomenon is related to the retrograde spread of the anaesthetic along the tibial nerve under the perineurium up to the sciatic bifurcation, and then anterograde along the peroneal nerve. In this case, it seems justified to be satisfied with a single tibial injection.

Comparison of the data between neurostimulation and ultrasound images can increase the precision of positioning of the needle, not only because that in the heterogenous popliteal environment of the nerves, their outline is sometimes difficult to determine (in particular for the common peroneal nerve), but also because it involves a nerve block with traditional neurostimulation, of prolonged time of onset in spite of "optimal" electrophysiological criteria. Observation of the method of spread of the local anaesthetic is an element that has a predictive potential for rapidity of onset and efficacy. The aim is to obtain more or less circumferential, subperineural spread, with a "rosette" around each of the two components.

Figures 7.63, 7.64, 7.65, 7.66, 7.67, 7.68, 7.69 and 7.70 show a sequence of a popliteal fossa block with a single

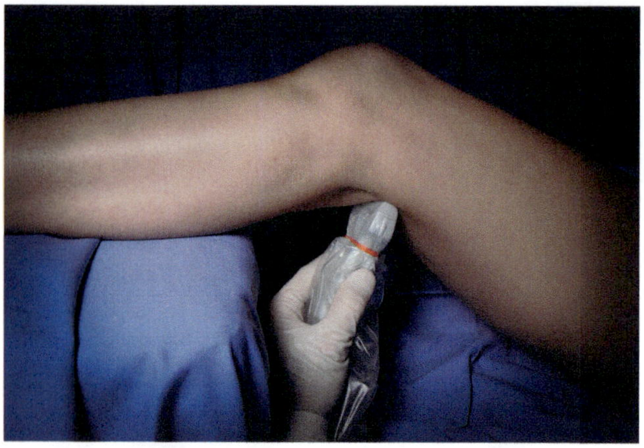

Fig. 7.60 Tilting of probe poorly showing nerve structures

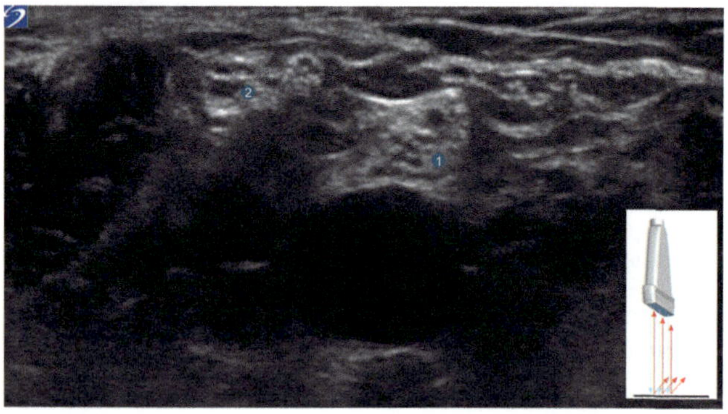

1. tibial n. 2. common peroneal (fibular) n.

Fig. 7.61 Tilting of probe more perpendicularly to the axis of nerves for better imaging

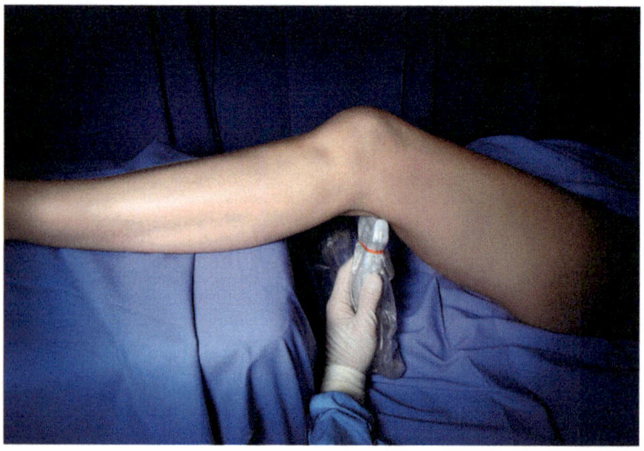

Fig. 7.62 Tilting of probe more perpendicular to the axis of nerves for better imaging

injection by lateral approach downstream of the sciatic nerve bifurcation. The initial injection is performed in contact with the tibial nerve, and then with the common peroneal nerve, lastly ending with a block of the terminal branch of the posterior cutaneous nerve of the thigh for anaesthesia of the entire length of the skin of the calf.

For Insertion of a Perineural Catheter

For placement of a perineural catheter, this can also be performed in plane (lateral or dorsolateral approach) or out of plane (posterior approach, with patient lying in the lateral or prone position):

- Posterior approach, out of plane (catheter placed along the axis of the nerve): The needle is advanced, with the aid of hydrolocalisation, and possibly with neurostimulation, up until the tip is placed alongside the sciatic nerve (generally, in contact with the tibial component, upstream of the sciatic bifurcation). It is probably necessary to insert the needle tip (and then the catheter) into the plane deep to the perineurium, that is, in direct contact with the epineurium, just as is done for blocks with a single injection. The catheter is then inserted a few centimetres in order to reduce the risk of subsequent displacement secondary to patient movement.
- Lateral approach, in plane (catheter placed perpendicular to the axis of the nerve): This time the needle is inserted perpendicularly to the axis of the sciatic nerve (or of its tibial and common peroneal branches if you are downstream of the bifurcation). According to the desired level of placement of the catheter (i.e. for analgesia which needs to be tibial, common peroneal, or both), the tip of the needle is placed immediately next to the desired nerve. Then the first phase of injection through the needle is performed, either directly with local anaesthetic or with a 5% dextrose solution, to enable the use of neurostimulation. When the injection appears correctly in contact with the epineurium (deep to the perineural sheath) (Figs. 7.64, 7.65 and 7.66), it then is possible to slide the catheter over/through the needle. Contrary to the approach with the catheter inserted in the axis of the nerve, the catheter must not be introduced much further than the very tip of the needle as this increases its distance from the nerve. With this tangential approach, the use of so-called pigtail catheters (e.g. PAJUNK® "Curl") may have an advantage. Once the peri-epineural space has been developed by the initial injection through the needle, the catheter can be introduced through the needle and inserted a further 3 cm, which corresponds to the length of deployment of the distal coil. Thus, the end of the catheter remains in immediate proximity to the nerve as the needle is withdrawn. When the injection through the catheter is observed, since the catheter tip is often difficult to visualise, one should ensure that the injectate spreads well in contact with the nerve, if necessary by using Doppler ultrasound as an aid. Insofar as the catheter is positioned perpendicularly to the axis of movement of the knee, the risk of secondary displacement of the catheter inserted in this manner is no greater, in our experience, than that of a catheter positioned in the axis of the nerve by posterior approach.

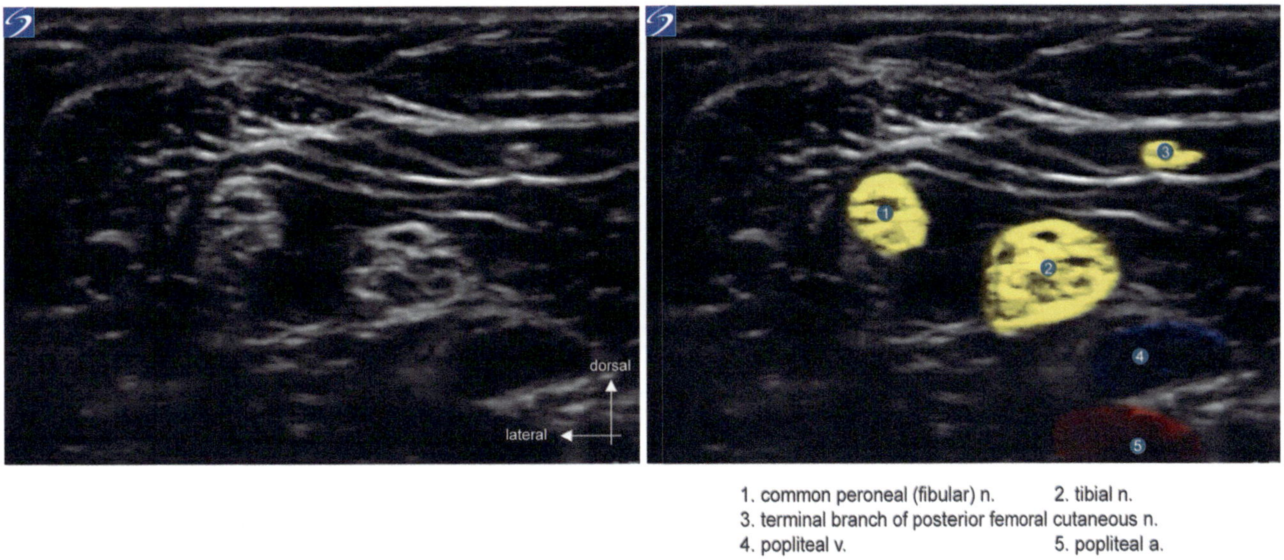

Fig. 7.63 Sciatic nerve block by a lateral popliteal approach. Transverse section with visualisation of the terminal branch of the posterior cutaneous nerve of thigh

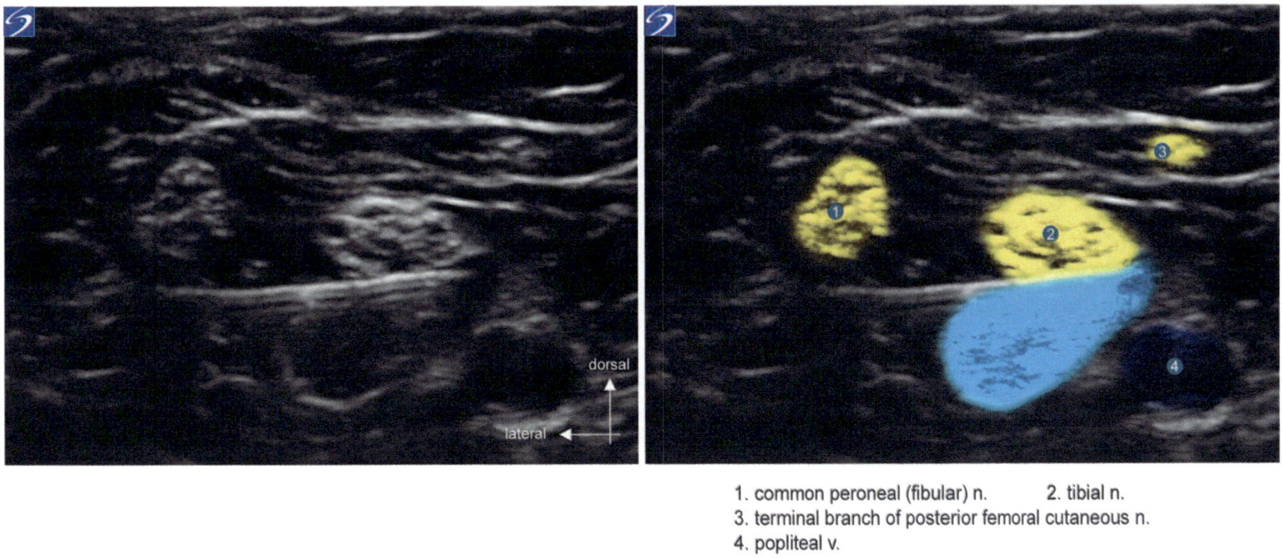

Fig. 7.64 Sciatic nerve block by a lateral popliteal approach. Injection of local anaesthetic in contact with the tibial nerve in its deep aspect

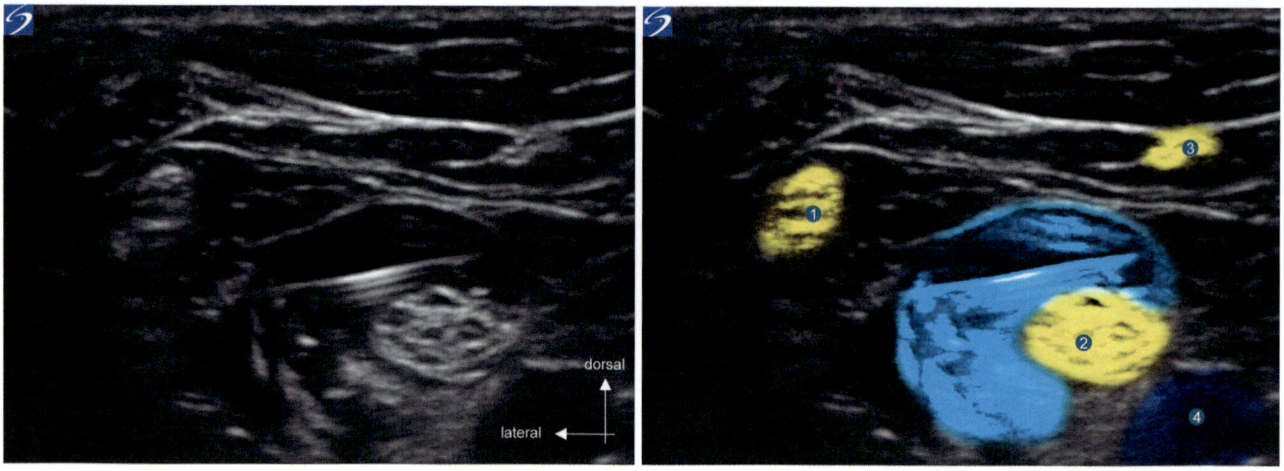

1. common peroneal (fibular) n. 2. tibial n.
3. terminal branch of posterior femoral cutaneous n.
4. popliteal v.

Fig. 7.65 Sciatic nerve block by a lateral popliteal approach. Injection of local anaesthetic in contact with the tibial nerve in its superficial aspect

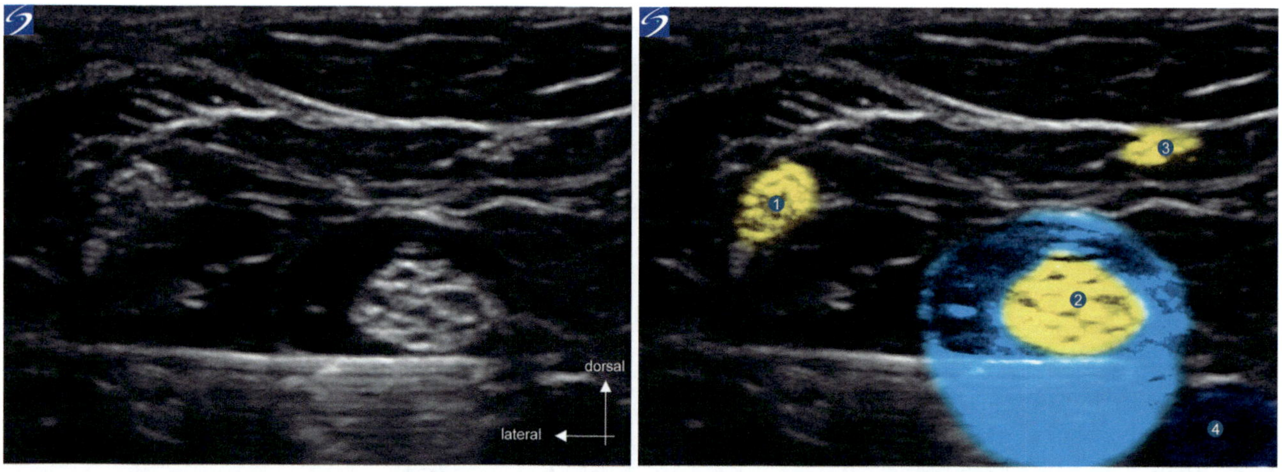

1. common peroneal (fibular) n. 2. tibial n.
3. terminal branch of posterior femoral cutaneous n.
4. popliteal v.

Fig. 7.66 Sciatic nerve block by a lateral popliteal approach. Circumferential spread of local anaesthetic around the tibial nerve following an additional injection in its deep aspect

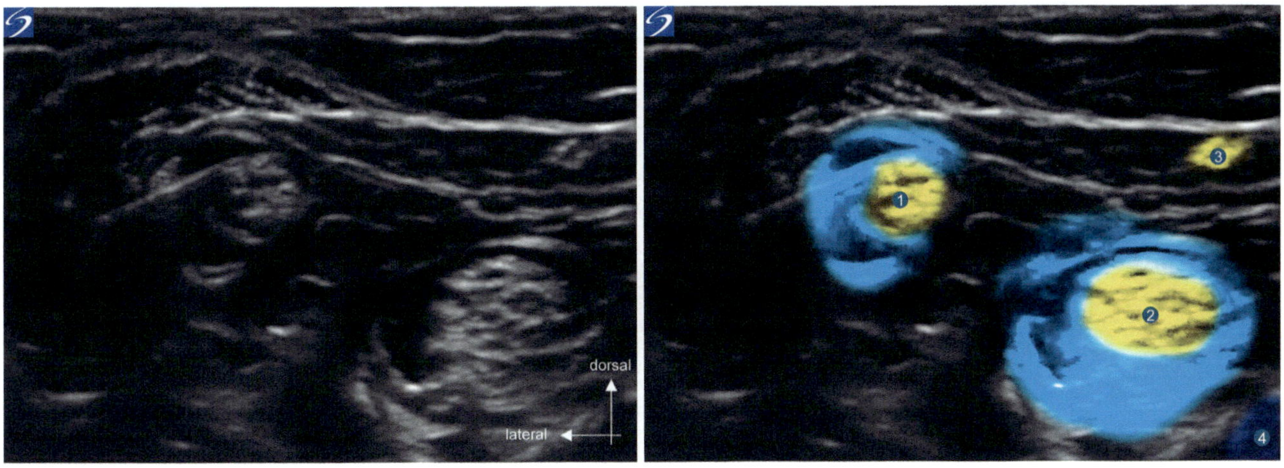

Fig. 7.67 Sciatic nerve block by a lateral popliteal approach. Injection of local anaesthetic in contact with common peroneal nerve, on its superficial aspect

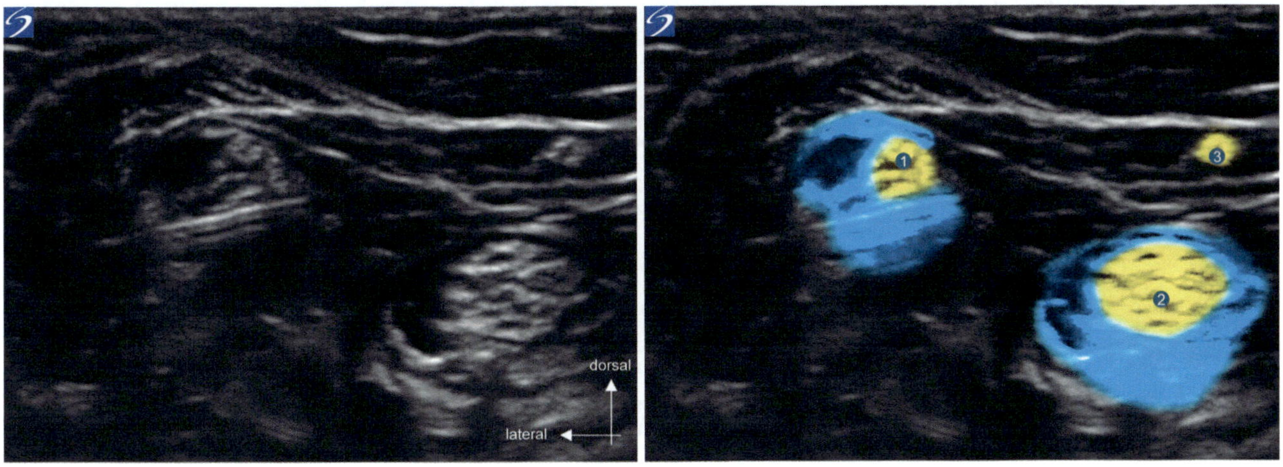

Fig. 7.68 Sciatic nerve block by a lateral popliteal approach. Injection of local anaesthetic in contact with common peroneal nerve, on its deep aspect

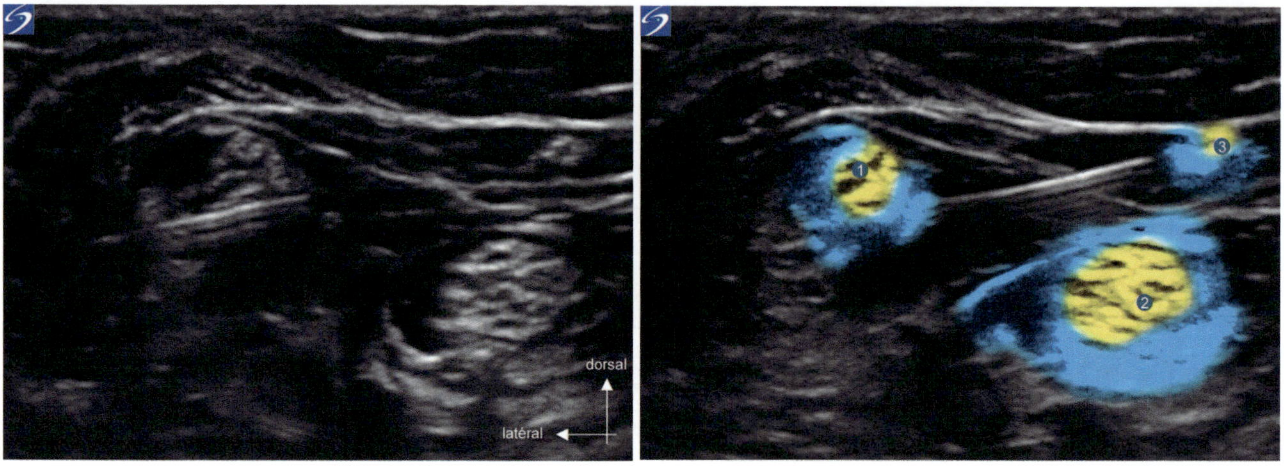

1. common peroneal (fibular) n. 2. tibial n.
3. terminal branch of posterior femoral cutaneous n.

Fig. 7.69 Sciatic nerve block by a lateral popliteal approach. Last step: injection of local anaesthetic in contact with the posterior cutaneous nerve of the thigh

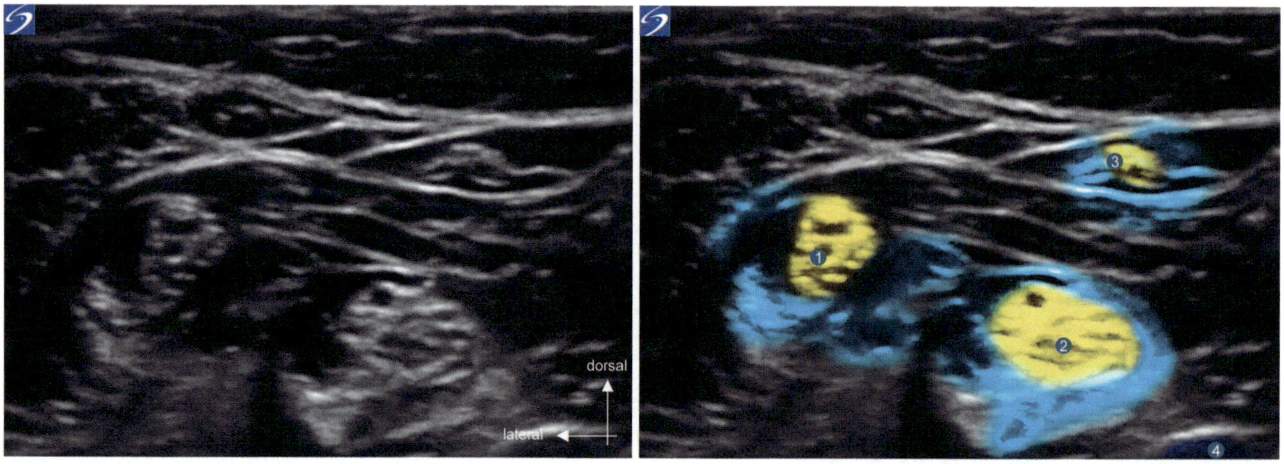

1. common peroneal (fibular) n. 2. tibial n.
3. terminal branch of posterior femoral cutaneous n.
4. popliteal v.

Fig. 7.70 Completed sciatic nerve block by a lateral popliteal approach. Spread of local anaesthetic around the tibial nerve, common peroneal nerve, and posterior cutaneous nerves of thigh

Tibial Nerve Block at the Ankle (Fig. 7.71)

Indications

This distal nerve block may be used along with other distal blocks around the ankle for the primary anaesthetic management of patients undergoing forefoot surgery or as a supplementary (or "rescue") block to enhance post-operative analgesia. It also is possible to insert a perineural catheter to prolong anaesthesia/analgesia.

Type of probe: Linear, 5 to 10 MHz or 6 to 13 MHz. The nerve is often very superficial, a probe with a small footprint is useful.
Axis of the probe: Transversal (Fig. 7.72).
Configuration: Nerve in short axis, needle out of plane.
Depth studied: 0.5 to 3 cm.
Neurostimulation: Enables further identification of the nerves visualised and can limit, by determination of an MIS > 0.3 mA (0.1 ms), the risk of accidental intrafascicular injection. Furthermore, it can supplement locating nerves when conditions of visibility are unfavourable.
Needle: 25 to 50 mm, 22 G.
Utility of Doppler ultrasound: Location of the posterior tibial artery and accompanying veins.

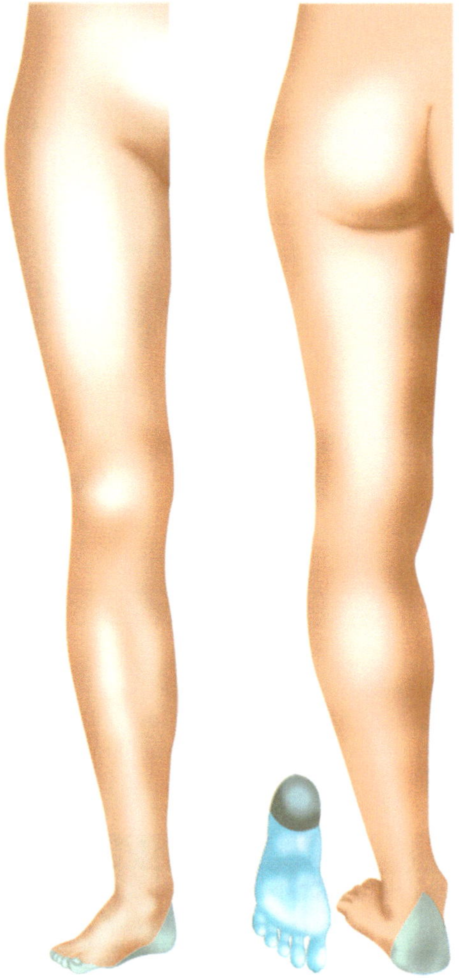

Fig. 7.71 Area of extension of tibial nerve block at the ankle

Echoanatomy

Although the study area is shallow, the nature of the underlying tissues can make it relatively complex in terms of ultrasound identification because it contains many interfaces (fascia, fat, aponeuroses, tendons etc.). It is essential to clearly identify the vascular structures—the posterior tibial artery and veins—to prevent vascular puncture but also to help positively locate the tibial nerve which, by its size, its shallow depth and heterogenous echogenicity of the area, is not always easy to visualise.

Approach with Ultrasound Guidance

The block area is small and shallow, and access to the nerve may be limited. Therefore, it may be preferable to use a high-frequency probe with a small footprint. The probe is positioned transversely 8–10 cm above the medial malleolus, in the continuation of the medial retromalleolar groove. The nerve is approached on the distal side of the probe (Fig. 7.73). Due to the presence of the tibia and the calcanean tendon, located immediately anterior and posterior to the target area respectively, an out-of-plane approach may be easier to perform. Control of the needle tip position can be performed by the coupled use of neurostimulation and hydrolocalisation.

Localisation

The tibial nerve lies medially and ventrally in relation to the flexor hallucis longus muscle/tendon, and usually dorsally in relation to the posterior tibial artery (Fig. 7.74). Note that pressure applied by the probe can inadvertently compress (and so obscure) the veins adjacent to the artery.

Puncture

After skin infiltration with local anaesthesia, the tip of the block needle is advanced towards the tibial nerve, with the aid of hydrolocalisation and, possibly neurostimulation, in the case of poor visualisation of the nerve.

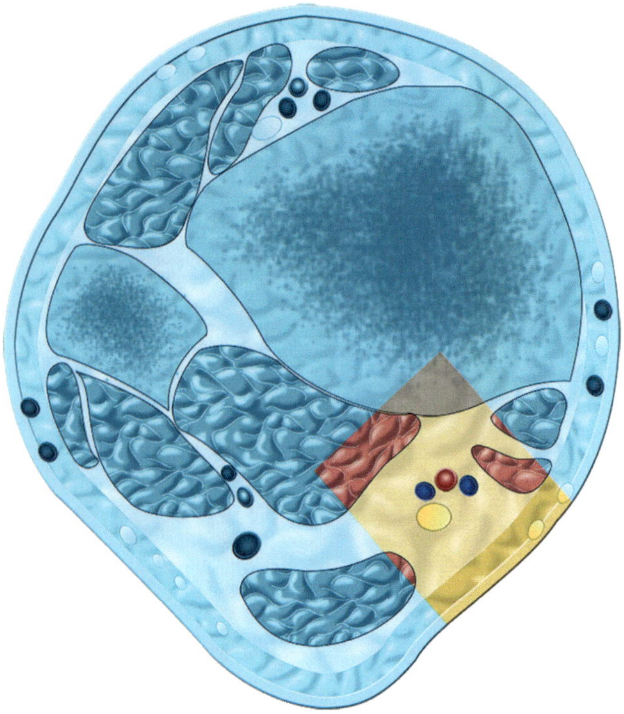

Fig. 7.72 Transverse section above the malleoli with materialisation of ultrasound beam

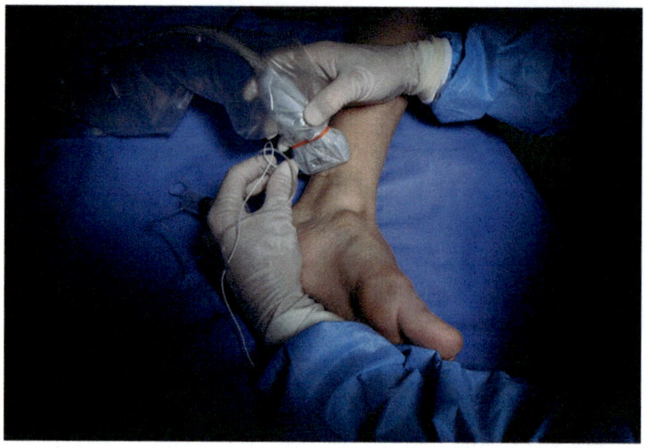

Fig. 7.73 Ultrasound-guided tibial block in ankle: position of probe. Injection out of plane

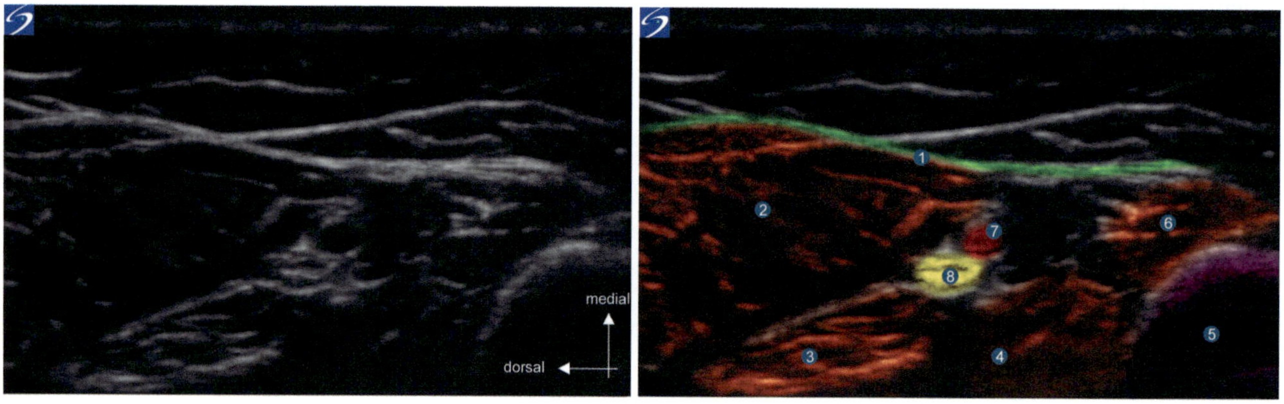

Fig. 7.74 Tibial nerve block at the ankle. Sonoanatomy

For placement of a perineural catheter, the injection technique is identical to that for the single-shot nerve block. After positioning the needle tip beside the tibial nerve, and after confirmation of its correct position by an initial injection through the needle, the catheter is then inserted a few centimetres further to secure its position.

Injection

The local anaesthetic should be injected immediately next to the tibial nerve, seeking to obtain, as far as possible, its circumferential distribution (Fig. 7.75).

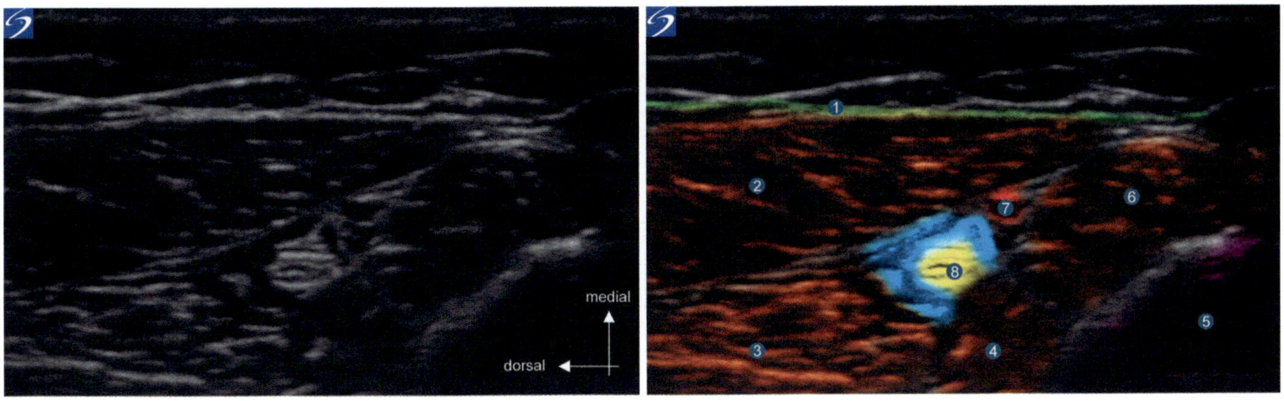

Fig. 7.75 Tibial nerve block at the ankle. Optimal spread of local anaesthetic

> **Paediatrics**
> This very superficial and distal nerve block requires a high frequency/small footprint probe to obtain precise visualisation, even more so in children. A more proximal block (tibial in the mid-leg or popliteal fossa) may be easier to perform.

Nerve Block of the Common Peroneal Nerve along the Biceps Femoris and Neck of the Fibula (Fig. 7.76)

Indications

This is a distal nerve block, used primarily as a supplementary block or combined with a distal tibial block (and possibly sural and saphenous block) for minor foot surgery.

> **Type of probe:** Linear, 5 to 10 MHz or 6 to 13 MHz.
> **Configuration:** Nerve in short or long axis, needle in plane.
> **Studied depth:** 0.5 to 3 cm.
> **Neurostimulation:** Enables further identification of the nerves visualised and can limit, by determination of an MIS > 0.3 mA (0.1 ms), the risk of accidental intrafascicular injection. Furthermore, it can supplement locating nerves when conditions of visibility are unfavourable.
> **Needle:** 25 or 50 mm, 22 G.
> **Utility of Doppler ultrasound:** None.

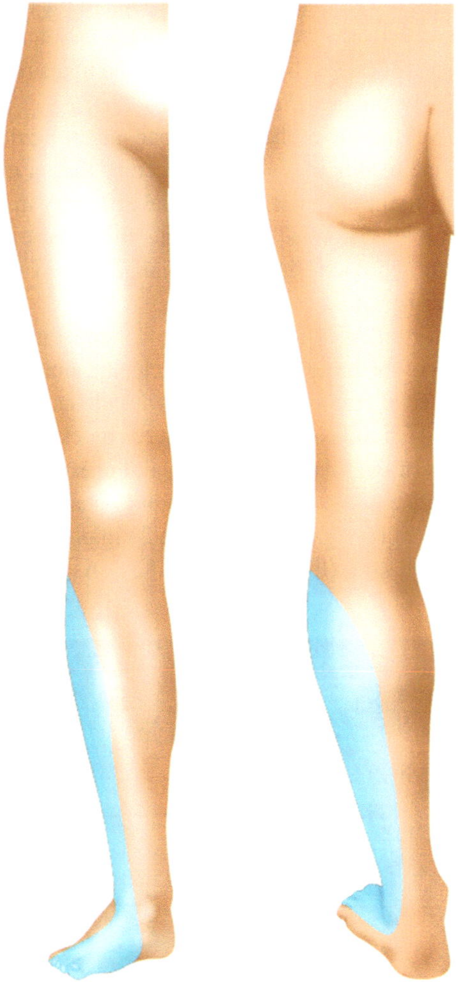

Fig. 7.76 Area of superficial extension of the common peroneal nerve block

Echoanatomy

The common peroneal nerve follows the medial border of the tendon of the biceps femoris muscle, then crosses the dorsal and then the lateral aspect of the lateral head of the gastrocnemius muscle. It then continues its path, lying on the periosteum of the neck of the fibula that it crosses obliquely, below and anteriorly, and finally divides into its two terminal branches.

Approaches and Location

It is possible to approach the common peroneal nerve selectively after its separation from the tibial nerve. This requires a high-frequency probe to obtain a precise image, essential in light of risk of intraneural injection in case of poor visibility of this very superficial nerve at this level. It is possible to block it along the dorsomedial border of the tendon of the biceps femoris muscle, between the flexion fold of the knee and the neck of the fibula (Fig. 7.77) or at the neck of the fibula (a probably more risky technique) (Figs. 7.78 and 7.79).

Whatever the site of the nerve block, inserting the needle out of plane results in a needle trajectory that is more perpendicular to the nerve, with the usual difficulty in identifying the tip of the needle. Inserting the needle in plane enables an approach that is more tangential to the nerve, which may be safer.

Thus, for a nerve block along the dorsomedial border of the tendon of the biceps femoris muscle, the aim is to visualise the nerve in **transverse** section, with the needle in plane (Fig. 7.80). On the contrary, if the nerve block is performed at the neck of the fibula where the nerve is extremely superficial and adherent to the periosteum, its visualisation in **longitudinal** section, again with the needle in plane, may make this approach safer (Figs. 7.81 and 7.82). (As with the distal tibial nerve block it may be necessary to use a small footprint probe due to access issues in this confined area.)

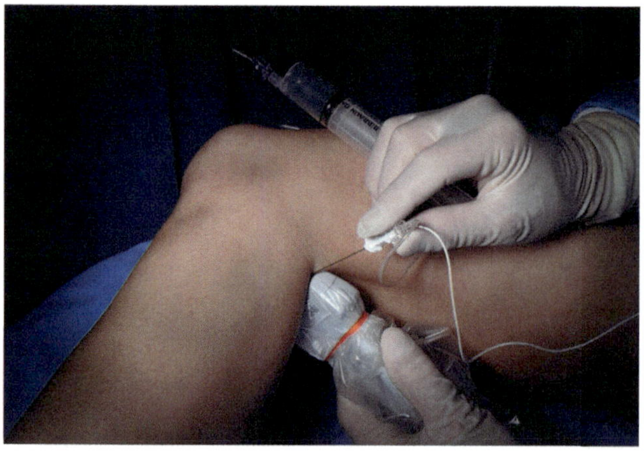

Fig. 7.77 Common peroneal nerve block along biceps femoris muscle

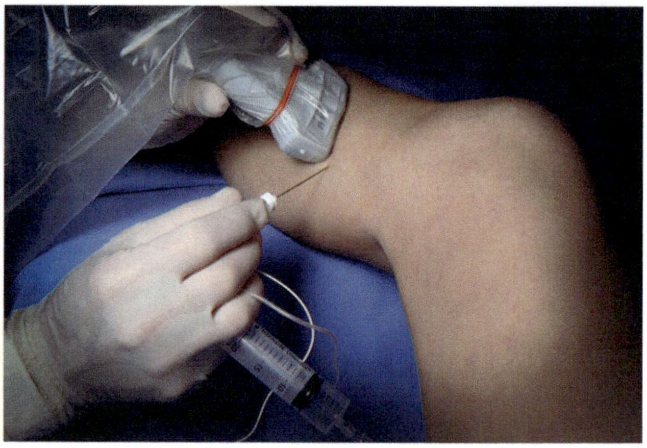

Fig. 7.78 Common peroneal nerve block at the neck of the fibula: injection in-plane by a cephalic approach

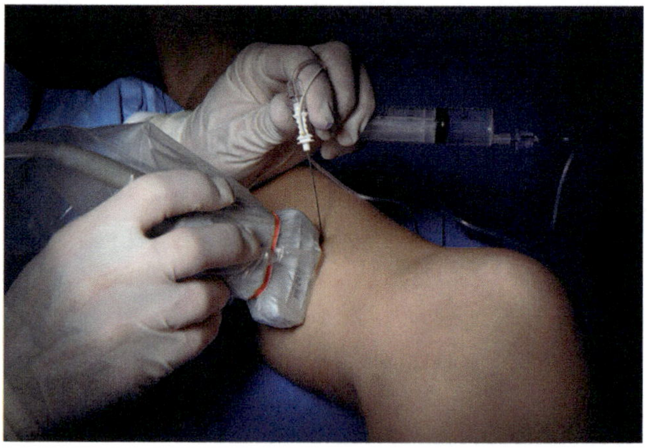

Fig. 7.79 Common peroneal nerve block at the neck of the fibula; needle in-plane by a caudal approach

Injection

Possibly aided by neurostimulation, the slow and fractionated injection of local anaesthetic makes it possible to limit the risk of an intrafascicular injection. Also, this risk is reduced by closely visualising the appearance of the nerve during injection and acknowledging any increase in size of the nerve. These signs are especially important to look for when, as is the case at the neck of the fibula, a nerve is relatively “fixed” in its position.

Paediatrics
This block might be avoided in paediatrics in light of the risks of nerve injury.

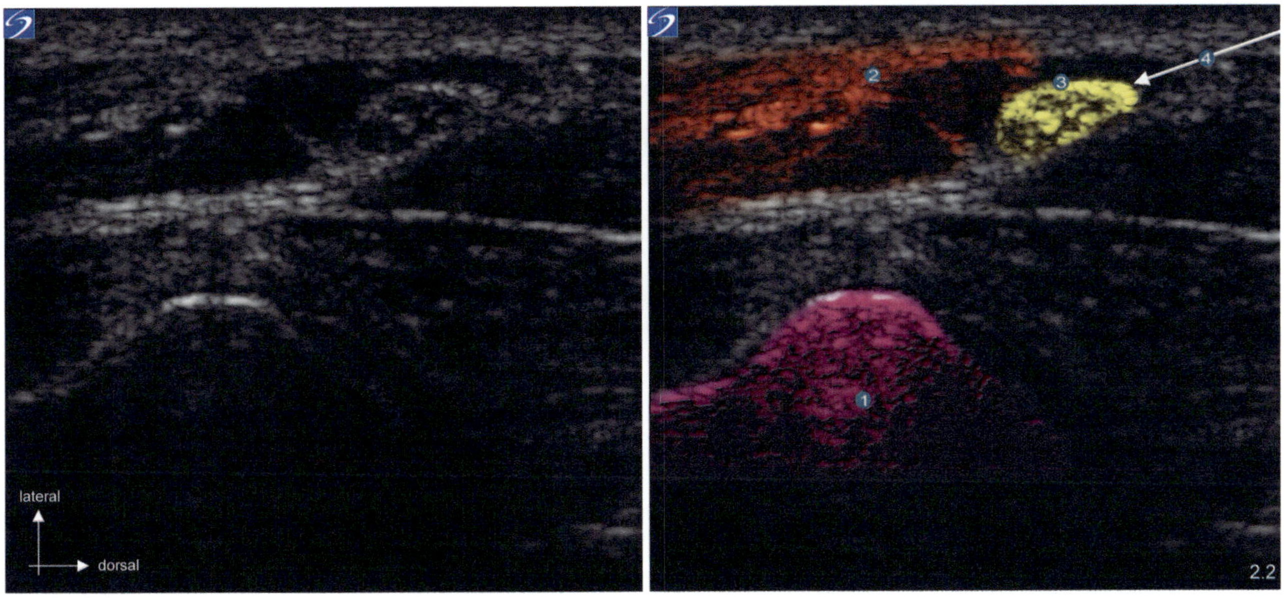

1. lateral condyle of femur
2. biceps femoris m.
3. common peroneal (fibular) n.
4. needle progression

Fig. 7.80 Transverse ultrasound section of the common peroneal nerve, along the dorsomedial border of the biceps femoris muscle

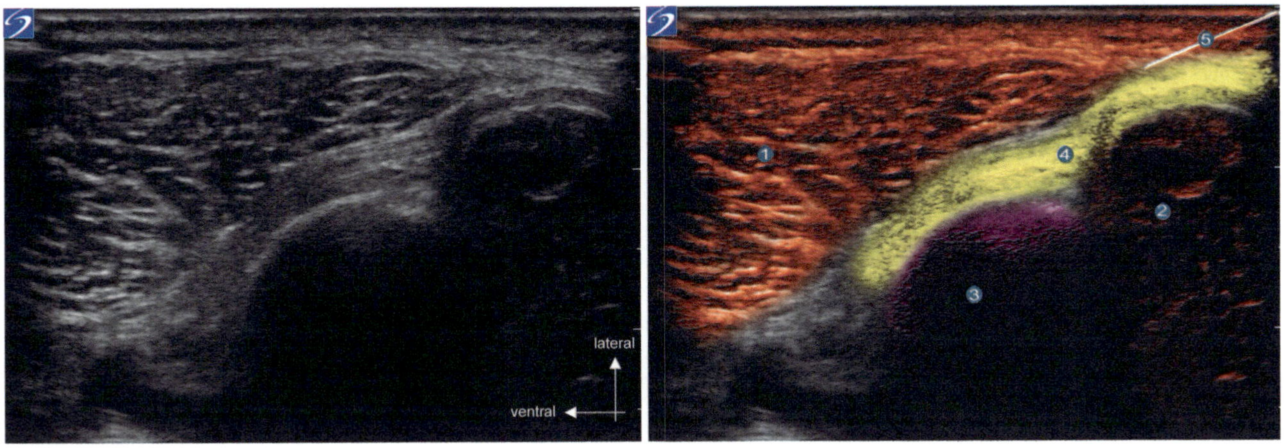

1. peroneus longus m.
2. gastrocnemius m. (lateral head)
3. fibula
4. common peroneal (fibular) n.
5. needle progression

Fig. 7.81 Common peroneal block at the neck of fibula. Longitudinal approach to the nerve, in plane, by a cephalic approach

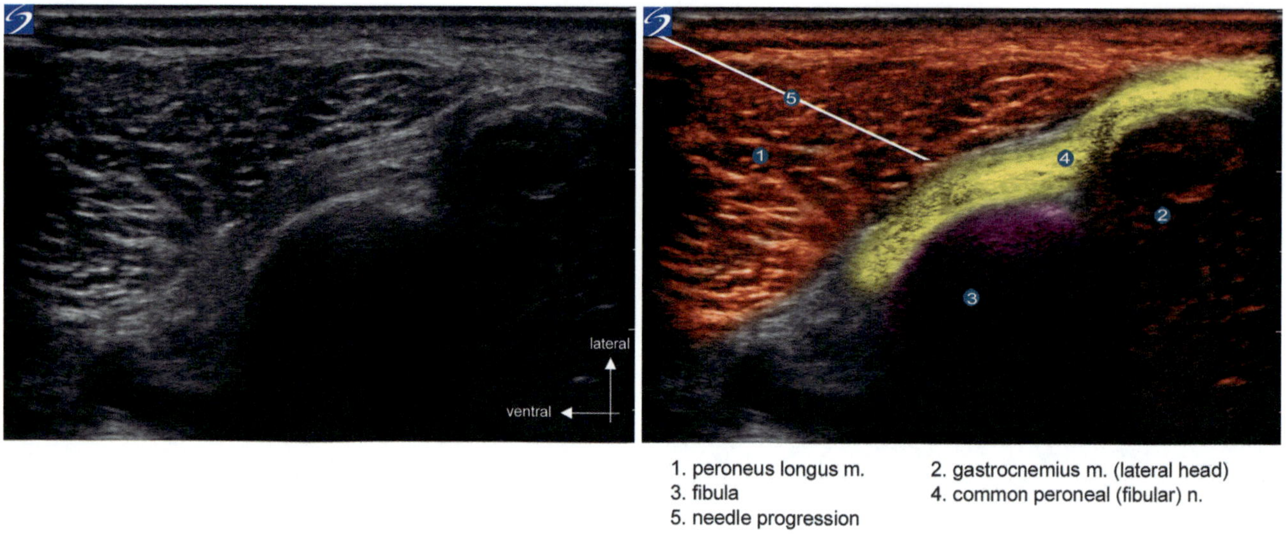

Fig. 7.82 Common peroneal block at the neck of fibula. Longitudinal approach to the nerve, in plane, by a caudal approach

References

1. Rouvière H, Delmas A. Anatomie humaine, descriptive, topographique et fonctionnelle. Tome III. Paris: Masson; 1984.
2. Beaton LE, Anson BJ. The relation of the sciatic nerve and of its subdivisions to the piriformis muscle. Anat Rec. 1937;70:1–5.
3. Chiby S. Multiple positional relationships of nerves arising from the sacral plexus to the piriformis muscle in humans. Kaibogaku-Zasshi. 1992;67:691–724.
4. Parsons FG, Keith A. Sixth annual report of the Committee of Collective Investigation of the anatomical Society of Great Britain and Ireland. J Anat Physiol. 1897;31:31–44.
5. Pokorny D, Jahoda D, Veigl D, Pinskerova V, Sosna A. Topographic variations of the relationship of the sciatic nerve and the piriformis muscle and its relevance to palsy after total hip arthroplasty. Surg Radiol Anat. 2006;28:88–91.
6. Cuvillon P, Ripart J, Jeannes P, et al. Comparison of the parasacral approach and the posterior approach, with single- and double-injection techniques, to block the sciatic nerve. Anesthesiology. 2003;98:1436–41.
7. Morris GF, Lang SA. Continuous parasacral sciatic nerve block: two case reports. Reg Anesth. 1997;22:469–72.
8. Gaertner E, Lascurain P, Venet C, et al. Continuous parasacral sciatic block: a radiographic study. Anesth Analg. 2004;98:831–4.
9. Mansour NY, Bennetts FE. An observational study of combined continuous plexus and single shot sciatic nerve blocs for post-knee surgery analgesia. Reg Anesth. 1996;21:287–91.
10. O'Connor M, Coleman M, Wallis F, Harmon D. An anatomical study of the parasacral block using magnetic resonance imaging of healthy volunteers. Anesth Analg. 2009;108:1708–12.
11. Ben-Ari AY, Joshi R, Uskova A, Chelly JE. Ultrasound localization of the sacral plexus using a parasacral approach. Anesth Analg. 2009;108:1977–80.
12. Bendtsen TF, Lönnqvist PA, Jepsen KV, Petersen M, Knudsen L, Børglum J. Preliminary results of a new ultrasound-guided approach to block the sacral plexus: the parasacral parallel shift. Br J Anaesth. 2011;107:278–80.
13. Taha AM. A simple and successful sonographic technique to identify the sciatic nerve in the parasacral area. Can J Anaesth. 2012;59:263–7.
14. Dillow JM, Rosett RL, Petersen TR, Vagh FS, Hruschka JA, Lam NC. Ultrasound-guided parasacral approach to the sciatic nerve block in children. Paediatr Anaesth. 2013;23:1042–7.
15. Mansour NY. Reevaluating the sciatic nerve block. Another landmark for consideration. Reg Anesth. 1993;18:322–3.
16. Raj PP, Parks RI, Watson TD, Jenkins MT. A new single-position supine approach to sciatic-femoral nerve block. Anesth Analg. 1975;54:489–93.
17. Di Benedetto P, Casati A, Bertini L, Fanelli G. Posterior subgluteal approach to block the sciatic nerve: description of the technique and initial clinical experiences. Eur J Anaesthesiol. 2002;19:682–6.
18. Chan VWS, Nova H, Abbas S, McCartney CJL, Perlas A. Xu Da Quan. Ultrasound examination and localization of the sciatic nerve: a volunteer study. Anesthesiology. 2006;104:309–14.
19. Karmakar MK, Kwok WH, Ho AM, Tsang K, Chui PT, Gin T. Ultrasound-guided sciatic nerve block: description of a new approach at the subgluteal space. Br J Anaesth. 2007;98:390–5.
20. Paqueron X, Bouaziz H, Macalou D, et al. The lateral approach to the sciatic nerve at the popliteal fossa: one or two injections? Anesth Analg. 1999;89:1221–5.
21. Hadzic A, Vloka J. A comparison of the posterior versus lateral approaches to the block of the sciatic nerve in the popliteal fossa. Anesthesiology. 1998;88:1480–6.
22. Andersen HL, Andersen SL, Tranum-Jensen J. Injection inside the perineural sheath of the sciatic nerve: direct comparison among ultrasound imaging, macroscopic anatomy, and histologic analysis. Reg Anesth Pain Med. 2012;37:410–4.
23. Missair A, Weisman RS, Suarez MR, Yang R, Gebhard R. A 3-dimensional ultrasound study of local anesthetic spread during lateral popliteal nerve block. What is the ideal end point for needle tip position? Reg Anesth Pain Med. 2012;37:627–32.
24. Sala Blanch X, de Riva N, Carrera A, López AM, Prats A, Hadzic A. Ultrasound-guided popliteal sciatic block with a single injection at the sciatic division results in faster block onset than the classical nerve stimulator technique. Anesth Analg. 2012;114:1121–7.
25. Perlas A, Wong P, Abdallah F, Hazrati LN, Tse C, Chan V. Ultrasound-guided popliteal block through a common perineural sheath versus conventional injection: a prospective, randomized, double-blind study. Reg Anesth Pain Med. 2013;38:218–25.

Cervical Plexus Blocks

8

Eryk Eisenberg, Elisabeth Gaertner, Philippe Clavert, and Olivier Choquet

Contents

Cervical plexus blocks are appropriate for neck surgery but are underused [1]. It may be that, as in many other types of nerve blocks using landmark-based approaches, the lack of precision was associated with block failure or an unacceptable level of risk of complications in this area. However, recently developed ultrasound-guided approaches are reliable and efficient [2].

E. Eisenberg (✉)
Département d'anesthésie et réanimation, Clinique Pôle Santé République, Clermont-Ferrand, France

E. Gaertner
Département d'anesthésie et réanimation chirurgicale, Hôpitaux Civils de Colmar, Colmar, France

P. Clavert
Service de chirurgie du membre supérieur, Centre de Chirurgie Orthopédique et de la Main (CCOM), Illkirch, France

O. Choquet
Département d'anesthésie-réanimation A, CHU de Montpellier, Hôpital Lapeyronie, Montpellier, France

Anatomy

Nerve Structures

Each cervical spinal nerve divides at its emergence into a dorsal and a ventral branch.

The **cervical plexus** is composed of **ventral branches** of the first 4 cervical spinal nerves; they emerge between the middle scalene, the levator scapulae muscle dorsally, and the longus capitis muscle ventrally. The ventral branch of C1 anastomoses with that of C2, circumscribing the atlas, thus forming the anastomotic arch of the atlas. The ventral branches of C2, C3, and C4 anastomose with each other by their ascending and descending branches, respectively, successively comprising the anastomotic arch of the Axis (loop of the Axis), and then a third arch joining C3 to C4. C4 also participates in formation of the brachial plexus by its anastomosis with the ventral branch of C5. Each nerve anastomoses with the sympathetic nervous system by one or two communicating branches that join the cervical sympathetic ganglion.

From this complex network, different sensory motor nerves arise and form the superficial cervical plexus and the deep cervical plexus (Fig. 8.1).

E. Eisenberg, E. Gaertner (eds.), *Ultrasound in Peripheral, Neuraxial and Perineuraxial Regional Anaesthesia*, https://doi.org/10.1007/978-3-031-08804-9_8

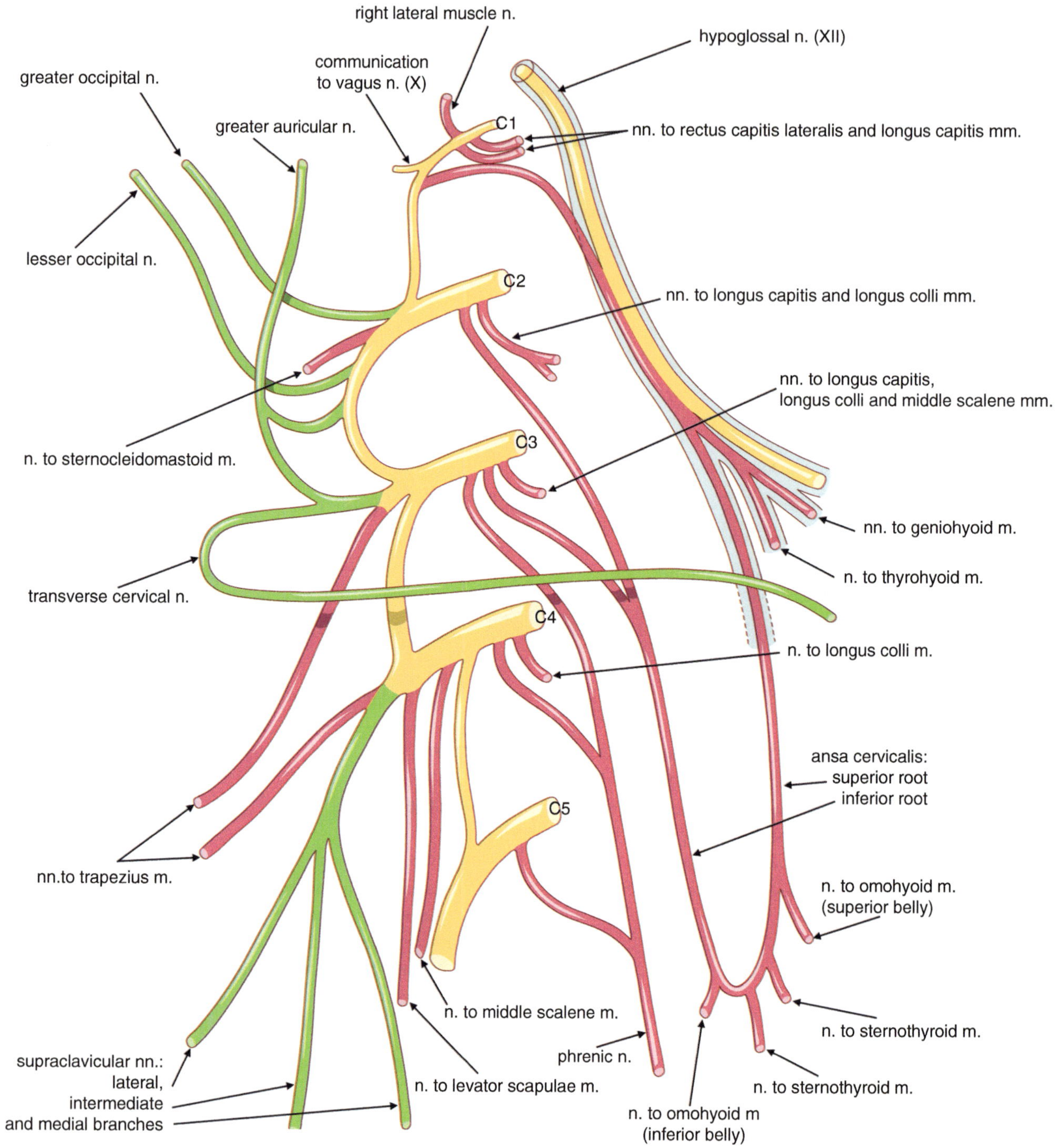

Fig. 8.1 Cervical plexus: overall organisation

The **dorsal cervical nerves** arise from the dorsal branches, especially the sub-occipital nerve (first branch of dorsal cervical nerve), the greater occipital nerve (second branch of cervical dorsal nerve), and third occipital nerve (third dorsal cervical branch).

The **deep cervical plexus includes** the roots, the anastomotic arches, and the motor branches. It ensures motor innervation of the majority of the muscles of the neck (sternocleidomastoid, levator scapulae, anterior and middle scalenes, trapezius, geniohyoid, thyrohyoid, sternohyoid, sternothyroid, omohyoid muscles) and roots C3 and C4 give rise to the phrenic.

The **superficial cervical plexus** is a purely sensory plexus; it provides sensation to the area of the neck, the shoulders, and the posterior part of the scalp in the manner of a nun's habit [3]. The area of sensory distribution is always subject to potential variations and overlaps with the adjacent territories (Figs. 8.2 and 8.3). It is composed of the following:

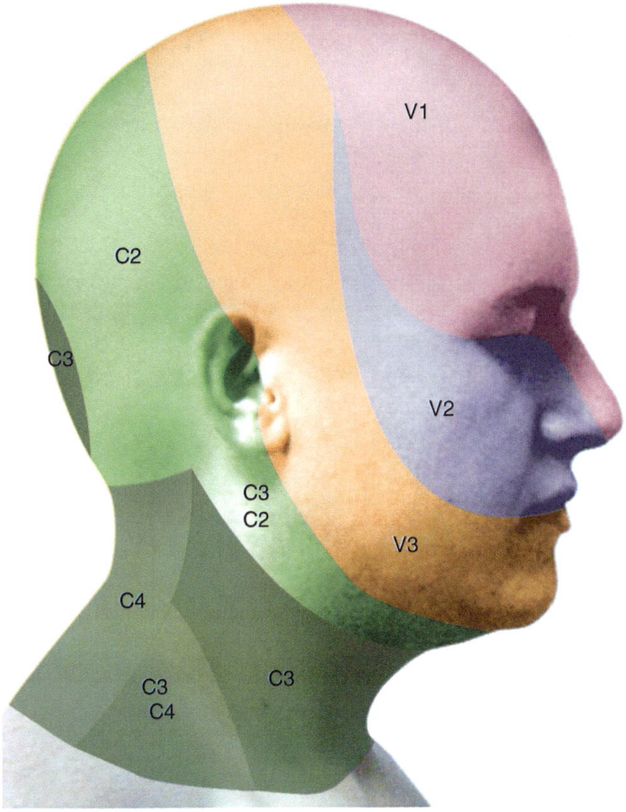

Fig. 8.2 Dermatomes of the head and neck

- The lesser occipital nerve (C2-C3): sensation of the skin of the lateral occipital area.
- The great auricular nerve (C2-C3): sensation of the skin of the mastoid, parotid area, and the outer ear.
- The transverse cervical nerve (C2-C3): innervates the skin of the supra- and subhyoid areas.
- The supraclavicular nerve (C3-C4): innervates the skin of the supra- and subclavicular area as well as that of the tip of the shoulder.

These four nerves pass around the posterior border of the sternocleidomastoid muscle in its middle part and emerge at a common point (Erb's point) to join the superficial fascia (Fig. 8.4).

Fascia and Planes of the Neck (Figs. 8.5 and 8.6)

The **cervical fascia** (or deep cervical fascia) demarcates the connective tissue spaces enabling swallowing and movements of the neck. It consists of three layers divided into three planes [4]:

- The superficial layer.
- The pretracheal layer.
- The prevertebral layer.

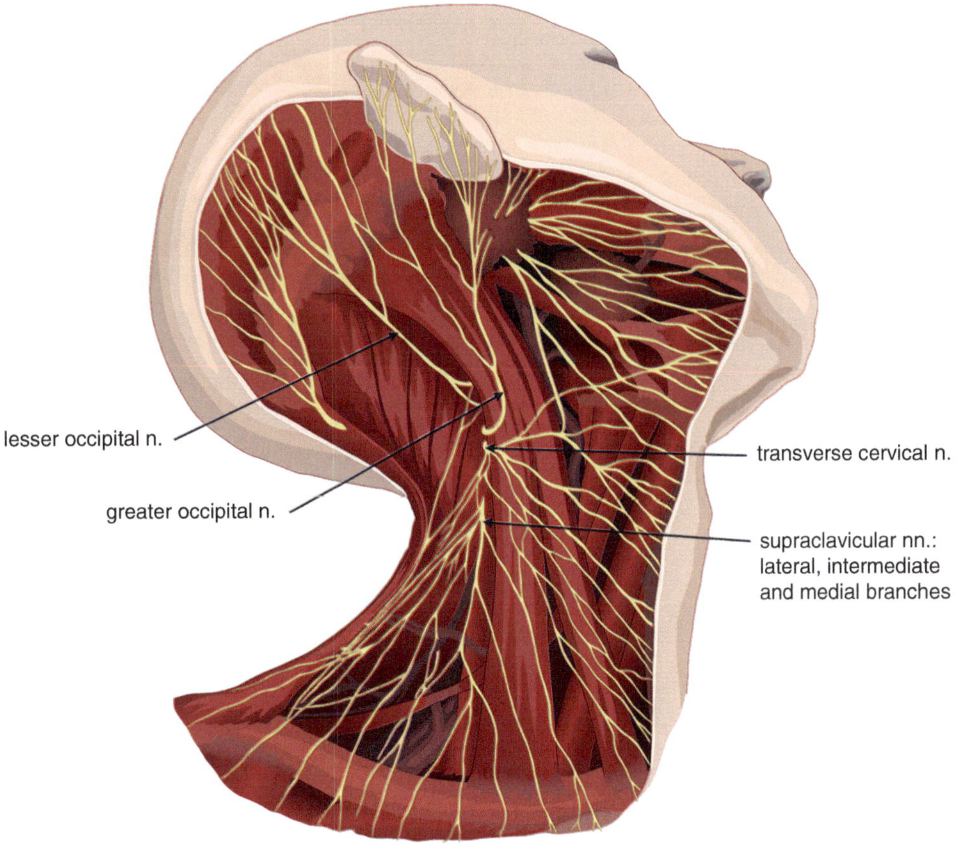

Fig. 8.3 Regional distribution of the branches of superficial cervical plexus

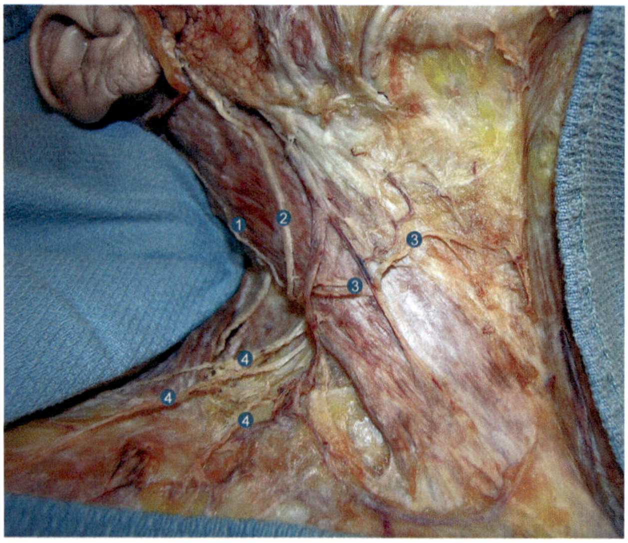

1. Lesser occipital n.
2. greater occipital n.
3. transverse cervical n.
4. supraclavicular nn.

Fig. 8.4 Emergence of nerves of superficial cervical plexus at Erb's point. Admir Hadzic iconography

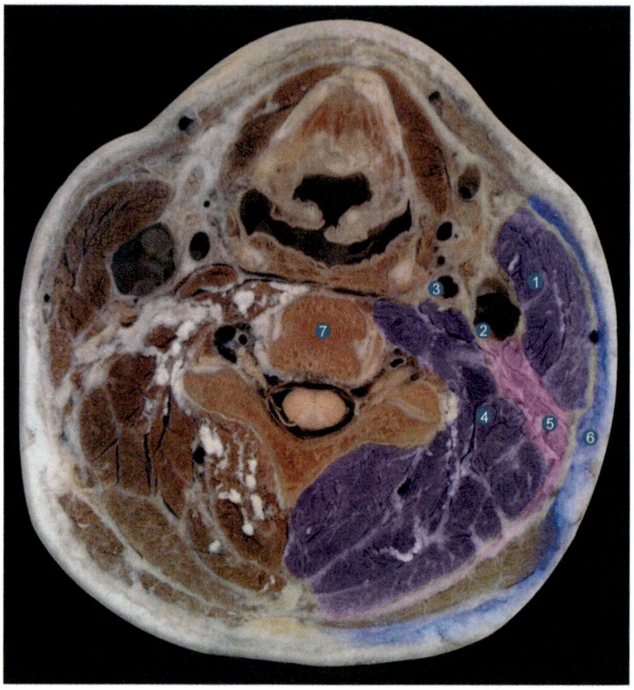

1. sternocleidomastoid m.
2. internal jugular v.
3. carotid a.
4. paravertebral space
5. posterior cervical space
6. subcutaneous space
7. C4 vertebra

Fig. 8.5 Anatomical section passing through vertebra C4

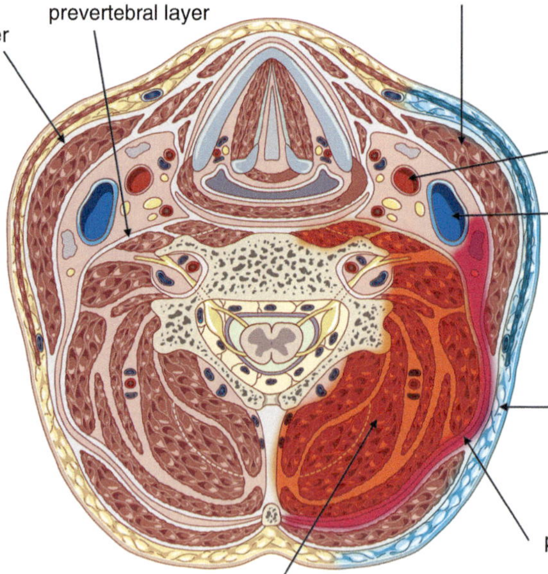

Fig. 8.6 Transverse section of the neck immediately upstream (caudal) of the carotid bifurcation (C4/C5 level)

The **superficial layer**, which fuses ventrally with its contralateral opposite number, is located under the skin and the platysma, extends from the mandible to the clavicle, and surrounds the sternocleidomastoid muscle and the trapezius muscle.

The **pretracheal layer** extends from the hyoid bone to the clavicle. It has two components:

- The **muscular** layer, which surrounds the infra-hyoid muscle.
- The **visceral** layer, which surrounds the viscera of the neck.

The **prevertebral layer** adheres to the anterior longitudinal ligament in the ventral aspect of the vertebral body, covers the prevertebral, scalene, levator scapulae, and rhomboid minor muscles. Posteriorly, it blends with the ligamentum nuchae in the midline.

The three layers described demarcate three anatomical spaces, important in locoregional anaesthesia (LRA):

- The subcutaneous space, which is **superficial**.
- The paravertebral space, which is **deep**.
- The posterior cervical space, which is **intermediate**.

The **subcutaneous space** lies between the dermis and the superficial layer of the cervical fascia. Injection of local anaesthetic in this space enables a superficial block of the **superficial cervical plexus**.

The paravertebral space in the periphery is contained by the prevertebral plate and extends from C2 to T1 [5]. Injection of a local anaesthetic spreads within it in a longitudinal manner and produces a deep nerve block of the cervical plexus.

The **posterior cervical space** superficially is limited by the superficial layer of cervical fascia, and more deeply by the prevertebral layer, in relation to the anterior and middle scalene, levator scapulae muscles medially, sternocleidomastoid muscle laterally, and trapezius muscles laterally and dorsally. It is crossed by the cutaneous branches of the cervical plexus and contains blood and lymphatic vessels. Injection of the local anaesthetic at this site defines an **intermediate** cervical block. It produces a superficial block of the cervical plexus; this space communicates with the paravertebral space which perhaps explains why this block may be sufficient for carotid surgery. Spread of local anaesthetic within this space will result in a block of the phrenic nerve.

Indications

- Carotid surgery (endarterectomy, stenting) is the principal indication of a superficial cervical plexus block [6]. It can be used as an alternative to general anaesthesia, thus enabling perioperative neurological monitoring [7]. In other cases (complicated surgery, an anxious patient etc.), it is used in combination with general anaesthesia: it promotes perioperative haemodynamic stability under general anaesthesia (perhaps less deep), as well as excellent post-operative analgesia. Occasionally, supplementary intraoperative perivascular injection of local anaesthetics may be necessary, in particular after a **subcutaneous** superficial cervical block. An **intermediate** ultrasound-guided cervical block seems to be as effective as a **subcutaneous** superficial cervical plexus block.
- Thyroid and parathyroid surgery can be performed with bilateral cervical block. This type of surgery is most often performed under general anaesthesia, with the combined block to ensure post-operative analgesia [8, 9].
- Other indications for this type of block include post-operative analgesia after cervical spine surgery or superficial soft tissue surgery (e.g. excisions and biopsies).
- A superficial cervical plexus block is also performed in addition to an interscalene block, when sensory testing finds insufficient spread of the anaesthetic to the skin medially, to the deltopectoral groove or to the tip of the shoulder. During painful procedures on the laterocervical area, it is useful to perform this superficial cervical plexus block in order to avoid multiple injections of a local anaesthetic (e.g. during placement of central venous catheters).
- Clavicle surgery benefits from analgesia obtained by a superficial cervical plexus block together with an interscalene block [10]. A clinical case of superficial block has been described recently for analgesia in a very painful clavicular fracture in the emergency department.
- In the ER, this block is useful for anaesthesia of wounds of the earlobe, superficial wounds of the neck area, and submandibular abscesses [11, 12].
- Lastly, bilateral block of the cervical plexus produces analgesia equivalent to morphine administered following anaesthesia, and to remifentanil in patients undergoing infratentorial or occipital craniotomy [13].

Type of probe: High frequency linear, 5 to 10 MHz or 6 to 13 MHz.
Axis of the probe: Transversal, across the sternocleidomastoid muscle (Fig. 8.7).
Configuration: Needle in plane, inserted at the posterior end of the probe (Fig. 8.15).
Depth studied: Up to about 5 cm depending on patient build.
Neurostimulation: enables confirmation of the existence of branches of the superficial cervical plexus by triggering electrical paraesthesiae in the relevant distribution.
Needle: 50 to 80 mm, isolated, 22 G.
Utility of Doppler ultrasound: Enables visualisation of the carotid artery, the internal and external jugular veins (often compressed by pressure of the probe), but also the vertebral vessels (Figs. 8.8, 8.9, and 8.10), as well as other vessels that can lead into the posterior cervical space of the paravertebral space.

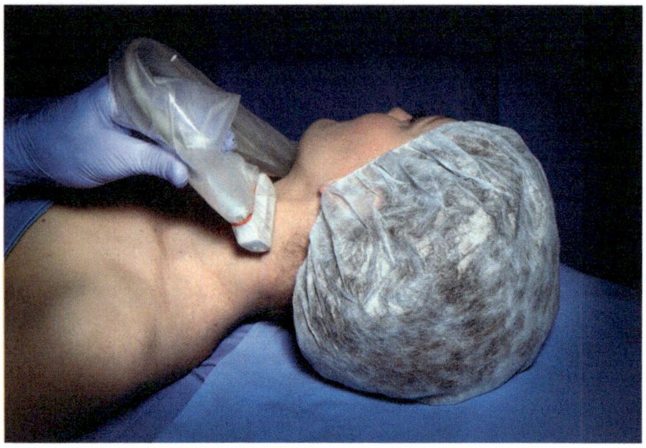

Fig. 8.7 Cervical plexus block. Position of the probe

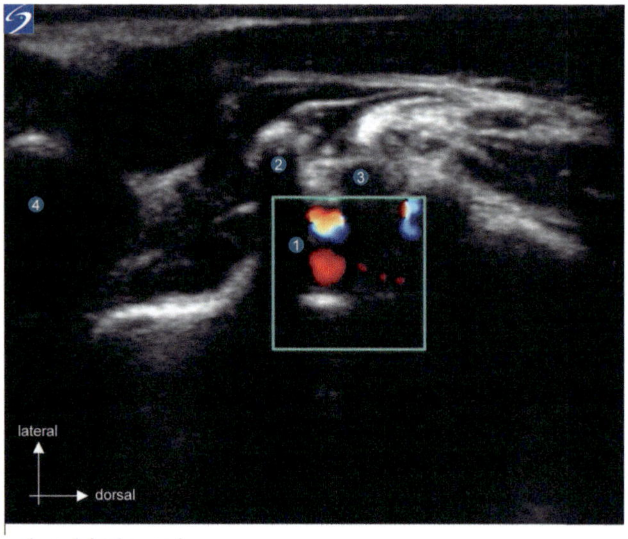

1. vertebral a. and v.
2. anterior tubercles of transverse process of C5 vertebra
3. posterior tubercles of transverse process of C5 vertebra
4. common carotid a.

Fig. 8.8 Vertebral blood vessels passing through the transverse foramen of C5

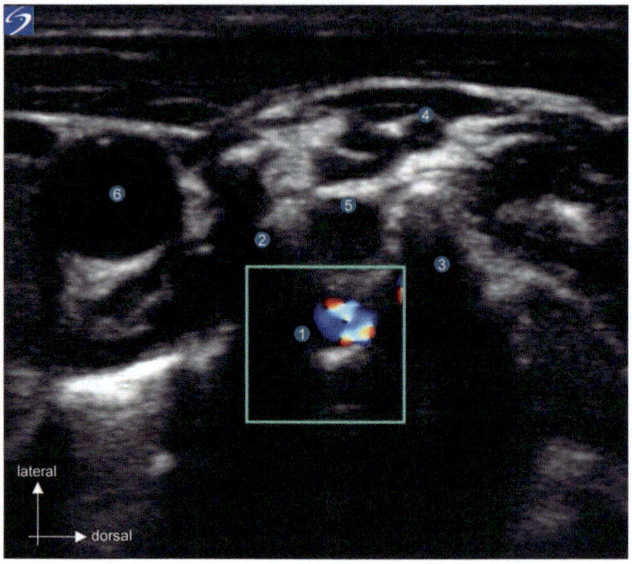

1. vertebral a.
2. anterior tubercles of transverse process of C6 vertebra
3. posterior tubercles of transverse process of C6 vertebra
4. C5 n. root
5. C6 n. root
6. common carotid a.

Fig. 8.9 Vertebral artery visible through the transverse foramen of C6

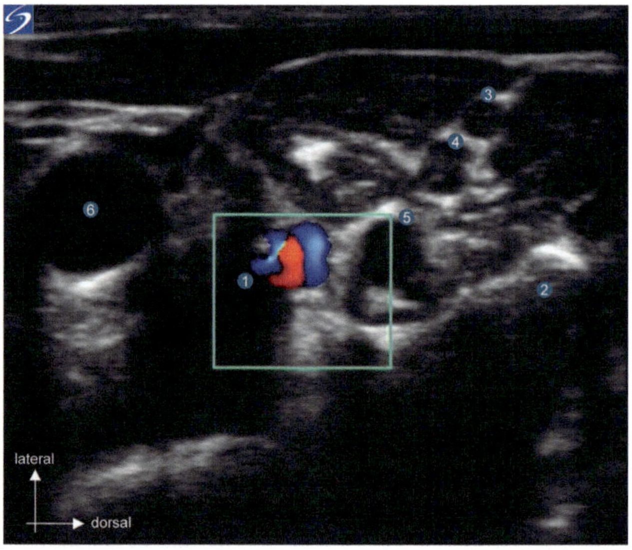

1. vertebral a. and v.
2. posterior tubercles of transverse process of C7 vertebra
3. C5 n. root
4. C6 n. root
5. C7 n. root
6. common carotid a.

Fig. 8.10 Vertebral vessels visible anterior to the transverse process of C7 (which does not have a transverse foramen); note root of C7 positioned immediately behind the vertebral vessels, possible source of confusion in the absence of use of Doppler ultrasound

Sonoanatomy and Location

The patient is supine with the head turned away from the side to be blocked. This reveals the posterior border of the sternocleidomastoid muscle (Fig. 8.7). The operator places himself on the side of the nerve block to be performed, with the ultrasound system facing opposite.

A high-frequency linear probe is placed transversely over the sternocleidomastoid muscle, at mid-point of its lateral border. Ventro-dorsal movement of the probe makes it possible to identify the principal anatomical components of the area:

- Ventrally, the trachea and blood vessels (the pulsatile common carotid artery and the internal/external jugular veins, although the latter may be compressed by pressure of the probe).
- Deeper, the anterior and middle scalene muscles.
- Superficially, the sternocleidomastoid muscle and its posterior border.

Craniocaudal movements of the probe enable observation of the cervical roots lying between the scalene muscles, at variable depths depending on their situation. It is possible to see the transverse processes of the cervical vertebrae from which they emerge. It is desirable to locate the vertebral artery: it passes through the successive transverse foramina between C6 and Atlas. It is generally visible with Doppler ultrasound, and relatively often with B mode ultrasound (Figs. 8.8, 8.9, and 8.10). The aim is to locate the bifurcation of the carotid artery in order to identify the vertebral level C4 (Fig. 8.11), the approximate level at which the nerves of the superficial cervical plexus emerge from the posterior border of the sternocleidomastoid muscle (Erb's point). A more in-depth examination of this area can locate the phrenic nerve whose principal origin is C4 (Fig. 8.12). It detaches from the ventral branch of C4 and continues on the surface of the anterior scalene muscle, whose lateral aspect it crosses and then the ventral aspect, as it runs caudally. It is often easier to locate distally and to follow it in the cranial direction up to the root of C4.

At the level of vertebra C4, it is generally possible to differentiate the branches of the superficial cervical plexus, in the form of a collection of hypoechoic rounded structures which continue into the connective tissue environment of the posterior cervical space. The large auricular nerve emerges from the posterior border of the sternocleidomastoid muscle, passing around it in an almost transverse plane. This makes it possible to obtain a characteristic ultrasound image [14] since it is visualised simultaneously on the surface of and deep into the sternocleidomastoid muscle (Figs. 8.13 and 8.14).

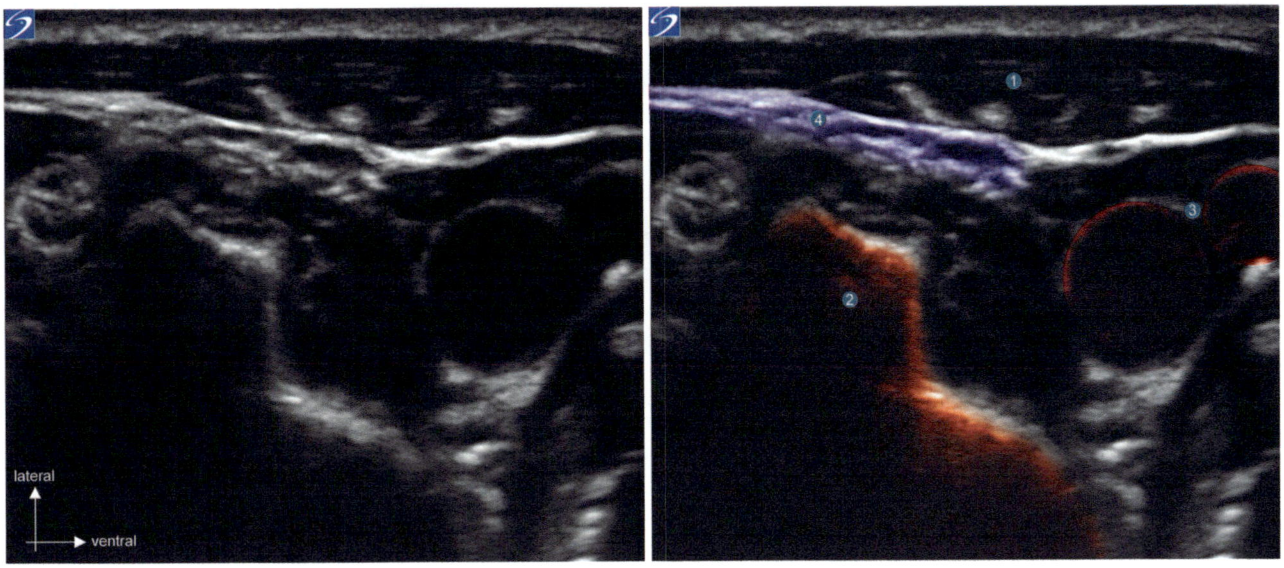

Fig. 8.11 Ultrasound section at the level of the C4 vertebra: carotid bifurcation

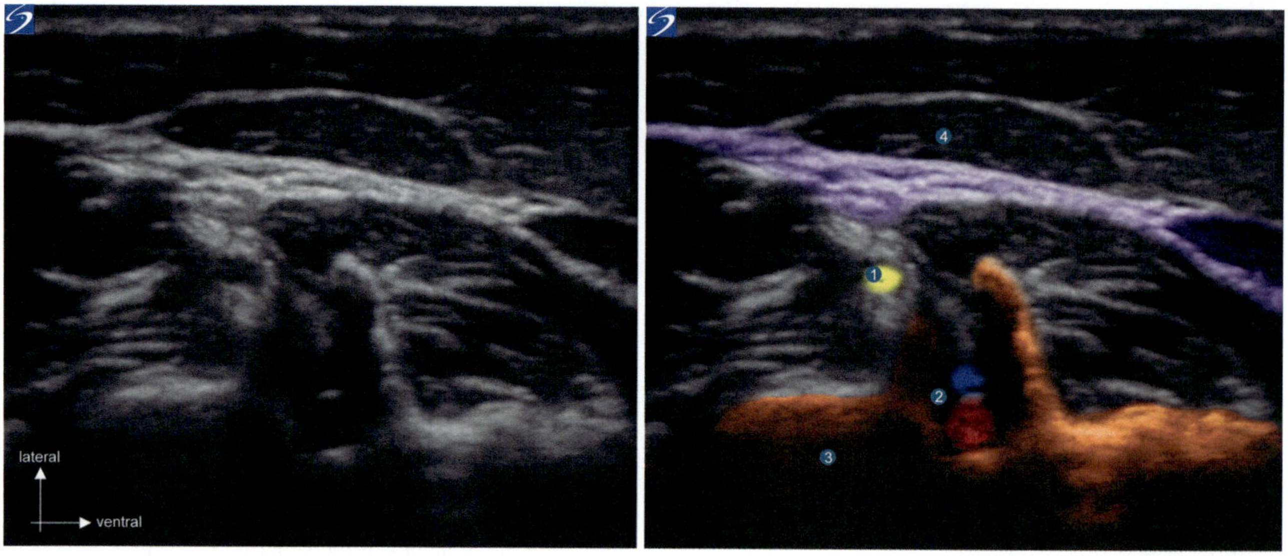

Fig. 8.12 Emergence of root of C4 at the level of the C4 transverse process

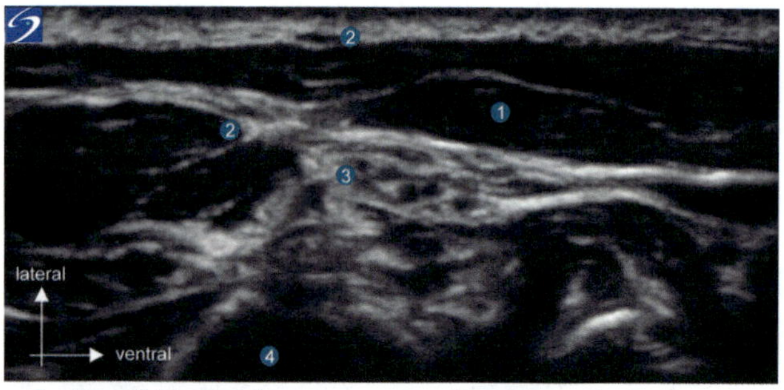

Fig. 8.13 Superficial cervical plexus in the posterior cervical space

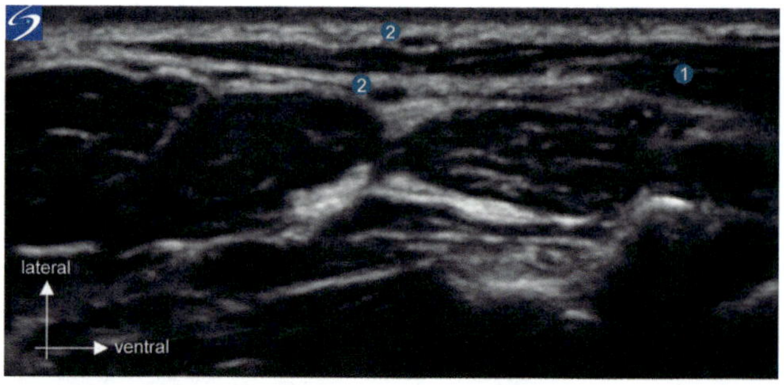

Fig. 8.14 Characteristic image of greater auricular nerve in proximity to Erb's point

Needle Puncture and Injection

Once the probe has been stabilised, anaesthesia of the skin is performed at its posterolateral end for an in-plane approach (Fig. 8.15).

- For a **superficial approach** to the cervical plexus:
  10 mL of local anaesthetic injected between the platysma muscle and the superficial layer of the cervical fascia (i.e. in subcutaneous position) are sufficient to block the 4 cutaneous sensory nerves (lesser occipital, greater auricular, transverse cervical, and supraclavicular).
- For an **intermediate approach** to the cervical plexus:
  The needle is inserted in the ultrasound plane, it crosses the superficial layer of the cervical fascia to the posterior border of the sternocleidomastoid muscle. Injection of the local anaesthetic then is started; the needle advances whilst hydrodissection is performed with the aid of the local anaesthetic in the posterior cervical space between the sternocleidomastoid muscle (superficially) and the prevertebral layer covering the anterior and middle scalene muscles (deeper) (Figs. 8.16 and 8.17). The injection can be continued until the needle tip is adjacent to the carotid sheath (and even up to the pericarotid area if echogenicity is favourable) (Fig. 8.18).

Injection of a total volume of 10 mL of local anaesthetic is sufficient.

- The **deep approach** to the cervical plexus possibly risks a higher rate of complications and should probably be avoided.
  The needle is inserted in plane. The local anaesthetic is injected into the superficial portion of the interscalene pathway, immediately under the prevertebral layer [15], at the level of root C4 (Fig. 8.19). Care should be taken to

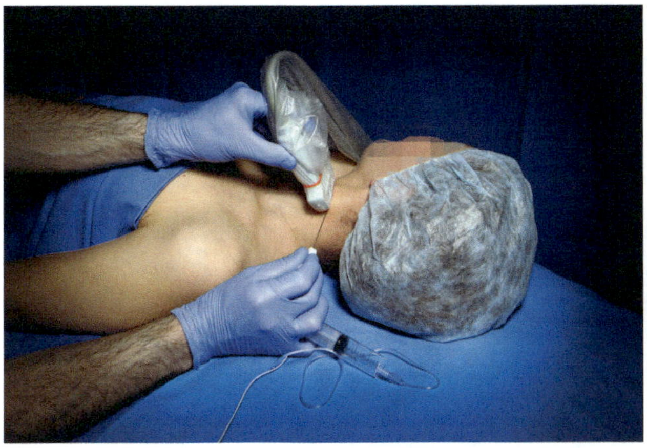

Fig. 8.15 Approach to the superficial cervical plexus in plane

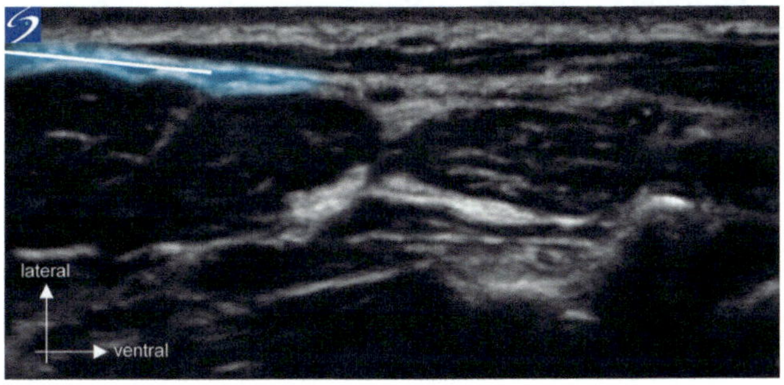

Fig. 8.16 Hydrodissection during progression of the needle for an intermediate cervical nerve block: stage 1

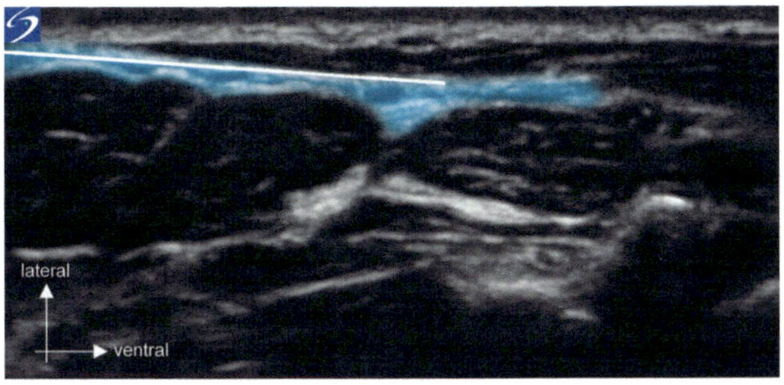

Fig. 8.17 Hydrodissection during progression of the needle for an intermediate cervical nerve block: stage 2

not inject in immediate proximity to emergence of the nerve root in order to avoid epidural or intrathecal spread. A maximum volume of 5–7 ml is injected.

Specific Points

Carotid Surgery

Hydrodissection performed during an intermediate block of the cervical plexus literally "opens up" access to the carotid sheath which is innervated by branches of cranial nerves X and XI. When the indication is precisely carotid surgery, it is

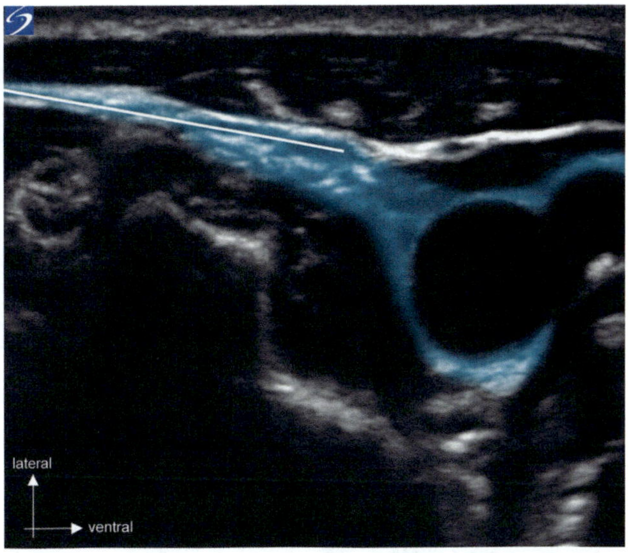

Fig. 8.18 Hydrodissection during progression of the needle for an intermediate cervical nerve block "pericarotid injection": stage 3

possible to combine an injection performed all along the pathway of the needle with an additional 5 mL bolus dose in immediate proximity of the carotid sheath [16] (Figs. 8.16, 8.17, and 8.18). Conditions for visualisation should be optimal. The procedure must be done very cautiously and by an experienced operator. A total volume of 15–20 mL of local anaesthetic is injected.

Surgery in Areas of Overlapping Innervation

At the ventromedial cervical level: A contralateral superficial cervical block or medial injection can be necessary.

In the subclavian area (implantable compartment, pacemaker, subclavian venous line): sensory participation of the first two thoracic ventral branches can require an additional local injection.

Complications and Side Effects

To date, no complication has been reported with superficial cervical plexus block (with or without ultrasound) [2]. It is with deep cervical plexus blocks, performed without ultrasound guidance, that all the serious complications have been reported [17]. The "blind" conduct of a cervical plexus block, by intermediate or deep approach, exposes patients to incidents precisely related to the limits of landmark-based techniques:

- Intraneural needle stick and injections.
- Vascular puncture and intravascular injection (immediate seizures in case of injection into the vertebral artery).
- Epidural spread or total spinal anaesthesia in case of injection in contact with a spinal nerve root or into the intervertebral foramen.
- Block failure due to incorrectly placed LA (e.g. intramuscular).

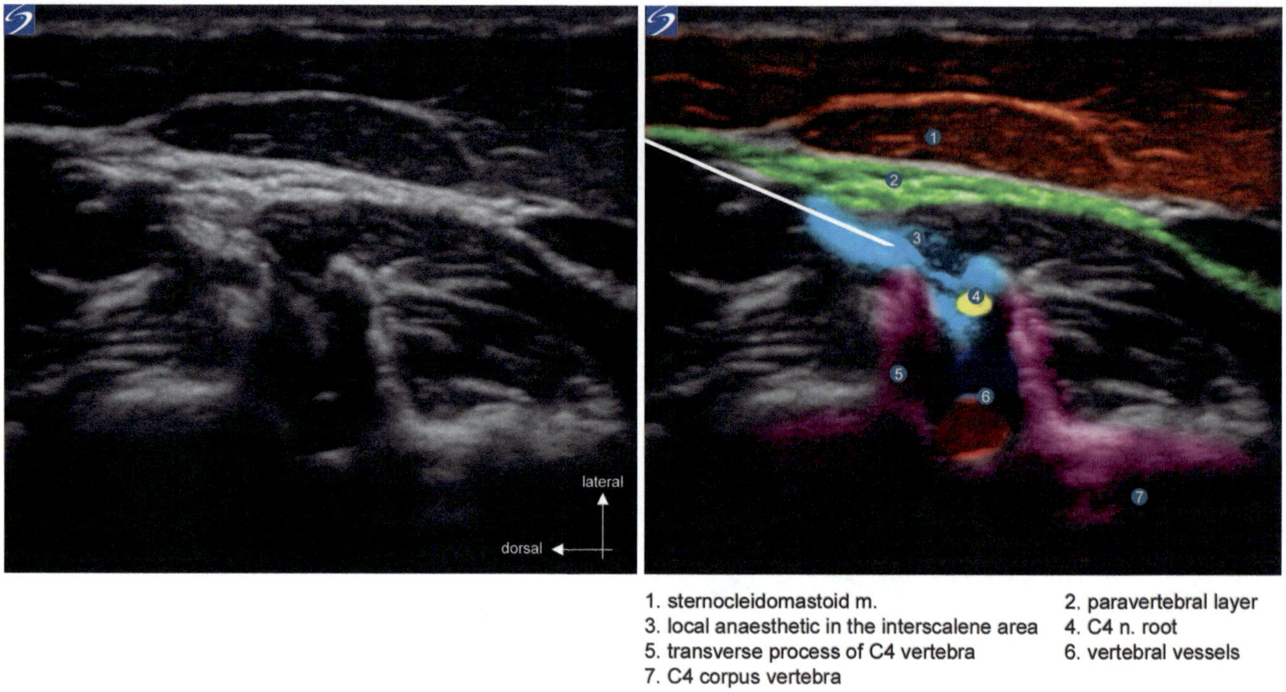

Fig. 8.19 Ultrasound-guided deep cervical block: injection of local anaesthetic (LA) in the paravertebral space, in proximity to the emergence of the root of C4

The anticipated benefits of ultrasound guidance used for blocks of the cervical plexus are a decrease in the incidence of significant complications and an improvement in the quality of the block.

Adverse reactions commonly encountered are: recurrent nerve anaesthesia, block of the vagus nerve, Claude Bernard-Horner syndrome, and anaesthetic spread to the brachial plexus. Transient cough, hoarse voice, facial numbness/paralysis, and dysphagia have been reported.

Precautions and Contraindications

In the presence of coagulation disorders, a superficial approach to the cervical plexus should be preferred.

In case of severe respiratory failure, diaphragmatic or contralateral recurrent nerve paralysis, it is inadvisable to perform an intermediate block of the cervical plexus and, even more so, a deep cervical block.

Bilateral intermediate and deep cervical plexus blocks are contraindicated due to the risk of bilateral phrenic nerve block.

References

1. Abbal B, Choquet O, Capdevila X. Blocs du plexus cervical. Mapar. 2013;
2. Choquet O, Dadure C, Capdevila X. Ultrasound-guided deep or intermediate cervical plexus block: the target should be the posterior cervical space. Anesth Analg. 2010;111:1563–4.
3. Pascal J, Navez M. Anesthésie locorégionale de la tête et du cou. In: Gautier-Lafaye P, Muller A Gaertner E. Anesthésie locorégionale et traitement de la douleur. Paris: Masson, 2009.
4. Telford RJ, Stoneham MD. Correct nomenclature of superficial cervical plexus blocks. Br J Anaesth. 2004;92:775.
5. Dhonneur G, Saidi NE, Merle JC, Asfazadourian H, Ndoko SK, Bloc S. Demonstration of the spread of injectate with deep cervical plexus block: a case series. Reg Anesth Pain Med. 2007;32:116–9.
6. Ramachandran SK, Picton P, Shanks A, Dorje P, Pandit JJ. Comparison of intermediate *vs* subcutaneous cervical plexus block for carotid endarterectomy. Br J Anaesth. 2011;107:157–63.
7. Markovic D, Vlajkovic G, Sindjelic R, Markovic D, Ladjevic N, Kalezic N. Cervical plexus block *versus* general anesthesia in carotid surgery: single center experience. Arch Med Sci. 2012;8:1035–40.
8. Shih ML, Duh QY, Hsieh CB, et al. Bilateral superficial cervical plexus block combined with general anesthesia administered in thyroid operations. World J Surg. 2010;34:2338–43.
9. Steffen T, Warschkow R, Brändle M, Tarantino I, Clerici T. Randomized controlled trial of bilateral superficial cervical plexus block *versus* placebo in thyroid surgery. Br J Surg. 2010;97:1000–6.
10. Choi DS, Atchabahian A, Brown AR. Cervical plexus block provides post-operative analgesia after clavicle surgery. Anesth Analg. 2005;100:1542–3.
11. Herring AA. The ultrasound-guided superficial cervical plexus block for anesthesia and analgesia in emergency care settings. Am J Emerg Med. 2012;30:1263–7.
12. Shteif M, Lesmes D, Hartman G, Ruffino S, Laster Z. The use of the superficial cervical plexus block in the drainage of submandibular and submental abscesses: an alternative for general anesthesia. J Oral Maxillofac Surg. 2008;66:2642–5.
13. Girard F, Quentin C, Charbonneau S, et al. Superficial cervical plexus block for transitional analgesia in infratentorial

and occipital craniotomy: a randomized trial. Can J Anaesth. 2010;57:1065–70.
14. Thallaj A, Marhofer P, Moriggl B, Delvi BM, Kettner SC, Almajed M. Great auricular nerve blockade using high resolution ultrasound: a volunteer study. Anesthesia. 2010;65:836–40.
15. Roessel T, Wiessner D, Heller AR, Zimmermann T, Koch T, Litz RJ. High-resolution ultrasound-guided high interscalene plexus block for carotid endarterectomy. Reg Anesth Pain Med. 2007;32:247–53.
16. Rössel T, Kersting S, Heller AR, Koch T. Combination of high-resolution ultrasound-guided perivascular regional anesthesia of the internal carotid artery and intermediate cervical plexus block for carotid surgery. Ultrasound Med Biol. 2013;39:981–6.
17. Pandit JJ, Satya-Krishna R, Gration P. Superficial or deep cervical plexus block for carotid endarterectomy: a systematic review of complications. Br J Anaesth. 2007;99:159–69.

Abdominal Wall Nerve Blocks

9

Eryk Eisenberg, Elisabeth Gaertner, Philippe Clavert, and Christophe Aveline

Contents

Note to the Reader

The emblematic technique of this family of nerve blocks is represented by "TAP block" (*transversus abdominis plane block*). Whether performed "blindly" by means of the loss of resistance approach to the needle crossing of the musculoaponeurotic planes or indeed with ultrasound guidance, its principle consists of injecting a local anaesthetic between the transverse abdominal muscle (TA) and the internal oblique muscle into the *transversus abdominis plane*. Since the French translation of "*transversus abdominis plane*" leaves much to be desired, because it is a source of anatomical confusion (transverse muscle plane most often designating the muscular layer and not the fascial plane located at the surface of the transverse muscle), as a convention we will continue to call this space the "TAP", and the corresponding technique of anaesthesia "TAP block".

The original version of the chapter has been revised. A correction to this chapter can be found at https://doi.org/10.1007/978-3-031-08804-9_13

Supplementary Information The online version contains supplementary material available at https://doi.org/10.1007/978-3-031-08804-9_9.

E. Eisenberg (✉)
Département d'anesthésie et réanimation, Clinique Pôle Santé République, Clermont-Ferrand, France

E. Gaertner
Département d'anesthésie et réanimation chirurgicale, Hôpitaux Civils de Colmar, Colmar, France

P. Clavert
Service de chirurgie du membre supérieur, Centre de Chirurgie Orthopédique et de la Main (CCOM), Illkirch, France

C. Aveline
Hôpital Privé Sévigné, Cesson Sévigné, France

E. Eisenberg, E. Gaertner (eds.), *Ultrasound in Peripheral, Neuraxial and Perineuraxial Regional Anaesthesia*,
https://doi.org/10.1007/978-3-031-08804-9_9

Anatomy of the Ventrolateral Wall of the Abdomen

Ventrolateral Wall of the Abdomen (Figs. 9.1, 9.2, 9.3 and 9.4)

It consists of muscles of the abdomen and of their aponeuroses, all of which are bordered on the surface by the skin and more deeply by the parietal peritoneum.

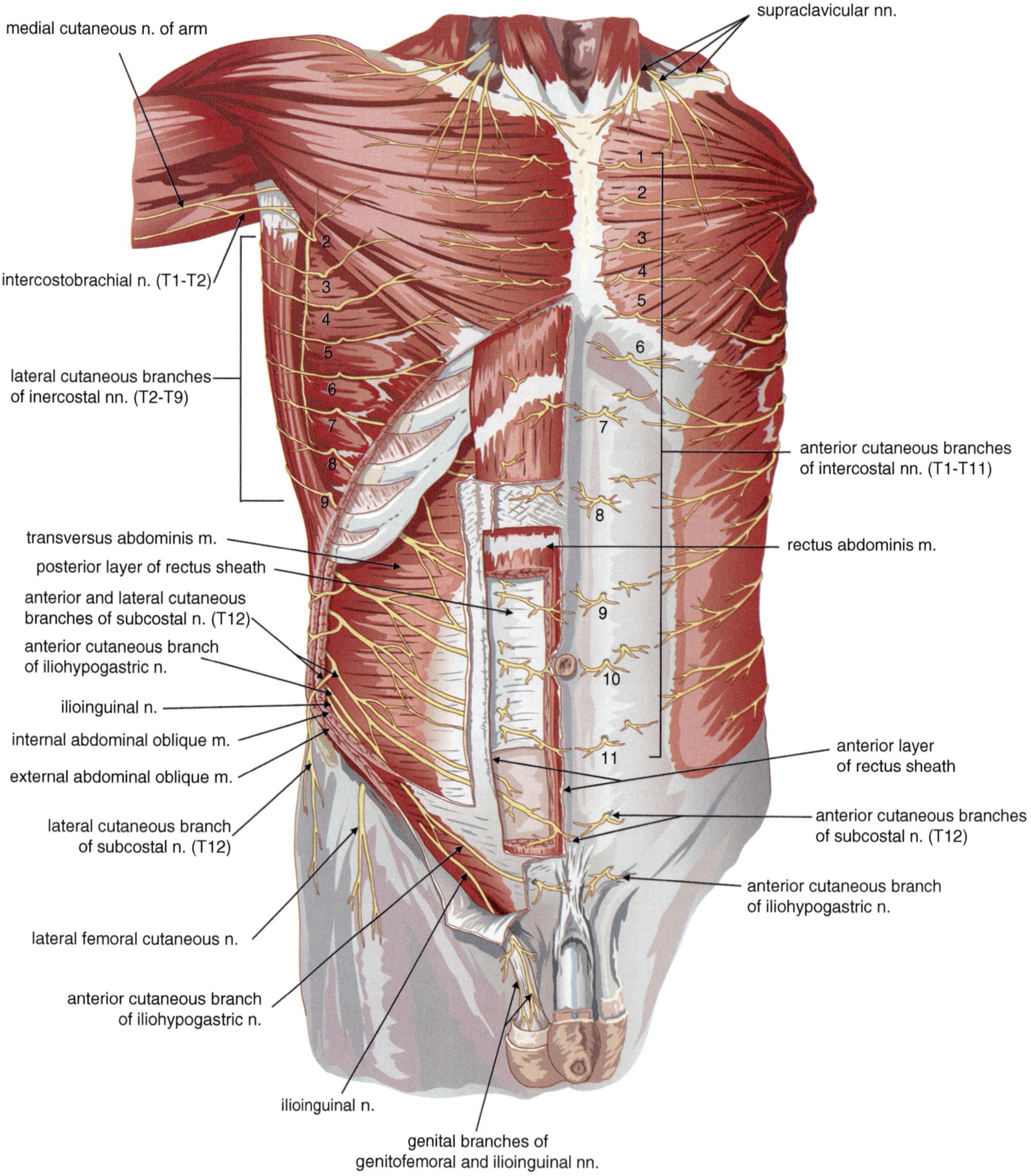

Fig. 9.1 Innervation of the thoracic and abdominal walls (from Netter FH. *Atlas of human anatomy*. third ed. Paris: Masson, 2007)

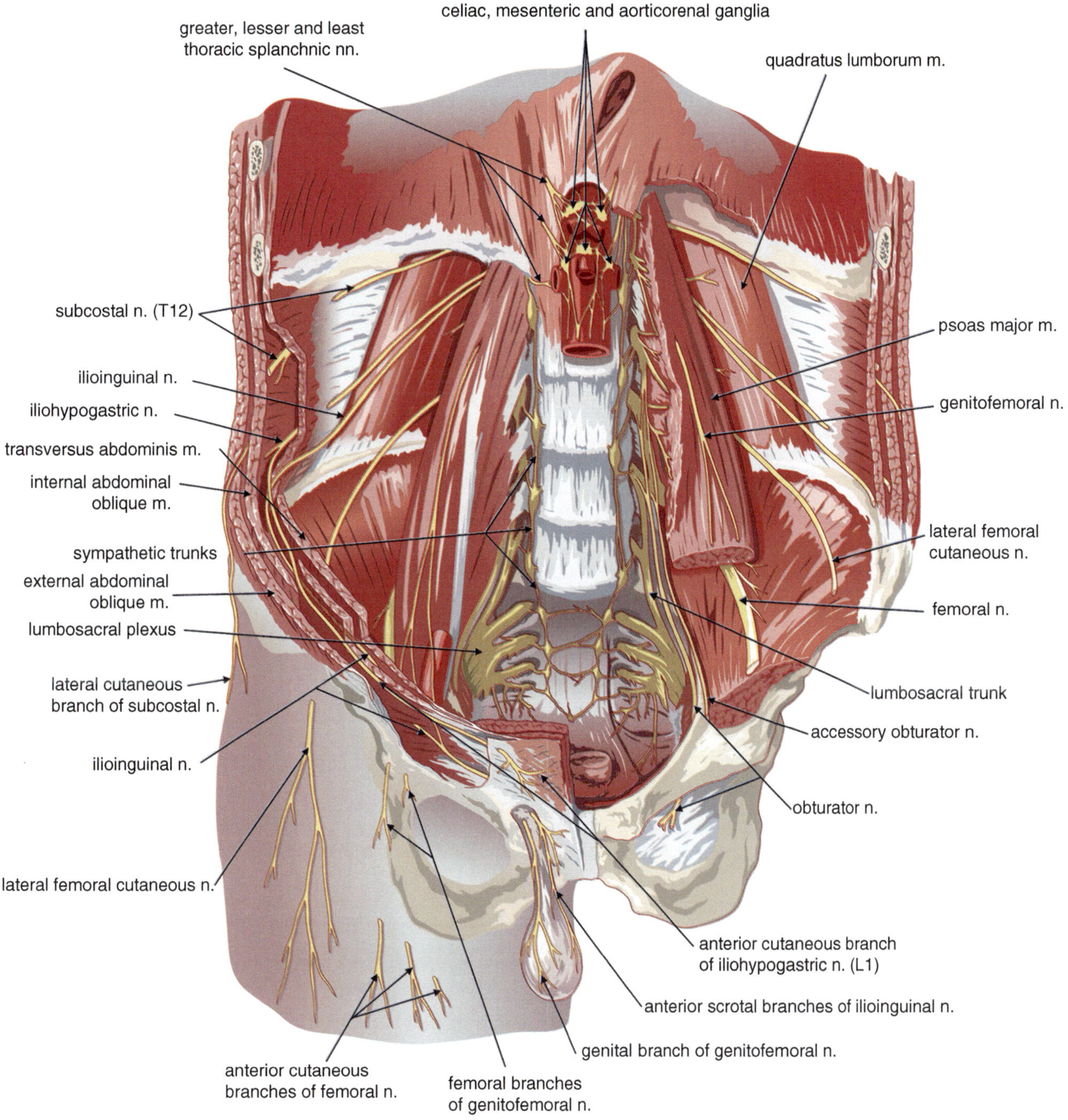

Fig. 9.2 Neuromuscular anatomy of the abdominal wall (from Netter FH. *Atlas of human anatomy*. third ed. Paris: Masson, 2007)

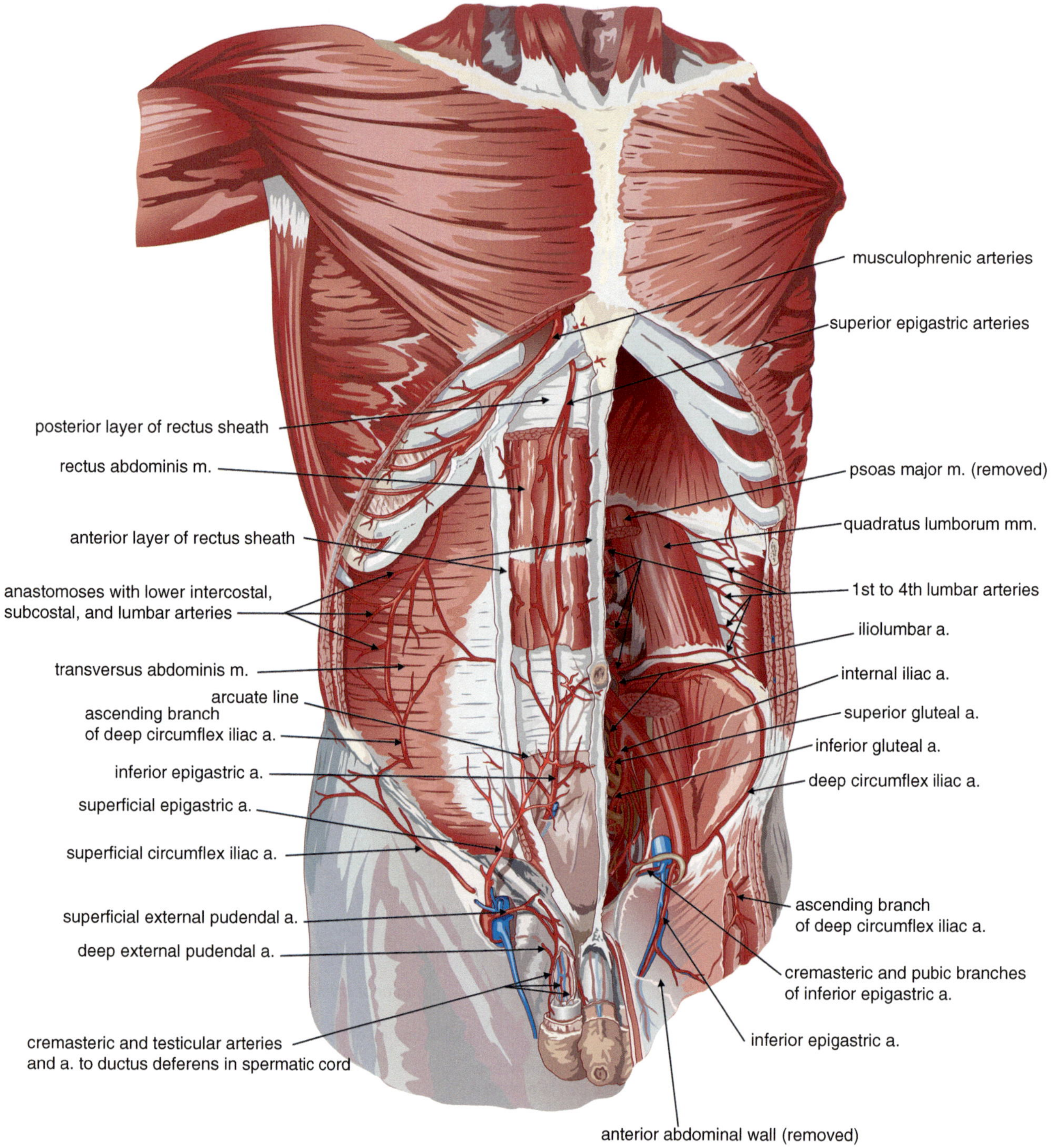

Fig. 9.3 Vascularisation of the abdominal wall (from Netter FH. *Atlas of human anatomy*. third ed. Paris: Masson, 2007)

The superficial fascia (superficialis facia) of the abdomen is located under the cutaneous lining of the abdominal wall, followed by the musculoaponeurotic components.

The latter are bounded by the following:

- Cranially: the ribs, the sternum, and the xiphoid process.
- Caudally: the pubis, the inguinal ligament and iliac crest.
- Dorsally and laterally: the thoracolumbar fascia and the transverse processes.
- Ventrally and medially: the linea alba.

On either side of the linea alba, which extends from the xiphoid process to the pubis, 4 muscle pairs primarily can be differentiated:

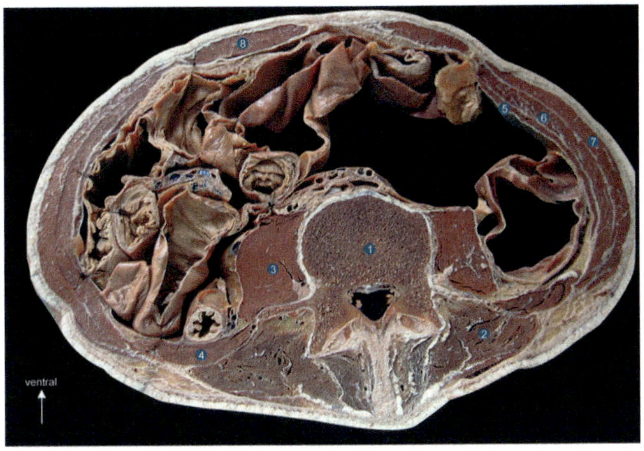

1. lumbar vertebra 2. erector spinae mm. 3. psoas major m.
4. quadratus lumborum m. 5. transversus abdominis m. 6. internal abdominal oblique m.
7. external abdominal oblique m. 8. rectus abdominis m.

Fig. 9.4 Transverse section of the abdomen at the lumbar level

- External oblique.
- Internal oblique.
- Transversus abdominis.
- Rectus abdominis.

The quadratus lumborum muscle participates in the posterior wall of the abdomen, and it is a key anatomical element in the method of spread of a local anaesthetic.

The so-called "broad" muscles of the abdomen (external oblique, internal oblique, and transverse muscles) are positioned in three successive layers from superficial to deep. They have more or less broad aponeuroses for insertion (aponeurotic tendons), which participate concurrently in particular in formation of the sheath of the rectus muscles and the linea alba. The fascia transversalis is the principal aponeurosis of lining of the ventrolateral wall of the abdomen. It covers the entire extent of the deep aspect of the transversus abdominis muscle, except its part located below the arcuate line, in front of the ventral and caudal aspect of the rectus abdominis muscle. In fact, the fascia transversalis here gives rise to the aponeurosis of the transverse muscle which passes behind the rectus muscle and remains the only structure separating it from the parietal peritoneum.

External Oblique Muscle

It is the most superficial muscular component of the abdominal wall. It inserts into the lateral aspect of the last 8 ribs, its fibres run ventrally and caudally, and it ends in the form of an aponeurosis: medially on the linea alba, caudally on the pubis, the inguinal ligament, and the anterior two-thirds of the iliac crest. It widens caudally and unites with the aponeurosis of the adjacent muscle (internal oblique). It has a natural orifice above the pubic tubercle corresponding to the superficial inguinal ring.

Its innervation is provided by the ventral branches of the thoracic nerves T6 to T11, as well as the subcostal nerve (T12).

Its principal motor actions are:

- Abdominal and pelvic compression and support of the internal organs in combination with the two other muscles (internal oblique and transversus abdominis muscles).
- Contralateral rotation of the trunk.
- Lateral flexion of the spine.
- Forced expiration.

Internal Oblique Muscle

It is located immediately deep into the external oblique muscle. It is separated from it by aponeurotic tissue. Its origin is on the anterior three-quarters of the iliac crest and the external one-third of the inguinal ligament, but also the posterior quarter of the iliac crest and the spinous process of L5 by a tendinous layer, which is combined with the thoracolumbar aponeurosis. It spreads out in a fan shape ventrally to end in a long line of attachment to the last 4 ribs (posterior fibres), the pubis (lower fibres which arise from the inguinal ligament), passing through the linea alba (middle fibres which give rise to the aponeurosis of the internal oblique muscle).

The lower fibres behave identically to the underlying transversus muscular fibres—they pass above and then behind the cord of the round ligament—and unite with each other, comprising the conjoint tendon which attaches to the pubis, to the pubic symphysis and to the pectineal line.

In the lateral border of the rectus muscle (thus participating in formation of the ipsilateral rectus muscle sheath), the aponeurosis of the internal oblique muscle divides into two layers: one anterior which passes in front of the ipsilateral rectus muscle and joins the aponeurosis of the external oblique muscle, and the other, posterior, passes behind the rectus abdominis by fusing with the anterior aponeurosis of the transverse muscle. This configuration is found only along the cephalic two-thirds or three-quarters of the rectus muscle. In effect, the arcuate line is the border below which the aponeurosis of the internal oblique, external oblique, and transverse muscles join together solely in front of the rectus muscle.

Innervation of the internal oblique muscle is provided by the ventral branches of T7–T12 with participation of L1.

Like the external oblique muscle, it provides:

- Abdominal and pelvic compression and support of the internal organs in combination with the two other muscles (external oblique and transversus muscles).
- Ipsilateral rotation of the trunk.
- Lateral flexion of the spine.
- Forced expiration.

Transversus Abdominis Muscle

It is the deepest muscle layer of the broad muscles of the ventrolateral wall of the abdomen. Between the transversus abdominis and the internal oblique muscles is found a relatively fine fascia, fused with the deep aspect of the muscular aponeurosis of the internal oblique muscle. Between this fascia and the superficial muscular aponeurosis of the transverse muscle are found nerve structures involved in a TAP block. At this level, vascular branches (deep iliac circumflex artery) are located that present a superolateral pathway above the iliac crest. The principal nerve trunks contained in this plane continue laterally to this artery.

The transversus abdominis muscle has its origins at the following levels:

- Costal: in the deep aspect of the last 6 costal arches.
- Lumbar: on the transverse processes of L1 to L4, and by the posterior aponeurosis of the transversus muscle (it occupies the entire space between the costal and iliac insertions).
- Iliac: the anterior two-thirds of the internal border of the iliac crest, and the lateral third of the inguinal ligament.

It projects ventrally to the linea alba where it ends in an aponeurosis located in the deep aspect of the ipsilateral rectus muscle in the cephalic two-thirds or three-quarters of this muscle, and at its ventral aspect in the most caudal portion, beneath the arcuate line.

The fibres that arise from the inguinal ligament behave identically to the muscular fibres of the internal oblique muscle placed more superficially—they pass above and then behind the cord or round ligament—and, by uniting to them, comprise the conjoint tendon which attaches to the pubis, to the pubic symphysis and to the pectineal line.

Innervation of the transverse muscle is provided by the ventral roots of T6–T12 and by that of L1. The transverse muscle provides compression of abdominal and pelvic internal organs.

Rectus Abdominis Muscle

It inserts by a short flat tendon on the ventro-cephalic part of the pubis, from the spine to the pubic symphysis and on the anterior aspect of the latter. Part of the tendinous fibres crisscross with those of the contralateral muscle. The fleshy body of the rectus muscle arises from the tendon, and widens from below upward to end in the xiphoid process and the fifth, sixth, and seventh costal cartilages. It has 2–5 transverse tendon intersections, often incomplete and partitioned [1]. Most often a first one is found at the point of the xiphoid process, a second one at the level of the umbilicus, a third one between the first two, sometimes with a fourth subumbilical one. The two rectus muscles of the abdomen are separated by the linea alba, which is a tendinous raphe. It extends from the xiphoid process to the pubic symphysis.

Its innervation arises from the ventral branches of the last 6 to 7 thoracic nerve roots (T6 or T7 to T12), as well as of L1 [2].

The rectus abdominis muscle flexes the spine and, in combination with the oblique muscles, maintains tension of the abdominal wall and compression of the abdominal internal organs.

Rectus Abdominis Muscle Sheath (Rectus Sheath)

This structure is demarcated by the aponeurosis of the external oblique, internal oblique, and transversus abdominis muscles. At the lateral border of the rectus muscle, the aponeurosis of the internal oblique muscle divides into two layers, passing, respectively, superficial to and deep to the muscle. The superficial layer is associated with the aponeurosis of the external oblique muscle to create the ventral part of the sheath, whilst the deep layer is associated with insertion of the transversus muscle to form the dorsal part of the sheath. The rectus sheath thus is demarcated as follows:

- Ventrally: by the aponeurosis of the external and internal oblique muscles.
- Dorsally: by that of the internal oblique and transversus muscles.

However, this configuration is present only along the cephalic two-thirds or three-quarters of the rectus muscles, the arcuate line (located a few centimetres below the umbilicus) demarcates its caudal limit. Beyond this, the aponeurosis of the broad muscles unites in front of the rectus muscle, whose deep aspect then is in direct contact with the fascia transversalis, which itself is fused with the extraperitoneal fascia and to the parietal peritoneum (Figs. 9.5, 9.6, and 9.7).

The linea semilunaris is the lateral border of the rectus muscle sheath, extending from the pubic tubercle to the ninth costal cartilage.

The rectus muscle adheres to its sheath only opposite the tendon intersections present on its ventral aspect. The fascial plane thus is compartmented ventrally, but is free dorsally.

Fascia Transversalis

The fascia transversalis lies deep into the transversus abdominis muscle. It is a lining aponeurosis separating the muscle from the peritoneal fat and the parietal layer of the peritoneum. Extending from the inguinal area to the diaphragm, it is in continuity in particular with the fascia iliaca with which it blends, with the fibrous sheath of the cord which it comprises and the vascular sheath around the external iliac vessels with which it participates. These communications can explain possible extension to the femoral nerve of an ilioin-

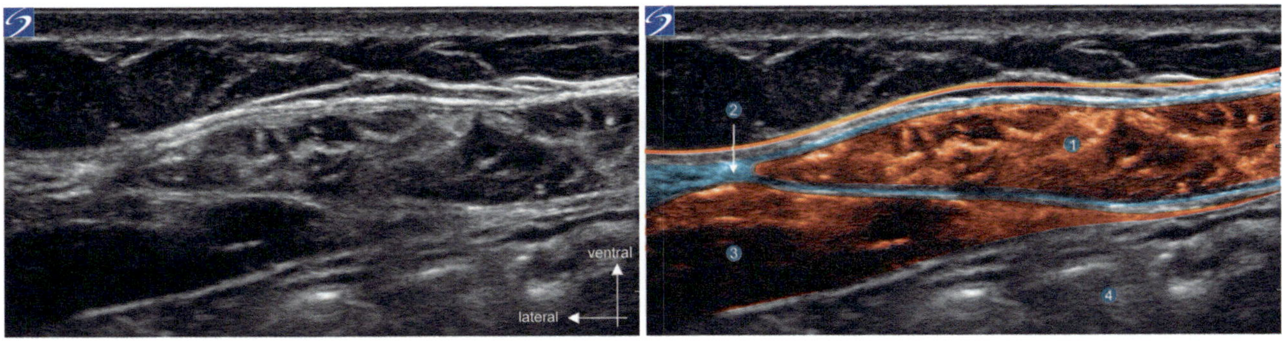

Fig. 9.5 Transverse ultrasound section of the rectus abdominis muscle in the subcostal area

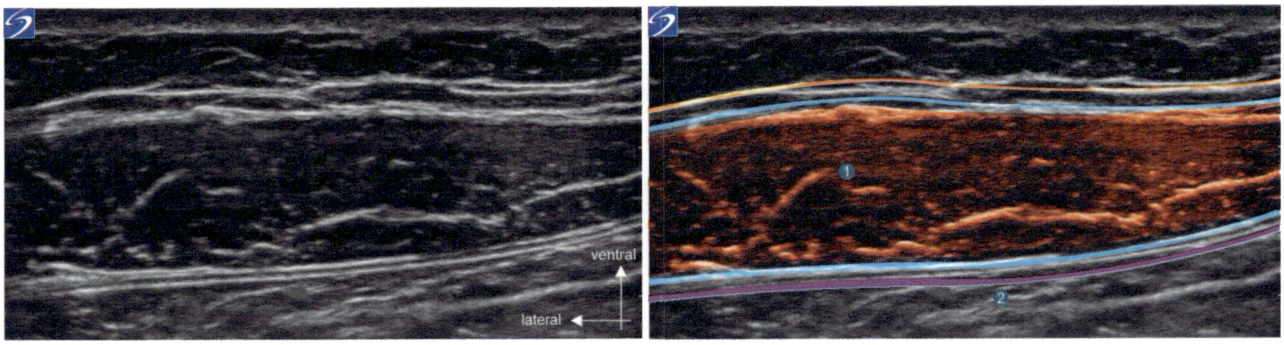

Fig. 9.6 Transverse ultrasound section of the rectus abdominis muscle at the supraumbilical level

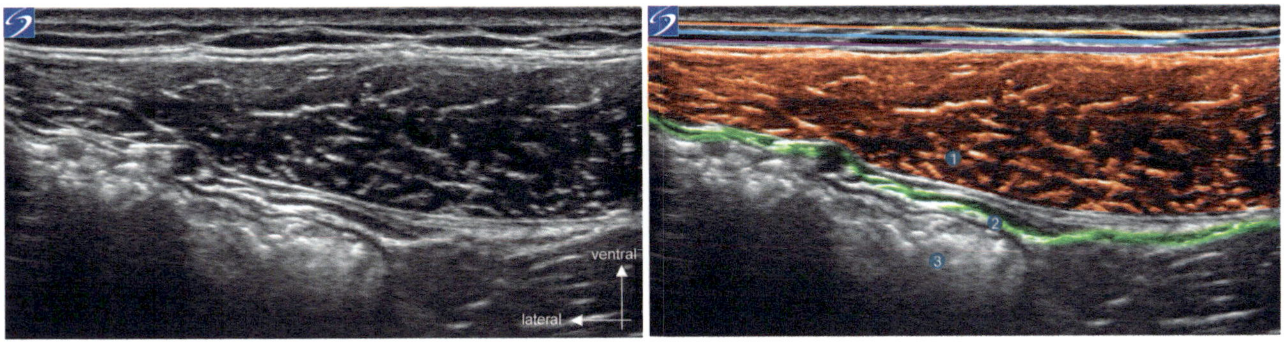

Fig. 9.7 Transverse ultrasound section of the rectus abdominis muscle below the arcuate line. The medial aponeurotic insertions of the transverse internal oblique and external oblique muscles are located at the ventral aspect of the rectus muscle. Deep to the rectus abdominis muscle are only fascia transversalis and parietal peritoneum

guinal block performed blindly, in particular, in children. Moreover, it forms the lateral part of the posterior aspect of the deep inguinal canal.

Quadratus Lumborum Muscle

This is a flat, quadrilateral muscle stretching from the iliac crest (where it is the widest) to the 12th rib and to the lumbar vertebral column. It is located in front of the spinal muscles from which it is separated by the aponeurosis of insertion of the transversus abdominis muscle of the abdomen [3]. It is contained within a strong aponeurotic sheath and presents three groups of bundles:

- The ilio-costal bundle: stretched caudally, between the internal border of the iliac crest and iliolumbar ligament, and cranially, the caudal border of the 12th rib.
- The ilio-transverse bundles: stretched between, caudally, the internal border of the iliac crest and the iliolumbar ligament, and medially, the apex of the transverse processes of the first 4 lumbar vertebrae.
- The costo-transverse bundles: these extend from the lower border of the 12th rib to the apex of the ventral aspect of the transverse processes of the lumbar vertebrae.

Innervation is provided by the ventral nerve roots of T12 and by that of L1, L2, and L3.

Contraction of the quadratus lumborum muscle provides ipsilateral inclination of the vertebral column and lowering of the 12th rib. When it takes its fixed point at the cranial level, it tilts the pelvis sidewards.

Nerve Structures (Figs. 9.1 and 9.2)

The ventral branches of the spinal nerves of T7 to L1 provide innervation to the skin and muscles of the ventrolateral abdominal wall, as well as the parietal peritoneum. The emergence of nerves in their respective pathways in the muscle layers are subject to many described variations. Creating an effective nerve block requires an injection of local anaesthetic with sufficient spread within the intermuscular spaces in order to reduce the risk of failure.

Intercostal Nerves from T7 to T11

These nerves continue anteriorly and inferiorly from their corresponding intercostal spaces, emerging from the inferior costal border before penetrating the transversus abdominis muscle. On emerging from this muscle the nerves lie immediately between transversus abdominis and the internal oblique. Continuing their ventromedial pathway, they anastomose widely between each other, penetrating the rectus muscle sheath and arriving at the linea alba from the deep aspect of this muscle (Fig. 9.8).

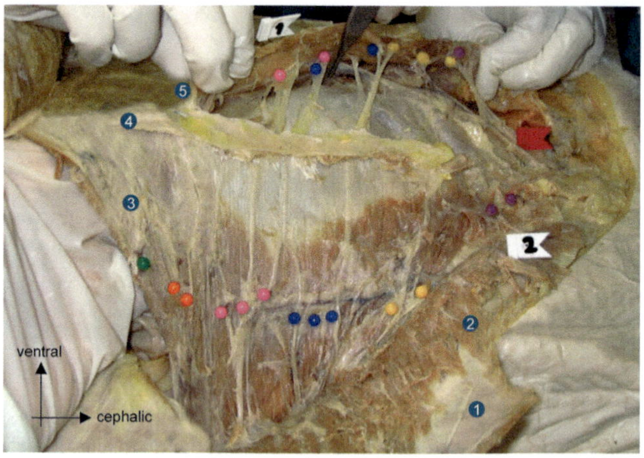

1. external abdominal oblique m. 2. internal abdominal oblique m.
3. transversus abdominis m. 4. linea semilunaris
5. rectus abdominis m.

Fig. 9.8 Dissection of the anterolateral aspect of the abdomen showing the nerves of the transverse muscle plane (TAP) and of the rectus sheath (from [2]) © Wiley. Captions for colour markers:– Green: branch issued from T8– Orange: branches issued from T9– Pink: branches issued from T10– Blue: branches issued from T11– Yellow: branches issued from T12– Purple: branches issued from L1Captions for white flags:– 1: level of the umbilicus– 2: anterosuperior iliac spine

Each of them distributes:

- Branches to the muscles of the thoracic wall during its intercostal pathway.
- A lateral cutaneous branch (which leaves the terminal nerve before it penetrates the transversus muscle, except for T11 [2]).
- A sensory branch that goes to the diaphragm, the peritoneum, and the pleura.
- Branches to the external oblique, internal oblique, and transversus muscles.
- Two ventral cutaneous branches:
  - One lateral which emerges from the lateral border of the rectus muscle.
  - One medial which often passes through the rectus muscle from its deep aspect before terminating in the skin overlying the linea alba.

Subcostal Nerve (12th Intercostal Nerve)

This is larger than the adjacent intercostal nerves. After crossing the ventral aspect of the quadratus lumborum muscle, the subcostal nerve penetrates the transversus abdominis muscle and then adopts the same abdominal position and pathway as the aforementioned intercostal nerves. Its lateral cutaneous branch arises from the lateral border of the quadratus lumborum muscle, becomes superficial above the middle of the iliac crest before being distributed to the skin of the gluteal area, sometimes as far as the greater trochanter. Its terminal branch lies at the craniomedial border of the iliohypogastric nerve to which it frequently provides an anastomosis before it continues into the subumbilical skin (Fig. 9.8).

Iliohypogastric Nerve

The most cranial terminal branch of the lumbar plexus, the iliohypogastric nerve is a mixed nerve which emerges from the ventral root of L1 with a variable component which arises from T12. It emerges laterally from the psoas major muscle to cross the ventral aspect of the quadratus lumborum muscle, lying caudally in relation to the 12th intercostal nerve, behind the posterior layer of the renal fascia. It then enters the transversus abdominis muscle at the lateral border of the quadratus lumborum muscle (or slightly more laterally), which places it in immediate proximity to the iliac crest, due to its oblique pathway. It then provides its lateral cutaneous branch which becomes superficial 2–3 cm laterally to the anterosuperior iliac spine after continuing between the internal and external oblique muscles above the iliac crest. This branch provides sensory innervation to the anterolateral gluteal area.

A little after the emergence of its lateral cutaneous branch, the iliohypogastric nerve, still placed between the transversus and the internal oblique muscles, divides into two branches, one abdominal and the other genital:

- The abdominal branch continues its pathway up to the rectus muscle sheath, sometimes penetrating the internal oblique muscle along the way. It ends with two perforating branches (medial and lateral) similar to the lower intercostal nerves. The medial nerve can cross the rectus muscle, to innervate the pyramidalis muscle.
- The genital branch crosses the internal oblique muscle near to the anterosuperior iliac spine, arrives at the deep orifice of the inguinal canal, and descends in ventrolateral position along the spermatic cord or the round ligament. It ends by medial branches which go to the skin of the pubis, the scrotum or the labia majora, and also, with lateral small branches for the skin of the proximal and medial part of the thigh.

These two terminal branches deliver along their passage many branches for the muscles of the abdominal wall.

The iliohypogastric nerve very frequently forms an anastomosis with the ilioinguinal nerve and the subcostal nerve. It can also be absent and replaced by a lateral branch of the 12th thoracic ventral branch, which results from the 11th ventral thoracic branch, or simply receive one of its branches.

Ilioinguinal Nerve

The second terminal branch of the lumbar plexus in the rostro-caudal plane, the ilioinguinal nerve is a sensory nerve. Its origins are in the ventral branch of L1, with possible contributions from T12 to L3 [4]. It follows the pathway of the iliohypogastric nerve along a variable distance, it has the same relations caudally located in relation to it. It penetrates the internal oblique muscle in the area adjacent to the anterosuperior iliac spine caudally and in front of the iliohypogastric nerve and it divides into two branches:

- The abdominal branch continues to terminate in the muscles of the abdominal wall.
- The genital branch crosses the external oblique muscle, possibly unites with that of the iliohypogastric nerve for which it has an analogous pathway and relations, accompanying the spermatic cord in the inguinal canal. It passes through the superficial inguinal ring to innervate the proximal medial aspect of the thigh, the upper part of the scrotum or of the labia majora.

Genitofemoral Nerve

This mixed nerve which arises from the ventral branches of L1 and L2, descends caudally and ventrally across the psoas major muscle from which it emerges opposite L3. Whilst descending under the fascia iliaca in the ventral aspect of the psoas major, it crosses dorsally the ureter and the gonadal vessels and continues along the lateral border of the external iliac vessels. In 20% of cases, the genitofemoral nerve upon its origin can be composed of two terminal branches [5]. However, traditionally, it divides only above the inguinal ligament into its two terminal branches:

- The genital branch passes in front of the external iliac artery (whose innervation it provides), and then across the internal inguinal ring and gives rise in males to the branch of the cremaster muscle, participates in the spermatic symptomatic plexus and vas deferens, as well as sensation to part of the skin of the scrotum. In the female, this branch follows the round ligament and ends in the pubis and in the labia majora.
- The femoral branch passes into the thigh under the inguinal ligament, laterally and ventrally to the femoral artery (which it innervates), and then crosses the fascia lata to innervate the skin of the superficial part of the femoral triangle (Fig. 6.24).

The intricate involvement of the sensory areas of the genitofemoral and ilioinguinal nerves is high, making it difficult to individualise them, respectively.

Vascular Supply of the Ventrolateral Wall of the Abdomen (Fig. 9.3)

The vascular supply of the deep layers of the ventrolateral abdominal wall depends primarily on the two collateral branches of the external iliac artery (deep iliac circumflex artery and inferior epigastric artery) and the superior epigastric artery (terminal branch of the internal thoracic artery).

Deep Iliac Circumflex Artery

It branches off the lateral border of the external iliac artery just above the inguinal ligament in the area around the emergence of the inferior epigastric artery which divides from it medially. It takes a craniolateral direction superficial to the transversalis fascia and divides into two terminal branches opposite the anterosuperior iliac spine:

- The ascending branch of the deep iliac circumflex artery, branches within the muscles of the abdominal wall (anastomosing with the branches of the inferior intercostal, subcostal, and lumbar arteries), some of whose branches frequently accompany the intercostal nerves of T9 to T11 in the plane of the transversus muscle.
- The iliac branch which continues along the iliac crest between the transversus and internal oblique muscles. It anastomoses with the lumbar and iliolumbar branches.

Inferior Epigastric Artery

Arising from the medial border of the external iliac artery, it gives rise to three small collaterals: one cremaster branch (or to the round ligament), a pubic branch and an anastomotic branch with the obturator artery. It then continues cranially and medially to the ventral aspect of the transversalis fascia, it continues under the arcuate line and then ascends to the umbilicus deep in the rectus muscle. The artery then penetrates into the muscle and its terminal branches anastomose with those of the superior epigastric artery. Other branches which go to the skin join the subcutaneous branches in particular from the lumbar arteries, the superior and subcostal epigastric arteries.

Superior Epigastric Artery

Emerging from the internal thoracic artery, it continues initially in the rectus sheath deep to the muscle, which penetrates only between the xiphoid process and the umbilicus.

Ventrolateral Abdominal Wall Nerve Blocks

General Data

All of the blocks described require antiseptic preparation in line with current recommendations [6]. Cleansing of the skin is performed with a 5% povidone-iodine or with 0.5% chlorhexidine, both of which are in alcohol solution. The practitioner should wear sterile gloves, and an appropriate sterile probe cover must be used. The cover should be checked before and after performing the nerve block. On completion of the block, an appropriate method of decontamination of the probe should be undertaken due to the possible contact between the probe and the patient's blood (a break in the protective covering, or an error made during its removal). Insertion of a catheter should be performed by using a strictly aseptic technique (i.e. use of a hat, mask, and gown). In the majority of cases, these nerve blocks are performed under general anaesthesia. However, in some cases they are done on patients who are awake, which makes it desirable first to perform local anaesthesia of the skin.

In the majority of cases, a high-frequency linear ultrasound probe (>10 MHz) is used, and in paediatrics a small footprint probe is preferred (with a hockey stick shape). In obese subjects, it may sometimes be necessary to use a low-frequency curvilinear probe (2–5 MHz) depending on the thickness of the abdominal wall.

Several techniques have been described to anaesthetise the ventrolateral wall of the abdomen (Figs. 9.9, 9.10 and 9.11). These are complementary, and their indications depend not only on the area to be anaesthetised, but also on patient factors (adult or a child, normal or elevated BMI, anatomical considerations etc.). Figures 9.12, 9.25, 9.32, 9.39, 9.47 and 9.57 and Table 9.1 summarise the patterns and extent of anaesthesia achieved by each of the described techniques. The block should be performed bilaterally in the context of midline or bilateral incisions.

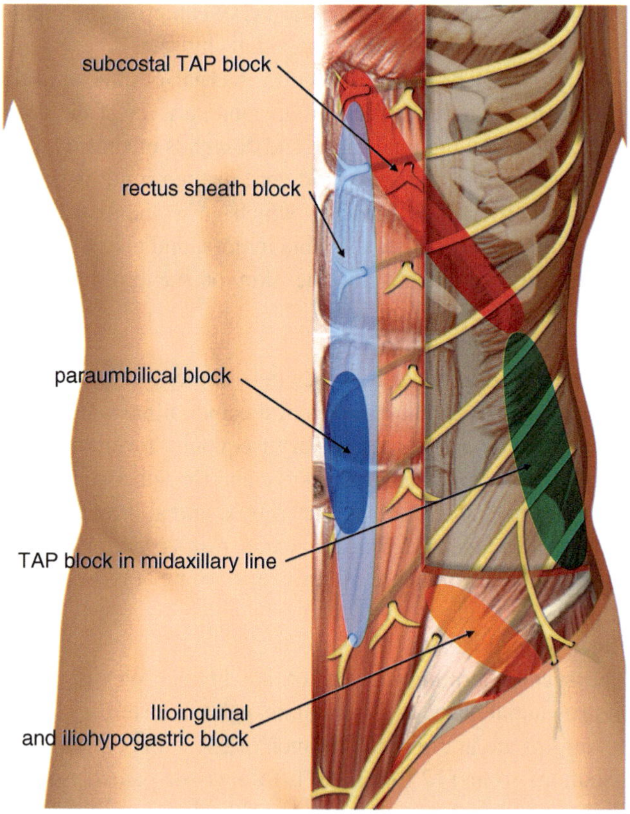

Fig. 9.9 Area where the injection of local anaesthetic should be administered, according to the type of block desired

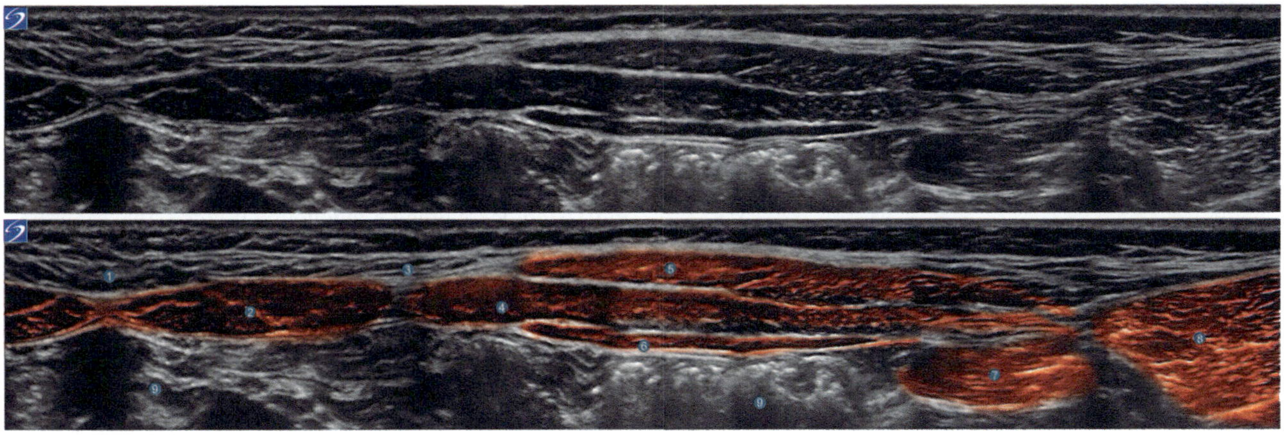

Fig. 9.10 Sonoanatomical section of muscles of the abdominal wall, from the linea alba to the paraspinal muscles

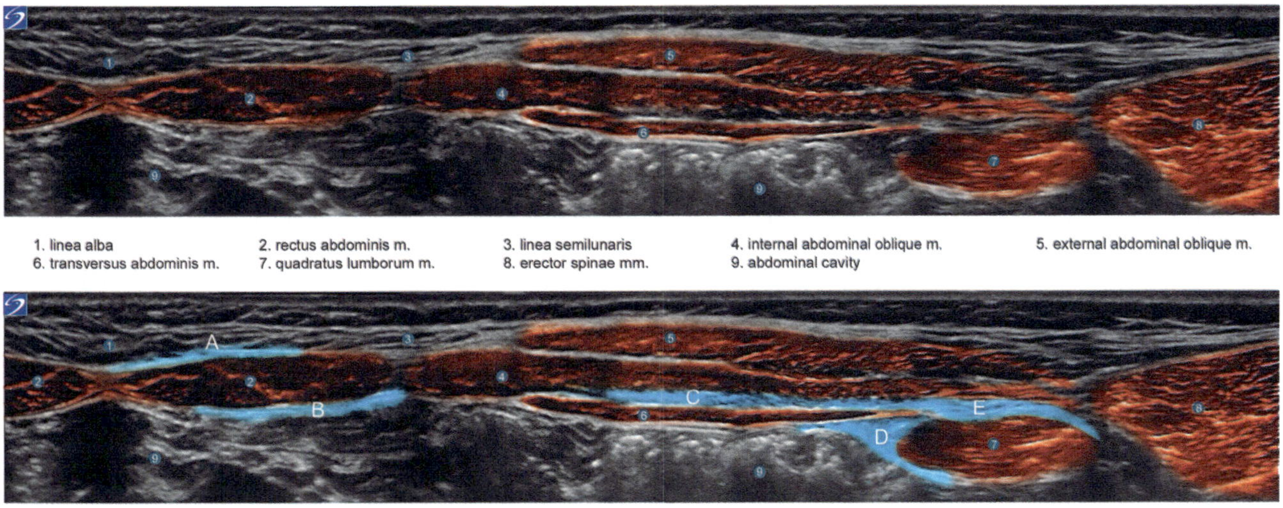

Fig. 9.11 Diffusion spaces for the abdominal wall nerve blocks

TAP Block

Indications

The bilateral TAP block appears to be an appropriate analgesic technique for supra- and subumbilical abdominal surgery [7]. It has taken the place of other techniques previously used for analgesia in abdominal surgery (ilioinguinal block for repair of an inguinal hernia, wound infiltration for caesarean delivery, peritoneal injection after colon surgery, and epidural anaesthesia).

It is performed in combination with conventional analgesia, in the setting of multimodal post-operative analgesia, for many abdominal procedures.

Performed in the mid-axillary line, the TAP block is indicated in cases of colorectal surgery, appendicectomy (laparotomy or laparoscopy), caesarean delivery, abdominal hysterectomy, abdominoplasty with subumbilical incision, and prostatectomy.

In spite of the extreme heterogeneity of studies, defects in methodology and the low number of patients cohorts, the results relating to this block overall show, during the first 24 hours post-operatively, pain scores at rest and on movement that are lower, an opioid-sparing effect, patients suffer fewer harmful effects of opioids (sedation, nausea, and post-operative vomiting). Patient satisfaction is improved. During abdominoplasty, TAP block enabled early ambulation of patients in the study group vs a control group. A bilateral TAP block significantly reduced the use of post-operative morphine after caesarean delivery [8], and resulted in fewer side effects such as nausea and vomiting [9].

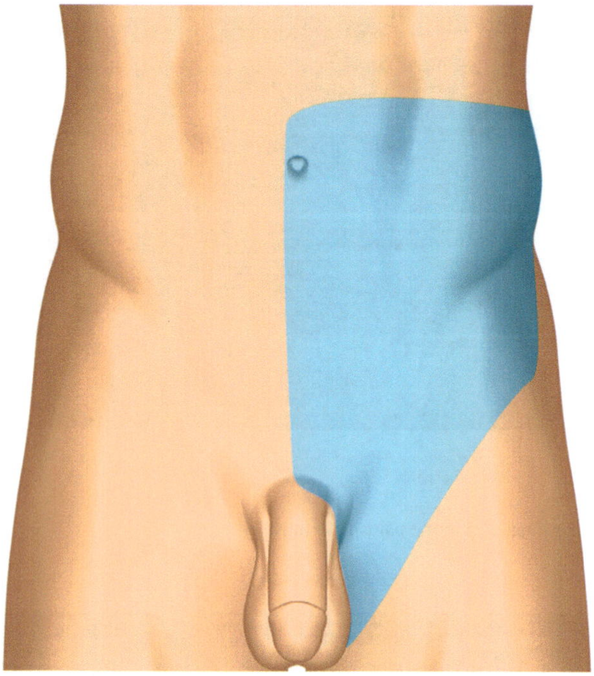

Fig. 9.12 Area of extension of a *TAP block* performed on the mid-axillary line

Table 9.1 Indication of the different techniques proposed

Type of block	Area involved	Indications
Rectus sheath block	T7 to T12 depending on site and number of injections	Median laparotomy
Paraumbilical block	T10	Umbilical hernia
Ilioinguinal/ iliohypogastric combined block	Branches of L1 (II-IH nerves)	Peroneal surgery inguinal hernia—Pfannenstiel procedure
TAP subcostal block	T7–T10	Supraumbilical surgery
TAP intercostal-iliac block	T10–T11–T12 often L1 and rarely T9	Abdominal-peroneal surgery

A study by Griffiths on analgesia after gynaecological cancer did not show any difference between the TAP block and conventional analgesia groups. In addition to heterogeneity of the groups, the variability of the surgery (according to the extension of cancer) which in many patients extends above the supraumbilical area can account for the negative result.

A TAP block may be a useful analgesic intervention when a laparoscopy is converted to a laparotomy.

In comparison to ilioinguinal and iliohypogastric blocks for repair of an inguinal hernia in children by Fredrickson et al. [10], a TAP block was less effective with increased use of analgesics in children with more pain.

The benefit of the TAP block seems significantly lower for analgesia when spinal analgesia with morphine is administered. The adverse events of opioids were, however, more common and more severe in the "intrathecal morphine" groups (post-operative nausea and vomiting, and pruritus). To extend the duration of the analgesic effect of the TAP block beyond a few hours, it is possible to insert a catheter.

Performed by subcostal approach, a TAP block is indicated for supraumbilical surgery (hepatobiliary, including laparoscopic [11], and renal). Compared to epidural analgesia, in a retrospective series, bilateral subcostal TAP blocks performed in the mid-axillary line were equivalent in terms of pain score, but have fewer disadvantages such as failure of the technique and hypotension. However, they require a higher use of opioids than the "thoracic epidural" groups. Niraj et al. [12] compared continuous thoracic epidural block (0.125% bupivacaine and 2 µg/mL fentanyl) vs bilateral subcostal TAP block with a catheter maintained by a bolus of 0.375% bupivacaine 1 mg/kg/8 h on each side, after upper abdominal surgery (hepatobiliary surgery and renal surgery). Pain scores at rest and during coughing were identical in the two groups up to 72 hours after surgery. However, the use of tramadol was higher in the TAP block group (p = 0.002). Barrington et al. [13] demonstrated more important extension of the contrast medium in a cadaver during multiple injections (hydrodissection by 5 mL bolus, associated with an injection of 5 mL between the rectus abdominis and the transverse abdominal muscle, total volume 20 mL) compared to a single injection of 20 mL with subcostal TAP block. In a more recent study, Albrecht et al. [14] demonstrated that subcostal TAP block does not improve gastric post-bypass analgesia if the trocar site is infiltrated with local anaesthetic, a finding also reported in another study on cholecystectomy [15].

TAP Block in the Middle Axillary Line (Intercostal-Iliac) (Fig. 9.12)

Echoanatomy

Ultrasound examination of the abdominal wall in the region of the mid-axillary line reveals the existence of three muscular layers separated by very echogenic structures that are their aponeuroses [16] (Fig. 9.13). The most superficial muscle is the external oblique, followed by the internal oblique, and lastly the transversus abdominis. The deep aspect of the transversus muscle (often the thinnest of the three) is in direct contact with the transversalis fascia, which itself lies on the parietal peritoneum which lines the abdominal cavity, where "mobile" digestive elements are visible. Under good echogenic conditions, it is possible to differentiate two fine

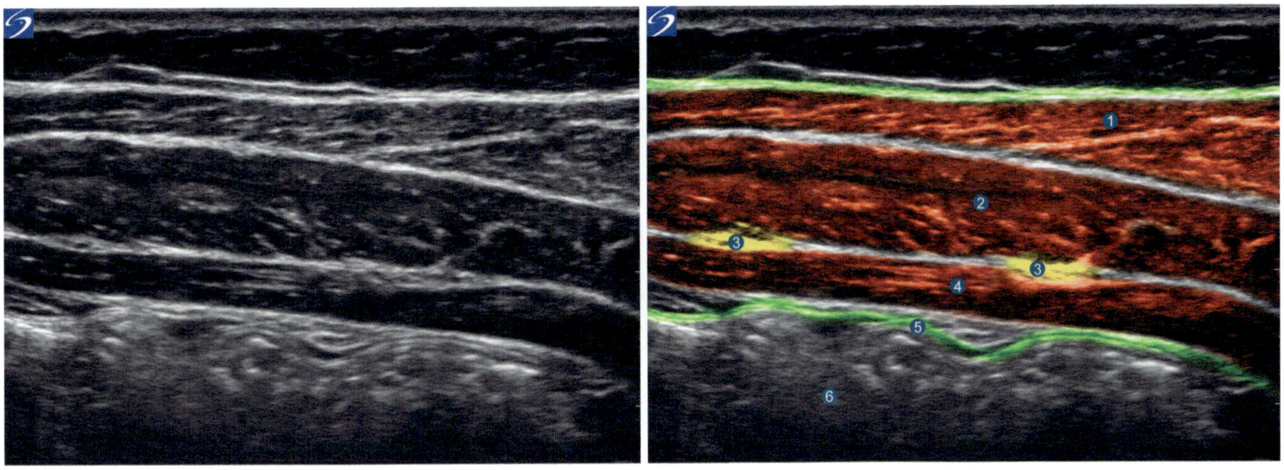

1. external abdominal oblique m. 2. internal abdominal oblique m.
3. nn. 4. transversus abdominis m.
5. peritoneum 6. abdominal cavity

Fig. 9.13 *TAP block* on the mid-axillary line: the three muscular layers are separated by strongly echogenic aponeuroses

"lamellar" structures corresponding to the deep aponeurosis of the transversus muscle and the parietal peritoneum which are superimposed, and separated by the transversalis fascia (Fig. 9.13). Between the surface of the probe and the external oblique muscle lie the skin and subcutaneous adipose tissue, of variable thickness. It is important not to confuse these tissues with the most superficial muscular layer (i.e. the external oblique). In case of doubt, it is useful to identify the muscles by starting with the deepest one, i.e. the transversus abdominis, since it always lies on parietal peritoneum independent of the patient's body habitus. In the intercostal-iliac space (between the lateral costal border and the iliac crest), the subcostal, iliohypogastric and ilioinguinal nerves lie close together before diverging and reaching the ventral area of the abdomen. Injection of the local anaesthetic into this more limited space may be useful in respect of this fact. By moving the probe slightly more dorsally, the lateral part of the quadratus lumborum muscle is visualised, which is in contact with the three major muscles of the abdominal wall (Fig. 9.14). This represents the ideal position of the probe to aid needle insertion and injection of local anaesthetic so that it can be distributed optimally in contact with the branches from T10 to T12 and extending perhaps to L1 (iliohypogastric and ilioinguinal) (Figs. 9.15 and 9.16). The postulate, defended by some authors, of constant anaesthesia of dermatomes T10 to L1 by means of an injection of 20–25 mL of local anaesthetic in this region is controversial. There are those who assert that it is even possible to extend the block effectiveness up to T7 by utilising greater volumes of LA. However, if such an extension is desired with confidence, it is probably necessary to combine a TAP block of the middle axillary line with subcostal TAP block (see below) (Fig. 9.17).

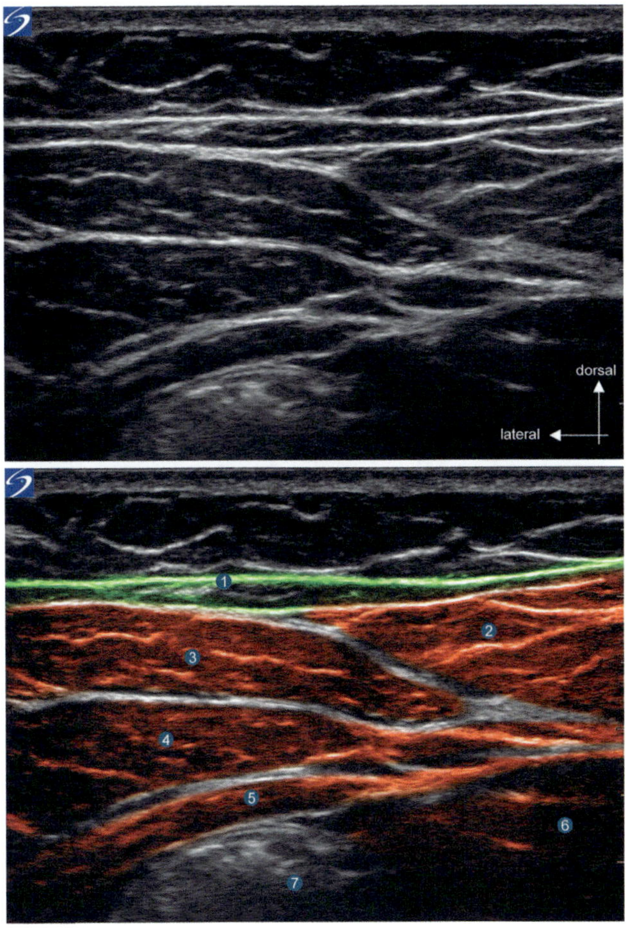

1. superficial fascia of abdomen 2. erector spinae mm.
3. external abdominal oblique m. 4. internal abdominal oblique m.
5. transversus abdominis m. 6. quadratus lumborum m.
7. abdominal cavity

Fig. 9.14 *TAP block* on the mid-axillary line. Probe moved dorsally: visualisation of quadratus lumborum muscle

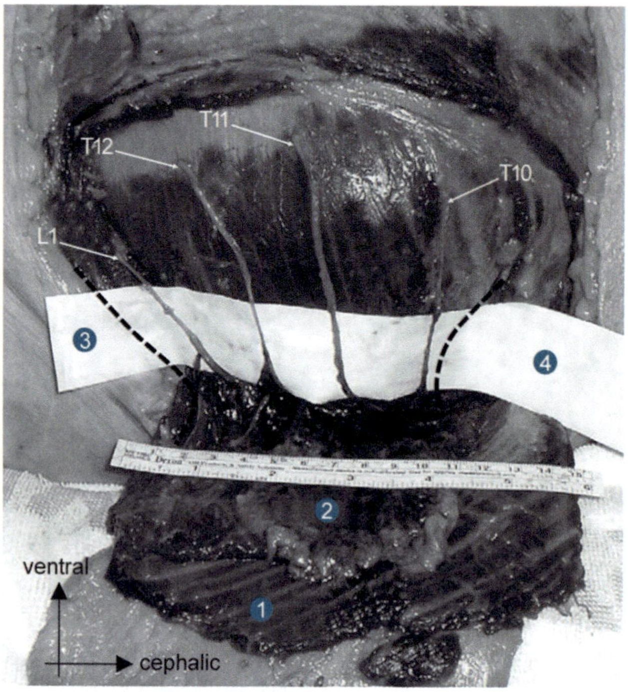

1. external abdominal oblique m.
2. internal abdominal oblique m.
3. iliac crest
4. costal margin

Fig. 9.15 Nerve branches from T10 to L1 (from [22]) © Oxford University Press

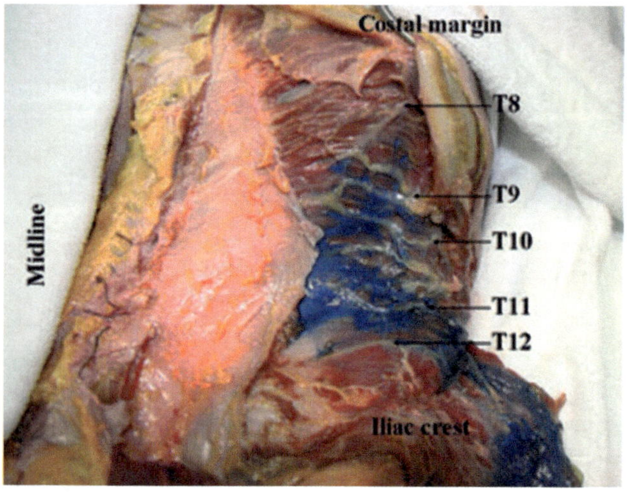

Fig. 9.16 Local anaesthetic diffusion space of T8 to T12 (from [13]) © Wiley

Location and Approaches.

> **Middle Axillary Line TAP Block.**
> **Type of probe:** Linear, high frequency 5–10 or 6–13 MHz. A small footprint probe is useful in paediatrics.
> **Axis of the probe:** Transversal, in the middle axillary line (Fig. 9.18).
> **Configuration:** Nerves inconsistently noticed, in transverse or oblique sections, needle in-plane (more rarely out of plane).
> **Depth studied:** Variable depending on body build and body mass index of the patient.
> **Neurostimulation:** Not very useful, but can possibly enable identification of the nervous structures observed.
> **Needle:** 80 to 100 mm, possibly isolated.
> **Utility of Doppler ultrasound:** Enables location of the arterial branches of the abdominal wall in order to avoid vascular puncture

The patient is placed in supine position. The skin is disinfected and the probe is prepared. The probe is initially positioned transversely, opposite the semilunar line (i.e. the lateral border of the ipsilateral rectus muscle). The probe then is slowly moved laterally/posteriorly in order to identify the three layers of the large abdominal muscles. Maintaining this view, the probe is moved further to reach the level of the mid-axillary line (or even slightly more posteriorly), at midpoint between the costal border of the 12th rib and the iliac crest (Fig. 9.18). The quadratus lumborum muscle is often relatively identifiable and is the dorsal limit of the area to be visualised (Fig. 9.14). If this muscle is not clearly visible, the practitioner must be satisfied with the typical pattern of the three muscle layers.

In-Plane Approach

A short bevel needle is inserted into the ultrasound plane in a ventrodorsal direction (Fig. 9.19). It successively crosses the skin, subcutaneous tissue, the external oblique muscle, and then the internal oblique muscle. At this point the needle tip is positioned in the TAP layer (Fig. 9.20). It is sometimes difficult to place the tip exactly in this intermuscular space, tending to be located either in the internal oblique muscle or

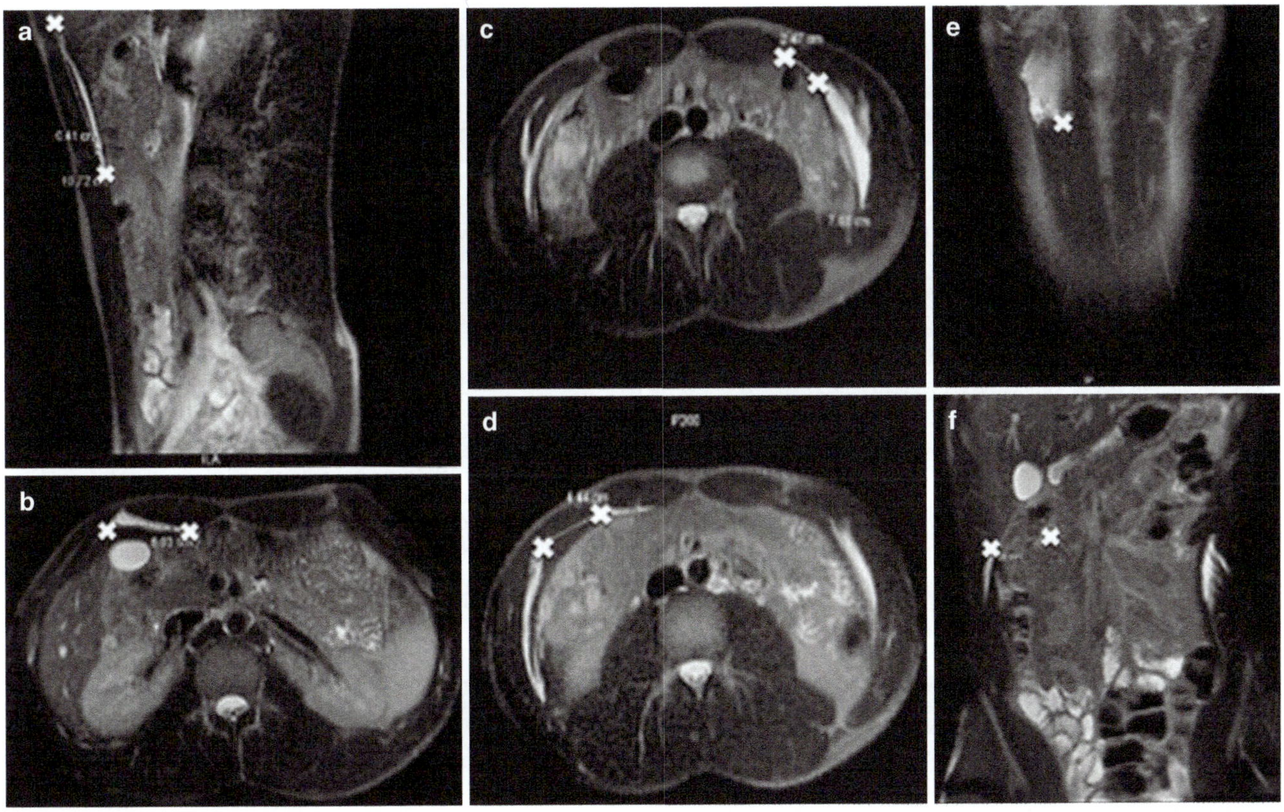

Fig. 9.17 Measurement by MRI of cephalocaudal, lateral, and coronal spread of the injection, in subcostal TAP block or performed along the mid-axillary line. From Borglum J, et al. Distribution patterns, dermatomal anaesthesia, and ropivacaine serum concentrations after bilateral dual transversus plane block. *Reg Anesth Pain Med* 2012; 37: 294–301. © Lippincott, Williams Wilkins

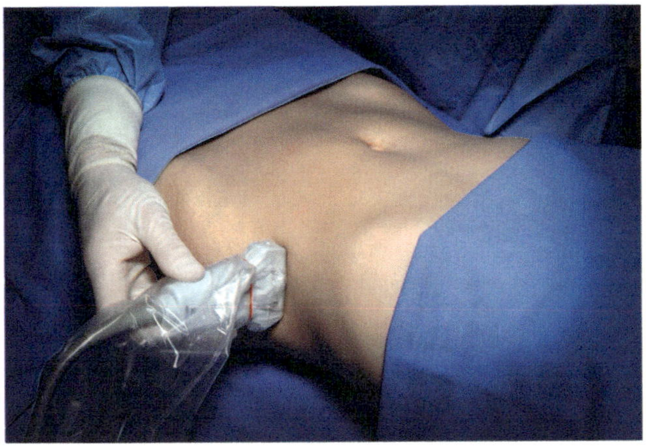

Fig. 9.18 TAP block on the mid-axillary line. Position of probe

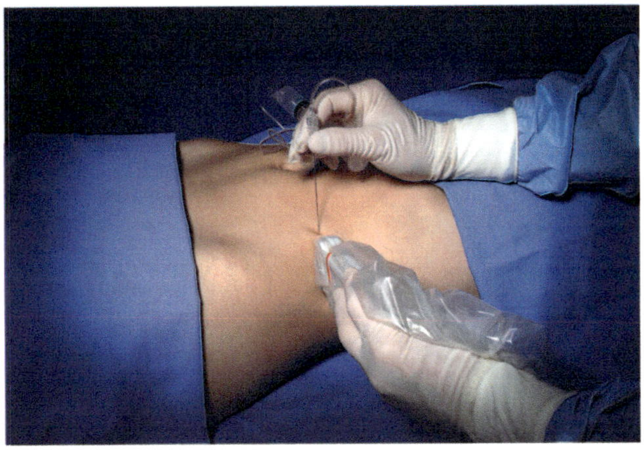

Fig. 9.19 TAP block on the mid-axillary line. Position of the probe. Needle inserted in plane

already within the transversus muscle. The use of small, repeated injections of 0.9% NaCl (or of a local anaesthetic) is useful for precise hydrolocalisation of the tip. Initially inserting the needle within the transversus muscle, then slowly withdrawing it whilst injecting fluid to "open up" the intermuscular space may result in a more accurate needle placement. Anatomically, a thin membrane exists between the deep aponeurosis of the internal oblique muscle and the one adjacent to the transverse muscle, difficult to differentiate but theoretically it should be crossed to enable ideal spread of the local anaesthetic within the TAP. It extends medially to the semilunar fold and covers all nerve structures which

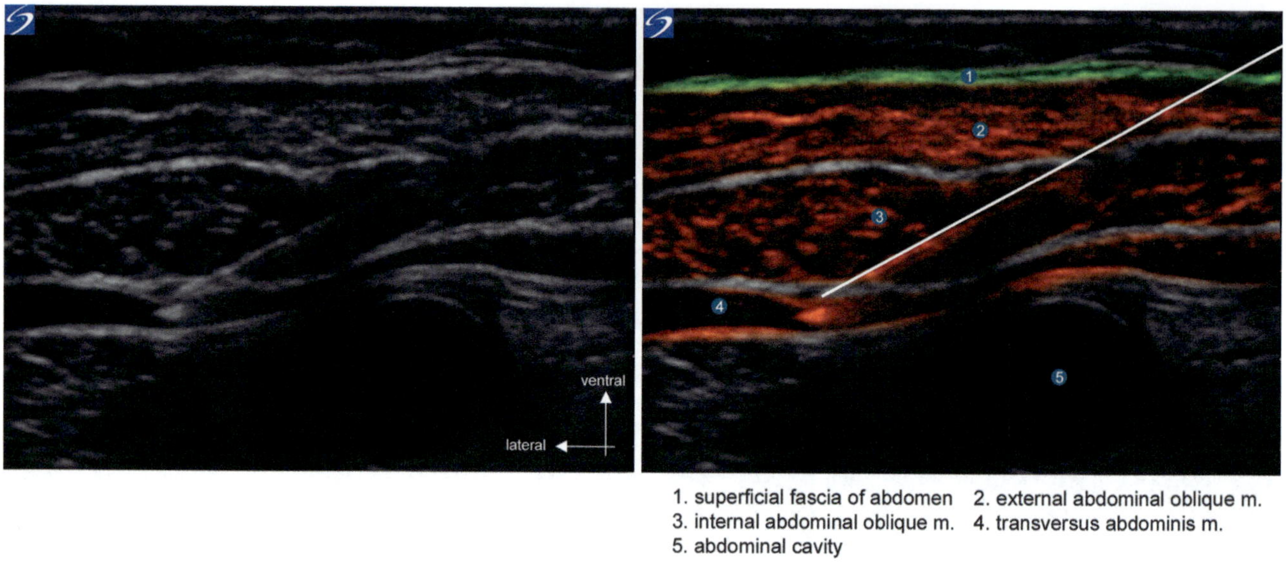

Fig. 9.20 TAP block on the mid-axillary line. Tip of needle positioned between internal oblique and transversus abdominis muscles

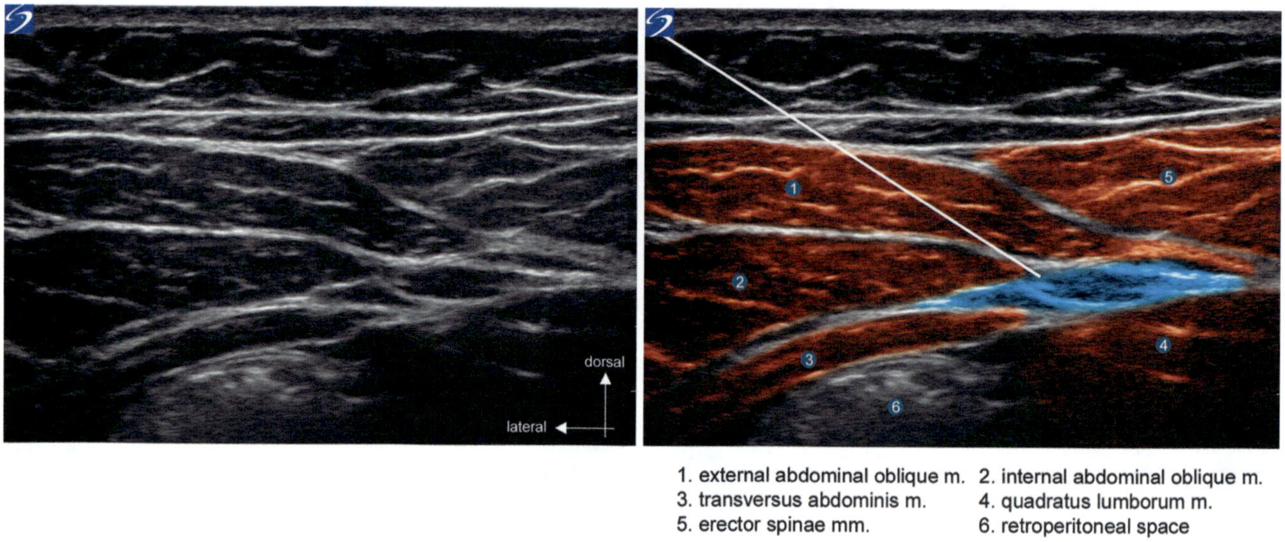

Fig. 9.21 TAP block on the mid-axillary line performed in immediate proximity to the quadratus lumborum muscle. Position of the tip of the needle and injection of local anaesthetic

lead into the TAP [2]. For a TAP block, it is not necessary to inject LA in direct contact with any visible structures that may (or may not) correspond to nerve branches. The anaesthetic solution is merely slowly deposited within the tissue plane. A sign of successful LA placement is the real-time formation of an intermuscular hypoechoic ellipse (in which the tip of the needle is identifiable), demarcated by the superficial and deep muscular aponeuroses (Figs. 9.21, 9.22 and 9.23). As the injection of the local anaesthetic is administered progressively, the hypoechoic ellipse has a natural tendency to spread towards each of its ends, with a protrusion of the transverse muscle into the abdominal cavity. To preserve patient safety, the injection should be performed slowly in a fractionated manner, with repeated aspiration tests.

Out-of-Plane Approach

The technique of performing TAP block out of plane can be useful particularly in obese patients due to the increased depth of the block target. The probe is orientated and positioned as per the in-plane approach (Fig. 9.18). After locating the relevant muscular and aponeurotic structures, the needle is inserted at a distance of 1–2 cm from the probe (Fig. 9.24) and is advanced out of plane to the appropriate depth. In order to control the real-time position of the needle,

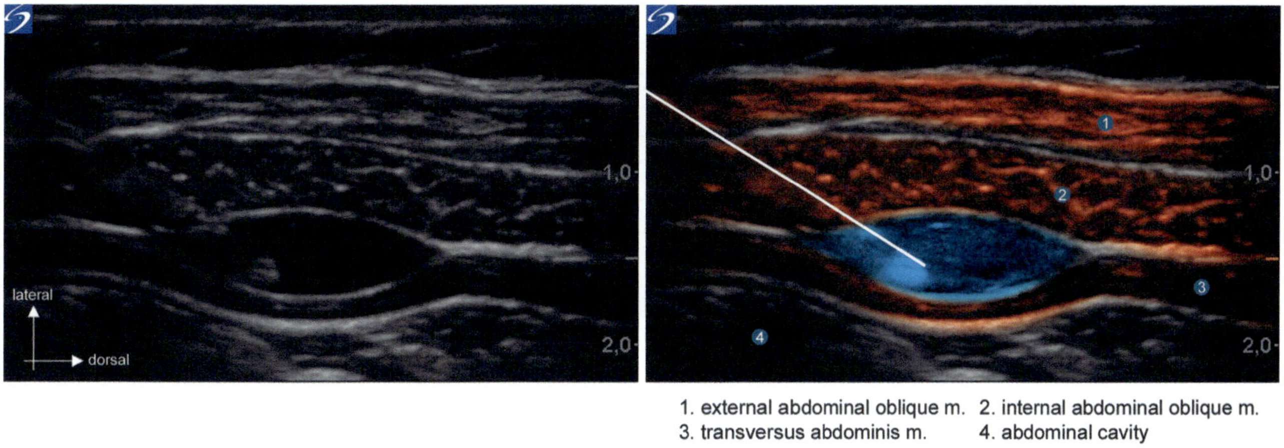

Fig. 9.22 TAP block on the mid-axillary line: formation of an intermuscular hypoechoic "lens" in which the needle is identifiable

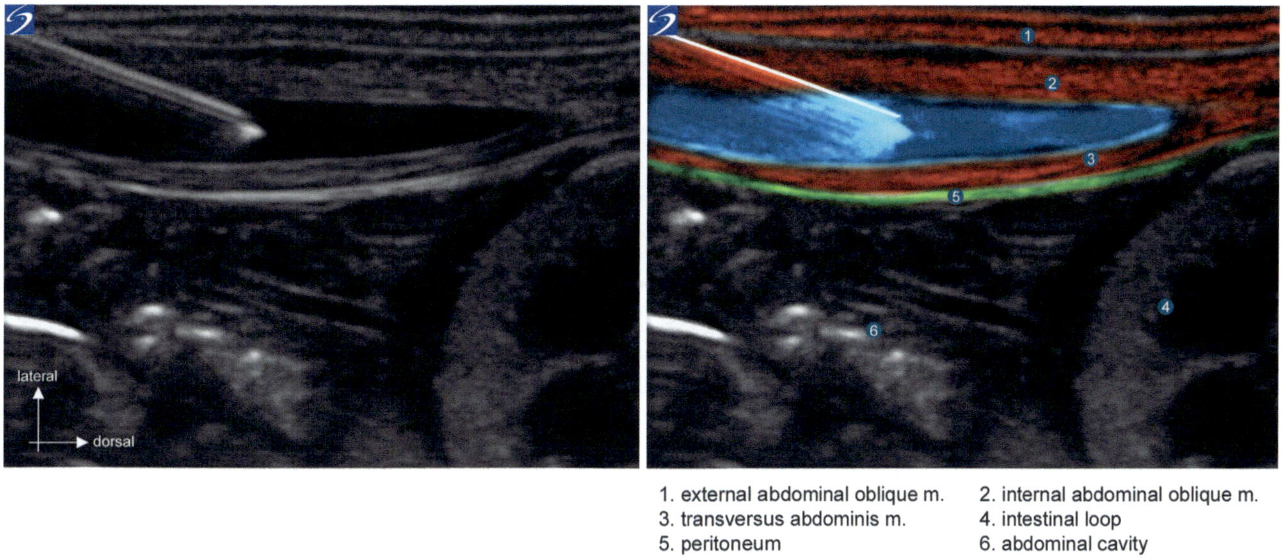

Fig. 9.23 TAP block on the mid-axillary line in paediatrics. During the injection, the muscular aponeuroses separate, materialising the "lens" of local anaesthetic

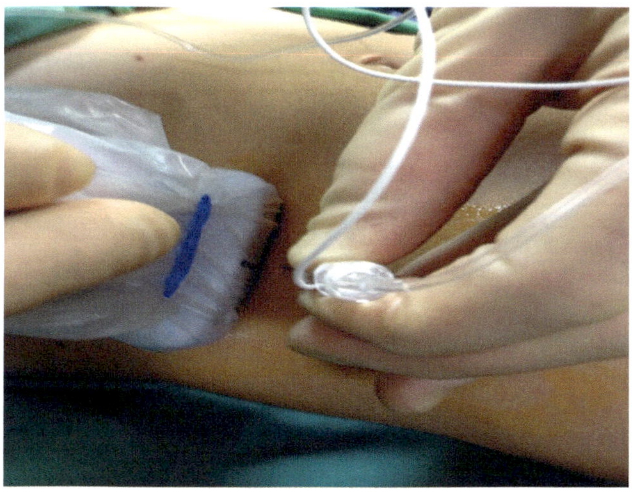

Fig. 9.24 TAP block performed on a child. Position of the probe. Injection out-of-plane

its progression should be slow and punctuated by small, "bouncing" movements, which are transmitted into visible movements within the tissues. Injection of small volumes of 0.9% NaCl (or of the local anaesthetic) during progression also aids in hydrolocalisation of the tip of the needle. Once the needle is positioned in the TAP, after an initial aspiration test, the local anaesthetic is injected slowly to obtain the image of the previously described ellipse.

Performed TAP block in or out of plane, the needle should not be advanced beyond the transversus abdominis muscle. This avoids risking puncture of the peritoneum [17] and also prevents the spread of the local anaesthetic between the transversus muscle and the transversalis fascia, which may result in an inadvertent femoral nerve block (particularly in children) [18, 19]. These complications have been described in "blind" TAP block (with loss of resistance) with a frequency between 5 and 15%. Ultrasound-guided injection can limit this risk [20]. Conduct of a TAP block is recommended under ultrasound guidance with grade B for the American Society of Regional Anaesthesia (ASRA) [21].

Continuous Injection by Catheter

With aseptic precautions, a perineural catheter can be inserted into the space created by the initial fluid bolus ellipse under ultrasound control. The catheter is introduced and then it is moved under ultrasound control to observe spread of the local anaesthetic between the internal oblique and transverse muscles.

TAP Subcostal Block (Fig. 9.25)

Echoanatomy and Approaches

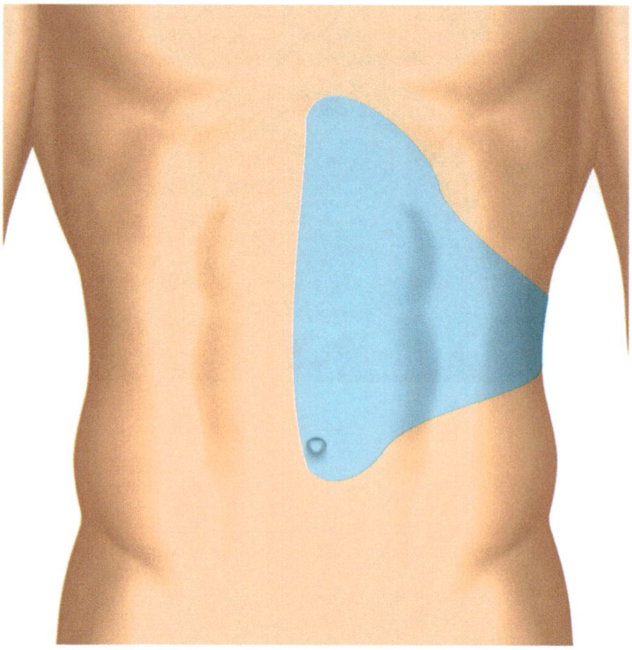

Fig. 9.25 Area of extension of the subcostal TAP block

> **TAP Subcostal Block.**
> **Type of probe:** High frequency, linear 5–10 or 6–13 MHz. A small footprint probe is useful in paediatrics.
> **Axis of the probe:** Oblique, parallel to the ventrolateral costal border, at about 2 cm from it (Figs. 9.26 and 9.27).
> **Configuration:** Nerves inconsistently noticed, on transverse or oblique section; needle in-plane, inserted at the medial or lateral extremity of the probe.
> **Depth studied:** Variable depending on patient build and body mass index.
> **Neurostimulation:** Minimally useful, can enable identification of nerve structures observed.
> **Needle:** 80–100 mm, possibly isolated.
> **Utility of Doppler ultrasound:** Enables locating the arterial branches of the abdominal wall in order to avoid vascular puncture.

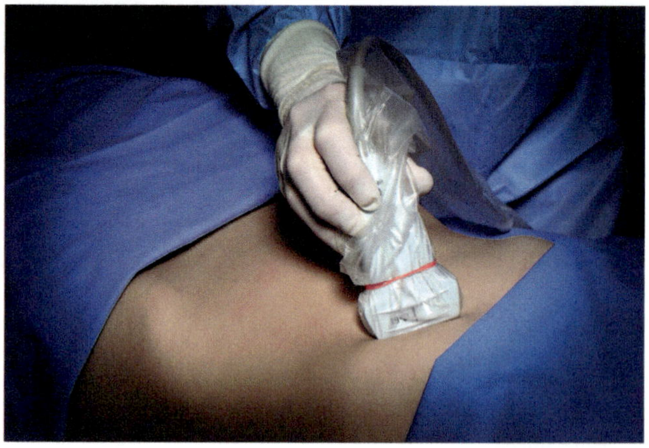

Fig. 9.26 Subcostal TAP block. Initial position of the probe about 2 cm from ventrolateral costal border

The patient is in the supine position. The high-frequency linear probe is positioned along a virtual line located under the ventrolateral costal border, about 2 cm below it. From this initial position, it is possible to study the abdominal wall obliquely from the midline just below the xiphoid process to the anterior axillary line laterally (Figs. 9.26 and 9.27). Success of the block requires spread of the local anaesthetic to the intercostal branches of T6–T9. These nerves are spared by the traditional approach to the TAP block when performed more caudally in the middle axillary line (from this "intercostal iliac" position, extension of the anaesthetic only usually involves dermatomes of T10 to L1, at best [22]) (Fig. 9.28). At the lateral border of the rectus muscle, the intercostal nerves of T6 to T9 are lying in the plane between the transversus abdominis muscle and the tendinous portion of the internal oblique muscle. To perform the block, a relatively long needle is required (80–100 mm in adults and

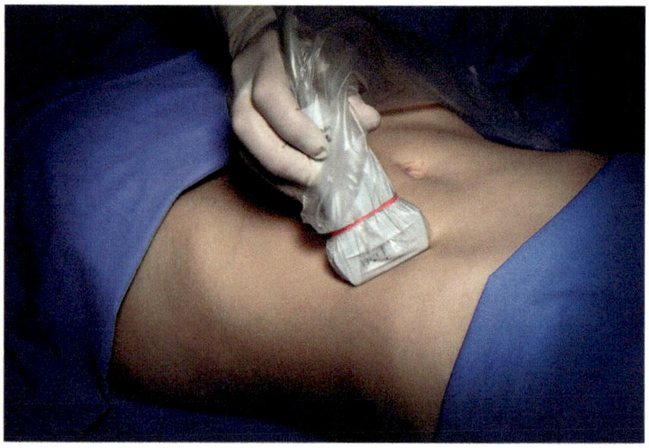

Fig. 9.27 Subcostal TAP block. Movement of probe up to the anterior axillary line

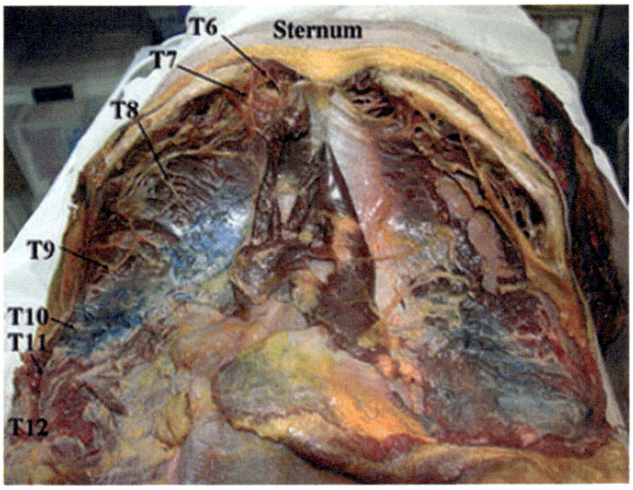

Fig. 9.28 Spread of anaesthetic from a subcostal TAP block (left side of picture). From [13] © Wiley

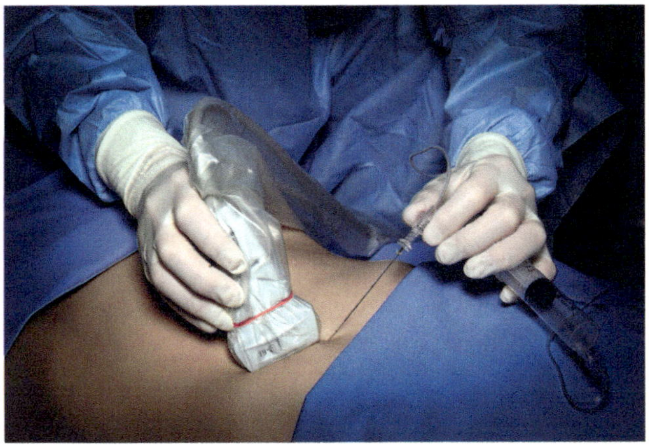

Fig. 9.29 Subcostal TAP block. Medial approach in-plane

50 mm in children) in order to inject the entire volume of local anaesthetic within the block plane via a single needle entry point. The block can be performed in plane from either the medial or lateral ends of the probe. When using either approach it is advisable to plan for the needle path which offers the lowest risk of puncturing the superior epigastric artery.

Medial Approach

The probe is placed under the ventral subcostal margin, with its medial end inside the lateral border of the rectus muscle. The visible underlying structures from this position are: the rectus muscle, the linea semilunaris, and the three abdominal muscle layers (from superficial to deep: the external oblique, internal oblique, and transversus abdominis). In order to avoid accidental vascular puncture, the superior epigastric artery should be sought and its position identified with Doppler ultrasound. The needle is inserted at the medial border of the probe, and is manoeuvred in plane (Fig. 9.29). The initial injection point starts within the lateral third of the rectus muscle. The axis of the needle is angled obliquely (lateralcaudal), and the TAP is targeted near the edge of the linea semilunaris (Fig. 9.30). The injection of the local anaesthetic is started, "opening up" the plane between the transversus abdominis and the rectus muscle, in exactly the same manner as during a TAP block at the mid-axillary line. As the hypoechoic ellipse of local anaesthetic develops, the needle is advanced further into the ultrasound plane in the latero-caudal direction, and the probe is moved along the costal border to follow, in real time, the progression of the needle and anaesthetic spread (Fig. 9.31). Thus, all the local anaesthetic is injected from a single needle entry point into the TAP, and across to the anterior axillary line.

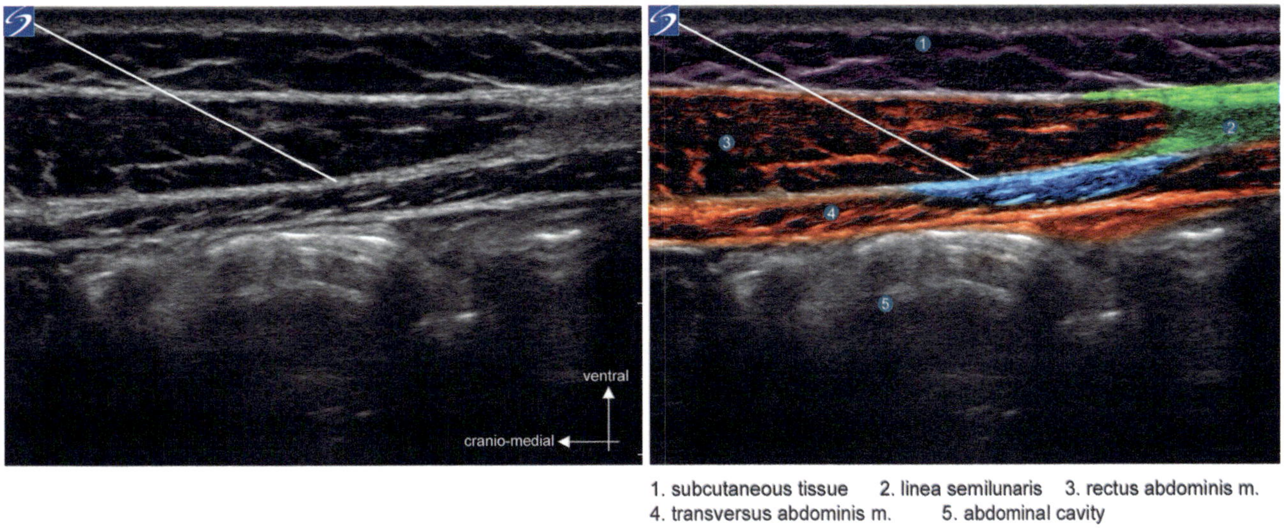

Fig. 9.30 Subcostal TAP block. Medial approach. Axis of needle in the direction of the semilunar line and injection of local anaesthetic starting at the deep aspect of the rectus abdominis muscle in the TAP (transversus abdominis plane)

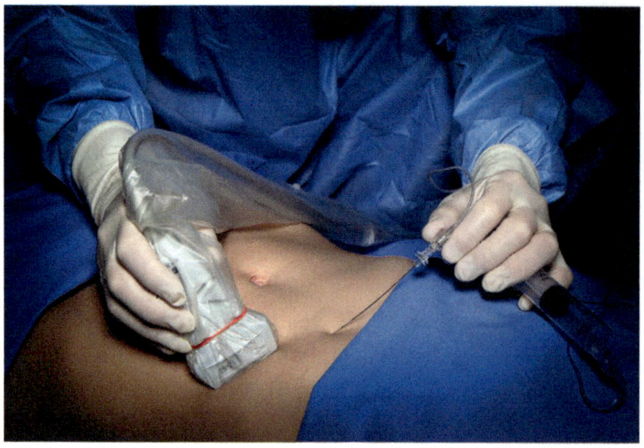

Fig. 9.31 Subcostal TAP block. Medial approach. Direction of needle

Lateral Approach

The probe is positioned on the subcostal line described above, with its lateral border at the anterior axillary line. The needle is inserted at the lateral edge of the probe, and is manoeuvred in plane. The tip is advanced into the TAP using hydrolocalisation. The injection of the local anaesthetic is started, "opening up" the plane between the transversus abdominis and the internal oblique muscles, in exactly the same manner as during a TAP block at the mid-axillary line. As the hypoechoic ellipse of local anaesthetic develops, the needle is advanced further into the ultrasound plane in an oblique (ventro-cephalic) direction. The probe is moved along the costal border to follow, in real time, the progression of the needle and spread of the local anaesthetic. Thus, all the local anaesthetic is injected from a single needle entry point into the TAP to the lateral border of the rectus muscle.

Iliohypogastric and Ilioinguinal Nerve Block (Fig. 9.32)

Iliohypogastric and Ilioinguinal Nerve Block

Type of probe: High-frequency linear 5–10 or 6–13 MHz. A small footprint probe is useful in paediatrics.

Axis of the probe: Oblique, on a line connecting the anterior superior iliac spine to the umbilicus (Fig. 9.33).

Configuration: Nerves are inconsistently seen in transverse or oblique sections, needle in- or out of plane.

Depth studied: Variable depending on body build and body mass index of the patient.

Neurostimulation: Minimally useful, can enable identification of nerve structures observed.

Aiguille: 80–100 mm, possibly isolated.

Utility of Doppler ultrasound: Enables locating the arterial branches in the abdominal wall in order to avoid vascular puncture.

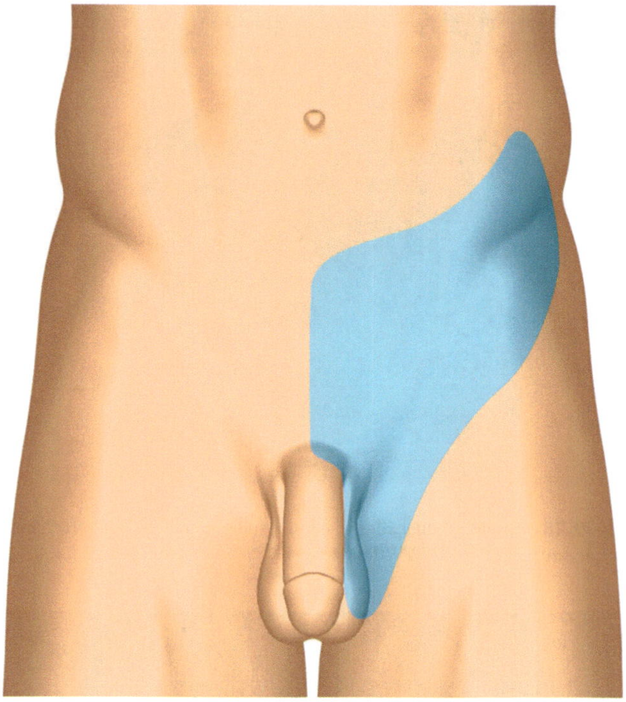

Fig. 9.32 Area of extension of iliohypogastric and ilioinguinal blocks

Indications

The territory for this block includes the hypogastric area, the inguinal fold, the craniomedial part of the thigh, the pubis, the base of the penis, and the anterior part of the scrotum or the ventral part of the labia majora. Major individual variations exist within these sensory territories.

Iliohypogastric and ilioinguinal nerve blocks are commonly performed for hernia surgery in children [23] and adults, but also for analgesia in many types of surgeries involving an ipsilateral inguinal incision [24–26]. The ilioinguinal block thus is used for analgesia in peritoneal-vaginal canal surgery, orchidopexy, spermatic cord surgery, or repair of a hydrocele. Ultrasound makes it possible to optimise the position of the needle between the internal oblique and the transversus abdominis muscles and to significantly reduce the dose of local anaesthetic injected to achieve analgesia. This block carries certain risks (e.g. puncture of the colon and haematoma), however, the use of ultrasound provides an additional level of safety. Post-operative pain after hernia surgery can be sometimes significant and prolonged, and carries the risk of the patient developing disabling chronic pain. When performed on the right side, this block is also indicated for analgesia in open appendicectomy by the Mc Burney point approach.

Echoanatomy

Iliohypogastric and ilioinguinal nerve block consist of TAP block sited in a more ventralcaudal position than traditional "intercostal iliac" TAP block, aiming to inject the local anaesthetic in immediate proximity to the nerves concerned. The patient is in the supine position. The high-frequency linear probe is positioned on a line connecting the anterior superior iliac spine (ASIS) to the umbilicus (Fig. 9.33). After observing the three muscle layers (from deep to superficial: the transversus abdominis, internal oblique, and external oblique), the aim is to identify the iliohypogastric and ilioinguinal nerves as they lie between the transversus muscle and the internal oblique muscle. Small craniocaudal movements of the probe may aid nerve identification and the use of Doppler ultrasound can reveal satellite arterial branches. Each nerve branch appears in the form of a small hypoechoic disc within a hyperechoic image in the shape of a "small boat" in intermuscular position (Fig. 9.34). The iliohypogastric nerve is in a more medial and cranial position than the ilioinguinal nerve. The iliac branch of the deep iliac circumflex artery is sometimes visible with colour Doppler ultrasound, as it lies on the iliac crest. Near to the ASIS, the iliohypogastric and ilioinguinal nerves are generally located between the transversus and internal oblique muscles. However, one (usually the iliohypogastric nerve) or both nerves are sometimes located between the external oblique and the internal oblique muscles. Moreover, in paediatrics, the external oblique muscle is often very thin or even aponeurotic at this level [24], and the ultrasound system may demonstrate only the two superimposed muscular layers of the internal oblique and transverse muscles (Fig. 9.35). In thin adults and in children, it is possible to observe a fourth

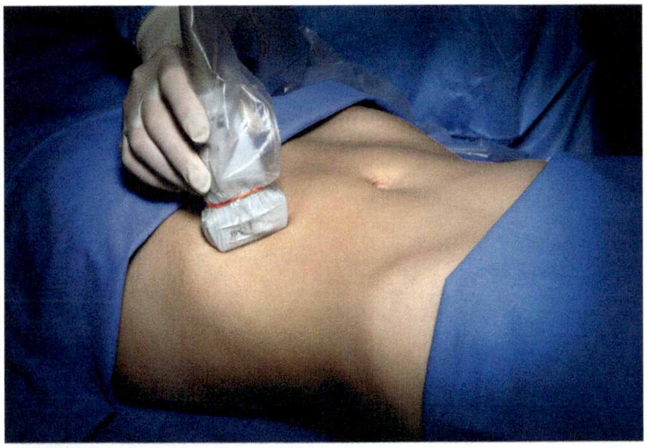

Fig. 9.33 Iliohypogastric and ilioinguinal blocks. Position of probe between anterosuperior iliac spine and umbilicus

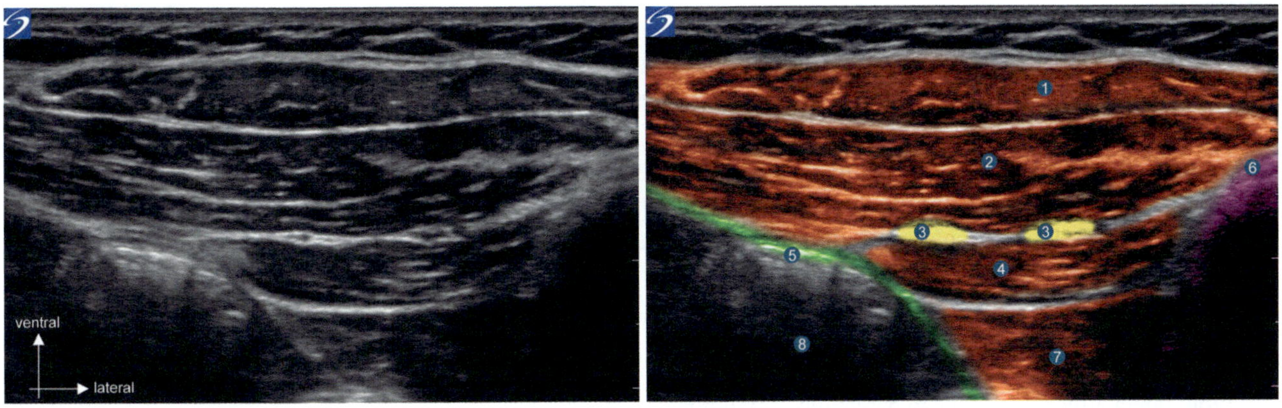

1. external abdominal oblique m.
2. internal abdominal oblique m.
3. iliohypogastric and ilioinguinal nn.
4. transversus abdominis m.
5. peritoneum
6. anterior superior iliac spine
7. iliacus m.
8. abdominal cavity

Fig. 9.34 Iliohypogastric and ilioinguinal blocks. Transverse ultrasound view in adult

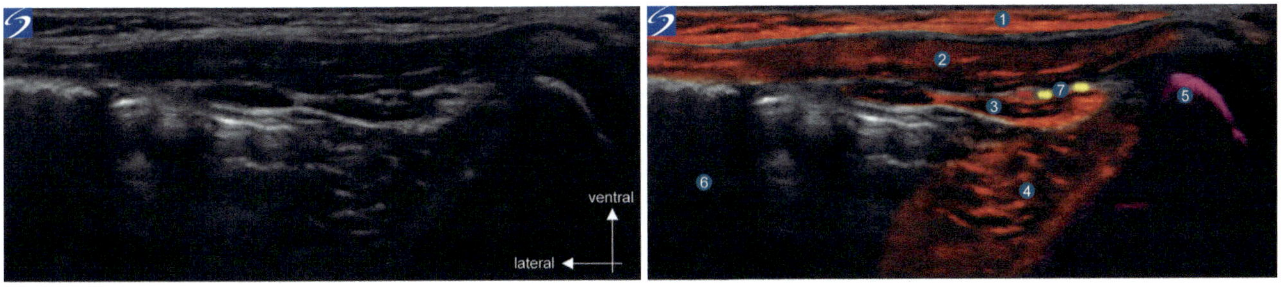

1. external abdominal oblique m.
2. internal abdominal oblique m.
3. transversus abdominis m.
4. iliacus m.
5. anterior superior iliac spine
6. abdominal cavity
7. ilioinguinal and iliohypogastric nn.

Fig. 9.35 Ultrasound view at the level of the iliac crest in a 4-year-old child (6–13 MHz linear probe) placed in a transverse orientation visualising the iliohypogastric nerve and the ilioinguinal nerve under the internal oblique muscle; the external oblique muscle is aponeurotic at this level

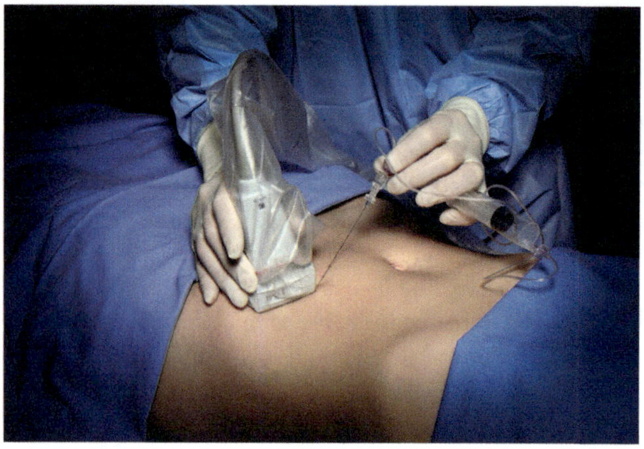

Fig. 9.36 Iliohypogastric/ilioinguinal block. Medial approach in-plane

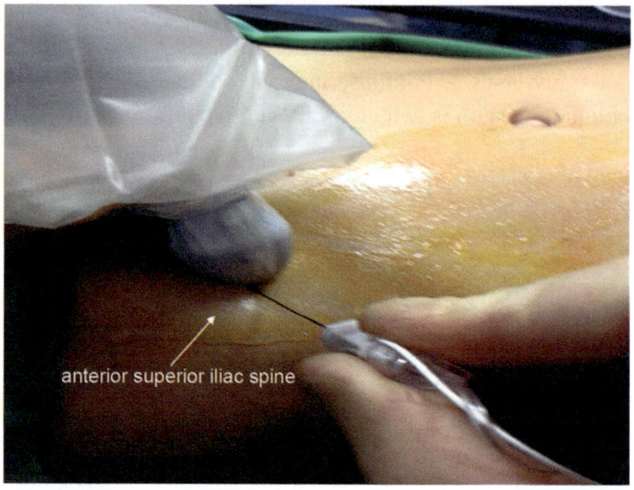

Fig. 9.37 Iliohypogastric block. Lateral approach in-plane

muscular layer placed on the iliac bone: this is the iliacus muscle, not to be confused with the transversus abdominis (Fig. 9.34). In paediatrics, the use of a small footprint probe is useful in performing this block.

Approach

The injection can be performed in three different manners.

Medial Approach, in Plane (Fig. 9.36)
In this approach, the needle is inserted at the medial border of the probe and is advanced in plane until reaching the two nerves. The initial injection of small volumes (1–2 mL) ensures the strictly intermuscular position of the tip (by noting the elliptical spread of liquid between the muscular aponeuroses). Thereafter, the local anaesthetic is deposited adjacent to the iliohypogastric and ilioinguinal nerves. Care must be taken to avoid an intraneural or intravascular injection: inject either side of the hyperechoic "small boat" containing the nerve branches (not directly into it), and perform an aspiration test before each injection. Puncture of the branch of the inferior epigastric artery may be avoided by use of Doppler ultrasound during this approach.

Lateral Approach, in Plane (Fig. 9.37)
The process for identification of the intermuscular space and the injection technique is exactly as described for the medial approach. This time the needle is inserted at the lateral end of the probe. In order to avoid being hindered by the ASIS, the clinician may decide to move the probe slightly cranially before commencing the injection.

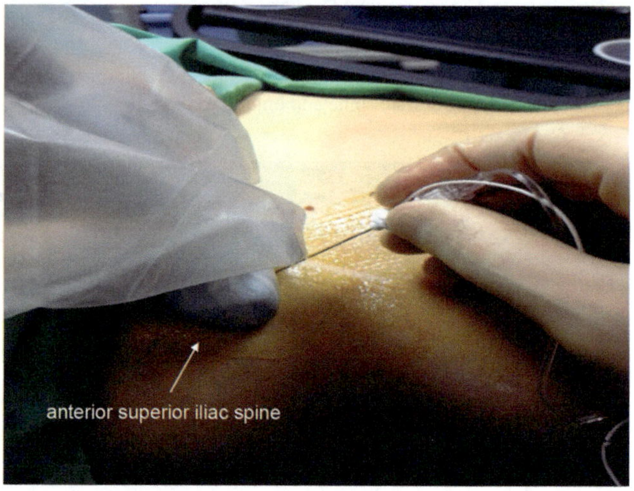

Fig. 9.38 Iliohypogastric/ilioinguinal block. Approach out-of-plane

Out-of-Plane Approach (Fig. 9.38)
The out-of-plane approach to this block is facilitated by appreciating changes in resistance felt during the passage of the needle, along with visualising movements within the tissues transmitted by the needle coupled with hydrolocalisation. This approach offers the advantage of a shorter and more direct pathway for the needle which can be useful, particularly in obese patients. The procedure for identification of the intermuscular space and the injection technique is exactly as described in the in-plane approach.

Nerve Blocks for the Quadratus Lumborum Plane (Fig. 9.39)

Nerve Block of Quadratus Lumborum Plane

Type of probe: High-frequency linear, 5–10 to 6–13 MHz in patients with standard body build. A low-frequency curvilinear probe 2–5 MHz may be required in patients with a high BMI.

Axis of probe: Transversal in dorsolateral position (Fig. 9.42).

Configuration: Nerves inconsistently seen, on transverse or oblique sections, needle in plane.

Depth studied: Variable depending on body build and body mass index of the patient.

Neurostimulation: Minimally useful, but may enable identification of nerve structures.

Needle: 80–100 mm, possibly isolated.

Utility of Doppler ultrasound: Enables locating the arterial branches of the abdominal wall in order to avoid vascular puncture.

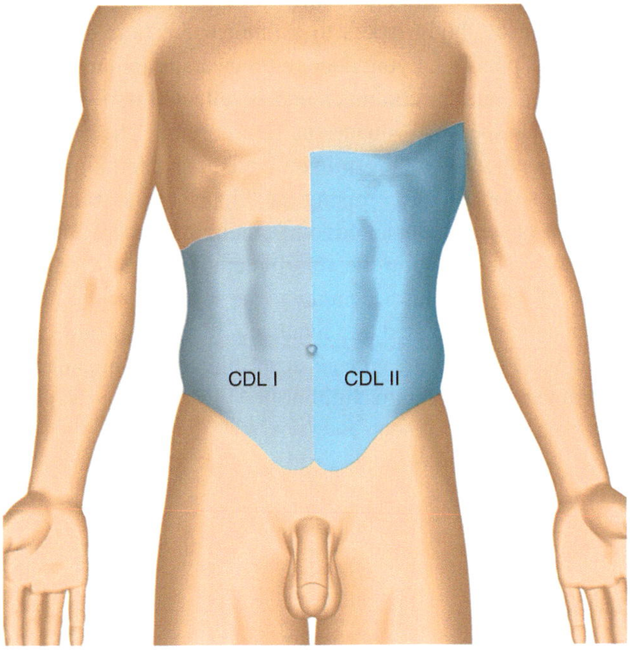

Fig. 9.39 Territories of extension of types I and II quadratus lumborum blocks

Definition of Type I and Type II Nerve Blocks

Two types of nerve blocks are defined depending on the injection site and spread of the local anaesthetic in relation to the quadratus lumborum muscle (Figs. 9.40 and 9.41):

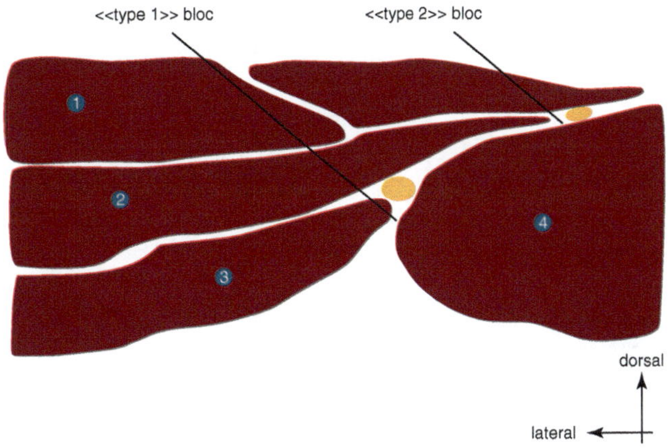

Fig. 9.40 Plane of quadratus lumborum block: in type I, the injection point lies at the lateral or ventrolateral border of the quadratus lumborum muscle; in type II, the injection is administered at the dorsal aspect of the muscle in order to obtain a paravertebral spread of the local anaesthetic

- A **type I nerve block of the quadratus lumborum plane** (original description) refers to an injection at the lateral border of the muscle, immediately caudal to the subcostal nerve.
- **A type II block of quadratus lumborum plane** (original description modified) describes an injection in the dorsal plane of the quadratus lumborum muscle. MRI examinations have confirmed that this approach can achieve spread of the anaesthetic into the paravertebral space.

Indications

The sensory block obtained extends from T8 to T12 with a type I injection. The effects of a type II injection can extend from T5 to T12 as the result of its paravertebral spread.

Thus, analgesia for a high number of surgical procedures can be achieved by these two types of nerve blocks: e.g. cholecystectomy, renal surgery, gynaecological surgery, and hernia repair (with possible supplement).

Sonoanatomy and Approach

After lateral positioning of the patient, application of monitoring, probe preparation, and asepsis of the skin, an echographic examination of the ventrolateral abdominal wall is performed. The probe is placed transversely superior to the most cranial part of the iliac crest in the mid- or posterior axillary line. Its position is adjusted to visualise the quadratus lumborum muscle (usually by moving slightly more posteriorly). The block is performed in plane with the needle inserted at the anterior end of the probe (Fig. 9.42).

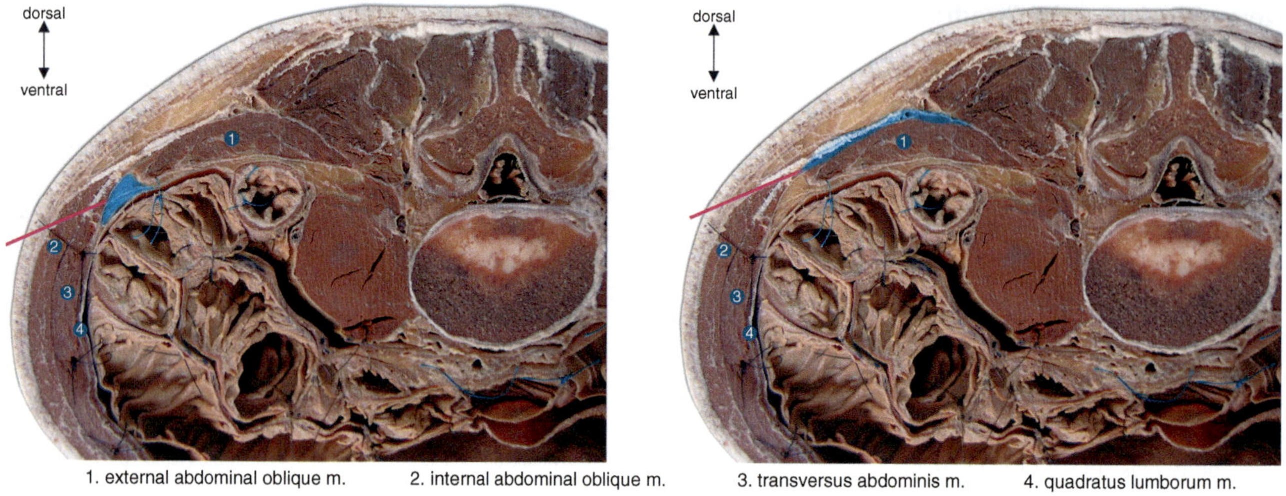

Fig. 9.41 Quadratus lumborum plane nerve block. Type I (image on left) and type II (image on right). Injection point and spread of local anaesthetic

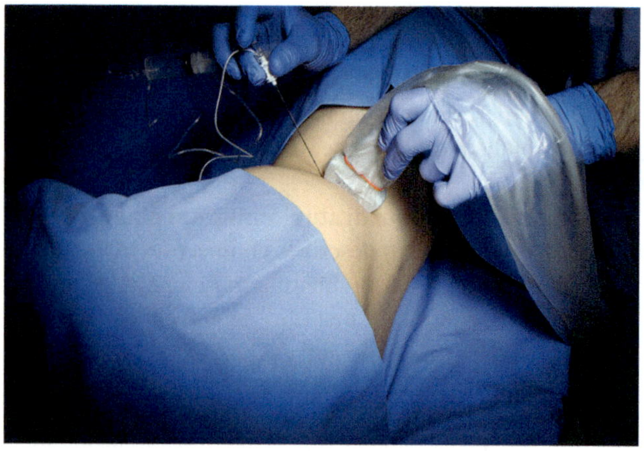

Fig. 9.42 Position of probe and insertion of the needle for types I and II quadratus lumborum blocks

Type I Technique

The three abdominal muscle layers (external oblique, internal oblique, and transversus abdominis) are followed dorsally until quadratus lumborum is identified. The sonographic appearance of the muscles is then seen to change. The transversus muscle body tapers to a fine hyperechoic layer (its tendinous portion). Lying deep to transversus abdominis, the fascia transversalis continues and is reflected onto the ventrolateral aspect of quadratus lumborum. The external oblique is then seen to terminate whilst the internal oblique continues dorsally (comprising the "floor" of the triangle of Petit) to cover the dorsal edge of quadratus lumborum. The needle is inserted up to the ventrolateral portion of the quadratus lumborum muscle, caudal to the subcostal nerve (T12) which can be visualised as a hyperechoic structure. Injection of the local anaesthetic spreads according to a cranial–caudal axis anterior to quadratus lumborum (Figs. 9.43 and 9.44).

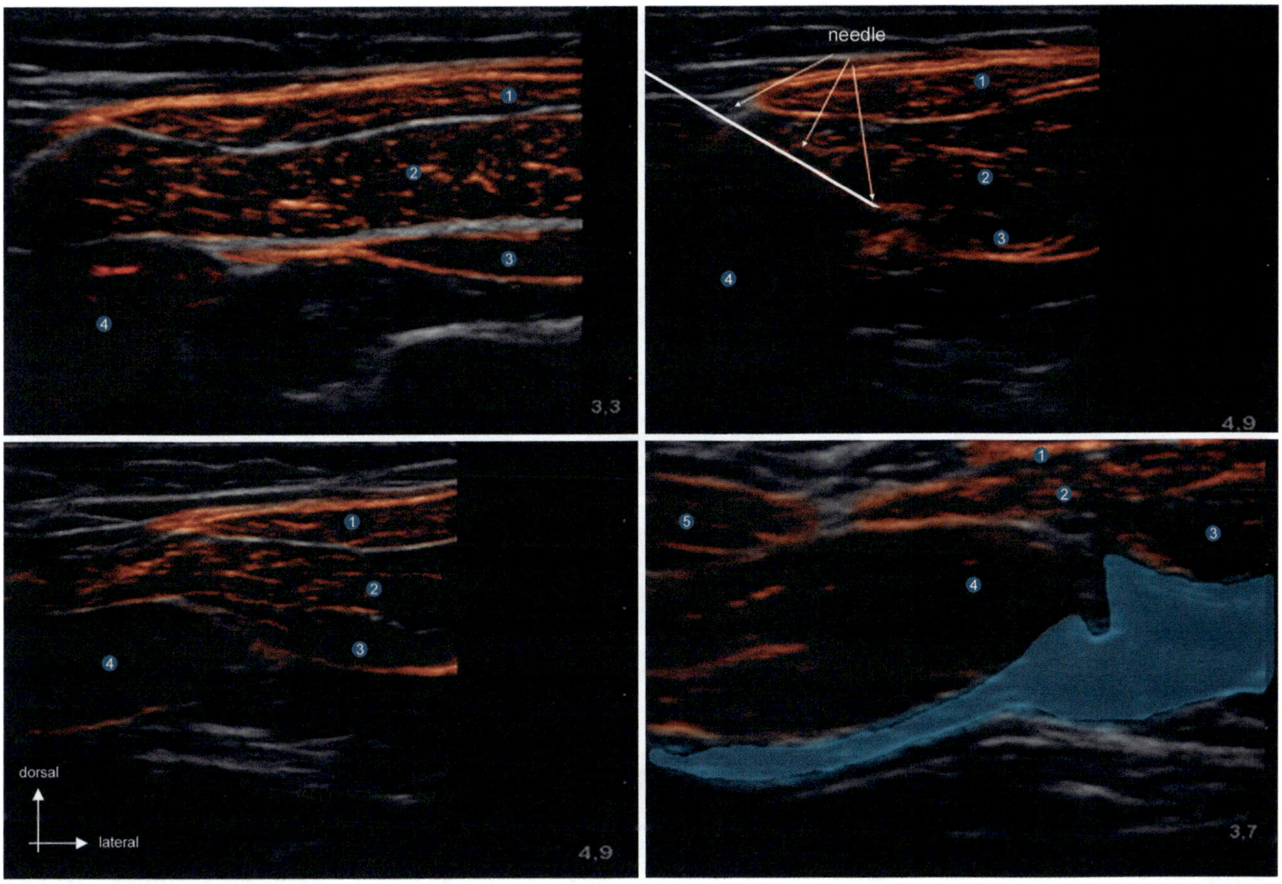

1. external abdominal oblique m. 2. internal abdominal oblique m. 3. transversus abdominis m.
4. quadratus lumborum m. 5. erector spinae mm.

Fig. 9.43 Type I quadratus lumborum block: injection at the lateral border of the quadratus lumborum muscle with ventrolateral spread of local anaesthetic. Note that in these images, exceptionally a dorsoventral injection pathway is used, contrary to that shown in Fig. 9.42, which is ventrodorsal

Type II Technique

The three abdominal muscle layers (external oblique, internal oblique, and transversus abdominis) are followed dorsally until quadratus lumborum is identified. In this approach, the needle is angled more posteriorly and directed towards the dorsal aspect of quadratus lumborum. Local anaesthetic is deposited in this plane—between the dorsal aspect of the muscle and its fascial covering (the aponeurotic insertion of the transversus and internal oblique muscles) (Figs. 9.40 and 9.45). Performing hydrodissection in a medial and cranial direction during injection of LA improves its spread within the plane. With MRI imaging, the extent of spread of the injection is observed, reaching the paravertebral space, possibly through the subcostal space T12 (Fig. 9.46).

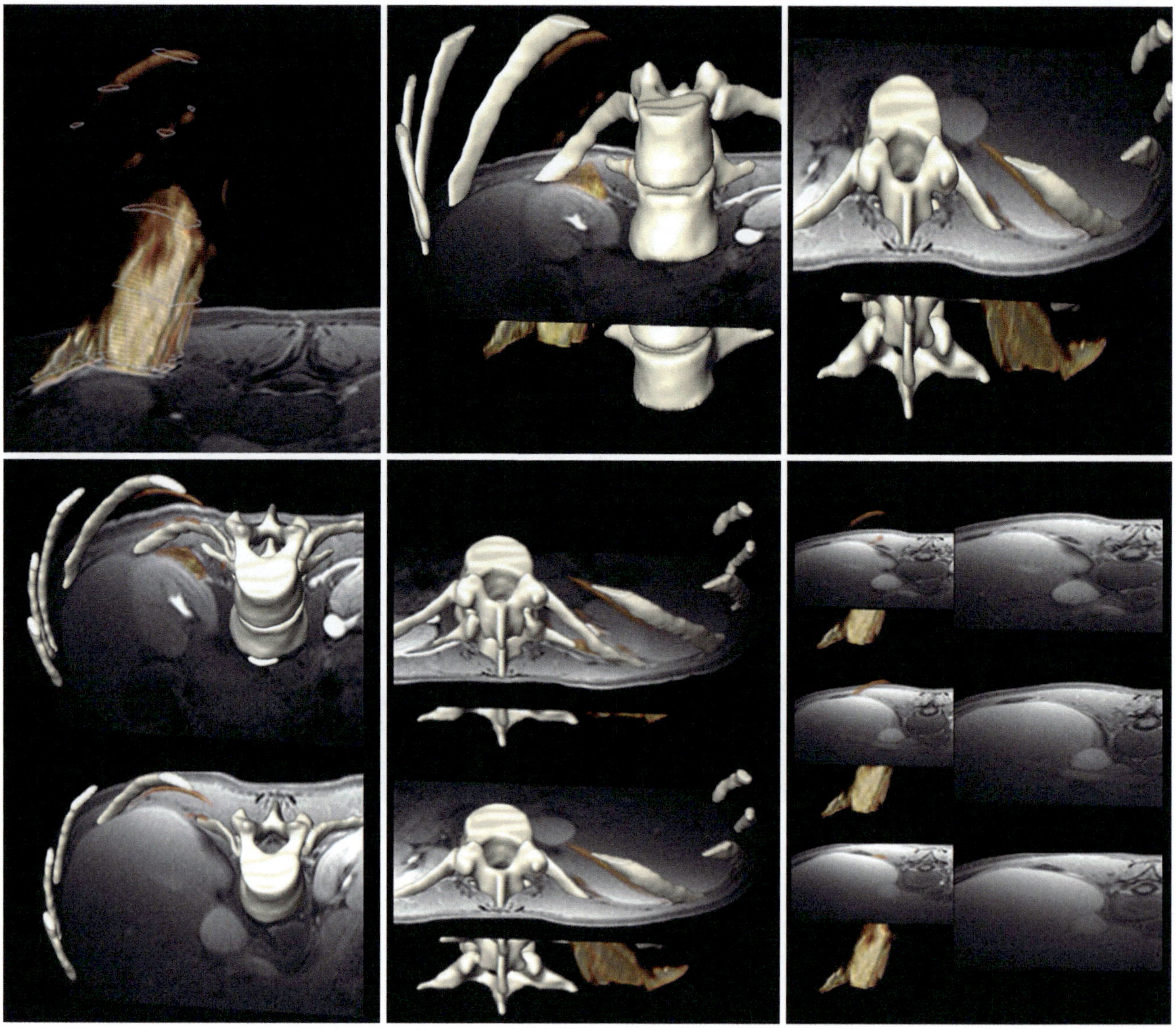

Fig. 9.44 Type I quadratus lumborum plane nerve block: minor spread of contrast medium into paravertebral space and T11 and T12 intercostal spaces. Rafael Blanco iconography

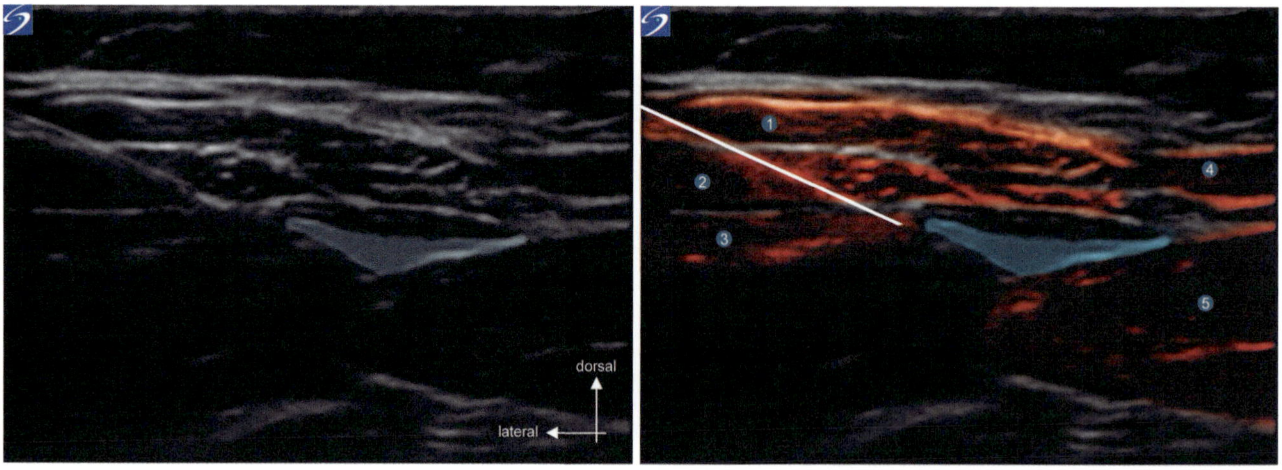

Fig. 9.45 Type II quadratus lumborum block. Injection into dorsal plane of quadratus lumborum muscle

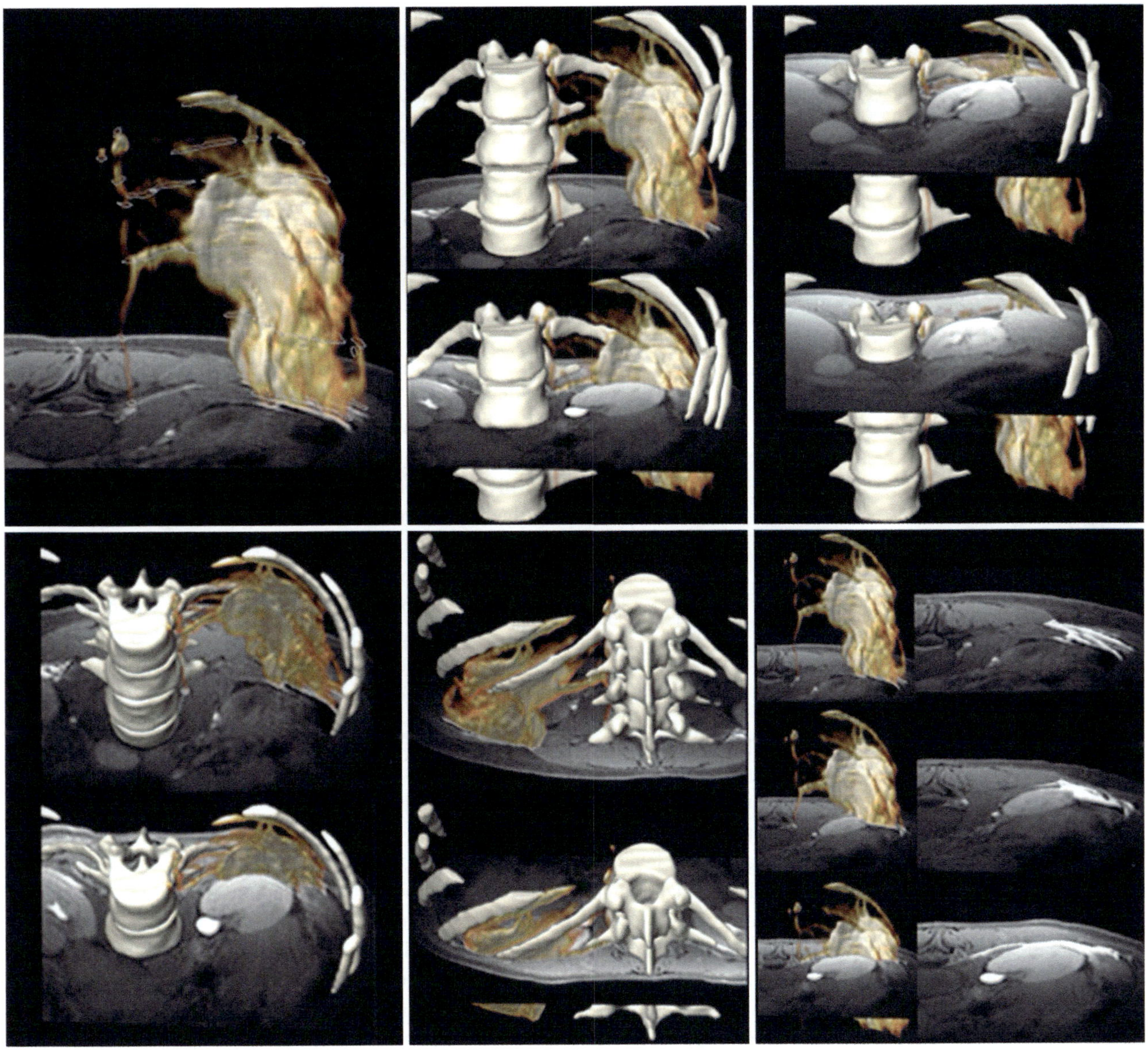

Fig. 9.46 Type II quadratus lumborum block. Spread of contrast medium into the paravertebral space in intercostal spaces much more pronounced than in a type I block (Fig. 9.44). Rafael Blanco iconography

Rectus Sheath Block (Fig. 9.47)

Block of the Rectus Sheath

Type of probe: High-frequency linear from 5–10 or 6–13 MHz. A small footprint probe is useful in paediatrics.

Axis of the probe: Transverse, at the lateral one-third of the rectus abdominis muscle and linea semilunaris (Figs. 9.53 and 9.54). A bilateral block is necessary to anaesthetise the midline.

Configuration: Nerves rarely seen; needle in-plane, lateral, or medial approach.

Depth studied: Variable depending on body build and body mass index of the patient.

Neurostimulation: Unnecessary.

Needle: 80–100 mm.

Utility of Doppler ultrasound: Enables locating the branches of the superior or inferior epigastric arteries.

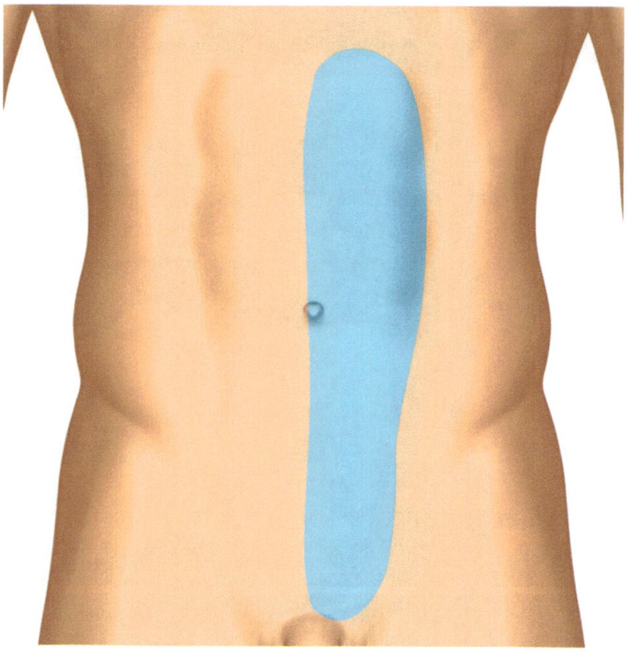

Fig. 9.47 Area of extension of a rectus abdominis muscle sheath block (unilateral block in this sketch)

General Considerations

The ventral portion of the rectus sheath is partitioned (compartmentalised by several tendinous intersections passing between the ventral aspect of the rectus muscle and the ventral layer of the sheath). However, the fascial plane between the posterior surface of the rectus muscle and deep layer of the sheath has no such partitions. Thus, if an injection of local anaesthetic is made within this plane it can (theoretically) spread along the entire length of the ipsilateral sheath. Since the linea alba presents a barrier to LA spread to the opposite side of the sheath it is necessary to perform bilateral rectus sheath blocks to fully anaesthetise the midline. Also note that despite the posterior rectus sheath being theoretically a single, continuous space, to ensure the optimal craniocaudal spread of LA it is advisable to perform several injections on each side (at 5–7 cm intervals), depending on the extent of the analgesia required.

Examples:

- For supraumbilical midline anaesthesia (T6-T7-T8-T9-T10): bilateral injections at two levels are required (i.e. 4 injections in total).
- For subumbilical midline anaesthesia (T10-T11): bilateral injections at a single level, to which it might be necessary to add blocks of the subcostal and iliohypogastric nerves to cover the hypogastric and median prepubic area (i.e. essentially a TAP block).
- For midline incisions extending both above and below the umbilicus: bilateral injections at three levels (i.e. 6 injections) are required, at least. However, in this case, it would be more logical to plan for bilateral TAP blocks in the mid-axillary line, performed with bilateral subcostal TAP blocks.

A few details need to be provided on what is called a paraumbilical block. This technique describes the "blind" injection of LA for analgesia for surgery involving the umbilicus and the periumbilical area (e.g. hernia repair). Traditionally, it involves inserting a needle into the skin and then appreciating the tactile sensation of the tip piercing the aponeurosis (the ventral layer of the sheath). The injection of local anaesthetic is given into the **ventral** compartment of the rectus sheath (i.e. in front of the muscle). This is contrary to a so-called "rectus sheath block" with the injection performed at the **dorsal** aspect of the muscle, between the latter and the dorsal layer of the sheath. A paraumbilical block can be performed under ultrasound guidance with better anatomical precision than the traditional technique.

Indications [27–29]

A rectus sheath block makes it possible to block the nerves located within the rectus abdominis sheath, i.e. the anterior cutaneous branches of the ventral rami of the thoracic nerves from T7 to T12. This block can be used alone for anaesthesia in hernia repair in the midline, but it is also most often performed in combination with general anaesthesia for post-operative analgesia after the same procedure, or after laparotomy (with midline or small transverse incisions) in gastrointestinal or gynaecological surgery.

A paraumbilical block is useful for post-operative analgesia after umbilical hernia repair, small midline herniae, gastrointestinal or gynaecological laparoscopy, pylorotomy (by the umbilical approach), and removal of tumours from the umbilical area. It is a block of the abdominal wall and it provides analgesia only to superficial structures. It was first performed by Schleich in 1899 who described a local anaesthetic injection made deep to the rectus abdominis muscle, in contact with the deep branch of the nerve, with risk of perforation of the peritoneum. In current practice, the local anaesthetic is deposited under the anterior aponeurosis of the rectus abdominis muscle. It can be used alone for anaesthesia in umbilical hernia repair, possibly combined with intraoperative infiltration and sedation.

Haematoma, injection site infection, and deep abscess are extremely rare complications, but have all been reported.

Use of ultrasound guidance is recommended with grade A evidence by the ASRA (level Ib) with a major reduction in risk of needle puncture of the peritoneum [21, 22].

Ultrasound guidance is essential for the precise conduct of the rectus sheath block. It enables the LA to be correctly deposited between the deep aspect of the rectus muscle and the deep aponeurosis of the sheath, thereby reducing the risk of peritoneal puncture. It should be kept in mind that below the arcuate line, the posterior layer of the rectus sheath is absent and the only structure between the muscle and the surface of the parietal peritoneum is the fascia transversalis. Ultrasound guidance also enables identification of vascular components present in this space [30]. Insertion of a catheter into the rectus sheath has also been described [31].

Using 0.1 mL/kg of 0.25% levobupivacaine during bilateral ultrasound-guided block of the rectus sheath in 20 children who underwent surgery for repair of umbilical hernia, a study observed satisfactory spread of the local anaesthetic in all cases and with analgesic success (based on the need for peri- and post-operative analgesia) in 100% (all) cases [29]. Dolan et al. observed the more frequent success by means of ultrasound guidance compared to a "loss of resistance" technique (89% vs 45%, respectively) [32]. The same study demonstrated a 21% incidence of intraperitoneal location of the needle with injections performed blindly versus none under ultrasound guidance. Conduct of diagnostic and therapeutic nerve blocks in chronic pain is an application, which is developing with recent success in a case of post-herpetic neuralgia [33].

Echoanatomy

With the patient in the supine position, a high-frequency linear ultrasound probe is initially placed in transverse position on the midline (Fig. 9.48), and then is slid laterally opposite the body of the rectus muscle. The rectus muscles appear as two ellipses separated by a white line (linea alba), lying beneath the skin and the superficial tissues. They are surrounded by the rectus sheath, easily visualised in the form of a hyperechoic envelope. Above the arcuate line, by moving the probe laterally, the ventral layer and the dorsal layer of the sheath are seen to join together at the lateral border of each rectus muscle (Fig. 9.49). In the midline of the abdomen, the two layers of the sheath unite with each other and with those of the contralateral sheath by inserting into the linea alba (Fig. 9.50). Below the arcuate line, the aponeuroses of the external oblique, internal oblique, and transverse muscles lie entirely ventral to the rectus muscle. Thus, the latter is separated from the parietal peritoneum only by the fascia transversalis (Fig. 9.7, p. 238).

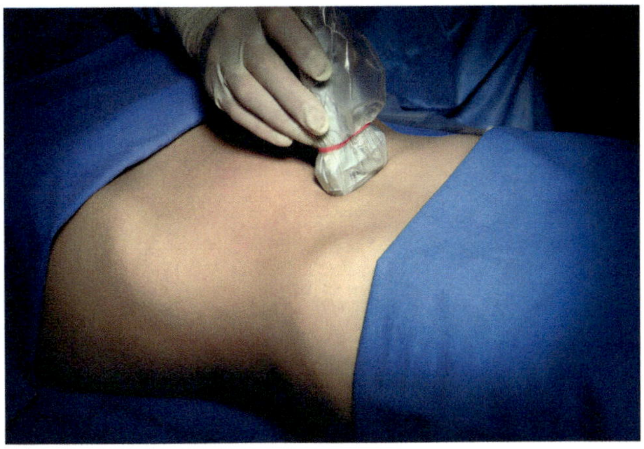

Fig. 9.48 Rectus abdominis muscles sheath block. Initial position of probe on the midline

Although it is very difficult to visualise them, the seventh, eighth, ninth, 10th, and 11th intercostal nerves are located between the rectus muscle and the dorsal layer of its sheath, before piercing and traversing the muscle to give rise to the anterior cutaneous branches at the skin surface. With Doppler ultrasound, it is also possible to visualise the epigastric arterial branches in the body of the muscle or in the sheath behind the rectus muscle (Figs. 9.51 and 9.52). Deep to the sheath are located: the preperitoneal fascia transversalis, the parietal peritoneum, and the intestines, whose peristaltic movements can be observed.

The sonoanatomical markers of paraumbilical block are more superficial. The probe is placed transversely, lateral to the umbilicus. It is positioned at the lateral part of the rectus muscle and the ventral layer of its sheath is identified. Since the injection, in this case, is to take place at the surface of the muscle, it is not essential to locate the branches of the inferior epigastric artery which usually lie in/deep into the muscle body at this level.

Approaches

In the literature, the described techniques are:

- In transverse ultrasound section:
  - In-plane approach (medial approach and lateral approach).
  - Out-of-plane approach.
- In paramedian sagittal ultrasound section:
  - In-plane approach.

Contrary to the situation of the TAP block where the transversus abdominis muscle is the layer that limits the risk

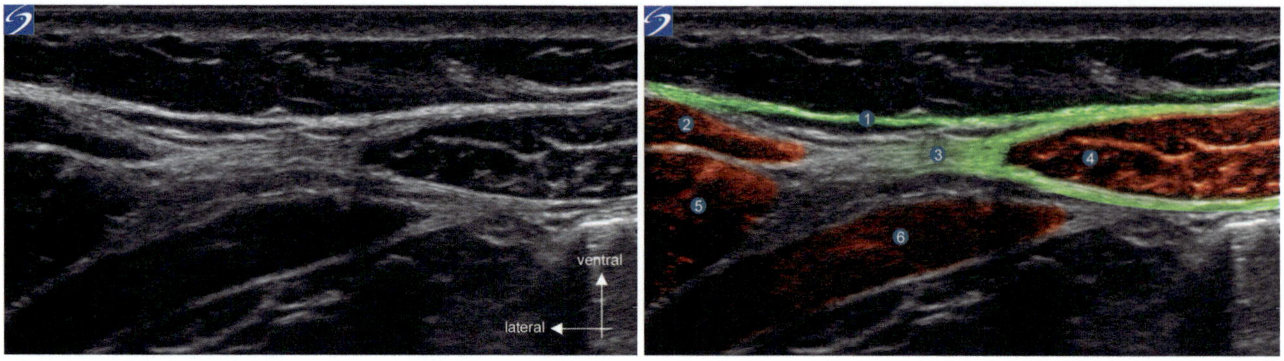

Fig. 9.49 Sonoanatomy: lateral border of the rectus abdominis muscle

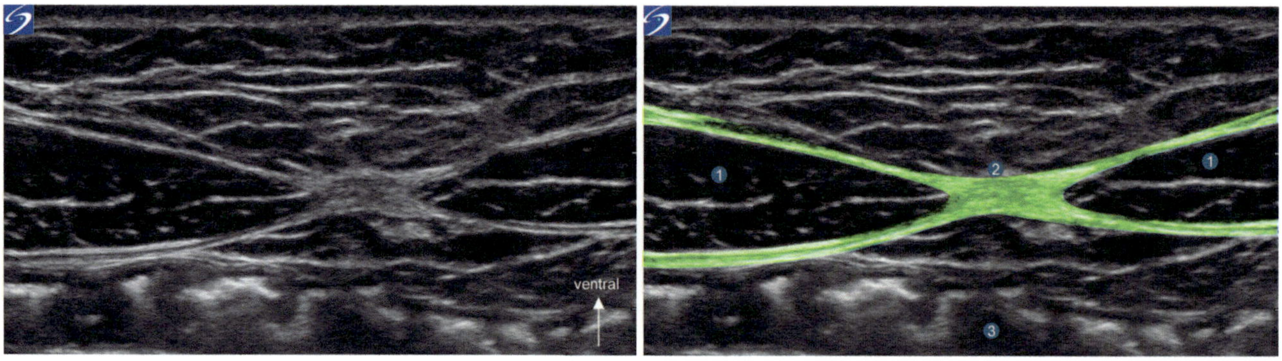

Fig. 9.50 Sonoanatomy: medial part of rectus abdominis muscles on the midline. Linea alba

of puncture of the peritoneum, in the rectus sheath block the deep aspect of the rectus muscle is in very close proximity to the peritoneum/peritoneal cavity. Therefore, an **out-of-plane** approach to this block may create a higher risk of needle puncture of the peritoneum than with an **in-plane** approach where the needle path is more tangential to the probe and offers better needle tip identification. Moreover, due to the presence of epigastric arterial branches within/deep into the muscle body, it may be more favourable to perform an in-plane approach from a **lateral starting point** and thereby avoid penetration of the rectus muscle. As previously described, for optimal analgesia the block is performed bilaterally, and may need to be repeated at several levels depending on degree of craniocaudal extension required.

Lateral Approach, in Plane

The probe is positioned transversely, initially on the midline of the abdomen, and then is moved laterally to the side where the nerve block is to be performed. When the lateral border of the rectus muscle is visible on the screen, its position is

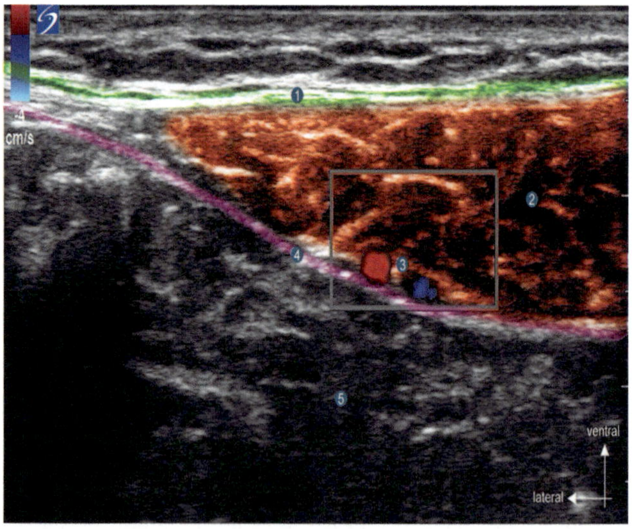

Fig. 9.51 Rectus abdominis muscle under the arcuate line. Branches of the inferior epigastric vessels

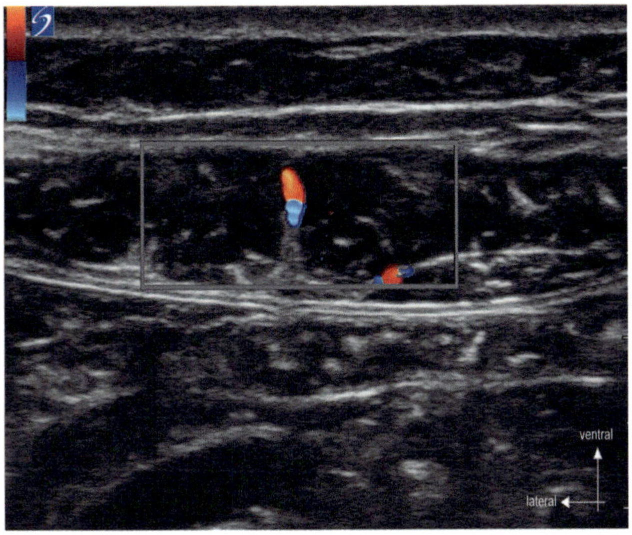

1. abdominal cavity 2. rectus abdominis m. 3. branches of superior epigastric a.

Fig. 9.52 Rectus abdominis muscle in the supraumbilical area. Branches of the epigastric artery

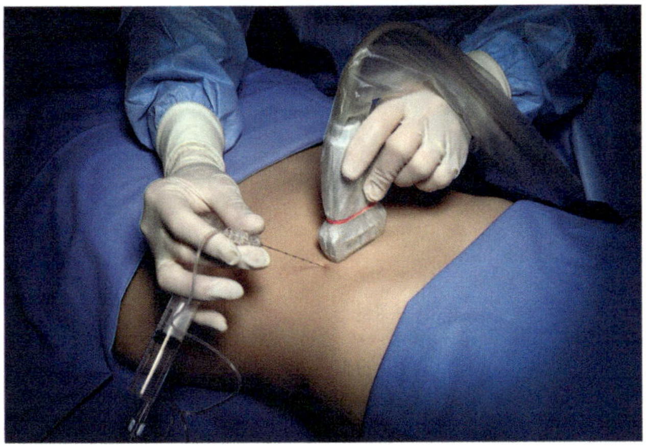

Fig. 9.53 Rectus abdominis sheath block. Lateral approach. Position of the probe, needle in-plane

anchored. The skin injection is performed at the lateral end of the probe (Fig. 9.53) over the linea semilunaris. The needle is slowly advanced and followed until the tip is positioned at the lateral border of the rectus muscle, between its dorsal aponeurosis and the deep layer of its sheath (Fig. 9.54). After verification of the correct needle tip position in the fascial plane (a 1–2 mL bolus dose) and a negative aspiration test, the local anaesthetic is injected between the dorsal aspect of the rectus muscle and the sheath (Fig. 9.55). By injecting small amounts of LA as the needle advances carefully more medially it is possible to promote medial spread.

Medial Approach, in Plane
The patient and probe are positioned as for the lateral approach. The skin is punctured at the medial end of the probe. The needle is introduced in plane and is aimed laterally towards to lateral edge of the rectus muscle, on its deep aspect. Since the needle path passes through the body of the rectus muscle care must be taken to observe the presence of arterial branches and to avoid inadvertent vascular puncture.

Paraumbilical Nerve Block With the probe placed transversely, lateral to the umbilicus, over the lateral portion of the rectus muscle, the needle is introduced in plane, from the lateral (or medial) end of the probe (Fig. 9.56). It is guided until it penetrates the ventral layer of the sheath opposite the lateral one-third of the rectus muscle. At this level, a small amount of local anaesthetic is injected after a negative aspiration test. Once the correct needle tip position is confirmed (the LA spreads between the ventral aspect of the rectus muscle and the ventral layer of its sheath), the remaining LA is injected under real-time ultrasound control.

Block of the Genital Branches of the Genitofemoral, Iliohypogastric, and Ilioinguinal Nerves (Fig. 9.57)

Type of probe: High-frequency linear probe for children of 5–10 years of age, 6–13 MHz. Small footprint probe useful in paediatrics.
Axis of probe: At lateral border of the pubic spine, the probe is placed according to a 2 h/8 h axis for the left side and 10 h/4 h for the right side.
Configuration: Needle in plane (medial or lateral approach), or out of plane with hydrolocalisation.
Depth studied: Variable according to body build and body mass index of the patient.
Neurostimulation: Can provide useful information in case of poor visualisation of the cord by motor stimulation of the genital nerve branches.
Needle: 50–80 mm.
Utility of Doppler ultrasound: Visualisation of the arterial branches contained in the cord to aid in its identification.

Indications
No description exists of ultrasound-guided genitofemoral nerve block in clinical practice for post-operative analgesia, and its utility has not been evaluated. An old study in children in inguinal hernia repair using a blind approach to the genitofemoral nerve [34] with injection under the aponeurosis of the external oblique muscle, laterally just above the pubic tubercle, showed a 60% reduction in haemodynamic signs during the traction on the cord with no effect on postoperative pain. Anatomical variability of this nerve, in par-

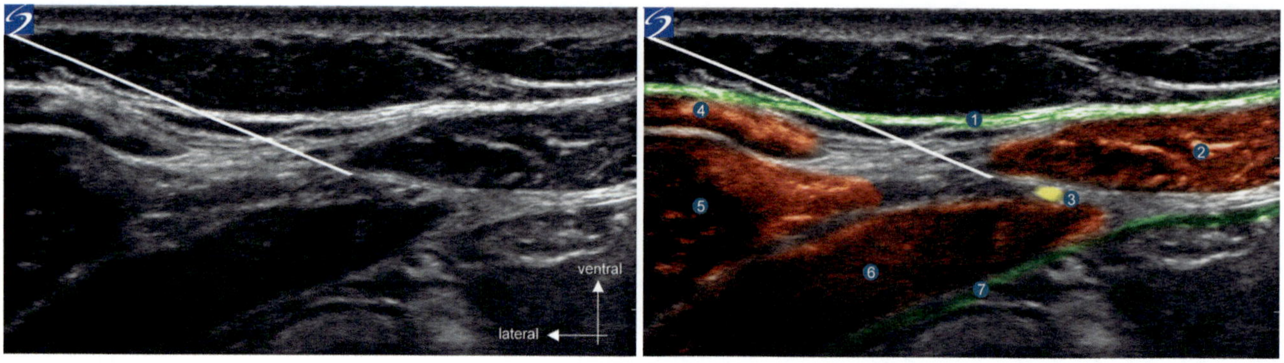

Fig. 9.54 Rectus abdominis sheath block. Lateral approach. Bevel of the needle positioned at the lateral border of rectus abdominis muscle, between dorsal aspect of muscle and deep layer of its sheath

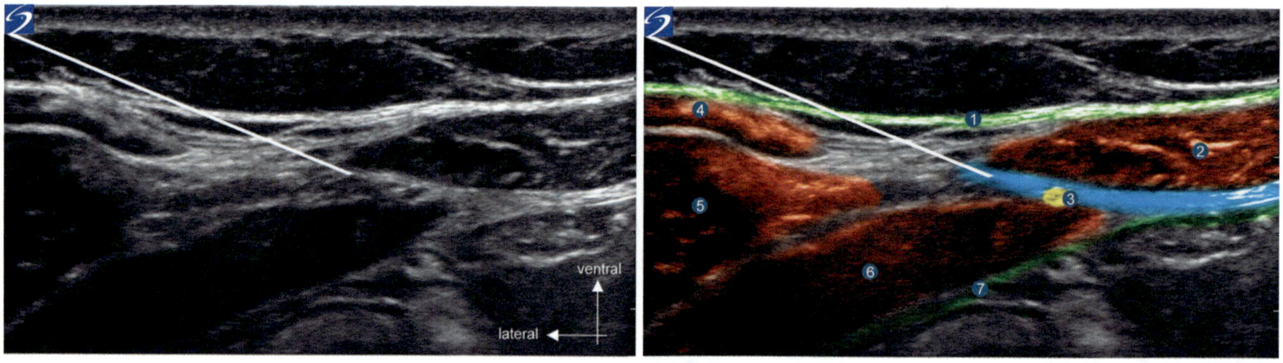

Fig. 9.55 Rectus abdominis sheath block. Injection of local anaesthetic and diffusion space during lateral approach in-plane

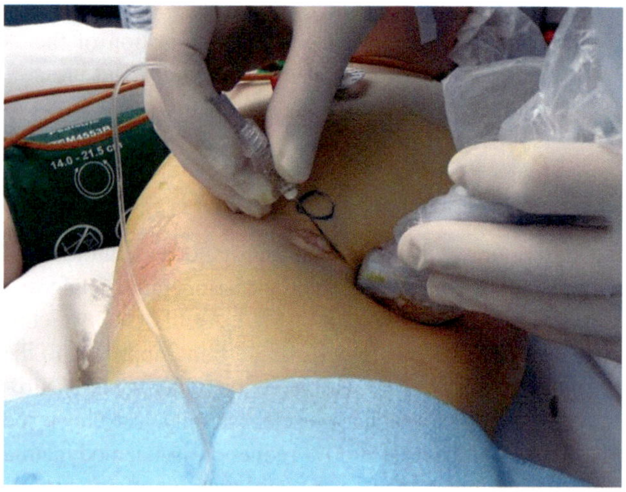

Fig. 9.56 Paraumbilical block. Medial approach, needle in-plane

ticular with the sensory branches of the ilioinguinal nerve, makes its use as a separate block uncommon, except in the setting of open abdomen hernia surgery under ilioinguinal/iliohypogastric nerve block only. However, an injection during handling of the hernia sac by the surgeon probably will be as effective even if this is not yet demonstrated. Lastly, the risk related to injection of the spermatic cord is not negligible, and distension of the sac can make individual identification of the structures difficult by the operator during surgical repair. In two clinical cases, ultrasound guidance was used to treat precise neuropathic sequelae [35, 36].

Echoanatomy

The patient is in supine position. The high-frequency linear probe is placed laterally to the pubic spine, in oblique position (a 2 h/8 h axis for the left side and a 10 h/4 h for the right

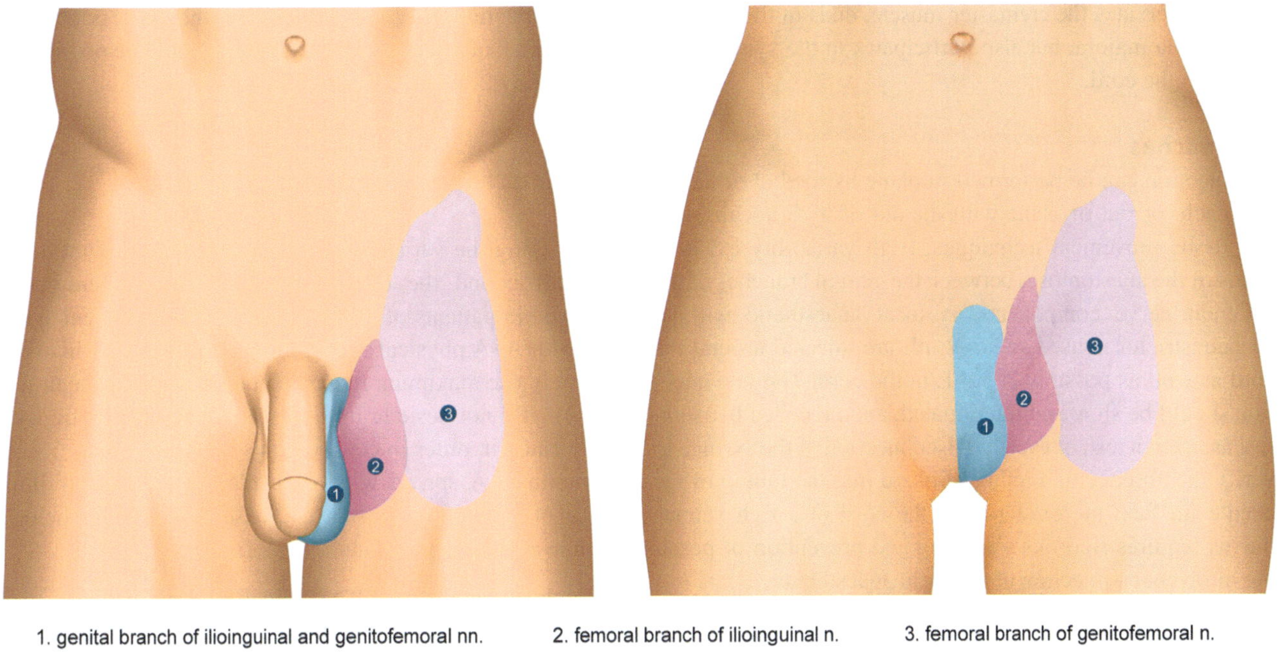

Fig. 9.57 Area of extension of nerve blocks of the genital branches of the genitofemoral, iliohypogastric, and ilioinguinal nerves

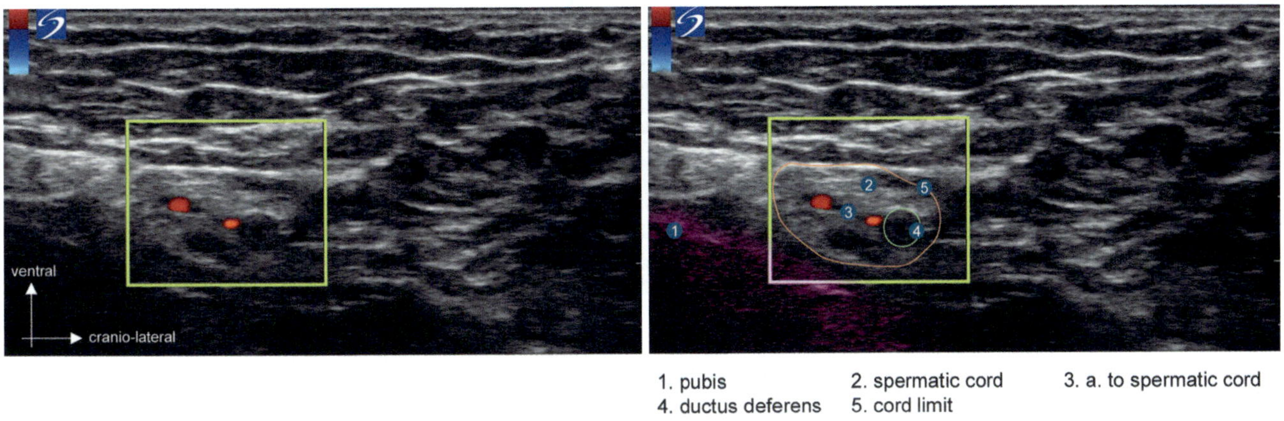

Fig. 9.58 Sonoanatomical view of spermatic cord

side). The spermatic cord can be visualised under the superficial fascia, medial to the femoral artery. In case of difficulty in locating it, it is also possible to position the probe initially in the inguinal fold (along its axis), and then to follow the femoral artery cranially up to the inguinal ligament; at this level, the cord is visible medially to the artery. It is then necessary to direct the probe obliquely (at about a 45° angle in relation to the transverse line, 2 h/8 h axis or 10 h/4 h axis according to the side approach) to obtain a more perpendicular section. It appears as a non-compressible, non-pulsatile "sac", but in which one or more arterial Doppler signals can be detected (testicular and cremaster arteries) (Fig. 9.58). In the cord, the existence of an anechoic rounded structure is seen with no corresponding Doppler signal in the vas deferens. Upon its emergence from the superficial inguinal ring, the coverings of the cord are comprised, from superficial to deep layer, by the following:

- The external spermatic fascia.
- The cremaster muscle and cremaster fascia.
- The internal spermatic fascia.

The genital branches of the iliohypogastric and ilioinguinal nerves anastomose with each other, continuing in the inguinal canal ventrally and medially to the spermatic cord (or to the round ligament), and then are distributed to the genital small branches which meander on the surface of the cord or in its envelopes.

The genital branch of the genitofemoral nerve descends into the inguinal canal behind the cord (or the round ligament), innervates the cremaster muscle, ends in the scrotum or the labia majora, but also participates in the sympathetic plexus in the cord.

Approaches

An injection can be performed in plane by medial or lateral approach, or out of plane with the aid of hydrolocalisation and tissue movement techniques. High variability exists in terms of the anastomoses between the genital branches of the different nerve components. A local anaesthetic without vasoconstrictor is used and 5–8 mL are injected around the cord as well as possibly 2–3 mL in the cord. The last injection should be slow and fractionated, and preceded by negative aspiration test, or even by disconnection of the syringe to verify the absence of a reflux of blood into the tubing of the needle. In fact, the existence in the cord of a rich venous plexus requires rigorous screening and prevention of possible intravenous injection of the local anaesthetic.

Local Anaesthetic: What Strategy, What Product, What Volume, What Concentration?

In the setting of post-operative analgesia, except in specific cases, long-acting local anaesthetics with lower toxicity (i.e. ropivacaine or levobupivacaine) are used. Recommended dosage is 3–4 mg/kg (respectively for superior and inferior limb nerve block), with a maximum dose admissible in adults in a single injection of 225 mg for ropivacaine and 150 mg for levobupivacaine. Short-acting products, such as lidocaine or mepivacaine, generally are used in the context of diagnostic blocks in the setting of chronic pain.

Abdominal wall blocks are performed within a fascial plane; their efficacy is related both to the volume injected (which affects spread) and the concentration of local anaesthetic used. At equal volume, duration of analgesia increases with increase in concentration of the local anaesthetic [37], but its reabsorption is highly variable. Maximum plasma concentrations above theoretical limits of toxicity can be found for total doses that would be considered subtoxic. It has been shown that strictly intermuscular injection of a local anaesthetic performed under ultrasound guidance is the guarantor of much better efficacy [26], and may be at the origin of enhanced plasma reabsorption in comparison to injections performed during "blind" procedures, which are more frequently intramuscular [38].

It is important to constantly look for the best volume/concentration compromise based on context and extent of the area to be anaesthetised (number and type of blocks to be performed): concentrations of local anaesthetic used should be adjusted to the total volume necessary to achieve the appropriate spread. Reduction of the total dose of local anaesthetic during a nerve block of the wall potentially reduces risk of systemic toxicity, and it should be possible to reduce it by 20 to 30%.

In Adults, a Few Proposals

Considering the variability of the reabsorption of the local anaesthetic and the unpredictable plasma concentrations achieved in patients of different sizes and in different states of health/ASA physical status, it is probably desirable to not routinely use maximum recommended doses of these drugs. One should not hesitate to dilute the local anaesthetic with 0.9% NaCl in order to conserve the "volume" effect (i.e. in promoting LA spread within the plane) and thereby limit doses of the active substance. These suggestions (below) assume an average size adult in good health:

- Exclusive unilateral **TAP block** in mid-axillary line: 20 mL of 0.5% levobupivacaine or 0.5% ropivacaine.
- Exclusive bilateral **TAP block** in mid-axillary line: 15 mL × 2 of 0.357% levobupivacaine or 0.5% ropivacaine.
- Exclusive unilateral **subcostal TAP block**: 20 mL of 0.5% levobupivacaine or 0.5% ropivacaine.
- Exclusive bilateral **subcostal TAP block**: 15 mL × 2 of 0.375% levobupivacaine or 0.5% ropivacaine.
- **TAP block** in the mid-axillary line + **subcostal TAP block** unilaterally: 15 mL × 2 of 0.375% levobupivacaine or 0.5% ropivacaine.
- **TAP block** in mid-axillary line + bilateral **subcostal TAP block**: 15 mL × 4 of 0.25% levobupivacaine or 0.20 or 0.25% ropivacaine.
- **Rectus sheath block (bilaterally only):**
  Supraumbilical midline: 2 levels of bilateral block, i.e.: 15 mL × 4 of 0.25% levobupivacaine or 0.20% or 0.25% ropivacaine.
  Supra- and subumbilical midline: 3 levels of bilateral block i.e.: 10 mL × 6 of 0.25% levobupivacaine or 0.20% or 0.25% ropivacaine.
  Subumbilical midline (as an alternative to bilateral TAP block): 15 mL × 2 of 0.375% levobupivacaine or 0.5% ropivacaine.
- **Analgesia via catheter:** 0.125% levobupivacaine or 0.2% ropivacaine at flow rate of 5–10 mL/h.
- **Iliohypogastric and ilioinguinal block:** 5–8 mL per nerve of 0.5% levobupivacaine or 0.5% ropivacaine.

In Children

Current Marketing Authorisations for ropivacaine and levobupivacaine are restricted in children.

- **Ropivacaine:**
  Children under 12 years of age: "Acute peri- and post-operative pain in the neonate, infants and children up to and including 12 years of age, by caudal epidural nerve block and by continuous epidural infusion".
- **Levobupivacaine:**
  Iliohypogastric and ilioinguinal nerve block by injection.

However, these products are routinely used in paediatric RA, under guidance by expert formal recommendations of the ADARPEF in 2010 [39]. This text suggests that it is preferable to use the levo-isomeric forms (ropivacaine and levobupivacaine) at the expense of bupivacaine because of their lower cardiac toxicity. Furthermore, the text specifies limitations in dosage and discusses the utility of adrenaline as an adjuvant to limit the risk of systemic toxicity:

"For peripheral nerve blocks or of the trunk or limbs, no more than 0.5 mL/kg of 2 mg/mL ropivacaine (= 1 mg/kg) or of 2.5 mg/mL levobupivacaine (= 1.25 mg/kg) probably should be injected" [39].

"It is possible to decrease the toxic risk of short-acting local anaesthetics by using solutions with adrenaline at maximum concentration of 5 mg/mL (1/200,000^e^)" [39].

"The combination of adrenaline with a local anaesthetic administered by caudal, perineural or local approach probably should not be used in combination to prolong analgesia" [39].

- **Iliohypogastric/ilioinguinal nerve block:** 0.2 mL/kg of 0.2% ropivacaine or 0.25% levobupivacaine per nerve block, without exceeding a total volume of 0.5 mL/kg. A clinical study showed the possibility of reducing the total dose of 0.25% levobupivacaine with 0.075 mL/kg producing perioperative analgesia and during the first 4 hours that is identical to an injection of 0.2 or 0.1 mL/kg.
- **Paraumbilical block:** 0.2 mL/kg of 0.2% ropivacaine or 0.25% levobupivacaine, without exceeding 0.5 mL/kg [39].

References

1. Rizk NN. A new description of the anterior abdominal wall in man and mammals. J Anat. 1980;131:373–85.
2. Rozen WM, Tran TM, Asqhton MW, Barrington MJ, Ivanisic JJ, Taylor GI. Refining the course of the thoracolumbar nerves: a new understanding of the innervation of the anterior abdominal wall. Clin Anat. 2008;21:325–33.
3. Rouvière H, Delmas A. Anatomie humaine descriptive, topographique et fonctionnelle 12e édition. Paris: Masson; 1985.
4. Papadopoulos NJ, Katritsis ED. Some observations on the course and relations of the iliohypogastrics and ilioinguinal nerves (based on 348 specimens). Anat Anz. 1981;149:357–64.
5. Anson BJ. Morris' human anatomy. 12th ed. New York: The Blakiston Division, McGraw-Hill Book Company; 1966.
6. http://www.sfar.org/article/906/echographie-en-anesthesie-locoregionale-rfe-2011
7. Petersen PL, Mathiesen O, Torup H, Dahl JB. The transversus abdominis plane: a valuable option for post-operative analgesia? Acta Anaesthesiol Scand. 2010;54:529–35.
8. Abdallah FW, Halpern SH, Margarido CB. Transversus abdominis plane block for post-operative analgesia after caesarean delivery performed under spinal anesthesia? A systematic review and meta-analysis. Br J Anaesth. 2012;109:679–87.
9. Kanazi GE, Aouad MT, Abdallah FW, Khatib MI, Adham AMBF, Harfoush DW, Siddik-Sayyid SM. The analgesic efficacy of subarachnoid morphine in comparison with ultrasound-guided transversus abdominis plane block after cesarean delivery: a randomized controlled trial. Anesth Analg. 2010;111:475–81.
10. Fredrickson MJ, Paine C, Hamill J. Improved analgesia with the ilioinguinal block compared to the transversus abdominis plane block after pediatric inguinal surgery: a prospective randomized trial. Ped Anesth. 2010;20:1022–7.
11. Petersen PL, Stjernholm P, Kristiansen VB, Torup H, Hansen EG, Mitchell AU, Moeller A, Rosenberg J, Dahl JB, Mathiesen O. The beneficial effect of transversus abdominis plane block after laparoscopic cholecystectomy in day-case surgery: a randomized clinical trial. Anesth Analg. 2012;115:527–33.
12. Niraj G, Kelkar A, Jeyapalan I, Graff-Baker P, Williams O, Darbar A, Maheshwaran A, Powell R. Comparison of analgesic efficacy of subcostal transverses abdominis plane blocks with epidural analgesia following upper abdominal surgery. Anesthesia. 2011;66:465–71.
13. Barrington MJ, Ivanusic JJ, Rozen WM, Hebbard P. Spread of injectate after ultrasound-guided subcostal transversus abdominis plane block: a cadaveric study. Anesthesia. 2009;64:745–50.
14. Albrecht E, Kirkham KR, Endersby RV, Chan VW, Jackson T, Okrainec A, Penner T, Jin R, Brull R. Ultrasound-guided transversus abdominis plane (TAP) block for laparoscopic gastric-bypass surgery: a prospective randomized controlled double-blinded trial. Obes Surg. 2013;23:1309–14.
15. Ortiz J, Suliburk JW, Wu K, Bailard NS, Mason C, Minard CG, Palvadi RR. Bilateral transversus abdominis plane block does not decrease post-operative pain after laparoscopic cholecystectomy when compared with local anesthetic infiltration of trocar insertion sites. Reg Anesth Pain Med. 2012;37:188–92.
16. Hebbard P, Fujiwara Y, Shibata Y, Royse C. Ultrasound-guided transversus abdominis plan (TAP) block. Anaesth Intensive Care. 2007;35:616–7.
17. Johr M, Sossai R. Colonic puncture during ilioinguinal nerve block in a child. Anesth Analg. 1999;88:1051–2.
18. Salib Y, Kukreja PK, Parikh MK. Prolonged femoral nerve palsy after Ilio-inguinal nerve block. Reg Anesth Pain Med. 2007;271
19. Rosario DJ, Jacob S, Luntley J, Skinner PP, Raftery AT. Mechanism of femoral nerve palsy complicating percutanous Ilio-inguinal nerve block. Br J Anaesth. 1997;78:314–6.
20. Wulf H, Behnke H, Vogel I, Schröder J. Clinical usefulness, safety, and plasma concentrations of ropivacaine 0,5% for inguinal hernia repair in regional analgesia. Reg Anest Pain Med. 2001;26:348–51.
21. Abrahams MS, Horn JL, Noles M, Aziz MF. Evidence-based medicine. Ultrasound guidance for truncal blocks. Reg Anesth Pain Med. 2010;35:S36–42.
22. Tran TM, Ivanusic JJ, Hebbard P, Barrington MJ. Determination of spread of injectate after ultrasound-guided transversus abdominis plane block: a cadaveric study. Br J Anaesth. 2009;102:123–7.
23. Ford S, Dosani M, Robinson AJ, Campbell GC, Ansermino JM, Lim J. Lauder GR defining the reliability of sonoanatomy identification by novices in ultrasound-guided pediatric ilioinguinal and iliohypogastric nerve blockade. Anesth Analg. 2009:1793–8.

24. Willschke H, Marhofer P, Bösenberg A, Jonhston S, Wanzel O, Cox SG, Sitzwohl C, Kapral S. Ultrasonography for ilioinguinal/iliohypogastric nerve blocks in children. Br J Anaesth. 2005;95:226–30.
25. Willschke H, Bösenberg A, Marhofer P, Johndton S, Kettner S, Eichenberger U, Wanzel O, Kapral S. Ultrasonographic-guided ilioinguinal/iliohypogastric nerve block in pediatric anesthesia: what is the optimal volume? Anest Analg. 2006;102:1680–4.
26. Weintraud M, Marhofer P, Bösenberg A, Kapral S, Willschke H, Felfening M, Kettner S. Ilioinguinal/iliohypogastric blocks in children: where do we administer the local anesthetic without direct visualization. Anesth Analg. 2008;106:86–93.
27. Rozen WM, Tran TM, Barrington MJ. Ashton MW avoiding denervation of the rectus abdominis muscle in DIEP flap harvest III: a functional study of the nerves to the rectus using anesthetic blockade. Plast Reconstr Surg. 2009;124:519–22.
28. Ludot H. Application of ultrasound in paediatric abdominal trunk blocks. Ann Fr Anesth Réanim. 2012;31:e21–4.
29. Willschke H, Bösenberg A, Marhofer P, Johnston S, Kettner SC, Wanzel O, Kapral S. Ultrasonography-guided rectus sheath block in paediatric anesthesia: a new approach to an old technique. Br J Anaesth. 2006;97:244–9.
30. Husain NKA, Ravalia A. Ultrasound-guided ilioinguinal and rectus sheath block. Anesthesia. 2006;61:1116.
31. Sandeman DJ, Dilley AV. Ultrasound-guided rectus sheath block and catheter placement. ANZ J Surg. 2008;78:621–3.
32. Dolan J, Lucie P, Geary T, Smith M, Kenny GN. The rectus sheath block: accuracy of local anesthetic placement by trainee anesthesiologists using loss of resistance on ultrasound guidance. Mag Anesth Pain Med. 2009;34:247–50.
33. Kato J, Ueda K, Kondo Y, Aono M, Gokan D, Shimizu M, Ogawa S. Does ultrasound-guided rectus sheath block reduce abdominal pain in patients with postherpetic neuralgia? Anesth Analg. 2011;112:740–1.
34. Sasaoka N, Kawaguchi M, Yoshitani K, Kato H, Suzuki A, Furuya H. Evaluation of genitofemoral nerve block, in addition to ilioinguinal and iliohypogastric nerve block, during inguinal hernia repair in children. Br J Anaesth. 2005;94:243–6.
35. Campos NA, Chiles JH, Plunkett AR. Ultrasound-guided cryoablation of genitofemoral nerve for chronic inguinal pain. Pain Physician. 2009;12:997–1000.
36. Peng PWH, Tumber PS. Ultrasound-guided interventional procedures for patients with chronic pelvic pain: a description of techniques and review of literature. Pain Physician. 2008;11:215–24.
37. Johansson B, Hallerbäck B, Stubberöd A, et al. Preoperative local infiltration with ropivacaine for post-operative pain relief after inguinal hernia repair. A randomised controlled trial. Eur J Surg. 1997;163:371–8.
38. Weintraud M, Lunblad M, Kettner SC, Willschke H, Kapral S, Lönqvist PA, Koppatz K, Turnheim K, Bösenberg A, Marhofer P. Ultrasound versus landmark-based techniques for ilioinguinal-iliohypogastric nerve blockade in children: the implications of plasma levels of ropivacaine. Anesth Analg. 2009;108:1488–92.
39. http://www.sfar.org/article/336/anesthesie-loco-regionale-en-pediatrie-rfe-2010-maj-mars-2010

Thoracic Paravertebral Block

10

Eryk Eisenberg, Elisabeth Gaertner, and Philippe Clavert

Contents

Performing a **thoracic paravertebral block** (TPVB) consists of injecting a local anaesthetic into the thoracic paravertebral space, with the objective of segmental, somatic, and sympathetic ipsilateral anaesthesia. Ultrasound guidance is becoming the gold standard for the majority of practitioners who perform this type of nerve block in routine practice [1].

The original version of the chapter has been revised. A correction to this chapter can be found at https://doi.org/10.1007/978-3-031-08804-9_13

Supplementary Information The online version contains supplementary material available at https://doi.org/10.1007/978-3-031-08804-9_10.

E. Eisenberg (✉)
Département d'anesthésie et réanimation, Clinique Pôle Santé République, Clermont-Ferrand, France

E. Gaertner
Département d'anesthésie et réanimation chirurgicale, Hôpitaux Civils de Colmar, Colmar, France

P. Clavert
Service de chirurgie du membre supérieur, Centre de Chirurgie Orthopédique et de la Main (CCOM), Illkirch, France

The **thoracic paravertebral spaces** are located on either side of the thoracic spine (Fig. 10.1a and b). Each of them contains in its fatty tissue environment primarily the following:

- The ipsilateral thoracic spinal nerves which emerge from the intervertebral foraminae.
- Their divisions: dorsal, ventral branch (intercostal nerve), recurrent meningeal branches etc.
- The ipsilateral thoracic sympathetic chain to which each spinal nerve is connected by white and grey communicating branches.
- The ipsilateral posterior intercostal vessels.
- The blood vessels which supply the spinal cord (radicular and anterior segmental spinal arteries, intervertebral veins).

Opposite each intervertebral foramen, is a "single" paravertebral space whose volume schematically is **wedge-shaped** [2] (Fig. 10.2), in direct continuation with the paravertebral spaces sited above and below, whose limits are as follows:

- Ventro-laterally: the endothoracic fascia and the parietal pleura.

E. Eisenberg, E. Gaertner (eds.), *Ultrasound in Peripheral, Neuraxial and Perineuraxial Regional Anaesthesia*,
https://doi.org/10.1007/978-3-031-08804-9_10

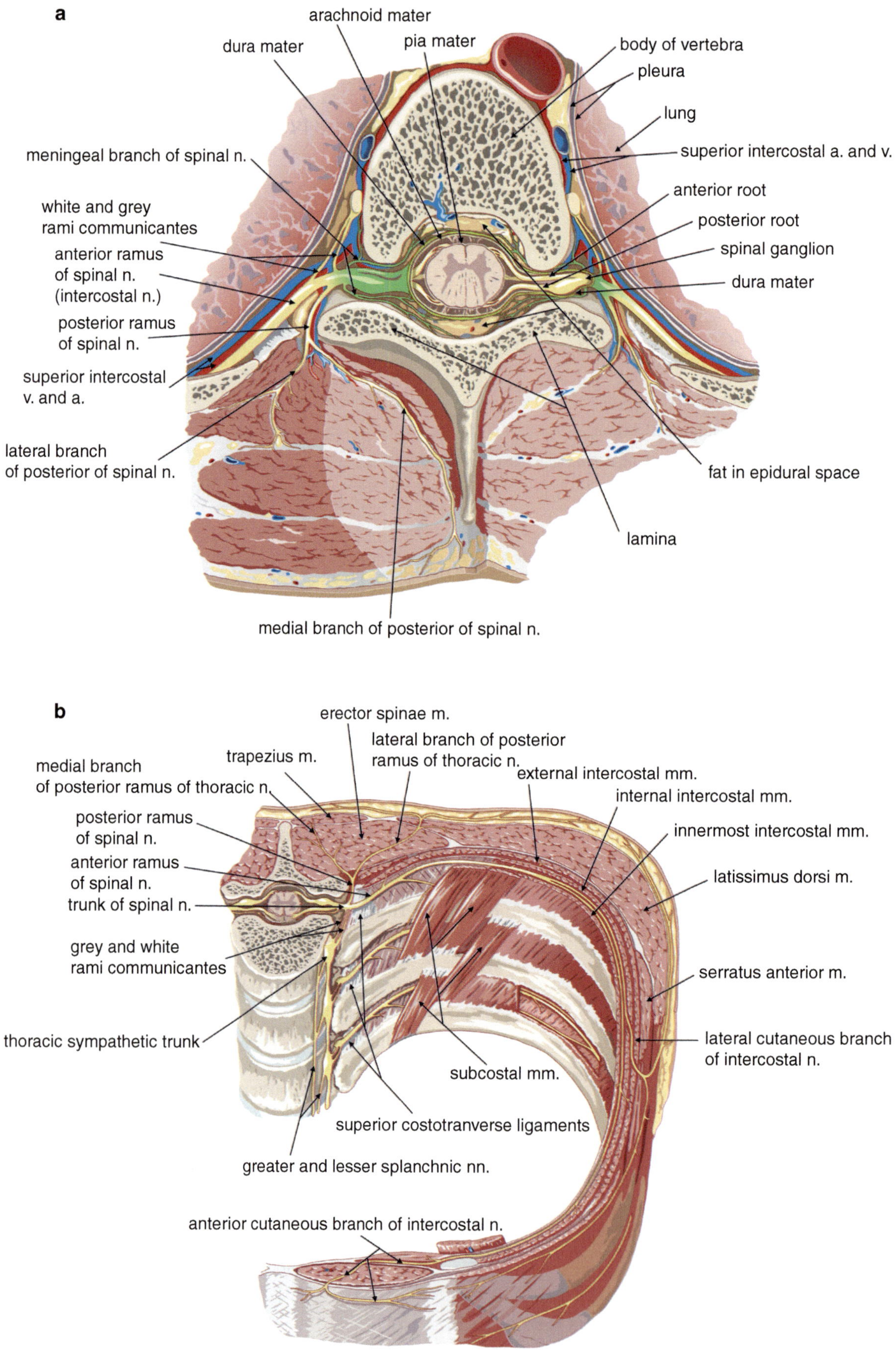

Fig. 10.1 (**a**) Anatomy of paravertebral space on both sides of the thoracic spine. (**b**) Thoracic spinal nerves

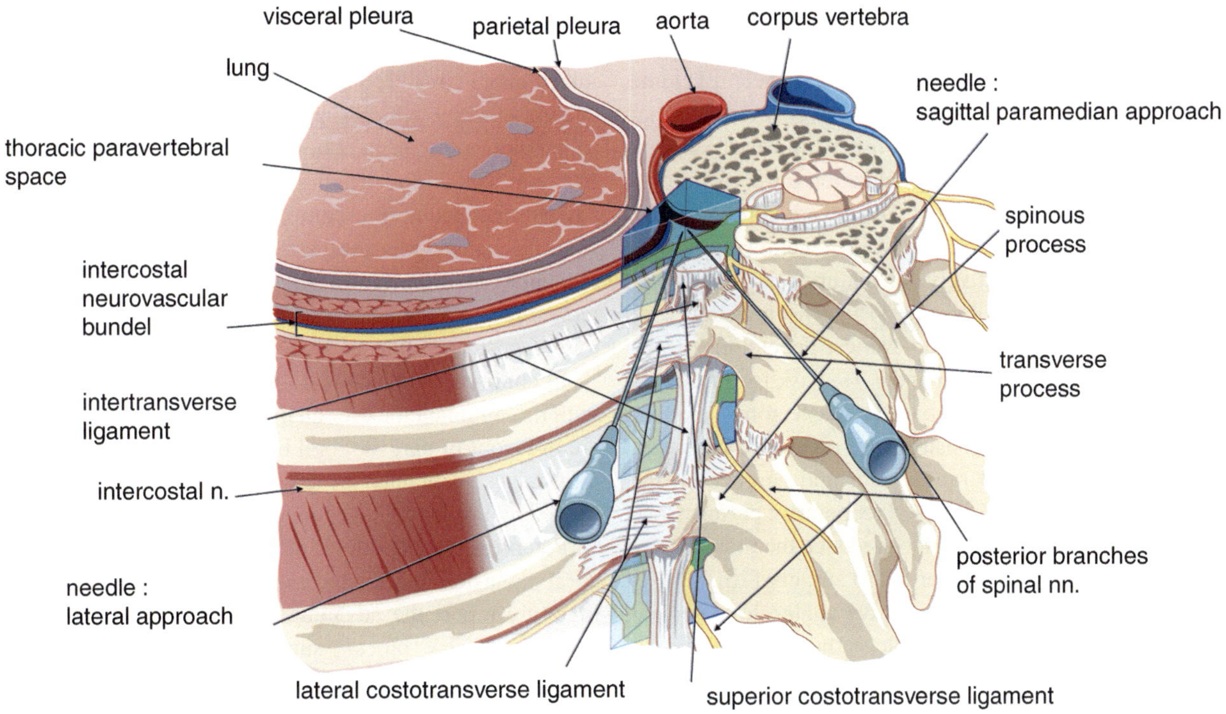

Fig. 10.2 Thoracic paravertebral spaces

- Dorsally: the superior costotransverse ligament (extending between the subadjacent rib and the supra-adjacent transverse process). This ligament "fuses" laterally with the internal intercostal membrane, the medial extension of the internal intercostal muscle.
- Medially: the intervertebral foramen, the lateral aspect of the vertebral body and the intervertebral disc. This medial aspect corresponds to the "base" of the wedge shape.

This TPV space is a longitudinal fascial plane which communicates with [2–7]:

- The epidural space through the intervertebral foramen, medially.
- The intercostal space, laterally.
- The contralateral TPV space via the epidural spaces or a prevertebral fascial plane.
- The retroperitoneal space (for the caudal part of the TPV space) via the lateral and medial arcuate ligaments.

The cranial limits of the TPV space are not clearly identified, but cervical paravertebral spread of LA has been observed after TPV injection.

The role of the endothoracic fascia in methods of spread of the local anaesthetic is unclear. It is not a sufficiently resistant physical barrier to restrain the progress of a catheter during its introduction, but on the other hand it seems to affect the pattern of spread of a local anaesthetic, and therefore its distribution in contact with the nerve structures.

Indications

The principal indication for a paravertebral block is **unilateral thoracic surgery:** breast surgery, thoracotomy, or video thoracoscopy [8, 9].

This block has demonstrated its unquestionable efficacy in **breast surgery** in improvements during the post-operative phase of the following [10, 11]:

- Decrease in post-operative pain scores.
- Decrease in the use of morphine.
- Decrease of post-operative nausea and vomiting.
- Very low incidence of complications.

In minor breast surgery, a paravertebral catheter compared to general anaesthesia certainly decreases VAS scores (12 vs 45 mm) during the initial post-operative phase, improving patient satisfaction, but the risk–benefit ratio in the context of this type of surgery was not in favour of a paravertebral block in 2002 [12]. In fact, the authors described the complications of epidural extension and pleural puncture amongst a small series of patients.

In comparison with local injection by the surgeon, a paravertebral block (a single injection at T4) improves the patient's cooperation, decreases the use of intraoperative propofol and improves pain scores at home for ambulatory patients [13].

The availability of **ultrasound** has re-established the use of this block even for minor surgery, in the hands of experienced practitioners. A single injection of a sufficient volume of local anaesthetic can extend the block over several dermatomes, and so it is appropriate to use it in breast surgery. Breast cancer surgery is a strong indication for performing a paravertebral block. For patients with severe comorbidities, this block makes it possible to avoid general anaesthesia [14]. Performed preoperatively for major breast surgery, it decreases length of hospital stay [15]. The prevalence of chronic pain after breast surgery varies, depending on the type of procedure, between 25 and 77%. Risk factors for its occurrence are young age of patient, mastectomy, axillary node dissection, breast reconstruction, previous breast surgery, intensity of post-operative pain, and surgery on the upper-outer quadrant of the breast. Preoperative performance of a TPV block has shown that it can reduce the incidence of post-operative chronic pain after breast surgery [16].

In **thoracic surgery**, post-operative analgesia generated by TPV block is at least comparable to that induced by thoracic epidural anaesthesia. Furthermore, it is often superior at rest and during movement, in comparison with parenteral and epidural analgesia, with lesser use of morphine [17, 18]. In addition to improved analgesia, TPV block seems to have a favourable post-operative impact compared to thoracic epidural block [19, 20] with the following:

- Decrease in pulmonary complications.
- Decrease in arterial hypotension.
- Decrease in nausea and vomiting.
- Decrease in urinary retention.

Continuous paravertebral block is indicated for multiple rib fractures [21].

Paravertebral block can also improve post-operative analgesia in certain types of abdominal surgery with a lateral incision (cholecystectomy, hepatic surgery, nephrectomy [22]). In the literature, **bilateral** thoracic paravertebral blocks are described for analgesia in certain types of major abdominal or vascular surgery (colectomies, ileostomies, stoma closure, splenectomy, cystectomy, abdominoplasty, repair of an abdominal aortic aneurysm etc.). Therefore, this block may be considered as an alternative to epidural analgesia in certain circumstances.

TPV block has been studied for inguinal hernia repair: Ozkan et al. [23] compared 2 injections (T10 and L1) vs 4 injections (T10, T11, T12, and L1) without finding a difference in efficacy, which suggests that it may be unnecessary to perform multiple injections. In ambulatory practice, this technique can diminish pain scores and reduce post-operative morphine consumption compared with general anaesthesia [24]; nausea and vomiting are also less frequent. Compared to spinal anaesthesia for inguinal hernia, this block can reduce the frequency of urinary retention and produce prolonged post-operative analgesia [25].

In children, studies which compared paravertebral block with ilioinguinal block for analgesia in inguinal hernia repair are in favour of paravertebral block [26].

Analgesia obtained by paravertebral block can be prolonged by insertion of a catheter.

Potential complications of paravertebral block are epidural extension and pleural puncture. Several articles in the literature report the advantages of using ultrasound whilst performing TPV blocks, suggesting improvements in the quality of analgesia with an improved risk-benefit profile [27, 28].

Mechanism of Action

The mechanism of action by which paravertebral block produces anaesthesia and segmental thoracic analgesia is not perfectly clear. The local anaesthetic injected acts by spreading into the specific space initially targeted, but also by spread into the supra- and subjacent contiguous spaces, into the intercostal space or the adjacent epidural space, or by a combination of these types of spread [2, 3, 29–31]. The role of epidural spread of a local anaesthetic, due to a modest fraction of total volume injected [29] and generally ipsilateral only [31], may be predominant in the mechanism of action of TPV. It is observed in 40–70% of cases and then produces major extension of an ipsilateral sensory block [5, 31].

Injection of 15–20 mL of a local anaesthetic on average spreads to the 4 vertebral levels with a trend to a more caudal than cephalic distribution in relation to the injection site [5, 6, 32–34]. The extent of sensory block, however, is superior in the majority of cases.

Ultrasound-Guided Thoracic Paravertebral Block

Ultrasound guidance is becoming the preferred technique to perform TPV block [1], which previously was performed by locating the surface landmarks and looking for loss of resistance. However, although reported in a rapidly increasing number of cases, there are still few published articles in this field [5, 21, 27, 34–48].

Although Pusch et al. [46] in 2000 validated the precision of the pre-block ultrasound measurement of depth of the transverse process and its usefulness before a "loss of resistance" technique, the first description of ultrasound-guided TPV block was reported by Hara et al. in 2009 [44]. They. performed 25 TPV blocks at T4 with real-time guidance of the injection out of plane up until contact with the transverse process. The end of the injection was done by "blindly" looking for loss of resistance and then injection of the local anaesthetic was observed in real time.

In 2009 also, Luyet et al. published a study on the placement of 20 TPV catheters between T4 and T8 in cadavers [43]. After determination of an ultrasound plane described as "optimum" (cranially and medially oblique) with a low frequency probe, they introduced the needle and catheter via an in plane approach and after confirming the correct position of the needle tip by injection of a bolus dose of 0.9% NaCl. CT scans performed after injection of 10 mL of a contrast medium into the catheter revealed 11 satisfactory opacifications (TPV and intercostal space), 1 intrapleural, 2 posterior mediastinal opacifications, and 6 epidural opacifications without contrast medium in the TPV space (30%). Performed on cadavers, this study nevertheless made it possible to emphasise the difficulty in managing the direction of the catheter once it emerges from the tip of the needle, even if the latter was correctly positioned. An additional study by the same author demonstrated similar findings: performing **ultrasound-guided** injection (out of plane, with the probe in the transverse position) or by looking for *loss of resistance*, and then insertion of a conventional "linear" catheter (non-preformed) frequently gave rise to ectopic placement of catheters [45].

Again in 2009, published reports by Shibata and Nishiwaki [41] and Ben-Ari et al. [42] described a transverse approach in plane with the posterior intercostal space as the target, at the end of the transverse process. This tangential approach to the pleura may make it possible to limit the risk to the pleura at the expense of an injection which may be more painful.

In 2010, Marhofer et al. [37] this time described a paramedial sagittal needle insertion in the same posterior intercostal space, out of plane, with an ultrasound probe in a transverse position. Although the technique used by Shibata and Nishiwaki [41] most likely is safe with respect to the pleura, this is not true for the one proposed by Marhofer et al. who approached it perpendicularly at a place where the posterior intercostal space is not very thick.

In 2010, O'Riain et al. [39] described a paramedian sagittal approach in plane. In spite of difficulties in visualisation and advancement of the needle inherent in its very angulated trajectory, compared to the probe, correct placement of catheters was possible in 80% of attempts on cadavers and in 100% of cases in the clinical setting.

In 2012, Karmakar et al. [48] published the first 3D ultrasound study on the thoracic paravertebral space. They showed the feasibility and underlined the utility of multiplane ultrasound for a more precise anatomical study than that permitted by 2D technology. However, computer processing power is currently insufficient to perform 4D ultrasound (i.e. real-time 3D). For its part, 3D imaging can enable a better understanding of the underlying anatomy, but its "non-real time" characteristic makes it less useful in clinical practice during conduct of the procedure itself.

In 2013, in a study on healthy volunteers, Marhofer et al. concluded that the probable involvement of individual anatomical parameters could explain the discrepancies between the anatomical paravertebral extension of the local anaesthetic and the much greater extent of the sensory block observed [34]. In fact, by performing a bilateral TPV block at T6 (but not US guided in real time) they noticed that the profile of paravertebral anatomical spread and sensory somatic extension were comparable on both sides in any given subject, even though they could be very different between subjects, independently of the technique used which was standardised.

General Considerations

Preparation

The patient is **prepared** as for conduct of any central or peripheral nerve block: placement of a peripheral venous cannula, ECG, blood pressure, and pulse oximetry monitoring. Nasal oxygen or by face mask is **highly recommended** because of risk of vasovagal symptoms, even more so if sedation/analgesia is administered to perform this block.

Aseptic Technique

It must be rigorous:

- Insertion of a PVB catheter mandates observing strict aseptic precautions: "4-phase" disinfection, mask, head covering, sterile gloves and gown, and peripheral drapes.
- For a single-shot block, it is recommended that the operator at least follow conditions identical for spinal anaesthesia: "4-phase" disinfection, mask, head covering, and sterile gloves.

Nerve Block with Patient Awake or under General Anaesthesia?

As in all regional anaesthesia procedures in adults, it is preferable to perform TPV block in the awake. "However, under conditions where the risk/benefit ratio is favourable and justified, it is possible to perform a block in a patient who has been anaesthetised or under light sedation. In this case, ultrasound guidance probably provides additional safety" [49].

The block can be performed with the patient seated, in lateral recumbent position, or in prone position in case of procedures that require fluoroscopy (in management of chronic pain). Once the probe has been positioned, local anaesthesia of the skin and deeper planes (subcutaneous layer and muscle tissue) is performed with ultrasound guidance using a long needle (intramuscular) and by replicating the pathway intended for the needle when performing the block.

Continuous Injection and Problem of Catheters

Depending on the indication, the block may be performed either with a single injection or via a catheter to provide continuing analgesia. Although ultrasound guidance enables correct positioning of the tip of the needle, the major problem with catheters is the difficulty in controlling the position of its distal end [43]. In fact, the direction of a traditional linear catheter cannot be controlled beyond its emergence from the needle tip, even when correctly placed in the TPV space. Instead of remaining within the paravertebral space (Fig. 10.3), its end may lodge itself in the epidural position (Figs. 10.4 and 10.5), prevertebral, intrapleural (Fig. 10.6), or intramuscular positions [43, 45, 50]. On the other hand, using a catheter with a preformed end which self-deploys in a coil upon its emergence from the tip of the needle (Fig. 10.7) may ensure more consistent positioning [38]. As can readily be observed in fig. 10.8, once the needle is correctly placed in the TPV space, deployment of the catheter (SonoLong

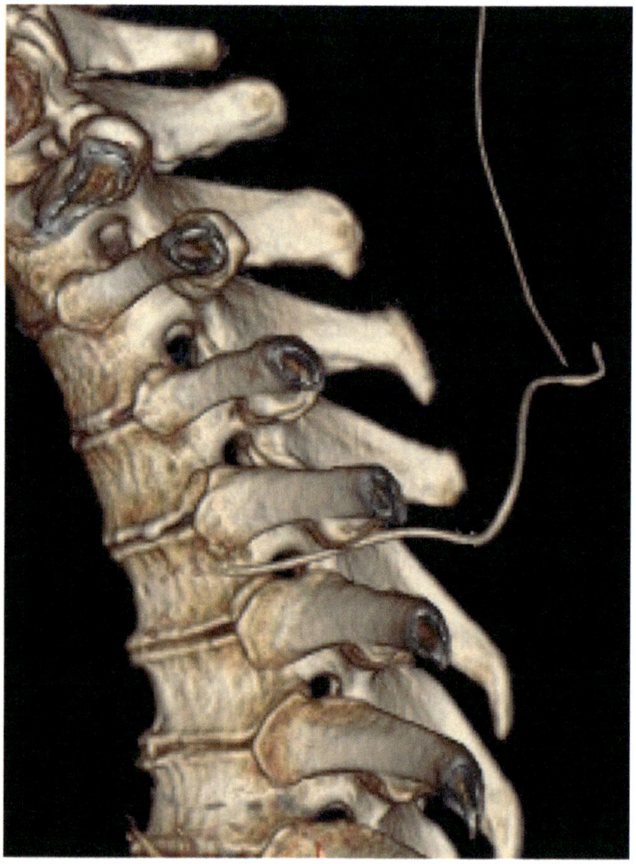

Fig. 10.3 Paravertebral catheter in place. Philippe Gautier iconography

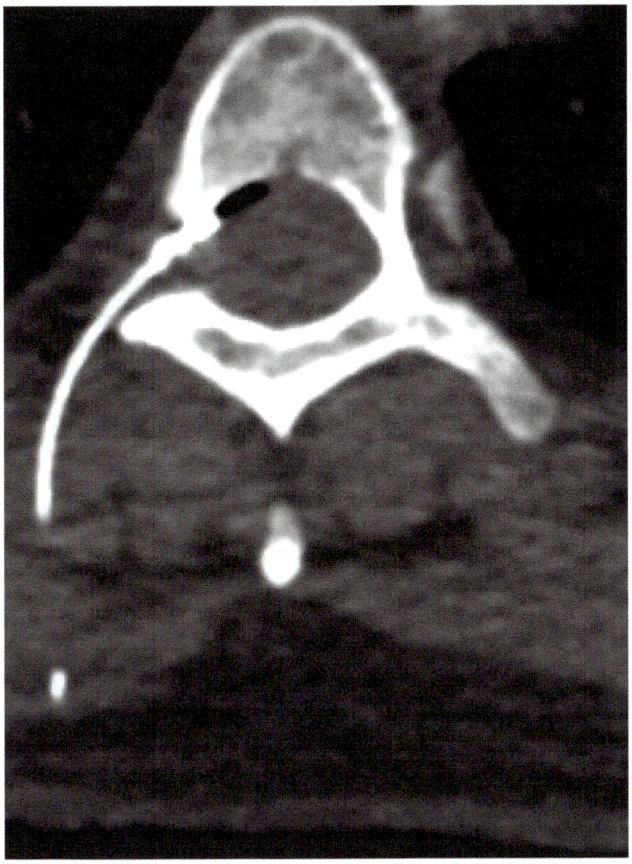

Fig. 10.4 Paravertebral catheter migrated into an epidural position. Philippe Gautier iconography

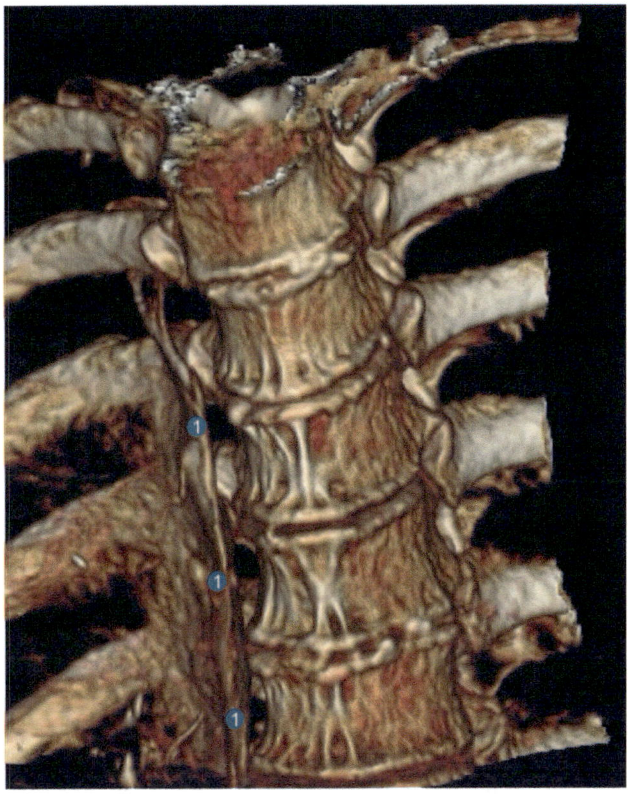

1. intrapleural diffusion

Fig. 10.6 Injection into a paravertebral catheter is unfortunately positioned in the pleural space. Philippe Gautier iconography

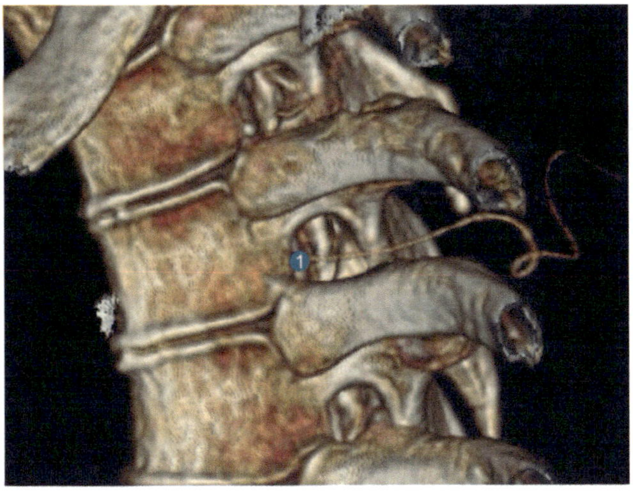

1. Catheter in intervertebral foramen

Fig. 10.5 Paravertebral catheter in an intervertebral foramen. Philippe Gautier iconography

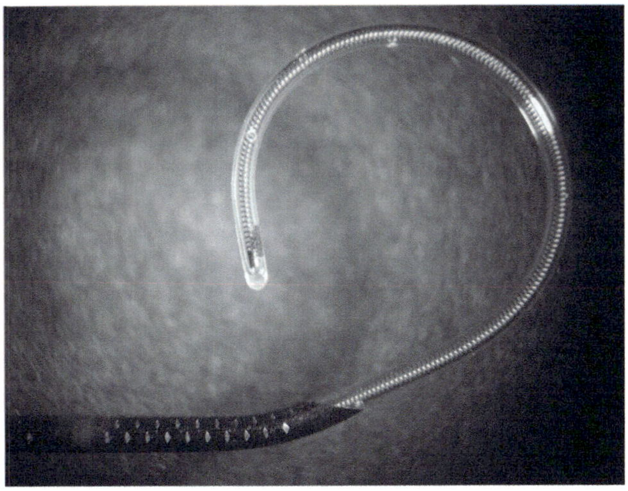

Fig. 10.7 "Pigtail" preformed catheter (Sonolong Curl Sono, Pajunk). Bertrand Fabre iconography

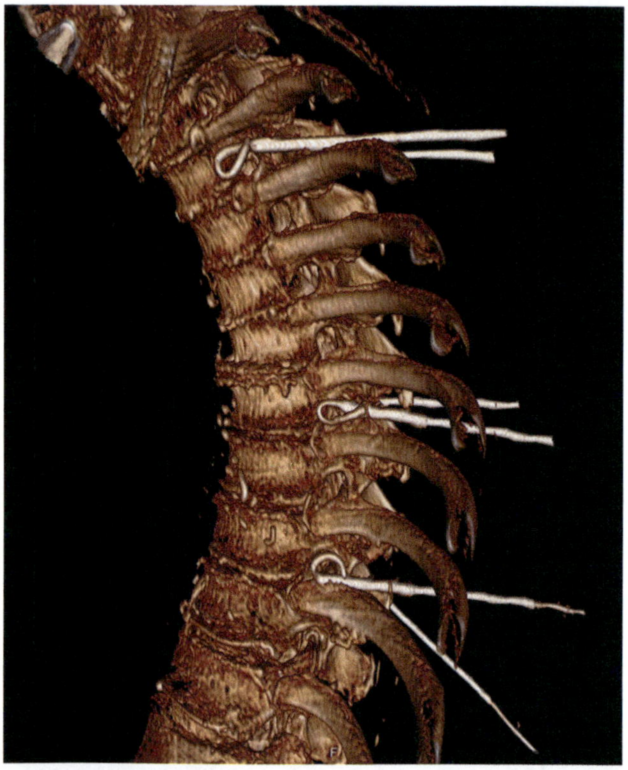

Fig. 10.8 "Pigtail" catheters (Sonolong Curl Sono, Pajunk) introduced into the TPV space at different levels. This 3D reconstruction confirms that end of this type of catheter is positioned in immediate proximity to the tip of the needle (Cedric Luyet, iconography)

Curl Echo, Pajunk®) happens in immediate proximity to the needle tip, without migration or advancement of its end. Logically, this could result in more predictable analgesic effects from the infusion of local anaesthetic via the catheter.

Needles

Whether it is a question of a single block or a continuous infusion via a catheter, it is preferable to use a "**Tuohy**" needle whose tip design is probably less traumatic with respect to the nerve, vascular, and other underlying structures, and which may not easily be visualised. However, the larger diameter and blunter characteristics of a Tuohy needle can make it less comfortable for the patient, so performing thorough skin and subcutaneous infiltration anaesthesia prior to its insertion is essential. The use of "**echogenic**" needles is also recommended in light of the depth of the space to be reached and the often unfavourable angle of the needle for its visualisation.

Expected Morphological Criteria

With Ultrasound (Figs. 10.9, 10.10 and 10.11) The position of the needle tip should be perfectly known and controlled throughout conduct of the TVP block. **Hydrolocalisation** can prove very useful when advancing the needle through the paravertebral muscle masses, aiding the identification of the

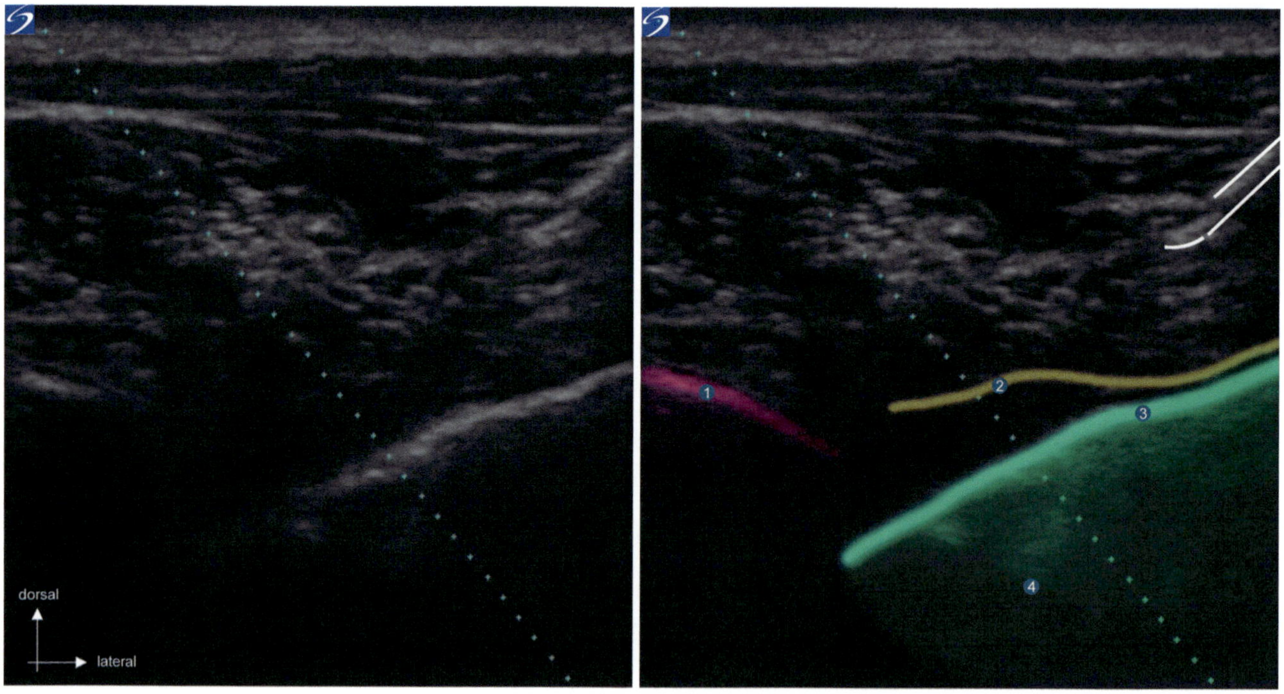

Fig. 10.9 Thoracic paravertebral block. Transverse section, in-plane approach. Beginning of the progression of the Tuohy needle

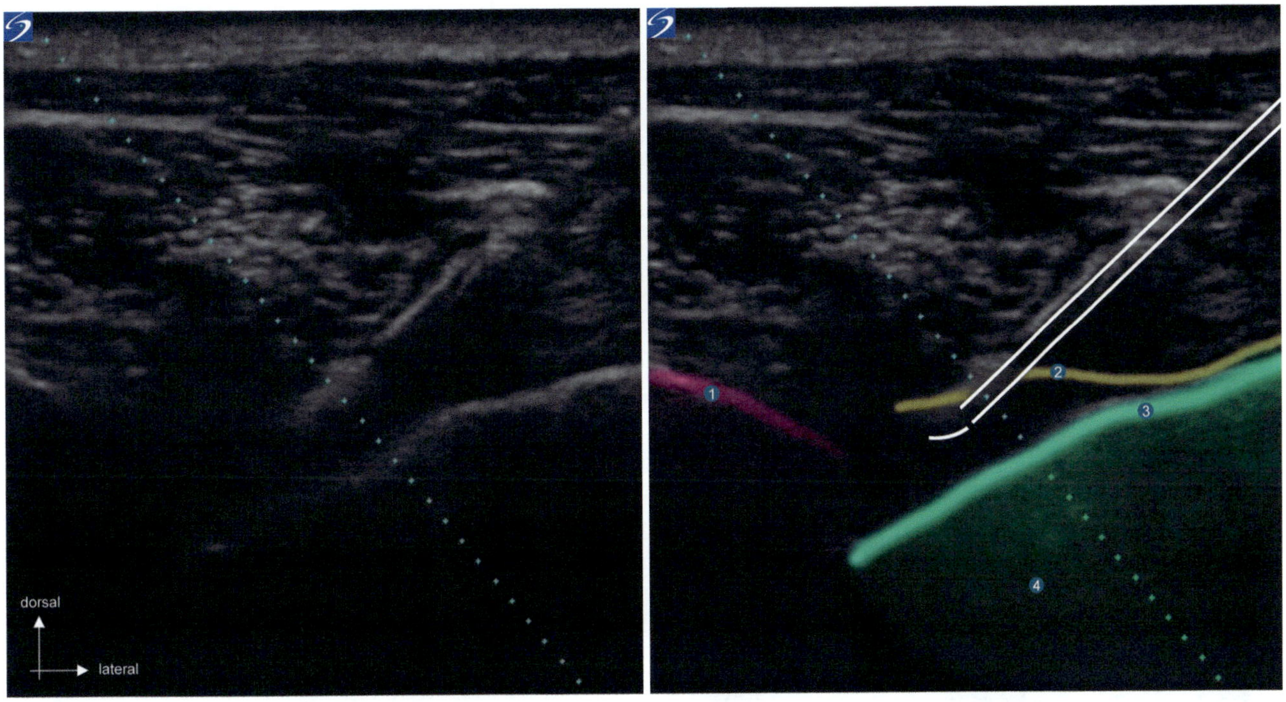

1. lamina and acoustic shadow
2. internal intercostal membrane
3. pleura
4. lung

Fig. 10.10 Thoracic paravertebral block. Transverse section, in-plane approach. Needle in place (no injection yet performed)

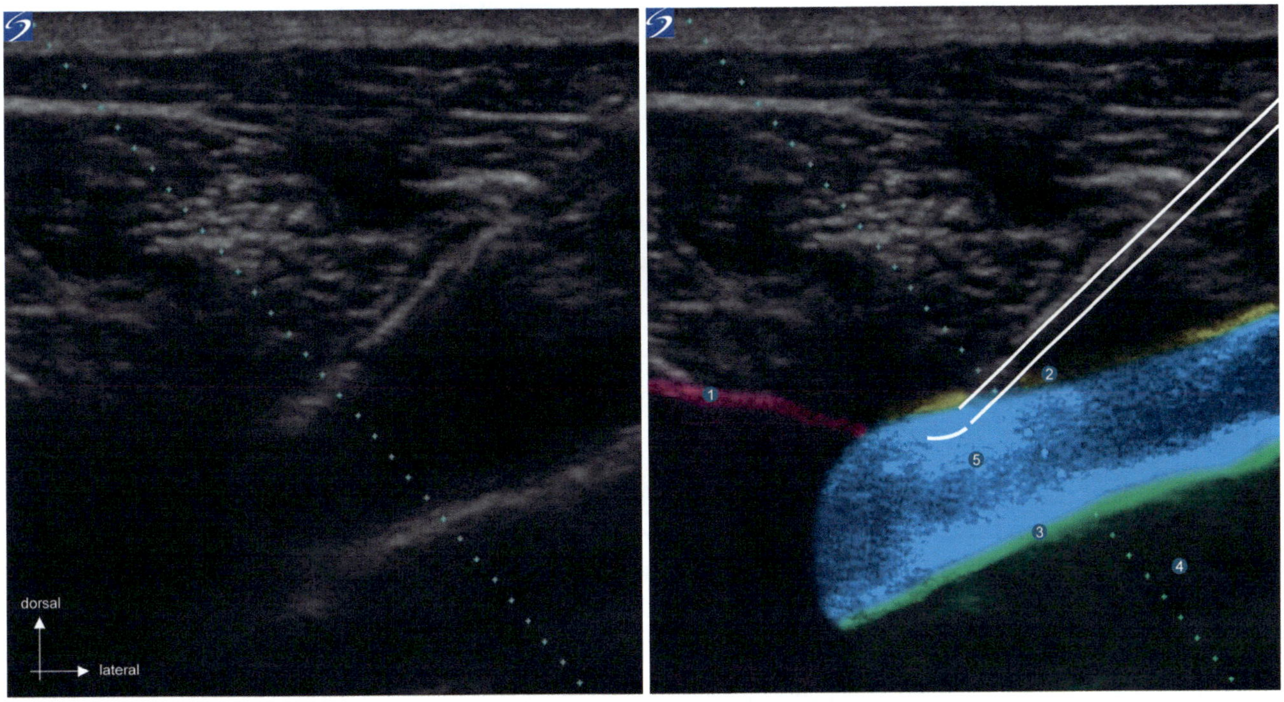

1. lamina and acoustic shadow
2. internal intercostal membrane
3. pleura
4. lung
5. local anaesthetic

Fig. 10.11 Thoracic paravertebral block. Transverse section, needle in place, authenticated by ventral movement of pleura during injection through needle LA: local anaesthetic needle tip as it approaches the TPV space. However, its position is only truly confirmed by the occurrence of the following during injection:

- Ventral movement of the parietal pleura (pathognomonic sign of paravertebral spread).
- Widening of the posterior intercostal space located between the parietal pleura and the internal intercostal membrane.

With Conventional Radiology If an injection with radiological contrast medium is performed, the correct positioning of the injection seen on the X-ray demonstrates longitudinal paravertebral spread (Figs. 10.12 and 10.13) whose extent varies depending on the volume injected. An image of intercostal spread (Fig. 10.14) or latero- or prevertebral spread (Fig. 10.15) may be associated with it.

With CT Scan 3D CT scan reconstruction reveals very precise images of spread of the contrast medium in TVP blocks (Fig. 10.16). The tendency to circumscribe the lateral border or the ventral border of the vertebral body is evidenced, and the adjacent intercostal spread may be noticed. Depending

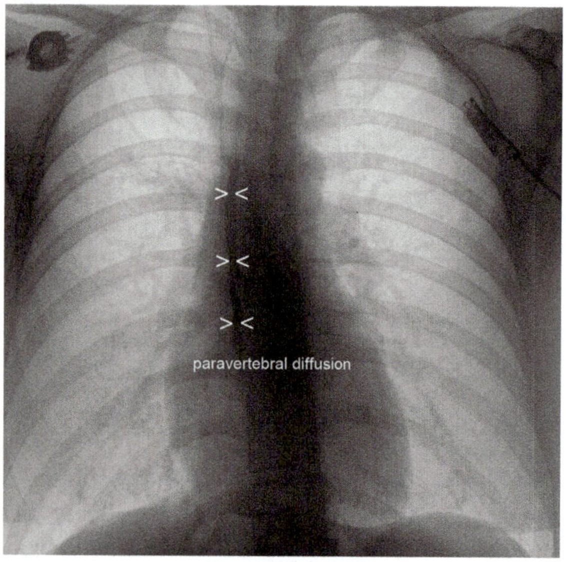

Fig. 10.12 Opacification of right TPV space through well-positioned catheter

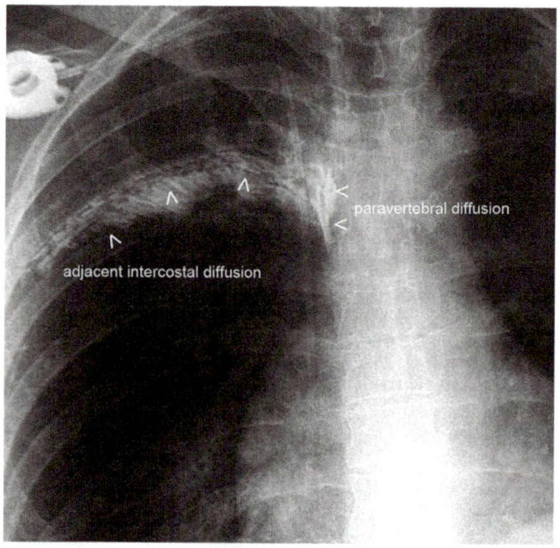

Fig. 10.14 Opacification of a TPV catheter placed at the right T5-T6 level. With only 2 mL injected, spread of the contrast medium is observed into the right T5-T6 TPV space and into the adjacent intercostal space

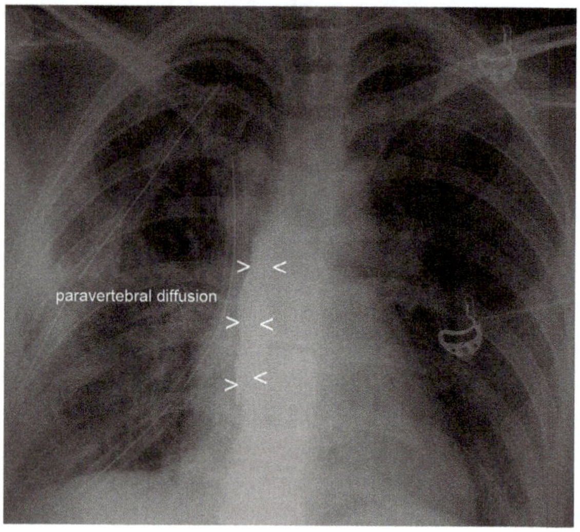

Fig. 10.13 Opacification of right TPV space through well-positioned catheter

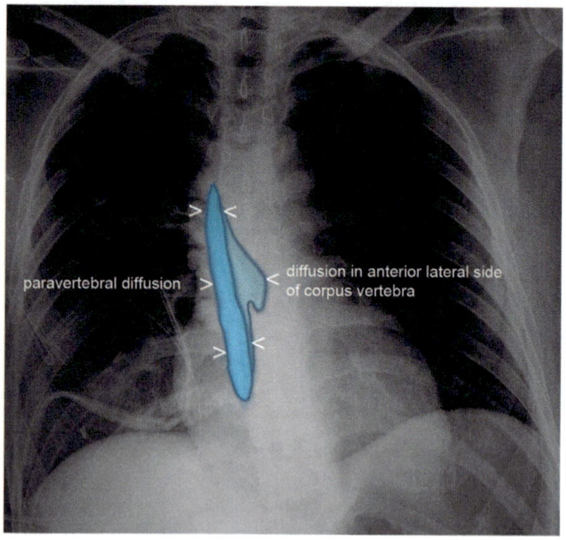

Fig. 10.15 Para and prevertebral spread of a paravertebral block

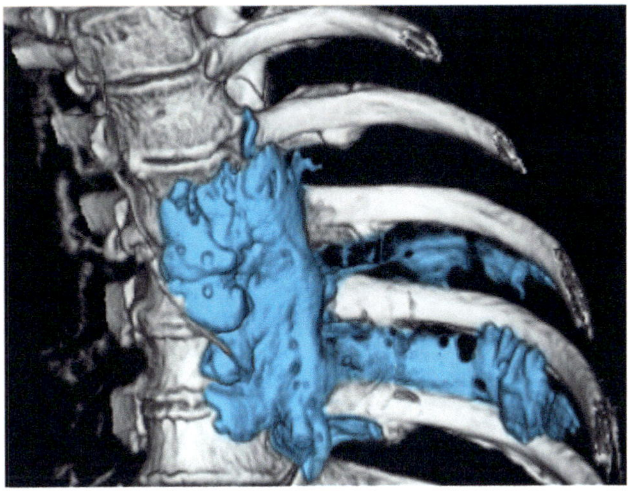

Fig. 10.16 3D CT scan reconstruction of the diffusion space of a paravertebral block. Philippe Gautier iconography

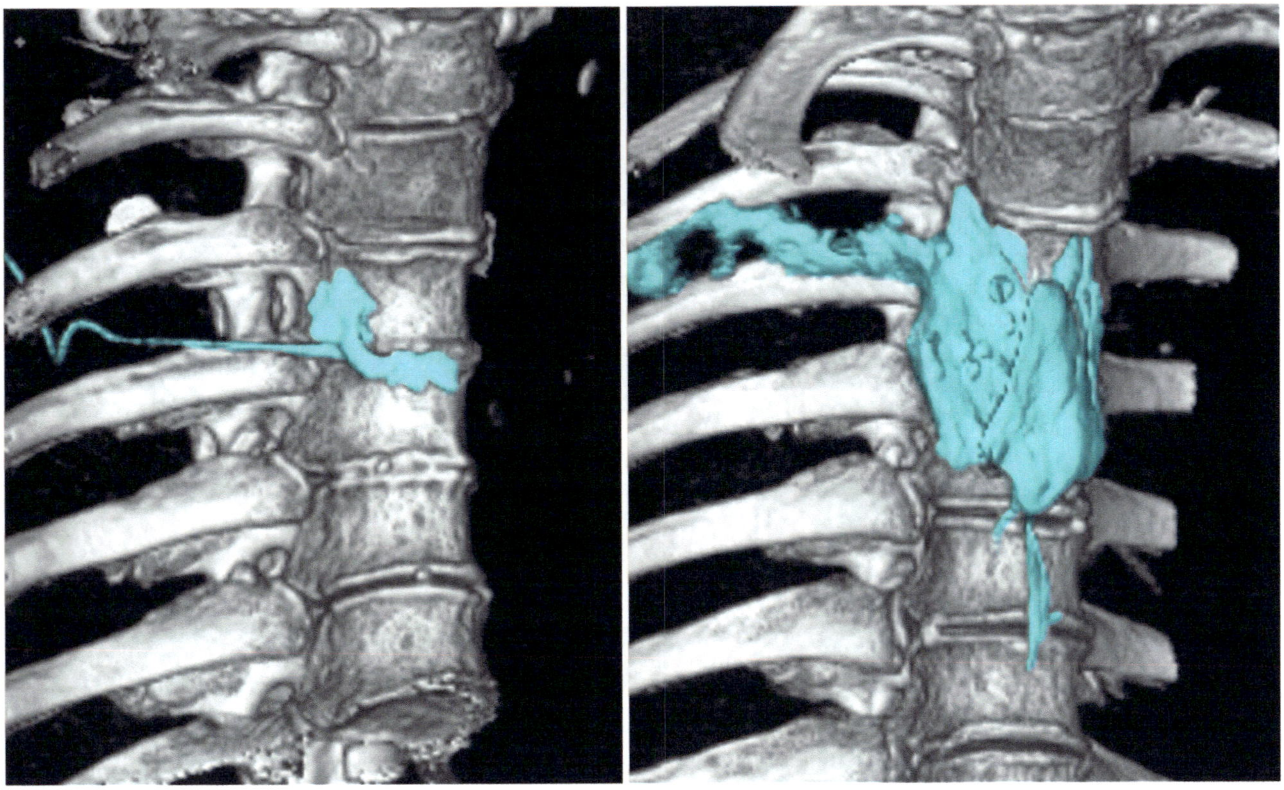

Fig. 10.17 Diffusion space of a paravertebral block depending on the volume injected. Philippe Gautier iconography

on the volume injected, the extension produced is clearly seen (Fig. 10.17).

With Thoracoscopic Vision Figure 10.18 shows, from the interior of the thoracic cavity, the protrusion of the parietal pleural after an injection into the thoracic paravertebral space.

Conduct

Prior to performing the TPV block, the paravertebral area is scanned. Depending on operator preference and patient body habitus and compliance, either a high-frequency linear probe (6–13 MHz) or a low-frequency curvilinear probe (2–5 MHz) may be used. Although these two types of probes are com-

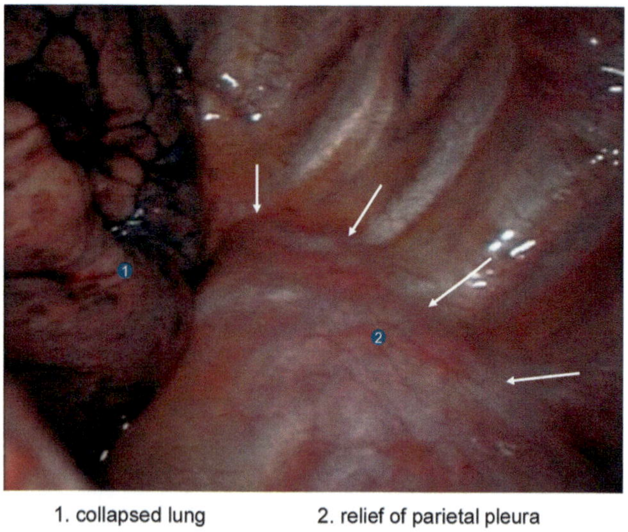

Fig. 10.18 Protrusion of pleura due to paravertebral injection. Fetracco P, Ori C. A new look at the paravertebral block: a percutaneous video-assisted technique. *Reg Anesth Pain Med* 2007; 32: 538–9. © Lippincott, Williams Wilkins

plementary in several regards, the high-frequency linear probe creates a more detailed image. Probe comparison:

- Low-frequency curvilinear probe:
  - "wider" sono-anatomical examination as result of a "fan-like" beam
  - A deeper examination.
- High-frequency linear probe:
  - Better spatial resolution.
  - Anatomical visualised in greater detail.
  - Smaller footprint than a low-frequency curvilinear probe.

The pre-procedural examination as well as the block itself can be performed with the patient seated, in the lateral decubitus or prone position (which tends to be reserved for procedures performed in a chronic pain context, often with US being used in combination with fluoroscopy). Whichever position is chosen, in order to create the space necessary for the ultrasound probe placement and the unimpeded manipulation of the needle, it is important to protract the shoulder on the block side. If the block is performed with patient in the lateral decubitus position, the patient should be lying on the non-block side and the ipsilateral arm positioned in order to free the scapula from the midline. The probe may be orientated in either the **transversal** or the **paramedian sagittal** ultrasound planes. Either probe position allows for the block to be performed either in plane or out of plane, with same the objective of placing the needle tip safely within the paravertebral space (Figs. 10.19, 10.20, 10.21 and 10.22).

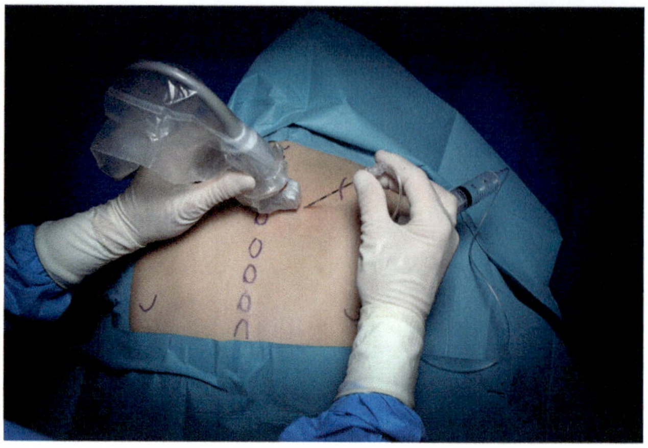

Fig. 10.19 TPV block. Transverse probe placement. Needle inserted in plane

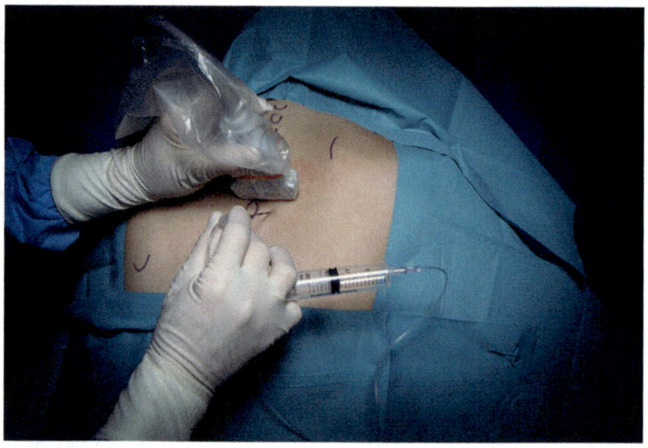

Fig. 10.20 TPV block. Transverse probe placement. Needle inserted out of plane

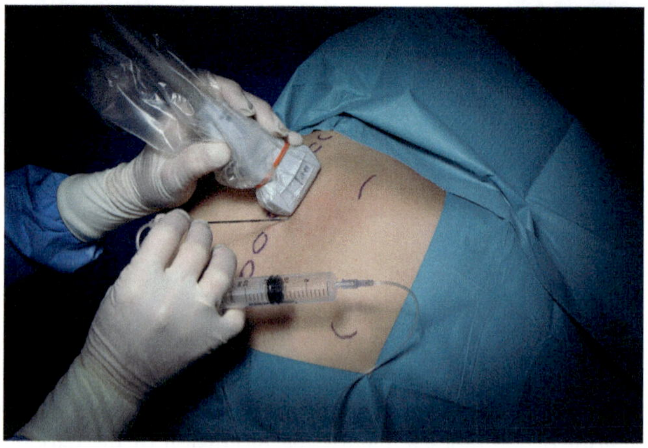

Fig. 10.21 TPV block. Paramedian parasagittal probe placement. Needle inserted in plane

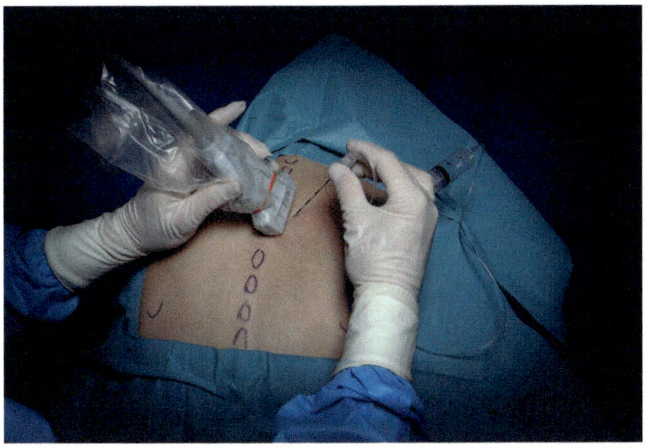

Fig. 10.22 TPV block. Paramedian parasagittal probe placement. Needle inserted out of plane

However, currently, no consensus nor scientific data exist that have determined the optimal approach to performing a TPV block.

Transverse Ultrasound Section

The image obtained with a **low-frequency curvilinear probe** in transverse position, centred on the midline, is especially instructive in understanding the overall anatomy of the thoracic spinal area. Muscle structures can be identified here (erector spinae muscles, intercostal muscles), along with bony structures (spinous and transverse processes, ribs), and the lungs located deep in the pleura. The pleura is visible as a hyperechoic line, characterised by sliding movements synchronous with respiration. Artefacts such as the "comet tail" can also be observed under the pleural line [51].

In Fig. 10.23, the ultrasound plane passes through the spinous and transverse processes (Fig. 10.24). As with all bony structures, the bone/periosteum interface creates a characteristic hyperechogenic surface, which casts an acoustic shadow, masking the paravertebral space located beneath it. The characteristic shape of the transverse processes of the thoracic vertebrae is noted, which have a dorsal concave curvature. In this image, beyond the lateral edge of the transverse process, the pleura is visualised (but not the corresponding rib), because a slight rostro-caudal shift exists between a rib and the transverse process with which it articulates.

In Fig. 10.25, the ultrasound plane this time is slightly shifted caudally, positioned between the adjacent transverse processes, to overlie the vertebral laminae (Fig. 10.26). This plane makes it possible to reveal the location of the paravertebral spaces, on each side of the vertebrae, in contact with the pleura. Note that a low-frequency probe does not enable the precise visualisation of the outline of the TPV spaces.

When a **high-frequency linear probe** is placed transversely to the spinal axis, it is useful to start the examination by a median transverse section (Fig. 10.27), and then to shift the probe laterally opposite the transverse process (Fig. 10.28). The operator then seeks to obtain three types of ultrasound sections:

- Section passing through the transverse process and the adjacent rib.
- Section passing through the transverse process and the pleura.
- Section passing through the vertebral laminae and the pleura.

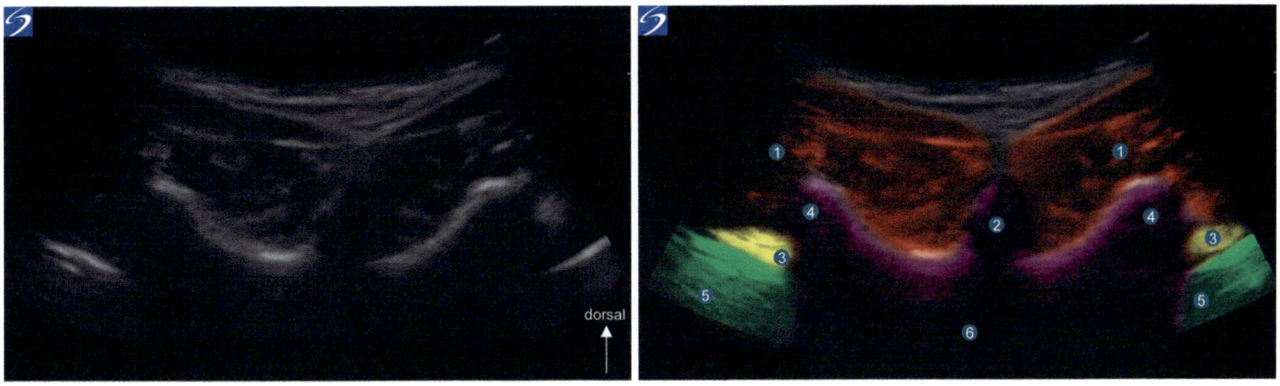

1. mm.
2. spinous process
3. lateral part of paravertebral space
4. tranverse process
5. lung
6. acoustic shadow

Fig. 10.23 Median transverse thoracic ultrasound view, cutting through the transverse processes (*see* fig. 10.24)

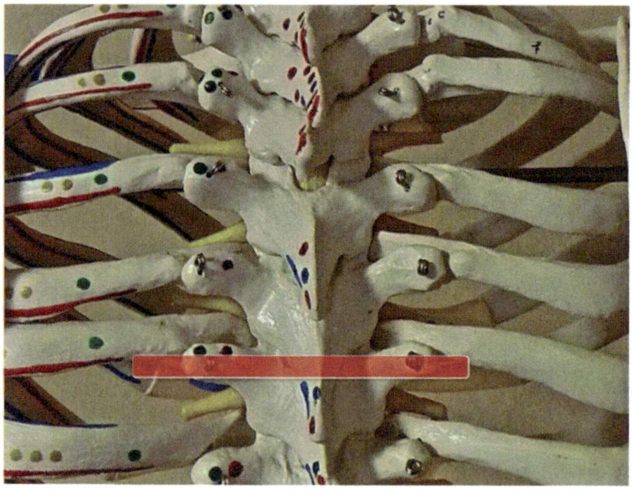

Fig. 10.24 Transverse position of ultrasound plane cutting through the transverse processes

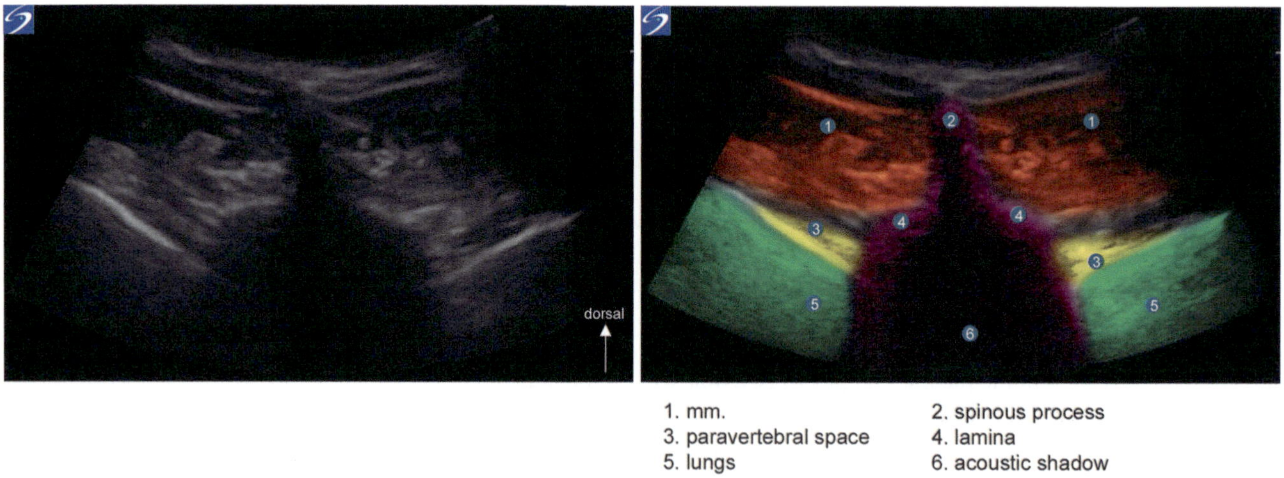

Fig. 10.25 Median transverse thoracic ultrasound section, not passing through transverse processes (*see* fig. 10.26)

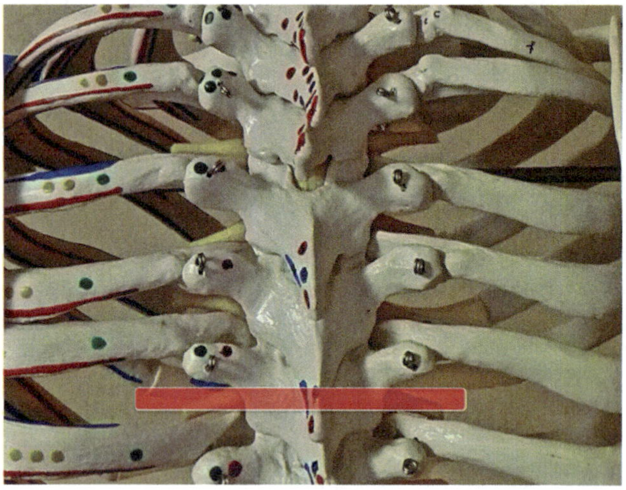

Fig. 10.26 Transverse position of ultrasound plane at the level of the lamina (not cutting through the transverse processes)

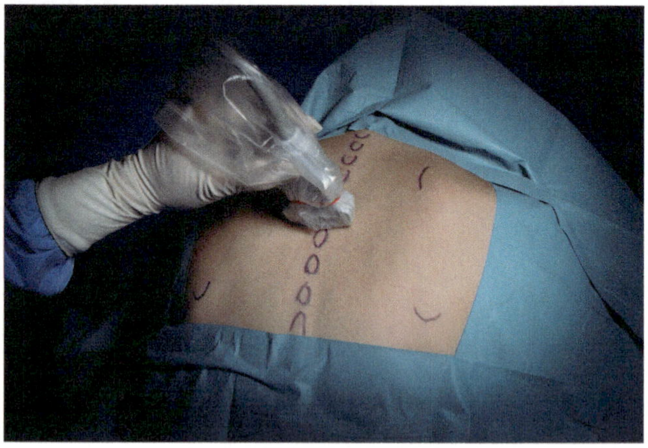

Fig. 10.27 Position of probe for median transverse view

Section Passing through the Transverse Process and the Adjacent Rib (Figs. 10.29 and 10.30)

This image is not usable for conduct of a TPV block. However, it makes it possible to identify the position of the costotransverse joint and is part of the pre-procedural scan to correctly understand all of the elements of the sonoanatomy of the paravertebral region.

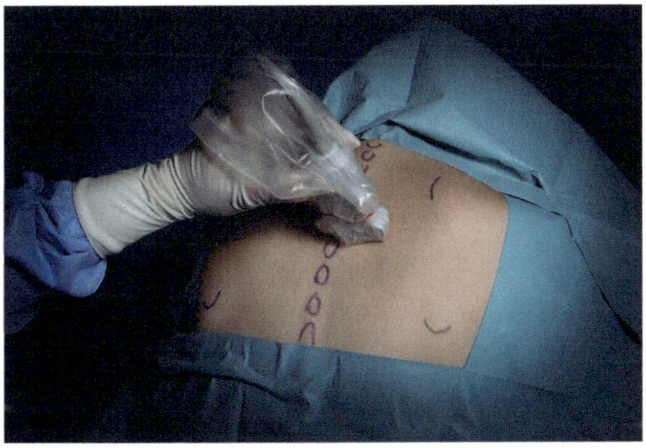

Fig. 10.28 Transverse view with lateral sliding of the probe

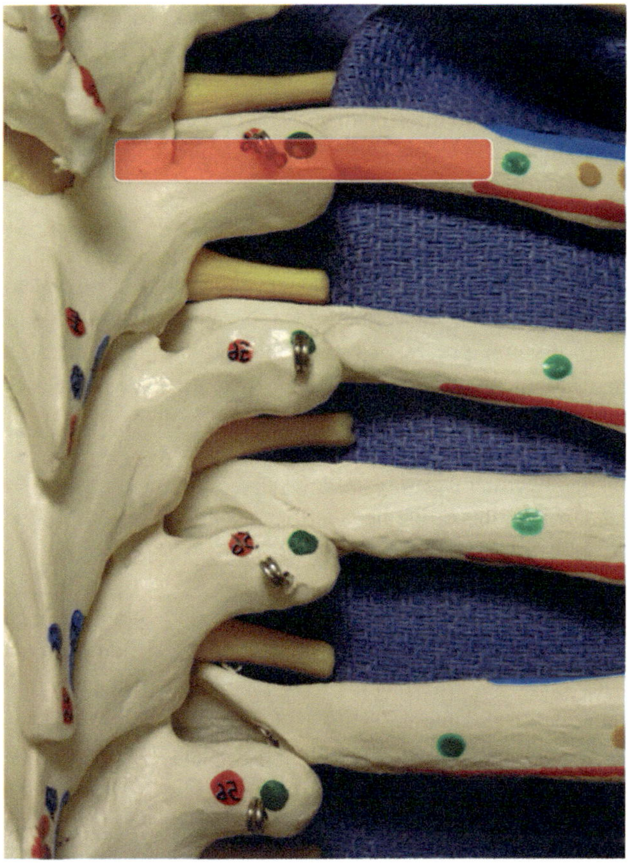

Fig. 10.30 Position of probe for view cutting through the transverse process and the adjacent rib

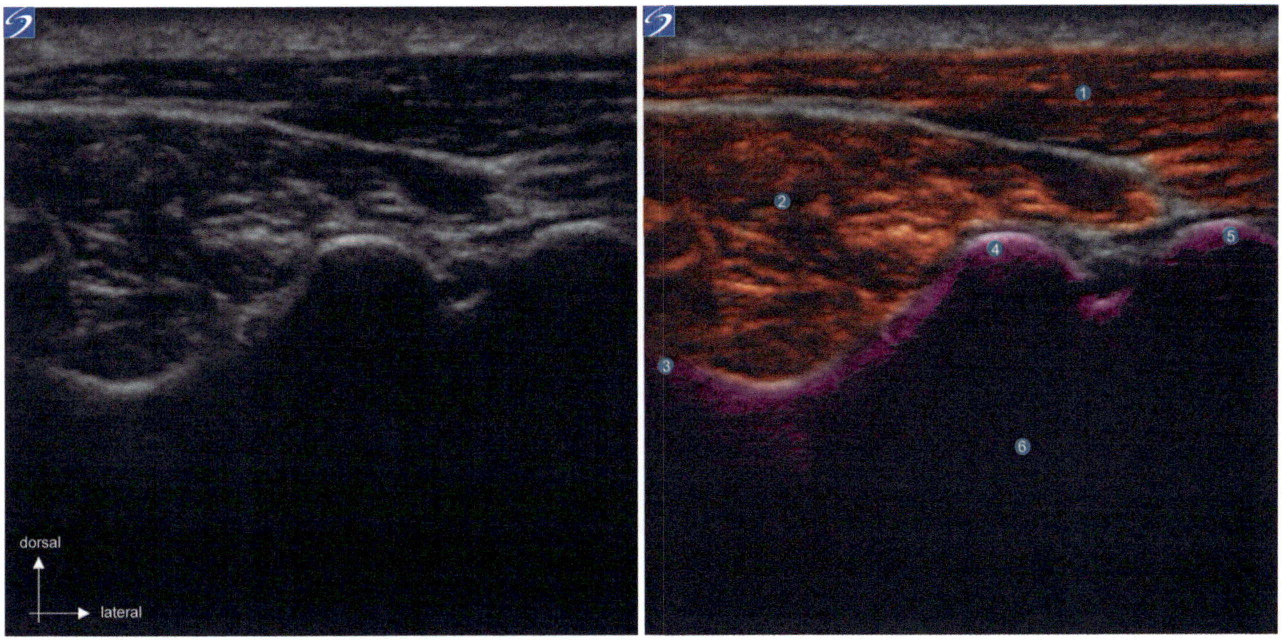

1. trapezius m.
2. erector spinae m.
3. lamina
4. transverse process
5. rib
6. acoustic shadow due to process and rib

Fig. 10.29 Transverse ultrasound view cutting through the transverse process and the adjacent rib (*see* fig. 10.30)

Section Passing through the Transverse Process and the Pleura (Figs. 10.31 and 10.32)

This produces an ultrasound view that is perfectly suitable to perform a TPV block. Moreover, this approach is described by Shibata [41] and Ben-Ari [42] when studying this block.

Approach in Plane (Fig. 10.19)

The needle is inserted into the ultrasound plane, and advanced into the posterior intercostal space seen at the lateral edge of the transverse process, between the parietal pleura and the internal intercostal membrane. The theoretical advantages of this approach are as follows:

- A tangential pathway to the pleura.
- The position of the needle tip at a safe distance from the intervertebral foramen.

However, the space where the needle tip is placed is narrow, which makes the procedure somewhat difficult to perform (Fig. 10.33). When the local anaesthetic is injected, the ventral movement of the parietal pleural must be visualised—this is a pathognomonic sign of appropriate spread into the intercostal space (Fig. 10.34). If this is not the case, the needle should be repositioned appropriately.

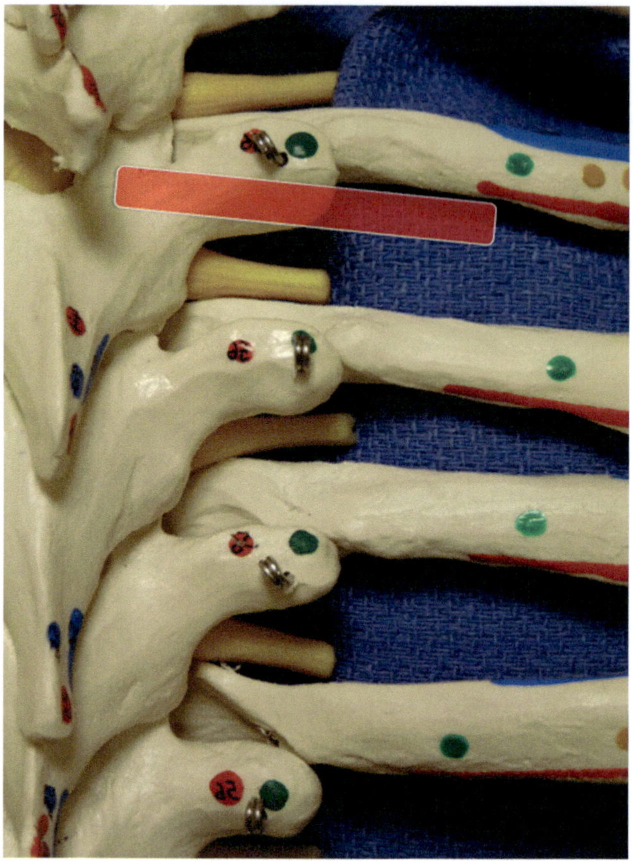

Fig. 10.32 Position of probe for view cutting through transverse process and pleura

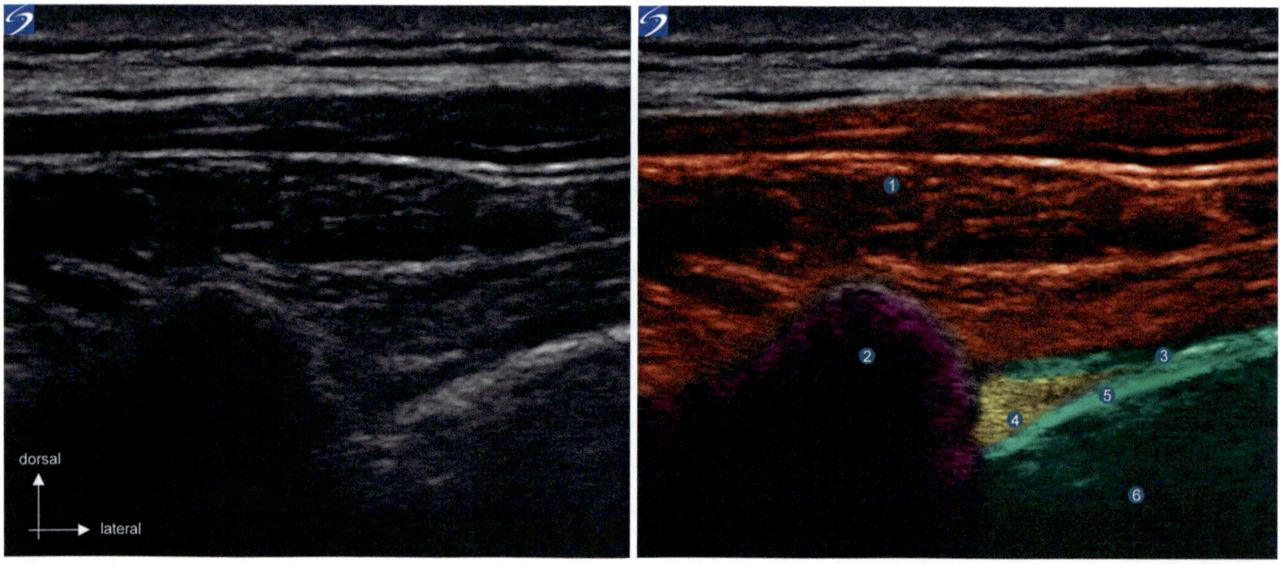

1. mm. 2. transverse process and acoustic shadow
3. internal intercostal membrane 4. posterior interscostal space
5. pleura 6. lung

Fig. 10.31 Transverse ultrasound view cutting through both transverse process and pleura (*see* fig. 10.32)

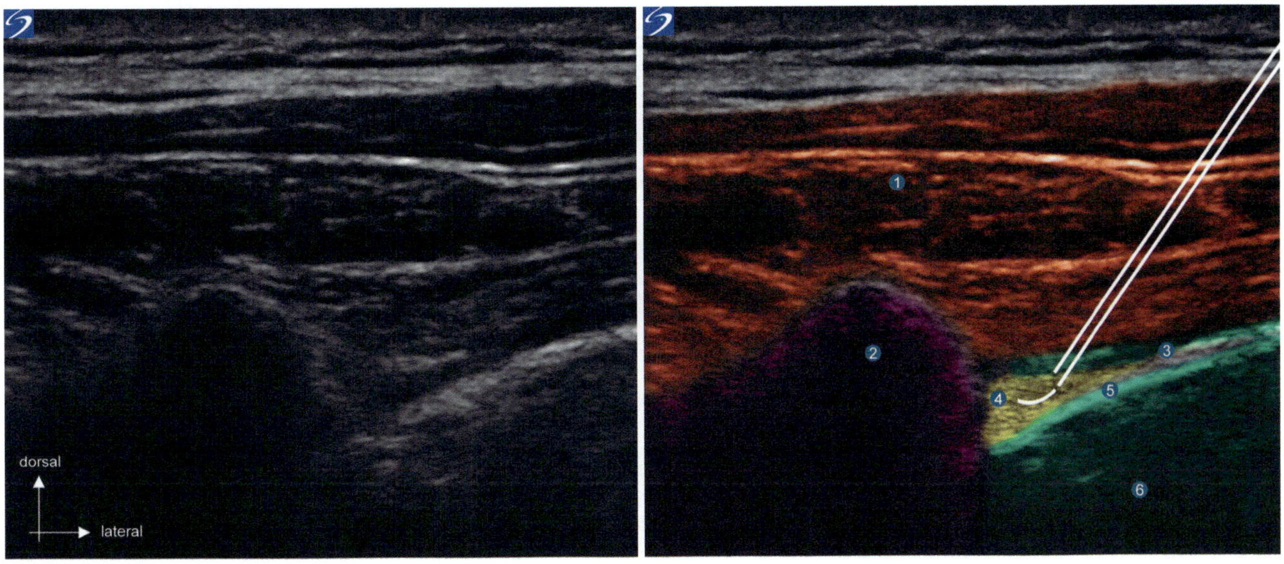

1. mm. 2. transverse process and acoustic shadow
3. internal intercostal membrane
4. needle in posterior interscostal space
5. pleura 6. lung

Fig. 10.33 Transverse ultrasound view cutting through transverse process and pleura. Theoretical position of needle in posterior intercostal space, inserted in plane

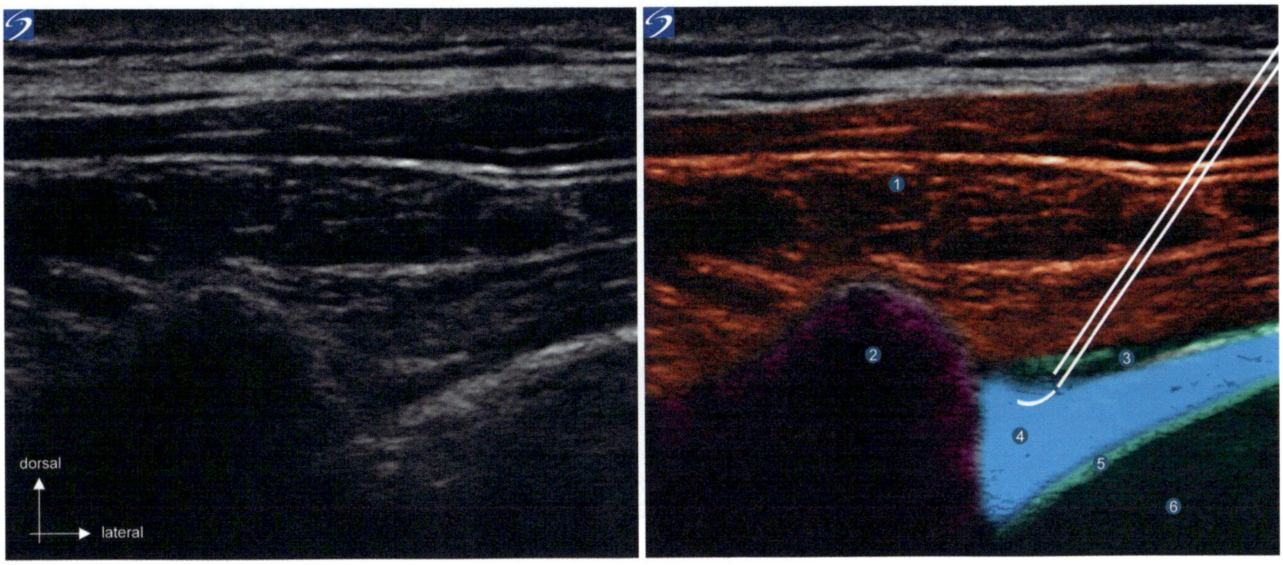

1. mm. 2. transverse process and acoustic shadow
3. internal intercostal membrane
4. local anaesthetic in posterior interscostal space
5. pleura (moving down due to the injection)
6. lung

Fig. 10.34 Transverse ultrasound view cutting through transverse process and pleura. Injection of local anaesthetic (LA) into the posterior intercostal space

Approach out of Plane (Figs. 10.20 and 10.35)

Due to the dorsal kyphosis and the slight rostro-caudal shift of the costotransverse joint (Fig. 10.32), generally, the injection is performed at the caudal border of the probe (Fig. 10.20). Insertion into the skin should be performed at a distance of at most 1–2 cm from the probe, in order to have an injection pathway that is relatively vertical and not hindered by the apex of the most caudal transverse process.

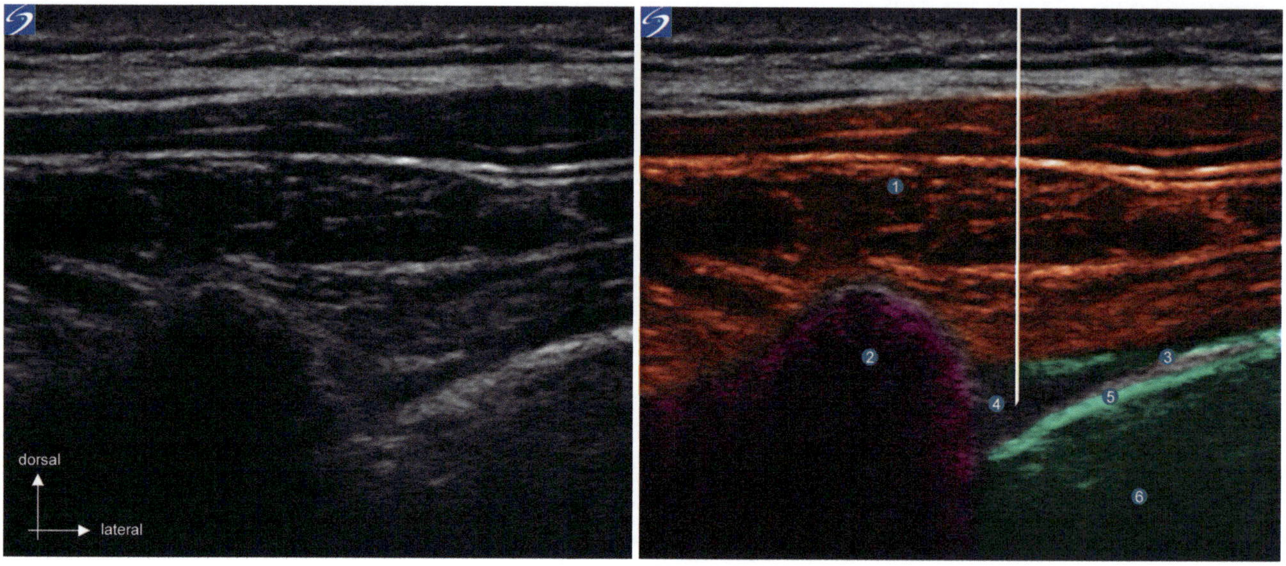

1. mm. 2. transverse process and acoustic shadow
3. internal intercostal membrane
4. posterior interscostal space 5. pleura 6. lung

Fig. 10.35 TPV block by injection into the posterior intercostal space. Theoretical trajectory of needle for out-of-plane approach

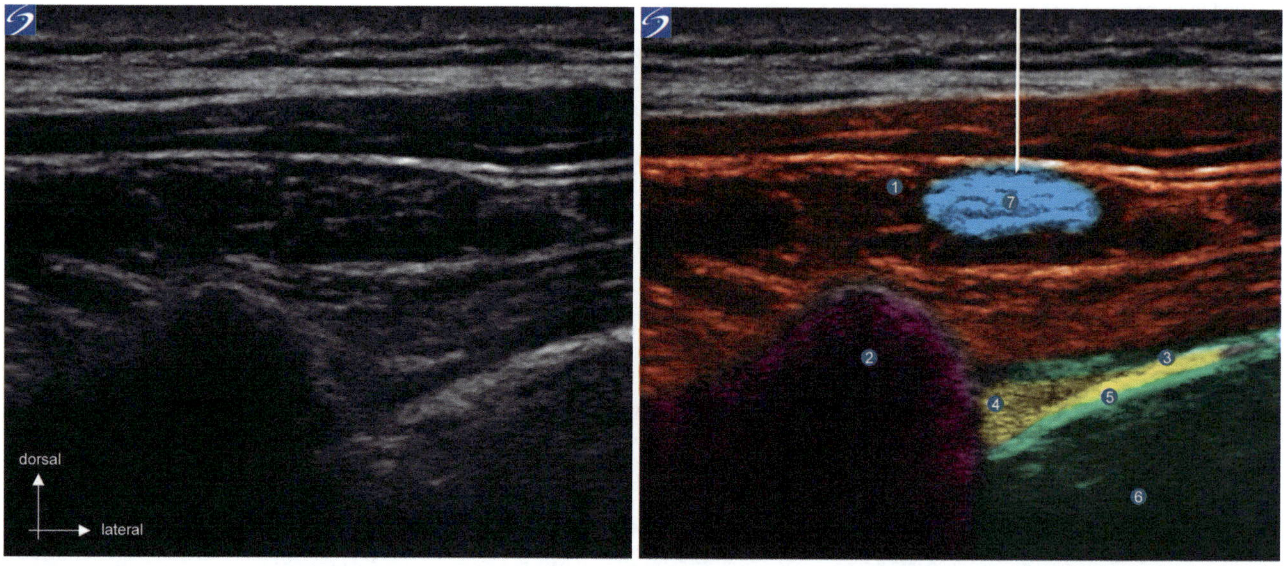

1. mm. 2. transverse process and acoustic shadow
3. internal intercostal membrane
4. posterior interscostal space 5. pleura
6. lung 7. bolus for hydrolocalization

Fig. 10.36 TPV block by injection into the posterior intercostal space. Out-of-plane approach with hydrolocation: first bolus visualised inside the erector spinae muscle

To facilitate the direction of the needle, it is useful to position the target (the tip of the transverse process) at the midpoint of the probe, and to insert the needle opposite this point also. Advancement of the needle is controlled by **hydrolocalisation** in order to know precisely the depth to which the needle tip is located. Up to the internal intercostal membrane (which is a lateral extension of the superior costotransverse ligament), bolus doses successively injected appear within the thickness of the spinal erector muscles (Figs. 10.35, 10.36, and 10.37). Once the internal intercostal membrane has been pierced, the injection then is accompanied by ventral pleural movement with widen-

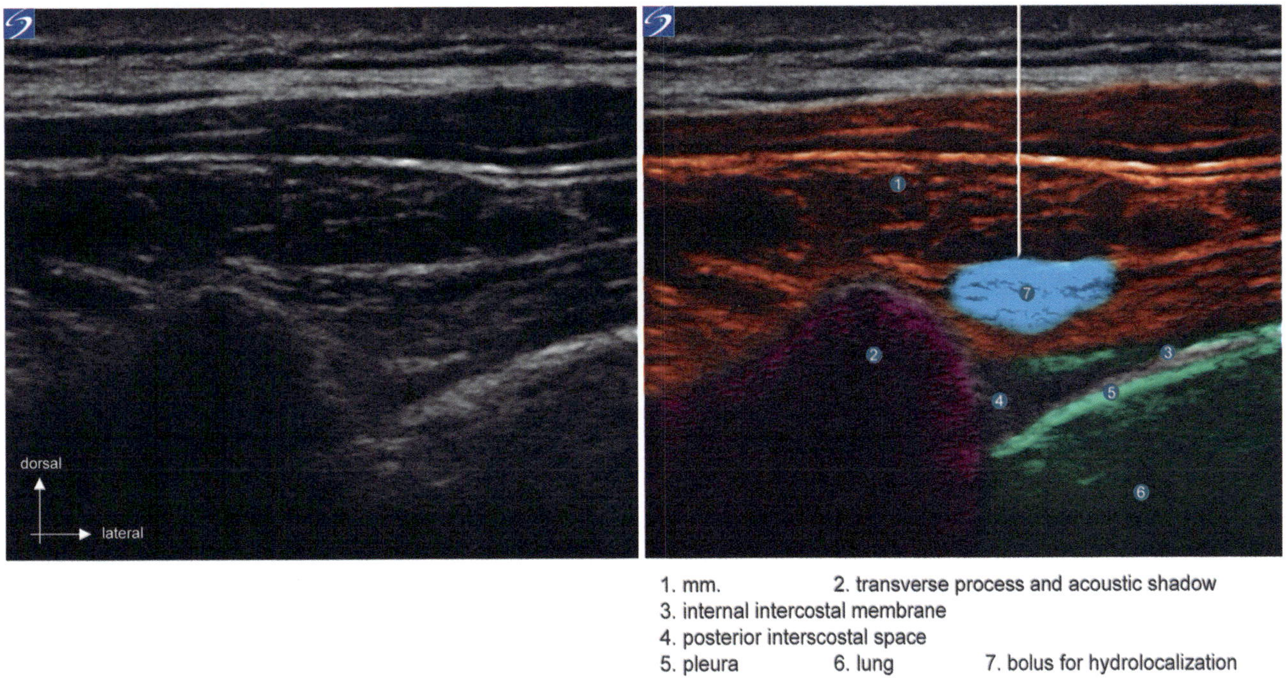

Fig. 10.37 TPV block by injection into the posterior intercostal space. Out-of-plane approach with hydrolocation: second bolus visualised inside the external intercostal muscles

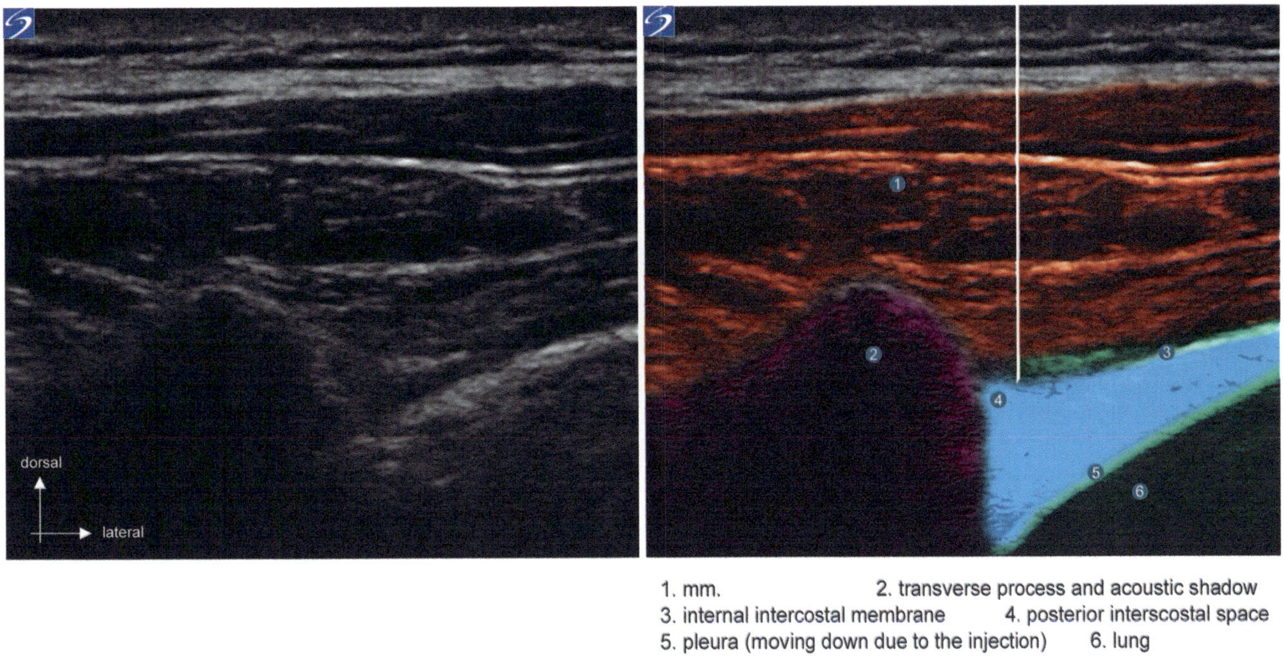

Fig. 10.38 TPV block by injection into the posterior intercostal space. Out-of-plane approach with hydrolocation: this last bolus is accompanied by a ventral displacement of the pleura, confirming that the needle has in fact crossed the internal intercostal membrane and that its tip is well-positioned in the posterior intercostal space

ing of the posterior intercostal space, confirming the correct position (Fig. 10.38). If this is not the case, the needle should not be considered as being correctly positioned, necessitating its repositioning. This technique described by Marhofer et al. [37] involves an approach with the needle **perpendicular** to the pleura, which may be a consideration when access to the target space (the posterior intercostal space) is especially limited**.**

It is also possible to use ultrasound guidance to place the needle tip in contact with transverse process, and then redirect it to complete its "blind" insertion into the TPV space by feeling a "loss of resistance" (about 1.5 cm more in depth).

Single or Continuous Injection?

- **Single injection block:** Once the needle tip has been correctly placed, the entire volume of local anaesthetic then can be injected slowly, in successive bolus doses of 5 mL, with repeated aspiration tests, under constant ultrasound visual control, and looking for signs of possible intravenous injection (maintaining verbal contact with the patient, monitoring of the ECG tracing etc.).
- **Continuous block:** Placement of a paravertebral perineural catheter is performed after injection of the first bolus dose of 5–7 mL, intended to "open" the paravertebral space. In light of current knowledge, it may be prudent to use a "Curl" preformed catheter (Sonolong Curl Echo, Pajunk®) for continuous TPV blocks. This product allows the distal coil of catheter to deploy in situ as it emerges from the needle tip, i.e. about 2.5–3 cm of catheter. If a traditional linear catheter is used, no more than 2–3 cm should be introduced after its emergence from the needle in order to reduce the risk of ectopic positioning (epidural, prevertebral etc.). In spite of this precaution, only by using X-ray imaging with radiological contrast medium is it possible to confirm correct positioning of the end of the catheter.

Section Passing through the Vertebral Laminae and the Pleura (Figs. 10.39 and 10.40)

When the echogenicity of the tissues permits, this approach offers the widest view of the TPV space making placement of the needle tip within it very feasible. The visible portion of the TPV space is demarcated as follows:

- Medially, by the acoustic shadow of the vertebral lamina.
- Ventrally, by the parietal pleura (and the endothoracic fascia).
- Dorsally, by the superior costotransverse ligament (whose outline is not always clearly visible), which continues laterally as the internal intercostal membrane.

This transverse ultrasound section through the lamina and the pleura offers a wider view of the TPV space (Fig. 10.39). A TPV block can be performed in this position either by an approach **in plane** (Fig. 10.41), or **out of plane** (Fig. 10.42). In both cases, the needle approach is very or relatively tangential to the pleura, which unquestionably provides a greater margin of safety. This is also the case for the approach described by Shibata and Nishiwaki (in plane, towards the tip of the transverse process) [41] (Fig. 10.33), but the view of the posterior intercostal space that their ultrasound section provides is much more limited.

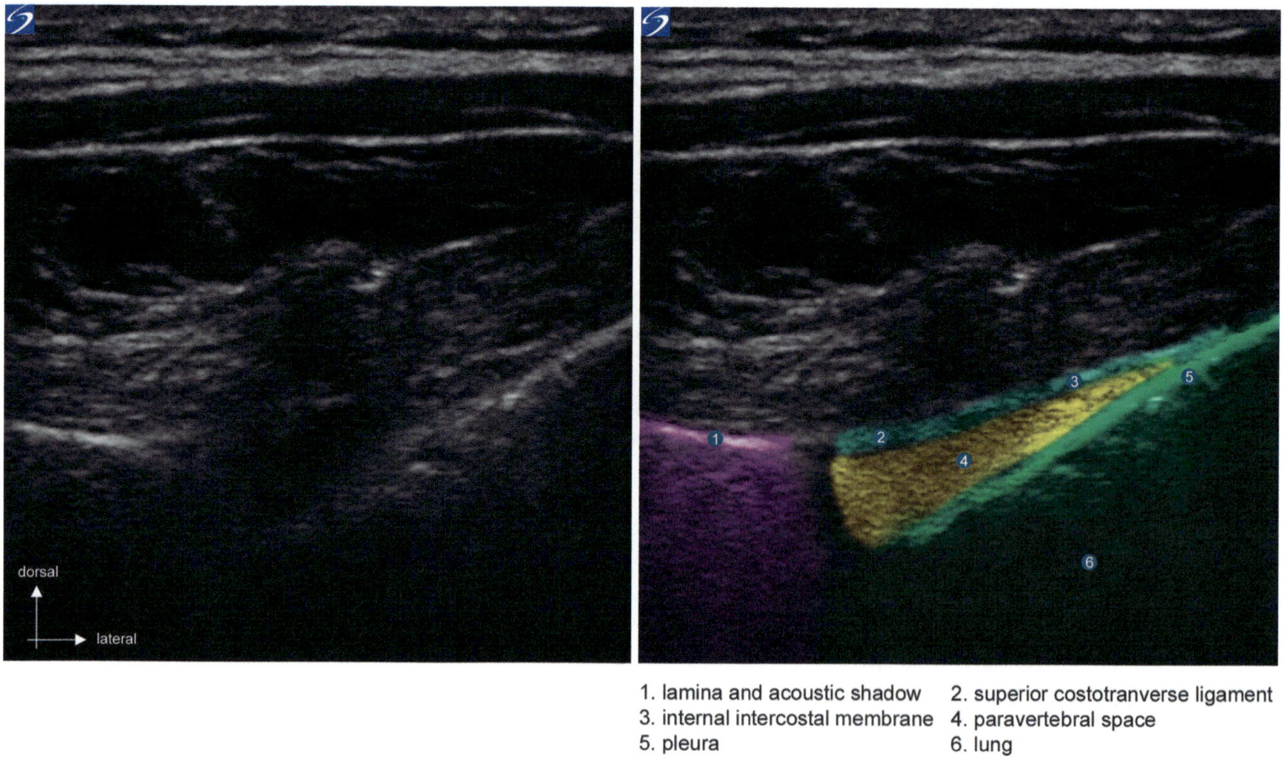

Fig. 10.39 Transverse ultrasound view cutting through lamina and pleura (see fig. 10.40)

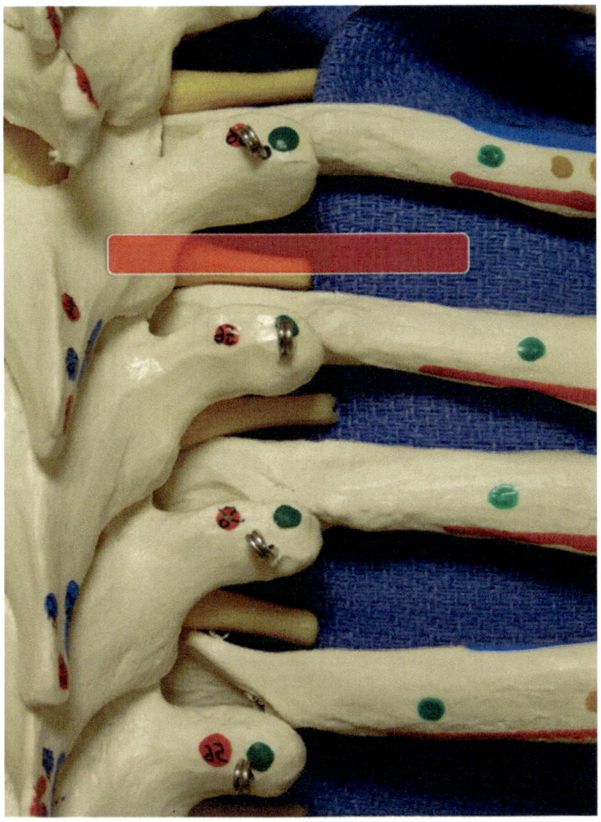

Fig. 10.40 Position of probe for view cutting through lamina and pleura

Under usual conditions of visualisation, the author considers this to be the most satisfactory approach to the TPV space (Figs. 10.39 and 10.40) in terms of both feasibility and safety.

Approach in Plane (Figs. 10.19 and 10.41)

The needle is inserted at the lateral end of the probe. It is inserted cautiously up to the TPV space, taking care to maintain visual contact with the position of the tip of the needle.

In case of poor needle handling, errors in positioning can have variable consequences. Intramuscular injection in the superficial plane in the TPV space (paravertebral muscles) leads only to block failure. An intrapleural or intrapulmonary injection will also result in block failure, but also leads to harmful respiratory consequences. Worse still, if the needle tip passes into the acoustic shadow of the vertebral lamina, which is medial to the paravertebral space visualised, it can come perilously close to midline neurological structures: risk of penetration of the intervertebral foramen or even puncture of the spinal cord etc. These worrisome consequences **cannot occur** if visual control of the needle tip is maintained within the ultrasound plane (it should always remain lateral to the lateral border of the lamina).

During its insertion, the needle successively crosses the skin, the plane of the spinal erector muscles (Fig. 10.43), and

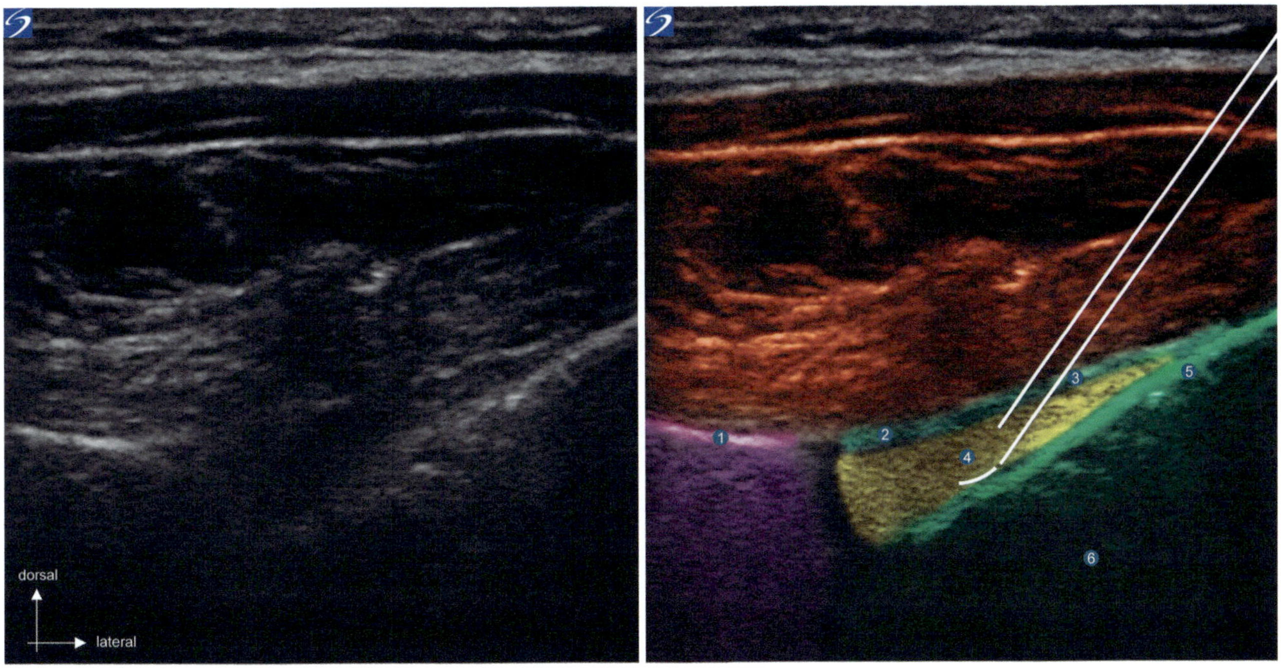

1. lamina and acoustic shadow 2. superior costotranverse ligament
3. internal intercostal membrane 4. paravertebral space
5. pleura 6. lung

Fig. 10.41 Transverse ultrasound view cutting through lamina and pleura. Injection in plane

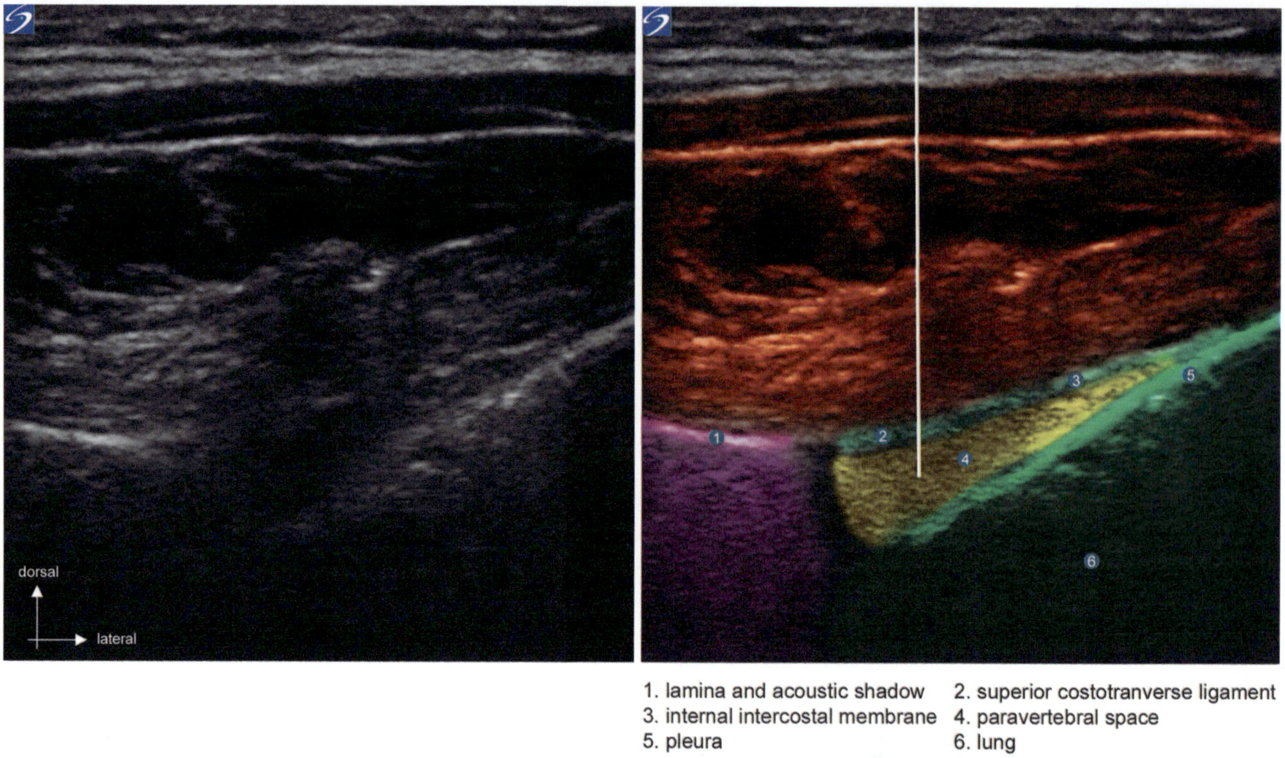

Fig. 10.42 Transverse ultrasound view cutting through vertebral plate and pleura. Injection out of plane

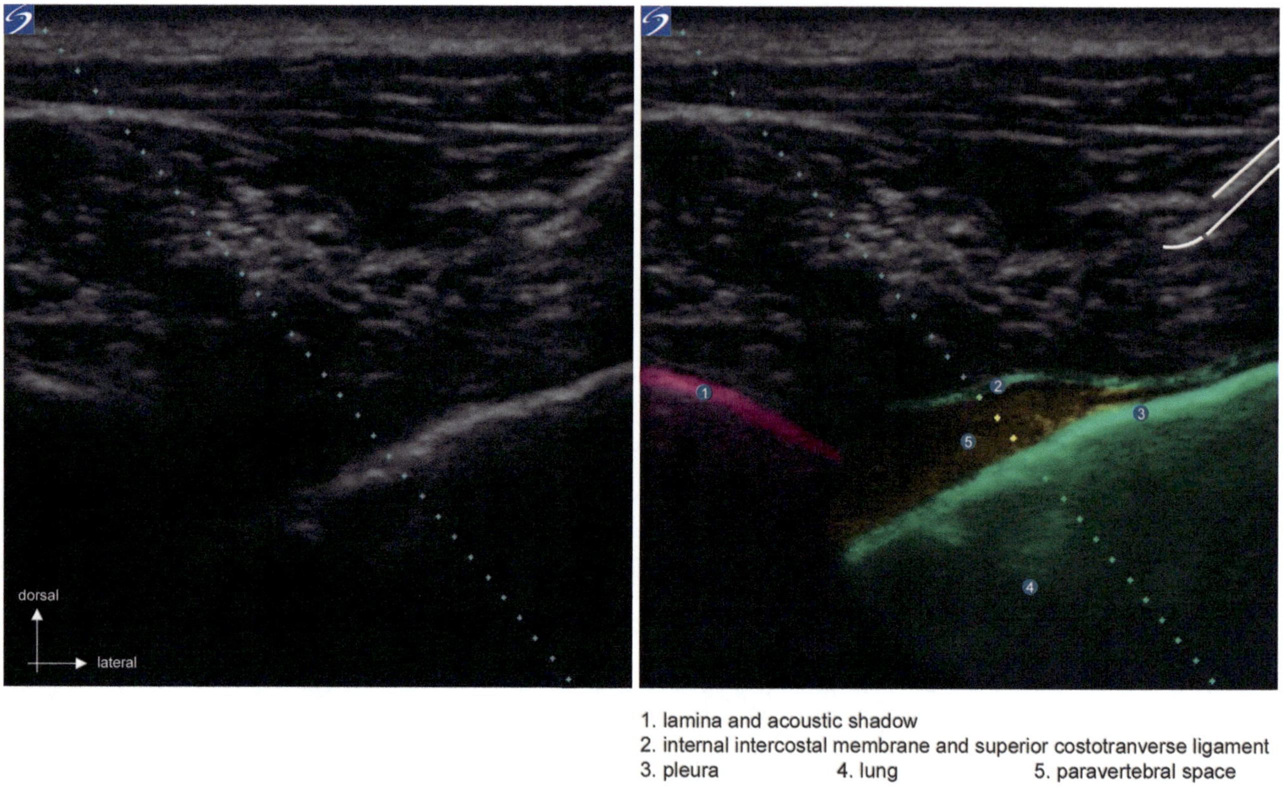

Fig. 10.43 Thoracic paravertebral block at end of transverse process: out of plane approach [37]

then the superior costotransverse ligament (which is not always distinguished clearly, but which continues as the internal intercostal membrane laterally, which is more visible) (Fig. 10.41). The tip of the needle is positioned in the "centre" of the paravertebral space, and the injection is started. Then, the ventral movement of the parietal pleura is observed with widening of the posterior intercostal space under the pressure related to the injection (Fig. 10.11). In the event that the injection is not accompanied by these signs, repositioning of the needle is essential.

Approach out of Plane (Figs. 10.20 and 10.42)

The needle is inserted at the mid-point of the caudal border of the probe (Fig. 10.47) at a maximum distance of 1–2 cm from the probe. This position creates a relatively vertical pathway for the needle and it is not hindered by the apex of the most caudal transverse process. Advancement of the needle is controlled by **hydrolocalisation** in order precisely identify the depth at which the needle tip lies. Up to the superior costotransverse ligament (for which the sensation of piercing it is frequent, but inconsistent), bolus doses successively injected appear within the substance of the erector spinae muscles. Once the ligamentary plane has been crossed, the injection then is accompanied by ventral pleural movement verifying the correct position of the needle tip, and is associated with a widening of the posterior intercostal space at the intervertebral level approached. If this is not the case, the needle should not be considered as being in the correct position and its repositioning is essential.

Single or Continuous Injection?

- **Single injection block:** Once the needle tip has been correctly placed, the entire volume of local anaesthetic then can be injected slowly, in successive bolus doses of 5 mL, with repeated aspiration tests, under constant ultrasound visual control, and looking for signs of possible intravenous injection (maintaining verbal contact with the patient, monitoring of the ECG tracing etc.).
- **Continuous block:** Placement of a paravertebral perineural catheter is performed after injection of the first bolus dose of 5–7 mL, intended to "open" the paravertebral space. In light of current knowledge, it may be prudent to use a "Curl" preformed catheter (Sonolong Curl Echo, Pajunk®) for continuous TPV blocks. This product allows the distal coil of catheter to deploy in situ as it emerges from the needle tip, i.e. about 2.5–3 cm of catheter. If a traditional linear catheter is used, no more than 2–3 cm should be introduced after its emergence from the needle in order to reduce the risk of ectopic positioning (epidural, prevertebral etc.). In spite of this precaution, only by using X-ray imaging with radiological contrast medium it is possible to confirm the correct positioning of the end of the catheter.

Paramedian Sagittal Ultrasound Section

The paramedian sagittal plane makes it possible to observe the ribs and the transverse processes in cross section. These bony structures have a hyperechogenic surface, which cast an acoustic shadow. The adjacent ribs and transverse processes are separated by an acoustic window enabling the underlying lung tissue to be seen. In their paramedian posterior thoracic part, the ribs articulate with the transverse processes. In sagittal section, the cross-sectional outline of the rib differs from that of the transverse process by its shape (the rib is more rounded) but in particular (principally) by its height above the surface of the pleura. The ribs generally terminate 0.7–1 cm above the pleural plane (Figs. 10.44 and 10.45), whilst the apex of the transverse process is often located 1.2 and 1.5 cm above the pleura (Fig. 10.46). Moreover, it is easy to see the transition between the costal section and the transverse section during ultrasound scanning from lateral to medial. A sagittal section also makes it possible to evidence a slight rostro-caudal shift between a rib and the transverse process with which it articulates (Fig. 10.47) (the rib is in a slightly more cranial position than the transverse process).

The sagittal ultrasound image of the TPV space is obtained by placing the probe parallel to the vertebral spinous process, on a line located about 2–3 cm from the midline. In this configuration, deep in the erector spinae muscles, the transverse processes are visible as rounded structures, with a hyperechoic surface and acoustic shadow in depth. The acoustic window located between two adjacent transverse processes should make it possible to observe the following from the superficial to deep level (Fig. 10.48):

- The erector spinae muscles.
- The superior costotransverse ligament and the intertransverse ligament.
- The thoracic paravertebral space.
- The pleura.
- The lung.

However, in light of the ventromedial direction of the pleura at this level, to correctly visualise it, it is necessary to incline the probe laterally (Figs. 10.49 and 10.50) so that the ultrasound beam approaches it according to a more favourable angle (i.e. more perpendicularly). In this case, the ultrasound beam generates a very lateral image of the paravertebral space, or even sometimes an image of a posterior intercostal space (in particular if the latter is especially deep, such as in obese patients). But it is as a result of this lateral inclination that the TPV space is more visible on this paramedian sagittal ultrasound section.

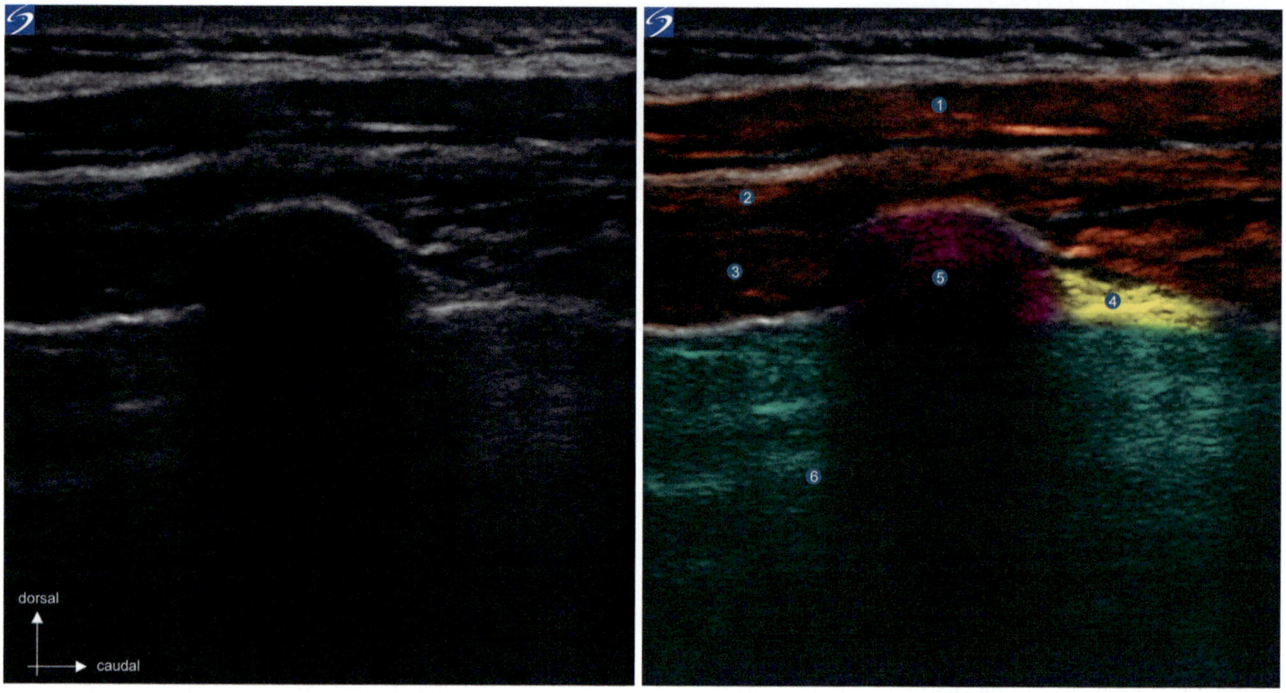

Fig. 10.44 Paramedian sagittal ultrasound view: rib view

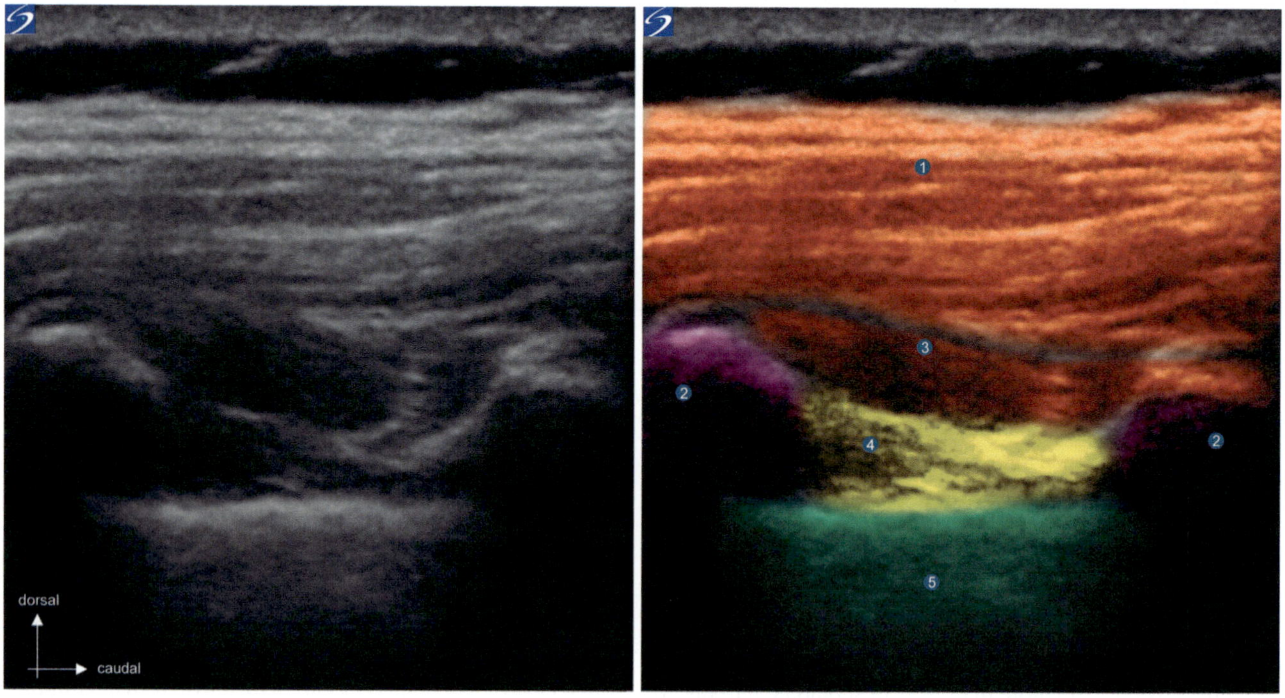

Fig. 10.45 Paramedian sagittal ultrasound view: rib view near tip of transverse process (not yet visualised)

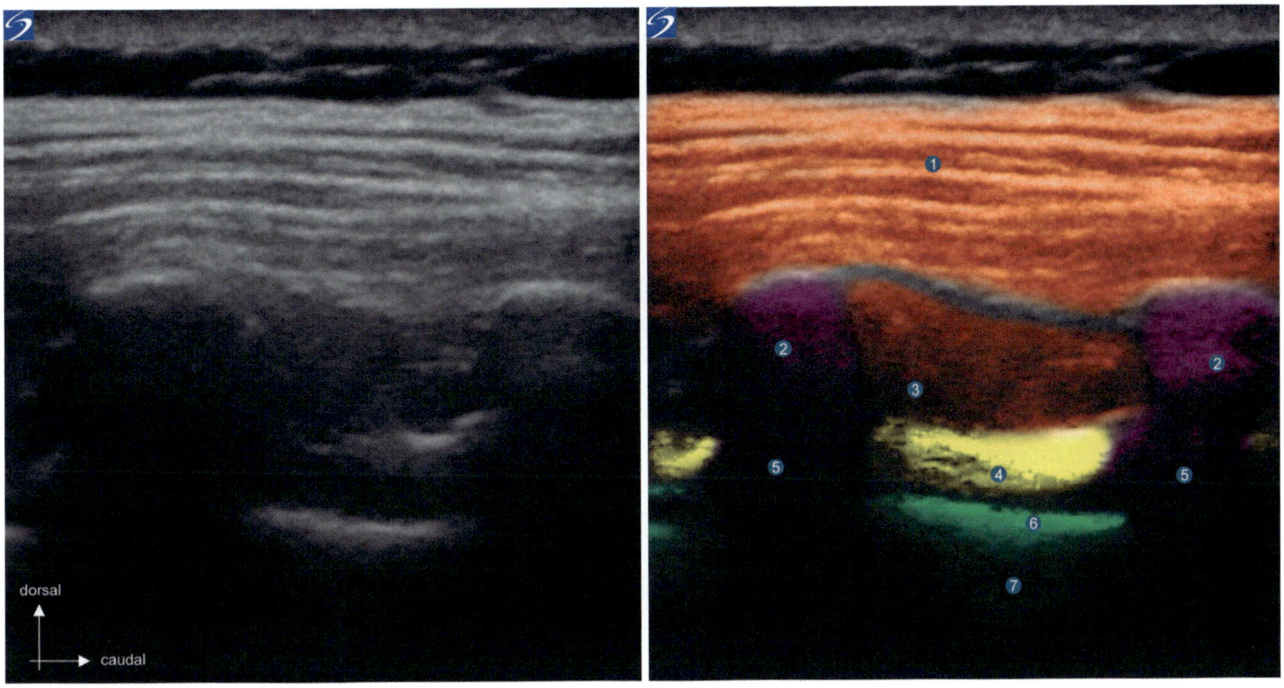

1. trapezius, rhomboid and erector spinae mm.
2. transverses process
3. superior costotransverse and intertransverse ligaments
4. lateral part of thoracic paravertebral space, next to the intercostal space
5. rib 6. pleura 7. lung

Fig. 10.46 Paramedian sagittal ultrasound view: view of transverse processes with ribs and costo-transverse joint

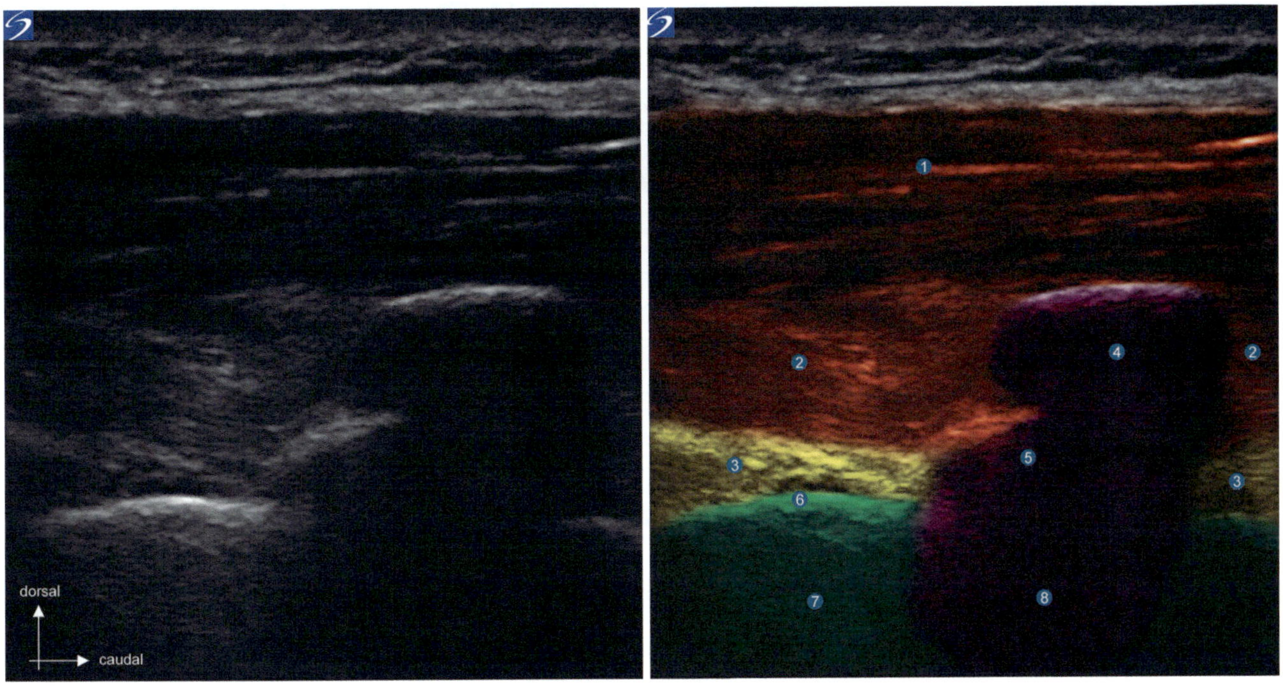

1. trapezius, rhomboid and erector spinae mm.
2. ligaments : posterior part of paravertebral space
3. paravertebral space 4. transverse process
5. rib 6. pleura
7. lung 8. acoustic shadow of bones

Fig. 10.47 Parasagittal ultrasound view of the costo-transverse joint showing rostro-caudal shift of these two bony structures

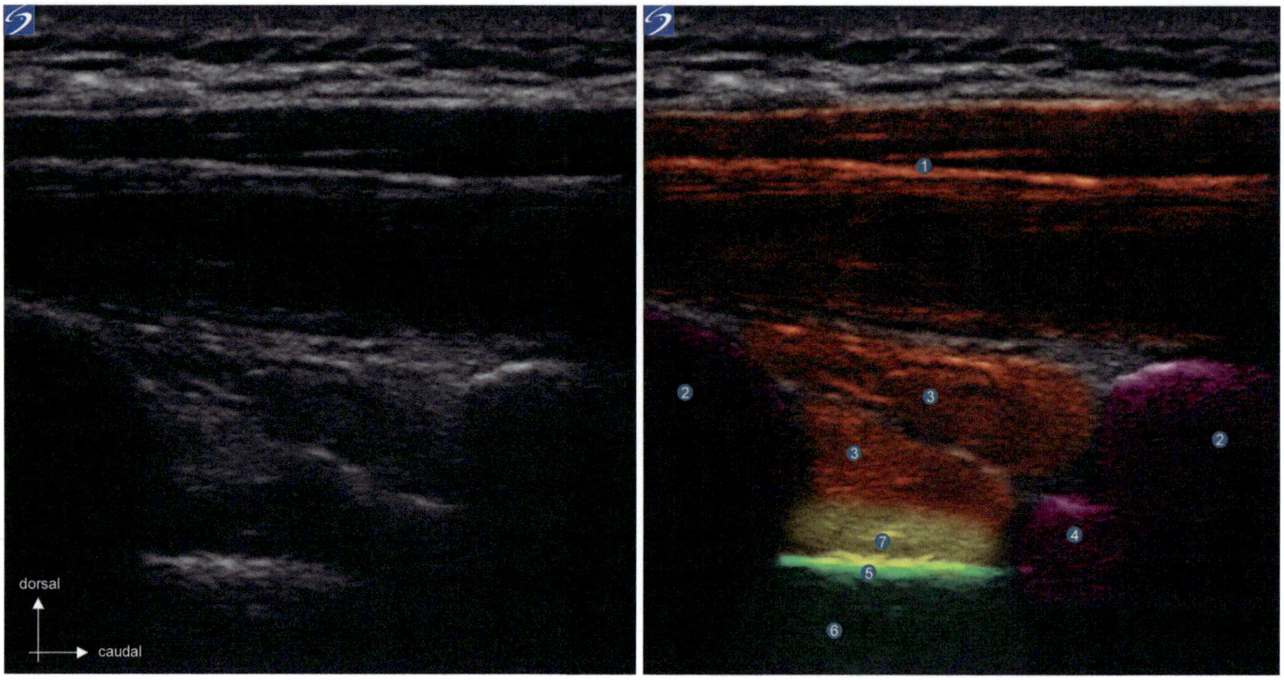

1. trapezius, rhomboid and erector spinae mm.
2. transverse process
3. superior costotransverse and intertransverse ligaments
4. rib 5. pleura 6. lung
7. lateral part of paravertebral space

Fig. 10.48 Sagittal ultrasound image of the paravertebral space

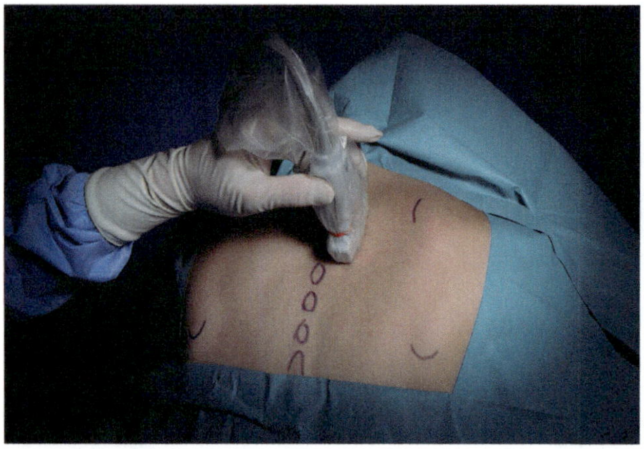

Fig. 10.49 Probe in strict parasagittal position not enabling to visualise pleura

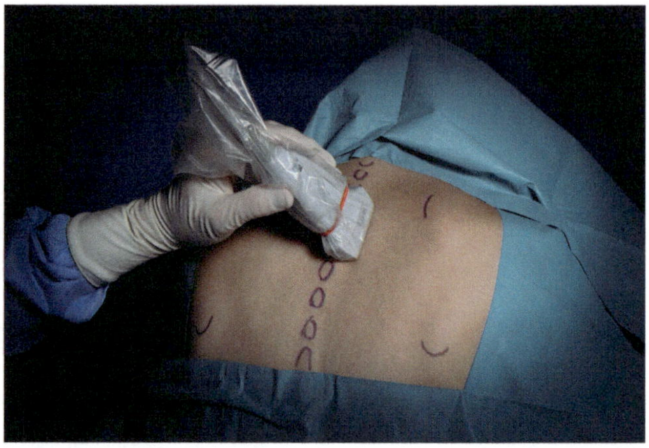

Fig. 10.50 Probe in parasagittal position with lateral tilt necessary to visualise pleura

Approach in Plane (Fig. 10.21)

As a result of the configuration of the superior costotransverse ligamentary plane, the PV space of a given intertransverse level is "thicker" in the caudal border of the cephalic transverse process than in the cephalic border of the most caudal transverse process. Consequently, the puncture is performed at the **caudal** end of the probe in order to target the TPV space at its thickest portion.

When the intervertebral level desired has been identified, the probe is positioned in a manner so that the intertransverse acoustic window is level with the centre of the screen, limited by the two adjacent transverse processes. An injection made at the immediate caudal end of the probe enables the needle to be advanced unhindered by the edge of the most caudal transverse process. The needle is carefully advanced successively through the skin, the paraspinal muscles, the intertransverse ligament, and the superior costotransverse ligament, up to the TPV space (Fig. 10.51). Throughout the injection process, it is essential to keep **strict visual control** of the position of the tip, possibly with the aid of hydrolocalisation. This is especially important as the needle pierces the ligaments that form the posterior limit of the TPV space. When the last ligamentary plane has been crossed, the injection of a small bolus of 0.9% NaCl (1–2 millilitre), will create ventral movement of the parietal pleura (Fig. 10.52). In the absence of this confirmatory sign, the needle cannot be considered as being correctly positioned and repositioning is required. With this approach, entry into the TPV space is often delicate or even difficult. This is due to the "narrow" view achieved, together with a more aggressive axis of injection towards the pleura.

Approach out of Plane (Fig. 10.22)

We have already seen that on paramedian sagittal ultrasound section, due to the ventromedial direction of the pleura, it is necessary to guide the probe slightly laterally to better visualise the TPV space (Fig. 10.50). Consequently, to perform a TPV block out the plane, the needle can be inserted only at its lateral border (Fig. 10.22). The needle should be inserted at a maximum distance of 1–2 cm from the probe so that the pathway of the needle is relatively vertical. Its entire advancement is controlled by **hydrolocalisation** in order to constantly identify the depth at which the tip of the needle is located. Bolus doses successively injected appear first in the substance of the erector spinae muscles (Figs. 10.53 and 10.54), and then the intertransverse and superior costotransverse ligaments (Fig. 10.55). A palpable sensation of piercing these ligaments, the dorsal border of the TPV space, is frequent but inconsistent. Once the ligamentary plane has been crossed, the injection is accompanied by ventral pleural movement confirming the correct position (Fig. 10.56). In the absence of this sign, the placement of the needle cannot be considered as correct and repositioning is required.

Single or Continuous Injection?

- **Single injection block:** Once the needle tip has been correctly placed, the entire volume of local anaesthetic then can be injected slowly, in successive bolus doses of 5 mL,

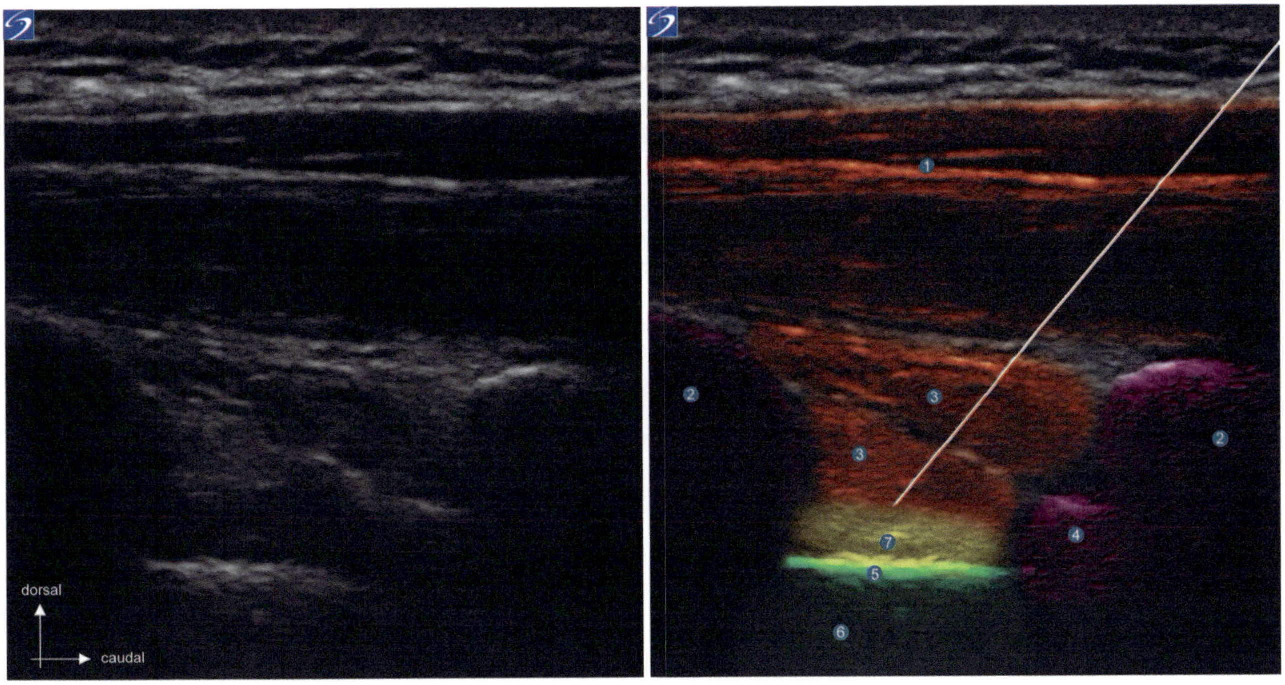

1. trapezius, rhomboid and erector spinae mm.
2. transverse process
3. superior costotransverse and intertransverse ligaments
4. rib 5. pleura 6. lung
7. lateral part of paravertebral space

Fig. 10.51 TPV block, paramedian sagittal view. Needle in plane

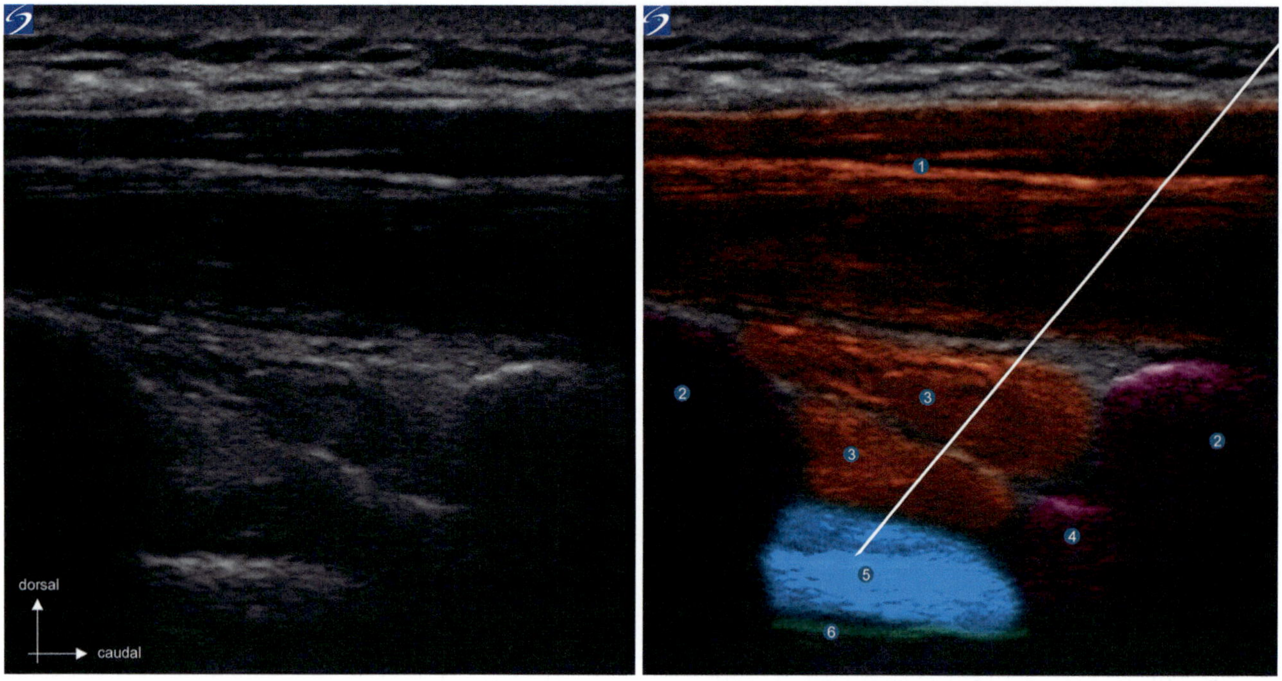

1. trapezius, rhomboid and erector spinae mm.
2. transverse process
3. superior costotransverse and intertransverse ligaments
4. rib
5. local anaesthetic in paravertebral space
6. pleura (moving down due to the injection)

Fig. 10.52 TPV block. Paramedian sagittal view, needle in plane. Ventral displacement of pleura during injection of local anaesthetic (LA) into the TPV space

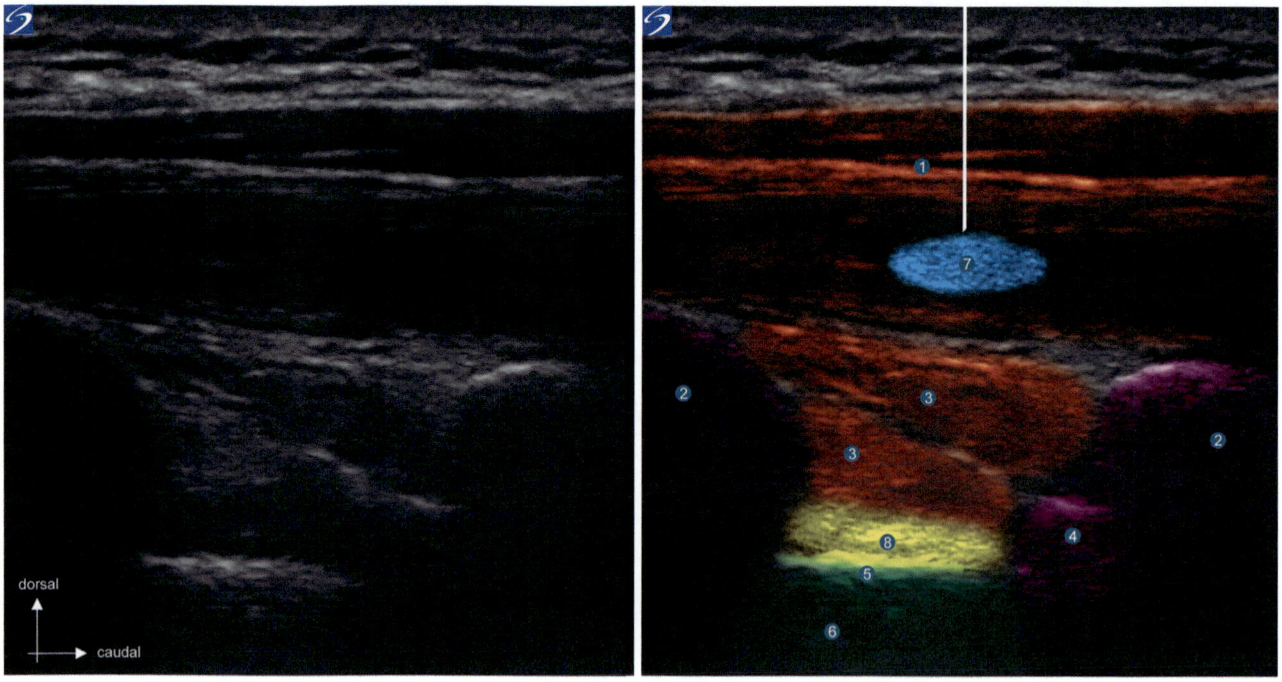

1. trapezius, rhomboid and erector spinae mm.
2. transverse process
3. superior costotransverse and intertransverse ligaments
4. rib 5. pleura 6. lung

Fig. 10.53 TPV block, paramedian sagittal view. Hydrolocation: local anaesthetic visualised inside the erector spinae muscle

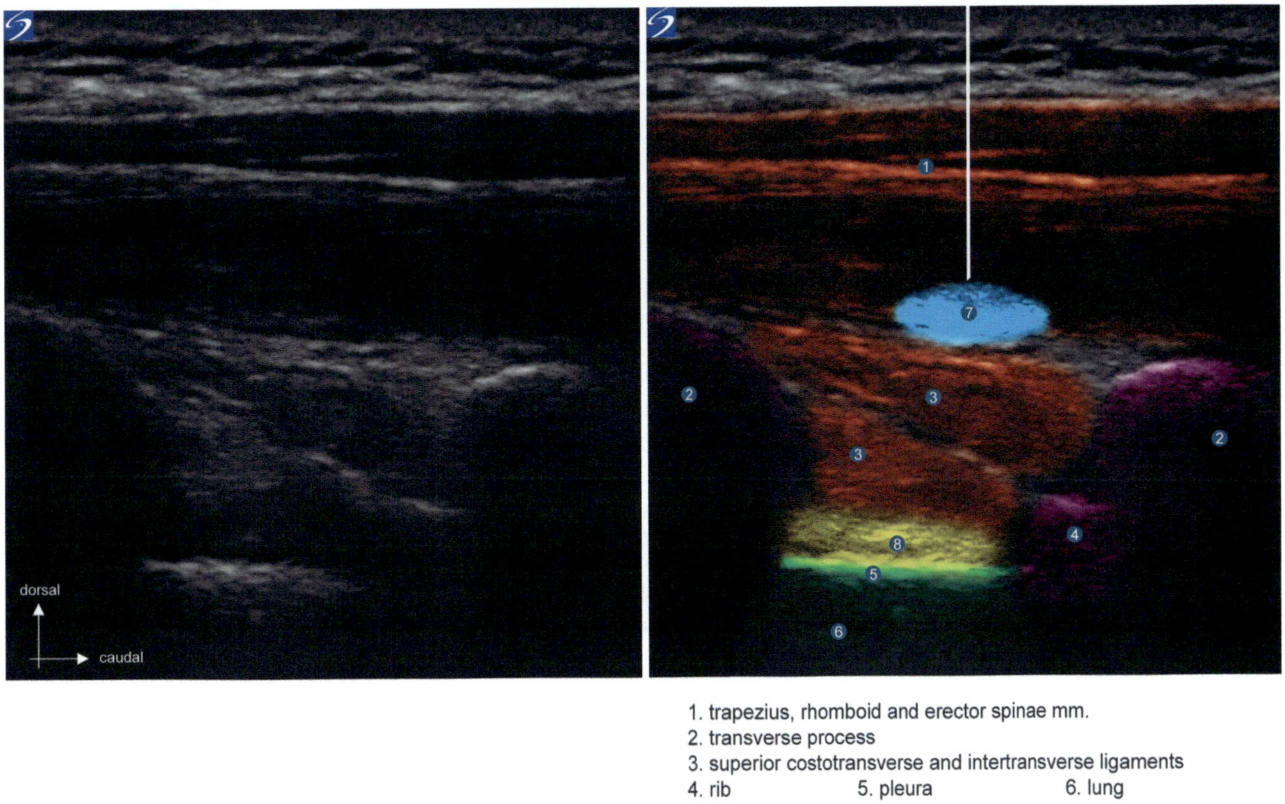

Fig. 10.54 TPV block, paramedian sagittal view. Hydrolocation: local anaesthetic block visualised in deep aspect of spinal erector muscles

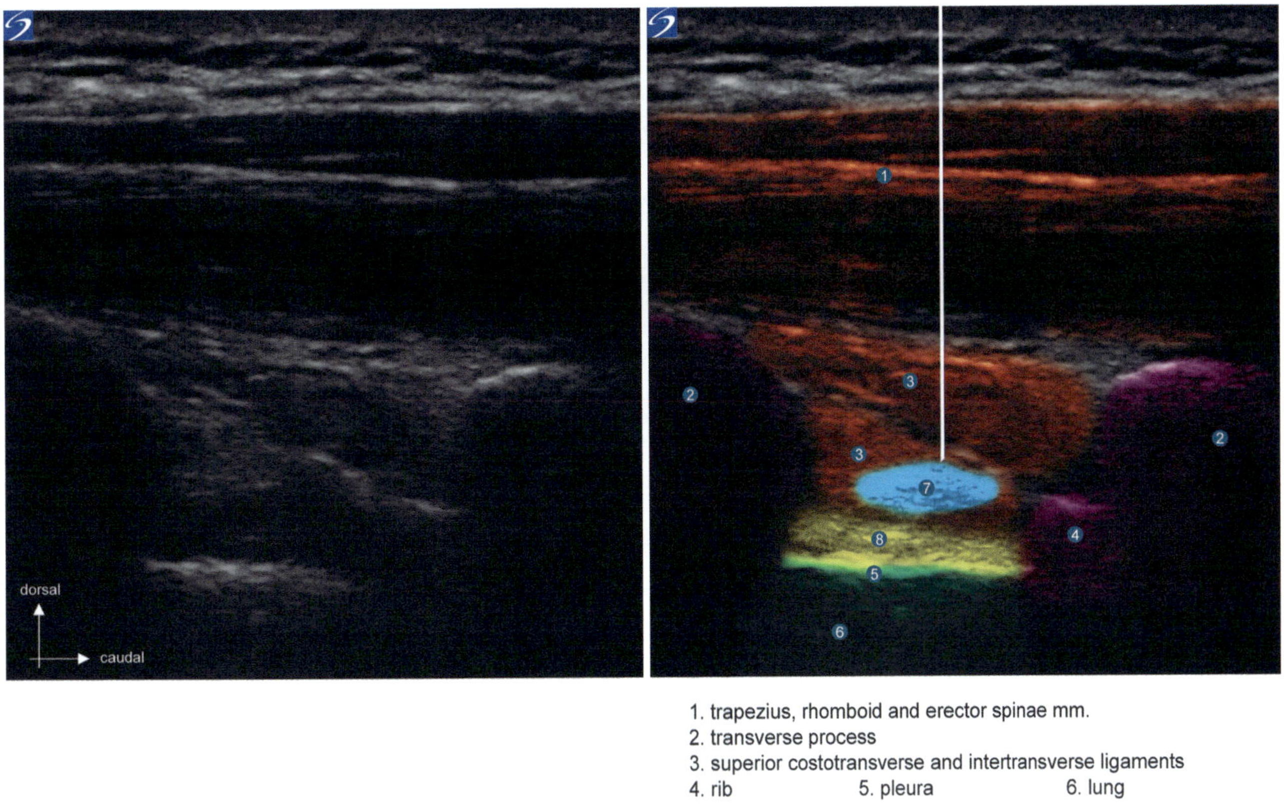

Fig. 10.55 TPV block, paramedian sagittal view. Hydrolocation: local anaesthetic visualised in intertransverse and superior costo-transverse ligaments

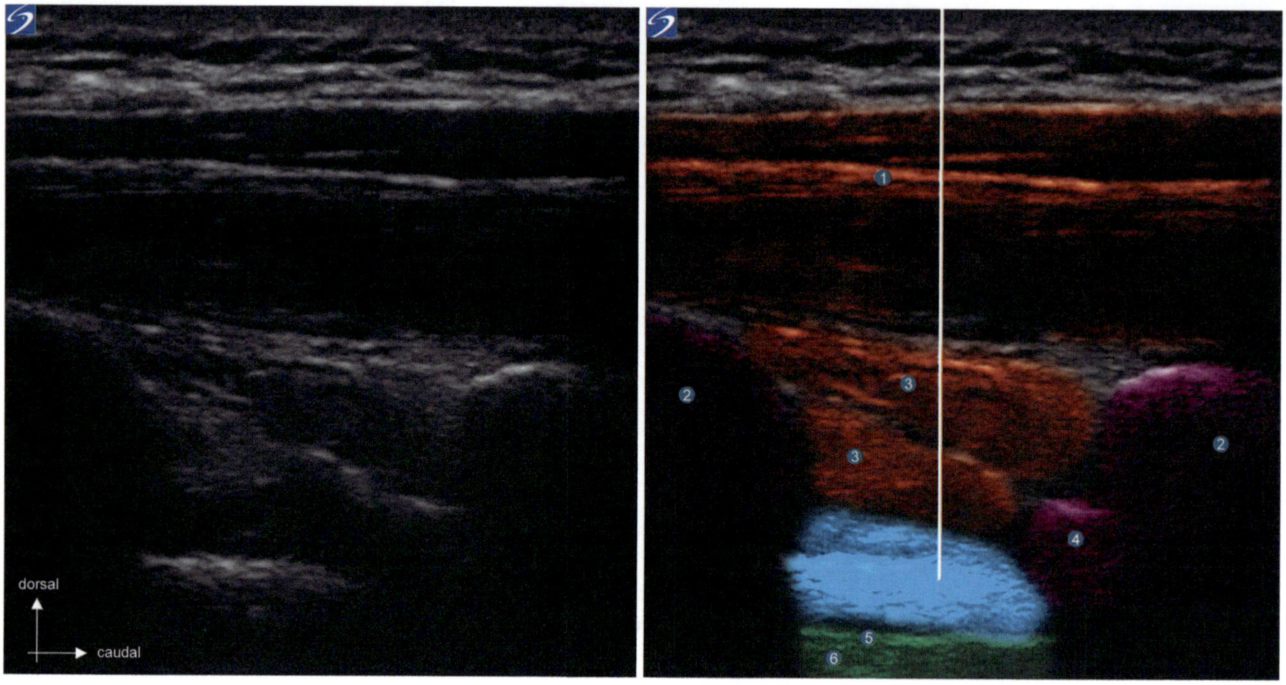

1. trapezius, rhomboid and erector spinae mm.
2. transverse process
3. superior costotransverse and intertransverse ligaments
4. rib 5. pleura (moving down due to the injection)
6. lung

Fig. 10.56 TPV block. Paramedian sagittal view, needle outside of ultrasound plane. Ventral displacement of pleura during injection of local anaesthetic in the TPV space

with repeated aspiration tests, under constant ultrasound visual control, and looking for signs of possible intravenous injection (maintaining verbal contact with the patient, monitoring of the ECG trace etc.).

- **Continuous block:** Placement of a paravertebral perineural catheter is performed after injection of the first bolus dose of 5–7 mL, intended to "open" the paravertebral space. In light of current knowledge, it may be prudent to use a "Curl" preformed catheter (Sonolong Curl Echo, Pajunk®) for continuous TPV blocks. This product allows the distal coil of catheter to deploy in situ as it emerges from the needle tip, i.e. about 2.5–3 cm of catheter. If a traditional linear catheter is used, no more than 2–3 cm should be introduced after its emergence from the needle in order to reduce the risk of ectopic positioning (epidural, prevertebral etc.). In spite of this precaution, only by using X-ray imaging with radiological contrast medium it is possible to confirm the correct positioning of the end of the catheter.

Conclusion

There are constant technological advances in the field of ultrasound with resultant improvements in our ultrasound systems. Ultrasound-guided techniques now allow for the precise determination of the thoracic paravertebral anatomy. By a simple, structured pre-procedure examination, it is easy to define the position and depth of the transverse processes and of the pleura and lungs. Therefore, a pre-block maximum safe depth of needle insertion can be measured and thereby reduce the risk of creating a pneumothorax during a "blind" (non-US guided) TVP block. However, by using ultrasound guidance throughout the performance of the block it is possible to position the needle tip precisely in the TPV space and to observe the real-time spread of the LA, either during a single-shot block or one performed with a perineural catheter for a continuous block. We have seen that schematically 4 configurations exist enabling to perform a TPV block:

- **Transverse** ultrasound section and approach **in plane.**
- **Transverse** ultrasound section and approach **out of plane**.
- **Paramedian sagittal** ultrasound plane and approach **in plane**.
- **Paramedial sagittal** ultrasound approach and approach **out of plane**.

However, although no consensus or scientific data exist to determine the optimum approach, some advantages in respect of **visibility** and **safety** support the first two of these techniques.

Advantages of the in-Plane Approach with Transverse Ultrasound Section (Figs. 10.19 and 10.41)

- Improved **visibility of the needle** both during needle insertion and injection of LA, enhanced by using echogenic needles and catheters.
- Improved **visibility of the paravertebral space,** particularly at the level of the vertebral lamina which offers a more ergonomic approach to the entire process of needle handling or catheter placement in the TVP.
- Improved **safety** (in respect of pneumothorax) due to the approach of the needle being tangential to the pleura and lung.

Advantages of the Out-of-Plane Approach, in Transverse Ultrasound Section (Figs. 10.20 and 10.42)

This approach to the TVP block offers the same advantages of visibility of the paravertebral anatomical structures as that described above. Although visibility of the needle in this orientation is often less obvious and requires a good level of expertise, careful use of hydrolocalisation makes it very safe. This approach is perhaps preferable when performing the block with a paravertebral catheter placed **close to the sagittal axis** (thoracic surgery with posterolateral incision, a flap of the latissimus dorsi muscle in reconstructive breast surgery).

The two configurations used with **paramedian sagittal ultrasound section** (i.e. with the needle directed in plane or out of plane) often result in poorer imaging of the structures in the TPV space. It is not easy to visualise, the pleura directly (which has a ventromedial direction) unless the ultrasound probe is inclined laterally. These approaches also create less access to the TPV space which is more confined in this orientation, and the needle trajectory is inclined more aggressively towards the pleura (in both approaches).

In terms of equipment, use of an **"echogenic" needles** whose surface is specifically made more reflective, is strongly recommended. Concerning **catheters,** the risk of ectopic positioning of the end of a catheter should be minimised. For a traditional linear catheter, insert no more than 2–3 cm beyond the tip of the needle. **Preformed ("pigtail") catheters,** which deploy a coil upon exiting from the needle tip (SonoLong Curl Echo, Pajunk®) may allow for a more predictable and reliable positioning of its end.

References

1. Chelly JE. Paravertebral blocks. Anesthesiol Clin. 2012;30:75–90.
2. Karmakar MK. Thoracic paravertebral block. Anesthesiology. 2001;95:771–80.
3. Karmakar MK, Kwok WH, Kew J. Thoracic paravertebral block: radiological evidence of contralateral spread anterior to the vertebral bodies. Br J Anaesth. 2000;84:263–5.
4. Karmakar MK, Gin T, Ho AM. Ipsilateral thoraco-lumbar anesthesia and paravertebral spread after low thoracic paravertebral injection. Br J Anaesth. 2001;87:312–6.
5. Cowie B, McGlade D, Ivanusic J, Barrington MJ. Ultrasound-guided thoracic paravertebral blockade: a cadaveric study. Anesth Analg. 2010;110:1735–9.
6. Eason MJ, Wyatt R. Paravertebral thoracic block-a reappraisal. Anesthesia. 1979;34:638–42.
7. Lönnqvist PA, Hildingsson U. The caudal boundary of the thoracic paravertebral space. A study in human cadavers. Anesthesia. 1992;47:1051–2.
8. Karmakar MK. Ultrasound-guided thoracic paravertebral block. Tech Reg Anesth Pain Manage. 2009;13:142–9.
9. Komatsu T, Sowa T, Takahashi K, Fujinaga T. Paravertebral block as a promising analgesic modality for managing post-thoracotomy pain. Ann Thorac Cardiovasc Surg. 2013;20(2):113–6.
10. Thavaneswaran P, Rudkin GE, Cooter RD, Moyes DG, Perera CL, Maddern GJ. Brief reports: paravertebral block for anesthesia: a systematic review. Anesth Analg. 2010;110:1740–4.
11. Schnabel A, Reichl SU, Kranke P, Pogatzki-Zahn EM, Zahn PK. Efficacy and safety of paravertebral blocks in breast surgery: a meta-analysis of randomized controlled trials. Br J Anaesth. 2010;105:842–52.
12. Terheggen MA, Wille F, Borel Rinkes IH, Ionescu TI, Knape JT. Paravertebral blockade for minor breast surgery. Anesth Analg. 2002;94:355–9.
13. Gardiner S, Rudkin G, Cooter R, Field J, Bond M. Paravertebral blockade for day-case breast augmentation: a randomized clinical trial. Anesth Analg. 2012;115:1053–9.
14. Bugelli GJ, Samra WA, Stuart-Smith K. Simple mastectomy and axillary lymph node biopsy performed under paravertebral block and light sedation in a patient with severe cardiorespiratory comorbidities: proposed management of choice in high-risk breast surgery patients. Clin Breast Cancer. 2013;13:153–5.
15. Coopey SB, Specht MC, Warren L, Smith BL, Winograd JM, Fleischmann K. Use of preoperative paravertebral block decreases length of stay in patients undergoing mastectomy plus immediate reconstruction. Ann Surg Oncol. 2013;20:1282–6.
16. Kairaluoma PM, Bachmann MS, Rosenberg PH, Pere PJ. Preincisional paravertebral block reduces the prevalence of chronic pain after breast surgery. Anesth Analg. 2006;103:703–8.
17. Daly DJ, Myles PS. Update on the role of paravertebral blocks for thoracic surgery: are they worth it? Curr Opin Anaesthesiol. 2009;22:38–43.
18. Fortier S, Hanna HA, Bernard A, Girard C. Comparison between systemic analgesia, continuous wound catheter analgesia and continuous thoracic paravertebral block: a randomised, controlled trial of postthoracotomy pain management. Eur J Anaesthesiol. 2012;29:524–30.

19. Davies RG, Myles PS, Graham JM. A comparison of the analgesic efficacy and side-effects of paravertebral *vs* epidural blockade for thoracotomy: a systematic review and meta-analysis of randomized trials. Br J Anaesth. 2006;96:418–26.
20. Powell ES, Cook D, Pearce AC, et al.. UKPOS InvestigatorsA prospective, multicentre, observational cohort study of analgesia and outcome after pneumonectomy. Br J Anaesth. 2011;106:364–70.
21. Murata H, Salviz EA, Chen S, Vandepitte C, Hadzic A. Case report: ultrasound-guided continuous thoracic paravertebral block for outpatient acute pain management of multilevel unilateral rib fractures. Anesth Analg. 2013;116:255–7.
22. Langille GM, Launcelott GO, Rendon RA. Access to the extrapleural space at the time of surgery for continuous paravertebral block after flank incision: description of the technique and case series. Urology. 2013;81:675–8.
23. Ozkan D, Akkya T, Comert A, et al. Paravertebral block in inguinal hernia surgeries: 2 segments or 4 segments? Reg Anesth Pain Med. 2009;34:312–5.
24. Hadzic A, Kerimoglu B, Loreio D, et al. Paravertebral blocks provide superior same-day recovery over general anesthesia for patients undergoing inguinal hernia repair. Anesth Analg. 2006;102:1076–81.
25. Akcaboy EY, Akcaboy ZN, Gogus N. Ambulatory inguinal herniorraphy: paravertebral block *versus* spinal. Minerva Anesthesiol. 2009;75:684–91.
26. Naja ZM, Raf M, El-Rajab M, et al. A comparison of nerve stimulator guided paravertebral block and Ilio-inguinal nerve block for analgesia after inguinal herniorrhaphy in children. Anesthesia. 2006;61:1064–8.
27. Kaur B, Vaghadia H, Tang R, Sawka. A real-time thoracic paravertebral block using an ultrasound-guided positioning system. Br J Anaesth. 2013;110:852–3.
28. Kus A, Gurkan Y, Gul Akgul A, Solak M, Toker K. Pleural puncture and intrathoracic catheter placement during ultrasound guided paravertebral block. J Cardiothorac Vasc Anesth. 2013;27:e11–2.
29. Conacher ID, Kokri M. Post-operative paravertebral blocks for thoracic surgery. A radiological appraisal. Br J Anaesth. 1987;59:155–61.
30. Conacher ID. Resin injection of thoracic paravertebral spaces. Br J Anaesth. 1988;61:657–61.
31. Purcell-Jones G, Pither CE, Justins DM. Paravertebral somatic nerve block: a clinical, radiographic, and computed tomographic study in chronic pain patients. Anesth Analg. 1989;68:32–9.
32. Richardson J, Jones J, Atkinson R. The effect of thoracic paravertebral blockade on intercostal somatosensory evoked potentials. Anesth Analg. 1998;87:373–6.
33. Cheema SP, Ilsley D, Richardson J, Sabanathan S. A thermographic study of paravertebral analgesia. Anesthesia. 1995;50:118–21.
34. Marhofer D, Marhofer P, Kettner SC, et al. Magnetic resonance imaging analysis of the spread of local anesthetic solution after ultrasound-guided lateral thoracic paravertebral blockade: a volunteer study. Anesthesiology. 2013;118:1106–12.
35. Ponde VC, Desai AP. Echo-guided estimation of formula for paravertebral block in neonates, infants and children till 5 years. Indian J Anaesth. 2012;56:382–6.
36. Renes SH, Bruhn J, Gielen MJ, Scheffer GJ, van Geffen GJ. In-plane ultrasound-guided thoracic paravertebral block: a preliminary report of 36 cases with radiologic confirmation of catheter position. Reg Anesth Pain Med. 2010;35:212–6.
37. Marhofer P, Kettner SC, Hajbok L, Dubsky P, Fleischmann E. Lateral ultrasound-guided paravertebral blockade: an anatomical-based description of a new technique. Br J Anaesth. 2010;105:526–32.
38. Luyet C, Meyer C, Herrmann G, Hatch GM, Ross S, Eichenberger U. Placement of coiled catheters into the paravertebral space. Anesthesia. 2012;67:250–5.
39. O'Riain SC, Donnell BO, Cuffe T, Harmon DC, Fraher JP, Shorten G. Thoracic paravertebral block using real-time ultrasound guidance. Anesth Analg. 2010;110:248–51.
40. Bouzinac A, Delbos A, Mazières M, Rontes O. Ultrasound-guided bilateral paravertebral thoracic block in an obese patient. Ann Fr Anesth Réanim. 2011;30:162–3.
41. Shibata Y, Nishiwaki K. Ultrasound-guided intercostal approach to thoracic paravertebral block. Anesth Analg. 2009;109:996–7.
42. Ben-Ari A, Moreno M, Chelly JE, Bigeleisen PE. Ultrasound-guided paravertebral block using an intercostal approach. Anesth Analg. 2009;109:1691–4.
43. Luyet C, Eichenberger U, Greif R, Vogt A, Szücs Farkas Z, Moriggl B. Ultrasound-guided paravertebral puncture and placement of catheters in human cadavers: an imaging study. Br J Anaesth. 2009;102:534–9.
44. Hara K, Sakura S, Nomura T, Saito Y. Ultrasound guided thoracic paravertebral block in breast surgery. Anesthesia. 2009;64:223–5.
45. Luyet C, Herrmann G, Ross S, et al. Ultrasound-guided thoracic paravertebral puncture and placement of catheters in human cadavers: where do catheters go? Br J Anaesth. 2011;106:246–54.
46. Pusch F, Wildling E, Klimscha W, Weinstabl C. Sonographic measurement of needle insertion depth in paravertebral blocks in women. Br J Anaesth. 2000;85:841–3.
47. Kaur B, Tang R, Vaghadia H, Sawka A. Ultrasound-guided thoracic paravertebral block using the SonixGPS® system in human cadavers. Can J Anaesth. 2013;60:331–2.
48. Karmakar MK, Li X, Li J, Hadzic A. Volumetric three-dimensional ultrasound imaging of the anatomy relevant for thoracic paravertebral block. Anesth Analg. 2012;115:1246–50.
49. Bouaziz H, Aubrun F, Belbachir AA, et al. Société française d'anesthésie et de réanimation. Locoregional anesthesia and echography. Ann Fr Anesth Réanim. 2011;30:e33–5.
50. Luyet C, Siegenthaler A, Szucs-Farkas Z, Hummel G, Eichenberger U, Vogt A. The location of paravertebral catheters placed using the landmark technique. Anesthesia. 2012;67:1321–6.
51. Lichtenstein DA, Menu Y. A bedsign ultrasound signruling out pneumothorax in the critically ill. Lung sliding. Chest. 1995;108:1345–8.

Peripheral Nerve Blocks of the Thoracic Ventrolateral Wall: Type I PECS Block, Type II PECS Block, and Serratus Plane Block

11

Eryk Eisenberg, Elisabeth Gaertner, Philippe Clavert, and Rafael Blanco

Contents

Innervation of the ventrolateral thoracic wall is provided primarily by the **intercostal nerves** (ribs, intercostal muscles, skin), the **lateral and medial pectoral nerves** (pectoralis major and minor muscles), the **long thoracic nerve** (serratus anterior muscle) and by the **supraclavicular nerves** (skin of the neck area, supra- and infraclavicular areas) (Figs. 4.1, 9.1, 11.1 and 11.2). Innervation of the breast and of the overlying skin is derived from the lateral and ventral cutaneous branches of the intercostal nerves.

For surgery performed in this area, anaesthesia and/or analgesia can be obtained by different anaesthetic techniques:

– Thoracic epidural anaesthesia provides extensive segmental and bilateral anaesthesia/analgesia.
– Thoracic paravertebral block provides segmental unilateral anaesthesia/analgesia.
– Peripheral nerve blocks which are: type I PECS block, type II PECS block and Serratus Plane Block. The sensory area blocked is less extensive than that of an epidural or thoracic paravertebral block, but, although these blocks have only recently been described and not yet subjected to rigorous investigation, they undoubtedly belong in the arsenal of relevant and effective techniques for a proportion of surgical procedures.

The choice between these **peripheral** blocks depends on the type and extent of surgery.

- For retropectoral prosthetics, the pectoralis major muscle is involved (lateral and medial pectoral nerves).
- For tumour excision (lumpectomy), mastectomy and sentinel lymph node excision, the intercostal nerves are the principal nerve structures which require anaesthesia.
- Lastly, in complex breast reconstruction, the thoracodorsal nerve and the thoracic long nerve are also involved.

E. Eisenberg (✉)
Département d'anesthésie et réanimation, Clinique Pôle Santé République, Clermont-Ferrand, France

E. Gaertner
Département d'anesthésie et réanimation chirurgicale, Hôpitaux Civils de Colmar, Colmar, France

P. Clavert
Service de chirurgie du membre supérieur, Centre de Chirurgie Orthopédique et de la Main (CCOM), Illkirch, France

R. Blanco
Anaesthetic Department, Corniche Hospital, Abu Dhabi, United Arab Emirates

E. Eisenberg, E. Gaertner (eds.), *Ultrasound in Peripheral, Neuraxial and Perineuraxial Regional Anaesthesia*,
https://doi.org/10.1007/978-3-031-08804-9_11

Fig. 11.1 Emergence of lateral and medial pectoral nerves at the deep aspect of pectoralis major muscle. Dissection Bertrand Fabre

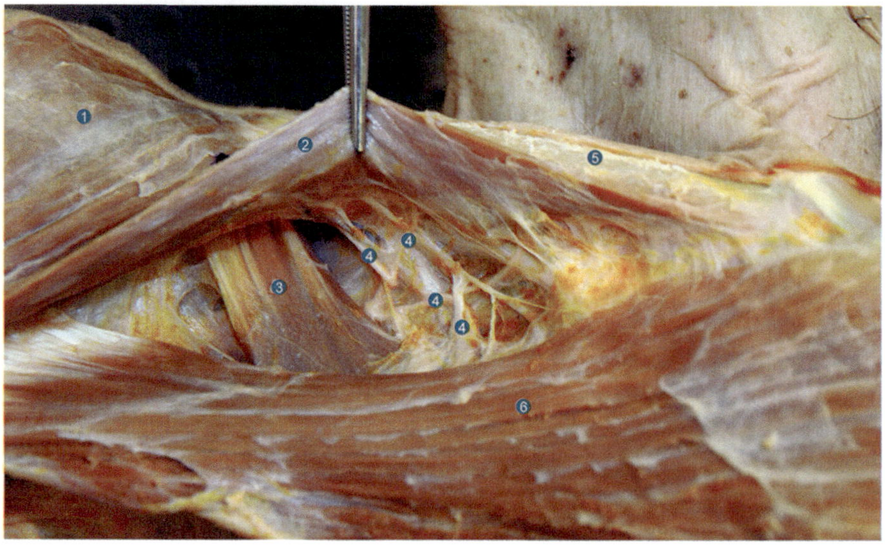

1. deltoid m.
2. pectoralis major m. (removed)
3. pectoralis minor m.
4. lateral and medial pectoral nn.
5. collarbone

Fig. 11.2 Pectoral region: lateral and medial pectoral nerves continue between pectoralis major and minor muscles. Dissection Bertrand Fabre

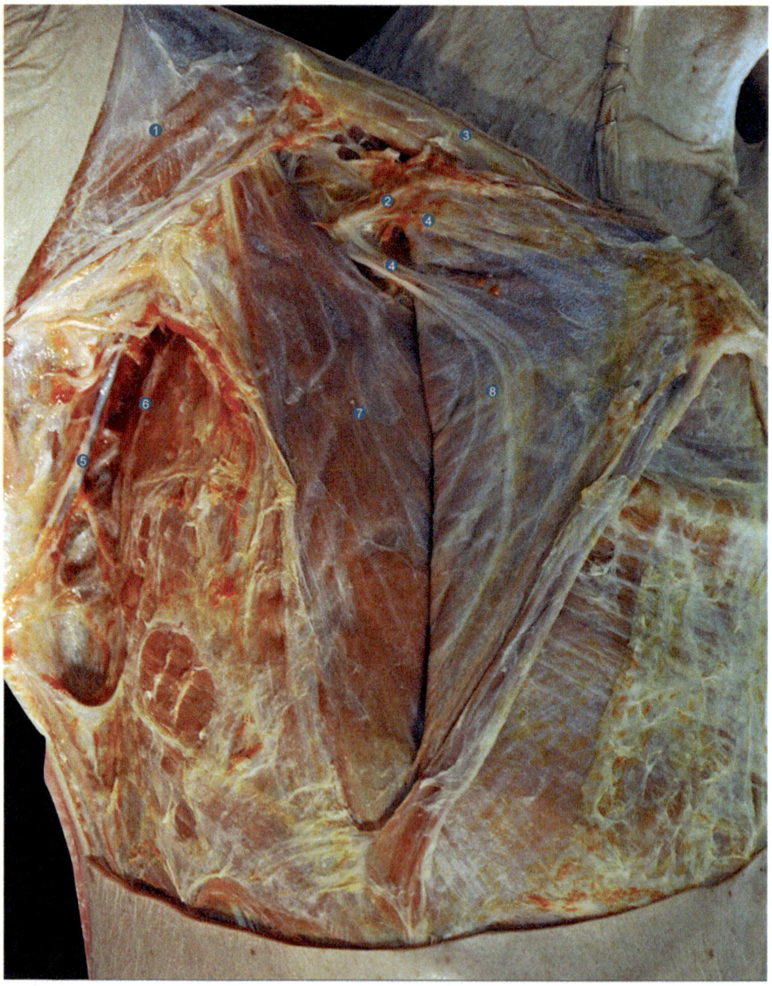

1. deltoid m.
2. pectoral branch of thoracoacromial a.
3. collarbone
4. branches of lateral and medial pectoral nn.
5. thoracodorsal vessels and nn.
6. long thoracic n.
7. pectoralis minor m.
8. pectoralis major m. (removed)

Rafael Blanco recently proposed three types of blocks:

- PECS I block (PECS I) [1].
- PECS II block (PECS II) [2].
- Serratus Plane Block [3].

Type I PECS Block

Type of probe: High-frequency linear 5–10 or 6–13 MHz.

Axis of probe: Oblique caudally and laterally, in an infraclavicular position (Fig. 11.3).

Configuration: Lateral and medial pectoralis nerves are not consistently identifiable, but the thoraco-acromial artery and its pectoral branch accompanying the nerves are usually visualised. The needle is inserted in plane, introduced at the cephalic/medial end of the probe, but an injection from the caudal end of the probe is possible (in plane, directed towards the pectoral nerves emergence (as in Fig. 11.4).

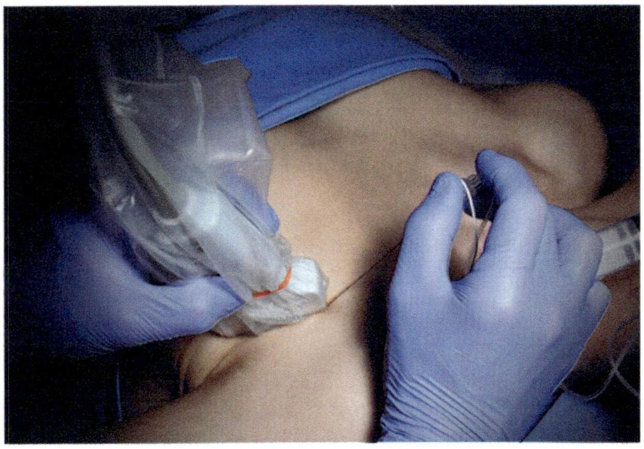

Fig. 11.3 Position of probe for performance of type I or II PECS blocks, needle inserted in-plane from the cephalad side of the probe

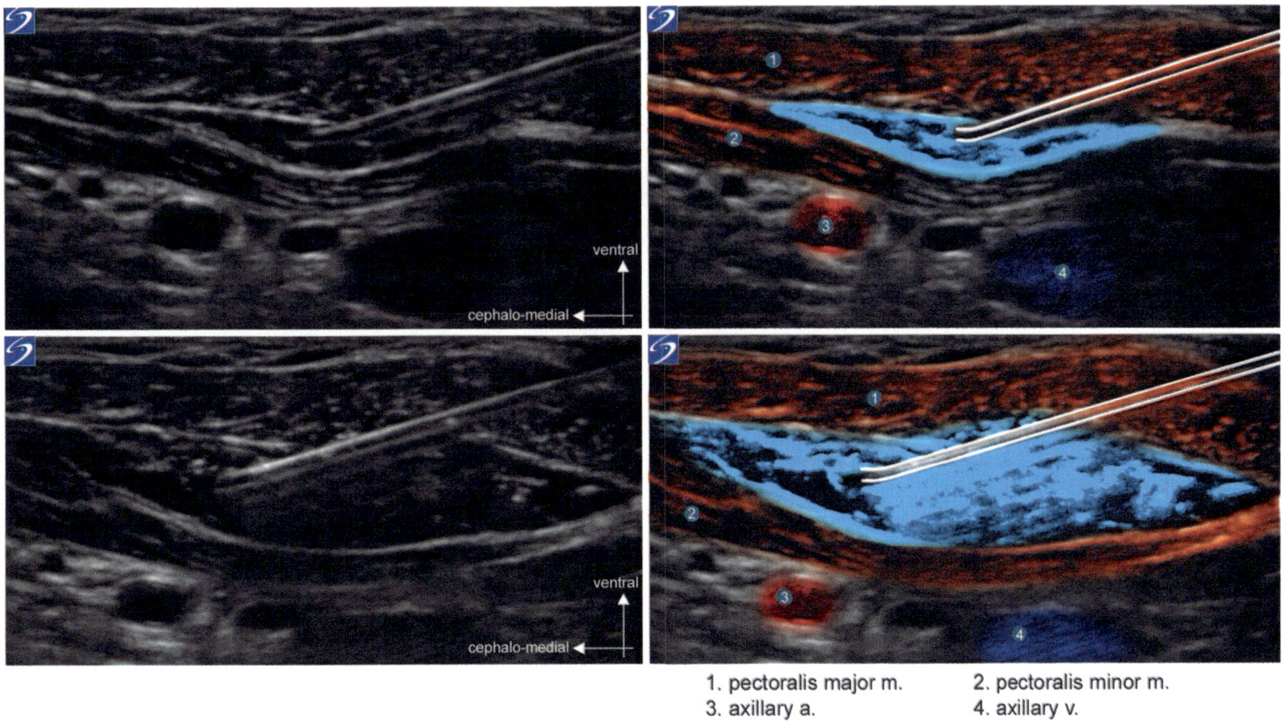

Fig. 11.4 Type I PECS block: spread of local anaesthetic between pectoralis major and minor muscles. Note that the needle here has been inserted in-plane but from the caudad side of the probe

Depth studied: 2 and 5 cm.

Neurostimulation: Not used.

Needle: 22 G, 80 mm.

Utility of Doppler ultrasound: Thoraco-acromial artery and its pectoral branch.

This is a simple block, which can be performed safely and quickly, with the intent of achieving anaesthesia/analgesia of the pectoral muscles and associated connective tissues. Performed with real-time ultrasound-guided technique, it consists of injecting local anaesthetic (e.g. 0.2 mL/kg of levobupivacaine, 1.25 mg/mL or 2 mg/mL ropivacaine) between the pectoralis major and minor muscles, aiming to block both the lateral and the medial pectoral nerves (Figs. 11.1 and 11.2).

Indications

The efficacy of this block is primarily only in relation to surgery involving the pectoralis major and minor muscles. PECS I will have no effect on the overlying skin in this area unless a significant spread of the local anaesthetic occurs towards the axilla, and therefore towards the lateral cutaneous branches of the upper intercostal nerves.

Amongst the procedures applicable for this analgesic block are: breast reconstruction, retropectoral implants, implantable compartments with pacemakers and shoulder surgery with deltopectoral groove incision. Its efficacy is incomplete where surgery involves axillary dissection.

Sonoanatomy and Approach

The patient in the supine position, with the arm abducted to 90°. A high-frequency linear probe is placed in infraclavicular position, in a caudal-lateral oblique axis (Fig. 11.3). The pectoralis major and minor muscles are identified along with the thoraco-acromial artery (or its pectoral branch) lying between the muscles. Deep to pectoralis minor lie the axillary artery and vein, seen in cross-section surrounded by components of the brachial plexus.

The needle is introduced in plane, at the cephalic end of the probe (also possible at its caudal end Fig. 11.4), with the needle tip directed towards the intermuscular plane between the pectoralis major and minor, avoiding the thoraco-acromial artery (or its pectoral branch). The local anaesthetic infusion is started opposite the second and third ribs after a negative aspiration test, ensuring that the needle is in fact between the two muscles (and not within one of them). The proper position of the needle tip is confirmed by the injected LA forming a hypoechoic intermuscular ellipse (Fig. 11.4).

This block induces motor and sensory anaesthesia of the pectoral muscles, as well as inconsistent anaesthesia of the lateral cutaneous branches of the intercostal nerves T2 to T4, depending on degree of spread of the local anaesthetic into the axilla.

Type II PECS Block

Type of probe: High-frequency linear of 5–10 or 6–13 MHz.

Axis of probe: Oblique caudally and laterally, in infraclavicular position (for interpectoral injection), and then in the same axis but more laterally opposite to ventral axillary line (for "axillary" injection).

Configuration: Needle in plane, introduced at the cephalic end of the probe. Once the initial interpectoral injection has been performed, the needle is advanced up to the lateral border of the pectoralis minor muscle under ultrasound control to perform an injection of the "axillary" component without using a second puncture point. This approach describes a "true" PECS II block. However, if only the axillary component is required (i.e. for surgery not requiring interpectoral injection), the needle is introduced at the cephalic end of the probe in a more lateral position (Fig. 11.5).

Depth studied: 2 and 5 cm.

Neurostimulation: Not used.

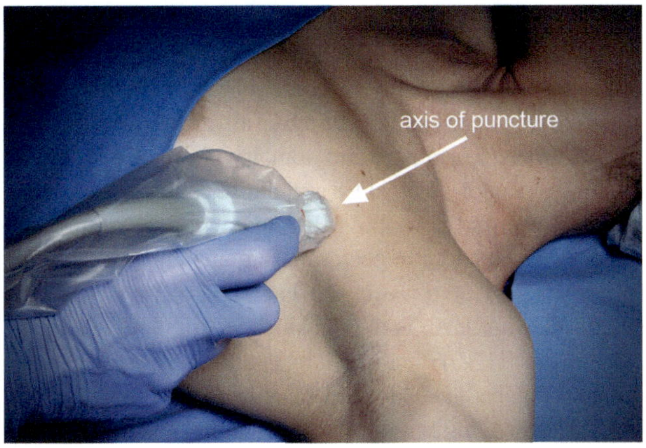

Fig. 11.5 Position of probe for performance of "axillary" phase in injection: needle will be inserted in-plane from the side of the probe

Needle: 22 G, 80–100 mm.

Utility of Doppler ultrasound: Thoraco-acromial artery and its pectoral branch.

It is a simple block that can be performed safely and quickly. R. Blanco described it as a modification of PECS I, intended to supplement its efficacy by an additional "axillary" injection. The latter in fact makes it possible to ensure anaesthesia/analgesia **of the axilla and of its contents,** as well as **of the skin of the ventrolateral aspect and of the upper part of the thorax**. It is performed in two complementary stages:

– The first phase consists of performing **PECS I** (0.2 mL/kg of levobupivacaine 1.25 mg/mL or ropivacaine 2 mg/mL) to produce anaesthesia of the pectoralis major and minor muscles ("interpectoral" injection).
– The second **"axillary"** phase is injection of local anaesthetic (0.4 mL/kg of levobupivacaine 1.25 mg/mL or ropivacaine 2 mg/mL), at the lateral border of the pectoralis minor muscle on the surface of (or even deep to) the serratus anterior muscle, opposite the cranial border of the fourth rib. This injection provides anaesthesia of the axilla and of the skin of the ventrolateral aspect of the upper part of the thorax (Fig. 11.12). Therefore, it may be sufficient to perform only the "axillary" component of PECS II when anaesthesia of the pectoralis muscle is not required.

Indications

- **Type II PECS (i.e. PECS I + "axillary" injection):** Breast reconstruction, radical (Halsted) mastectomy, wide thoraco-axillary surgery etc.
- **Axillary injection only:** Simple axillary dissection, lumpectomy, wedge resection etc.

By injecting the local anaesthetic at the costal insertions of the serratus anterior muscle, this block makes it possible to obtain anaesthesia of the long thoracic nerve, the intercostal brachial nerve (from the lateral cutaneous branches of T1 and T2, ± T3) and the lateral cutaneous branches of T2 up to T4, or even more caudally (Fig. 11.6). It is required for analgesia after insertion of a prosthesis, particularly in the retropectoral position, as a result of the surgical approach

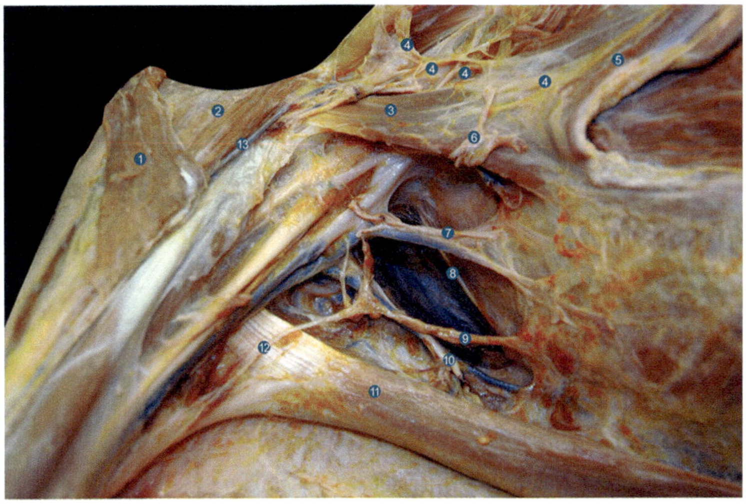

1. pectoralis major m. removed
2. clavicular head of deltoid m.
3. pectoralis minor m.
4. branches of lateral and medial pectoral nn.
5. pectoralis major m. removed
6. pectoral branch of thoracoacromial a.
7. lateral cutaneous branch of 2nd intercostal n.
8. long thoracic n.
9. lateral cutaneous branch of 3rd intercostal n.
10. thoracodorsal n. and neurovascular bundel
11. latissimus dorsi m.
12. intercostobrachial n.
13. cephalic v.

Fig. 11.6 Axillary fossa: lateral intercostal cutaneous branches and intercostobrachial nerve. Long thoracic and thoracodorsal nerves. Dissection Bertrand Fabre performed for the muscular dissection. Although the pectoralis major muscle is involved mostly in this type of surgery, there also seems to exist a pain component involving the serratus anterior muscle. Dissection of the pectoral fascia at the lateral border of the pectoral muscles promotes the spread of local anaesthetic into the axilla and therefore promoting block of the nerves which are located in it (Fig. 11.7). If this block is performed after surgery, the aponeurotic planes are already dissected and so promote rapid and effective distribution of the local anaesthetic into this area.

Sonoanatomy and Approach

As previously seen, the PECS II is the combination of an initial injection of local anaesthetic between the pectoralis muscles (PECS I) and a second one performed at the lateral border of the pectoralis minor muscle at the third and fourth ribs, affecting the T2 to T4 dermatomes (at least).

If PECS II is performed in its totality (PECS I + "axillary" injection), once the injection of PECS I has been performed by an injection in plane at the cephalic end of the probe, the probe is moved in an oblique axis, caudally and laterally (Fig. 11.5). This enables both components of the PECS II block to be made via a single skin insertion point. If the interpectoral injection is not necessitated by surgery and only the "axillary" component is required, it then is possible to proceed as follows: place the probe initially in the infraclavicular fossa opposite the lateral one-third of the clavicle in an oblique axis, caudally and laterally. After identification of the subclavius muscle, the axillary artery and vein, and pectoralis major (Fig. 11.8), the probe is progressively moved towards the axilla, whereupon pectoralis minor is visualised lying deep to pectoralis major (Fig. 11.9). Moving the probe further towards the lateral border of pectoralis minor, the fourth rib is then positioned in the middle of the screen (Fig. 11.10) (confirmed by counting from the first rib, which may be located deep to the axillary artery, as in Fig. 11.8, or lying within the acoustic shadow of the clavicle). The approach is performed in plane by injection at the cephalic end of the probe (Fig. 11.5). Note that the visible ribs (third, fourth and 5th) are covered by the serratus anterior muscle, which itself is covered by the suspensory ligament of the axilla (Gerdy's ligament). At this level, at the lateral border of pectoralis minor (Fig. 11.10) opposite the third rib (which marks the entry to the axilla), it is necessary to inject the local anaesthetic onto the surface of the serratus anterior muscle—an additional injection into the deep plane of the muscle is also possible (Fig. 11.11).

The cutaneous area affected by the "axillary" injection of PECS II is relatively extensive (up to T4/T5) (Fig. 11.12), sometimes reaching up to T9 (personal observation).

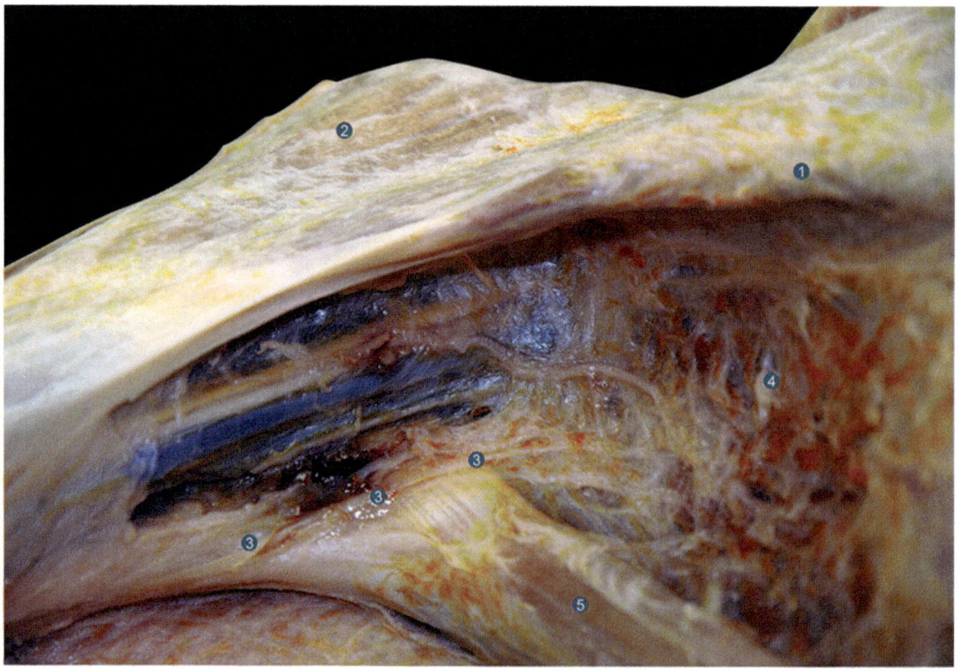

Fig. 11.7 Axillary space: axillary fascia in continuity with pectoral fascia. Dissection Bertrand Fabre

1. pectoralis major m. with pectoral fascia and axillary fascia laterally
2. deltoid m.
3. intercostobrachial n.
4. axilla covered by axillary fascia
5. latissimus dorsi m.

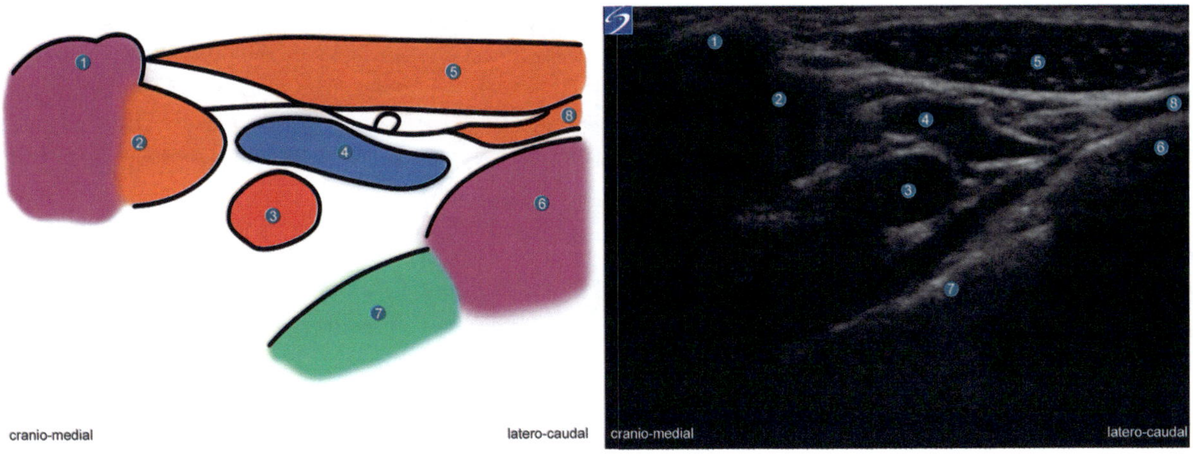

1. collarbone
2. subclavius m.
3. axillary a.
4. axillary v.
5. pectoralis major m.
6. 2nd rib
7. pleura and lung
8. pectoralis minor m.

Fig. 11.8 Stage 1: infraclavicular ultrasound view

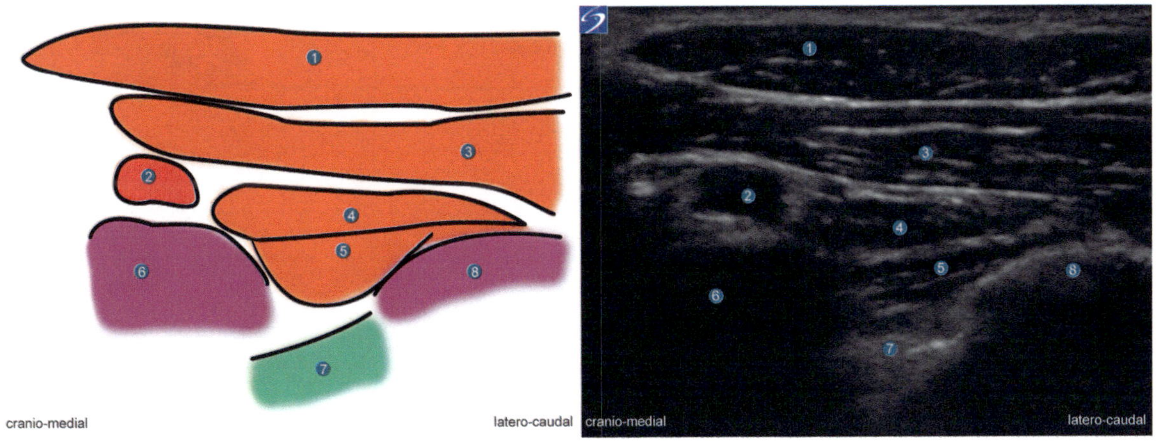

1. pectoralis major m.
2. axillary a.
3. pectoralis minor m.
4. serratus m.
5. intercostal m.
6. 2nd rib
7. pleura and lung
8. 3rd rib

Fig. 11.9 Stage 2: progressive movement of probe in direction of axillary area: identification of pectoralis major and minor muscles

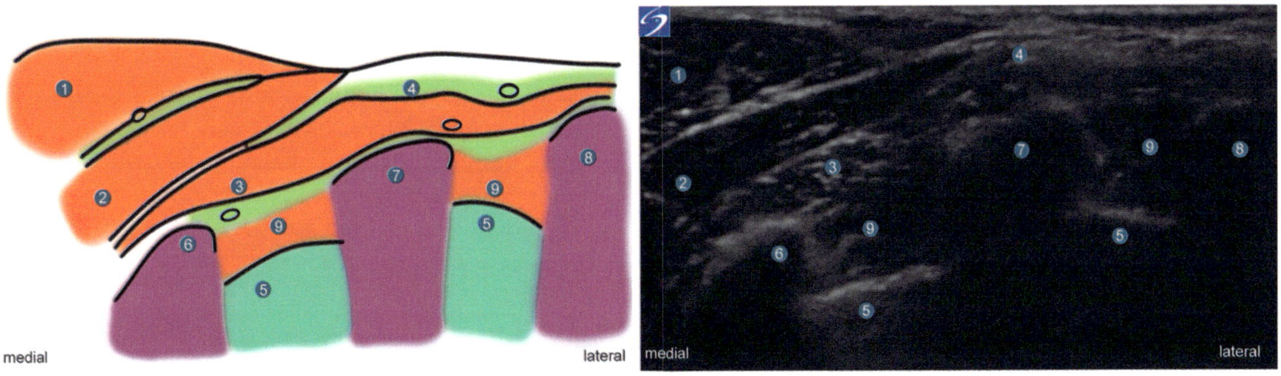

1. pectoralis major m.
2. pectoralis minor m.
3. serratus m.
4. suspensory ligament of axilla
5. pleura
6. 3rd rib
7. 4th rib
8. 5th rib
9. intercostal mm.

Fig. 11.10 Stage 3: visualisation of lateral border of pectoralis minor muscle facing the fourth rib

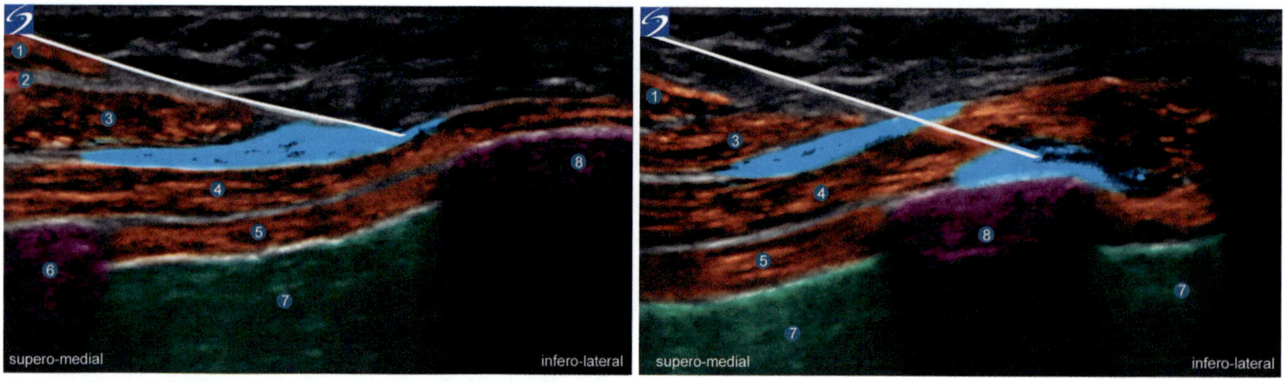

Fig. 11.11 "Axillary" phase of PECS II: injection into superficial plane (image on left) ± deep plane (image on right) of the serratus anterior muscle

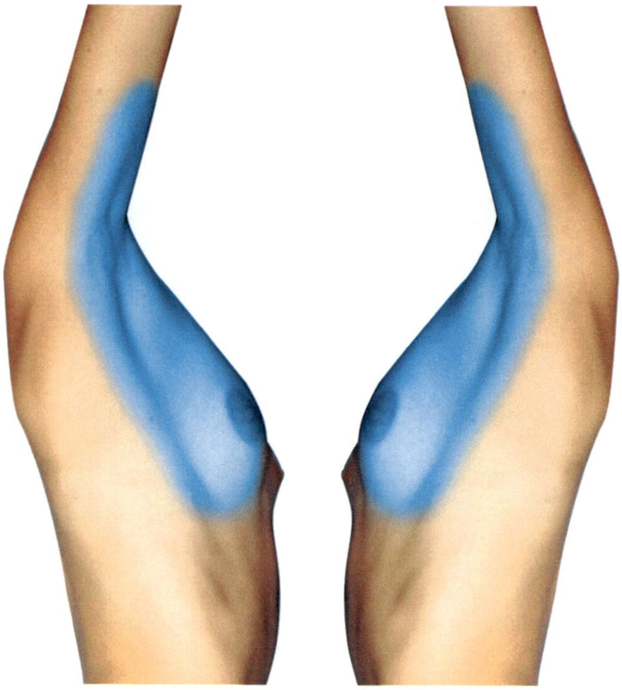

Fig. 11.12 Cutaneous area blocked by the "axillary" injection of PECS II

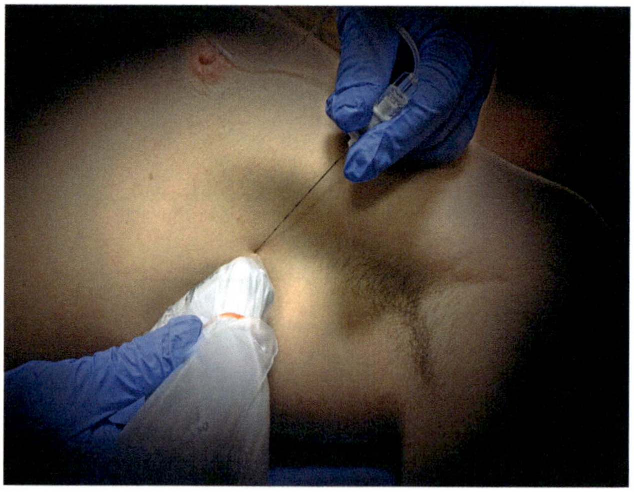

Fig. 11.13 Position of probe for performance of a Serratus Plane Block: injection in-plane

Serratus Plane Block (SPB)

Type of probe: High-frequency linear 5–10 or 6–13 MHz.

Axis of the probe: Oblique caudally and laterally (Fig. 11.13).

Configuration: The probe is positioned obliquely in caudal and lateral direction, in the mid-axillary line, with the fifth rib placed in the centre of the ultrasound image. The needle in plane, introduced at the cephalic end of the probe, the injection is performed in the plane deep to the latissimus dorsi muscle onto the surface of the serratus anterior muscle.

Depth studied: 2 and 5 cm.

Neurostimulation: Not used.

Needle: 22 G, 80–100 mm.

Utility of Doppler ultrasound: Thoracodorsal artery.

This block, which is simple to perform, is intended to provide anaesthesia and/or analgesia to an extensive part of the ventrolateral thoracic wall and of the axilla. The local anaesthetic is deposited between the latissimus dorsi and the serratus anterior in the mid-axillary line, opposite the fifth rib.

Volume and dosage: 0.6 mL/kg levobupivacaine 1.25 mg/mL or 2 mg/mL ropivacaine.

Indications

Mastectomies, lumpectomies, wide excisions, axillary dissection, breast reconstruction with a latissimus dorsi flap, prosthetic breast reconstruction, multiple rib fractures, lateral and ventrolateral thoracotomies, post-radiotherapy chronic pain etc.

Sonoanatomy and Approach

The patient is in the supine position with the arm abducted to 90°.

The high-frequency linear probe is initially positioned as per the second "axillary" stage of PECS II. It is then moved further laterally and caudally to the mid-axillary line and the fifth rib is identified (Fig. 11.13).

Latissimus dorsi (superficial and dorsal), teres major (cranial) and the serratus anterior (deep and caudal) are easily identifiable, located in the superficial plane of the ribs (including the 5th). It is also possible to locate the thoracodorsal artery which aids in identification of the superficial plane of serratus anterior. The needle is inserted at the cephalic end of the probe and advanced in plane. Injection of the local anaesthetic is performed opposite the fifth rib, deep to latissimus dorsi, onto the surface of serratus anterior (Fig. 11.14). R. Blanco has compared injections made at the surface of serratus anterior to those which are performed deep to it. 3D reconstructions have demonstrated that when injection of the local anaesthetic takes place at its surface, it produces a more extensive spread, greater than that achieved by an injection made deep to it (Figs. 11.15 and 11.16).

The **area** anaesthetised by this block is extensive, and although little different according to whether the injection takes place on either aspect of serratus anterior, Rafael Blanco observed that the **duration** of effect was two and a half times longer in case of injection made superficial to serratus anterior muscle than with injection on its deep aspect.

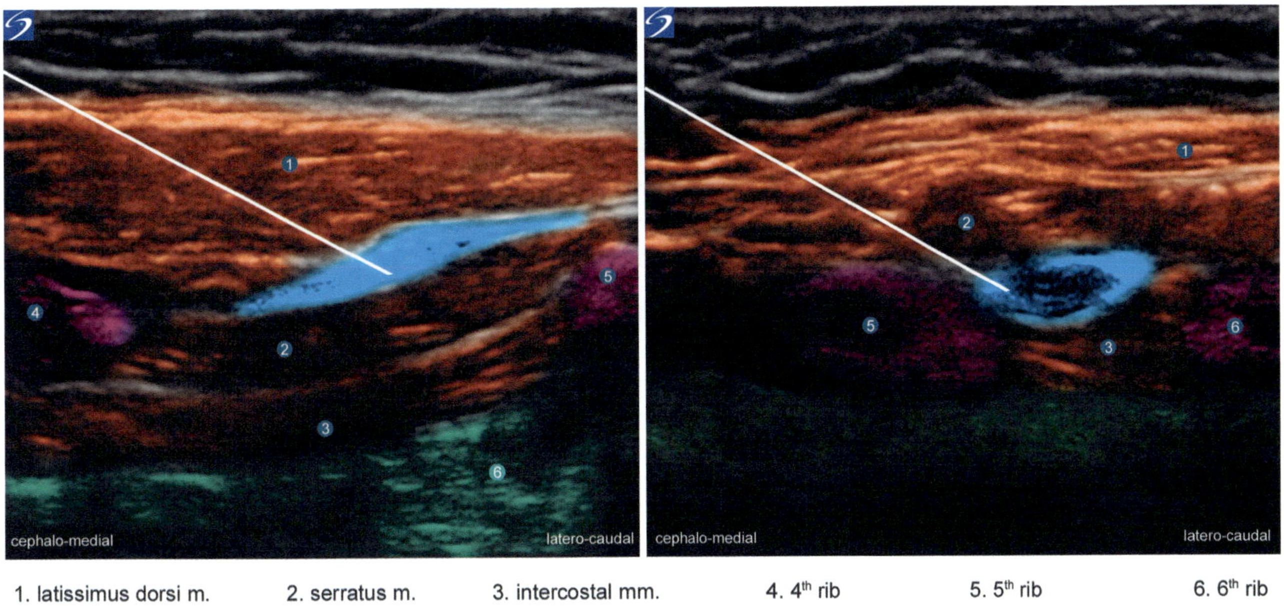

Fig. 11.14 Serratus Plane Block: injection between latissimus dorsi muscle and serratus anterior muscle (image on the left). In the image on the right, the injection is performed into the deep plane of the serratus anterior muscle

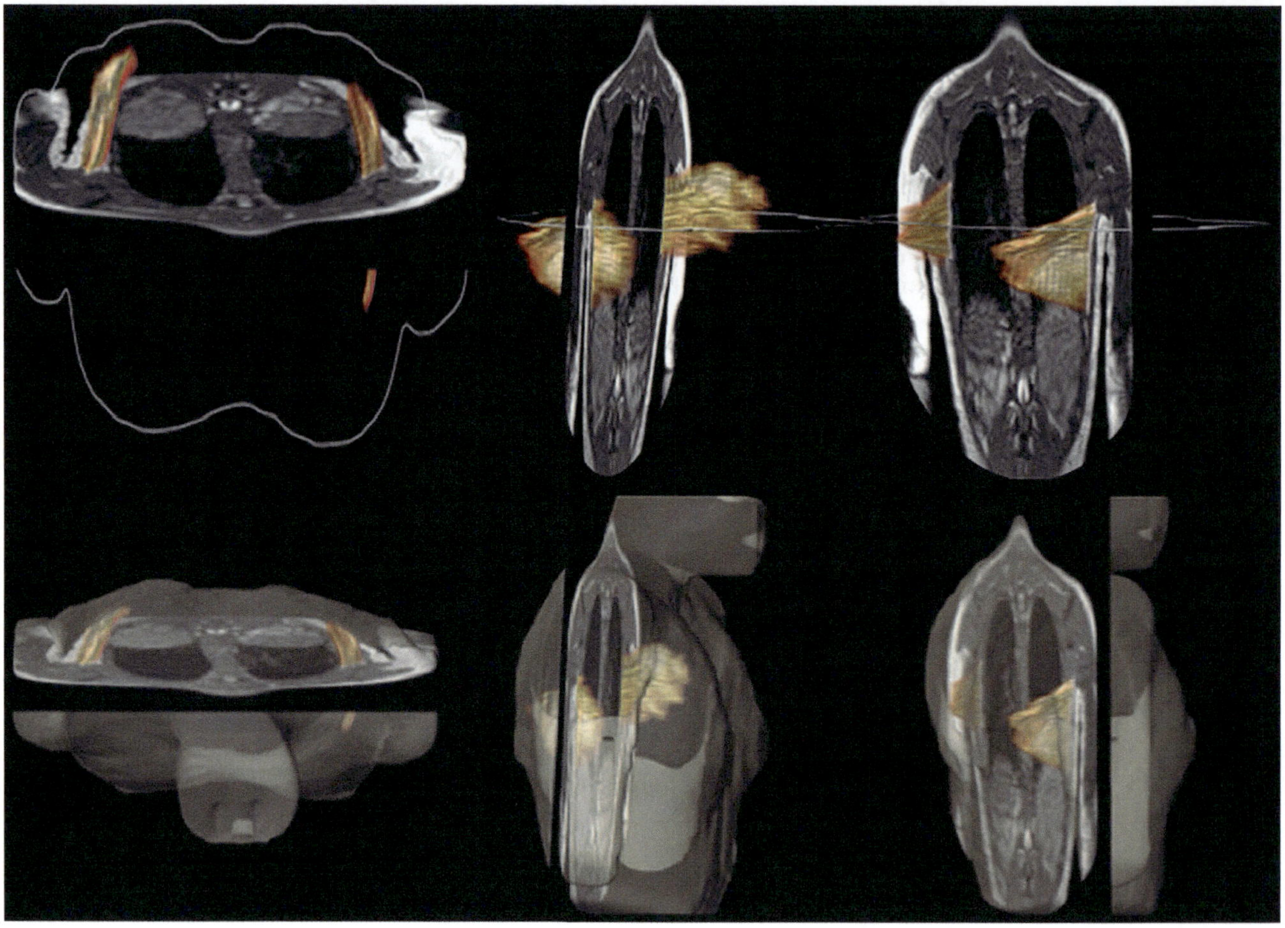

Fig. 11.15 3D reconstruction of spread of the local anaesthetic after conduct of Serratus Plane Block. Rafael Blanco iconography

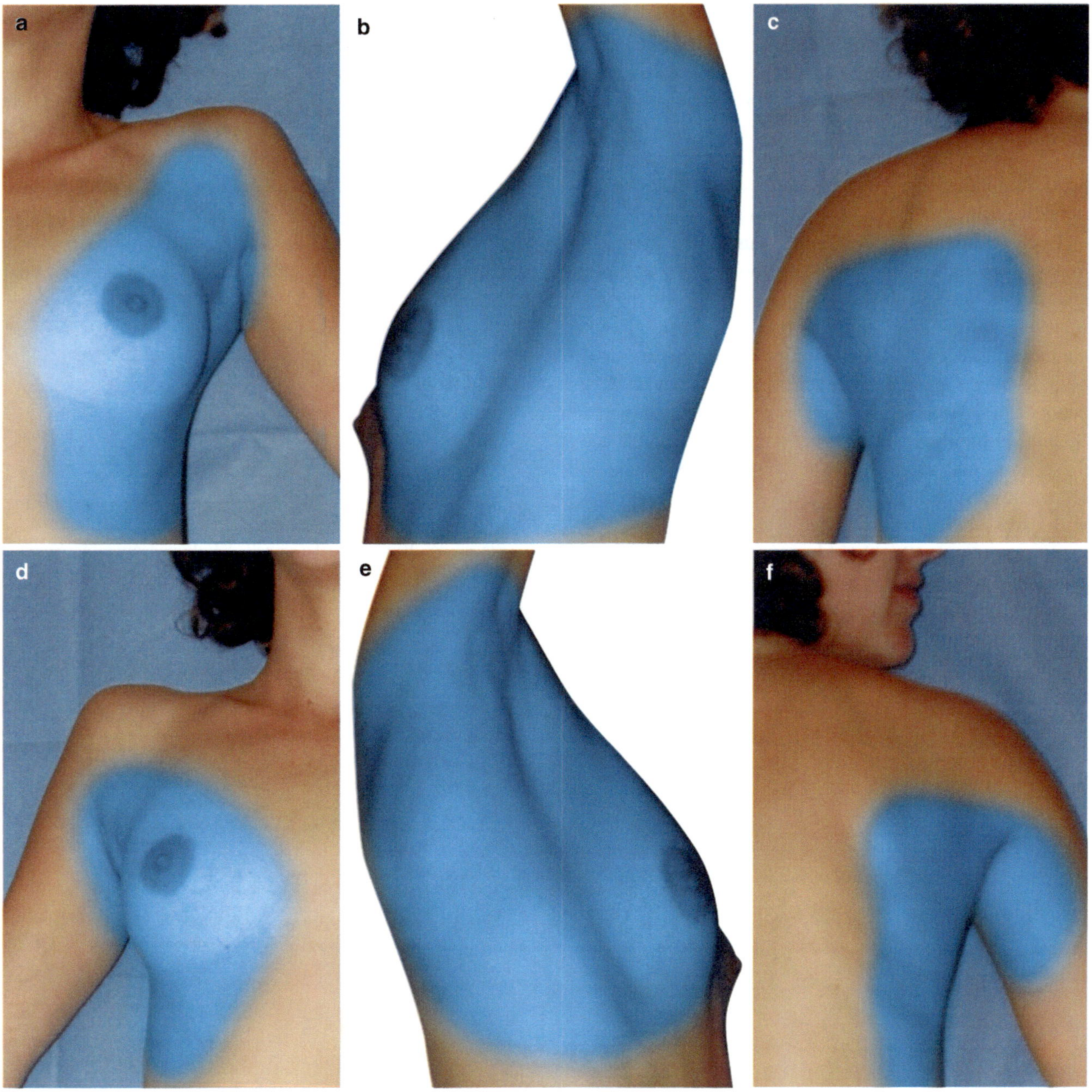

Fig. 11.16 Area of coverage of Serratus Plane Block: injection into the superficial plane of serratus anterior muscle (**a**, **b**, and **c**) or its deep plane (**d**, **e**, and **f**)

In Summary

Types I and II PECS blocks, as well as Serratus Plane Block are **peripheral** nerve blocks, performed by injection into fascial planes involving only peripheral nerves:

- The **lateral and medial pectoral** nerves (collateral branches of brachial plexus) for PECS I.
- The **lateral cutaneous branches of the intercostal nerves** as well as the **long thoracic and thoracodorsal** nerves for "axillary" injection of PECS II and Serratus Plane Block.

They are effective analgesic or anaesthetic techniques, recently described in clinical use and to date have been little studied, whose effective area is more limited than that observed in the case of epidural and thoracic paravertebral techniques.

TPV blocks, which are to be considered as **central** blocks, can anaesthetise all of the segmental spinal nerves of the thoracic region. Additionally, the TPV approach achieves a block of the **entire** intercostal nerve, that is, both **lateral and ventral** cutaneous branches. Such is not the case for PECS II and for Serratus Plane Block where local anaesthesia only achieves its effect on the **lateral** cutaneous branches of the intercostal nerves. The lack of efficacy of the peripheral blocks in the ventral cutaneous branches makes them, in a good number of surgical cases, more appropriate for post-operative analgesia than solely for anaesthesia.

Specific case of axillary dissection: Innervation of the axilla is provided mainly by nerve root of T2. To produce anaesthesia, a TPV block must be performed at the inter-transverse level T2-T3, or T3-T4 (that is, opposite the inter-spinous space T1-T2, or T2-T3, respectively, due to the oblique direction of the thoracic spinous process). These cephalic approaches are often very difficult to perform because of the increased depth of the paravertebral space at these levels and due to its poor visibility. TPV blocks thus are often performed at lower level (e.g. T4-T5 or T5-T6), with the associated risk of failure at the root of T2.

This is why the PECS II block (or at least its "axillary" injection on the surface of serratus anterior at the lateral border of pectoralis minor) is truly a technique of choice for analgesia in axillary dissection, and is preferable compared to TPV block. And although axillary dissection is performed during surgery for more complicated breast surgery for which an analgesic TPV block may have been performed, an "axillary" injection for PECS II block can be used in combination with the latter for optimum analgesia of the axilla. Also, a TPV can be combined with PECS II (with its two injections) if one also wishes to ensure analgesia of the pectoral muscles (e.g. a "Halsted" procedure), which is not achieved by TPV blocks alone (the pectoral nerves being collateral branches of the brachial plexus).

References

1. Blanco R. The "pecs block": a novel technique for providing analgesia after breast surgery. Anesthesia. 2011;66:847–8.
2. Blanco R, Fajardo M, Parras MT. Ultrasound description of PECS II (modified PECS I): a novel approach to breast surgery. Rev Esp Anestesiol Reanim. 2012;59:470–5.
3. Blanco R, Parras T, McDonnell JG, Prats-Galino A. Serratus plane block: a novel ultrasound-guided thoracic wall nerve block. Anesthesia. 2013;68:1107–13.

Ultrasound in Neuraxial and Perineuraxial Blocks

12

Eryk Eisenberg and Elisabeth Gaertner

Contents

The original version of the chapter has been revised. A correction to this chapter can be found at https://doi.org/10.1007/978-3-031-08804-9_13

Supplementary Information The online version contains supplementary material available at https://doi.org/10.1007/978-3-031-08804-9_12.

E. Eisenberg (✉)
Département d'anesthésie et réanimation, Clinique Pôle Santé République, Clermont-Ferrand, France

E. Gaertner
Département d'anesthésie et réanimation chirurgicale, Hôpitaux Civils de Colmar, Colmar, France

Note to the Reader

In the French language literature, spinal anaesthesia (SA), epidural anaesthesia (EDA), and caudal anaesthesia (CA) are commonly grouped under the heading "anesthésies périmédullaires" (perispinal anaesthesia) or are still called "central nerve blocks", compared to so-called "peripheral nerve blocks". This definition leaves some doubt about the place of thoracic paravertebral (TPV) and lumbar plexus (LP) blocks: are they "proximal peripheral blocks" or "peripheral central blocks"? The English language literature combines spinal anaesthesia, epidural anaesthesia, and caudal anaesthesia under the label "*central neuraxial blocks*", whilst TPV and LP blocks are combined under the term "*perineuraxial blocks*". This very rightly promotes the concept of the close anatomical relationship between the thoracic paraver-

E. Eisenberg, E. Gaertner (eds.), *Ultrasound in Peripheral, Neuraxial and Perineuraxial Regional Anaesthesia*, https://doi.org/10.1007/978-3-031-08804-9_12

tebral space and the neuraxis for TPV, and between the psoas compartment and the neuraxis for LP block. This relationship is confirmed by the functional consequences of these "perineuraxial" blocks: the commonplace and undesirable epidural spread of local anaesthetics observed during TPV and LP blocks. Perhaps in the French language literature, SA, EDA, and CA could be categorised under the term **anesthésies médullaires** ("spinal anaesthesia" with reference to *central neuraxial blocks*) and TPV block and LP block considered as **anesthésies périmédullaires** ("perispinal anaesthesia" with reference to *perineuraxial blocks*)? By analogy, the extremely low incidence of epidural spread during conduct of an interscalene block (ISB) leads us to classify it amongst peripheral nerve blocks. However, it could also be argued that ISB belongs under the heading "*perineuraxial blocks*" as the result of its injection site being so centrally located, and the proximity of the tissue planes which can lead to the direct injection or diffusion of local anaesthetic into the CSF. The rarity of this especially undesirable event is, without a doubt, due to the rigour with which ISB is practised: performed without ultrasound guidance (meticulous use of landmarks and safe axis of injection, depth of insertion limited by use of 25 mm needles etc.). The performance of TPV and LP blocks may suffer more from the lack of such precision as a result of their greater depth, their less common use and the fact that they are less well controlled (when compared to ISB).

Introduction

As with peripheral nerve blocks (PNB), the application of ultrasound to **central neuraxial** and **perineuraxial** anaesthesia has been the subject of increasing interest during the last few years [1, 2]. The constant advances in ultrasound imaging and the development of a better understanding of spinal sonoanatomy, account for much of this interest. There is no doubt that the use of ultrasound is a valid and effective alternative/adjunct to traditional surface landmarks for many procedures in medicine; so the "why not?!" attitude which initially motivated its use in spinal anaesthesia, has now been legitimised by a growing, evidence-based argument. From now on, safety, reliability, efficacy, and comfort are all mentioned when using ultrasound to guide the performance of spinal and epidural anaesthesia [3], as is the case with peripheral regional anaesthesia.

Ultrasound in central neuraxial anaesthesia can be seen in two different ways:

- In a **"preprocedural"** setting: the practitioner performs an ultrasound-guided examination before the conduct of anaesthesia. Its use aids the identification of a number of anatomical structures and thus facilitates the process, perhaps making its conduct safer. Subsequently, the procedure is performed according to the "usual" technique but modified in the light of new information provided by the initial ultrasound. In this context, we then can refer to an "**ultrasound-assisted**" procedure**.**
- In **"real-time ultrasound-guided"** mode: the practitioner uses ultrasound not only as a tool to study the sonoanatomy of the region, but also as an instrument for guidance of the needle during the injection. In this case, we refer to the procedure as **"ultrasound-guided"** [4–6].

In light of our current practices, the role and conduct of these techniques (ultrasound localisation and/or ultrasound guidance) for central neuraxial blocks are not yet perfectly established. Even the most experienced practitioners acknowledge that real-time ultrasound guidance is not part of their routine practice *during* central neuraxial blocks, unlike perineuraxial blocks such as TPV [7] and LP [8] blocks. Consistent imaging of deep structures, located within confined "ultrasound windows" scattered with bony obstacles require particularly high levels of expertise and dexterity. Also, integrating real-time guidance of the needle is disadvantaged by the current relative lack of availability of specific, echogenic needles.

Indications for Central Neuraxial Anaesthesia

Central blocks, spinal anaesthesia, and epidural anaesthesia/analgesia are the techniques most widely used in regional anaesthesia (RA). Conventional spinal anaesthesia is widely used in orthopaedics, abdominal surgery, urology and gynaecology, and produces consistent anaesthesia with a rapid onset. Epidural anaesthesia is utilised routinely for labour analgesia and commonly in the provision of post-operative analgesia for many abdominal, urological, thoracic, and orthopaedic surgeries. Certainly, it decreases the VAS score at rest and during movement, and may also promote the accelerated post-operative recovery of the patient.

Developments in these techniques during the last century concentrated mainly on equipment (e.g. needle tip design), local anaesthetics used, LA adjuvants, and evaluating the optimum technique in generating patient benefits in the context of different surgical categories. Haemodynamic impacts are better controlled, complications are very rare. Ultrasound localisation (by preprocedural ultrasound examination) and ultrasound guidance represent the next step in this development process: identification of the midline, location of the level of the injection, the depth of specific structures (e.g. ligamentum flavum and posterior dura) [9, 10], or the real-time guidance of the needle, visualisation of the LA spread or position of the catheter.

Spinal Anaesthesia

Unilateral Spinal Anaesthesia

With low doses of local anaesthetic, it is possible to anaesthetise a single lower limb or single side of the groin/abdomen.

The technique promotes better haemodynamic stability and is associated with a reduced incidence of urinary retention. The time to patient discharge is reduced when compared with "classical" spinal anaesthesia and so is especially well-suited to ambulatory anaesthesia [11]. The better haemodynamic profile produced by this technique may offer advantages over bilateral spinal anaesthesia or general anaesthesia in "high risk" cases: in frail patient, patients with severe hypertension and/or heart failure. Continuous spinal anaesthesia (allowing for repeated, low dose "top-ups") has also been proposed in this context but single-shot unilateral SA is technically easier to perform with reproducible results without having the disadvantages associated with spinal catheter insertion and management. In practice, the most common indication for unilateral SA described in the literature is in patients with a fractured neck of femur [12, 13].

Saddle Block Anaesthesia

Spinal anaesthesia is performed in low lumbar area with the patient seated. With the needle tip directed caudally, a small dose of hyperbaric local anaesthetic solution is injected slowly. The hyperbaric LA "sinks" in the CSF to become localised within the sacral canal. The patient is kept in the seated position for a further 10 minutes to promote sacralisation of the anaesthesia. The most common side effect of this technique is prolonged bladder atony (leading to urinary retention), but the haemodynamic stability gained is considerably better than with traditional spinal anaesthesia. Its principal indication is anorectal and urological surgery [14–16].

Spinal Analgesia

Used as an analgesia technique for abdominal, thoracic, and complex orthopaedic surgeries, an intrathecal injection of an opioid (+/− LA) can produce prolonged spinal analgesia with few harmful effects. However, certain caveats must be observed: "*Caring an intrathecal injection of morphine can be performed in a traditional hospitalization, if the dose of morphine injected without preservative is less than or equal to 0.1 mg in the ASA I or II subject*" [17].

Continuous Spinal Anaesthesia

It is useful in the frail elderly, particularly if they have major associated comorbidities. Inserting an intrathecal catheter enables the slow titration of spinal anaesthesia whilst maintaining haemodynamic stability [18, 19]. Initially, the spinal needles and intrathecal catheters used were of an extremely small diameter leading to technical difficulties in their handling and management. The catheters could not be easily injected nor aspirated. In attempting to extend the effect of the anaesthesia by repeated injections of LA, this led to the sacral accumulation of a large quantity of local anaesthetic. The LA reached neurotoxic concentrations in the dural "cul-de-sac", which caused serious, permanent neurological consequences for the patients involved. As a result, these items were withdrawn from the market.

Continuous spinal anaesthesia-analgesia in obstetrics has been reviewed by C.M. Palmer who estimated that it is underused due to lack of adequate equipment [20]. His article summarises all the advantages/disadvantages of this technique in this indication.

Epidural Anaesthesia

SFAR experts who set criteria for safety in epidural practice state: "It is recommended after major thoracic or intra-abdominal surgery (gastric, pancreatic, colic, small bowel surgery, oesophageal surgery, and cystectomy), in order to improve analgesia, to reduce duration of post-operative ileus and to shorten the time to extubation. For epidural analgesia it is recommended to use low concentrations of local anaesthetics with an opioids, and to insert the catheter at the midpoint of the dermatomes to be blocked".

Lumbar Epidural Anaesthesia

Lumbar epidural anaesthesia was traditionally performed to enhance post-operative analgesia in complicated orthopaedic surgery of the lower limb. It has largely been given up for peripheral nerve blocks which provide good quality analgesia with fewer adverse events (better haemodynamic stability, less urinary retention, fewer dysaesthesiae). However, it remains an indication in patients undergoing bilateral arthroplasties (e.g. TKR and THR). Lumbar EDA produces good quality analgesia in this context but certain commentators advise against it because of the difficulty of early mobilisation of the patient. Some commentators promote local infiltrative anaesthesia (LIA) as an alternative to lumbar EDA.

"It is recommended in major pelvic surgery, prostate, bladder, abdominal, renal, and vascular surgery. It may be used alone or in combination with general anaesthesia or sedation" [17].

Thoracic Epidural Anaesthesia

Combined with general anaesthesia in thoracic surgery, it makes it possible to reduce the doses of anaesthesia and to provide good quality postoperative analgesia, enabling opioid-sparing and fewer harmful adverse events related to opioids, early rehabilitation, a decreased rate of complications (especially respiratory ones) and a decreased hospital stay. Thoracic EDA reduces the development of chronic pain [21].

Epidural anaesthesia is also widely used in treatment of chronic pain and this technique is used at the cervical, thoracic, lumbar, and sacral levels.

Contraindications to central nerve blocks (spinal and epidural anaesthesia) include general contraindications for RA procedures (e.g. sepsis, patient's unwillingness), but also include hypovolemia and intracranial hypertension. The many debates regarding coagulation disorders and the risk of spinal haematoma have given rise to French and International recommendations [22] and it is essential to know and to comply with these. Some contraindications are controversial: patients with tattoos, conduct of epidural anaesthesia under general anaesthesia or in a patient who is generally uncommunicative.

Anatomy of the Spine

General Considerations

A vertebra is symmetrical around a sagittal plane. Overall, it consists of the union of the **vertebral body** (placed ventrally) and of the **vertebral arch** (located dorsally) (Fig. 12.1a). The latter consists laterally of two pedicles which form its most anterior parts, and then the transverse

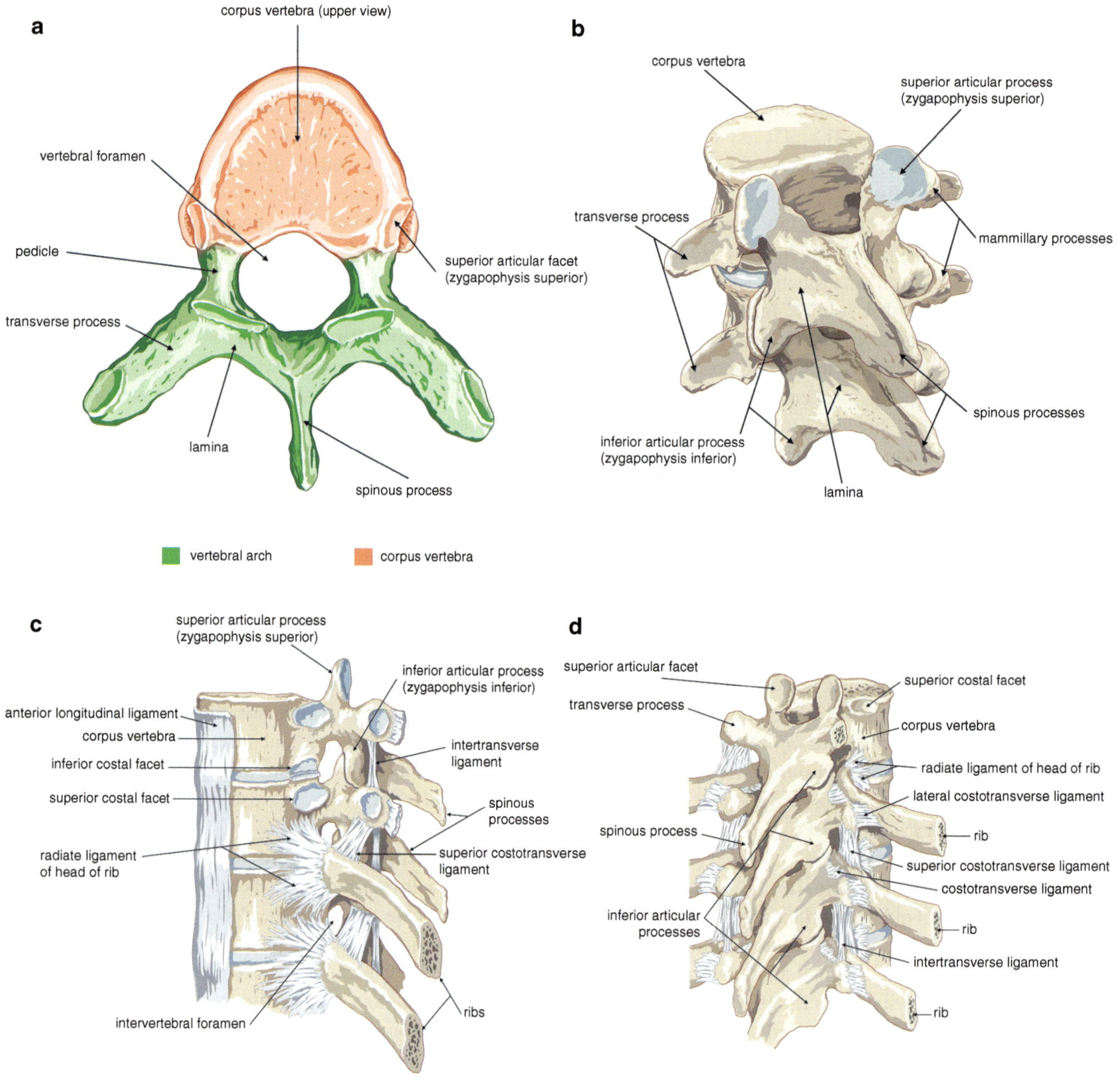

Fig. 12.1 (**a**) Superior view of a thoracic vertebra. (**b**). Oblique view of two lumbar vertebrae. (**c**) Lateral view of the osteoligamentary network of the thoracic vertebrae (from Netter FH. *Atlas of human anatomy*. third ed. Paris: Masson, 2007). (**d**) View of osteoligamentary network of the thoracic vertebrae (from Netter FH. *Atlas of human anatomy*. third ed. Paris: Masson, 2007)

processes, the articular processes, and the laminae which unite sagittally at the base of the spinous process, the only unpaired bony structure of the arch, which closes it posteriorly. The spinal canal, containing the neuraxis, is bounded in front by the posterior aspect of the vertebral body, laterally by the pedicles, and rearwards by the laminae and the spinous process.

Access to the spinal canal is via the intervertebral foraminae laterally (from which the spinal nerves emerge) and the interlaminar and interspinous spaces dorsally. On each side of the midline, the interlaminar space is closed by the corresponding ligamentum flavum ("yellow ligament"), which unites the upper and lower laminae of adjacent vertebra [23]. At each intervertebral level, the ligamentum flavum from each side usually fuses in the midline, thus forming a continuous "ligamentous wall".

Medially, the adjacent spinous processes are linked by the interspinous ligament, which is thin and not especially dense. On the contrary, the supraspinous ligament extending between the apices of the spinous processes and which covers them dorsally is dense and very tough.

In the spinal canal are found the spinal cord, the cauda equina, and the cerebrospinal fluid (CSF), surrounded by the dura mater and the arachnoid space which comprise a true sac enveloping these structures (the thecal sac). The epidural space is located between this meningeal covering and the bony spinal canal. It is not uniformly structured and its anatomy is more complex than is generally described [24]. Its posterior segment, which is of interest to us in particular in spinal anaesthesia, is **discontinuous:** practically non-existent opposite the laminae to which the posterior dura mater (PDM) adheres, but takes shape opposite the interlaminar spaces where it is filled with fatty tissue and the internal posterior venous plexus network. The lateral epidural spaces at the level of the intervertebral foraminae contain the spinal nerves, the radicular vessels and fat. The anterior epidural space contains fat and the important anterior internal vertebral venous plexus.

Lumbar Spine (Fig. 12.1b)

Unlike in the thoracic region where they overlap, the lumbar laminae lie distinctly separate from each other and the interlaminar spaces (and deeper structures) may be visualised with ultrasound. The spinous processes, wide and flattened sagittally, are projected rearward with a slightly oblique angle caudally. This allows for a needle to be inserted towards the spinal canal, whether it is in the midline (interspinous) or paramedian (interlaminar), which can be facilitated by anterior flexion of the trunk. However, a lumbar injection can be made more difficult (particularly with the sagittal approach) by the existence of calcifications, hypertrophy of the articular processes, or simply because of age-related changes in spinal anatomy.

Thoracic Spine (Fig. 12.1c and d)

Three distinct regions exist in the thoracic spine; the vertebrae have different morphology:

- The most **cephalic vertebrae (T1 to T4)** have "cervical" characteristics: articular processes are oriented dorsally and almost vertically; the spinal process is directed dorsally.
- The most **caudal vertebrae (T9 to T12)** have characteristics similar to lumbar vertebrae: articular processes directed laterally, spinous processes flat and thick, directed dorsally.
- The **intermediate vertebrae (T5 to T8)** have spinous processes whose caudal angulation is very pronounced, whose extremity surpasses the middle of the laminae of the adjacent vertebrae. The laminae of these adjacent vertebrae overlap to the point of not allowing them to separate from each other, despite altering the patient's position. In the best of cases, there is only a small interlaminar gap, which makes it more difficult to approach/enter with a needle.

General Data from the Literature

Central neuraxial anaesthesia (collectively: spinal or epidural anaesthesia, combined or sequential spinal and epidural anaesthesia, caudal epidural anaesthesia) is routinely used during the perioperative period, as in analgesia and anaesthesia for obstetrics, and in the management of chronic pain disorders [25]. Traditionally performed based on surface anatomical landmark guidance, tactile sensation during advancement of the needle (passing through tissues with different densities, with the sensation of "loss of resistance") and/or of evidencing the presence of CSF within the needle, the technique suffers from the lack of objectivity of these methods, and due to their variable and unpredictable characteristics. It is often impossible for the practitioner to predict the simplicity or the difficulty of the technique prior to performing it, due to its "blind" nature.

Ultrasound imaging is a simple, hazard-free technique, with no side effects and which allows for images to be generated in real time. The practitioner can perform a sonoanatomical examination of the patient and then optimise the performance of the spinal/epidural in light of the ultrasound-guided observations. Furthermore, the ultrasound technique makes it possible to limit the use of conventional radiology and therefore patient (and personal) exposure to ionising radiation (especially useful in repeated procedures in **chronic pain** practice). Its role in obstetrical spinal/epidural anaesthesia has been most widely studied to date. These studies have mainly explored the "ultrasound-assisted" techniques (i.e. preprocedural location), but also include some evaluations of real-time ultrasound guidance.

As a result of the ability to perform a precise, personalised, and interactive anatomical analysis, ultrasound imaging promotes the overall reliability of the anaesthetic technique for the following reasons:

- Precisely identifying the vertebral level.
- Pre-evaluation and measurement of the needle path.
- Determining the site and spread of the local anaesthetic.

History

The first practitioners who described the use of ultrasound in spinal interventional procedures (lumbar punctures) were Bogin and Stulin in 1971. In 1978, Porter et al. used it in diagnostic radiology to measure the size of the lumbar canal [26]. It is in 1980 that the first team of anaesthesiologists (Cork et al.) studied and described the relevant ultrasound landmarks for conduct of epidural analgesia (EDA) [27]. Relatively quickly, other studies followed describing spinal anatomy and the metrics of the region (measurements from the skin to the laminae and to the epidural space) [28, 29]. The series of publications by Grau et al. between 2001 and 2004 [4, 30–35] unquestionably marked a turning point that demonstrated the utility of ultrasound in spinal anaesthesia. These studies comprise the basis of our current understanding of spinal ultrasonography. The first published report of spinal anaesthesia (combined spinal and epidural), visualised in real time with ultrasonography, was by Grau et al. in 2004 [4]. The technological advances achieved since then have dramatically improved the quality of ultrasound imaging described in the majority of recent seminal studies, such as those published by Karmakar et al. [5, 36, 37], and in particular in procedures performed with real-time ultrasound guidance [4–6].

What Does Ultrasound Provide?

The difficulties and failures encountered in central neuraxial anaesthesia performed in a "traditional" manner, without the aid of imaging, should not be underestimated.

The apparent simplicity surrounding **spinal anaesthesia** compared to PNB should be seen in a relative context.

- De Filho et al. noted a first successful attempt at a successful injection only in 61.5% of cases in a study conducted on 1481 patients [38].
- In 1–3% of cases, an injection can be impossible [39, 40].
- Out of 300 cases of spinal anaesthesia, 15% have been considered as technically difficult, and 10% have required more than 5 attempts at injection [41].
- Six percent of anaesthetic failures in a cohort of 3224 patients where CSF in the needle was observed [42].
- Concerning **EDA**:
- Complete failures in efficacy range from 20 to 50% when used for perioperative analgesic purposes [43–45], 5 to 20% in an obstetric analgesic context [46].
- Failures after a second attempt at insertion were 6%.
- Insufficient/inappropriate analgesia was around 20% [47], independently of the indication.
- The incidence of a breach of the dura mater (resulting from poor needle handling) is between 0.16 and 3%, and was symptomatic in 16–86% of cases [48].

Part of the difficulties, failures, complications, and/or errors revealed in the literature during central spinal anaesthesia performed with a conventional "blind" technique are manifestly related to **the discrepancy between the technique performed and the patient's specific anatomical configuration**. Therefore, imaging can be of assistance.

Ultrasound Facilitates Identification of the Intervertebral Approach

The ultrasound study of the spine makes it possible to determine the level of injection [10, 49, 50] for which the surface landmarks ("Tuffier's", or the intercristal line) very often provide only an approximation to the position and direction of the spinous process, the depth of the ligamentum flavum, of the epidural space and of the posterior dura mater. As the result of the low calcification of the spine in neonates and infants, ultrasound imaging of the spine is especially helpful in this context [51] (Fig. 12.2). Although it is possible to visualise the dura mater, the epidural space, spread of the local anaesthetic and even the presence of an epidural catheter with relative ease in children [52–54], with modern US systems, improvements in sonoanatomical knowledge, combined with a level of expertise, it is possible to visualise many of these elements in adults also. Thus, ultrasound proves very useful in populations that are technically problematic, such as obese subjects [55], in whom palpation of bony landmarks is imprecise or impossible, patients presenting with spinal disorders such as scoliosis [56] or a surgical history with or without implanted surgical materials (laminectomies, metalwork etc.) [57, 58].

Even in the absence of spinal abnormalities, the literature mentions frequent problems with a purely clinical estimate of the intervertebral level approach [9, 59–63]. In a popula-

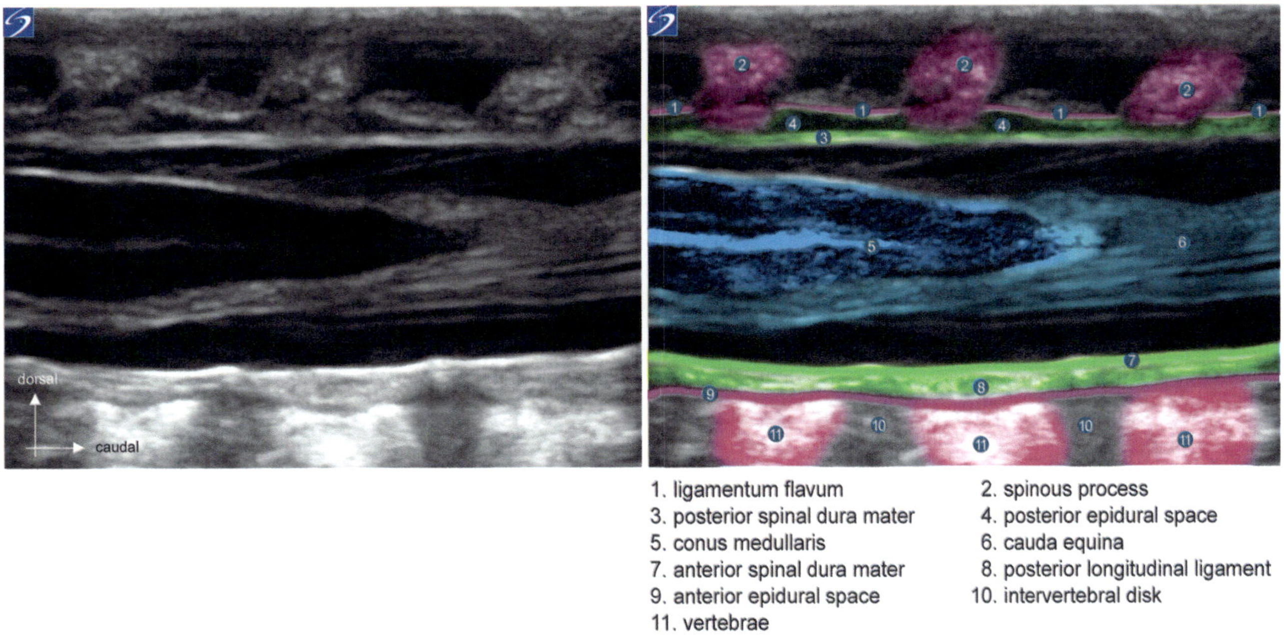

Fig. 12.2 Spinal ultrasonography imaging in children: conus medullaris and origin of cauda equina

tion of pregnant women where the median level of "Tuffier's" line lies just below L2-L3 [63], a study conducted in 2008 by Schlotterbeck [50] on 99 obstetric patients who underwent perispinal anaesthesia demonstrated that the estimated level of injection corresponded to reality only in 36% of cases, and in 50% of cases, the injection had been performed above the desired level. This type of error can result in inappropriate injection between one to three levels above that estimated [64, 65]. Some studies comparing other imaging techniques to ultrasound show that the latter makes it possible to identify a spinous process or an intervertebral space with accuracy only in 68 to 76% of cases [9, 49, 66]. Whilst a clinically identified space generally is more cephalic than the anatomical reality, a study by Locks et al. [59] suggests the contrary. But divergent opinions concerning the extent of error of the level and of the more cephalic or caudal tendency of the estimate are sometimes to be gauged; experience, as well as a rigorous training, makes it possible to obtain more than 90% of accurate assessment of the level [66], in particular by avoiding errors in identification of the junction L5-S1 [67]. Anomalies of the lumbosacral junction exist in about 12% of the general population [68] and are one of the causes of error. Sacralisation of L5 is the most common abnormality. It ranges from a variable degree of fusion between L5 and sacrum (involving the transverse process or both) up to complete fusion of L5, resulting in the existence of only 4 lumbar vertebrae. Conversely, complete lumbarisation of S1, which is much rarer, results in visualisation of 6 lumbar vertebrae.

Improper assessment of the intervertebral level accessed carries a real risk of a spinal injury [69, 70]. In fact, although the end of the conus medullaris traditionally lies opposite the vertebral body of L1, its position in the general population follows a "normal" distribution curve ranging from the middle of T12 up to the upper border of L3 [71], and below L1-L2 in about 42% of adults. Yet, in adults, the conus medullaris and the outline of the spinal cord are not visible, even though they may be in paediatric patients, in the form of hyperechoic lines marking their outlines [54, 72].

Ultrasonography Decreases the Technical Difficulty of Central Neuraxial Anaesthesia

Preprocedural localisation can reduce by half the number of attempted injections [3, 4, 30, 31] and increases significantly the success rate of the first attempt (75% vs 20%) [32]. It also makes it possible to improve the success rate of novice practitioners [33, 73]. Arzola et al. have observed that when spinal ultrasonography for localisation had been performed, in 92% of cases it was not necessary to reinsert the needle in the epidural space, in 74% of cases it was not even necessary to redirect the needle, and in 97% of cases identification of the epidural space was successful after two redirections or less of the needle [74]. In obese parturient patients, Balki et al. noted that three redirections of the needle or less were necessary in 93% of cases to correctly position the epidural catheter after ultrasound determination of the "ideal" insertion point [75]. In a randomised control study on 72 patients anticipated to pose difficulty in the conduct of a lumbar EDA approach, Grau et al. had already shown in 2001 the significant contribution of ultrasound location in the reduction of the number of injections and number of spaces approached [30]. Chin et al. also recently confirmed these observations in a study evaluating the preprocedural ultrasound location before spinal anaesthesia in a population considered difficult (unreliable anatomical landmarks, spinal deformities and/or a history of spinal surgery etc.): they concluded that there was a significant improvement in the success rate of the first attempt of injection and a reduction of the mean number of injections resulting in success of the procedure [76]. This was corroborated in 2013 by Shaikh et al. [3] who showed, in a meta-analysis combining 14 randomised studies on 1334 patients (674 underwent ultrasound), that the use of ultrasound enabled a significant reduction in the total number of injections, in the number of redirections of the needle and in the number of traumatic injections (failure rate divided by 7).

Even though the majority of published reports describe preprocedural localisation, Grau et al., in 2004, identified 30 parturients who were to undergo combined lumbar EDA anaesthesia in three groups: "loss of resistance", "preprocedural ultrasound", and "real-time ultrasound guidance". Although all the anaesthesias were effective, the number of injection attempts was significantly lower in the two groups where ultrasound was used [4].

Can Use of Ultrasound Decrease the Risk of Complications?

Ultrasound screening to detect a failure of the medial fusion of the ligamentum flavum (LF) may make it possible to limit the risk of an accidental breach of the dura mater. This particular anatomical variation (Fig. 12.3) is frequently found and can be present at the cervical, thoracic, and lumbar levels. In their anatomical studies, Lirk et al. [77, 78] described this particularity with the following frequencies: 4.4% in T6-T7, 2.1% in T7-T8, 4.4% in T8-T9, 17.6% in T9-T10, 35.2% in T10-T11, 28.5% in T11-T12, 15.8% in T12-L1, 22% in L1-L2, 11.4% in L2-L3, 11.1% in L3-L4, 9.3% in L4-L5, and 0% in L5-S1.

In a retrospective study of a population of parturients who had already experienced an accidental dural puncture during epidural anaesthesia, Lee et al. showed that an ultrasound-specific features could suggest a **problem of median continuity of the LF.** An "abnormality of the LF" was found in 72% of cases who had experienced an accidental dural puncture versus 17% in the control group [34].

The anatomical reality of these ultrasound abnormalities of the LF, however, remains to be specified. However, in the

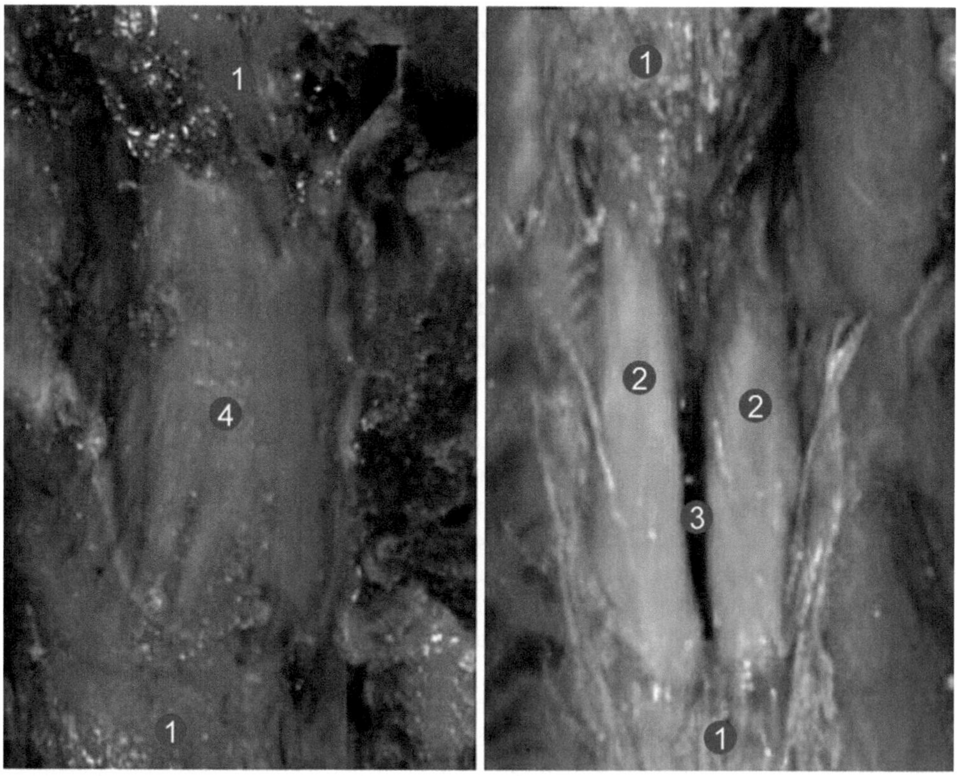

1. vertebral arch
2. ligamentum flavum
3. problem of continuity between the two "hemi ligaments"
4. the two fused "hemi ligamentum flavum"

Fig. 12.3 Anomaly of median fusion of ligamentum flavum (from [79] © Oxford Journals) case of a US image suggesting an underlying problem, the possibility of choosing an alternative spinal level where a "traditional" sonogram is found perhaps provides additional safety.

Can Use of Ultrasound Improve Clinical Efficacy of Central Neuraxial Anaesthesia?

A preprocedure ultrasound examination seems to have had a favourable influence on the efficacy of central neuraxial anaesthesia in four randomised control studies [4, 30, 31, 73]. Although in the three oldest studies, efficacy was evaluated by the practitioner who himself had performed the procedures, this was not the case in the one published in 2010 by Vallejo et al. [73] who showed that prior ultrasound localisation enabled a significant decrease in the number of times epidural catheters required repositioning due to insufficient analgesia in a randomised population of 370 parturients.

Central Neuraxial Anaesthesia at the Thoracic Level

To date, little has been published on the contribution of ultrasound imaging in thoracic central neuraxial anaesthesia. The study by Grau et al. in 2002 suffers from a limited size cohort (20), comprised solely of thin, young subjects, with no spinal abnormality, examined only at the level T5-T6 [79]. However, as is the case at lumbar level, it demonstrated the possibility of identifying structures as relevant as the LF and PDM, as well as the **anterior complex** (hyperechoic image generated by the superimposition of anterior dura mater (ADM), the posterior longitudinal ligament (PLL), and the posterior aspect of the vertebral body located immediately ventrally to the spinal canal). Thoracic spinal imaging, however, proved more difficult than lumbar imaging [2, 80] as the result of the shape and inclination of the spinous process and of the laminae, which limit access of ultrasound to the spinal canal.

The possibilities for obtaining useful images change depending on the level at which one positions oneself. Although between T9 and T12 vertebral morphology is very similar to the lumbar spine, it is much less so above T9 where the spinous processes become very oblique, the laminae increasingly flat with a tendency to overlap, creating increasingly narrow spaces (and so smaller **acoustic windows**). In this unfavourable region, it is the **paramedian sagittal oblique (PSO)** ultrasound plane *(see below)* which is the most useful for conduct of spinal anaesthesia: determination of the intervertebral level, visualisation of the LF and of the PDM (or of the **posterior complex**, superimposed images of the LF and PDM). The anterior complex of the spinal canal is visualised inconsistently. Although the **median transverse (MT)** plane *(see below)* visualises fewer elements, they are however crucial: the position of the midline, existence/extent of possible axial vertebral rotation (scoliosis), degree of sagittal inclination required to demonstrate a potential needle trajectory to the epidural space.

In the prospective observational study conducted by Avramescu et al. [80] on 61 healthy volunteers, although 98% of ultrasound examinations of the PSO and MT ultrasound planes were considered "conclusive" [sic] in the lower thoracic part, overall the PSO ultrasound plane proved to be the most "conclusive" at all levels of the thoracic spine (74% vs 37.5%), as a result of the narrowness of the acoustic windows in the transverse probe view and upper thoracic limitations.

Just as in the lumbar levels, one of the principal advantages of ultrasound in the thoracic region is in evaluating the anatomically abnormal spine. US studies of morphological parameters before introduction of an epidural catheter in a scoliotic spine have been described by Pandin et al. [81] (the most favourable interlaminar window, the skin–epidural space distance) as well as by McLeod et al. [82] (degree of axial vertebral rotation) and seem to have been beneficial in conduct of the procedures.

The opinion of Chin et al. is that "even if the vertebral canal is not clearly visible, a preprocedural scan may provide information that will facilitate thoracic epidural catheter insertion. Apart from determining axial rotation (as described by McLeod *et al.*), the depth to the lamina may be measured (as a surrogate marker of depth to the epidural space), the levels of the thoracic interspaces may be determined more accurately, and the locations of the midline and interlaminar spaces can be marked on the skin. Triangulation using this information will facilitate estimation of the appropriate needle insertion site and trajectory for a paramedian or midline approach. Currently, no published data support or refute these assertions" [2].

Techniques

Basic Notions for Ultrasound in Central Neuraxial Anaesthesia

If we are in the preprocedural setting (**ultrasound-assisted** technique), the patient should be placed in the position in which anaesthesia will subsequently be administered. The question obviously does not arise when it involves an **ultrasound-guided** technique, since ultrasound and the anaesthetic procedure are simultaneous.

The direction of the probe is described based on the elementary anatomical nomenclature (**sagittal**, **transverse,** and **coronal** planes) (Fig. 12.4).

Therefore, we will mention the following ultrasound planes.

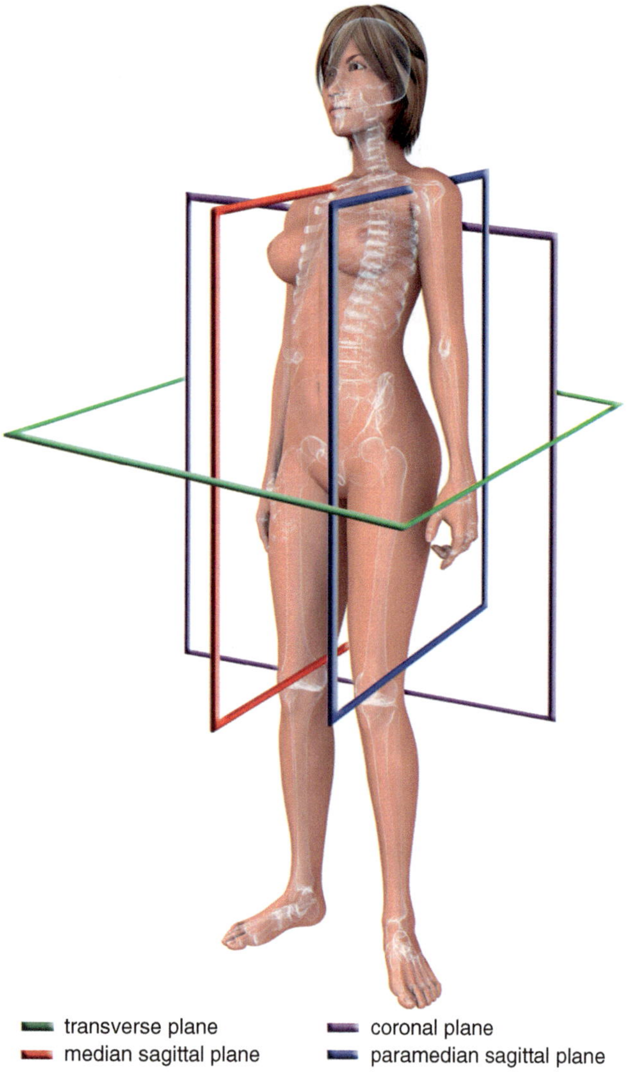

Fig. 12.4 Basic anatomical nomenclature

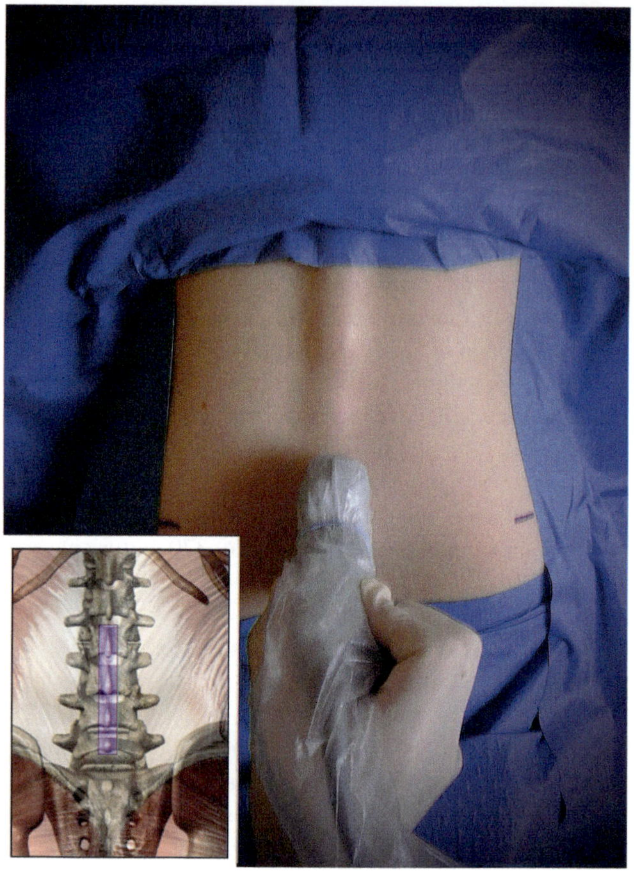

Fig. 12.5 Position of probe for median sagittal plane

Median Sagittal

The probe is positioned parallel to the patient's craniocaudal axis, placed on the line of the spinous process (midline) and directed perpendicularly to the coronal anatomical plane (Fig. 12.5). The Median Sagittal (MS) plane often is little informative because of the very superficial apices of the spinous process generate major acoustic shadows, making even their simple counting difficult. On the contrary, in an obese subject, the deeper spinous processes are more easily identifiable (Fig. 12.6).

Paramedian Sagittal

It is sagittal but does not pass through the line of the spinous process. It can be Paramedian Sagittal (PS) straight right or left, placed at a variable distance from the line of the spinous process and to enable visualisation of different structures (laminae, articular processes, transverse processes etc.) (Fig. 12.7).

Paramedian Sagittal Oblique

The probe initially in PS position is directly slightly in the direction of the midline (Fig. 12.8). It can be obtained from the right or left PS plane. As will be shown later, in case of vertebral rotation in the transverse plane (scoliosis), the right and left Paramedian Sagittal Oblique (PSO) planes do not provide identical information due to the direction of the sagittal axis of the specific vertebrae (plane of symmetry of the vertebrae), in relation to the plane of the patient's back. The PSO plane on the side of the **convexity** of the scoliosis is more informative and easier to acquire (Fig. 12.9).

Median Transverse

Synonymous with "median **axial**" or simply "transverse" view, this view is generated by positioning the probe parallel to the transverse anatomical plane and in the midline (Fig. 12.10). Where vertebral axial rotation exists, it is this ultrasound plane which most easily demonstrates its existence and allows its significance to be evaluated.

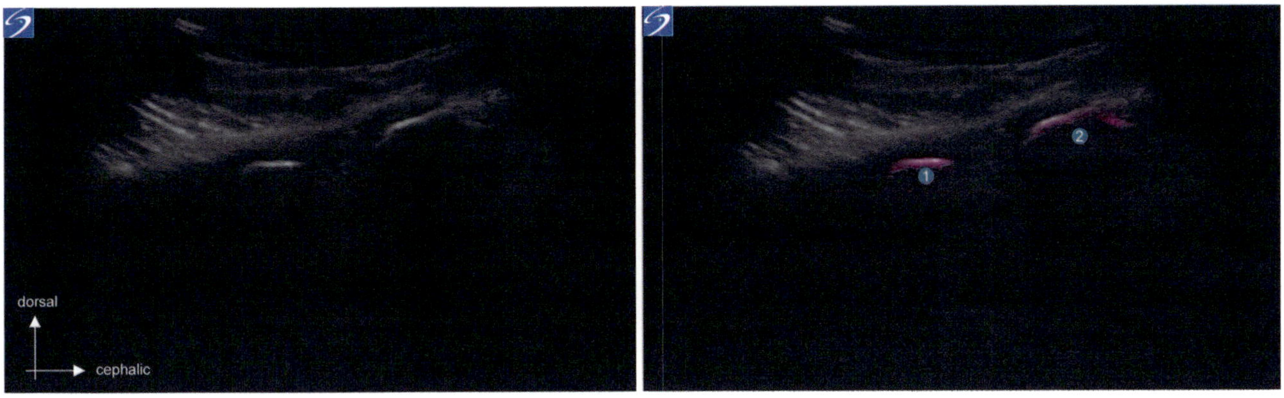

Fig. 12.6 Spinous processes in an obese patient in median sagittal view

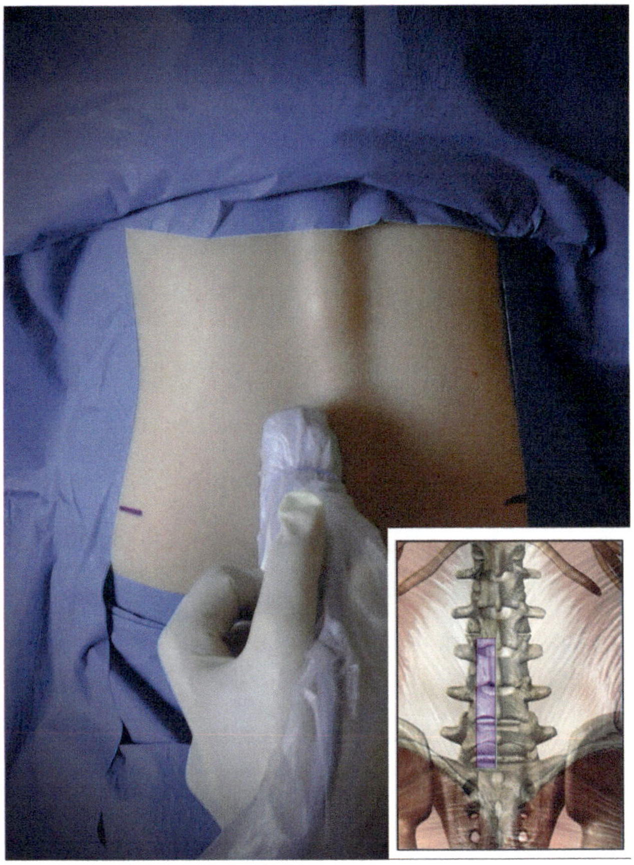

Fig. 12.7 Position of probe for paramedian sagittal plane

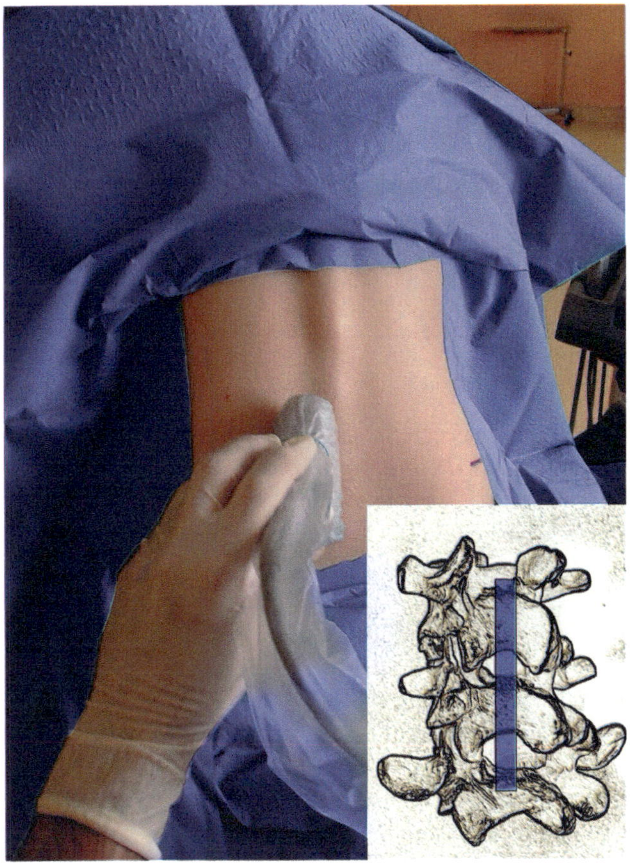

Fig. 12.8 Position of probe for oblique paramedian sagittal plane

Equipment

- **The ultrasound system** used is the same apparatus usually dedicated to RA.
- **A low-frequency convex probe (2–5 MHz)** is necessary for adults, its advantages are as follows:
  - A wide ultrasound field in a "fantail" enabling an advantageous anatomical overall view.
  - Lower frequency ultrasound facilitates a deeper examination.
- **The ultrasound gel** can be used for preprocedural localisation, but physiological saline should be used in preference in the case of ultrasound guidance [4], to avoid causing the gel to come into direct contact with the spinal/perispinal structures (safety unknown). Although the use of gel is preferable because it creates a better coupling

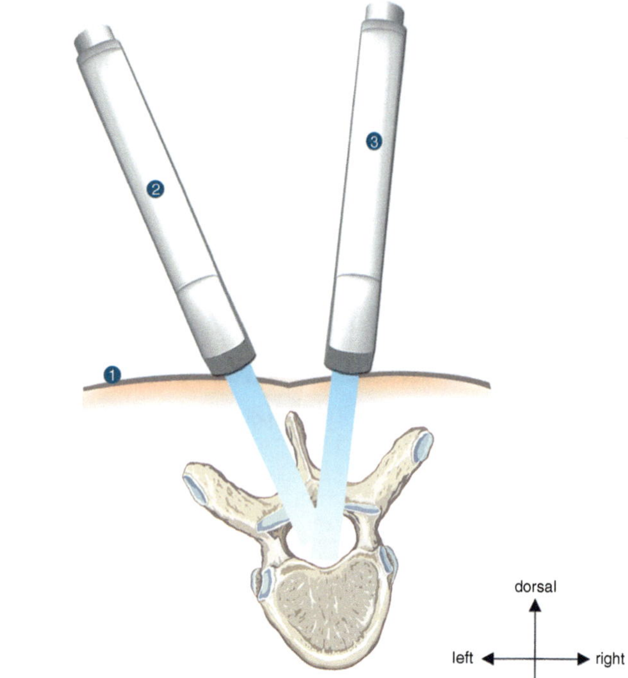

Fig. 12.9 Comparison of right and left OPS plane in case of vertebral rotation

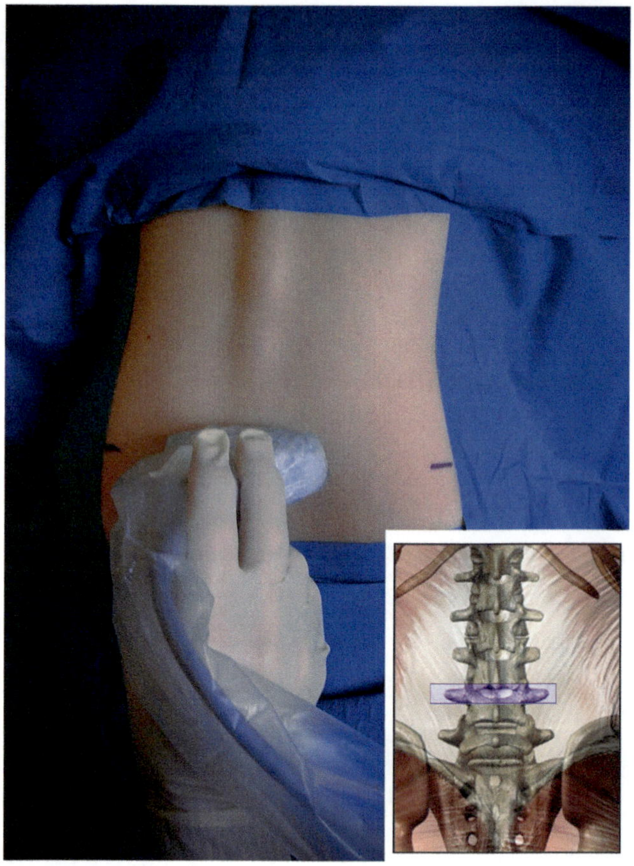

Fig. 12.10 Position of probe for median transverse plane

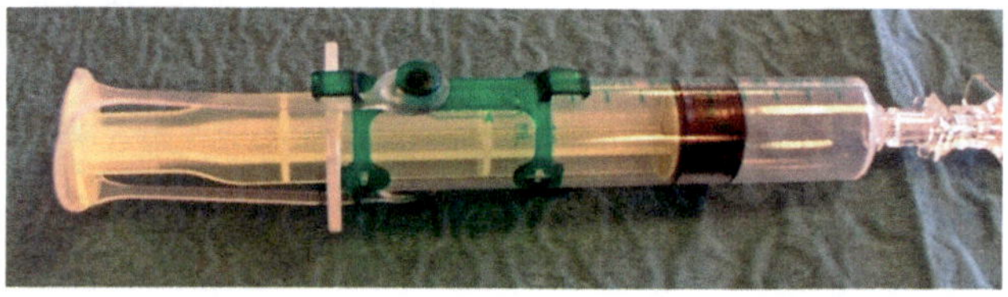

Fig. 12.11 Special syringe to identify the epidural space ("Épimatic" syringe, Vygon®)

interface between the probe and skin (and thereby produces better US images), great care must be taken to wipe it off the skin prior to commencing the needle puncture.

- **An appropriate sterile sheath and surgical asepsis** are necessary for real-time guidance procedures.
- **Needles:**
  - For **spinal anaesthesia**: the usual pencil-point needles are used (27 G, 25 G, or 22 G). Their echogenicity is variable, and often poor in the case of real-time ultrasound guidance; however, echogenic needles for spinal anaesthesia are now available (Pajunk®).
  - For **epidural anaesthesia**: If it is performed **after prior ultrasound localisation,** the usual kits are used (needle, catheter, filter, loss of resistance syringe).
  - To screen for loss of resistance: in case of **real-time ultrasound guidance,** the need to release one hand to hold the ultrasound probe necessitates the use of specific syringes to verify entering the epidural space. These devices maintain a constant force on the plunger of the syringe means of either a rubber band (Fig. 12.11), or an internal spring [83].
- **Routine monitoring** of the patient is mandatory [12].

Preprocedural Ultrasound Localisation

Localisation prior to perispinal anaesthesia has the purpose of determining:

- The level of injection.
- The ideal approach of the needle (skin injection site, sagittal and transverse angulation).
- The optimum site of injection of the local anaesthetic or insertion of the catheter.

Visualisation of the sacrum, the vertebral laminae and the interlaminar spaces, the LF and PDM (together = the posterior complex), the anterior complex, the spinous processes, and the interspinous spaces, provides information necessary for these objectives.

Chin et al. [2] propose a 7-successive step procedure:

I. Preparation with ultrasound localisation.
II. Paramedian sagittal plane (PS): with visualisation of the transverse processes.
III. Paramedian sagittal plane (PS): with visualisation of the articular processes.
IV. Paramedian sagittal oblique plane (PSO).
V. Identification and marking of the intervertebral levels.
VI. Interlaminar median transverse plane (MT).
VII. Marking of the skin at the injection site for a midline approach.

This process may be further simplified depending on the practitioner's experience, the procedure planned, and characteristics of the patient. To determine the intervertebral level, it is not always necessary to go through steps I–III, but instead to directly position the probe in PSO (stage IV), and then look for the level of the lumbosacral junction (L5-S1 interspace) and count in ascending manner the intervertebral levels.

I. **Preparation for ultrasound localisation.**

The patient is either seated or lying lateral, with the trunk and hips flexed in order to "open" the interspinous and interlaminar spaces. Additionally, such flexion moves the spinal nerves into a more anterior position within the spinal canal [35, 84], which theoretically may limit the risk of accidental nerve root injury during injection (Figs. 12.12 and 12.13).

The depth of the ultrasound field (generally 7–10 cm), focal point, and gain are adjusted accordingly.

II. **Paramedian sagittal plane (PS): visualisation of the transverse process.**

The probe placed in PS position at 3–4 cm from the midline makes it possible to visualise the "trident sign" consisting of the acoustic shadows of the adjacent transverse process [85] (Fig. 12.14).

III. **Paramedian sagittal plane (PS): visualisation of the articular processes.**

The probe is slid medially whilst maintaining its SP direction.

A pattern resembling a "camel's back" is observed, which is a continuous hyperechoic line generated by the succession of adjacent articular processes (Fig. 12.15).

Although strict PS section avoids the acoustic shadows of the spinous process, however, it does not make it possible to easily visualise the spinal canal as it intersects it more laterally, almost tangentially (Fig. 12.16).

IV. **Paramedian sagittal oblique plane (PSO).**

After obtaining the image of the articular processes of the SP plane, a slightly medial inclination of the probe obtains a more useful viewing angle, the PSO plane [5, 86] (Fig. 12.8).

It is in this section that the interlaminar space (i.e. between the laminae of adjacent vertebrae) opens to provide a wider acoustic window to the deeper structures. The succession of adjacent laminae generates a characteristic "sawtooth" pattern, with each lamina having a profile likened to a "horses head" (according to Karmakar) (Fig. 12.17). The LF and the PDM (whose depth can be measured), the spinal canal, and the anterior complex are all visible between the lamina, but are masked at regular intervals by the acoustic shadow cast by each successive lamina.

According to a study of the thoracic spine for conduct of epidural anaesthesia, the PSO plane is particularly useful as the obliquity of the spinous process masks the view in the median sagittal plane. However, since the laminae here are wider than in the lumbar spine, the interlaminar space is correspondingly narrower. The interlaminar acoustic window which gives access to the structures present in the spinal canal is also narrower than that seen at the lumbar level. Nevertheless, the PSO plane is the preferred plane for real-time ultrasound-guided procedures, both at the lumbar level as well as at the thoracic level (Fig. 12.18).

In case of axial vertebral rotation (present in scoliosis), the most favourable PSO view is gained on the side of rotation (Fig. 12.9), i.e. with the probe placed on the side of the convexity of the scoliosis. Apart from these cases, it is often advisable to compare the quality of visualisation of both sides in order to choose which could offer the best image quality.

V. **Identification and marking of intervertebral levels.**

By placing the ultrasound probe over the lumbosacral junction, in the PSO position (Fig. 12.19), a continuous hyperechoic line is observed caudally which corresponds to the sacrum. Moving the probe in a cephalic direction allows the laminae of L5, and then L4, to be identified in succession etc. (Fig. 12.20), up to the interlaminar level desired, which is then centred in the image. The corresponding interspinous space is identified either by palpating it opposite the middle of the probe (provided that it is palpable), or by moving the probe medially (up to the **median sagittal** position) to observe the image of the spinous process beneath. (Note that in some cases it is possible to perform a **descending count** of the intervertebral spaces from T12 which can be identified by its articulation with the 12th rib. A combination of the two methods may be useful to dispel doubt.)

Once the desired level of injection has been identified, its position is marked by an indelible pen. In the

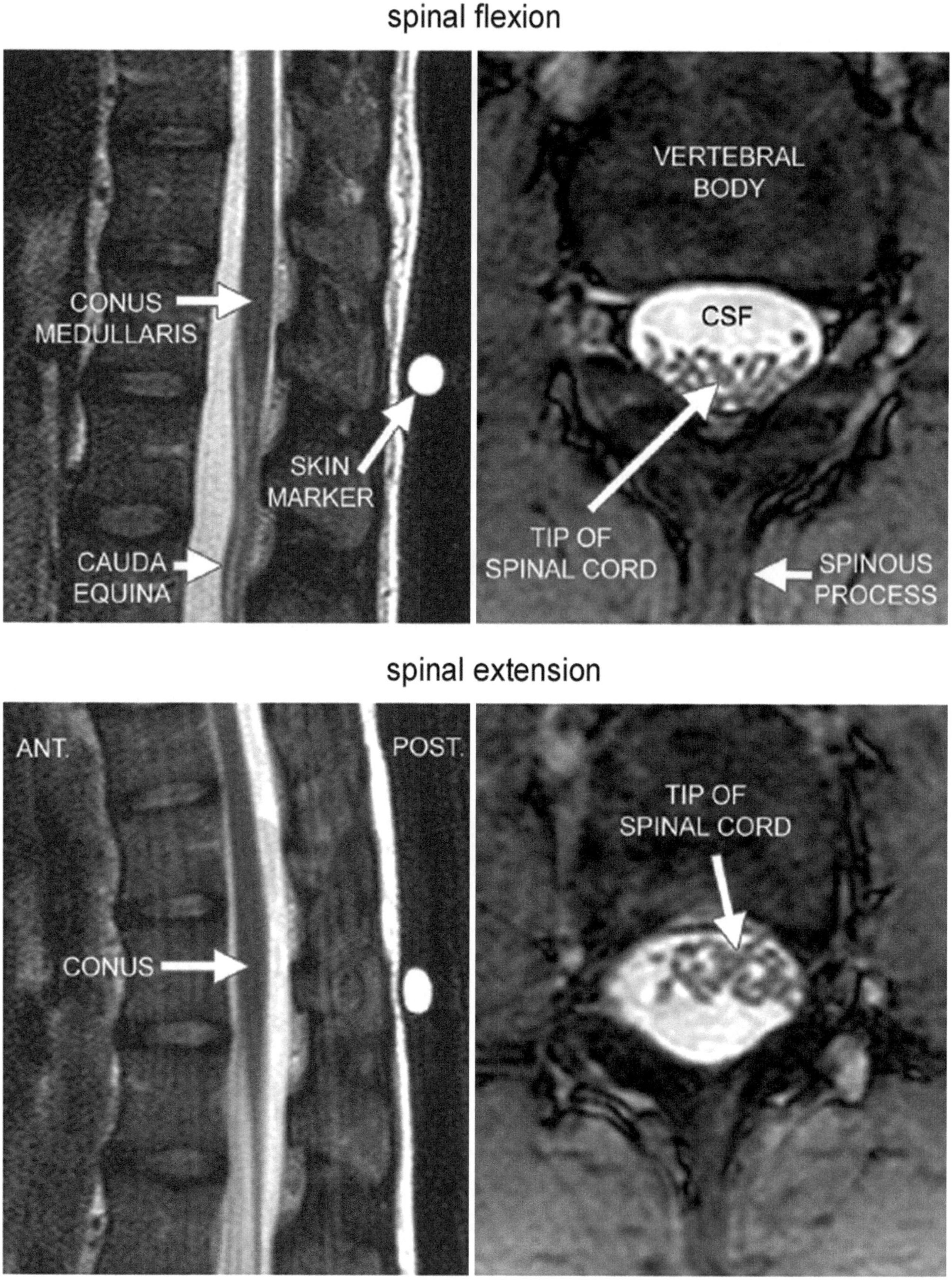

Fig. 12.12 Change to position of roots of the cauda equina depending on dorsoventral spinal flexion (from [85] © Oxford Journals). *Conus medullaris*; *skin marker*; *cauda equina*; *vertebral body*; *CSF: cerebrospinal fluid*; *Tip of spinal cord*; *spinous process*

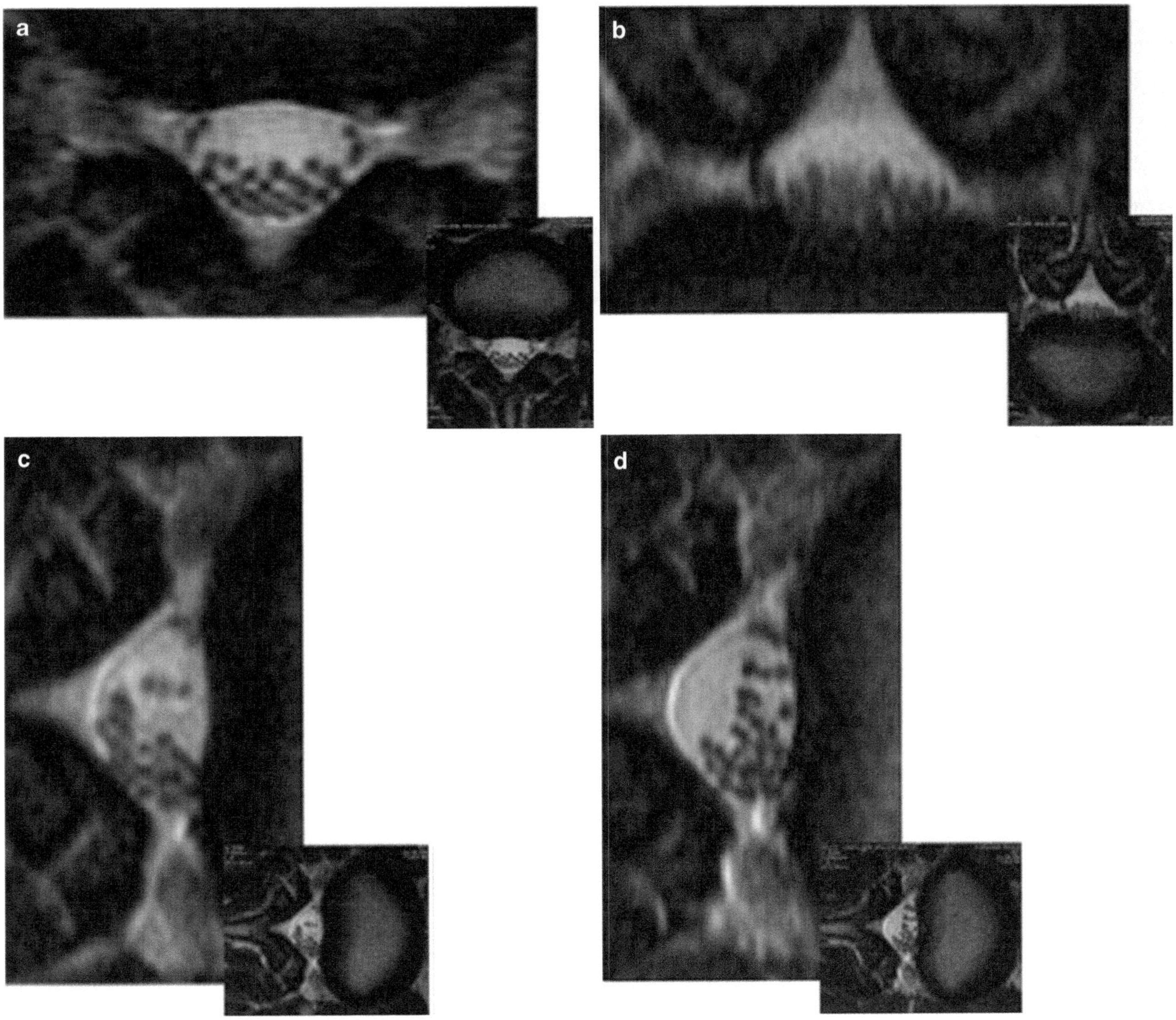

Fig. 12.13 Change to position of roots of the cauda equina depending on patient position: supine position (**a**), ventral (**b**), or lateral (**c**). D: lateral recumbent + flexion of spine (from [35] © Lippincott Williams Wilkins)

majority of cases, the technique of median sagittal injection will be chosen. To obtain the optimum skin marking, a 90° rotation of the probe is performed. The MT section obtained gives an interspinous view which allows for the point of injection and its axis/trajectory to be confirmed (in case of a median sagittal approach *(see steps VI and VII in the following)*.

In some cases it is preferred to perform a paramedian injection. Then the usual preprocedural localisation is undertaken, but with marking of the skin appropriate for a paramedian injection. It is the PSO plane (and not the MT) that facilitates this (Fig. 12.21). In the same manner as described above, the landmarks of the skin injection are marked with an indelible pen when the ideal image is obtained: this involves intersection of the line, passing through the middle of the probe longitudinally and transversely (Fig. 12.22). Then, by directing the needle in the same direction as that of the probe during location and marking (median orientation and more or less cephalic), perispinal anaesthesia is performed whilst taking account of the pre-measured depths of the LF/PDM (Fig. 12.23).

VI. **and VII. Interspinous median transverse (MT); marking of the skin of the point of skin puncture for median approach.**

All the informations provided by prior ultrasound localisation are extremely reliable and allow to determine the ideal injection site [73]. When the interspinous level desired has been identified in the PSO plane, the probe is rotated 90° to lie transversely to the median line of the spinous process (Fig. 12.10) (i.e. theoretically in the interspinous space). However, dur-

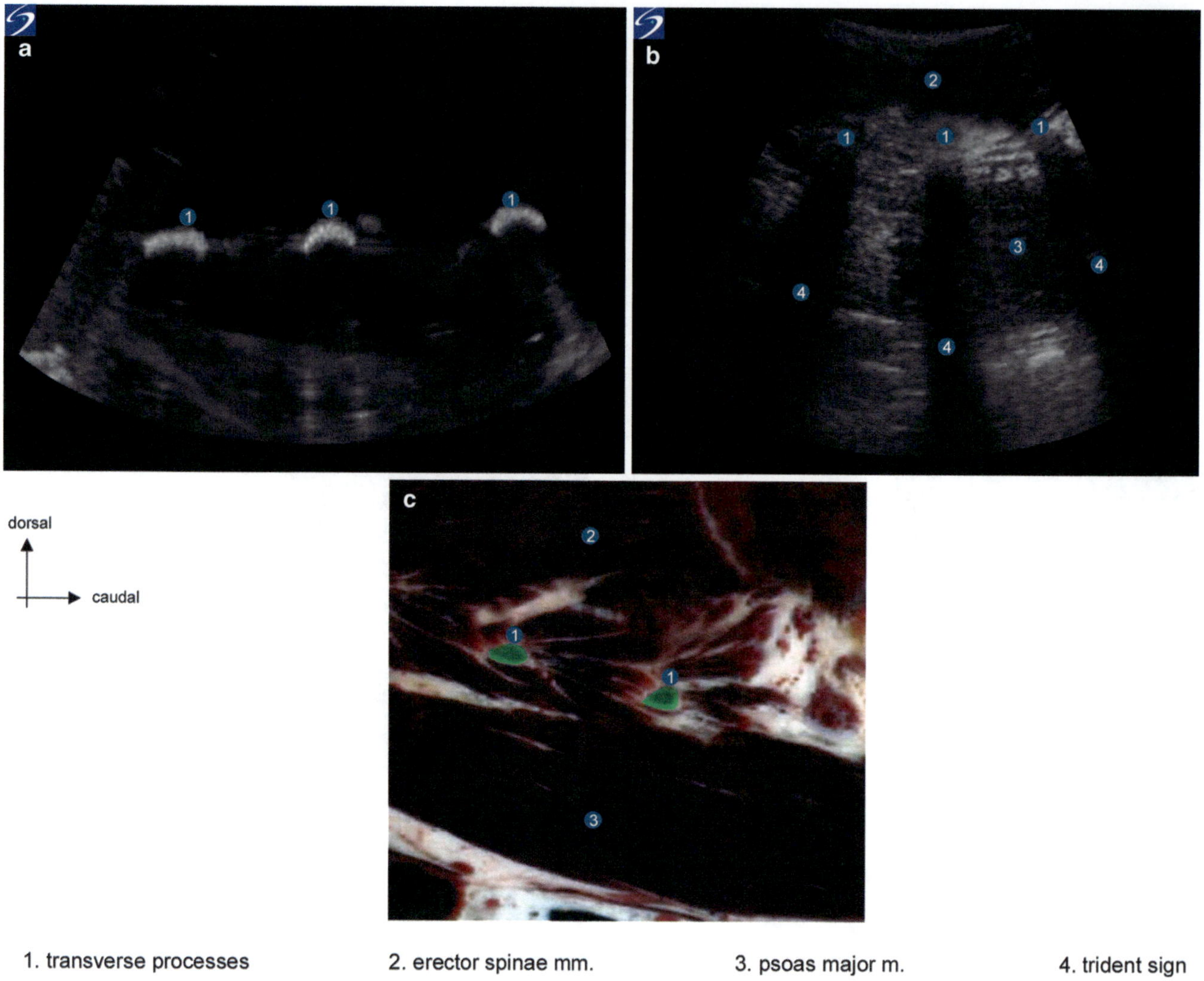

Fig. 12.14 Paramedian sagittal plane: visualisation of transverse processes *(Trident sign)* (from [86] © British Institute of Radiology). (**a**) ultrasound image obtained on a phantom; (**b**) ultrasound image obtained on model; C: anatomical section in cadaver

ing this procedure, often the clinician's position is slightly shifted, which results in an MT plane more or less intercepting the cranial or caudal spinous process. It appears in the form of an acoustic shadow (shadow cone) starting superficially, which may be enhanced by a hyperechoic signal on the surface (Fig. 12.24). The laminae appear in this ultrasound section as hyperechoic lines in ventrolateral position, on either side of the base of the spinous process, and they also generate acoustic shadows.

At this stage of localisation, the possible existence of a **transverse vertebral rotation** (axial) should be determined, a finding which is very common in scoliosis (Fig. 12.25). This is revealed when the probe is placed in median transverse position strictly perpendicular to the plane of the back, and the underlying vertebra does not appear centred and symmetrical on the screen (Fig. 12.26). To determine the extent of vertebral rotation, it is necessary to direct the probe laterally (generally in the same direction as the convexity of scoliosis) until a centred and symmetrical view of the vertebra is obtained in the middle of the screen. The spinous process appears in the form of a vertical acoustic shadow in the centre of the screen, and then, on either side, fusing with the vertebral laminae placed symmetrically beneath. The direction of the probe thus predicts the transverse angulation that should be applied to the needle during its insertion and advancement (Fig. 12.27). The positions of the spinous processes then are outlined on the skin with an indelible pen, demonstrating their alignment along the **superficial sagittal spinal axis**.

At present, to visualise the LF, the epidural space, and the PDM optimally, it is necessary to move the probe within the interspinous space and to make small movements of translation, usually with a slight sagittal

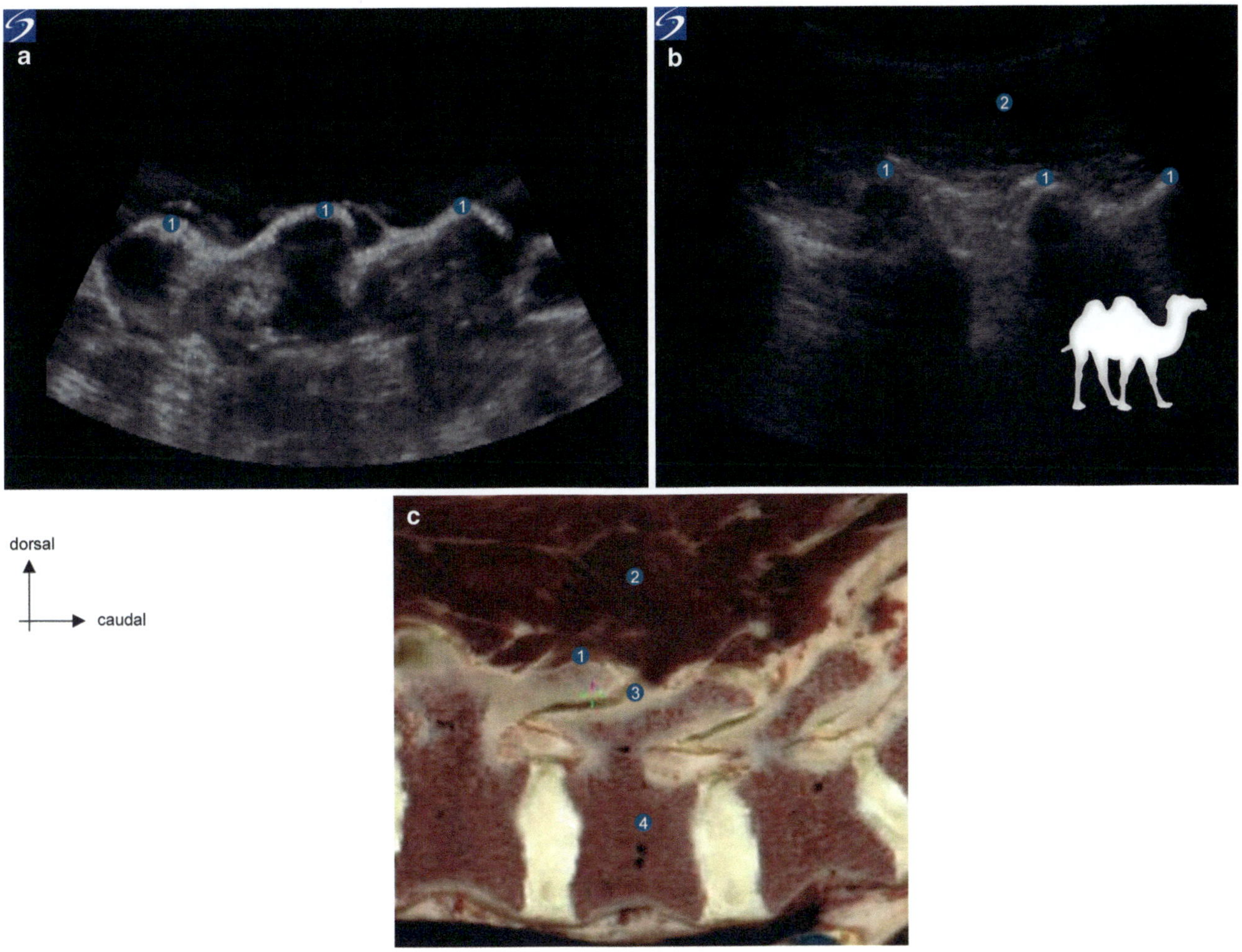

1. articular processes 2. erector spinae mm. 3. joint space 4. L4 vertebral body

Fig. 12.15 Paramedial sagittal plane: visualisation of joint process ("camelback" presentation) (from [86] © British Institute of Radiology). (**a**) ultrasound image obtained on a phantom; (**b**) ultrasound image obtained on model; C: anatomical section in cadaver

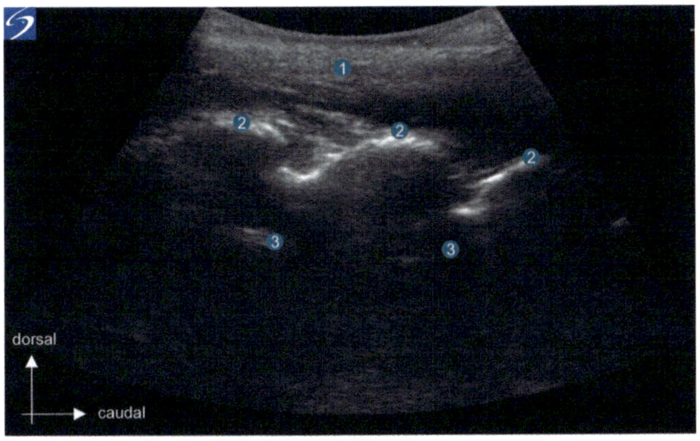

1. erector spinae mm. 2. articular processes
3. parasagittal section, tangential to spinal canal, seen in acoustic windows.

Fig. 12.16 Strict paramedian sagittal section (PS), almost tangential to spinal canal

inclination of the probe. Sagittal tilting should be performed whilst taking care **to minimise lateral angulation of the probe.** The correct position is confirmed by holding the symmetrical image of the vertebra in the centre of the screen. The LF and the PDM may appear as two separate hyperechoic linear structures which demarcate the epidural space (hypoechoic) lying between them. The hyperechoic fine line of the dura mater can be visualised frequently, whilst the LF, which is slightly more superficial and thicker, is less echoic (Fig. 12.28). Often, particularly in the elderly, only a single linear hyperechoic image is discerned, a type of "fusion" of the two previous images: **the posterior complex**. Note that the depth from the skin to the LF, PDM, or posterior complex should be measured and used to facilitate the safe performance of the spinal/epidural anaesthesia.

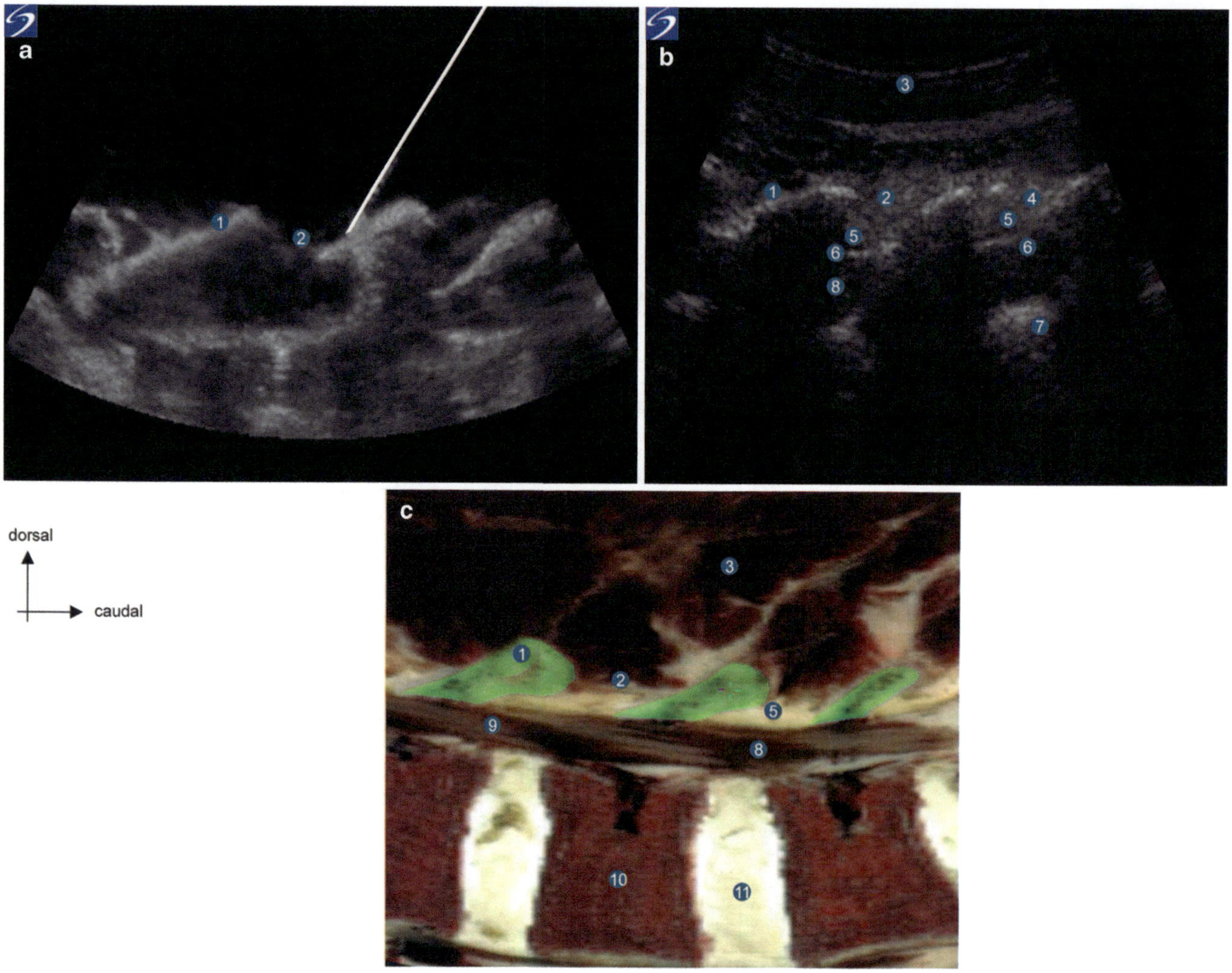

Fig. 12.17 Sagittal paramedian oblique plane ("horses' heads" presentation) (from [86] © British Institute of Radiology). (**a**) ultrasound image obtained on a phantom; (**b**) ultrasound image obtained on model; C: anatomical section in cadaver

In front of the posterior complex, we can see the spinal canal that is almost anechoic. In front of the canal, **the anterior complex** is visualised, sometimes with possibility of differentiating hyperechoic lines corresponding to the ADM, the anterior epidural space, and then a very hyperechoic signal corresponding to the posterior longitudinal ligament and the posterior aspect of the vertebral body (Fig. 12.28). If the intervertebral space chosen does not make it possible to obtain a conclusive image or it predicts an especially difficult procedure, the examiner should inspect the adjacent spinal levels in order to choose an alternative space with an easier to approach.

At the thoracic level when the optimum interspinous view is obtained, the sagittal angulation of the probe is more pronounced (Fig. 12.29), as the result of the increased obliquity of the spinous process. The position of the probe is marked with an indelible pen at each of its lateral ends, at its mid-point (Fig. 12.30). The intersection of this transverse level and of the superficial sagittal spinal axis makes it possible to define the point of puncture of the skin (Fig. 12.31).

Once the markings on the skin have been made, the clinical procedure is commenced, keeping in mind the depth of the epidural space which has been measured. **The direction of the needle path is indicated by the transverse and sagittal directions that are made with the probe when obtaining the best image.**

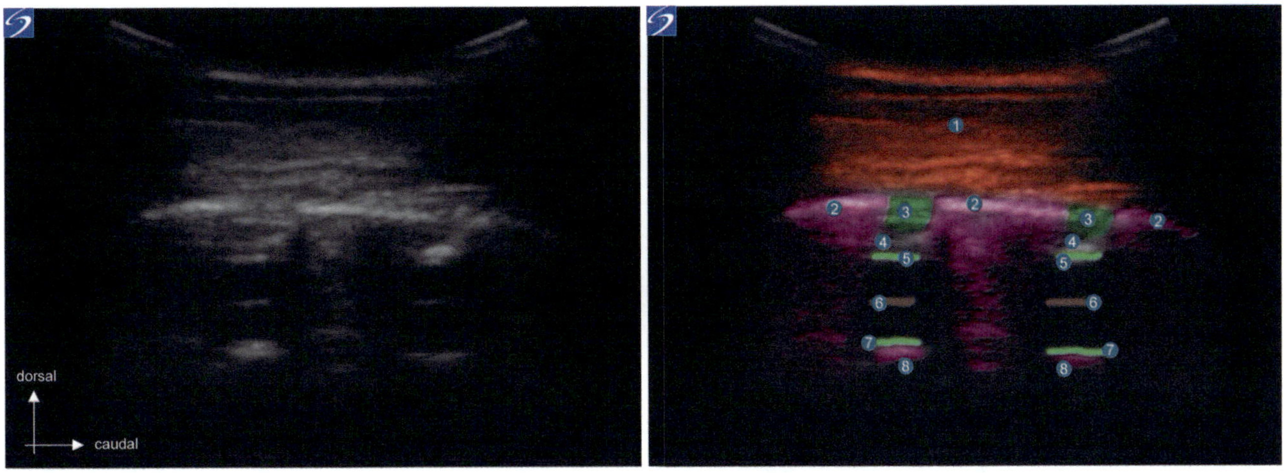

Fig. 12.18 Sagittal paramedian oblique plane at the thoracic spinal level

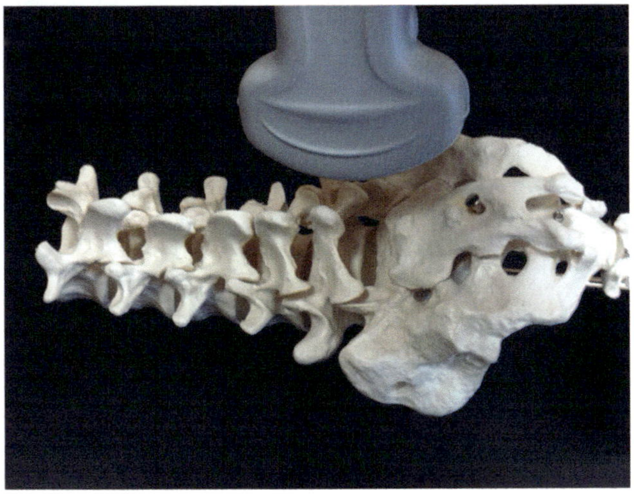

Fig. 12.19 Position of probe in SPO at level of lumbosacral junction at start of procedure for evaluation of spinal level

Real-Time Ultrasound Guidance

Real-time ultrasound guidance in central neuraxial anaesthesia is of a higher level of difficulty than for all other RA techniques described previously. The reasons relate to the footprint of the low-frequency probe (size and thickness), its position with respect to the needle, the possible need to enlist the help of a colleague (e.g. to hold the probe), and often the poor visibility of the needle.

The PSO approach is the best option to perform ultrasound-guided central neuraxial anaesthesia in plane [5], with the needle in longitudinal view in the ultrasound beam. Just as for peripheral nerve blocks, a real-time technique does not dispense with the need for a prior ultrasound study of the region examined.

Spinal Anaesthesia

After location of the sacrum, the probe in the PSO position is moved cranially up to the interlaminar level intended for spinal anaesthesia (generally L3/L4 or L4/L5, sometimes L2/L3 or L5/S1). After looking for the best view of the LF/PDM complex, local anaesthesia of the skin is performed at the caudal end of the probe. The injection is performed in plane with the spinal needle introducer, with the latter itself being inserted (Fig. 12.32). The difficulty of the procedure lies on one hand in the requirement for perfect immobility of the probe and, on the other hand, in advancing the needle strictly in plane through the erector muscles of the spine, towards the upper border of the lamina which borders the interlaminar space caudally. The needle then crosses the LF, the epidural space, and then lastly penetrates the PDM. Reflux of CSF through the needle confirms the correct placement of the needle tip. On piercing the dura mater, its initial anterior movement is sometimes observed, and then its return to normal position after penetration, but this process is not constant. Injection of the local anaesthetic into the CSF may be observed with colour Doppler or power Doppler ultrasound (Fig. 12.33).

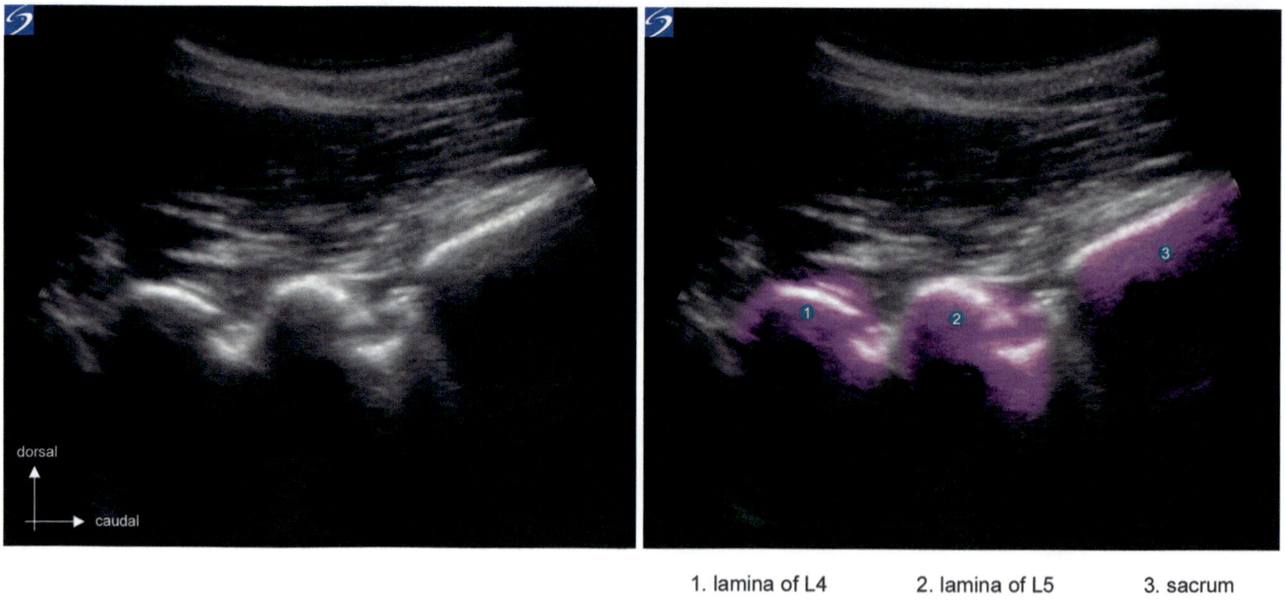

Fig. 12.20 Sagittal paramedial oblique plane with visualisation of sacrum and of lamina of vertebrae L5 and L4

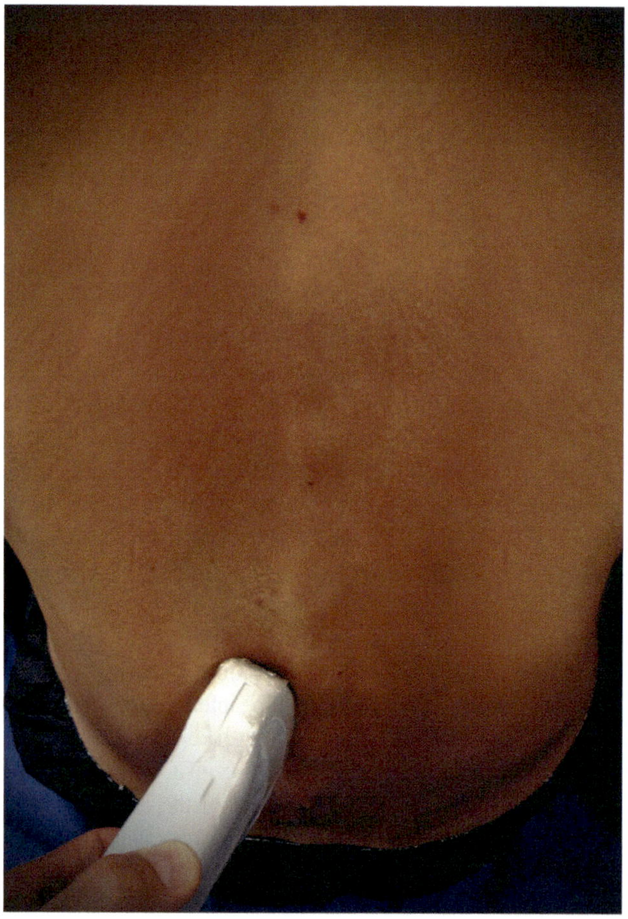

Fig. 12.21 Position of probe for sagittal paramedian oblique section (SPO)

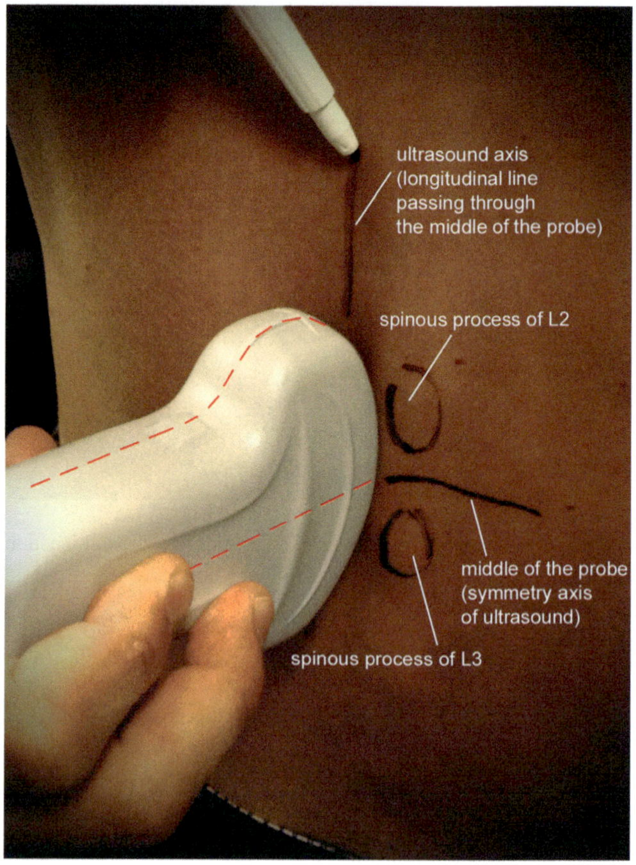

Fig. 12.22 Marking of skin during preprocedural scanning for parasagittal approach

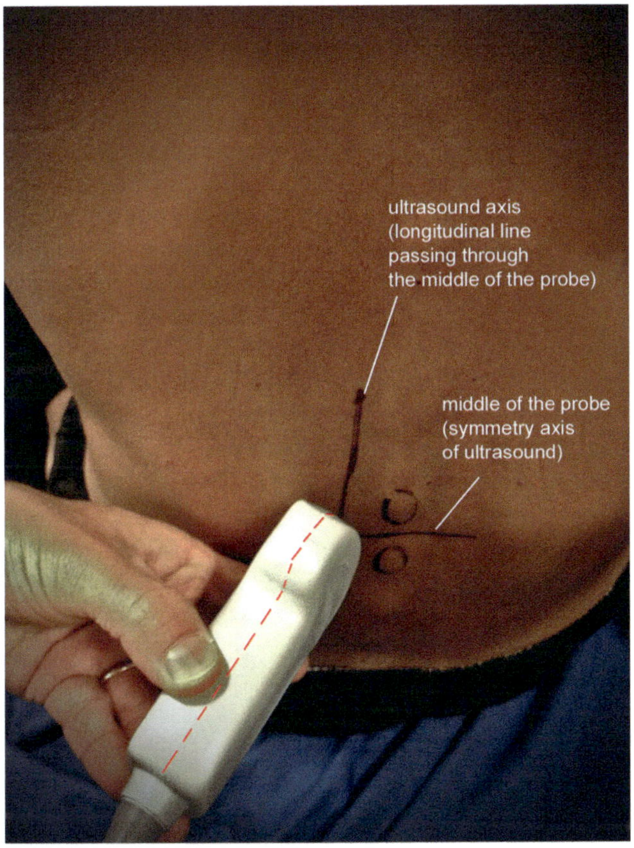

Fig. 12.23 Parasagittal injection: the direction of the needle will be that given by the probe during preprocedural scanning

Epidural Anaesthesia

The initial process (locating the level, the PSO axis of the probe, optimisation of the image etc.) is the same as that for spinal anaesthesia up to the skin infiltration with LA at the caudal end of the probe. M.K. Karmakar [5] has described conduct of ultrasound-guided lumbar epidural anaesthesia in 15 patients according to this approach. The needle for epidural puncture is introduced and followed under visual control in plane, from the caudal end of the probe up to the upper border of the lamina caudal to the interlaminar space chosen. The practitioner holds the probe with one hand and the needle with the other hand. The automatic/self-deploying loss of resistance syringe is attached to the Tuohy needle only **after** its positioning in the LF (the densest structure). This may avoid any false occurrences of loss of resistance. The direction and depth of the needle make its visualisation a subtle procedure. During entry into the epidural space, the author has observed in 53.3% of cases the anterior movement of the dura mater and widening of the epidural space, generated by injection of 0.9% NaCl contained in the syringe. The success rate in this study was 93.3%.

The epidural catheter then is inserted, and it is sometimes possible to observe it at the upper intervertebral level, even to follow its advancement (Fig. 12.34).

The concomitant deployment of the plunger of the syringe, sensation of loss of resistance, and of visualisation of the needle tip between the LF and the PDM all may validate the correct epidural position of the needle tip.

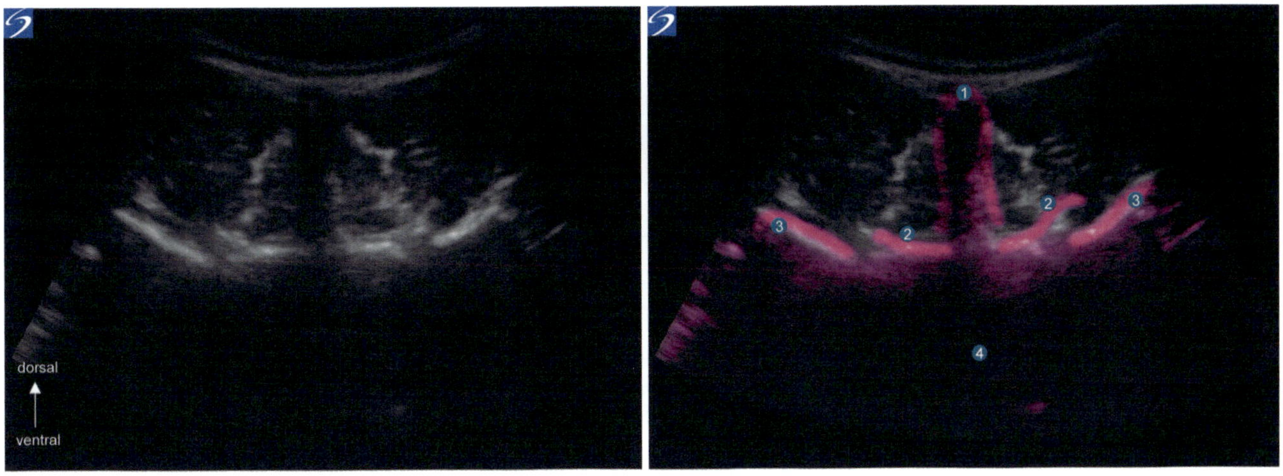

Fig. 12.24 Median transverse view cutting through the spinous process

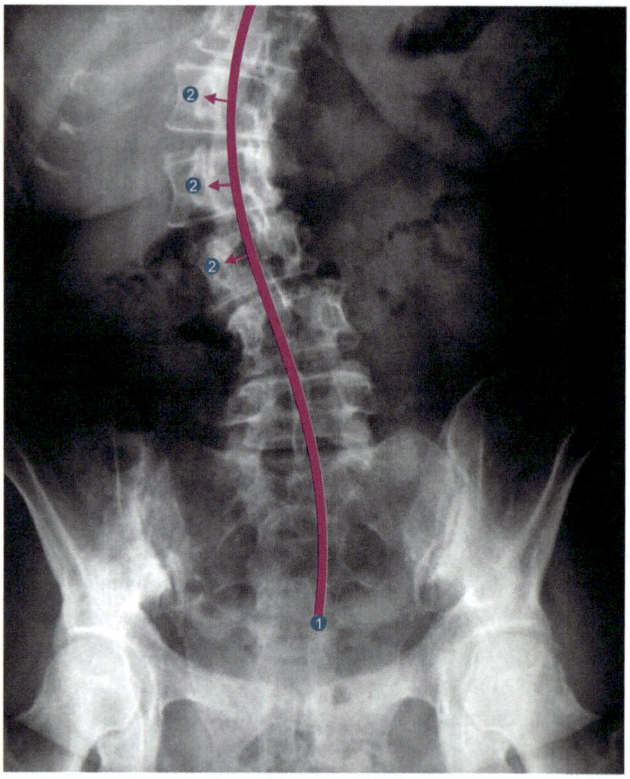

1. superficial sagittal spinal axis (line of spinous processes)
2. vertebral rotation : arrows on the X-ray.
Maximal rotation at the top of spinous convexity.

Fig. 12.25 X-ray of scoliotic spine. Drawing the projection of the "line of the spinous processes" allows evidencing concomitant vertebral rotation

As has also been shown [6], the use of an injection guide is helpful during the procedure of real-time guidance of the needle, and it may even become a compulsory aid for this technique.

Contrary to spinal anaesthesia for which the theoretical level of puncture is located below L2 so as not to accidentally injure the spinal cord, an epidural puncture can be performed at a more cephalic level, depending on the surgical site. In thoracic approaches, the PSO section is also the most useful and probably the only one usable for real-time ultrasound guidance. It offers a narrower acoustic window than at the lumbar level considering the flattened and overlapping characteristic of the laminae. The management of the trajectory of the needle under ultrasound control in the epidural space requires a particular level of dexterity.

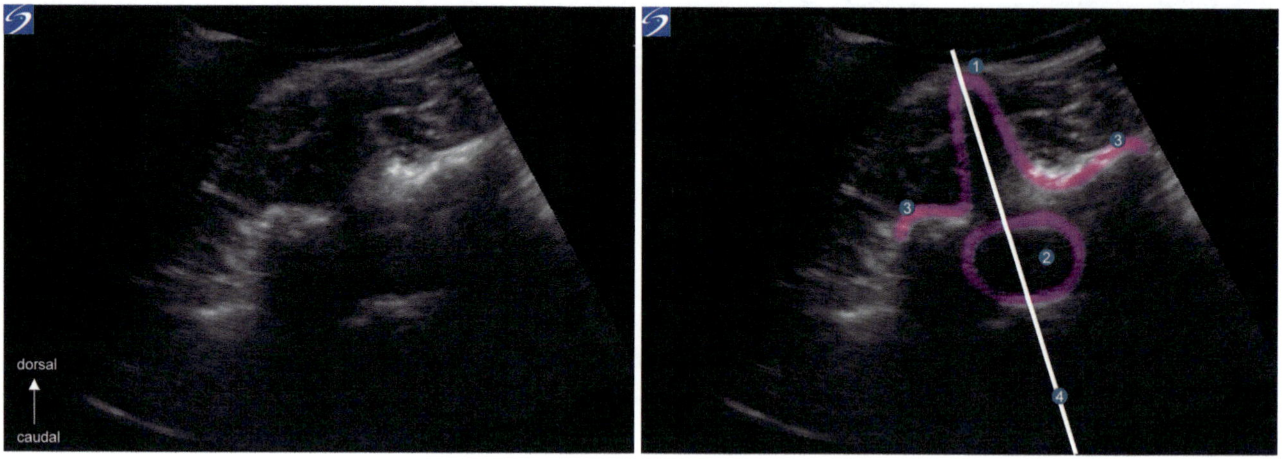

1. spinous process
2. spinal canal
3. lamina
4. sagittal vertebral axis

Fig. 12.26 Axial vertebral rotation in scoliosis. The sagittal vertebral axis is "off-centred" when the probe is in neutral transverse position

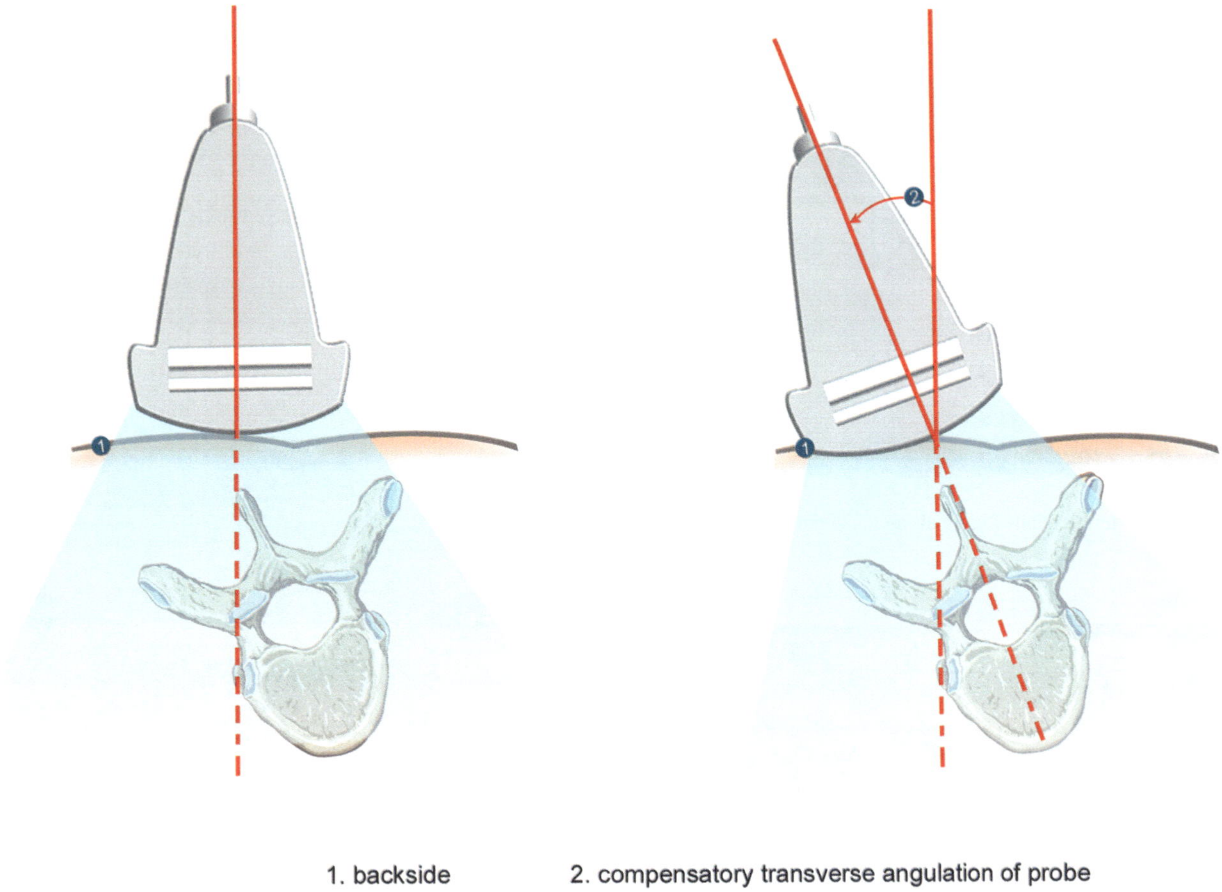

Fig. 12.27 Compensatory transverse angulation of the probe in case of scoliosis

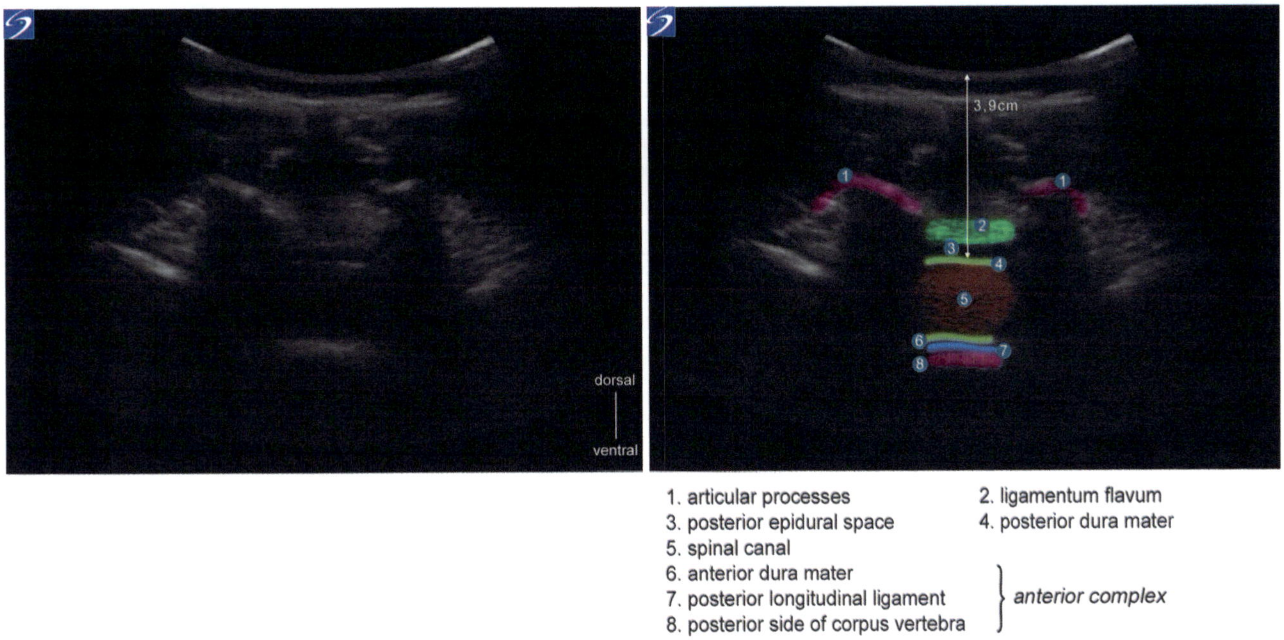

Fig. 12.28 Transverse plane: visualisation of major structures

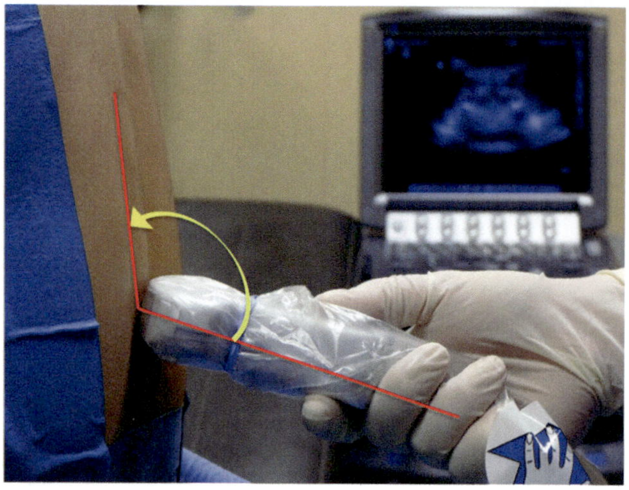

Fig. 12.29 Determination of optimal sagittal angulation of probe

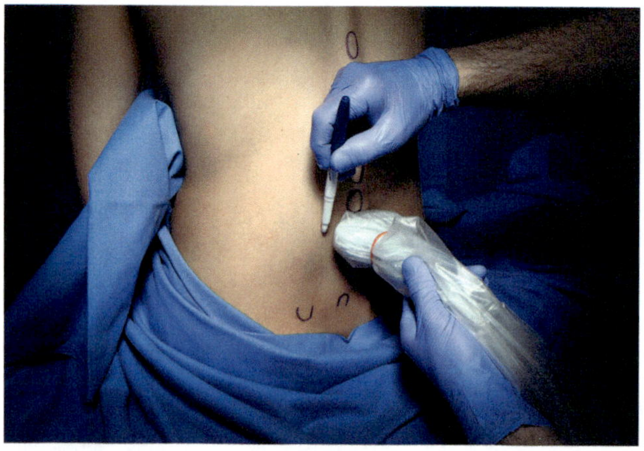

Fig. 12.30 Precise marking of the skin at level of the chosen interspinous space

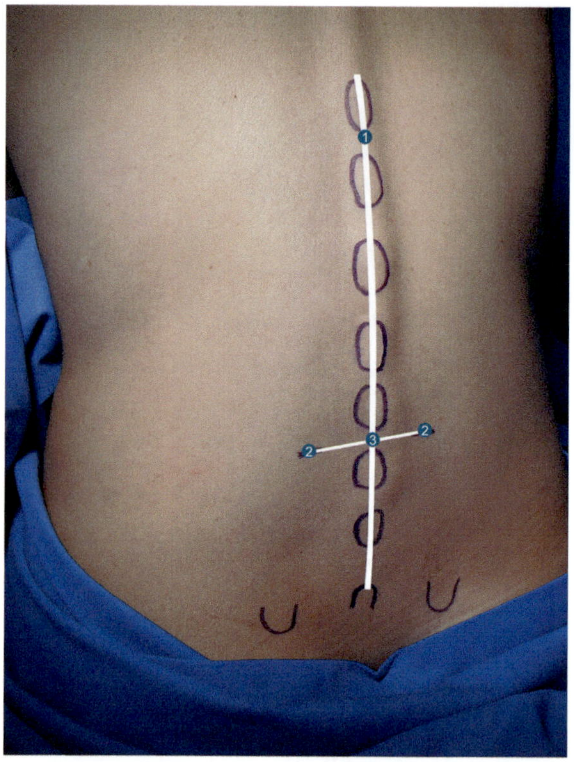

1. superficial sagittal spinal axis 2. interspinous space (L3-L4)
3. puncture point

Fig. 12.31 Determination of needle insertion site

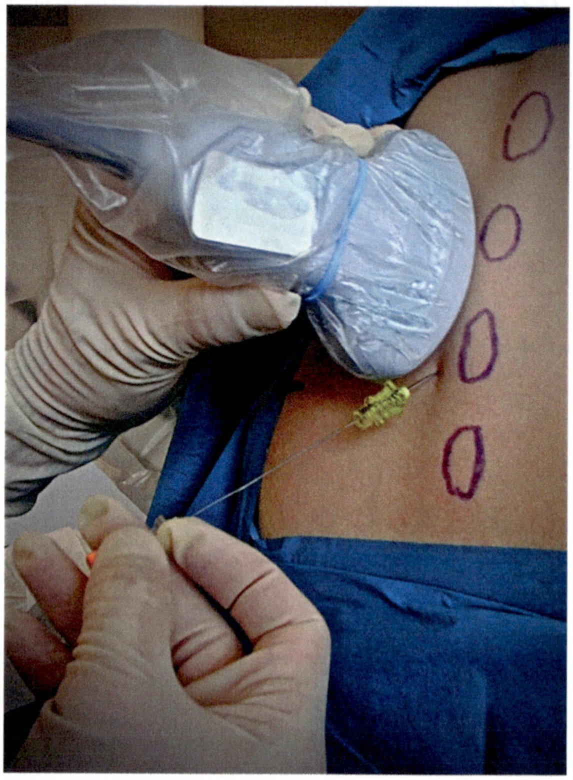

Fig. 12.32 Performance of ultrasound-guided spinal anaesthesia by OPS approach

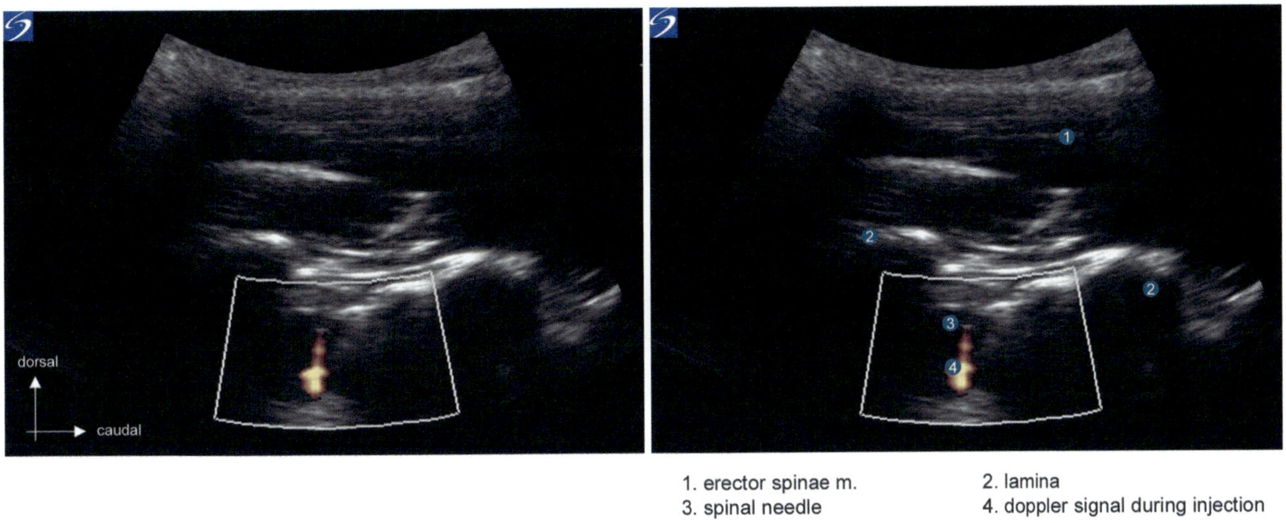

Fig. 12.33 Injection of local anaesthetic observed with Doppler power ultrasound during ultrasound-guided spinal anaesthesia

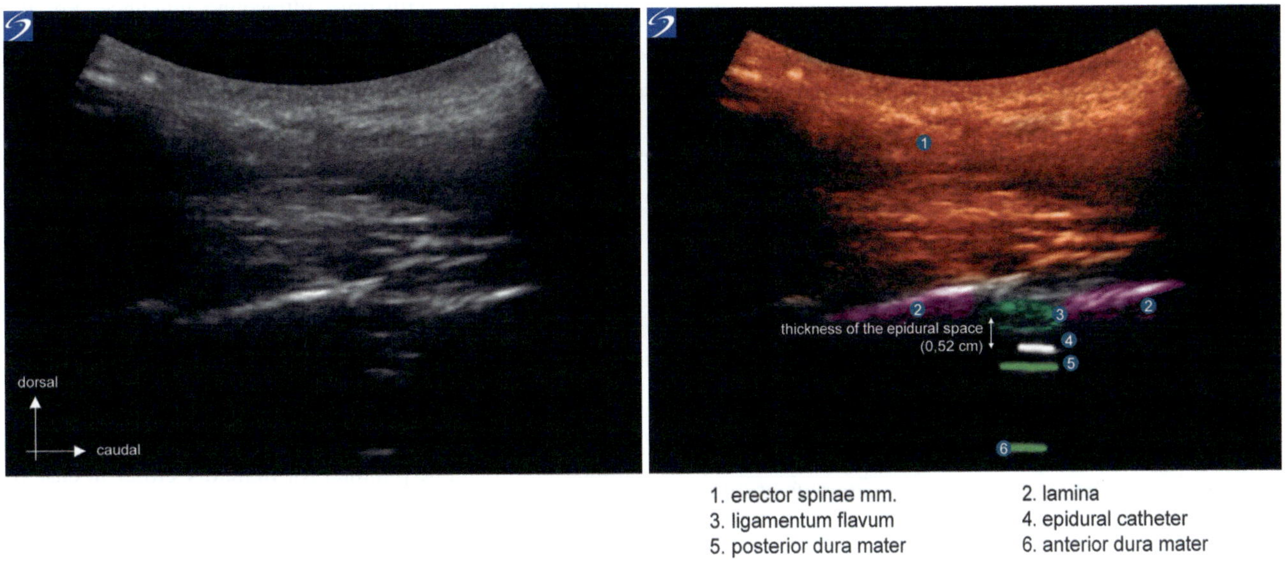

Fig. 12.34 Visualisation of epidural catheter in the interlaminar acoustic window just cephalad to the needle insertion level

Conclusions

Use of ultrasound by an experienced practitioner can make it possible to predict the level of difficulty of spinal anaesthesia [87]. Thus, its use promotes **objectivity** in technical and medical decision-making (anaesthetic strategy) and facilitates conduct of the procedure.

At the end of a structured procedure which is performed often more rapidly and simply than is described verbally, the preprocedural ultrasound location enables the practitioner to obtain **crucial information**:

- Identification of level of puncture.
- Ideal position of puncture site.
- Existence and extent of possible transverse vertebral rotation.
- Sagittal inclination of the interspinous space chosen.
- Depth of the epidural space.

Knowing this information enables the practitioner at the outset to position and direct the needle precisely, to limit uncertainty during the procedure, to reduce multiple redirections, and to anticipate objectively the depth to which the injection should be made.

Table 12.1 Ultrasound and spinal anaesthesia: summary of evidence

	Level of evidence	Jadad scores
Ultrasound-guided spinal anaesthesia		
Ultrasound performed at the patient's bedside enables the determination of a lumbar space with more precision than palpation only	IIa	N/A
Ultrasound localisation enables the very precise prediction of the depth of the epidural space	Ib	2
Ultrasound-guided epidural injection performed in obstetrics by experienced anaesthesiologists results in a similar success rate as well as a decrease in number of attempts at injection and in the number of intervertebral spaces punctured	Ib	2
Ultrasound-guided epidural puncture can improve the success rate of practitioners who are in training	Ib	2
Ultrasound-guided epidural puncture in young children		
An ultrasound-guided epidural injection performed by experienced anaesthesiologists leads to a similar success rate and furthermore can reduce the procedure duration and limit the number of "bony contacts" compared to the standard technique	Ib	2

The majority of published studies have been performed by very experienced teams but few in number. Few controlled randomised studies exist; almost all have been conducted in paediatrics and obstetrics. However, in light of these studies, it is now possible to outline much more than trends [10, 88] (Table 12.1).

N/A: not applicable.

In expert hands, use of ultrasound for spinal anaesthesia has demonstrated its ability:

- To estimate more reliably the intervertebral level approached [10, 49, 50].
- To reduce the overall number of attempts at injection, needle redirections and of failures [3, 4, 30, 31], increasing the success rate with the first attempt at epidural needle insertion by 30–60% compared to the "blind" traditional technique.
- To improve the success rate in approach to the epidural space at the first attempt [32].
- To reduce the number of intervertebral spaces punctured [4, 31, 32].
- To decrease the number of traumatic injections [10].
- To improve overall patient satisfaction [63].

Obese patients and/or patients with spinal abnormalities (who have undergone surgery or not) can obtain a real benefit from the use of spinal ultrasound [30, 55–58, 81, 82, 88–97]. In these populations where it proves, especially difficult to perform and/or interpret, there are still very few studies but they are already persuasive [30, 58, 76].

There are not yet prospective studies, comparing in terms of safety, central neuraxial anaesthesia with ultrasound imaging (real-time ultrasound or preprocedural ultrasound examination) to "standard" techniques based on surface landmarks.

However, by improving the preprocedural estimate of the position and depth of the epidural space [4, 23, 27–29, 31, 34, 74, 76, 98], and of the detection of possible abnormalities of the LF which compromise the safe performance of "blind" technique (by searching for loss of resistance) [34, 77], it is possible to choose a needle with an appropriate length and to include a component of **objectivity** into the planning of the procedure along with the perception of the tactile sensations felt during it.

The preprocedural ultrasound examination of the spine takes on average 2–3 minutes [4, 31]; considering the reduction in number of injection attempts, potentially there is no prolongation of the time required for conduct of epidural anaesthesia with prior ultrasound guidance. For the time being, this parameter has not been measured in comparative studies, but it is the impression which emerges from teams familiar with this procedure. In the specific case of obstetrics where the "time constraint" in insertion of the epidural catheter is perhaps a more sensitive concept than in other surgical indications, one may also ask the following question: "If we do not have 3 minutes for a preprocedural spinal ultrasound examination, do we truly have the time to perform epidural anaesthesia?"

Real-time ultrasound-guided procedure is a technique which is still relatively uncommon, performed only by the experts and which certainly needs to be validated by larger studies. Its role has not been scientifically determined, but it can be planned for logically either in case of failure of an injection in spite of preprocedural location conducted optimally, or at the outset where there are serious criteria predictive of difficulty in the conduct of anaesthesia.

In January 2008, in the United Kingdom, the NICE *(National Institute of Clinical Excellence)* recommended, based on bibliographical data then available, the routine use of ultrasound for insertion of epidural catheters [99]. However, it appears that the lack of specific training makes this practice difficult [100]. Amongst the principal obstacles to implementation of the technique, the lack of ultrasound equipment available on the site where a central neuraxial anaesthesia is performed may also participate (with ultrasound systems undoubtedly more concentrated in orthopaedic anaesthesia units). However, since these systems used for obstetric examination in the delivery room are perfectly appropriate for spinal ultrasound, part of the perceived difficulty may be circumvented.

In their publication "Ultrasonography of the adult thoracic and lumbar spine for central neuraxial blockade" [2], Chin et al. concluded, without differentiating ultrasound-assisted and ultrasound-guided techniques, by "[...] we do not believe the technique should supplant the traditional surface landmark-based techniques of spinal and epidural anaesthesia; these are simple, safe, and effective in most patients. Instead, the utility of the ultrasound-guided approach is most evident in patients in whom technical difficulty is expected because of poor surface anatomic landmarks (*e.g.*, in obesity or after spinal surgery) or distorted spinal anatomy (*e.g.*, scoliosis)".

If applied to the conduct of peripheral nerve blocks, this reasoning would result in use of ultrasound solely in cases where the surface anatomy predicts technical difficulty... Yet, in spite of the simplicity, safety, and efficacy for which neurostimulation had demonstrated in the performance of PNB [101], the use of ultrasound for **all patients** receiving RA is increasingly unquestionable.

Contrary to the apparent simplicity surrounding central neuraxial anaesthesia, its conduct is marked by well-documented drawbacks and difficulties which should not be underestimated. The literature reports that spinal anaesthesia performed "traditionally", by using the surface landmarks only, is less safe and less effective than PNB with neurostimulation only. Apart from those issues which are related to drugs injected and their possible consequences, problems encountered generally are of a technical and anatomical nature.

Ultrasound, therefore, is proposed as an **essential** tool whose **systematic** use can become the norm in spinal anaesthesia: it provides reliable and precise information on many critical factors not accessible by the clinical examination alone but which are necessary for the conduct of appropriate and personalised central neuraxial anaesthesia.

But customs have a long life...

Take-Home Message

- Conduct of central neuraxial anaesthesia based only on surface landmarks carries documented drawbacks and difficulties which should not be underestimated.
- The spinal ultrasound procedure provides objective findings on the anatomy of the patient, helping to define the anaesthetic strategy to be implemented.
- Use of ultrasound in central neuraxial anaesthesia can be viewed in two different manners: preprocedural examination and real-time ultrasound guidance.
- Preprocedural ultrasound (simple ultrasound examination prior to conduct of anaesthesia) is a simple and rapid technique. It is sufficient and very useful in a vast majority of cases.
- Contrary to a simple analysis of anatomical surface landmarks, preprocedural ultrasound makes it possible to identify precisely the level of injection, the depth of the epidural space or of the posterior complex, and the axis of puncture appropriate for the patient's underlying anatomy. During needle insertion, it provides some objectivity in the critical analysis of length of the needle introduced and of the sensations perceived.
- The use of ultrasound guidance makes it possible to reduce the overall number of puncture attempts, to improve significantly the success rate of the first attempt, to reduce the number of spaces punctured in order to improve patient satisfaction.
- Real-time ultrasound guidance of the needle is planned logically either in the case of failure of injection in spite of preprocedural examination conducted optimally, or at the outset in case of "serious" factors predictive of the difficulty in conduct of the anaesthesia (significant anatomical anomalies, history of spinal surgery, especially reduced acoustic window).
- Ultrasound is especially useful in obese patients or patients presenting with spinal anomalies.

References

1. Carvalho JC. Ultrasound-facilitated epidurals and spinals in obstetrics. Anesthesiol Clin. 2008;26:145–8.
2. Chin KJ, Karmakar MK, Peng P. Ultrasonography of the adult thoracic and lumbar spine for central neuraxial blockade. Anesthesiology. 2011;114:1459–85.
3. Shaikh F, Brzezinski J, Alexander S, et al. Ultrasound imaging for lumbar punctures and epidural catheterisations: systematic review and meta-analysis. BMJ. 2013;346:f1720.
4. Grau T, Leipold RW, Fatehhi S, Martin E, Motsch J. Real-time ultrasound observation of combined spinal-epidural anesthesia. Eur J Anaesthesiol. 2004;21:25–31.
5. Karmakar MK, Li X, Ho MH, Kwok WH, Chui PT. Real-time ultrasound-guided paramedian epidural access of a novel in-plane technique. Br J Anaesth. 2009;102:845–54.
6. Tran D, Kamani AA, Al-Attas, et al. Single-operator real-time ultrasound-guidance to aim and insert a lumbar epidural needle. Can J Anaesth. 2010;57:313–21.
7. O'Riain SC, Donnell BO, Cuffe T, Harmon DC, Fraher JP, Shorten G. Thoracic paravertebral block using real-time ultrasound guidance. Anesth Analg. 2010;110:248–51.
8. Karmakar MK, Vandepitte C. Ultrasound of the lumbar paravertebral space and considerations for lumbar plexus block. In: Hadzic A, editor. Hadzic's peripheral nerve blocks and anatomy for ultrasound-guided regional anesthesia. McGraw-Hill; 2012. p. 503–8.
9. Furness G, Reilly MP, Kuchi S. An evaluation of ultrasound imaging for identification of lumbar intervertebral level. Anesthesia. 2002;57:277–80.
10. Lee AJ, Ranasinghe JS, Chehade JM, et al. Ultrasound assessment of the vertebral level of the intercristal line in pregnancy. Anesth Analg. 2011;113:559–64.
11. Nair S, Abrishami A, Lermitte J, Chung F. Systematic review of spinal anesthesia using bupivacaine for ambulatory knee arthroscopy. Br J Anaesth. 2009;102:307–15.
12. Les blocs périmédullaires chez l'adulte. Recommandations pour la pratique clinique. SFAR 2006, www.sfar.org.

13. Dufeu N, Buntinx J. La rachianesthésie unilatérale (chapitre Blocs centraux), www.i-alr.com 2013.
14. Obi AO, Okafor VU, Nnodi PI. Prospective randomized trial of spinal saddle block *versus* periprostatic lignocaine for anesthesia during transrectal prostate biopsy. Urology. 2011;77:280–5.
15. Kazak Z, Ekmekci P. Kazbek K hyperbaric levobupivacaine in anal surgery: spinal perianal and spinal saddle blocks. Anaesthesist. 2010;59:709–13.
16. Schmittner MD, Schreiber H, Janke A, et al. Randomized clinical trial of perianal surgery performed under spinal saddle block *versus* total intravenous anesthesia. Br J Surg. 2010;97:12–20.
17. Recommandations formalisées d'experts. Prise en charge de la douleur post-opératoire chez l'adulte et l'enfant. Ann Fr Anesth Réanim. 2008;2008:1035–41.
18. Turc J, Lamblin A, Klack F, Wey PF, Martinez JY. Pulmonary hypertension and femoral neck fracture: interest of continuous spinal anesthesia. Ann Fr Anesth Réanim. 2012;31:557–9.
19. Biboulet P, Jourdan A, Van Haevre V, et al. Hemodynamic profile of target-controlled spinal anesthesia compared with 2 target-controlled general anesthesia techniques in elderly patients with cardiac comorbidities. Reg Anesth Pain Med. 2012;37:433–40.
20. Palmer CM. Continuous spinal anesthesia and analgesia in obstetrics. Anesth Analg. 2010;111:1476–9.
21. Andreae MH, Andreae DA. Local anaesthetics and regional anesthesia for preventing chronic pain after surgery. Cochrane Database Syst Rev. 2012;10:CD007105.
22. Horlocker TT, Wedel DJ, Rowlingson JC, et al. *Regional anesthesia in the patient receiving antithrombotic or thrombolytic therapy american society of regional anesthesia and pain medicine evidence-based guidelines* (3rd edition). Reg Anesth Pain Med. 2010;35:64–101.
23. Cramer GD, Darby SA. The lumbar region, basic and clinical anatomy of the spine, spinal cord and ANS. 2nd ed. St Louis MO: Mosby; 2005. p. 242–307.
24. Hogan QH. Epidural anatomy: new observations. Can J Anaesth. 1998;45:R40–8.
25. Cook TM, Counsell D, Wildsmith JA. Major complications of central neuraxial block: report on the third national audit project of the royal college of anaesthetits. Br J Anaesth. 2009;102:179–90.
26. Porter RW, Wicks M, Ottewell D. Measurement of the spinal canal by diagnostic ultrasound. J Bone Joint Surg Br. 1978;60-B:481–4.
27. Cork RC, Kryc JJ, Vaughan RW. Ultrasonic localization of the lumbar epidural space. Anesthesiology. 1980;52:513–6.
28. Currie JM. Measurement of the depth of the epidural space using ultrasound. Br J Anaesth. 1984;56:345–7.
29. Wallace DH, Currie JM, Gilstrap LC, Santos R. Indirect sonographic guidance for epidural anesthesia in obese pregnant patients. Reg Anesth. 1992;17:233–6.
30. Grau T, Leipold RW, Conradi R, Martin E. Ultrasound control for presumed difficult epidural puncture. Acta Anaesthesiol Scand. 2001;45:766–71.
31. Grau T, Leipold RW, Conradi R, Martin E, Motsch J. Efficacy of ultrasound imaging in obstetric epidural anesthesia. J Clin Anesth. 2002;14:169–75.
32. Grau T, Leipold RW, Conradi R, Martin E, Motsch J. Ultrasound imaging facilitates localization of the epidural space during combined spinal and epidural anesthesia. Reg Anesth Pain Med. 2001;26:64–7.
33. Grau T, Bartusseck E, Conradi R, Martin E, Motsch J. Ultrasound imaging improves learning curves in obstetric epidural anesthesia: a preliminary study. Can J Anaesth. 2003;50:1047–50.
34. Lee Y, Tanaka M, Carvalho JC. Sonoanatomy of the lumbar spine in patients with previous unintentional dural punctures during labor epidurals. Reg Anesth Pain Med. 2008;33:266–70.
35. Takiguchi T, Yamaguchi S, Tezuka M, Kitajima T. Measurement of shift of the cauda equina in the subarachnoid space by changing position. Reg Anesth Pain Med. 2009;34:326–9.
36. Karmakar MK. Ultrasound for central neuraxial blocks. Techniques Reg Anesth Pain Manage. 2009;13:161–70.
37. Karmakar MK, Li X, Li J, Hadzic A. Volumetric three-dimensional ultrasound imaging of the anatomy relevant for thoracic paravertebral block. Anesth Analg. 2012;115:1246–50.
38. De Filho GR, Gomes HP, da Fonseca MH, Hoffman JC, Pederneiras SG, Garcia JH. Predictors of successful neuraxial block: a prospective study. Eur J Anaesthesiol. 2002;19:447–51.
39. Tarkkila PJ, Kaukinen S. Complications during spinal anesthesia: a prospective study. Reg Anesth. 1991;16:101–6.
40. Pan PH, Bogard TD, Owen MD. Incidence and characteristics of failures in obstetric neuraxial analgesia and anesthesia: a retrospective analysis of 19,259 deliveries. Int J Obstet Anesth. 2004;13:227–33.
41. Tarkkila P, Huhtala J, Salminen U. Difficulties in spinal needle use. Insertion characteristics and failure rates associated with 25-, 27- and 29-gauge Quincke-type spinal needles. Anesthesia. 1994;49:723–5.
42. Kinsella SM. A prospective audit of regional anesthesia failure in 5,080 caesarean deliverys. Anesthesia. 2008;63:822–32.
43. Andersen G, Rasmussen H, Rosenstock C, et al. Post-operative pain control by epidural analgesia after transabdominal surgery. Efficacy and problems encountered in daily routine. Acta Anaesthesiol Scand. 2000;44:296–301.
44. Blanche R, Eisenach JC, Tuttle R, Dewan DM. Previous wet tap does not reduce success rate of labor epidural analgesia. Anesth Analg. 1994;79:291–4.
45. Ready LB. Acute pain: lessons learned from 25,000 patients. Reg Anesth Pain Med. 1999;24:499–505.
46. Le Coq G, Ducot B, Benhamou D. Risk factors of inadequate pain relief during epidural analgesia for labour and delivery. Can J Anaesth. 1998;45:719–23.
47. Burstal R, Wegener F, Hayes C, Lantry G. Epidural analgesia: prospective audit of 1,062 patients. Anaesth Intensive Care. 1998;26:165–72.
48. Tanaka K, Watanabe R, Harada T, Dan K. Extensive application of epidural anesthesia and analgesia in a university hospital: incidence of complications related to technique. Reg Anesth. 1993;18:34–8.
49. Whitty R, Moore M, Macarthur A. Identification of the lumbar interspinous spaces: palpation *versus* ultrasound. Anesth Analg. 2008;106:538–40.
50. Schlotterbeck H, Schaeffer R, Dow WA, Touret Y, Bailey S, Diemunsch. Ultrasonographic control of the puncture level for lumbar neuraxial block in obstetric anesthesia. Br J Anaesth. 2008;100:230–4.
51. Marhofer P, Bosenberg A, Sitzwohl C, Willschke H, Wanzel O, Kapral S. Pilot study of neuraxial imaging by ultrasound in infants and children. Paediatr Anaesth. 2005;15:671–6.
52. Roberts SA, Guruswamy V, Galvez I. Caudal injectate can be reliably imaged using portable ultrasound. A preliminary study. Paediatr Anaesth. 2005;15:948–52.
53. Rapp HJ, Folger A, Grau T. Ultrasound-guided epidural cathéter insertion in children. Anesth Analg. 2005;101:333–9.
54. Willschke H, Bosenberg A, Marhofer P, et al. Epidural catheter placement in neonates: sonoanatomy and feasibility of ultrasonographic guidance in term and preterm neonates. Reg Anesth Pain Med. 2007;32:34–40.
55. Peng PW, Rofaeel A. Using ultrasound in a case of difficult epidural needle placement. Can J Anaesth. 2006;53:325–6.
56. Prasad GA, Tumber PS, Lupu CM. Ultrasound-guided spinal anesthesia. Can J Anaesth. 2008;55:716–7.
57. Costello JF, Balki M. Caesarian delivery under ultrasound-guided spinal anesthesia, in a parturiant with poliomyelitis and Harrington instrumentation. Can J Anaesth. 2008;55:606–11.
58. Chin KJ, Macfarlane AJR, Chan V, Brull R. The use of ultrasound to facilitate spinal anesthesia in a patient with previous lumbar laminectomy and fusion: a case report. J Clin Ultrasound. 2009;37:482–5.

59. Locks Gde F, Almeida MC, Pereira AA. Use of the ultrasound to determine the level of lumbar puncture in pregnant women. Rev Bras Anestesiol. 2010;60:13–9.
60. Render CA. The reproducibility of the iliac crest as a marker of lumbar spine level. Anesthesia. 1996;51:1070–1.
61. Kettani A, Tachinante R, Tazi A. Evaluation of the iliac crest as anatomic landmark for spinal anesthesia in pregnant women. Ann Fr Anesth Reanim. 2006;25:501–4.
62. Chakraverty R. Which spinal levels are identified by palpation of the iliac crests and the posterior superior iliac spines? J Anat. 2007;210:232–6.
63. Margarido CB, Mikhael R, Arzola C, Balki M, Carvalho JC. The intercristal line determined by palpation is not a reliable anatomical landmark for neuraxial anesthesia. Can J Anaesth. 2011;58:262–6.
64. Broadbent CR, Maxwell WB, Ferrie R, et al. Ability of anaesthetists to identify a marked lumbar interspace. Anesthesia. 2000;55:1122–6.
65. Holmaas G, Frederiksen D, Ulvik A, et al. Identification of thoracic intervertebral spaces by means of surface anatomy: a magnetic resonance imaging study. Acta Anaesthesiol Scand. 2006;50:368–73.
66. Halpern SH, Banerjee A, Stocche R, Glanc P. The use of ultrasound for lumbar spinous process identification: a pilot study. Can J Anaesth. 2010;57:817–22.
67. Margarido CB, Arzola C, Balki M, Carvalho JC. Anesthesiologists' learning curves for ultrasound assessment of the lumbar spine. Can J Anaesth. 2010;57:120–6.
68. Bron JL, van Royen BJ, Wuisman PI. The clinical significance of lumbosacral transitional anomalies. Acta Orthop Belg. 2007;73:687–95.
69. Reynolds F. Damage to the conus medullaris following spinal anesthesia. Anesthesia. 2001;56:238–47.
70. Hamandi K, Mottershead J, Lewis T, Ormerod IC, Ferguson IT. Irreversible damage to the spinal cord following spinal anesthesia. Neurology. 2002;59:624–6.
71. Saiffudin A, Burnett JJ, White J. The variation of position of the conus medullaris in an adult population. A magnetic resonance imaging study. Spine. 1998;23:1452–6.
72. Dick EA, Patel K, Owens CM, De Bruyn R. Spinal ultrasound in infants. Br J Radiol. 2002;75:384–92.
73. Vallejo MC, Phelps AL, Singh S, Orebaugh SL, Sah N. Ultrasound decreases the failed labor epidural rate in resident trainees. Int J Obstet Anesth. 2010;19:373–8.
74. Arzola C, Davies S, Rofaaeel A, Carvalho JC. Ultrasound using transverse approach to the lumbar spine provides reliable landmarks for labor epidurals. Obst Anesth. 2007;104:1188–92.
75. Balki M, Lee Y, Halpern S, Carvalho JC. Ultrasound imaging of the lumbar spine in the transverse plane: the correlation between estimated and actual depth to the epidural space in obese parturients. Anesth Analg. 2009;108:1876–81.
76. Chin KJ, Perlas A, Chan V, Brown-Shreves D, Koshkin A, Vaishnav V. Ultrasound imaging facilitates spinal anesthesia in adults with difficult surface anatomical landmarks. Anesthesiology. 2011;115:94–101.
77. Lirk P, Moriggl B, Colvin J, et al. The incidence of lumbar ligamentum flavum midline gaps. Anesth Analg. 2004;98:1178–80.
78. Lirk P, Colvin J, Steger B, et al. Incidence of lower thoracic ligamentum flavum midline gaps. Br J Anaesth. 2005;94:852–5.
79. Grau T, Leipold RW, Delorme S, Martin E, Motsch J. Ultrasound imaging of the thoracic epidural space. Reg Anesth Pain Med. 2002;27:200–6.
80. Avramescu S, Arzola C, Tharmaratnam U, Chin KJ, Balki M. Sonoanatomy of the thoracic spine in adult volunteers. Reg Anesth Pain Med. 2012;37:349–53.
81. Pandin P, Haentjens L, Salengros JC, Quintin J, Barvais L. Combined ultrasound and neurostimulation-guided thoracic epidural catheter placement for analgesia following anterior spine fusion in scoliosis. Pain Pract. 2009;9:230–4.
82. McLeod A, Roche A, Fennelly M. Case series: ultrasonography may assist epidural insertion in scoliosis patients. Can J Anaesth. 2005;52:717–20.
83. Habib AS, George RB, Allen TK, Olufabi AJ. A pilot study to compare autodetect syringe with the glass syringe for identification of the epidural space in parturients. Anesth Analg. 2008;106:541–3.
84. Ranger MRB, Irwin GJ, Bunbury KM, Peutrell JM. Changing body position alters the location of the spinal cord within the vertebral canal: a magnetic resonance imaging study. Br J Anaesth. 2008;101:804–9.
85. Karmakar MK, Li X, Kwok WH, Ho AM, Ngan Kee WD. Sonoanatomy relevant for ultrasound-guided central neuraxial blocks via the paramedian approach in the lumbar region. Br J Radiol. 2012;85:e262–9.
86. Grau T, Leipold RW, Horter J, Conradi R, Martin EO, Motsch J. Paramedian access to the epidural space: the optimum window for ultrasound imaging. J Clin Anesth. 2001;13:213–7.
87. Chin KJ, Chan V. Ultrasonography as a preoperative assessment tool: predicting the feasibility of central neuraxial blockade. Anesth Analg. 2010;110:252–3.
88. Perlas A. Evidence for the use of ultrasound in neuraxial blocks. Reg Anesth Pain Med. 2010;35:S43–6.
89. Peterson MA, Abele J. Bedside ultrasound for difficult lumbar puncture. J Emerg Med. 2005;28:197–200.
90. O'Donnell D, Prasad A, Perlas A. Ultrasound-assisted spinal anesthesia in obese patients. Can J Anaesth. 2009;56:982–3.
91. Whitty RJ, Maxwell CV, Carvalho JC. Complications of neuraxial anesthesia in an extreme morbidly obese patient for cesarean section. Int J Obstet Anesth. 2007;16:139–44.
92. Kawaguchi R, Yamauch M, Sugino S, Tsukigase N, Omote K, Namiki A. Two cases of epidural anesthesia using ultrasound imaging. Masui. 2007;56:702–5.
93. Yamauchi M, Honma E, Mimura M, Yamamoto H, Takahashi E, Namiki A. Identification of the lumbar intervertebral level using ultrasound imaging in a post-laminectomy patient. J Anesth. 2006;20:231–3.
94. Minville V, Lavidalle M, Bayoumeu F, Parrant O, Fourcade O. Ultrasound-assisted spinal anesthesia in obstetric patient with corrected scoliosis. Ann Fr Anesth Réanim. 2010;29:501–2.
95. Somers R, Jacquemyn Y, Sermeus L, Vercauteren M. Corrected scoliosis, cholinesterase deficiency, and cesarean section: a case report. Case Report Med. 2009;2009:957479.
96. Yeo ST, French R. Combined spinal-epidural in the obstetric patient with Harrington rods assisted by ultrasonography. Br J Anaesth. 1999;83:670–2.
97. Chin KJ, Chan VW, Ramlogan R, Perlas A. Real-time ultrasound-guided spinal anesthesia in patients with a challenging spinal anatomy: two case reports. Acta Anaesthesiol Scand. 2010;54:252–5.
98. Tran D, Kamani AA, Lessoway VA, Peterson C, Hor KW, Rohling RN. Preinsertion paramedian ultrasound guidance for epidural anesthesia. Anesth Analg. 2009;109:661–7.
99. National Institute for Health and Clinical Excellence (NICE). Ultrasound-guided epidural catheterization of the epidural space: understanding NICE guidance. January 2008. http://guidance.nice.org.uk/IPG249/guidance/pdf/English
100. Mathieu S, Dalgleish DJ. A survey of local opinion of NICE guidance on the use of ultrasound in the insertion of epidural catheters. Anesthesia. 2008;63:1146–7.
101. Auroy Y, Benhamou D, Bargues L, et al. Major complications of regional anesthesia in France: the SOS-ALR anesthesia hotline service. Anesthesiology. 2002;97:1274–80.

Correction to: Ultrasound in Peripheral, Neuraxial and Perineuraxial Regional Anaesthesia

Eryk Eisenberg and Elisabeth Gaertner

Correction to:
E. Eisenberg, E. Gaertner (eds.), *Ultrasound in Peripheral, Neuraxial and Perineuraxial Regional Anaesthesia*, https://doi.org/10.1007/978-3-031-08804-9

This book was inadvertently published without the video files and links. This has been corrected.

The updated original version of these chapters can be found at
https://doi.org/10.1007/978-3-031-08804-9_5
https://doi.org/10.1007/978-3-031-08804-9_6
https://doi.org/10.1007/978-3-031-08804-9_9
https://doi.org/10.1007/978-3-031-08804-9_10
https://doi.org/10.1007/978-3-031-08804-9_12

E. Eisenberg, E. Gaertner (eds.), *Ultrasound in Peripheral, Neuraxial and Perineuraxial Regional Anaesthesia*,
https://doi.org/10.1007/978-3-031-08804-9_13

GPSR Compliance

The European Union's (EU) General Product Safety Regulation (GPSR) is a set of rules that requires consumer products to be safe and our obligations to ensure this.

If you have any concerns about our products, you can contact us on ProductSafety@springernature.com

In case Publisher is established outside the EU, the EU authorized representative is:

Springer Nature Customer Service Center GmbH
Europaplatz 3
69115 Heidelberg, Germany

Batch number: 10755966

Printed by Printforce, the Netherlands